Presurgical Evaluation of Epileptics

Basics, Techniques, Implications

Edited by H. G. Wieser and C. E. Elger

With 191 Figures, 65 Tables, and a Clinical Questionnaire
in a Pocket Inside the Back Cover

Springer-Verlag
Berlin Heidelberg New York
London Paris Tokyo

Priv.-Doz. Dr. med. HEINZ GREGOR WIESER
Department of Neurology
University Hospital
Frauenklinikstr. 26
8091 Zürich, Switzerland

Priv.-Doz. Dr. med. CHRISTIAN E. ELGER
Department of Neurology
University Hospital
University of Münster
Albert-Schweitzer-Str. 33
4400 Münster
Federal Republic of Germany

ISBN-13: 978-3-642-71105-3 **e-ISBN-13: 978-3-642-71103-9**
DOI: 10.1007/978-3-642-71103-9

Library of Congress Cataloging-in-Publication Data. Presurgical evaluation of epileptics. Based on the International Symposium on Methods of Presurgical Evaluation of Epileptic Patients: Basics, Techniques, and Implications for Epileptology and Surgical Epilepsy Therapy, held in Zürich, Sept. 12-15, 1985. Dedicated to: Rudolf M. Hess. Includes bibliographies and index. 1. Epilepsy - Surgery - Congresses. 2. Epilepsy - Diagnosis - Congresses. 3. Brain - Examination - Congresses. 4. Electrophysiology - Congresses. I. Wieser, Heinz Gregor. II. Elger, C. E. (Christian Erich) III. Hess, Rudolf M., 1913- . IV. International Symposium on Methods of Presurgical Evaluation of Epileptic Patients: Basics, Techniques, and Implications for Epileptology and Surgical Epilepsy Therapy (1985: Zurich, Switzerland) [DNLM: 1. Epilepsy - diagnosis - congresses. 2. Epilepsy - surgery - congresses. WL 385 P942 1985] RD594.P66 1987 616.8'53 87-2275 ISBN 0-387-16344-1 (U.S.)

Softcover reprint of the hardcover 1st edition 1987

Typesetting, printing, and bookbinding by Appl, Wemding
2125/3130-543210

Honorary Address for Professor Rudolf Hess delivered on 12 September 1985

Dear Rector, My Dear Akert,
Dear Presidents, My Dear Wieser,
Dear Colleagues and Friends, and, above all,
My Dear Rudi and Silvia

It is with great pleasure that I deliver tonight's address in honor of Prof. Rudolf Hess. He may have accepted words of praise in his local circles on the occasion of his retirement. Now it falls to me to pay him due international tribute. This symposium was chosen for such a purpose because it is dedicated to his field of interest, it takes place in his native town and he himself is contributing personally to the meeting, which unites around 150 participants of 19 countries representing 5 continents.

I appear to be the appropriate speaker to honor Rudi Hess because both of us passed through an almost identical career with only a few years difference: promotion as a medical doctor and specialization in neurology in the early forties; subsequent engagement in electroencephalography, which gave occasion for our meeting each other for the first time at the first international EEG congress in London, 1947; educational stay at the National Hospital, Queen Square, London, where Rudi Hess stayed for a longer period under the care and influence of William A. Cobb, whereas I stayed longer than he at the Burden Neurological Institute, Bristol, exposed to the unique mental emanations of W. Grey Walter. Some years later, both of us stayed for 6 months at the Montreal Neurological Institute, under the auspices of Wilder Penfield and Herbert Jasper, where each of us tried his hand at experimental neurophysiology, neither with much success for that matter. Thereafter, we both returned to clinical neurophysiology, specifically to EEG and epileptology, to which we devoted the remainder of our professional lives - Rudi Hess as the head of the EEG department of the Kantonsspital Zürich, myself as the head of the EEG department of the Hôpital de la Timone, Marseille. Both departments were first attached to the respective neurological clinics, which implied administrative complications and disputes about competencies, so that later they were rendered independent. In these positions we pursued, over roughly 30 years, our trifold tasks: as physicians at the service of our patients; as researchers, at the service of science; and as teachers, at the service of our students. Such similar endeavors led in the parallel courses of our academic careers, to election to the presidency of our respective national EEG Societies and Leagues Against Epilepsy. They also naturally led to numerous personal meetings, on which occasions our firm friendship developed.

There may still be another reason for my feeling qualified to deliver this speech: the astonishing difference in our personalities, contrasting with the so similar curricula. This did not in any way affect our mutual friendship; rather, it let us appreciate each other's qualities. As a Latin of modest descent, enthusiastic and ambitious as I am, I can, better than anyone else, esteem the calm and unassuming Alemannic Swiss, descended from a famous father (a Nobel prize winner in 1949) from whom he adopted a certain reserve and learned to sacrifice futile honors in the skilful accomplishment of his assignments.

RUDOLF M. HESS, Dr. med., Professor Emeritus, University Zürich

Rudi Hess received the Hans Berger prize of the German EEG Society of which he is an honorary member. He is also honorary member of the French and Swiss EEG Societies, corresponding member of the German League, and honorary president of the Swiss League Against Epilepsy. He was treasurer of the International Federation of EEG Societies, but he refrained from standing as candidate for its presidency, a decision which was surprising to many; he chose instead to invest his time and energy in his clinical work.

Although he was convinced that, for many neurophysiological problems, animal experimentation was an indispensible approach – and he had repeatedly engaged in such – Hess's personal inclination decidedly was for clinical research. Independently from the research of the Gibbs', he found the EEG correlate of malignant infantile spasms and the even stronger correlation with age of these anarchic patterns and the more organized slow spikes and waves. The benign rolandic sharp waves of school age children were identified independently from the French authors. The spatial distribution of sleep potentials was recognized, before Brazier's classic paper appeared in the EEG Journal. He contributed to the pathogenetic explanation of intermittent frontal δ-rhythms and to the still intriguing and much debated cross-connections between headaches and epilepsy. One of Rudi Hess's outstanding interests was the EEG expression of cerebral tumors, which he examined in great numbers in Krayenbühl's neurosurgical unit. Interesting facts and correlations emerged from this endeavor of many years, and were published in a monograph; the sober overall conclusion was, however, that the EEG, for the diagnosis of cerebral tumors, was less reliable than many earlier authors had presumed.

To sum up, Rudi Hess never thought of himself as a great scientist, nor as a pioneer, called up to lead EEG to new horizons. He was rather the cautious type who saw his task and his merits in "preventing over-optimistic and unwarrantable conclusions and ideas which, at that time, were blossoming."

This self-judgement reflects his main quality: his honest objectivity. It will not prevent future generations from regarding him as one of the pioneers of clinical EEG. It also explains the appreciation of those who worked with him for his incorruptible character. His friends also know him as a lover of the mountains, whose preferred recreations are hiking and skiing. And they admire his adorable wife Silvia, with her four children, three of whom are themselves engaged in the neuropsychiatric fields in which their father and grandfather gained their reputations.

Professor Henri Gastaut

Preface

From the 12th to the 15th September 1985 the International Symposium on Methods of Presurgical Evaluation of Epileptic Patients: Basics, Techniques, and Implications for Epileptology and Surgical Epilepsy Therapy was held in Zürich. This symposium was a consequence of the increasing recognition by Europeans, especially from German-speaking countries, of a growing need for surgical therapy of epileptics. The main aim was to provide a venue for critical review and lively discussion of presurgical evaluation protocols, with special emphasis on the electrophysiological aspects, including invasive techniques. To provide a necessary background, some basic aspects along with postsurgical results had to be dealt with by leading experts in the different fields of experimental and clinical epileptology and neurosurgery. It was intended to be an international but moderately scaled meeting. Finally, however, in addition to the European contingent, there were participants from Australia, Brazil, Canada, Israel, Japan, Korea, and the United States of America, including pioneers from the surgically active centers.

This international discussion was surely facilitated by the close temporal relationship to the International Congresses (XIIIth World Congress of Neurology and XVIth Epilepsy International Congress) which took place in Hamburg. And, probably most important, is the fact that the Zürich symposium was dedicated to Prof. Rudolf Hess and his lifelong devotion to epilepsy diagnosis and treatment in Switzerland, as outlined in the Honorary Address by his eminent colleague and personal friend, Prof. Henri Gastaut.

This resulting volume follows closely the scientific program, covering five main topics:

(1) Basic considerations relevant to seizure origin and spread
(2) Electrophysiological and other diagnostic procedures
(3) Electroclinical semiology of the epilepsies
(4) The surgical approach: preconditions, rationale, efficacy
(5) Consequences for the practice of epilepsy therapy

Each topic was divided into several subtopics, and each subtopic was dealt with by three to five leading representatives, who were specially requested to emphasize the more generally valid aspects.

The presentations were extremely well-discussed, and the book unfortunately cannot fully reflect this. However, the editors have tried to incorporate the main points of the discussions within the Editors' Commentaries following each subtopic. We hope that, in this way, the volume provides a logical sequence of information to the reader and satisfies the criteria for a competent overview in this field of growing importance. The chapters are intended to summarize the present state of knowledge and development of the principal, well-established facets of preconditions and techniques of surgical therapy of the epilepsies.

We gratefully acknowledge the generous financial contributions of many European sponsors, the editorial assistance of Dr. phil. Valerie Hylton, and the editorial and production assistance of the staff of Springer-Verlag.

Heinz Gregor Wieser
Christian E. Elger

Contents

Contributors

Asterisks indicate individuals who lectured at the Symposium*, chaired sessions**, or contributed with posters*** only.

Akert, K.**, Professor, Dr. med., Rector of Zürich University, Former Director of the Zürich Brain Research Institute, Rämistr. 71, 8006 Zürich, Switzerland

Bancaud, J., Dr. med., Director de Recherche, Unité de Recherches sur l'Epilepsie, INSERM U 97, Centre Paul Broca, 2 ter rue d'Alésia, 75014 Paris, France

Baumgartner, G.**, Professor, Dr. med., Department of Neurology, University Hospital, Frauenklinikstr. 26, 8091 Zürich, Switzerland

Black, S. E., M. D., F. R. C. P. (C), Research Associate, Department of Clinical Neurological Sciences, St. Joseph's Hospital, Research Institute, 268 Grosvenor Street, London, Ontario N6A 4V2, Canada

Blume, W. T., Professor, Clinical Neurological Sciences, University Hospital, 339 Windermere Road, London, Ontario N6A 5A5, Canada

Brunet, P., Technicien de Recherche, Unité de Recherches sur l'Epilepsie, INSERM U 97, Centre Paul Broca, 2 ter rue d'Alésia, 75014 Paris, France

Bruton, C. J., M. D., Department of Neuropathology, Runwell Hospital, Wickford, Essex SS117QE and Division of Psychiatry, C. R. C. Northwick Park Hospital, Harrow, Middlesex, United Kingdom

Cahan, L. D., M. D., Associate Professor, Division of Neurosurgery, University of California, Irvine Medical Center, Orange, CA 92668, USA

Carr, T., Clinical Assistant Professor, University of Western Ontario, Department of Nuclear Medicine, St. Joseph's Hospital, London, Ontario N6A 4V2, Canada

Chauvel, P., M. D., Unité de Recherches sur l' Epilepsie, INSERM U 97, Centre Paul Broca, 2 ter rue d'Alésia, 75014 Paris, France

Chodkiewicz, J. P., Professor, Centre Hospitalier Sainte-Anne, 1 rue Cabanis, 75674 Paris, France

Comte, P., Dr. ès. sciences, Avenue des Cerisiers 41, 1009 Pully, Switzerland

Corrie, W. S., M. D., Associate Professor of Neurology, Department of Neurology and Department of Neurological Surgery, The Medical College of Ohio, 3000 Arlington Avenue, C. S. 1008, Toledo, OH 43699-0008, USA

Crandall, P. H., Professor of Surgery and Neurology, UCLA Medical Center and Brain Research Institute, University of California, 760 Westwood Plaza, Los Angeles, CA 90024, USA

Creutzfeldt, O. D., Professor, Dr. med., Department of Neurobiology, Max-Planck-Institute for Biophysical Chemistry, P. O. Box 2841, 3400 Göttingen, Federal Republic of Germany

De Rooij, N. F., Professor, Dr., Institute of Microtechnology, University of Neuchâtel, Rue A.-L.-Breguet 2, 2000 Neuchâtel, Switzerland

Delgado-Escueta, A. V., M. D., Professor of Neurology, Comprehensive Epilepsy Program, Neurology Department, UCLA School of Medicine, Los Angeles, CA 90024, USA

Elger, C. E., Privatdozent, Dr. med., Neurological Clinic, University Hospital, University of Münster, Albert-Schweitzer-Str. 33, 4400 Münster, Federal Republic of Germany

Engel, J., Jr., M. D., Professor of Neurology and Anatomy, Reed Neurological Research Center, UCLA School of Medicine, Los Angeles, CA 90024, USA

Ferguson, S. M., M. D., Professor of Psychiatry and Neurological Surgery, Section of Neuropsychiatry, Department of Psychiatry, Medical College of Ohio, 3000 Arlington Avenue, C. S. 1008, Toledo, OH 43699, USA

Gastaut, H., Professor, WHO Collaborating Centre for Teaching and Research in Neurology, Institut de Recherches Neurologiques, Faculté de Médecine, 13385 Marseille, Cedex 5, France

Giallonardo, A. T., M. D., Va Clinica Neurologica Università degli Studi "La Sapienza", Viale dell'Università 30, 00100 Rome, Italy

Gloor, P., Professor of Clinical and Experimental Neurophysiology, Montreal Neurological Institute and Hospital, Department of Neurology and Neurosurgery, McGill University, 3801 University Street, Montréal, Québec H3A 2B4, Canada

Haas, H. L., Professor, Dr. med., Institute of Physiology, University of Mainz, Saarstr. 21, 6500 Mainz, Federal Republic of Germany

Halgren, E., Ph. D., Associate Professor of Psychiatry, Department of Psychiatry, UCLA Brain Research Institute, Los Angeles, CA 90024, USA

Hess, R. M., Professor, Dr. med., Rietholzstr. 34, 8125 Zollikerberg, Switzerland

Janz, D., Professor, Dr. med., Department of Neurology, Free University of Berlin, Klinikum Charlottenburg, Spandauer Damm 130, 1000 Berlin 19, Federal Republic of Germany

Jensen, I., Professor, Dr. med., Rigshospitalet, University of Copenhagen, Blegdamsvej 9, 2100 Copenhagen, Denmark, and Vejlesøevej 122, 2840 Holte, Denmark

Kausel, W., Dr. med., Department of Neurology, University Hospital, Frauenklinikstr. 26, 8091 Zürich, Switzerland

Kertesz, A., M. D., F. R. C. P. (C), Professor, Department of Clinical Neurological Sciences, St. Joseph's Hospital, Research Institute, 268 Grosvenor Street, London, Ontario N6A 4V2, Canada

Kleihues, P., Professor, Dr. med., Institute of Neuropathology, University Hospital, Schmelzbergstr. 12, 8091 Zürich, Switzerland

Koella, W. P., Professor emer., Dr. med., Buchenstr. 1, 4104 Oberwil, Switzerland

Liegeois-Chauvel, C., Ph. D., Unité de Recherches sur l'Epilepsie, INSERM U 97, Centre Paul Broca, 2 ter rue d'Alésia, 75014 Paris, France

Lopes da Silva, F. H., Dr., Professor of Animal Physiology, Department of Zoology, University of Amsterdam, Kruislaan 320, 1098 SM Amsterdam, The Netherlands

Maggauda, A., M. D., Ph. D., WHO Collaborating Centre for Teaching and Research in Neurology, Institut de Recherches Neurologiques, Faculté de Médecine, 13385 Marseille, Cedex 5, France

Maldonado, H. M., M. D., Ph. D., Comprehensive Epilepsy Program, Neurology Department, UCLA School of Medicine, Los Angeles, CA 90024, USA

Mariani, E., Dr., WHO Collaborating Centre for Teaching and Research in Neurology, Institut de Recherches Neurologiques, Faculté de Médecine, 13385 Marseille, Cedex 5, France

Mars, N. J. I., Professor, Dr., Department of Computer Science, University of Twente, P. O. Box 217, 7500 AE Enschede, The Netherlands

Mayanagi, Y., M. D., D. M. Sc., Chief Neurosurgeon, Neurological Center, Tokyo Metropolitan Police Hospital, 2-10-41, Fujimi, Chiyodaku, Tokyo, Japan 102

McLachlan, R. S., Assistant Professor, Departments of Clinical Neurological Sciences and Physiology, University of Western Ontario, Epilepsy Unit, University Hospital, London, Ontario N6A 5A5, Canada

Müller, R. U., Dr. med., Department of Neurology, University Hospital, Frauenklinikstr. 26, 8091 Zürich, Switzerland

Munari, C., M. D., Unité de Recherches sur l'Epilepsie, INSERM U 97, Centre Paul Broca, 2 ter rue d'Alésia, 75014 Paris, France

Musolino, A., M. D., Hôpital Sainte Anne, Service de Neurochirurgie, 1 rue Cabanis, 75014 Paris, France

Nadig, T., Dr. phil., Department of Neurology, Neuropsychology Unit, University Hospital, Frauenklinikstr. 26, 8091 Zürich, Switzerland

Nicholson, R. L., M. D., R. R. C. P. C., Associate Professor (Clin.), University of Western Ontario, Department of Nuclear Medicine, St. Joseph's Hospital, London, Ontario N6A 4V2, Canada

Oana, Y.***, M. D., Department of Neuropsychiatry, Tokyo Medical College Hospital, 6-7-1 Nishishinjuku, Shinjukuku, Tokyo 160, Japan

Ojemann, G. A., M. D., Professor of Neurological Surgery, Department of Neurological Surgery, RI-20, University of Washington, School of Medicine, Seattle, WA 98118, USA

Pandya, D. N., M. D., Professor of Anatomy, Edith Nourse Rogers Memorial Veterans Hospital, 200 Springs Road, Bedford, MA 01730, USA

Perret, E., Professor, Dr. phil., Neuropsychological Institute, Promenadengasse 18, 8001 Zürich, Switzerland

Petsche, H., Professor, Dr. med., Institute of Neurophysiology, University of Vienna, and Brain Research Institute, Austrian Academy of Sciences, Währinger Str. 17, 1090 Vienna, Austria

Phelps, M. E., Ph. D., Laboratory of Nuclear Medicine and Biophysics, UCLA School of Medicine, Los Angeles, CA 90024, USA

Pockberger, H., Dr. med., Institute of Neurophysiology, University of Vienna, and Brain Research Institute, Austrian Academy of Sciences, Währinger Str. 17, 1090 Vienna, Austria

Quesney, L. F., M. D., Ph. D., Chief Department of EEG and Clinical Neurophysiology, Montreal Neurological Institute and Hospital, McGill University, 3801 University Street, Montréal, Québec H3A 2B4, Canada

Rand, R. W., M. D., Ph. D., Professor of Neurological Surgery, Department of Surgery, UCLA Center for the Health Sciences, Los Angeles, CA 90024, USA

Rappelsberger, P., Universitätsdozent, Dipl. Ing., Dr., Institute of Neurophysiology, University of Vienna, and Brain Research Institute, Austrian Academy of Sciences, Währinger Str. 17, 1090 Vienna, Austria

Rasmussen, T., M.D., Professor of Neurology and Neurosurgery, Montreal Neurological Institute and Hospital, McGill University, 3801 University Street, Montréal, Québec H3A 2B4, Canada

Rausch, R., Ph.D., Associate Professor, Reed Neurological Research Center, University of California, 710 Westwood Plaza, Los Angeles, CA 90024, USA

Ravagnati, L., M.D., Associate Professor of Neurosurgery, Institute of Neurosurgery, University of Milan, Policlinico, Via F. Sforza 35, 20122 Milan, Italy

Rayport, M., M.D., C.M., Ph.D., Professor and Chairman, Department of Neurological Surgery, Medical College of Ohio, 3000 Arlington Avenue, C. S. 10008, Toledo, OH 43699-0008, USA

Robitaille, Y., M.D., Associate Professor, Departments of Neuropathology, Neurosurgery, and Neurology, McGill University, Montreal Neurological Institute, 3801 University Street, Montréal, Québec H3A 2B4, Canada

Schmidt, D., Professor, Dr. med., Department of Neurology, Klinikum Grosshadern, Ludwig-Maximilian University, Marchioninistr. 15, 8000 München 70, Federal Republic of Germany

Silfvenius, H., M.D., Ph.D., Department of Neurosurgery, University Hospital, 901 85 Umeå, Sweden

Speckmann, E.-J., Professor, Dr. med., Department of Experimental Epileptology, Institute of Physiology - Neurophysiology, University of Münster, Robert-Koch-Str. 27a, 4400 Münster, Federal Republic of Germany

Stefan, H., Professor, Dr. med., Neurological Clinic, University of Erlangen-Nürnberg, Schwabachanlage 6, 8520 Erlangen, Federal Republic of Germany

Stodieck, S.R.G.*, Dr. med., Department of Neurology, University Hospital, Frauenklinikstr. 26, 8091 Zürich, Switzerland

Sutherling, W.W., M.D., Assistant Professor of Neurology, Reed Neurological Research Center, UCLA School of Medicine, Los Angeles, CA 90024, USA

Swartz, B.E., M.D., Ph.D., Assistant Professor of Neurology, Wadsworth VA Medical Center and UCLA School of Medicine, Department of Neurology, Los Angeles, CA 90024, USA

Talairach, J., Professor, Dr. med., Unité de Recherches sur l'Epilepsie, INSERM U 97, Centre Paul Broca, 2 ter rue d'Alésia, 75014 Paris, France

Vetsch, M., lic. phil., Neuropsychological Institute, Promenadengasse 18, 8001 Zürich, Switzerland

Vieth, J., Professor, Dr. med., Department of Experimental Neuropsychiatry, Neurological Clinic, University of Erlangen-Nürnberg, Schwabachanlage 6, 8520 Erlangen, Federal Republic of Germany

Vignal, J.P., Dr., Service de Neurochirurgie, Hôpital Sainte-Anne, 1 rue Cabanis, 75014 Paris, France

Walden, J., Dr. med., Dipl.-Psych., Institute of Physiology, University of Münster, Robert-Koch-Str. 27a, 4400 Münster, Federal Republic of Germany

Walsh, G.O., M.D., Comprehensive Epilepsy Program, Department of Neurology, University of California, Center for the Health Sciences, Los Angeles, CA 90024, USA

Wieser, H.G., Privatdozent, Dr. med., Department of Neurology, University Hospital, Frauenklinikstr. 26, 8091 Zürich, Switzerland

Wyler, A. R., M. D., Professor of Neurosurgery, Anatomy, and Neurobiology, University of Tennessee, Memphis and Semmes-Murphy-Clinic, 956 Court Avenue, Memphis, TN 38163, USA

Yaşargil, M. G., Professor, Dr. med., Department of Neurosurgery, University Hospital, Rämistr. 100, 8091 Zürich, Switzerland

Yeterian, E. H., Ph. D., Associate Professor of Psychology, Department of Psychology, Colby College, Waterville, ME 04901, USA

Zifkin, B., M. D., C. M., F. R. C. P. (C)., State University of New York, Health Science Center at Brooklyn, Institute for Basic Research in Developmental Disabilities, Epilepsy Unit, Box 75, 450 Clarkson Avenue, Brooklyn, NY 11203, USA

Acknowledgements

The editors gratefully acknowledge the generous support of the following donators, who sponsored the symposium.

Chemische Fabrik Promonta GmbH, 2000 Hamburg 26, Federal Republic of Germany

Ciba-Geigy GmbH, 7867 Wehr/Baden, Federal Republic of Germany

Desitin-Werk Carl Klinke GmbH, 2000 Hamburg 63, Federal Republic of Germany

German Branch of International League Against Epilepsy

German EEG Society

Grünenthal GmbH, 8756 Mitlödi, Switzerland

ICI-Pharma, 6002 Luzern, Switzerland

Interpharma (Ciba-Geigy AG, Hoffmann-La Roche & Co. AG, Sandoz AG), 4002 Basel, Switzerland

Janssen Pharmaceutica AG, 6340 Baar, Switzerland

Lucenta-Cosmetic, 7505 Ettlingen, Federal Republic of Germany

Merz & Co. GmbH, 6000 Frankfurt 1, Federal Republic of Germany

Nordmark Arzneimittel GmbH, 2082 Uetersen, Federal Republic of Germany

Oxford Instruments Deutschland GmbH, 6200 Wiesbaden, Federal Republic of Germany

Sanofi Pharma AG, 4009 Basel, Switzerland

Sapos AG, 1211 Genève 1, Switzerland

Schürholz-Arzneimittel GmbH, 8000 München 60, Federal Republic of Germany

Schwarz Pharma AG, 4410 Liestal, Switzerland

Schwarzer-Picker International GmbH, 8000 München 60, Federal Republic of Germany

Spectralab Bruno Fricker, 8802 Kilchberg, Switzerland

"Stiftung Michael", 5300 Bonn 1, Federal Republic of Germany

Swiss Branch of International League Against Epilepsy

Swiss EEG Society

Wander AG-Pharma Schweiz, 3001 Bern, Switzerland

Basic Considerations Relevant to Seizure Origin and Spread

Hodology of Limbic and Related Structures: Cortical and Commissural Connections*

D. N. Pandya[1, 2, 3] and E. H. Yeterian[1, 4]

Introduction

A number of theories have been suggested to account for the genesis of seizures. For example, seizures could arise as a result of disinhibition of reciprocal cortical pathways. Alternatively, it could be that an overactivation of afferent and efferent cortical pathways is the basis of seizure phenomena. Another notion proposes that seizures arise from antidromic stimulation of cortical axons of passage which course through deeper structures. Finally, seizures could result from an abnormal reorganization of higher pathways during development. Thus, it seems that seizures could arise from any point in the cortex as a result of any one of these possible mechanisms. Nevertheless, the most common site of seizure genesis is in the deep structures of the temporal and frontal lobes. In this chapter, we wish to present an anatomical viewpoint on how these cortical structures, especially in the temporal lobe, may underlie various seizure phenomena.

Cortical Architecture and Connections

Seizure phenomena may be characterized as involving disturbances of the sensorimotor and emotional, or limbic, spheres. In other words, seizure disorders have a broad range of manifestations primictally and postictally as well as during the ictal period. In order to understand the symptom complex attached to seizure disorders, it may be helpful to have an overall understanding of cortical organization, in terms of both architecture and connections. As the basis of our presentation, we would like to draw upon the evolutionary approach to the anatomy of the cerebral cortex as developed by Sanides (1969). Sanides' notion has been referred to as the "evolutionary architectonic concept." We will briefly describe this concept in order to try to understand possible mechanisms underlying epilepsy originating in particular from the temporal lobe. We would like to emphasize from the outset, however, that this is an anatomical perspective, and that functional verification remains to be obtained.

As shown in Fig. 1 A, the evolutionary architectonic concept proposes that the cortex has evolved from two prime moieties, namely the archicortex (hippocampus) and the paleocortex (olfactory cortex). From these two moieties, stepwise progressions of cortical architecture can be traced leading to the development of periallocortex, which in turn leads to the proisocortex. From proisocortex, true six-layered isocortex evolves (Fig. 1 B). The further differentiation of the isocortices of each moiety leads to the somatosensory and motor areas from the archicortical moiety, on the one hand, and to the sensory cortices from the paleocortical moiety on the other. Likewise, the frontal lobe shows a progressive development from two moieties that is, ventrolateral cortex from the paleocortical moiety, and medial and dorsolateral cortex from that archicortical moiety (Fig. 1 C).

Let us now examine how each system has evolved from the proisocortex. As shown in Fig. 2 A–C, the auditory-related cortical system can be traced in a successive manner from the proisocortex up to the primary auditory area (Galaburda and Pandya 1983; Pandya and Yeterian 1985). This progression occurs in stages, with each stage including a core line and a belt line of architectonic regions. The core line leads to the primary auditory region, whereas the belt line contains association areas at each stage. Figure 2 D

[1] Edith Nourse Rogers Memorial Veterans Hospital, 200 Springs Road, Bedford, MA 01730, USA
[2] Departments of Anatomy and Neurology, Boston University School of Medicine, Boston, MA 02118, USA
[3] Harvard Neurological Unit, Beth Israel Hospital Boston, MA, USA
[4] Department of Psychology, Colby College, Waterville, ME 04901, USA

* This project is supported by the Veterans Administration, Edith Nourse Rogers Memorial Veterans Hospital, Bedford, Massachusetts 01730, NIH grant 16841, and Colby College social science grant A22143.

Presurgical Evaluation of Epileptics
Edited by H. G. Wieser and C. E. Elger

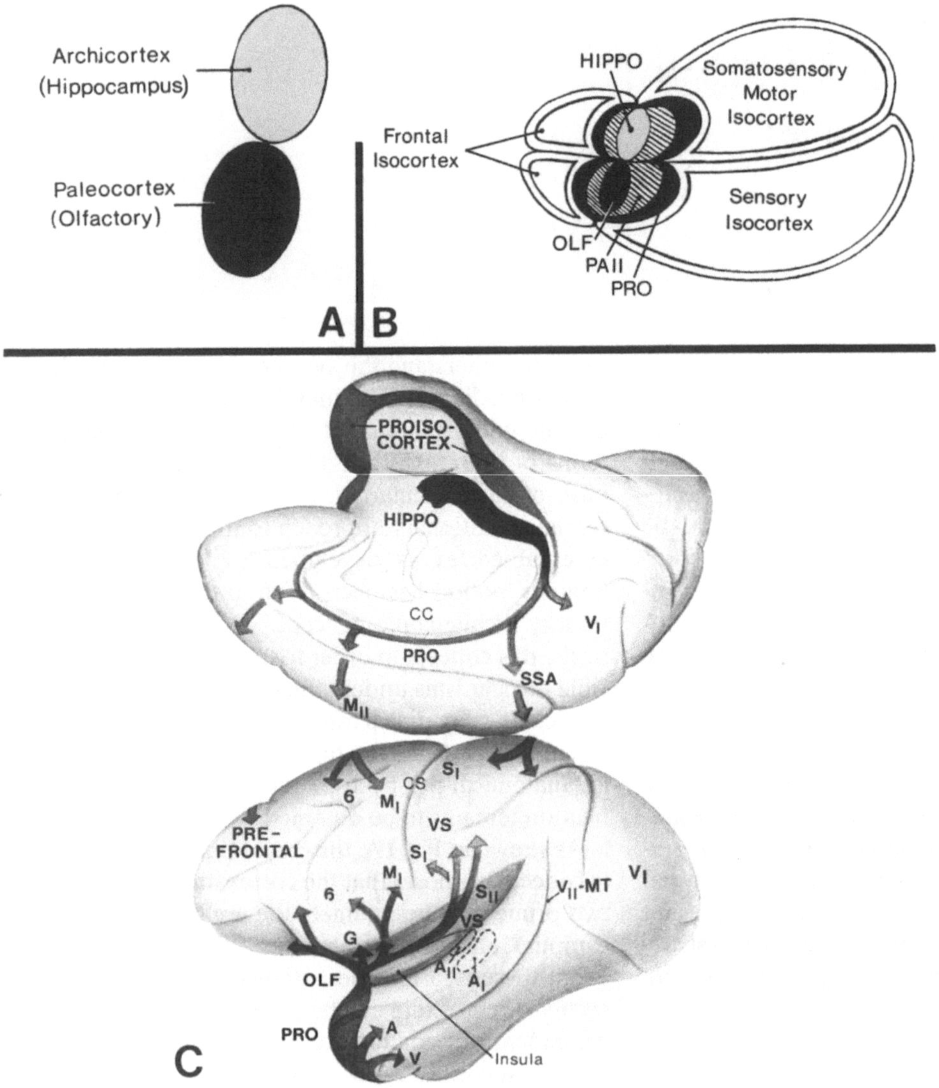

Fig. 1 A–C. Schematic diagrams showing the progessive development of cortical areas from the two primordial moieties (archicortical and paleocortical) through successive steps: periallocortex *(PALL)* to proisocortex *(PRO)* to isocortex, and culminating in pre- and post-Rolandic sensorimotor and association areas. *A*, auditory areas; A_I, primary auditory area; A_{II}, second auditory area; *CC*, corpus callosum; *CS*, central sulcus; *G*, gustatory area; M_I, primary motor cortex; M_{II}, supplementary motor cortex; *MT*, middle temporal visual area; *OLF*, olfactory cortex; *PRO*, proisocortex; S_I, primary somatosensory cortex; S_{II}, second somatosensory cortex; *SSA*, supplementary sensory area; *V*, visual areas; V_I, primary visual area; V_{II}, visual area *MT* in the superior temporal sulcus; *VS*, vestibular area

shows a similar architectonic progression in the visual system (Rosene and Pandya 1983; Pandya and Yeterian 1985). Thus, here one can trace the progression from the proisocortical region to TE1 to TE2 to TE3 to visual areas V4, V3, V2, and, ultimately, V1, the primary visual cortex. This comprises the progression for the central visual field. Similar steps relating to the peripheral visual field begin in the archicortical moiety and continue through the periallocortex to the proisocortex, progressing through the parahippocampal gyrus and ultimately to the medial surface of the occipital lobe. Likewise, from the paleocortical moiety other sensory systems can be traced, including the gustatory, vestibular, and head- and face-related portion of the somatosensory system (Fig. 1 C). The archicortical moiety similarly progresses in sequential steps, through the proisocortex of the cingulate gyrus to the supplementary sensory and motor areas, and eventually to the somatosensory and motor cortices (Fig. 1 C). In summary, then, it is the paleocortical or olfactory moiety that leads to cortical sensory systems, whereas the archicortical, or hippocampal, moiety leads to somatosensory and motor cortices.

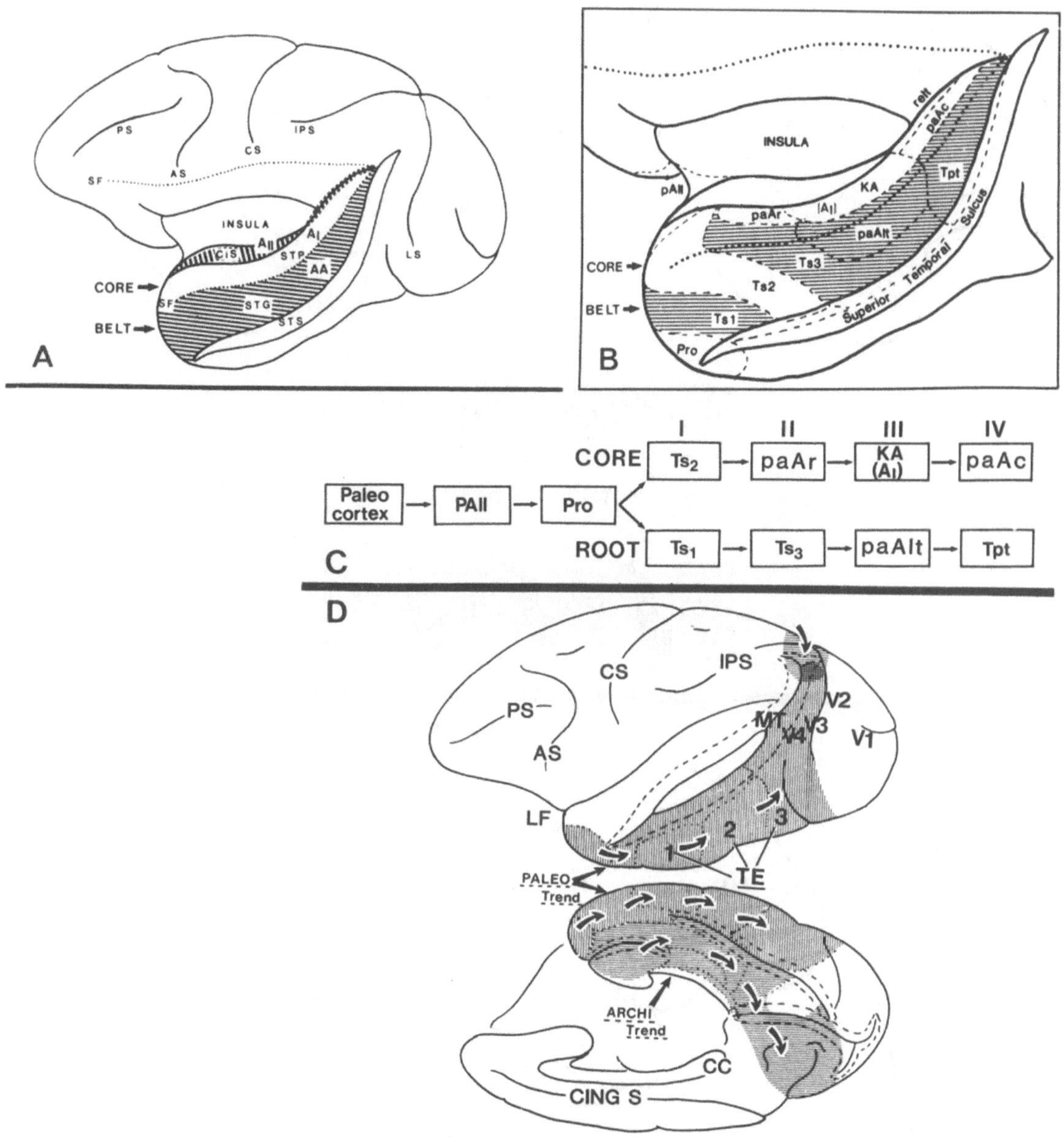

Fig. 2. **A** The lateral surface of the cerebral hemisphere showing two architectonic lines, core and belt, in the superior temporal region of the rhesus monkey. The Sylvian fissure is opened up to expose the insula and banks. The core line is in the supratemporal plane, and the belt line is in the superior temporal gyrus. **B** Enlarged view of the superior temporal region showing architectonic areas within the core and belt lines. **C** Block diagram depicting four successive architectonic stages beginning in proisocortex, each containing core and belt regions. **D** Diagrammatic representation of the two trends in the cortical visual system involving the inferotemporal area (paleocortical trend) and the parahippocampal gyrus, occipital lobe, and posterior parietal region (archicortical trend). *AS*, arcuate sulcus; *CING S*, cingulate sulcus; *CiS*, circular sulcus; *IPS*, intraparietal sulcus; *KA*, auditory koniocortex; *LF*, lateral fissure; *LS*, lunate sulcus; *paAc*, caudal parakoniocortex; *paAlt*, lateral parakoniocortex; *paAr*, rostral parakoniocortex; *Pro*, proisocortex; *PS*, principal sulcus; *relt*, retroinsular temporal cortex; *STG*, superior temporal gyrus; *STP*, supratemporal plane; *STS*, superior temporal sulcus; *Tpt*, temporoparietal cortex; *Ts1*, *Ts2*, *Ts3*, temporalis superior, 1, 2, 3; *V1*, *V2*, *V3*, *V4*, and *MT*, visual areas

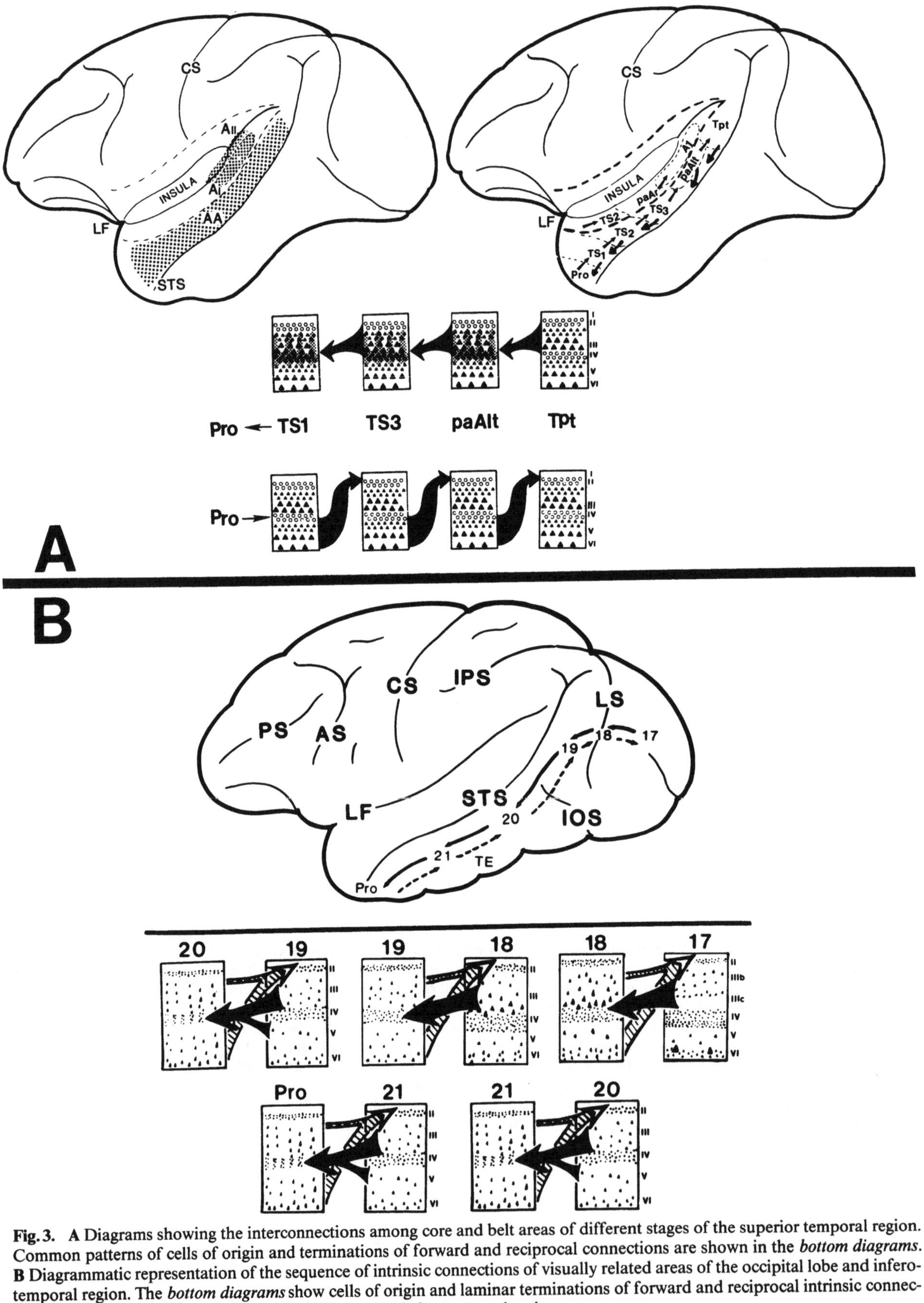

Fig. 3. **A** Diagrams showing the interconnections among core and belt areas of different stages of the superior temporal region. Common patterns of cells of origin and terminations of forward and reciprocal connections are shown in the *bottom diagrams*. **B** Diagrammatic representation of the sequence of intrinsic connections of visually related areas of the occipital lobe and inferotemporal region. The *bottom diagrams* show cells of origin and laminar terminations of forward and reciprocal intrinsic connections of visually related areas of the occipital lobe and inferotemporal region

In recent years, we have traced the connectivity of the subregions of these sensory and motor cortices. These types of connections are termed "intrinsic," since they interrelate adjacent architectonic areas within single modalities. As shown in Fig. 3 A, each region in the auditory system is interconnected in a stepwise manner. Thus, a series of sequential connections extends from the primary area all the way to the proisocortex in the temporal pole. Also, reciprocal connections exist beginning in the proisocortex and leading to the primary region (Galaburda and Pandya 1983; Pandya and Yeterian 1985). The other remarkable feature is in terms of laminar specificity for cells of origin and terminations of these connections. Thus, the forward connections originate from the third layer neurons, and terminate in a columnar manner in rostrally adjacent regions. The reciprocal connections originate from the neurons in layers V and VI and terminate in layer I of caudally adjacent regions. This laminar specificity is evident between all adjacent areas from the proisocortex to the primary cortices. This sequence is present not only in auditory-related areas, but, as shown in Fig. 3 B, exists also as a bidirectional system in visually related areas (Rosene and Pandya 1983; Pandya and Yeterian 1985). Preliminary evidence suggests that a similar sequence occurs in the cortical somatosensory system as well. In summary, then, the architectonic progression which can be followed in each sensory system of the cerebral cortex is paralleled by a systematic bidirectional flow of intrinsic connectivity (Pandya and Yeterian 1985).

Another structure which has a significant role in seizure disorders is the amygdala. Its close ties with several cortical regions would seem to implicate it in different aspects of seizure disorders (Halgren et al. 1978; Wieser 1979; Gloor et al. 1982). In terms of connections, as shown in Fig. 4 A, the amygdaloid nuclear complex receives input from the anterior temporal region, such as the rostral superior temporal gyrus and the inferotemporal cortex, which are related, respectively, to the auditory and visual systems. Several paralimbic regions such as the temporal pole, posterior orbital cortex, anterior cingulate gyrus, and insula also send connections to the amygdala. In addition, the amygdala receives projections from the lateral prefrontal cortex as well as the parahippocampal gyrus, including the perirhinal and entorhinal cortices (Van Hoesen 1981). Many of these connections are reciprocal in nature. Thus, as shown in Fig. 4 B, Amaral and Price (1984) have described widespread projections from the amygdala back to the frontal lobe: areas 24, 25, and 32 on the medial surface; areas 14, 13, and 12 on the orbital surface; and areas 45, 46, 6, 9, and 10 on the dorsolateral surface. The amygdala also projects to certain somatosensory fields, such as areas 3, 1, and 2, and SII, in the pericentral operculum. Other projections of the amygdala are to the temporal pole, area TA of the rostral superior temporal gyrus, and areas TE and TEO of the inferotemporal region. The amygdala also has projections to the prestriate cortex and the depths of the intraparietal sulcus. It should be pointed out that, just as the reciprocal intrinsic connections terminate in superficial cortical layers, these amygdaloid efferent projections terminate in layers I and II of the different cortical regions.

According to recent observations by Saunders and Rosene (1987) and Amaral (1986), the amygdala and the hippocampus are interconnected. Thus, the amygdala projects directly to the hippocampus, and likewise the hippocampus projects back to the amygdala (Fig. 5 A). These investigators have also shown that amygdalohippocampal connectivity could occur indirectly via the entorhinal cortex. Furthermore, as shown in Fig. 5 B, the entorhinal cortex receives input from sensory association areas, frontal lobe association areas, as well as the paralimbic region via the parahippocampal gyrus (Van Hoesen 1982). Considered together, these studies demonstrate that the amygdala and hippocampus are interconnected, and that both have close ties with the neocortices.

Each sensory system relates to limbic areas via long connections (Pandya and Kuypers 1969; Jones and Powell 1970; Van Hoesen et al. 1972; Seltzer and Van Hoesen 1979; Amaral et al. 1983; Mishkin et al. 1983; Pandya and Yeterian 1985). Thus, as shown in Fig. 6 A, the visual association areas of the inferotemporal cortex relate to the parahippocampal gyrus as well as the perirhinal region, with a small contingent to the cingulate gyrus. The rostal temporal visual areas, as mentioned, have systematic connections with the amygdala. Likewise, the auditory association areas are related to the cingulate gyrus, on the one hand, and to the parahippocampal gyrus on the other (Fig. 6 B). Finally, as shown in Fig. 6 C, the somatonsensory system also relates to the limbic region. These connections are to the cingulate gyrus as well as to the parahippocampal gyrus and the presubiculum.

Thus, information from each sensory system ultimately relates to limbic regions via two major routes; one, a "short route", consisting of what we call intrinsic connections within each modality, and another, a "long route," which includes connections between association cortices and paralimbic regions such as the cingulate and parahippocampal gyri.

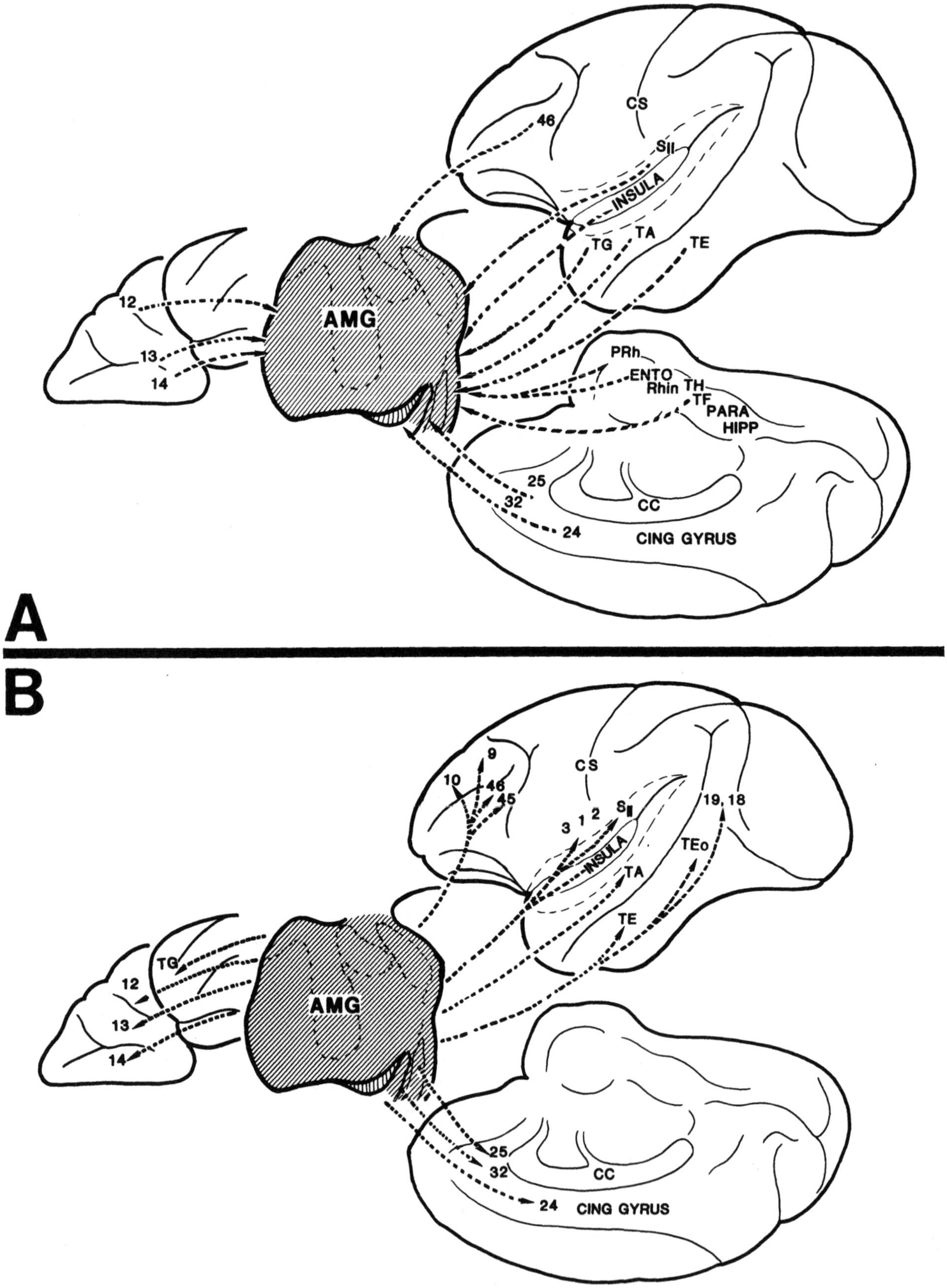

Fig. 4. Summary diagram showing cortical afferent (**A**) and efferent (**B**) connections of the amygdala. *AMG*, amygdala; *CING GYRUS*, cingulate gyrus; *ENTORhin*, entorhinal cortex; *PARAHIPP*, parahippocampal gyrus; *PRh*, perirhinal cortex

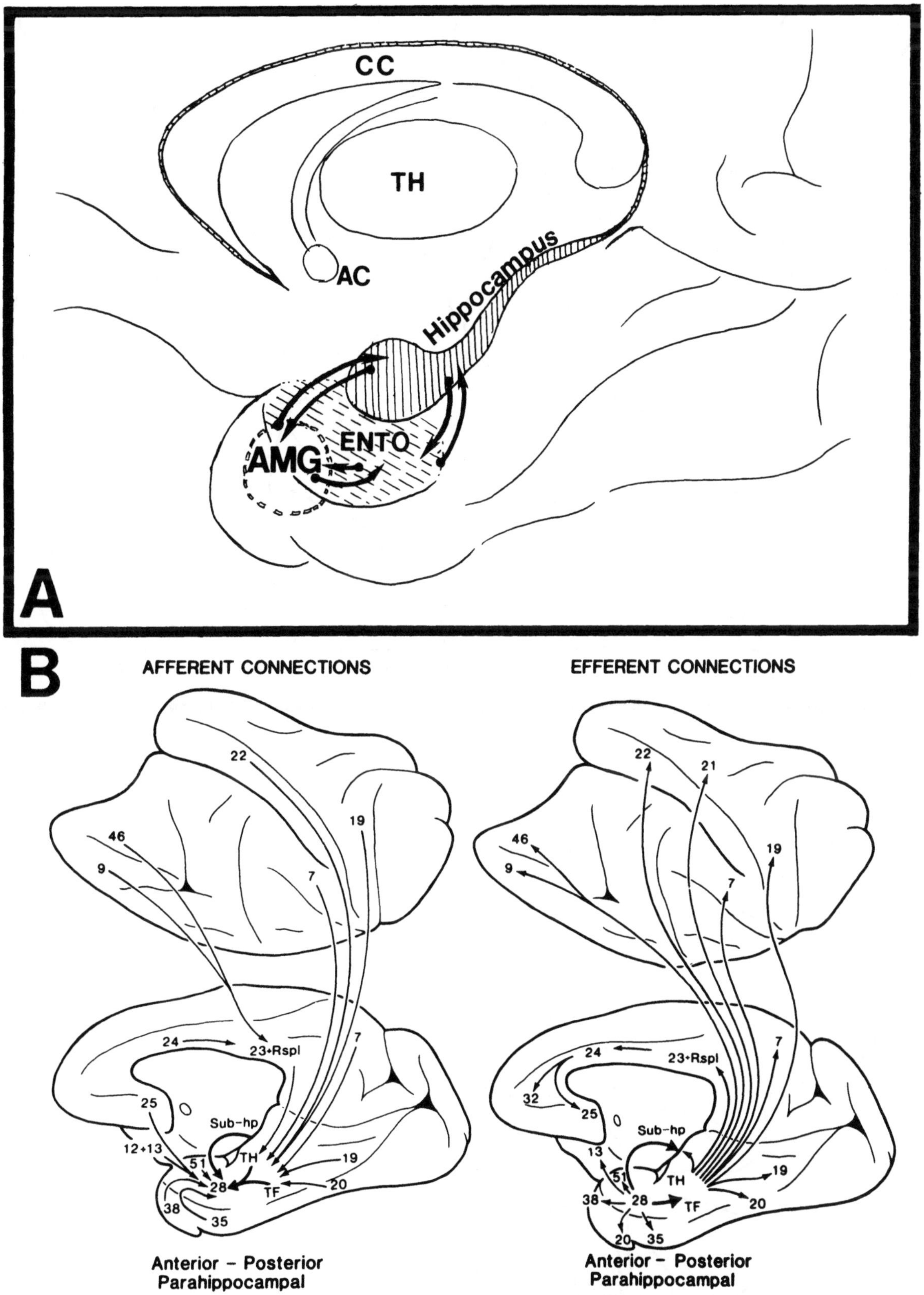

Fig. 5. **A** Diagram of the ventromedial region of the cerebral hemisphere showing interconnections between amygdala *(AMG)* and hippocampus, hippocampus and entorhinal cortex *(ENTO),* and entorhinal cortex and amygdala. **B** Summary diagram of afferent and efferent connections of the parahippocampal gyrus (Van Hoesen 1982). *AC,* anterior commissure; *Rspl,* retrosplenial cortex; *Sub-hp,* subiculum-hippocampus; *TH,* thalamus

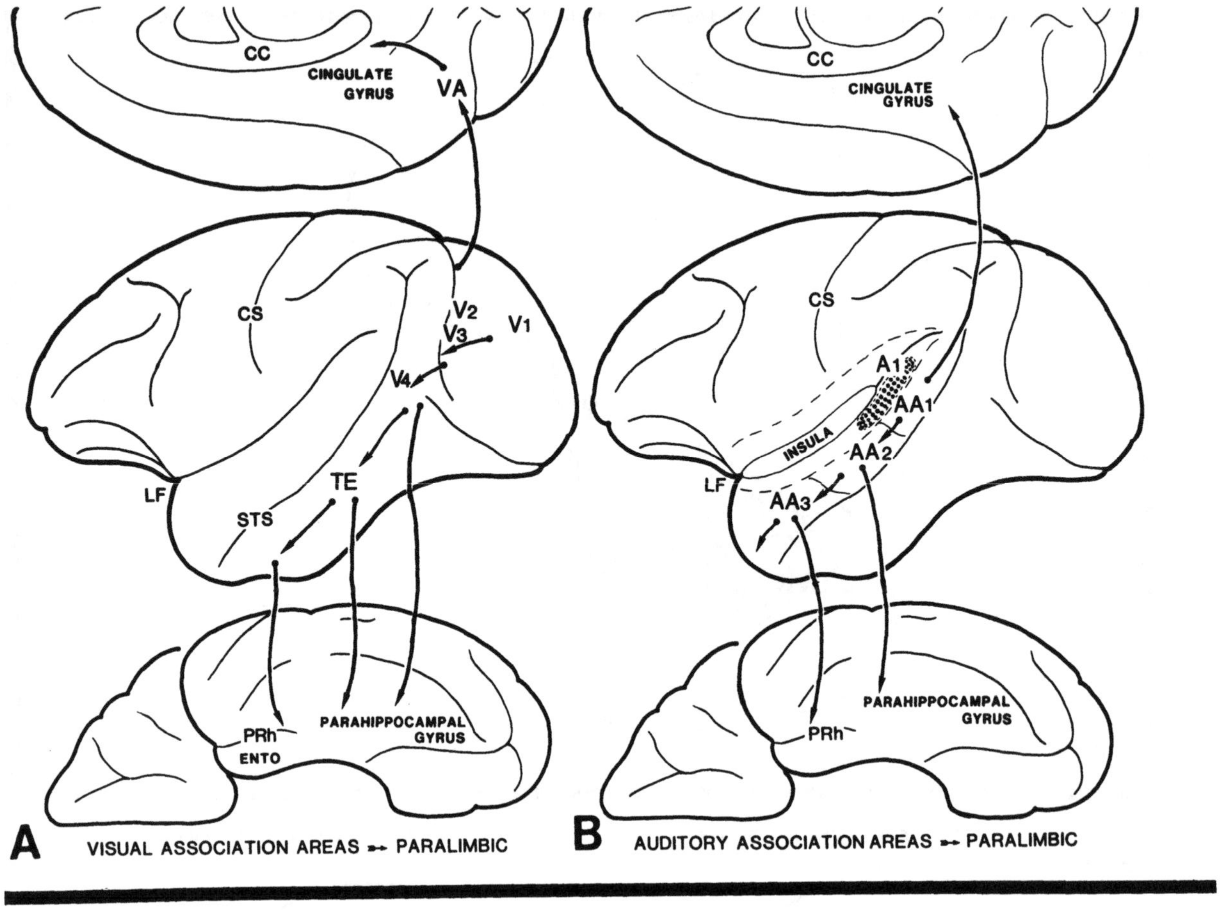

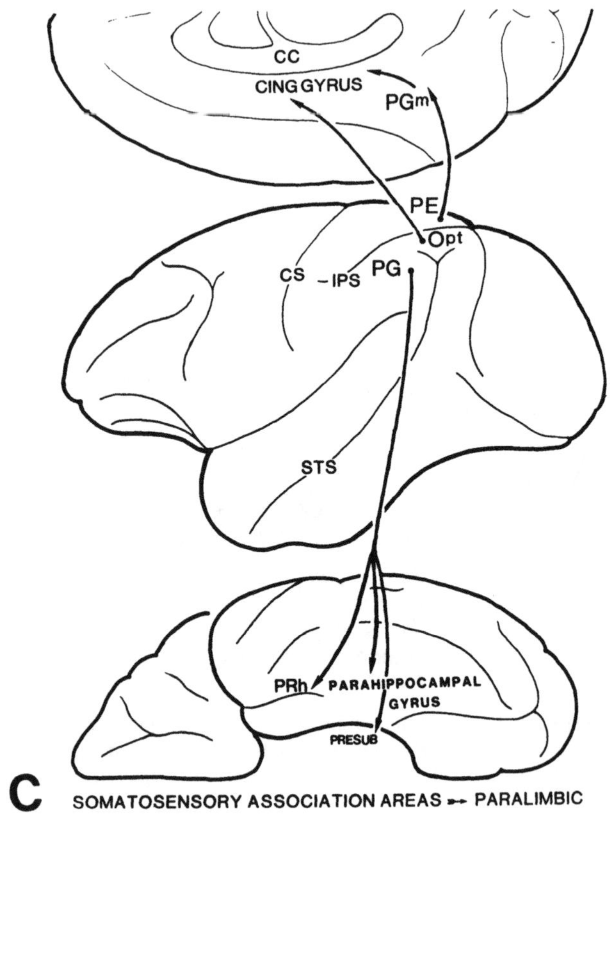

Fig. 6. Diagrams of the cerebral hemispheres showing paralimbic connections of visual (**A**), auditory (**B**), and somatosensory (**C**) association areas

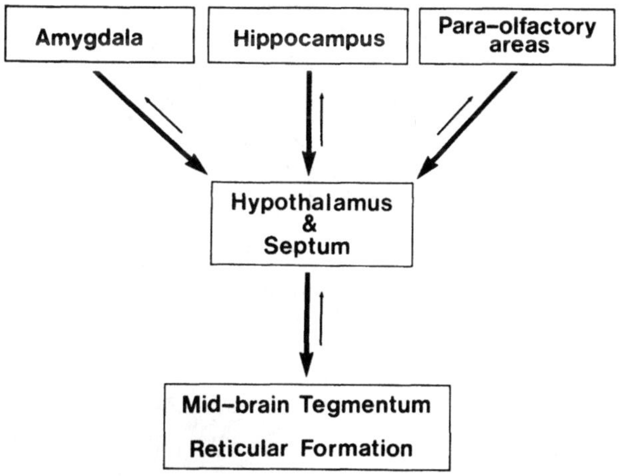

Fig. 7. Summary diagram showing the interconnections between cortical limbic areas (amygdala, hippocampus, and paraolfactory areas) and brain stem regions (mid-brain tegmentum and reticular formation) via the hypothalamus and septum

Moreover, certain cortical regions, in particular the rostral temporal, opercular, and medial and orbital frontal cortices, as well as the cingulate gyrus have a reciprocal relationship with the amygdala (Fig. 4).

It is beyond the scope of this chapter to elaborate upon the intrinsic circuitry of the hippocampus or the relationship of temporal lobe structures with the brain stem. Nevertheless, it is important to note that the amygdala as well as the structures related to the archicortical and paleocortical moieties, namely the hippocampus and paraolfactory regions, relate to the septum and hypothalamus (Nauta and Haymaker 1969). Furthermore, it has been shown that these structures in turn relate to other brain stem regions, including the reticular formation (Fig. 7) (Brodal 1981). Thus, the brain stem structures, in an evolutionary framework, can be considered the precursors of the cortical sensory and motor regions with which they are systematically interconnected.

Discussion

We now wish to examine various ways in which the phenomena of epilepsy may be correlated with the anatomy just described. We would emphasize again that this is primarily an anatomical viewpoint and is a first attempt at relating structures to the dynamic aspects of epilepsy.

It is our view that structurally or otherwise dysfunctional foci in the temporal lobe could disrupt the forward as well as the reciprocal flow of information normally taking place in the cortex. Thus, interictal behavioral manifestations such as viscosity, circumstantiality, paranoia, hyperreligiosity, hypergraphia, sexual dysfunction, and altered emotionality, may all be manifestations of abnormal cortical-limbic associative function. That is, due to a disturbance of the normal flow of information into and out of limbic structures, abnormal associations between stimuli and motivational and emotional states may occur, with resultant changes in personality (Bear 1979; Geschwind 1983). It should be pointed out, however, that the existence of a specific interictal syndrome remains controversial (Rodin and Schmaltz 1984). Nevertheless, such a syndrome would be consistent with the notion of abnormal sensory-limbic interactions.

The auras which may occur just prior to the motor manifestations of epilepsy may be viewed as phenomena related to the described pattern of bidirectional intrinsic connectivity. Thus, without the presence of peripheral sensory input, deep temporal foci may abnormally activate memories stored in the association cortices, leading to various types of auras. As we have pointed out, each modality has intrinsic connections which flow from the proisocortical regions out to the primary sensory areas. These proisocortical areas are connected to both the hippocampus and the amygdala. Therefore, the experiential phenomena observed in temporal lobe epilepsy, such as visual, auditory, olfactory, and epigastric hallucinations, déjà vu, and fear, among others, may be viewed as representing a disruption of the normal relationship between limbic structures and neocortical systems. In fact, Halgren et al. (1978), Wieser (1979), and Gloor et al. (1982) have shown that perceptual (auditory and visual) hallucinations or illusions, memory flashbacks, illusions of familiarity, forced thinking, and emotions can occur with seizures or electrical stimulation of limbic structures.

As we have already shown, the long association connections and the intrinsic connections reach the hippocampus via the entorhinal cortex, on the one hand, and the amygdala on the other. In addition, the amygdala and hippocampus themselves are interconnected. In the case of seizure disorders, the abnormal activity of these structures, especially that generated by the hippocampus, could be transmitted to the cingulate gyrus and ultimately to the supplementary and primary motor cortices via sequential cortical connections. This could result in the manifestation of seizure activity at the motor level.

The role of the brain stem in various seizure phenomena has been amply documented (Penfield and Jasper 1954; Lüders et al. 1984). We wish to point out that both paleo- and archicortical structures are related to the septohypothalamic regions, as is the

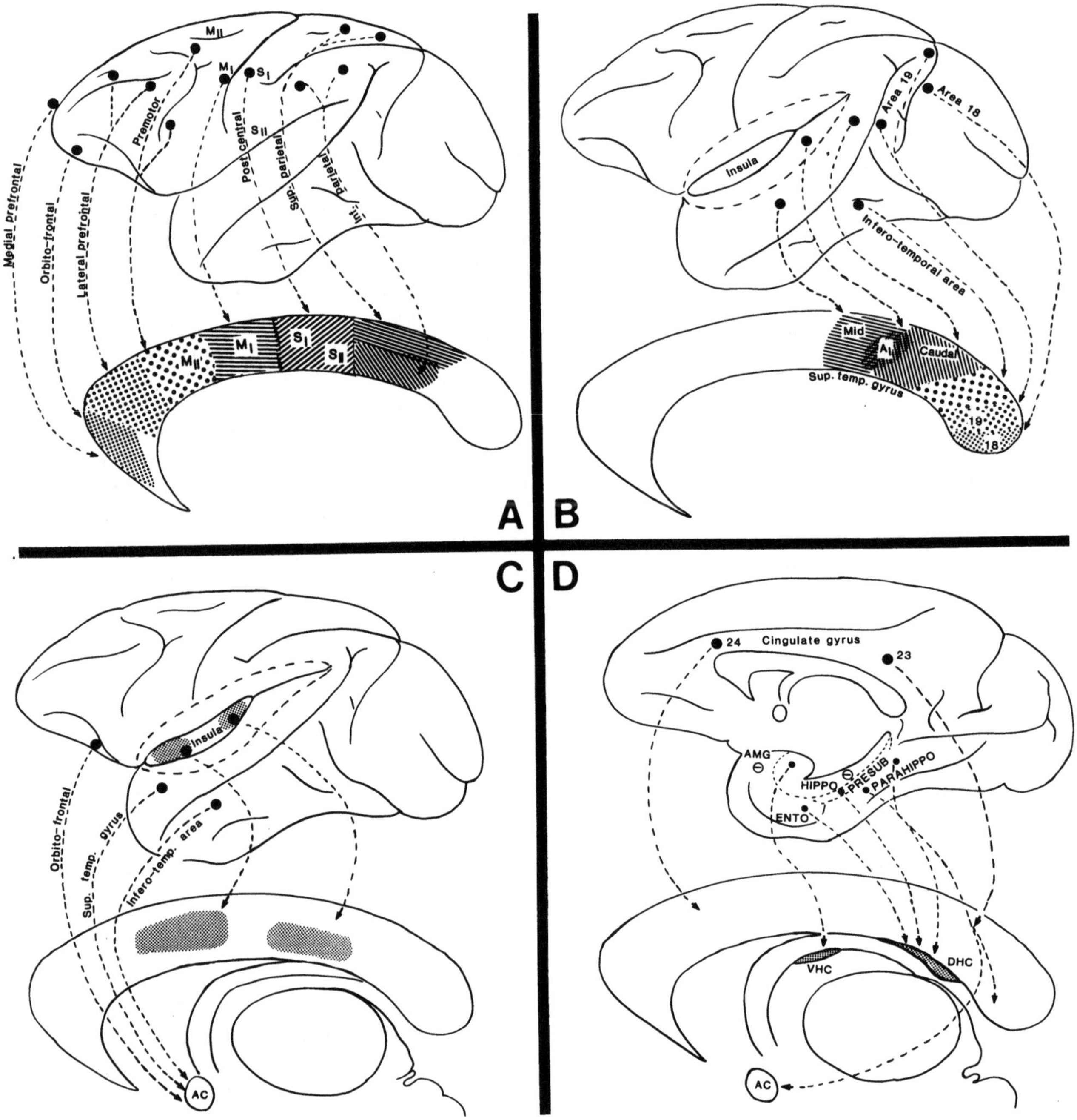

Fig. 8 A–D. Summary diagrams showing the topography and location of the commissural fibers of the corpus callosum, anterior commissure (*AC*), and hippocampal commissures (*DHC* and *VHC*, dorsal and ventral hippocampal commissures)

amygdala. Abnormal activity within this limbic circuitry could produce several of the autonomic phenomena associated with epilepsy.

Commissural Connections

Let us now turn to another aspect of forebrain anatomy. Since our theme here is the evaluation of patients prior to surgical intervention, of which commissural section is an important approach, it is of interest to present the topography of interhemispheric fibers. The cerebral hemispheres are interconnected via three commissures – the corpus callosum, the anterior commissure, and the hippocampal commissures (Demeter et al. 1985; Pandya and Rosene 1985). However, the corpus callosum is by far the most significant, both in size and in terms of the number of cortical areas it interconnects.

The rostral half of the corpus callosum carries fibers from various subdivisions of the frontal lobe (Fig. 8 A). Prefrontal areas send projections through the genu and the most rostral portion of the body of the corpus callosum. In contrast, fibers from the premotor and motor areas cross more caudally, but still in the rostral half of the body. Anterior insular and anterior cingulate fibers appear to overlap with those coming from the premotor and motor areas (Fig. 8 C, D).

The caudal half of the corpus callosum carries fibers orginating in the parietal, temporal, and occipital lobes (Fig. 8 B). Just behind the fibers from the motor cortex are those emanating from the primary and second somatosensory areas (Fig. 8 A). Posterior parietal association fibers are located more caudally in the body of the corpus callosum. Superior parietal fibers are more dorsal, whereas inferior parietal fibers are more ventral (Fig. 8 A). Overlapping these posterior parietal fibers are commissural fibers coming from the caudal superior temporal gyrus (Fig. 8 B). Those from the primary auditory area also cross within the caudal part of the body. Fibers from the posterior insular and posterior cingulate gyrus overlap with those coming from the parietal and temporal lobes (Fig. 8 C, D). Also in the caudal part of the body of the corpus callosum, immediately behind the parietal and superior temporal fibers, is the trajectory of interhemispheric connections originating in the caudal portion of the inferotemporal region. These fibers extend into the splenium (Fig. 8 B). Finally, fibers from the occipital lobe travel through the splenium, with those from area 18 and the juxtastriate region located ventral to fibers from area 19 (Fig. 8 B). Interhemispheric connections from the lateral and caudal portions of the parahippocampal gyrus also pass through the splenium (Fig. 8 D). Thus, the corpus callosum contains topographically organized fibers coming from different cortical regions. However, there is some overlap in the fiber trajectories of the different zones. This is especially true for fibers originating in the posterior superior temporal, insular, and cingulate regions.

The anterior commissure carries fibers from the caudal orbitofrontal cortex and the rostral portions of the superior temporal (including the temporal pole) and the inferotemporal areas (Fig. 8 C). Likewise, fibers from the lateral and rostral sectors of the posterior parahippocampal gyrus travel through this commissure (Fig. 8 D). Thus, unlike the corpus callosum, which contains interhemispheric fibers for all lobes, the anterior commissure appears to contain fibers only from the caudal orbitofrontal and rostral temporal cortices. Regardless of their cortical origins, the fibers in the anterior commissure show considerable overlap.

The dorsal hippocampal commissure (DHC) contains fibers from the presubiculum and entorhinal cortex, as well as the medial portion of the parahippocampal gyrus (Fig. 8 D). According to our observations, no significant commissural fibers arise from the amygdala in primates. With regard to the hippocampus, only the rostral, or uncal, portion appears to give rise to commissural fibers. These fibers reach the opposite hemisphere via the ventral hippocampal commissure (VHC) (Fig. 8 D). Thus, the corpus callosum, anterior commissure, and hippocampal commissures contain fibers originating from allocortex (hippocampus), periallocortex (entorhinal cortex and presubiculum), proisocortex (area TH), and even isocortex (area TF).

At the present time, it is not possible to correlate precisely these anatomical data on the topography of commissural connections with specific functional roles. Nevertheless, these anatomical observations may help in interpreting the effects of experimental and clinical commissural transections as well as clinical observations on lesions that involve the trajectories of commissural fibers. It is of particular interest to consider this topography in the light of the use of callosal sections for the control of intractable seizures. Accordingly, it may be useful to consider partial callosal transection when discrete epileptogenic foci can be identified. Moreover, in cases with seizure foci in the rostral temporal neocortex, section of the anterior commissure should be considered. Thus, by selective transection of the cerebral commissures, it may be possible to reduce the frequency and severity of undesirable side effects.

Acknowledgements. We are highly grateful to Mr. Brian Butler and Ms. Valerie Knowlton for excellent technical assistance, and to Ms. Karen Bourassa for typing the manuscript. We are also thankful to Dr. Douglas L. Rosene and Dr. David G. Amaral for providing pre-publication data on amygdaloid and hippocampal connections.

References

Amaral D (1986) Amygdalohippocampal and amygdalocortical projections in the primate brain. In: Schwarcz R, Ben-Ari Y (eds) Excitatory amino acids and epilepsy. Plenum, New York, pp 3-17

Amaral DG, Price JL (1984) Amygdalo-cortical projections in the monkey *(Macaca fascicularis).* J Comp Neurol 230: 465-496

Amaral DG, Insausti R, Cowan WM (1983) Evidence for a direct projection from the superior temporal gyrus to the entorhinal cortex in the monkey. Brain Res 275: 263-277

Bear D (1979) Temporal lobe epilepsy: a syndrome of sensory-limbic hyperconnection. Cortex 15: 357-389

Brodal A (1981) Neurological anatomy in relation to clinical medicine, 3rd edn. Oxford University Press, Oxford

Demeter S, Rosene DL, Van Hoesen GW (1985) Interhemispheric pathways of the hippocampal formation, presubiculum, and entorhinal and posterior parahippocampal cortices in the rhesus monkey: the structure and organization of the hippocampal commissures. J Comp Neurol 233: 30-47

Galaburda AM, Pandya DN (1983) The intrinsic architectonic and connectional organization of the superior temporal region of the rhesus monkey. J Comp Neurol 221: 169-184

Geschwind N (1983) Interictal behavioral changes in epilepsy. Epilepsia 24 [Suppl 1]: 523-530

Gloor P, Olivier A, Quesney LF, Andermann F, Horowitz S (1982) The role of the limbic system in experiential phenomena of temporal lobe epilepsy. Ann Neurol 12: 129-144

Halgren E, Walter RD, Cherlow DG, Crandall PH (1978) Mental phenomena evoked by electrical stimulation of the human hippocampal formation and amygdala. Brain 101: 83-117

Jones EG, Powell TPS (1970) An anatomical study of converging sensory pathways in the cerebral cortex of the monkey. Brain 93: 793-820

Lüders H, Lesser RP, Dinner DS, Morris HH III (1984) Generalized epilepsies: a review. Cleve Clin Q 51: 205-226

Mishkin M, Ungerleider LG, Macko KA (1983) Object vision and spatial vision: two cortical pathways. Trends Neurosci 6: 414-417

Nauta WJH, Haymaker W (1969) Hypothalamic nuclei and fiber connections. In: Haymaker W, Anderson E, Nauta WJH (eds) The hypothalamus. Thomas, Springfield, pp 136-209

Pandya DN, Kuypers HGJM (1969) Cortico-cortical connections in the rhesus monkey. Brain Res 13: 13-36

Pandya DN, Rosene DL (1985) Some observations on trajectories and topography of commissural fibers. In: Reeves AG (ed) Epilepsy and the corpus callosum. Plenum, New York, pp 21-39

Pandya DN, Yeterian EH (1985) Architecture and connections of cortical association areas. In: Peters A, Jones EG (eds) Cerebral cortex, vol 4 Association and auditory cortices. Plenum, New York, pp 3-61

Penfield W, Jasper H (1954) Epilepsy and the functional anatomy of the human brain. Little, Brown, Boston

Rodin E, Schmaltz S (1984) The Bear-Fedio personality inventory and temporal lobe epilepsy. Neurology 34: 591-596

Rosene DL, Pandya DN (1983) Architectonics and connections of the posterior parahippocampal gyrus in the rhesus monkey. Soc Neurosci Abstr 9: 222

Sanides F (1969) Comparative architectonics of the neocortex of mammals and their evolutionary interpretation. Ann NY Acad Sci 167: 404-423

Saunders RC, Rosene DL (1987) A comparison of the efferents of the amygdala and the hippocampal formation in the rhesus monkey. I. Convergence in the entorhinal, prorhinal and perirhinal cortices. J Comp Neurol, (in press)

Seltzer B, Van Hoesen GW (1979) A direct inferior parietal lobule projection to the presubiculum in the rhesus monkey. Brain Res 179: 157-161

Van Hoesen GW (1981) The differential distribution, diversity and sprouting of cortical projections to the amygdala in the rhesus monkey. In: Ben-Ari Y (ed) The amygdaloid complex. Elsevier/North Holland Biomedical, Amsterdam, pp 77-90 (INSERM symposium no 20)

Van Hoesen GW (1982) The parahippocampal gyrus. Trends Neurosci 5: 345-350

Van Hoesen GW, Pandya DN, Butters N (1972) Cortical afferents to the entorhinal cortex of the rhesus monkey. Science 175: 1471-1473

Wieser HG (1979) Ictal "psychical phenomena" and stereoelectroencephalographic findings. In: Lechner E, Aranibar A (eds) EEG and clinical neurophysiology: proceedings of the second European congress on EEG and clinical neurophysiology, Salzburg, Austria. Excerpta Medica, Amsterdam, pp 62-76 (International congress series no 526)

The Role of Limbic and Related Structures in the "Making" of Behavior: The Lessons from Animal Experimentation

W. P. Koella

Introduction and Methodology

Proper recognition and interpretation of the behavioral symptoms characterizing seizures originating in, and often confined to, the various components of the limbic system is not possible without at least some insight into, and understanding of, the role played by these structures in the organization of behavior as a whole. Yet, before the mid-thirties, there was little evidence available about the physiology of *Broca's grand lobe limbique*. It was eventually the pioneering work of Klüver and Bucy (1937, 1939) that gave the impetus for an ever-expanding series of investigations that produced the evidence which enables us today to make at least some educated guesses, if not even some safe statements, about the physiological functions of this previously so mystical conglomeration of nuclei, fiber tracts, and cortical structures.

All these investigations, including Klüver and Bucy's original work, made use of one or more of the traditional methods of neurophysiology, i.e., *lesioning, stimulation,* and *recording*. These were further amended by the more modern techniques of neuropharmacology and -biochemistry.

With the *lesion method* one reduces or completely eliminates activity in, and thus output from, a particular smaller or larger structure. Or, by transection of a particular connection, transmission from one structure to the other is eliminated. After making the lesion, one proceeds by observing changes in the various behaviors belonging to the (behavioral) armamentarium of a given species. Thus, one obtains insight into the necessary - yet certainly not sufficient - part played by the lesioned, i. e., "de-efferented and -afferented," structure in the making of a particular extroversive or an indirectly revealed internal, i.e., covert, behavior. In the "classical" era of neurophysiology, such lesions were made (rather indiscriminately) through electrolysis or, for the interruption of pathways, with the knife, as well as by local cooling, infusion of local anesthetics, or induction of spreading depression. In the modern times of mainly "wet" - i. e., neurotransmitter- (NT-)related - neurophysiology, lesions involving more or less specifically only one NT system or pathway can be produced by, e. g., local infusion of neurotoxins or, in a more transitory fashion, through local or systemic application of (post-synaptic) NT-receptor-blocking drugs, or by agents inhibiting NT-release and/or-synthesis.

With the *stimulation method* one enhances artificially the activity in, and thus output from, a particular structure. Again one observes, whenever possible in freely behaving animals, changes in, or the induction of, a particular behavior. Local activation was, and still is, achieved by electrical stimulation through implanted electrodes. For the activation of particular NT-pathways within, leading to, or emerging from a given structure, one can apply locally the neurotransmitter (to interact with its specific receptors); one can administer (systemically or locally) its precursor and thus drive NT-synthesis; one can enhance, through block of presynaptic receptors, NT-release (as in the case of noradrenergic transmission); or one can inhibit NT re-uptake.

Recording can be done by locally picking up the EEG, single cell activity, or, as done mainly in the case of pathway activity, one can use the evoked potential method. While recording the activity of a given structure or a multitude of them, one observes again the animals and is able to correlate certain behaviors (or blocking of such) with a particular (e. g., electrical) activity in a given structure or a multitude thereof. Applying the "wet" techniques, one can put the animal (quickly) to death while it exhibits a certain behavior and study in the excised CNS the activity of a variety of NTs in a variety of structures. By the use of push-pull cannulae, or of the recently developed method of voltametry, one is able to study NT-activity while the animal performs certain behaviors or "nonbehaviors."

We recently reviewed, in a rather extended fashion, these experimental findings (Koella 1982). In

Buchenstr. 1, 4104 Oberwil, Switzerland

Presurgical Evaluation of Epileptics
Edited by H. G. Wieser and C. E. Elger

this chapter, the same material (supplemented by some recent citations) is presented, though in a somewhat shortened manner. We hope we are forgiven this flagrant case of autoplagiarism.

Experimental Findings

Changes in Aggressive Behavior, Fear, Sexual Behavior, and Oral Exploration: The Klüver-Bucy Syndrome

In their classical experiments, Klüver and Bucy (1937, 1939) placed bilateral lesions in the temporal lobe area of rhesus monkeys. These lesions involved the uncus, the amygdala, parts of the hippocampus and, through anterograde and retrograde degeneration, a number of efferent as well as afferent pathways from and to these structures. All animals developed in the postoperative period (often extending over several years) a set of cardinal symptoms: *tameness* (i. e., loss of fear and aggressiveness), *hypersexuality* (i. e., excessive sexual drive and indiscriminate sexual behavior), and *excessive oral exploratory activity.* In addition, the operated animals revealed changes in their *dietary habits,* signs of *visual agnosia,* and *excessive reactions to visual stimuli.* Quite evidently, the structures affected in Klüver and Bucy's animals can be assumed to be directly or indirectly involved in the making and control of such behavioral components as fear, aggression, sexual drive and activity, and the oral aspects of exploratory patterns as well as, probably, visual recognition and taste for food.

Later investigators, intent on delineating in more detail and in a more selective manner the nuclei, cortical areas, and pathways responsible for the organization of these behavioral components, used not only the lesioning but also the stimulation and recording techniques. Schreiner and Kling (1956) bilaterally destroyed the whole of the amygdalar complex in a mountain lion and a series of domestic cats. They found tameness to be one of the main effects of this surgical intervention. Wood (1958), however, observed that bilateral lesions restricted to the central amygdalar nucleus resulted in an increase of aggressive behavior of cats. In the study of Green et al. (1957), cats became more "placid" after lesions restricted to the amygdala proper; however, such lesions were found to produce aggressive behavior ("rage") when they expanded onto the hippocampus. Furthermore, hippocampectomized rats were found to be impaired in their association of a neutral stimulus with the response elicited by a noxious stimulus (Olton and Isaacson 1969). Rats can be made aggressive also by septal lesions (King and Meyer, 1958) or by ablation of the olfactory bulb (Vergnes and Karli 1969). According to Rosvold et al. (1954), amygdalectomy leads to a drop in *social dominance* of monkeys. Jonason et al. (1973) placed sequential septal-amygdalar or amygdalar-septal lesions in rats. When the septal lesion was done first, social cohesiveness increased and then dropped following the amygdalar lesion. In turn, cohesiveness was reduced by an initial amygdalar lesion and then was brought back to control level after the septum was destroyed. Hess (1949) electrically stimulated in freely moving cats the perifornical area of the diencephalon and invariably produced a behavior having all the characteristics of *"rage,"* including growling, hissing, attack or flight, and such autonomic "symptoms" as piloerection, extreme mydriasis, tachycardia, and sweating on the foot pads. Hunsperger (1956), through electrical stimulation of a vast variety of areas, delineated a neural system, extending from the midbrain to the anterior diencephalon, which seems to be involved in the initiation and elaboration of emotional behavior - the *affective defense reaction* - including rage, aggression, and flight. On the basis of his experience with more than 500 such experiments, Hunsperger still is of the opinion that the perifornical area is the site with the lowest threshold if it comes to elicit such behaviors by electrical stimulation. Wood (1958) found that electrical stimulation of the lateral amygdaloid nucleus of the cat failed to induce emotional behavior, whereas stimulation in the basal nucleus was followed by responses signaling fear and/or anger. Adamec (1976) showed that electrical stimulation of the medial and ventromedial hypothalamus of cats stops predatory attack.

Evidently, fear, aggression, tameness, and social cohesiveness can be manipulated through lesioning or stimulation of a variety of sites in the limbic system. In fact, there cannot be a single "center" that is involved in the organization of these moods or drives, and one must assume that a particular level of affective behavior is the result of activity in, and of interaction between, a number of widely spaced structures and substructures. This assumption is well substantiated by the behaviors produced by simultaneous (or sequential) manipulation of two or more structures. The study of Jonason et al., mentioned above, can be seen as an example of such interaction. Furthermore, Egger and Flynn (1963) noted that electrical stimulation of the basal amygdala, and of the anterior and the medial portion of the lateral amygdala, suppressed the attack behavior produced in cats by electrical stimulation of the hypothalamus and that lesions in these areas facilitated this behavior. In turn, stimulation of the dorsolateral part of the posterolateral nucleus enhanced the hypothalamically induced attack behavior, whereas lesions in these structures inhibited such behavior. Hypothalamically induced attack behavior also can be inhibited by simultaneous electrical stimulation of the ipsilateral prefrontal cortex (Siegel et al. 1975).

Wood (1958) noted that in male cats *hypersexuality* can be produced by lesions restricted to the lateral area of the amygdala. Similar observations were made by Kling and Schwartz (1961). Green et al. (1957) found that removal of the piriform cortex (another limbic structure) of cats leads to changes in sexual behavior. Kling and Schwartz (1961) observed that hypersexuality induced in male cats by amygdalar lesions can be abolished by lesioning the septal area. This suggests that the latter structure is specifically involved in the activation of sexual behavior. MacLean and Ploog (1962) showed that in the squirrel monkey penile erection is produced by electrical stimulation of the septal area. This very same region seems to play a role too in the sexual behavior and activity of female animals; Sawyer (1962) noted electrographic signs of excitation in the septal region of female rabbits coincident with orgasmic activity.

Klüver and Bucy's monkeys also exhibited, as already mentioned, signs of *exaggerated oral exploratory behavior.* They tended to examine objects, even inedible ones, directly by mouth, instead of picking them up first with their hands. Hess (1949) noted oral tendencies, such as licking, chewing, salivation, forced eating (bulimia), and excessive sniffing in response to electrical stimulation in the preoptic, the septal, and the anterior hypothalamic regions. Ursin and Kaada (1960) remarked on oral signs, such as licking and chewing, as stimulation-induced effects in the amygdala.

A Limbic Reward System

In their "classical" experiments, Olds and Milner (1954) demonstrated that rats given the opportunity to switch on the stimulating current to excite certain of their brain structures quickly learned to "push the lever" with little fatigue for many hours. The animals appeared to experience pleasure from this kind of self-inflicted stimuli. For the rewarding experience of *intracranial self-stimulation* (ICSS) the animals were found to be ready to forfeit food and water, to forget about sex, and to accept pain - the crossing of an electrified grid in the path to the lever. In the initial experimental series, Olds and Milner noted that the stimulating electrodes had to be placed in, or in the neighborhood of, limbic structures - in particular in the medial forebrain bundle, the septal area, or the amygdala - to be "ICSS -active." When the electrodes had been placed outside these structures no reinforcement could be obtained. Therefore these authors in fact suggested that ICSS involves Nauta's mesencephalic limbic system with its rostral projecting extension (1958). An additional indication that the limbic system participates in ICSS was found by Ito and Olds (1971); they noted that self-stimulation periods were always attended by reduced firing of units in the hippocampus, but enhanced firing of units in the cingulum.

In a critical review, Routtenberg (1975) suggested that the "brain stimulation reward system may function as a memory consolidation system." According to Goddard (1964) the amygdala is involved in the suppression of motivated approach behavior; amygdalectomized animals have difficulties in associating fear and avoidance behavior with previous neutral stimuli. Along with this view, Jacobs and McGinty (1972) found unit activity in the basolateral amygdala to be positively correlated with behavioral inhibition. Ursin and Kaada (1960) suggested that the amygdala plays an important role for the organization of attentive behavior.

Concerning "wet" transmission instruments for the organization of reward behavior, the following suggestions and observations may be mentioned; Crow (1972, 1977), Routtenberg (1975), Routtenberg and Malsbury (1969), German and Bowden (1974), and Wise (1981) made a strong point for the participation of limbopetal *catecholaminergic pathways* in the organization of ICSS behavior. Stein and Wise (1969) and Arbuthnott et al. (1971) observed enhanced release of norepinephrine in such areas as the preoptic region and the amygdala during self-stimulation behavior. In a later article, Wise (1978) critically reviewed the many pieces of evidence supporting the notion of an important role of catecholaminergic pathways in the organization of self-stimulation. Collecting the already substantial evidence, Stein and Belluzzi (1979) discussed the possibility that (limbic) opiate pathways may also be involved in ICSS behavior and, as a logical consequence, in the organization of reward.

The Limbic System, Memory, Extinction, Conditioning

Grave defects in the *memory functions* have been observed after resection of the medial temporal lobe of monkeys (Orbach et al. 1960). Grastyán and Karmos (1962), after having placed bilateral lesions in the hippocampus of cats, noted a marked impairment of the performance of delayed reactions. Kaada et al. (1962) observed slowing in maze learning after bilateral lesions had been placed in the hippocampus of cats. Grastyán et al. (1959) found the 4–7/s hippocampal waves of the cat to be more pronounced in the early stages of conditioning, whereas a desynchronized pattern rather attended the later stages of the learning process. It is also well recognized that *theta-activity* in the hippocampus attends the approach phase of conditioned responses (Adey and Walter 1963). Niki (1967) noted that the ablation of the hippocampus of rats is followed by a disinhibition of a previously extinguished bar pressing and by a decrease in the rate of response to nonreinforced stimuli. Furthermore, such ablations were also found to impair position reversal learning, to produce a severe deficit in single alternation learning, and to slow down habituation of the cardiac response to unconditioned stimuli.

Olton et al. (1978) observed that the destruction of the entorhinal area, of the fimbria-fornix system, of the septum, or of the postcommissural fornix is followed by a marked and persistent impairment of the *spatial discrimination* performance. According to these authors, the hippocampus with its various afferent systems plays an "important role in processing information about spatial location." O'Keefe and Dostrovsky (1971) and Black et al. (1977) inferred from their investigations with hippocampal lesions that "animals with such lesions cannot process information about places, and therefore cannot employ place strategies in avoidance learning and punishment situations." O'Keefe and Black (1978) with their recordings of single cell activity in the hippocampus, in response to sensory signals, produced additional evidence supporting the notion of a *hippocampal cognitive map*.

The effect of various lesions in the septo-hippocampal system upon extinction, after continous or partial reinforcement, was investigated by Gray et al. (1978). In the view of those authors, the septo-hippocampal complex constitutes a *behavioral inhibitory system*. Landfield (1977) studied in rats the effect of driving or blocking of the hippocampal theta-rhythm (by low- and high-rate stimuli, respectively, applied to the septum) on retention. He found that theta-driving improved retention performance in a one-way active avoidance task; this in contrast to animals receiving high-rate stimulation or nonstimulated controls. In a one-way passive avoidance task, theta-driven animals performed better than blocked ones. Landfield suggested those results to indicate that theta activity is associated with memory storage processes.

Using the microelectrode recording technique, Rolls (1981) found neurons in the monkey's extreme anterior thalamic area to respond to visual stimuli *only* when these were familiar. *Recognition memory* seems to involve this particular diencephalic region which, in fact, represents a way station within Papez' (1937) limbic circuit. Rolls' experimental results may well constitute a connecting piece of information toward better understanding of the visual agnosia in Klüver and Bucy's monkeys.

Effects in the Autonomic-Vegetative Sphere

Kaada (1951), in his extensive work on the limbic forebrain, observed manifold effects in the autonomic sphere in response to stimulation of these structures. These included changes in blood pressure, heart rate, respiration, intestinal activity, pupillary size, and piloerection (the latter two already mentioned also in connection with the stimulation-induced "rage"). Hess (1949) had noted such effects in response to electrical stimulation in the preoptic and septal areas. Of particular interest in the present connection are such "autonomic" effects – also demonstrated by Hess – as defecation and micturition in adequate posture and accompanied by such behavioral activities as "securing" the surrounding space against possible danger. Kawakami et al. (1973) have found that through electrical stimulation of the amygdala and the dorsal hippocampus, one can affect the serum concentrations of luteinizing and follicle-stimulating hormones, LH and FSH.

Behavior of Animals During Limbic Seizures

Of special interest in the context of the present volume are the behavioral manifestations of experimentally induced seizures in, and confined to, the limbic structures. According to Kaada (1951), such seizures – if, for instance, triggered in the hippocampus by an electrical stimulus – are usually restricted to that structure or, at most, may invade the limbic cortex. They occur, as MacLean (1954) writes, "without discernible alteration of the neocortical activity." In our own work with experimentally induced hippocampal seizures in cats (employed as a "chronic" experimental model for the testing of anticonvulsant agents), we always noted that such seizures may eventually generalize if the stimulating electrical current markedly exceeds the a priori low threshold value. Liberson and Akert (1955) also stressed the low seizure threshold of the hippocampus. Those authors noted that the behavior of the cats or guinea pigs under such attacks (if not spreading to other parts of the brain) was rather "unimpressive." In our own experiments in cats (Schmutz et al. 1981), carefully controlled by electrical recordings for "non-spread," we noted the following "spontaneous behavioral symptomatology": during stimulation and up to the end of the limbic seizure, the animals often showed an *arrest reaction* accompanied by a "staring, absence-like" look, the pupils being widely opened. Sometimes the cats turned their heads as though they were looking for the source of an unknown stimulus. Some seconds after cessation of the attack, the behavior of the animals returned to normal, often preceded by meowing and/or grooming movements. After repeated stimulations in the same area, licking, salivation, and some facial twitches may occur.

MacLean's (1954) description of the behavioral phenomenology - particularly in response to manipulations - deserves to be cited verbatim:

More dramatic changes are seen during the afterdischarge which follows (hippocampal) stimulation of greater strength. Any turning movements that were present during the initial stimulation appear to reverse themselves with the onset of the afterdischarge, and the pupils may suddenly dilate. Purring, if previously present, may cease and be replaced by occasional meows or yowls. Concurrently the animals assume attitudes that strike one at first as being rapt attention or fearful alerting for the unexpected. Further examination indicates, however, that the animal is poorly in contact with his environment. Although the pupils react to light, the animal will not avoid the light, nor will he cringe when one pretends to strike the face. If one forcefully blows smoke at him, he will withdraw a little, but will not avoid the smoke. A burning cigarette may be brought up to his nose with no response; but if there is momentary contact with the lighted end, the animal may lunge about wildly. The emotional changes that one may induce during an afterdischarge are most interesting: if one takes an animal which, prior to stimulation, simply meows when the tail is pinched, one finds that the same noxious stimulus will cause it to hiss, spit, and strike out during the afterdischarge. A prolonged or repeated stimulus may lead to states of wild excitement, accompanied by spitting and hissing. I have held on to the tip of the tail of such an animal and observed a state of rage in which he would not properly orient his attack, but rather would viciously bite at the stump of the tail. It is striking that these excited states may suddenly terminate in rather prolonged catatonic-like stances. A meow or series of meows usually signals the end of the afterdischarge.

Interpretation

This brief review of the neuropsychophysiology of the limbic system was not intended to be complete. We rather selected, from a large body of experimental studies, a (relatively) small number of typical results, sufficiently specific to allow characterization of the behavioral (and neural) effects obtained by "manipulations" of activity in well-defined structures of this system. With such data, supplemented in part by results from recording experiments, we attempted to collect the information necessary to facilitate, based on proper interpretative procedures, our understanding of the *principles of operation*, as well as the *functional role* played in terms of behavioral organization, by these long ill-understood structures of the brain. And it is hoped that by an improved insight into the "normal" neuropsychophysiology of the limbic system, we are able to better interpret the *pathoneuropsychophysiology* of these structures; for instance, the symptomatology and a number of pathogenetic aspects of temporal lobe epilepsy.

Many, yet not all, of the effects produced by manipulations of single limbic structures involve *higher functions*; i.e., a variety of "internal behaviors," manifested by respective extroversive activities. They signal such changes in mood as anger, loss of fear, anxiety, reward, pleasure; changes in what may be referred to as "motivation," and in attention; loss of contact with the surrounding space; changes in the various memory functions, in cognitive abilities, and in recognition; and changes in the various aspects of orienting behavior. But they also seem to involve such *lower functions* as sexual drive and behavior, micturition, and defecation. And they often signal changes in *autonomic activities*, as evidenced by variations in heart rate, respiration, pupillary size, piloerection, salivation, etc.

Most of the effects of limbic stimulation are complex in nature. Thus, for instance, the "rage" produced by electrical stimulation of the perifornical area includes, besides the obvious "internal anger", such well-coordinated motor activities as vocalization, specific stance, attack, striking out, or flight (depending on the momentary situation); and such signs of heightened sympathoadrenal activity as piloerection, mydriasis, and accelerated pulse. The pleasure-seeking activity induced by intracranial self-stimulation requires not only adequate and repeated motor performance (key pressing, approach to the lever), but also active suppression of other behaviors, e.g. satiation of thirst, hunger and sexual drive, and avoidance of pain, that potentially could reduce reward. Micturition and defecation, when induced by limbic stimulation, involve not only the autonomic effector organs (contraction of the smooth muscles in the wall of the respective recepticles and relaxation of the smooth and striated muscular sphincters), but also, as under natural conditions, a variety of postural sensorimotor mechanisms, and a heightened vigilance in those cognitive systems that provide proper protection from external danger.

Evidently, activation of even well-confined foci within the limbic system is capable of producing, to some extent, a *whole package of activities*. Unlike, for instance, the cerebral motor cortex, from which one can influence single muscles or, at best, small groups of muscles, in the sense of a point-to-point projection, the limbic structures almost invariably control "complete" (or close to complete) behaviors, consisting of a set of behavioral components with the necessary activation or inactivation of a vast variety of effector organs.

Furthermore, a variety of natural behaviors per se, or components thereof, are eliminated by *lesioning* certain parts of the limbic system. This again strongly indicates - and supports the notion derived from the stimulation experiments - that these parts are intimately involved in the organization and activation of whole behaviors or, at least components thereof. In turn, one notes that surgical or pharmacological elimination of well-delineated limbic structures is often followed by the appearance of "new" behaviors or behavioral components of a somehow exaggerated quantity and quality. This suggests that such structures normally exert an inhibitory influence on such behavioral components; the emerging symptoms of such lesions then must be interpreted as being the manifestation of a release mechanism.

Of considerable interest also is the experience that in many instances a particular effect is inducible, with only minor variations in quality and quantity, by manipulation of not only one anatomically well-defined area, but by stimulation or lesioning of *several different* and often spatially widely separated structures. Aggressive behavior can be triggered by lesioning the septum, the olfactory lobe, or the amygdala (together with parts of the hippocampus). Furthermore, a behavior, e.g., "rage" elicited by, e.g., stimulation of the perifornical area, can be up- and downmodulated by simultaneous stimulation of a variety of amygdalar sites, as well as by strategically placed lesions. One can indeed postulate that under normal conditions the overall organization, and the putting into proper action at the proper time, of a particular behavior is handled not by a single, but rather by a multitude of limbic structures. In fact, it is not unlikely that a given structure assumes responsibility for the coarse organization of such a behavior or behavioral component, while a series of additional networks are capable, through interaction with this "primary center" and dependent upon a vast variety of internal and external conditions, of quantitatively and qualitatively modifying the emerging behavior.

Finally, one has to be aware - relating what has been said so far to the central topic of this volume - that a good part of the stimulation- and, even more so, the lesion-induced limbic effects appear to manifest some degree of *abnormality*, if not psychoneuropathology. Or, to introduce the terms of a novel concept of behavioral organization as a whole (Koella 1982b, 1984, 1985), the majority of these symptoms can be viewed as being the manifestation of abnormally high or low *vigilance* in the respective "behavioral systems"; and this would be the consequence of abnormally high and low, respectively, *reactivity* in the limbic neuronal networks, responsible for the organization of the said behaviors. This concept holds true, certainly, for such phenomena as the (unfounded) anger induced by electrical stimulation in the diencephalon, the aggressive tendencies of rats with lesioned septa, the undiscriminative sexual behavior following limbic lesions, and the "insatiable" seeking of pleasure in the self-stimulation experiments where all caution and consideration for other needs (and dangers) is completely forgotten. This holds true too for the excessive use of oral exploration and for the obvious disturbances in the memory functions, as well as the signs of visual agnosia after limbic lesions. It is of considerable interest to note that virtually all these "symptoms" produced by manipulation of activity (or reactivity) in limbic structures can be "rediscovered" again as components of man's psychopathology. In fact, some of the changes and behavioral abnormalities produced by experimental intervention into the activity pattern of the limbic system of animals imitate, to some degree, the symptoms we are used to seeing in man suffering from schizophrenia, depression, mania, anxiety, psychopathy, and complex partial epilepsy. This should give us at least some hints as to where to look for the pathogenic focus of human psychopathology.

References

Adamec RE (1976) Hypothalamic and extrahypothalamic substrates of predatory attack. Suppression and the influence of hunger. Brain Res 106: 57-69

Adey WR, Walter DO (1963) Application of phase detection and averaging techniques in computer analysis of EEG records in the cat. Exp Neurol 7: 186-209

Arbuthnott GW, Fuxe K, Ungerstedt U (1971) Central catecholamine turnover and self-stimulation behavior. Brain Res 27: 406-413

Black AH, Nadel L, O'Keefe J (1977) Hippocampal function

in avoidance learning and punishment. Psychol Bull 84: 1107-1129
Crow TJ (1972) A map of the rat mesencephalon for electrical self-stimulation. Brain Res 36: 265-273
Crow TJ (1977) The neuroanatomy of intracranial self-stimulation: a general catecholamine hypothesis. Neurosci Res Program Bull 15: 195-204
Egger MD, Flynn JP (1963) Effects of electrical stimulation of the amygdala on hypothalamically elicited attack behavior in cats. J Neurophysiol 26: 705-720
German DC, Bowden DM (1974) Catecholamine systems as the neural substrate for intracranial self-stimulation: a hypothesis. Brain Res 73: 381-419
Goddard GV (1964) Functions of the amygdala. Psychol Bull 62: 89-109
Grastyán E, Karmos G (1962) The influence of hippocampal lesions on simple and delayed instrumental conditioned reflexes. In: Physiologie de l'Hippocampe. Colloques internationaux du Centre national de la Recherche Scientifique. CNRS 107: 225-239
Grastyán E, Lissak K, Madarasz I, Donhoffer H (1959) Hippocampal electrical activity during the development of conditioned reflexes. Electroenceph Clin Neurophysiol 11: 409-430
Gray J, Feldon J, Rawlins JNP, Owen S, McNaughton N (1978) The role of the septo-hippocampal system and its noradrenergic afferents in behavioral responses to non-reward. In: Elliot K, Whelan J (eds) Functions of the septo-hippocampal systems. Ciba foundation symposium 58 (new series). Elsevier, Amsterdam, pp 275-300
Green JD, Clemente CD, de Groot J (1957) Rhinencephalic lesions and behavior in cats. J comp Neurol 108: 505-545
Hess WR (1949) Das Zwischenhirn. Schwabe, Basel
Hunsperger RW (1956) Affektreaktionen auf elektrische Reizung im Hirnstamm der Katze. Helv Physiol Acta 14: 70-92
Ito M, Olds J (1971) Unit activity during self-stimulation behavior. J Neurophysiol 34: 263-273
Jacobs BL, McGinty DJ (1972) Participation of the amygdala in complex stimulus recognition and behavioral inhibition: evidence from unit studies. Brain Res 36: 431-436
Jonason KR, Enloe NJ, Contrucci J, Meyer PM (1973) Effects of simultaneous and successive septal and amygdaloid lesions on social behavior of the rat. J Comp Physiol Psychol 83: 54-61
Kaada BR (1951) Somato-motor, autonomic, and electrocorticographic responses to electrical stimulation of "rhinencephalic" and other structures in primates, cat, and dog. Acta Physiol Scand 24 [Suppl 83]: 1-285
Kaada BR, Rasmussen EW, Kveim O (1962) Impaired acquisition of passive avoidance behavior by subcallosal, septal, hypothalamic, and insular lesions in rats. J Comp Physiol Psychol 55: 661-670
Kawakami M, Terasawa E, Kimura F, Wakabayashi K (1973) Modulating effect of limbic structures on gonadotrophin release. Neuroendocrinology 12: 1-16
King FA, Meyer PM (1958) Effects of amygdaloid lesions upon septal hyperemotionality in the rat. Science 128: 655-656
Kling A, Schwartz NB (1961) Effects of amygdalectomy on sexual behavior and reproductive capacity in the male rat. Fed Proc 20: 335
Klüver H, Bucy PC (1937) "Psychic blindness" and other symptoms following bilateral temporal lobectomy in rhesus monkeys. Am J Physiol 119: 352-353
Klüver H, Bucy PC (1939) Preliminary analysis of functions of the temporal lobe in monkeys. Arch Neurol Psychiatry 42: 979-1000
Koella WP (1982a) The functions of the limbic system - evidence from animal experimentation. Adv Biol Psychiatry 8: 12-39
Koella WP (1982b) A modern neurobiological concept of vigilance. Experientia 38: 426-437
Koella WP (1984) The organization and regulation of sleep; a review of the experimental evidence and a novel integrated model of the organizing and regulating apparatus. Experientia 40: 309-338
Koella WP (1985) Local vigilance, the vigilance profile, and psychiatric disease; a new general theory on an old notion and its potential application in biological psychiatry. Integrative Psychiatry 3: 185-198
Landfield PW (1977) Different effects of post-trial driving or blocking of the theta rhythm on avoidance learning in rats. Physiol Behav 18: 439-445
Liberson WT, Akert K (1955) Hippocampal seizure in guinea pigs. Electroenceph Clin Neurophysiol 7: 211-222
MacLean PD (1954) The limbic system and its hippocampal formation. Studies in animals and their possible application to man. J Neurosurg 11: 29-44
MacLean PD, Ploog DW (1962) Cerebral representation of penile erection. J Neurophysiol 25: 29-55
Nauta WJH (1958) Hippocampal projections and related neural pathways to the mid-brain in the cat. Brain 81: 319-340
Niki H (1967) Effects of hippocampal ablation on learning in the rat. Prog Brain Res 27: 305-317
O'Keefe J, Black AH (1978) Single unit and lesion experiments on the sensory inputs to the hippocampal cognitive map. In: Elliot K, Whelan J (eds) Functions of the septo-hippocampal systems. Ciba foundation symposium 58 (new series). Elsevier, Amsterdam, pp 179-182
O'Keefe J, Dostrovsky J (1971) The hippocampus as a spatial map. Preliminary evidence from unit activity in the freely-moving rat. Brain Res 34: 171-175
Olds J, Milner P (1954) Positive reinforcement produced by electrical stimulation of the septal area and other regions of the rat brain. J Comp Physiol Psychol 47: 419-427
Olton DS, Isaacson RL (1969) Fear, hippocampal lesions, and avoidance behavior. Commun Behav Biol 3: 259-262
Olton DS, Walker JA, Gage FH (1978) Hippocampal connections and spatial discrimination. Brain Res 139: 295-308
Orbach J, Milner B, Rasmussen T (1960) Learning and retention in monkeys after amygdala-hippocampus resection. Arch Neurol 3: 230-251
Papez JW (1937) A proposed mechanism of emotion. Arch Neurol Psychiatry 38: 725-743
Rolls T (1981) Processing beyond the inferior temporal visual cortex related to feeding, memory, and striatal function. In: Katzuki Y, Norgren R, Sato M (eds) Brain mechanisms of sensation. Wiley, New York, pp 241-269
Rosvold HE, Mirsky AF, Pribram KH (1954) Influence of amygdalectomy on social behavior in monkeys. J Comp Physiol Psychol 47: 173-178
Routtenberg A (1975) Intracranial self-stimulation: catecholamine brain pathways and memory consolidation. In: Cole J, Sonderegger E (eds) Nebraska symposium on motivation. University of Nebraska Press, Lincoln, pp 161-182
Routtenberg A, Malsbury C (1969) Brainstem pathways of reward. J Comp Physiol Psychol 68: 22-30
Sawyer CH (1962) Triggering of the pituitary by the central nervous system. In: Bullock TH (ed) Physiological triggers. Waverly, Baltimore, pp 164-174
Schmutz M, Bürki H, Koella WP (1981) Electrically induced hippocampal afterdischarge in the freely moving cat: an animal model of focal (possibly temporal lobe) epilepsy. In:

Dam M, Gram L, Penry JK (eds) Advances in epileptology: XII epilepsy international symposium. Raven, New York, pp 59–65
Schreiner L, Kling A (1956) Rhinencephalon and behavior. Am J Physiol 184: 486–490
Siegel A, Edinger H, Dotto M (1975) Effects of electrical stimulation of the lateral aspects of the prefrontal cortex upon attack behavior in cats. Brain Res 93: 473–484
Stein L, Belluzzi JD (1979) Brain endorphins: possible role in reward and memory formation. Fed Proc 38: 2468–2472
Stein L, Wise CD (1969) Release of norepinephrine from hypothalamus and amygdala by rewarding medial forebrain bundle stimulation and amphetamine. J Comp Physiol Psychol 67: 189–198
Ursin H, Kaada BR (1960) Functional localization within the amygdaloid complex in the cat. Electroenceph Clin Neurophysiol 12: 1–20
Vergnes M, Karli P (1969) Effets de l'ablation des bulbes olfactifs et de l'isolement sur le développement de l'aggressivité interspécifique du rat. Soc Biol 163: 2704–2707
Wise RA (1981) Intracranial self-stimulation: mapping against the lateral bounderies of the dopaminergic cells of the substantia nigra. Brain Res 213: 190–194
Wise RS (1978) Catecholamine theories of reward: a critical review. Brain 152: 215–247
Wood CD (1958) Behavioral changes following discrete lesions of temporal lobe structures. Neurology 8: 215–220

I. Anatomical and Physiological Data

Physiology of the Frontal Cortex

O. D. Creutzfeldt[1] and H. G. Wieser[2]

The frontal association cortex comprises the homotypical granular cortex of the frontal lobe, i. e., Brodmann's areas 9–12, 45, and 46 (Fig. 1). These areas are also labelled the "prefrontal" cortex in order to separate them from the agranular frontal cortex consisting of the premotor areas 6 and 8 and the motor speech area (Brodmann's area 44) in the language-dominant hemisphere. Phylogenetically, the frontal granular and agranular cortex and their connections are already present in lower mammals but achieve prominence only in primates, exceeding other neocortical areas in relative growth, with the dorsolateral prevailing over the ventromedial aspect (Bianchi 1923; Feuchtwanger 1923; Luria 1969, 1973; Nauta 1971; Warren 1972).

Afferent Connections

The specific thalamic afferents to the frontal regions are topographically layed out (Akert 1964; Warren and Akert 1964): the primary motor area receives afferents from the ventrolateral nucleus of the thalamus; the premotor regions are reciprocally connected with the medial ventroanterior nucleus; and the prefrontal cortex is connected with the mediodorsal thalamic nuclei (MD). These projections are topically ordered: the most medial parts of MD (pars magnocellularis) project to the orbital, and the more laterally situated pars parvocellularis projects to the dorsal prefrontal cortex. The most lateral part of MD (pars paralamellaris) projects to the zone between the premotor and prefrontal cortex, i. e., Brodmann's area 8.

The medial magnocellular portion of MD receives its main input from the amygdaloid nuclei and from the temporo-basal cortex including the hippocampus, in part directly, in part through the septum.

Thus the thalamic input into the frontal orbital cortex derives principally from the limbic and olfactory systems. It is, however, not known to what extent medial and lateral portions of MD are different with respect to the specificity of their inputs.

The medioposterior extension of MD in the nucleus pulvinaris medialis projects into the transition zone between the prefrontal and premotor cortex. It can be assumed that through these connections visual and auditory afferents from the medial pulvinar complex have direct access to the prefrontal and premotor cortex. Afferents from the cortical association areas derive from the limbic cortex (temporobasal and anterior cingulate cortex, Brodmann's area 24). These pathways thus could provide an anatomical basis for information the prefrontal cortex

[1] Department of Neurobiology, Max-Planck-Institute for Biophysical Chemistry, Postfach 2841, 3400 Göttingen, Federal Republic of Germany

[2] Department of Neurology, University Hospital, Frauenklinikstr. 26, 8091 Zürich, Switzerland

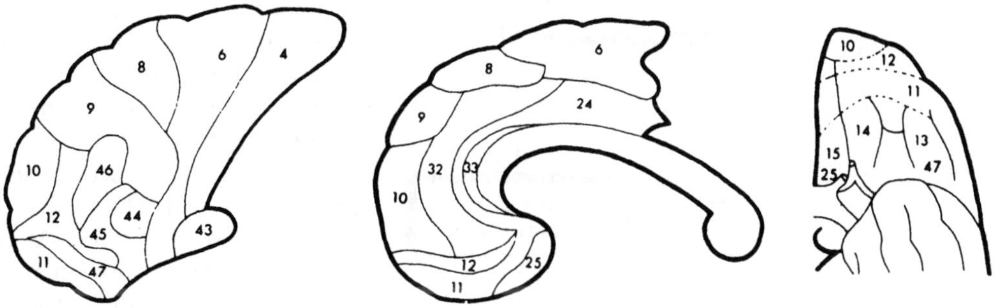

Fig. 1. Brodmann's cytoarchitectural designations of the lateral, medial, and orbital surfaces

Presurgical Evaluation of Epileptics
Edited by H. G. Wieser and C. E. Elger

receives about cerebral representations of the "inner world" (Flechsig 1896), i.e., from the limbic system through the basolateral and the medial limbic circuit (Nauta 1972), as well as from neocortical representations of the "outer world."

Efferent Connections

In addition to the reciprocal connections between the frontal cortex and the mediodorsal thalamus, the amygdala, and various neocortical areas (except the occipital lobe), the main efferents from the prefrontal cortex are towards the head of the caudate nucleus (and in part also to putamen and globus pallidus), the claustrum, the midbrain, and the hypothalamus including the preoptic region (Fig. 2).

The projections into the striatum are topographically organized, with the orbital cortex projecting to the ventrolateral, and the dorsolateral cortex projecting to the anterodorsal portion of the head of the caudate nucleus. The projection to the midbrain concentrates on the ventral portions of the periaqueductal grey, the nucleus centralis tegmenti superior (Bechterev), and the regions of nucleus ruber and substantia nigra.

Moreover, there is a projection from the dorsolateral prefrontal cortex to the most medial nuclei of the pons, and from area 8 (frontal eye field) and from the prefrontal dorsal association cortex to the deep layers of the tectum opticum and pretectum, which play a substantial role in the control of eye movements.

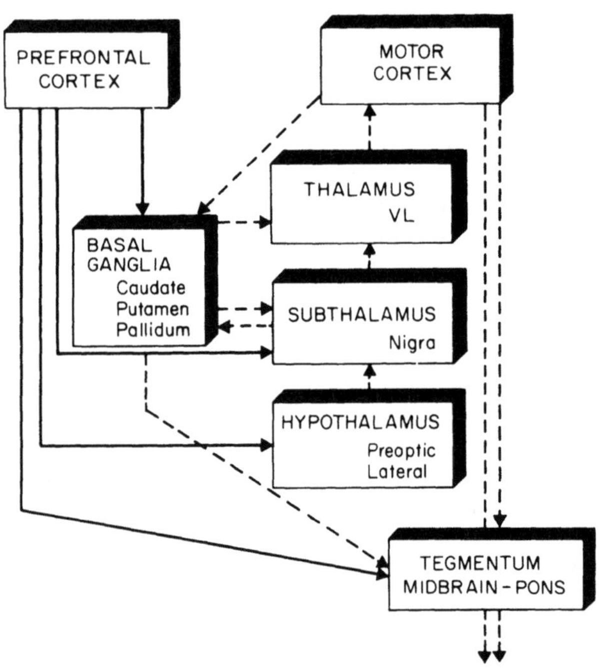

Fig. 2. Efferents of the prefrontal cortex (from Fuster 1980)

Functional Considerations

Neurophysiological Findings

In the *prefrontal cortex,* neurons responsive to visual, somatosensory, auditory (predominantly in the dorsal convexity), and olfactory (predominantly in the ventrobasal part) stimuli have been described. Combined behavioral and neurophysiological studies have shown, however, that responses can vary considerably with the behavioral significance of the stimulus.

A remarkable feature of some prefrontal neurons is the activation during the delay period of a delayed response task (for further reference see Fuster 1980) (Fig. 3).

This neuronal activation is accompanied by a slow negative DC potential, which is also recorded over the dorsal prefrontal cortex in a go-no go behavioral situation. It can also be recorded in men during a similar task, e.g. following a conditional stimulus while waiting for the unconditional signal (expectancy wave or contingent negative variation (CNV), Walter et al. 1964). It should be noted, however, that the delayed response activation of the neuronal discharge rate is not restricted to the cortex, but can also be recorded in the mediodorsal nucleus, i.e., the thalamic afferent relay to the prefrontal cortex (see Fuster 1980). After bilateral prefrontal ablation in monkeys, delayed response behavior is disturbed (see Rosenkilde 1979).

The *nonprimary, frontal motor cortex can be divided* into two functionally separate areas: the premotor cortex, which receives its thalamic input from the cerebellum, and the supplementary motor area, which receives its thalamic input mainly from the basal ganglia. The activities of neurons in the *premotor cortex* show complex interrelations with motor tasks. One aspect is the activation of some neurons preceding visually guided or triggered movements, suggesting function in sensory guidance and in a representation of the motor set for future movements (Wise 1985). One may also remember the classical observations of cortical electrical stimulation in monkeys and men, which elicit complex "movement synergies" (Foerster 1936). After a lesion, incoordination of synergistic movements may be observed.

The function of the *supplementary motor area (SMA)* is predominantly seen as being involved in

Fig. 3. Activity of a single unit in the prefrontal cortex during five delayed-response trials. The *horizontal bar* marks the cue period, and the *arrow* marks the end of the delay. Note the activation during the delay (from Fuster 1980)

the initiation of movements. This, however, is only one aspect. It also seems to be involved in functions related to the "motor set" or "conception" of movements as well as in the continuous flow of movement synergies (Goldberg 1985; Wise and Strick 1984). In self-paced movements, neurons discharging up to 1-2 s before onset of movement are found in both the premotor and in the supplementary motor areas, but more frequently in the latter (Okano and Tanji, to be published).

Thus, both premotor areas appear to function together with the primary motor cortex in a "cooperative control" (Foerster 1936) of self-paced and visually guided movements and in the coordination of movement sequences.

The eye movement-related activity in the *frontal eye field* is not simply a "command" activity. Spontaneous saccades as well as those elicited by optokinetic and vestibular stimuli appear without premovement activity in the frontal eye fields. During visually conditioned eye movements some neurons are activated and others inhibited; however, most of these activities do not precede the eye movements (Bizzi and Schiller 1970). On the other hand, recent investigations have demonstrated distinct classes of neurons which discharge before purposeful and visually guided eye movements, suggesting a trigger function of the prefrontal eye fields for this type of activity (Bruce and Goldberg 1985). Eye movements to the contralateral side can be elicited not only from the frontal eye field but also from a rather large region including the premotor cortex and the dorsal prefrontal cortex.

Thus, the frontal eye fields as well as widespread regions of the premotor and prefrontal cortex appear to be involved, in a complex manner, in directing the gaze and in focussing visual attention, making visual perception a dynamically active process. This is consistent with the observation that frontal eye field lesions can lead to a temporary deviation of the eyes to the side of the lesion, and/or to a temporary visual field neglect (first reported by Munk in 1881; see Latto and Cowey 1971), involving a lack of attention to objects and persons despite an intact visual field perimeter, and disproportionately long searching times for correct stimuli located in the visual field contralateral to the lesion. With bifrontal lesions patients suffer from "inertia of gaze," i. e., attention focusses persistently on only one aspect of the visual field. In addition patients suffering from massive bifrontal lesions fail to arrive at logical hypotheses on a thematic pictures test.

Neuropsychological Symptoms after Prefrontal Lesions

Lesions anterior to Broca's area in the language-dominant hemisphere result in a deterioration in verbal fluency (as measurable on the Thurstone Word Fluency Test). By correlating results obtained from verbal fluency tests with those from a modified version of the Stroop test it was concluded that the fluency decay may be only the surface manifestation of a more general dysfunctioning best described by general response spontaneity and initiative, or flexibility in switching cognitive sets (Perret 1974). Inability to use error feedback is a relatively characteristic feature of prefrontal lesions in the language-dominant hemisphere and consistent with findings of high perseverative error rates on the relatively nonverbal Wisconsin Card Sorting Test.

Successful problem-solving behavior involves several stages of processing. According to Lhermitte et al. (1972) they can be reduced to four: (1) systematic orientation to, and analysis of, initially presented task features; (2) generation of several alternative hypothetical solutions; (3) flexible formulation of appropriate motor programs for each of those solutions; and (4) evaluation of adequacy of results by comparison of the programs with higher-order error feedback. Some patients with frontal lesions fail problem-solving tasks because they cannot extract information-rich details from initially given task features. Other patients with prefrontal lesions appear to suffer a dysfunctioning in the process of comparing response feedback information with the planned action of the program. The goal-oriented action programs lose their regulating influence on behavior because of a breakdown in neuronal systems utilizing error feedback information in comparator processes. Luria et al. (1964) have suggested that both the "susceptibility to distraction by irrelevant stimuli" and the "perseverative tendencies in responding" result from the subject's "inability to organize and reorganize behavior when flexibility is demanded." Pribram (1973) theorizes that, in situations in which appropriate behaviors are incompletely specified by the environmental situation, the frontal lobes normally function to keep interference at a minimum through a context-dependent "flexible noticing order" system regulating information flow and processing. The flexible noticing order system functions by "assigning processing time and space priorities to a multitude of simultaneously noticed and weighted sensory events so that each event impinging on the system is processed in turn of importance while other events are held back to avoid overloading." The CNV (Walter et al. 1964), or expectancy wave, is quoted as a neurophysiological support for this hypothesis.

Similarly, Fuster (1980) considers the frontal cortex as a "timing device" which helps to keep action and behavior processes in correct temporal order. In a more general context one may argue that "the activities of the prefrontal cortex represent internal states of the forebrain, i.e., essentially emotional and goal-directed motivational drives in a social and cognitive context, which are thus incorporated through the efferent pathways of the prefrontal and frontal cortex, into the neocortical control of behavior. ... One may then speculate that the strong relative and absolute increase of the frontal cortex in primates and man became necessary, because a neocortical representation of internal states and their social implications was needed when behavior control came increasingly under cortical control. In lower mammals with less functional corticalization, a cortical representation of such states in less needed, as the respective subcortical structures determine behavior directly as drives, mediated through direct subcortical efferents" (Creutzfeldt 1983).

In the context of *epilepsy*, it has to be kept in mind that the prefrontal cortex is essentially an "electrically unexcitable region" of the brain. This means that circumscribed electrical stimulation has no, or only little, apparent behavioral effects. Thus, confined focal epileptic activity may not produce marked clinical symptoms. These become more obvious when the premotor cortex, i.e., areas 6 and 8, including the frontal eye fields and/or the SMA, are involved. The poor symptomatology of prefrontal focal seizure activity is not the only reason, however, why frontal foci typically appear to lead to generalized seizures without noticeable aura. There seem to exist some as yet unknown anatomical or functional mechanisms responsible for the fact that frontal foci lead to primarily generalized seizures more frequently than foci in other parts of the neocortex.

Finally the participation of the frontal cortex in photically elicited myoclonic seizures in the baboon should be mentioned (Silva-Comte et al. 1982). Although this light-induced epileptic activity in the premotor and motor areas may disappear after ablation of the occipital cortex, the latency and time course of the photically elicited potentials suggest that they depend on a visual projection parallel to the geniculocortical pathway rather than being mediated through association connections via the prestriate cortex.

References

Akert K (1964) Comparative anatomy of frontal cortex and thalamofrontal connections. In: Warren JM, Akert K (eds) The frontal granular cortex and behavior. McGraw-Hill, New York, pp 372-396

Bianchi L (1923) The mechanisms of the brain and the function of the frontal lobes. Livingstone, Edinburgh

Bizzi E, Schiller PH (1970) Single unit activity in the frontal eye fields of unanesthetized monkeys during eye and head movements. Exp Brain Res 10: 151-158

Bruce CJ, Goldberg M, (1985) Primate frontal eye fields: I. Single neurons discharging before saccades. J Neurophysiol 53: 603-635

Creutzfeldt OD (1983) Cortex cerebri. Springer, Berlin Heidelberg New York

Feuchtwanger E (1923) Die Funktionen des Stirnhirns, ihre Pathologie und Psychologie. Springer, Berlin

Flechsig P (1896) Gehirn und Seele. Verlag Veit, Leipzig

Foerster O (1936) Motorische Felder und Bahnen. In: Bumke O, Foerster O (eds) Handbuch der Neurologie, vol 6. Springer, Berlin, pp 1-357

Fuster JM (1980) The prefrontal cortex. Anatomy, physiology and neuropsychology. Raven, New York

Goldberg G (1985) Supplementary motor area structure and function: review and hypothesis. Behav Brain Sci 8: 567-615

Jouandet M, Gazzaniga MS (1979) The frontal lobes. In: Gazzaniga MS (ed) Handbook of behavioral neurobiology, vol 2: Neuropsychology. Plenum, New York, pp 25-59

Latto R, Cowey A (1971) Visual field defect after frontal eyefield lesions in monkeys. Brain Res 30: 1-24

Lhermitte F, Derouesne J, Signoret JL (1972) Analyse neuropsychologiques du syndrome frontale. Rev Neurol 172: 415-440

Luria AR (1969) Frontal lobe syndromes. In: Vinken PJ, Bruyn GW (eds) Handbook of clinical neurology, vol 2: localization in clinical neurology. North Holland, Amsterdam, pp 725-757

Luria AR (1973) The frontal lobes and the regulation of behaviour. In: Pribram KH, Luria AR (eds) Psychophysiology of the frontal lobes. Academic, New York, pp 3-26

Munk H (1881) Über die Funktion der Großhirnrinde. Hirschwald, Berlin

Nauta WJH (1971) The problem of the frontal lobe. A reinterpretation. J Psychiatr Res 8: 167-187

Nauta WJH (1972) Neural association of the frontal cortex. Act Neurobiol Exp (Warsz) 32: 125-140

Perret E (1974) The left frontal lobe of man and the suppression of habitual responses in verbal categorical behavior. Neuropsychologia 12: 323-330

Pribram KH (1960) The intrinsic system of the forebrain. In: Field J (ed) Handbook of physiology, neurophysiology, Sect 1, vol II. American Physiological Society, Washington DC, pp 1323-1344

Pribram KH (1973) The primate frontal cortex - executive of the brain. In: Pribram KH, Luria AR (eds) Psychophysiology of the frontal lobes. Academic, New York, pp 293-314

Rosenkilde CE (1979) Functional heterogeneity of the prefrontal cortex in the monkey: a review. Behav Neural Biol 25: 301-345

Silva-Comte C, Velluti J, Ménini C (1982) Characteristics and origin of frontal paroxysmal responses induced by light stimulation in *Papio papio* under allylglycine. Electroencephalogr Clin Neurophysiol 53: 479-490

Walter WG, Cooper R, Aldridge VJ, McCallum WC, Winter AL (1964) Contingent negative variation: an electric sign of sensory motor association and expectancy in the human brain. Nature 23: 380-384

Warren JM (1972) Evolution, behavior and the prefrontal cortex. Acta Neurobiol Exp (Warsz) 32: 581-593

Warren JM, Akert K (1964) (ed) The frontal granular cortex and behavior. McGraw Hill, New York

Wise SP (1985) The primate premotor cortex: past, present and preparatory. Ann Rev Neurosci 8: 1-19

Wise SP, Strick PL (1984) Anatomical and physiological organization of the non-primary motor cortex. Trends Neurosci 7: 442-446

Editors' Commentary 1

The purpose of these first chapters is to improve overall understanding of cortical architecture and connections, as well as their functional implications. Pandya and Yeterian review aspects related to the temporal lobe and the limbic system. They point out "how these cortical structures, especially in the temporal lobe, may underlie various seizure phenomena." Koella elaborates on the functional role of limbic and related structures by reviewing results of animal experiments. Creutzfeldt and Wieser's review on the "Physiology of the Frontal Cortex" concentrates on various aspects related to its function, pathology, and role in epilepsy.

The *discussions* dealt with two main categories of questions. The first impinged on the problem of extrapolation from experimental animal findings to human brain functions. With respect to the Klüver-Bucy syndrome (Akert et al. 1961), Akert made the point that inference from lesioning experiments in "small" animals as to the functions of the much "bigger" and more complex human brain with its considerable redundancy in terms of circuitry is limited. For understanding epileptic seizures he prefered using stimulation experiments. Gloor emphasized the important insights into brain function obtained by carefully analyzing spontaneous and induced seizure accompaniments because - unlike in animals - "the conscious content is accessible." In general, he found the correspondence between animal and human data "surprisingly good."

The other discussion dealt with more specific questions. One centered around the role of the supplementary motor cortex (SMA). Pandya emphasized its input from the sensory regions via corticocortical connections. Creutzfeldt, however emphasized its thalamic input from the basal ganglia as well as its efferent connections directly into the subcortical motor systems of the forebrain, the midbrain, and even the spinal cord. Also, the efferents of the prefrontal cortex to the basal ganglia project back, via the pallidum and the ventrolateral thalamus, to the SMA. Yeterian commented on the limbic lobe connections to the striatum. Akert underlined that the insular cortex receives quite a lot of olfactory and gustatory information and asked about the insular connections to the frontal lobe. From Pandya's response it became clear that the anterior insula has solid bidirectional connections with the ventral part of the frontal lobe, whereas the posterior part is related to the parietal and somatosensory systems. Creutzfeldt pointed out that responses to gustatory stimuli are found in the lateral prefrontal cortex, and they appear to arrive directly rather than through insular association fibers.

The question as to the role of the uncinate fasciculus in seizure propagation phenomena from temporal to frontal regions and vice versa was raised but could not be answered definitively. Likewise some participants questioned the existence and in particular the specificity of the "interictal personality and behavioral changes" associated with temporal lobe epilepsy (Geschwind 1977).

References

Akert K, Gruesen RA, Woolsey CN, Meyer DR (1961) Klüver-Bucy syndrome in monkeys with neocortical ablations of the temporal lobe. Brain 84: 480-498

Geschwind (1977) Interictal behavioral changes in epilepsy. Epilesia 24 (suppl 1): S23-S30

Basic Epileptology

C. E. Elger

An epileptic seizure is a paroxysmally occurring functional disturbance of the CNS. Prior to electrophysiological investigations Jackson (1870) suggested: "An epileptic seizure is a state produced by an abnormal excessive neuronal discharge within the central nervous system." More than 100 years of clinical and experimental investigations in the field of epileptology support and confirm this speculative consideration. The huge amount of experimental data has been made possible by the existence of a large variety of different animal models with focal and generalized epilepsy. Based on these data this brief review will mainly take into account experience from investigations dealing with focal epilepsy. To be within the scope of this volume, special emphasis will be put on the cellular mechanisms during epileptic activity, forming the first section. A second part describes the processes initiating and suppressing the outburst of a seizure. Finally an attempt will be made to elucidate the phenomena terminating a seizure. Other aspects, especially the generation of field potentials during epileptic activity and the mechanisms initiating the kindling process, will be described by Speckmann and Walden (this volume). Furthermore, a number of recent reviews cover other aspects of the basic epileptology in more detail (Conners and Gutnick 1984; Prince 1985).

Cellular Mechanisms

Seizure activity runs in parallel with characteristic, steep, high-amplitude potentials on the EEG. This indicates a synchronous activity of a large number of neurons in the area surrounding the electrode. Intracellular recordings of neocortical neurons performed during focal seizure activity revealed the unique behavior of the neurons involved in the epileptic process. It consisted of a high frequency discharge of action potentials initiating a steep depolarization, during which the action potential generation stopped. The rectangular depolarization is terminated by a more or less steep repolarization, sometimes followed by a hyperpolarization (Fig. 1). The event runs in parallel with the epileptiform field potential, thus indicating that a considerable number of neurons behave in the same manner. The typical membrane potential change in the neurons was firstly described by Goldensohn and Purpura (1963) and shortly there after named "paroxysmal depolarization shift" (PDS) by Matsumoto and Ajmone-Marsan (1964). Investigations on the various models of epileptic activity confirmed that PDS can be regarded as the cellular equivalent of the seizure discharges (Fig. 1; c.f. Prince 1978).

Approaching the question of the processes underlying the generation of PDS, several aspects have to be considered. First the question arises as to whether or not PDS is due to an excessive synaptic bombardment inducing a "giant" postsynaptic potential or to altered intrinsic properties of the neuronal membrane during the epileptic event. Whereas early intracellular studies demonstrated a similarity between postsynaptic potentials and PDS, recent investigations using invertebrate and slice preparations favor the latter concept (Ayala et al. 1970, Speckmann and Caspers 1973; Prince 1978; Schwartzkroin and Wyler 1982). As can be seen from part C of Fig. 1 molluscan neurons are capable of generating PDS which are quite similar in their properties to mammalian nerve cells. The generation of PDS after synaptic isolation of these cells is still possible (Speckmann and Caspers 1973). Experiments using slice preparation revealed that the PDS is at least in part the result of an interaction between voltage-dependent calcium and sodium inward currents and calcium-dependent potassium outward currents (Wong and Prince 1978, 1979). In vivo recordings from cortical neurons during penicillin-induced focal seizure activity confirmed further the proposal that nonsynaptic mechanisms are of central importance for the generation of PDS. It could be demonstrated that the intracellular application of a calcium antagonist

Neurological Clinic University Hospital, University of Münster, Albert-Schweitzer-Str. 33, 4400 Münster, Federal Republic of Germany

Presurgical Evaluation of Epileptics
Edited by H. G. Wieser and C. E. Elger

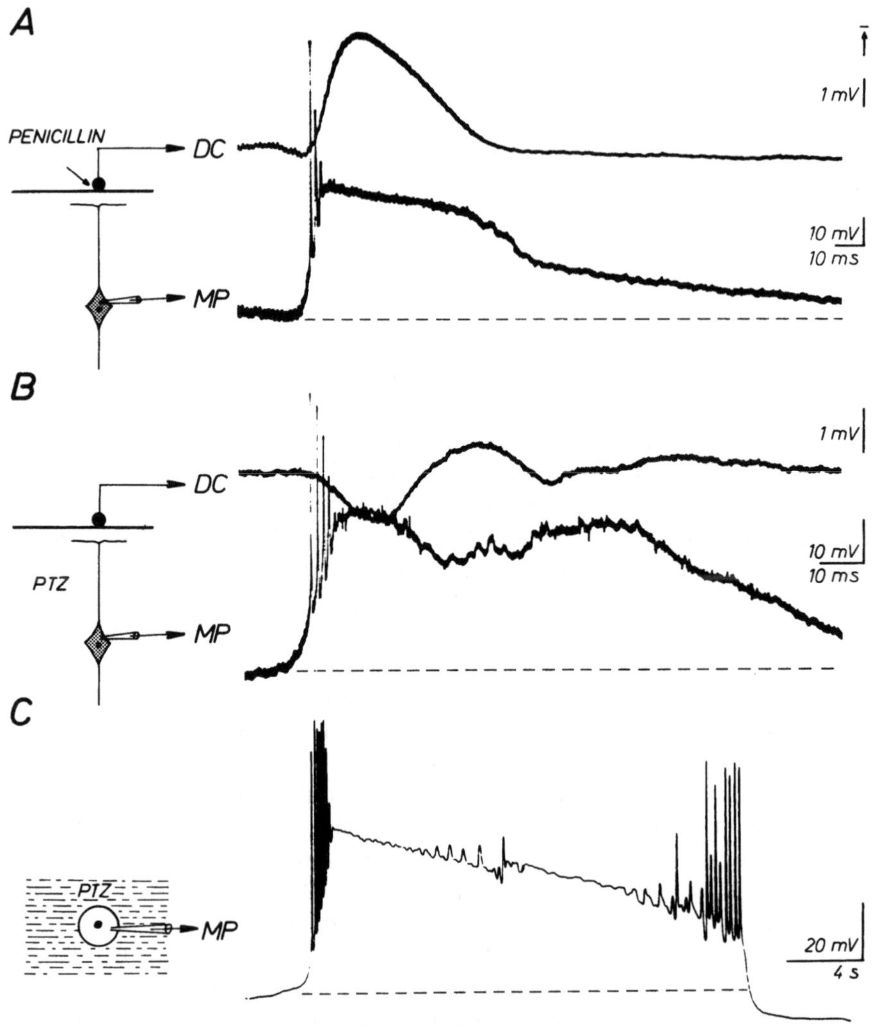

Fig. 1 A-C. Epileptiform potentials at the cortical surface *(DC)* and corresponding fluctuations of the membrane potential *(MP)* of neurons - (**A** rat, cortex; **B** cat, cortex; **C** *Helix pomatia*, buccal ganglion) under seizure conditions. **A** Focal activity induced by topical application of penicillin to the cortical surface. **B** Tonic-clonic seizures induced by systemic application of pentylenetetrazole *(PTZ)*. **C** Application of pentylenetetrazole to the bath solution. (Experiments together with E.-J. Speckmann; modified after Elger and Wieser 1984)

results in a significant reduction of the depolarization shift (Witte et al. 1983).

As a whole, the majority of data is in favor of the concept that PDS is the result of altered intrinsic properties of neurons directly involved in the epileptic process. The characteristic depolarization is thought to be due to a voltage-dependent, calcium inward current.

Spread and Restriction of Epileptic Activity

It is obvious that epileptic activity is linked to synchronized activity in a considerable number of neurons. From the high-frequency discharge initiating PDS it was speculated that a few neurons producing this type of burst activity are able to synchronize a large population of nerve cells and thus an epileptic event (Calvin 1972). As the cortical network consists of powerful inhibitory mechanisms, in most instances a development of seizure seems to be depressed by postsynaptic inhibition. Obviously, the development of the depression of seizure activity depends on a disturbance of the balance of excitatory and inhibitory postsynaptic potentials within the neuronal structure in question.

PDS is, as a rule, triggered by excitatory postsynaptic potentials (c.f. Prince 1985). If the triggering of epileptic neuronal events is initiated transsynaptically, the question then arises concerning which way

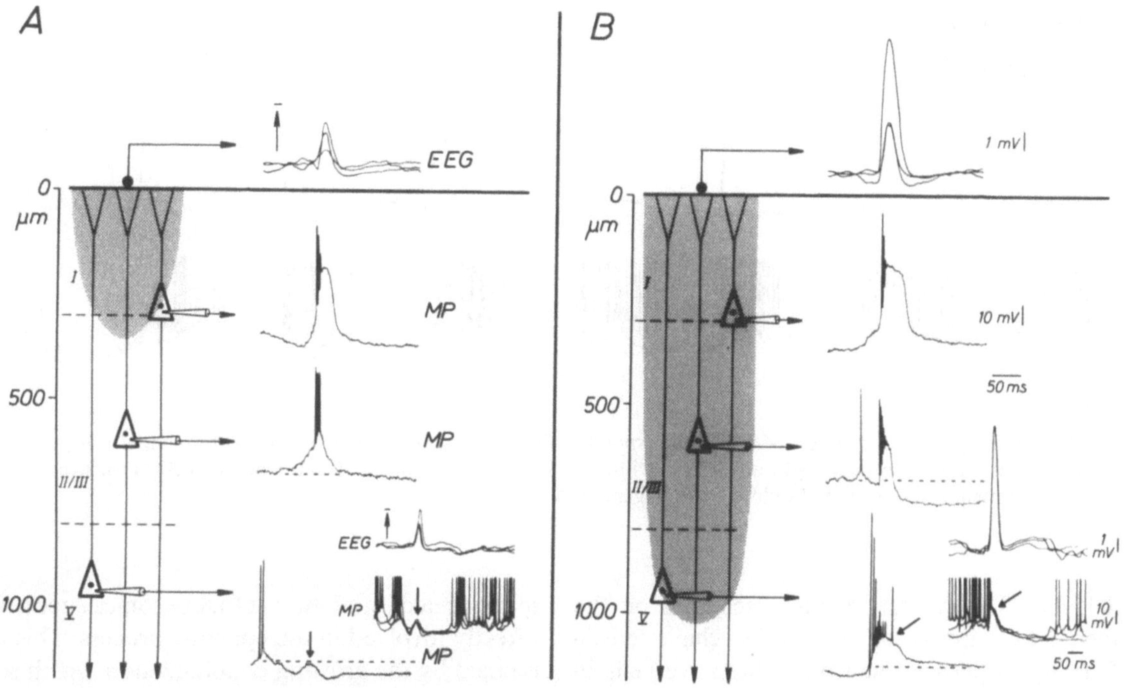

Fig. 2 A, B. Membrane potential changes *(MP)* of cortical neurons in different cortical laminae and corresponding EEG signals during focal seizure activity induced by topical application of penicillin to the cortical surface. The foci in **A** and **B** have different vertical extensions, indicated by the *shaded area* in the schematic drawing. The *numbers* refer to the depth below the cortical surface, and *roman numbers* to the cortical layers. The embedded depolarization is indicated by an *arrow*. (Rat, motor cortex; experiments together with E.-J. Speckmann; modified from Elger and Speckmann 1983)

the presynaptic discharges gain the necessary temporal and spatial synchronization. One can speculate that the propagation of excitatory impulses in recurrent connections of pyramidal cells or in the axonal arborization of thalamocortical neurons due to spontaneous axonal discharges serves this purpose (Gutnick and Prince 1974). However, recordings performed during the interictal state of a developing experimental focus showed that by the period 400-500 ms prior to the epicortical seizure potential, a continuous increase of the mean neuronal discharge frequency develops until the PDS is initiated (Elger and Speckmann 1983). A result of this increase could be a progressively positive interference finally initiating PDS in a large number of neurons. The answer to the question upon the origin of the early synaptic activity is still unsolved (Elger and Speckmann 1983).

It is beyond question that other mechanisms further support the spread of epileptic activity, especially alterations of the ionic milieu which can influence the bioelectric processes of the neuronal network retroactively. In this context the increase in potassium ion activity and the decrease in calcium ion activity are of relevance (Lux 1974; Heinemann et al. 1977). Furthermore, the highly active nerve cell population causes a significant release of transmitter. Acetylcholine, for example, may increase the neuronal excitability considerably, when large amounts appear in the extracellular space (Prince 1985). Much has been speculated about the role of electrotonically coupled neurons in the epileptic process since they have been demonstrated in the hippocampus and neocortex (Dudek et al. 1983). However, the contribution of these mechanisms and of large extracellular field potentials exciting neurons is still unclear.

From the number of processes supporting the spread of epileptic activity mentioned above, it seems surprising that epileptic activity of the brain is the exception and normal signal processing is the rule. As already described, the cortical and hippocampal neuronal networks consist of a powerful inhibitory neuronal network which serves under physiological conditions as an effective control mechanism.

The analysis of mechanisms underlying the restriction of focal seizure activity was possible on the basis of data derived from neurons surrounding an epileptic focus. An example is shown in Fig. 2. Part A of this figure shows an epileptic focus induced by penicillin, restricted to the superficial cor-

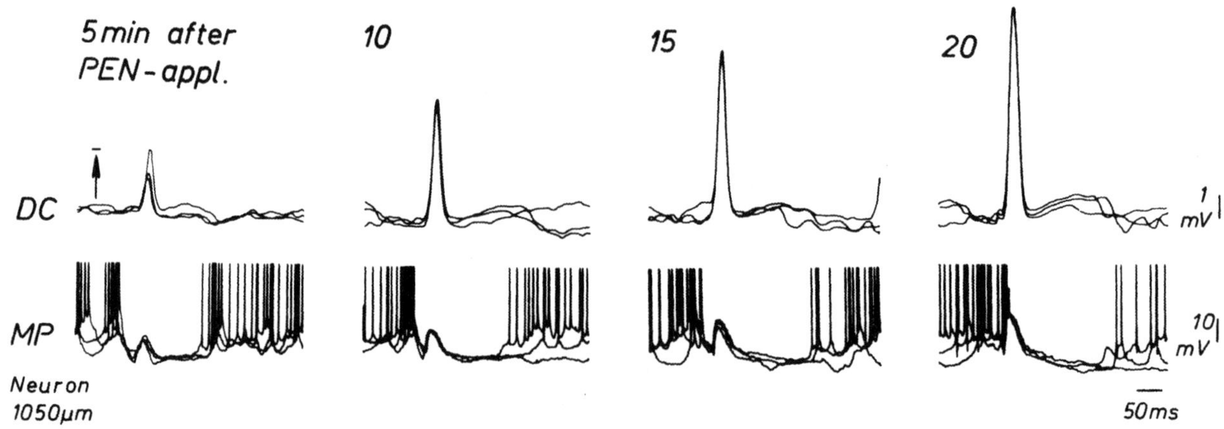

Fig. 3. Membrane potential changes *(MP)* and corresponding fluctuations of the epicortical field potential *(DC)* at different points of time after the epicortical application of penicillin. Pyramidal tract neuron; 1050 µm below cortical surface; rat, motor cortex. Action potentials truncated. (Modified after Elger and Speckmann 1983)

tical layers. The typical seizure discharges on the surface EEG are accompanied by characteristic PDS in the superficial neurons. The nerve cells in deeper laminae exhibited a powerful, time-linked inhibition, in which a small depolarization not reaching the threshold for action potential generation is embedded. This phenomenon was described as "vertical inhibition" (Elger and Speckmann 1983). With a further spread of the epileptic process within the cortex, the hypolarization vanishes and the depolarization increases, finally turning into a PDS (Fig. 2 B). In addition Prince and Wilder (1967) and Dichter and Spencer (1969) demonstrated that nerve cells in the neocortex and the hippocampus surrounding the focus in the horizontal plane exhibited prominent inhibitions, linked to PDS in the center, which were labelled "surround inhibition." From these findings it can be seen that at least the experimental epileptic focus is enveloped hemispherically by a zone of inhibition which is assumed to be the basis of the spatial restriction of focal seizure activity (c.f. Elger and Speckmann 1983; Traub 1983).

Considering the transition form focal restriction to spread of a seizure event, two possibilities have to be taken into account. In the first case neurons beyond the focus and its inhibiting zone are capable of generating PDS. If then the excitatory synaptic input increases, caused by focal epileptic activity, PDS will be triggered and the spread of pathological activity will go on supported by the processes already described. In the second case cells outside the focus are "nonepileptic" at first. Such a case is illustrated in more detail in Fig. 3. A neuron in the deep cortical layers shows shortly after penicillin application to the cortical surface the characteristic long-lasting inhibition with embedded depolarization. The neuron in the area of inhibition at first becomes successively directly involved in the epileptic process. This is indicated by the growing depolarization which seems to run in parallel with a diminution of the first part of the inhibition. Finally PDS is generated with its characteristic high-frequency discharge. The increase of the depolarization is thought to be a dendritic conductance change, finally reaching the soma regions (Elger and Speckmann 1983; Wong and Prince 1979). A number of studies indicate that voltage-dependent calcium currents play an important role. The process initiating the alteration in dendritic ion fluxes is quite unknown (c.f. Prince 1985).

Beside the phenomena described above, inhibitory and excitatory postsynaptic potentials play a major role during the spread of seizure activity. In particular, the interaction of some convulsant drugs with receptors of inhibitory transmitters led to the assumption that development and spread of the epileptic activity are at least in part due to a disinhibition (Wong and Prince 1979). Other studies showed, however, that PDS-generation has already been started while the cell is still capable of generating a powerful inhibition (Fig. 4; Elger and Speckmann 1983). Epicortical stimulation leads to a long-lasting inhibition in cortical neurons (Fig. 4 a). After the epicortical application of penicillin, PDSs develop which are triggered by the epicortical stimulation. Parallel to this, some stimuli are still able to initiate an inhibitory postsynaptic potential resembling the one under control conditions (Fig. 4 b, c). These observations reveal that a significant reduction or even an abolition of synaptic inhibition is not a prerequisite for the development of PDS. Similarly, a reduction of postsynaptic inhibition is tolerated before seizure initiation (Tribble et al. 1983).

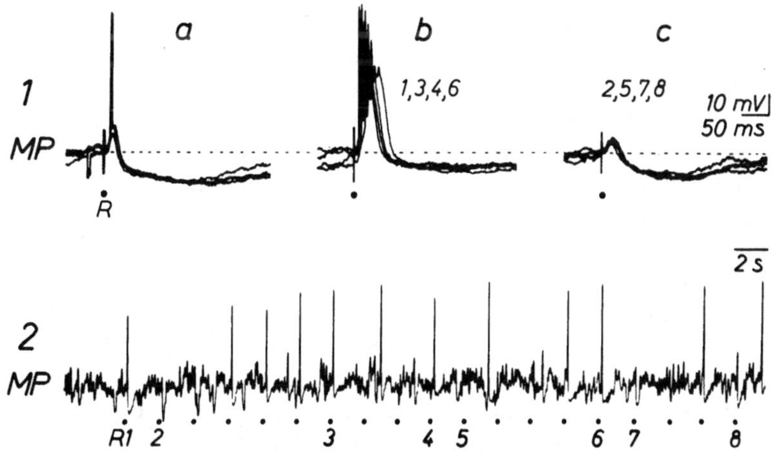

Fig. 4 a–c. Inhibitory postsynaptic potentials and paroxysmal depolarization shifts triggered by epicortical stimulation of the cortical surface (stimulation = *dots*) in a superficial cortical neuron (rat, motor cortex). **1 a** Control condition; **b** and **c** after epicortical penicillin application. Superposition of several responses (original tracings); *MP*, membrane potential; *dashed line*, resting MP. **2** Inkwriter tracing of MP. The recordings in **1** and **2** are related by numbers. (Modified after Elger and Speckmann 1983)

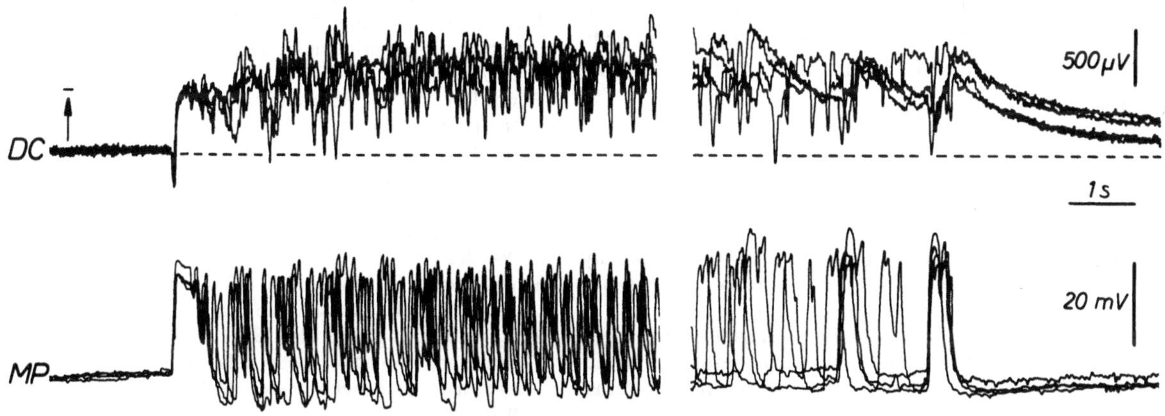

Fig. 5. Epicortical field potential *(DC)* and corresponding membrane potential changes *(MP)* during generalized tonic-clonic seizures induced by systemic application of pentylenetetrazole. Cat, pyramidal-tract neuron; superposition of four events. *Dashed line*, control value of the DC potential. (Modified after Speckmann and Caspers 1979)

Termination of Seizure Activity

Since seizure discharges are easily induced in the CNS under experimental conditions, a large number of studies have contributed to our understanding of various aspects of this pathological condition. Compared to this, a surprisingly small number of investigators devoted attention to the processes terminating a seizure. This might be due to the obvious assumption that metabolic exhaustion causes termination. Excessive neuronal activity results in a critical consumption of energy-supplying substances. Hypoxia, hypercapnia, and acidosis will lead to a suppression of neuronal activity. This view is further supported by neuropathological findings in epileptic patients showing alterations in the temporal lobe which indicate that hypoxia has taken place. Experimental studies are not in favor of this concept (Caspers and Speckmann 1972; Kreisman et al. 1981). Within a series of tonic-clonic seizures, at first an increase of the local PO_2 due to an increase of local blood flow occurred with each seizure of the series. During the further course of the experiment the single seizure ran in parallel with a drop of local tissue PO_2. Every seizure had a similar duration, no matter whether it happened at the beginning or the end of the whole series (Caspers and Speckmann 1972). Form these experiments the conclusion can be drawn that hypoxia is not a prerequisite for the termination of a seizure.

More recent studies indicate that neuronal processes contribute to the termination of the epileptic event. Figure 5 shows the membrane potential fluctuations in a cortical pyramidal tract cell during several tonic-clonic seizures induced by pentylenetetrazole. Each seizure ends with a steep re- and hyperpolarization of the neuron, demonstrating that no hypoxia, which would result in a depolarization, is present. As to the basic processes underlying this phenomenon, the activation of a voltage-dependent calcium conductance might provide an explanation (Wong and Prince 1978; Witte et al. 1983). The increase of intracellular calcium leads to an activation of a calcium-activated potassium outward current, which, like in pacemaker cells of molluscans, results in a suppression of the burst activity and in a re- and hyperpolarization of the neuronal membrane (Lux and Hofmeier 1982). Explanations of the high degree of synchronization of seizure termination, as indicated in the field potential recording of Fig. 5, are not available as yet.

As a whole, only a limited number of studies concerning the termination of seizure activity are available. They strongly indicate that metabolic processes, as a rule, play only a minor role. Intrinsic mechanisms of the single neuron play a more important part.

References

Ayala GF, Matsumoto H, Gumnit RJ (1970) Excitability changes and inhibitory mechanisms in neocortical neurons during seizures. J Neurophysiol 33: 73-85

Calvin WH (1972) Synaptic potential summation and repetitive firing mechanisms: input-output theory for the recruitment of neurons into epileptic bursting firing patterns. Brain Res 39: 71-94

Caspers H, Speckmann E-J (1972) Cerebral PO_2, PCO_2 and pH: changes during convulsive activity and their significance for spontaneous arrest of seizures. Epilepsia 13: 699-725

Conners BW, Gutnick MJ (1984) Cellular mechanisms of neocortical epileptogenesis in an acute neuronal model. In: Schwartzkroin P, Wheal H (eds) Electrophysiology of epilepsy. Academic, London, pp 79-105

Dichter M, Spencer WA (1969) Penicillin-induced interictal discharges from the cat hippocampus: II. Mechanisms underlying origin and restriction. J Neurophysiol 32: 663-687

Dudek FE, Andrew RD, MacVicar BA, Snow RW, Taylor CP (1983) Recent evidence for and possible significance of gap junctions and electrotonic synapses in the mammalian brain. In: Jasper HH, van Gelder NV (eds) Basic mechanisms of neuronal hyperexcitability. Liss, New York, pp 31-70

Elger CE, Speckmann E-J (1983) Penicillin induced epileptic foci in the motor cortex: vertical inhibition. Electroencephalogr Clin Neurophysiol 56: 604-622

Elger CE, Wieser HG (1984) Pathophysiologie der Epilepsie. Schweiz Med Wochenschr 114: 1278-1288

Goldensohn ES, Purpura DP (1963) Intracellular potentials of cortical neurons during focal epileptogenic discharges. Science 139: 840-842

Gutnick MJ, Prince D (1974) Effects of projected cortical epileptiform discharges on neuronal activities in cat VPL: I. Interictal discharges. J Neurophysiol 37: 1310-1327

Heinemann U, Lux HD, Gutnick MJ (1977) Extracellular free calcium and potassium activity during paroxysmal activity in the cerebral cortex of the cat. Exp Brain Res 27: 237-243

Jackson JH (1870) A study of convulsions. Trans St Andrews Med Grad Ass 3: 1-45. In: Taylor J (ed) (1931-1932) Selected writings of John Hughlings Jackson. Hodder and Stoughton, London

Kreisman NR, Lamanna JC, Rosenthal M, Sick TJ (1981) Oxidative metabolic responses with recurrent seizures in rat cerebral cortex: role of systemic factors. Brain Res 218: 175-188

Lux HD (1974) The kinetics of extracellular potassium: relation to epileptogenesis. Epilepsia 15: 375-393

Lux HD, Hofmeier G (1982) Activation characteristics of the calcium-dependent outward potassium current in *Helix*. Pflügers Arch 394: 61-69

Matsumoto H, Ajmone-Marsan C (1964) Cortical cellular phenomena in experimental epilepsy: interictal manifestations. Exp Neurol 9: 305-326

Prince DA (1978) Neurophysiology of epilepsy. Ann Rev Neurosci 1: 395-415

Prince DA (1985) Physiological mechanisms of focal epileptogenesis. Epilepsia 26 (suppl): 3-14

Prince DA, Wilder BJ (1967) Control mechanisms in cortical epileptogenic foci: "surround" inhibition. Arch Neurol 16: 194-202

Schwartzkroin PA, Wyler AR (1982) Mechanism underlying epileptiform burst discharge. Ann Neurol 7: 95-107

Speckmann E-J, Caspers H (1973) Paroxysmal depolarization and changes in action potentials induced by pentylenetetrazole in isolated neurones of *Helix pomatia*. Epilepsia 14: 397-408

Speckmann E-J, Caspers H (1979) Cortical field potentials in relation to neuronal activities in seizure conditions. In: Speckmann E-J, Caspers H (eds) Origin of cortical field potentials. Thieme, Stuttgart, pp 205-213

Traub RD (1983) Cellular mechanisms underlying the inhibitory surround of penicillin epileptogenic foci. Brain Res 216-277-284

Tribble GL, Schwindt PC, Crill WE (1983) Reduction of postsynaptic inhibition tolerated before seizure initiation: Spinal cord. Exp Neurol 80: 288-303

Witte OW, Walden J, Speckmann E-J, Elger CE (1983) Reduction of penicillin-induced paroxysmal depolarization shifts of rat cerebral cortex by intracellular injection of a calcium channel blocker. Neurosci Lett [Suppl] 14: 405

Wong RKS, Prince DA (1978) Burst generation and calcium spikes during intrinsic burst firing in hippocampal neurons. Brain Res 159: 385-390

Wong RKS, Prince DA (1979) Dendritic mechanisms underlying penicillin induced epileptiform activity. Science 204: 1228-1231

Special Features of the Hippocampal Formation with Respect to Seizure Conditions*

H. L. Haas

Introduction

The hippocampus has long been known as a region particularly prone to epileptiform discharges (Kandel et al. 1961). Connections and physiology of this archaic cortical structure are relatively well characterized and a wealth of information on features favoring exaggerated neuronal activity has emerged in recent years. The lamellar organization of the hippocampus (Andersen et al. 1971) may be one of these features; it certainly has facilitated their investigation. Tissue slices cut along the lamellae, perpendicular to the axis of the structure, contain a relatively undisturbed chain of neurons which can be rigorously investigated in vitro. The results from such experiments have allowed modelling and imitation of hippocampal seizures on a computer (Traub et al. 1985). The properties of hippocampal neurons and their connections are discussed here successively for didactic reasons although they overlap functionally.

Institute of Physiology, University of Mainz, Saarstr. 21, 6500 Mainz, Federal Republic of Germany

* This study was supported by the Swiss National Science Foundation (3.002.0.84) and Stiftung für Wissenschaftliche Forschung an der Universität Zürich.

Thus, the hallmark of epileptic activity, the paroxysmal depolarization shift (PDS), could equally well be considered a synaptic (Johnston and Brown 1981) or an intrinsic event (Schwartzkroin and Prince 1980).

Synaptic Connnections

The hippocampus consists of principal neurons, densely packed in one layer of a three-layered structure, which form the members in an excitatory chain (Fig. 1). These granule cells in the dentate area and pyramidal cells in Ammon's horn are surrounded by inhibitory interneurons (basket cells) with short axons. The latter receive excitatory inputs from and are inhibitory to the former. The function of other interneurons is not yet well understood.

Synaptic Excitation

Excitatory transmitters are amino acids, glutamate and aspartate being the most prominent candidates. An exaggerated excitatory transmission, either increased release or increased postsynaptic sensitivity, is one attractive hypothesis for epileptogenesis. It

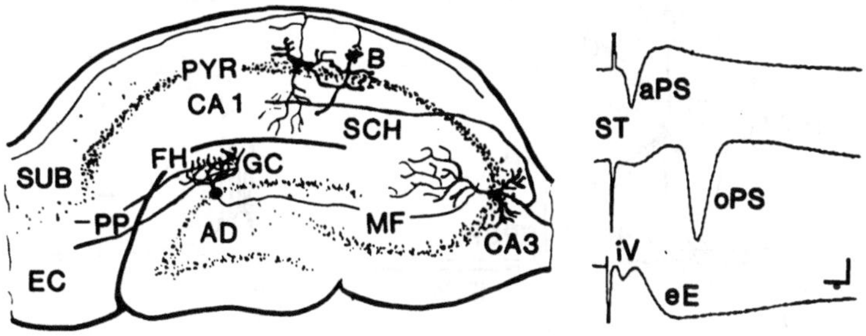

Fig. 1. Schematic representation of the hippocampal architecture and extracellularly recorded field potentials. *AD*, area dentata; *CA*, cornu ammonis; *SUB*, subiculum; *EC*, entorhinal cortex; *FH*, fissura hippocampi; *PP*, perforant path; *MF*, mossy fibers; *SCH*, Schaffer collaterals; *PYR*, pyramidal cell layer; *GC*, granule cell; *B*, basket cell; *ST*, stimulus artifacts; *aPS*, population spike in the CA1 cell body layer after alveus (antidromic) stimulation; *oPS*, synaptically evoked population spike in the same location after stimulation of the Schaffer collateral/commissural fibers; *iV*, input volley; *eE*, extracellular EPSP, synaptic field potential recorded in the apical dendritic region of CA1 after afferent stimulation. Negative is downward in this and all other figures. Calibration: 2 ms, 5 mV

Presurgical Evaluation of Epileptics
Edited by H. G. Wieser and C. E. Elger

has been suggested that the transmitters could act as excitotoxins mediating the anoxic brain damage (Meldrum 1985). CA3 pyramidal cells fire normally in bursts, and epileptiform activity in the intact structure arises from CA3 (Wong and Traub 1983). These pyramids have excitatory connections which can account for the triggering of a synchronous discharge of many cells from a single one (Miles and Wong 1983).

Electrical Synapses

Electrotonic (gap) junctions have been demonstrated between CA3 pyramids but their role in regulating neuronal excitability or synchronization is probably not important (McVicar and Dudek 1981).

Synaptic Inhibition

Classical experimental models of epilepsy like the penicillin paradigm depend on a loss of a GABAergic inhibition mediated by interneurons and a chloride conductance change (Dichter and Spencer 1969; Schwartzkroin and Prince 1977; Dingledine and Gjerstad 1980; Alger and Nicoll 1982a; Swann and Brady 1984; and author's Fig. 2). Recurrent and feed-forward inhibition is found on soma and dendrites. GABA hyperpolarizes the somata but has a mainly depolarizing effect on dendrites. The dendritic effect is nevertheless inhibitory as a large conductance change shunts the incoming excitation (Andersen et al. 1980; Alger and Nicoll 1982b). Working on (B-receptors, GABA can also activate a potassium conductance (Newberry and Nicoll 1984; Gähwiler and Brown 1985; Haas et al. 1985) which may account for pre- and postsynaptic inhibition. The inhibitory interneurons prevent CA1 pyramidal cells and dentate granule cells from expressing their intrinsic bursting properties (Wong and Prince 1979; Hablitz 1984). Transmitters or modulators that depress interneuron activity selectively are therefore epileptogenic in the hippocampus. Thus morphine has anticonvulsant properties on the whole organism but causes marked epileptiform activity in this structure.

Synaptic Plasticity

Short, subconvulsive, tetanic activation of afferent fibers leads to a persistent enhancement of excitatory transmission in the hippocampus. This long-term potentiation (LTP) was first described by Bliss et al. (1973) and has since attracted much attention as a possible learning paradigm. Repeated administration of similar tetani leads to a progressively epileptiform response with afterdischarges and finally generalized seizures. The cellular mechanisms for this kindling (Goddard et al. 1969) and LTP are probably the same although the hippocampus is not the most sensitive structure for kindling. Kindling and certain forms of epilepsy (e.g., the mirror focus) may thus be considered as a perversion of basic learning mechanisms (Fig. 3). Increased transmitter release and an enhanced postsynaptic response are likely to be involved in these phenomena. Interneuron-mediated disinhibition is not responsible, but an intrinsic disinhibition involving a reduced calcium-activated potassium conductance may operate pre- and postsynaptically (Haas and Rose 1984).

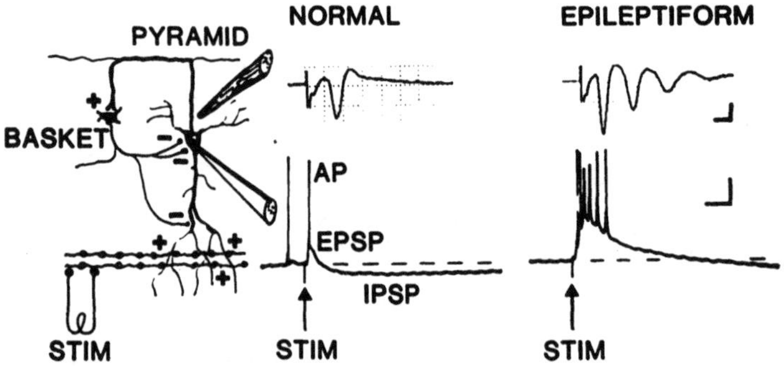

Fig. 2. Epileptiform responses to orthodromic stimulation after blockade of GABAergic basket cell inhibition in vitro. *Left*, schematic drawing of experimental situation. *Upper traces*, extracellularly recorded population spikes before and after addition of penicillin (2000 iU). *Lower traces*, intracellularly recorded EPSP - IPSP sequence before and after addition of bicuculline (10 μmol/liter) to the bathing solution. The IPSP is blocked and a large EPSP is unmasked which fires six (instead of one) action potentials. A late hyperpolarizing component of the synaptic potential is preserved. This is presumably mediated by GABA B receptors and due to a potassium conductance. (Calibrations: *upper traces*, 1 mV, 4 ms; *lower traces*, 20 mV, 50 ms)

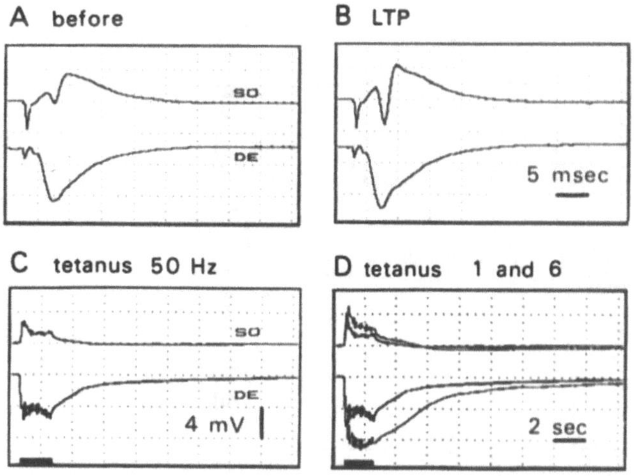

Fig. 3 A–D. Long-term potentiation and "kindling" in somatic and dendritic areas of CA1. **A, B** Field potentials recorded after (orthodromic) stimulation of the Schaffer collateral/commissural input. **B** 10 min after a tetanic stimulation (2 s, 50 Hz). The extracellularly recorded EPSP and population spike are enhanced. **C** Recording of the tetanus used to elicit LTP in the same experiment (indicated by *black bar below traces*). Further tetani were given every 10 min and the response to the sixth tetanus (after 1 h) is superimposed on the first one in **C** (DC recording). This field reflects depolarization of the cells and duration of calcium inflow, which is much larger in the "kindled" situation (**D**) especially in the dendritic region. **C** and **D** show only an envelope potential, not the single population spikes (in vitro, rat)

Field Interactions

After prolonged exposure to low-Ca high-Mg media, in the absence of synaptic transmission, the pyramidal cells in the CA1 area synchronize their firing spontaneously (Jefferys and Haas 1982; Haas and Jefferys 1984; Taylor and Dudek 1984). These field bursts or spreading excitation have been shown to depend on electrical field effects (ephaptic transmission). Structures with a laminar organization, such as the hippocampus, generate large field potentials during synchronous activity. Figure 4 illustrates that population spikes are associated with a depolarization of the transmembrane potential sufficient to trigger action potentials. Although field interactions may only play a minor role in normal function, they contribute significantly to synchronization in an excited population of densely packed neurons such as the CA1 pyramids.

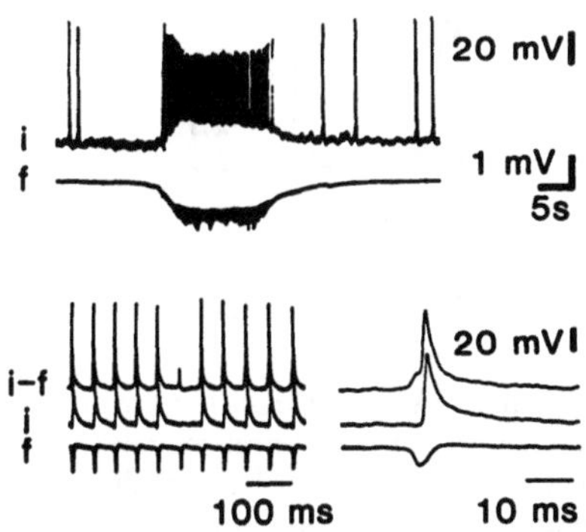

Fig. 4. Spreading excitation or field burst in the CA1 area. *i*, intracellular record; *f*, field recorded with a separate, nearby electrode; *i-f*, transmembrane potential. Lower traces show action potentials beginning with a negative (downward) potential. This reflects the electric field from the synchronous discharge of neighboring neurons, which is sufficient to entrain the recorded cell (in vitro, rat)

Extracellular Ion Concentrations

Epileptiform discharges are accompanied by a fall in Ca and an increase in K ions in the extracellular space (Heinemann et al. 1977). Both effects reinforce or cause excitation. The K accumulation is counteracted by the bufferring function of glia but if this is overrun, a wave of high K (ca. 30 mmol/liter) can move through the tissue, ignoring anatomical borders. This event is called spreading depression (Leao 1944; and author's Fig. 5). Such excitatory waves, followed by a marked and much longer lasting depression, may occur during seizures and have been seen in the retina during migraine attacks. As the transmembrane potential of the neurons approaches zero, action potentials are inactivated, and the input resistance is near zero for about 30 s. Later on, for many minutes the cell is hyperpolarized probably through an action of electrogenic pumps and perhaps an accumulation of adenosine (Greene and Haas 1985).

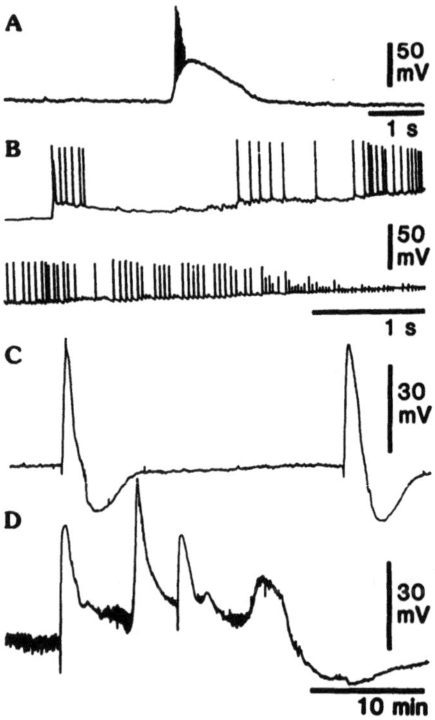

Fig. 5 A–D. Intracellular recording of abnormal excitation in hippocampal slices. **A** Paroxysmal depolarization shift; **B** continuous record of a spontaneous sudden irreversible depolarization in a CA1 pyramid recorded with a cesium chloride-filled electrode (Cs has partially blocked potassium conductances and prevents repolarization). As the potential shift continues, neighboring cells become entrained; at the end of the record there is no more action potential generation but field potentials are apparent. Recording was from a completely submerged slice in which spontaneous spreading depression (SD) did not occur. **C** Two phases of spreading depression registered with a pen writer in a half-submerged slice (fast potentials are not visible); **D** same cell after (transient) treatment with ouabain (1 μmol/liter). The hyperpolarizing phase of SD has disappeared; it only recovers at the end of the record, 20 min after ouabain washout. Note the (downward) field component at the beginning of some SD phases

Intrinsic Properties of Hippocampal Neurons

Every single neuron has a certain repertoire of mechanisms which governs its response to changes in voltage or neurotransmitters. These include electrogenic pumps and ion channels which have, through their specific activation-inactivation properties and their location on the neuron (soma, dendrites), quite selective and marked influences on the firing pattern.

Voltage-Dependent Inward Currents

The TTX-sensitive Na current responsible for the rising phase of the action potential has a transient and a noninactivating component (Hotson et al. 1979). Slower transient and persistent inward currents are blocked by Cd and are most probably Ca currents (Schwartzkroin and Slawsky 1977; Johnston et al. 1980; Halliwell 1983; Brown and Griffith 1983). These can generate Ca spikes and sustained depolarizations but are normally curtailed by the action of outward potassium currents (see below). Abnormalities in the time course and extent of inactivation of inward currents may be relevant for the production of epileptic potentials such as the PDS.

Voltage-Dependent Outward Currents

Several different potassium currents have been identified in pyramidal cells. The differences between them are in the voltage ranges and the time courses of activation and inactivation. The delayed rectifier, I_K (Segal and Barker 1984) is involved in the repolarization of action potentials and an early afterhyperpolarization. I_C (Brown and Griffith 1983), a Ca-dependent current, is also rapidly activated. Tetraethylammonium (TEA) blocks these currents. Another transient current, I_A (Gustafsson et al. 1982), is blocked by 4-aminopyridine (4-AP). Both TEA and 4-AP are convulsants and increase excitatory postsynaptic potentials (EPSPs). I_M (Halliwell and Adams 1982) is probably involved in the accommodation of firing and is blocked by muscarinic agonists. A slowly activating and inactivating Ca-dependent current which is responsible for accommodation of firing (Madison and Nicoll 1984), and the long-lasting afterhyperpolarization (AHP, author's Fig. 6; Alger and Nicoll 1980; Hotson and Prince

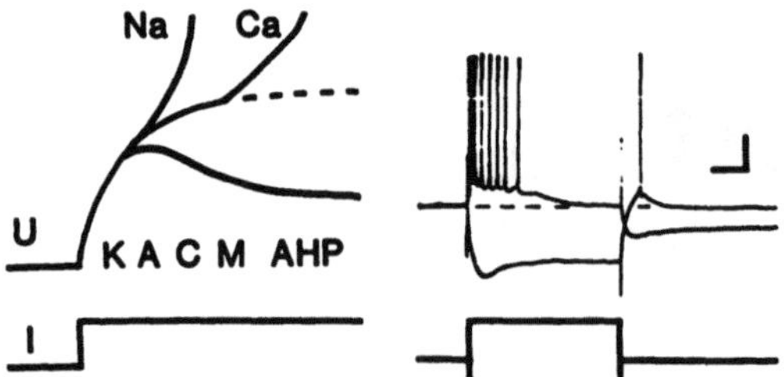

Fig. 6. Intrinsic properties of a CA1 cell. *On the left,* the deviation from the passive voltage response *(U)* of the membrane to depolarizing current injection *(I)* is shown schematically. Passive response: *exponential, broken line.* Sodium and calcium inward currents further depolarize the cell, while several potassium outward currents (I_K, I_A, I_C, I_M, I_{AHP}, see text) repolarize or prevent the neuron from reaching the firing threshold. *On the right,* response of a CA1 pyramid to ± 1-nA current injection. The depolarizing pulse evokes firing and a marked accommodation and afterhyperpolarization. The hyperpolarizing pulse evokes also a nonpassive response (I_Q)

1980) following single, or more pronounced, bursts of action potentials and Ca spikes, has been termed I_{AHP} (Adams and Lancaster 1985). This AHP is blocked by histamine (Haas and Konnerth 1983), noradrenaline (Madison and Nicoll 1982), corticotropin-releasing factor (Aldenhoff et al. 1983), cyclic AMP, and some phorbolesters (Baraban et al. 1985), and is enhanced by adenosine (Haas and Greene 1984). It seems that these substances act on the intracellular availability (sequestration) of Ca rather than directly on the K or the Ca channels. In this way the cellular excitability is linked to the genetic setup, the energy metabolism, and other biochemical events. Two of the potassium currents are sensitive to endogenous neuroactive substances (I_M and I_{AHP}) and it is likely that release of those from nerve endings normally regulates the response to excitatory signals. The frequent occurrence of seizures just after awakening may be explained by the known increased firing of locus coeruleus neurons which provide a diffuse noradrenergic innervation of the forebrain.

Conclusion

Dysregulations of all the above-mentioned mechanisms could lead to epileptogenesis. Excitatory and inhibitory events, synaptic or intrinsic, are closely interrelated and we are just beginning to understand the relative importance of the several contributing factors. At least in the hippocampus, postsynaptic regulation of excitability seems to occur mainly through a primary interaction with inhibitory events. Thus Ca spikes (and Ca currents) which are generally assumed to play a major role in epileptogenesis are not specifically and directly influenced by any drug or neuromodulator so far studied, but are profoundly modulated by K currents. Only divalent cations like Cd, Mn, Mg, Ni, and Co block Ca currents directly. These are not anticonvulsants; in fact Co ions are often used to elicit experimental epilepsy (Buchert-Rau and Sonnhof 1982). The dense packing of hippocampal principal neurons and the orientation of their dendrites, as well as the recurrent inhibition through GABAergic interneurons, predisposes the structure to synchronization and rhythmic activity. The capability of the cells to produce epileptiform potentials, large depolarization shifts, is normally under control of synaptic and intrinsic inhibitory mechanisms. Although the GABAergic inhibition has predominantly been related to epileptogenesis in the past, intrinsic inhibitions, in particular the voltage- and Ca-dependent K currents, have been attracting more attention recently.

References

Adams PR, Lancaster B (1985) Components of Ca-activated K current in rat hippocampal neurones. J Physiol 361: 23 P

Aldenhoff JB, Gruol DL, Rivier J, Vale W, Siggins GR (1983) Corticotropin releasing factor decreases postburst hyperpolarizations and excites hippocampal neurons. Science 221: 875-877

Alger BE, Nicoll RA (1980) Epileptiform burst afterhyperpolarization: calcium-dependent potassium potential in hippocampal CA1 pyramidal cells. Science 210: 1122-1124

Alger BE, Nicoll RA (1982a) Feed-forward dendritic inhibition in rat hippocampal pyramidal cells studied in vitro. J Physiol 328: 105-123

Alger BE, Nicoll RA (1982b) Pharmacological evidence for two kinds of GABA receptor on rat hippocampal pyramidal cells studied in vitro. J Physiol 328: 125-141

Andersen P, Bliss TVP, Skrede KK (1971) Lamellar organization of hippocampal excitatory pathways. Exp Brain Res 13: 222-238

Andersen P, Dingledine R, Gjerstad L, Langmoen IA, Mosfeldt-Laursen A (1980) Two different responses of hippocampal pyramidal cells to application of gamma-amino butyric acid. J Physiol 305: 279-296

Baraban JM, Snyder SH, Alger BE (1985) Protein kinase C regulates ionic conductance in hippocampal pyramidal neurons: electrophysiological effects of phorbol esters. Proc Natl Acad Sci USA 82: 2538-2542

Bliss T, Lomo T, Gardner-Medwin A (1973) Long-lasting potentiation of synaptic transmission in the dentate area of the anaesthetized rabbit following stimulation of the perforant path. J Physiol 232: 331-374

Brown DA, Griffith WH (1983) Calcium-activated outward current in voltage-clamped hippocampal neurones of the guinea-pig. J Physiol 337: 287-301

Buchert-Rau B, Sonnhof U (1982) An analysis of the epileptogenic potency of Co^{2+} - its ability to induce acute convulsive activity in the isolated frog spinal cord. Pflügers Arch 394: 1-11

Dichter M, Spencer WA (1969) Penicillin induced interictal discharges from the cat hippocampus: 1. Characteristics and topographical features. J Neurophysiol 32: 649-663

Dingledine R, Gjerstad L (1980) Reduced inhibition during epileptiform activity in the in vitro hippocampal slice. J Physiol 305: 297-313

Gähwiler BH, Brown DA (1985) GABA B receptor-activated K^+ current in voltage-clamped CA3 pyramidal cells in hippocampal cultures. Proc Natl Acad Sci USA 82: 1558-1562

Goddard CV, McIntyre DC, Leech CK (1969) A permanent change in brain function resulting from daily electrical stimulation. Exp Neurol 25: 295-330

Greene RW, Haas HL (1985) Adenosine actions on CA1 pyramidal neurons in rat hippocampal slices. J Physiol 366: 119-127

Gustafsson B, Galvan M, Grafe P, Wigström H (1982) A transient outward current in a mammalian central neurone blocked by 4-aminopyridine. Nature 299: 252-254

Haas HL, Greene RW (1984) Adenosine enhances afterhyperpolarization and accommodation in hippocampal pyramidal cells. Pflügers Arch 402: 244-247

Haas HL, Jefferys JGR (1984) Low calcium field burst discharges of CA1 pyramidal neurones in rat hippocampal slices. J Physiol 354: 185-201

Haas HL, Konnerth A (1983) Histamine and noradrenaline decrease calcium-activated potassium conductance in hippocampal pyramidal cells. Nature 302: 432-434

Haas HL, Rose G (1984) The role of inhibitory mechanisms in hippocampal long term potentiation. Neurosci Lett 47: 301-306

Haas HL, Jefferys JGR, Slater NT, Carpenter DO (1984) Modulation of low calcium induced field bursts in the hippocampus by monoamines and cholinomimetics. Pflügers Arch 400: 28-33

Haas HL, Greene RW, Olpe H-R (1985) Stereoselectivity of l-baclofen in hippocampal slices of the rat. Neurosci Lett 55: 1-4

Hablitz JJ (1984) Picrotoxin-induced epileptiform activity in hippocampus: role of endogenous versus synaptic factors. J Neurophysiol 51: 1011-1027

Halliwell JV (1983) Calcium-loading reveals two distinct Ca-currents in voltage-clamped guinea-pig hippocampal neurons in vitro. J Physiol 341: 10-11

Halliwell JV, Adams PR (1982) Voltage-clamp analysis of muscarinic excitation in hippocampal neurons. Brain Res 250: 71-92

Heinemann U, Lux HD, Gutnick MJ (1977) Extracellular free calcium and potassium during paroxysmal activity in cerebral cortex of the cat. Exp Brain Res 27: 237-243

Hotson JR, Prince DA (1980) A calcium-activated hyperpolarization follows repetitive firing in hippocampal neurons. J Neurophysiol 43: 409-419

Hotson JR, Prince DA, Schwartzkroin PA (1979) Anomalous inward rectification in hippocampal neurons. J Neurophysiol 42: 889-895

Jefferys JGR, Haas HL (1982) Synchronized bursting of CA1 hippocampal pyramidal cells in the absence of synaptic transmission. Nature 300: 448-450

Johnston D, Brown TH (1981) Giant synaptic potential hypothesis for epileptiform activity. Science 211: 294-297

Johnston D, Hablitz JJ, Wilson WA (1980) Voltage clamp discloses slow inward currents in hippocampal burst firing neurones. Nature 286: 391-393

Kandel ER, Spencer WA, Brinley FJ (1961) Electrophysiology of hippocampal neurons. J Neurophysiol 24: 225-285

Leao AAP (1944) Spreading depression of activity in the cerebral cortex. J Neurophysiol 7: 359-390

Madison DV, Nicoll RA (1982) Noradrenaline blocks accommodation of pyramidal cell discharge in the hippocampus. Nature 299: 636-638

Madison DV, Nicoll RA (1984) Control of the repetitive discharge of rat CA1 pyramidal neurones in vitro. J Physiol 354: 319-331

McVicar BA, Dudek FE (1981) Electrotonic coupling between pyramidal cells: a direct demonstration in rat hippocampal slices. Science 213: 782-784

Meldrum B (1985) Excitatory amino acids and anoxic/ischaemic brain damage. TINS 8: 47-48

Miles R, Wong RKS (1983) Single neurones can initiate synchronized population discharge in the hippocampus. Nature 306: 371-373

Newberry NR, Nicoll RA (1984) Direct hyperpolarizing action of baclofen on hippocampal pyramidal cells. Nature 308: 450-452

Schwartzkroin PA, Prince DA (1977) Penicillin-induced epileptiform activity in the hippocampal in vitro preparation. Ann Neurol 1: 463-469

Schwartzkroin PA, Prince DA (1980) Changes in excitatory and inhibitory synaptic potentials leading to epileptogenic activity. Brain Res 183: 61-76

Schwartzkroin PA, Slawsky M (1977) Probable calcium spikes in hippocampal neurons. Brain Res 135: 157-161

Segal M, Barker JL (1984) Rat hippocampal neurons in culture: potassium conductances. J Neurophysiol 51: 1409-1433

Swann JW, Brady RJ (1984) Penicillin-induced epileptogenesis in immature rat CA3 hippocampal pyramidal cells. Dev Brain Res 12: 243-254

Taylor CP, Dudek FE (1984) Excitation of hippocampal pyramidal cells by an electrical field effect. J Neurophysiol 52: 126-142

Traub RD, Dudek FE, Taylor CP, Knowles WD (1985) Simulation of hippocampal afterdischarges synchronized by electrical interactions. Neuroscience 14: 1033-1038

Wong RKS, Prince DA (1979) Dendritic mechanisms underlying penicillin-induced epileptiform activity. Science 204: 228-231

Wong RKS, Traub RD (1983) Synchronized burst discharge in disinhibited hippocampal slice. 1. Initiation in CA2-CA3 region. J Neurophysiol 49: 442-458

The Kindling Model of Epilepsy of the Hippocampus in the Rat: Electrophysiology and Behavior*

F. H. Lopes da Silva

Repeated tetanization at regular intervals, i.e., kindling of a brain structure, leads ultimately to the development of an epileptogenic focus. This is the so-called kindling model of epilepsy, described first by Goddard (1967). Several aspects of the phenomenology of the kindling model have been investigated, as reviewed by Racine (1978) and McNamara et al. (1980). In this study, we have tried to reply to two different questions regarding the development of an epileptogenic focus due to kindling: (a) which electrophysiological changes are responsible for this process and (b) can the kindling procedure offer a model not only for motor seizures but also for interictal behavioral deficits?

In this chapter, we will review briefly these two aspects (parts A, B) of our current research, without entering into detailed technical descriptions which are published elsewhere (Wadman et al. 1985; Kamphuis et al. 1985).

Part A: Electrophysiological Processes Underlying Kindling Epileptogenesis

In order to study such processes, it is important to choose a suitable brain area where detailed neurophysiological analysis can be carried out. For this purpose, we chose the hippocampus of the rat for a number of reasons: (a) the hippocampus has a laminated structure which offers the possibility of determining accurately the neuronal sources of field potentials; this is of importance since we wish to find out which synaptic processes are affected by the kindling form of stimulation; (b) a good deal of information is available regarding the basic physiology of hippocampal networks of the rat (Leung 1979; Leung et al. 1982; Wadman et al. 1983, 1985) and the cellular properties of hippocampal cells (Andersen et al. 1980; Schwartzkroin 1975; Knowles et al. 1984); (c) there is also a wealth of information concerning the histological structure of the hippocampus, not only at light microscopic level but also at the ultrastructural level (Ribak and Anderson 1980; Ribak 1985), and in relation to different systems of neurotransmitters (Storm-Mathisen and Ottersen 1984).

Material and Methods

Wistar male rats were chronically implanted under pentobarbitone anesthesia with stainless-steel electrodes (diameter 100 μm, cut sharp) in the dorsal hippocampus for recording and stimulation. The electrodes were placed in such a way that the Schaffer collaterals in CA_1 fields were optimally stimulated. Single-pulse stimulation of this pathway evoked a field potential which reached maximal negativity (corresponding to the extracellular EPSP) at the level of stratum radiatum and became positive toward the stratum pyramidale. Usually, a bundle consisting of three electrodes was used for stimulating and at least two electrodes were used for recording: one was placed in stratum radiatum and the other above it in stratum pyramidale. Reference electrodes were screws placed on the skull. The electrodes were cemented in place and connected to a socket which was fixed to the skull. On the plug FET amplifiers were mounted, functioning as pre-amplifiers and connected via a cable to a commutator and standard amplifiers. Recordings were performed about 1 week after implantation. The signals were sampled by a microcomputer which was programmed for obtaining averages.

The kindling procedure consisted in stimulating the Schaffer collaterals with 1-s train pulses (50 Hz, 0.2 ms duration), two or three times a day at intervals of at least 3 h. The procedure was carried out every

Department of Zoology, University of Amsterdam, Kruislaan 320, 1098 SM Amsterdam, The Netherlands

* This work was in part subsidized by CLEO (Commissie Landelijke Epilepsie Onderzoek).

Presurgical Evaluation of Epileptics
Edited by H. G. Wieser and C. E. Elger

day until generalized convulsions occurred. The stimulus intensity for kindling was usually the same as that necessary to obtain a field potential with an amplitude equal to half of the saturation level.

At the end of an experiment, which usually lasted 4 or 6 weeks, the rats were put to death and the brain was perfused for further histological investigations. The localization of the electrodes was confirmed histologically.

Results

The kindling stimulus leads to the occurrence of afterdischarges (ADs) which tend to increase in length in the course of time. Another EEG phenomenon which is characteristic of the development of an epileptogenic focus is the occurrence of interictal epileptiform transients of different types, as already published elsewhere (Wadman et al. 1983). In order to probe the state of excitability of the neural network within and around the area where the epileptogenic focus becomes established, we used the method of paired-pulse stimulation (cf. Creager et al. 1980). We pay particular attention here to the changes in the corresponding field potentials evoked by Schaffer stimulation as they take place during the kindling process. In Fig. 1 a comparison of the field potentials to paired-pulse stimulation obtained during the control period (session 1), the middle period (session 9), and the final period (session 17) is shown. The main characteristics of paired-pulse responses in the control period were as follows:

- The response to the first pulse was characterized by an initial negative-going deflection at the level of the stratum radiatum which lasted for 10–12 ms and was followed by a "late" positive-going slower deflection, if the stimulus intensity was above a certain level.
- The response to the second pulse, which was usually given 20 ms after the first, also showed a negative-going deflection followed by a slower component of opposite polarity. At low intensities, the amplitude of the second negative wave was larger than that of the first; this phenomenon is known as paired-pulse facilitation (Creager et al. 1980). At high intensities, however, the second was clearly smaller than the first.

Before examining the changes in these field potentials, we have to understand the basic physiology of these responses. Even at low intensities, one stimulus evokes an EPSP at the level of the synapses formed by the Schaffer collaterals and the apical dendrites of the pyramidal cells (Lopes da Silva et al. 1984). However, this weak stimulus may not be sufficient to provoke cell firing and thus the inhibitory interneurons which are normally excited by collaterals of the axons of CA_1 pyramidal cells may not be activated (Andersen et al. 1964). The "late" positive component occurs at the time where intracellularly the early IPSP can be recorded (Knowles et al. 1984). This inhibitory component will not occur in the case of weak stimuli. The second stimulus may elicit a larger field EPSP probably due to the mobilization of neurotransmitter, as is currently assumed (Creager et al. 1980). In case the strength of the first stimulus is sufficient to elicit cell firing, the inhibitory pathway may be activated and thus a field IPSP may occur. This inhibitory pathway may be recurrent and/or feed-forward (Buszaki and Eidelberg 1982).

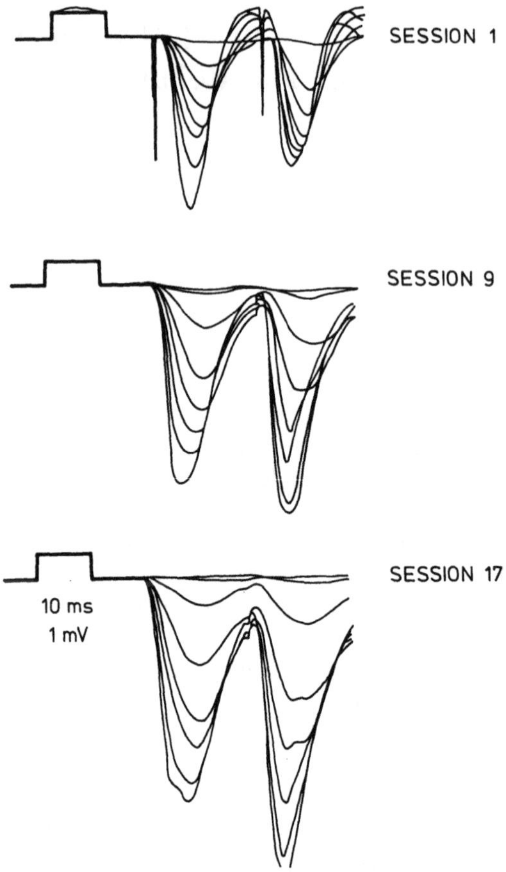

Fig. 1. CA field potentials to paired-pulse stimulation of the Schaffer collaterals. Recording is from one electrode placed at the level of the apical dendrites (stratum lacunosum-moleculare) and ipsilaterally to the site of kindling stimulation. Different traces correspond to field potentials evoked by stimuli of different strengths. Note the changes in field potentials along kindling sessions (1st, 9th and 17th): (a) the change in the positive-going wave occurs after the field EPSP; (b) the increase in amplitude of the field EPSPs; (c) the change in the ratio between the amplitude of the field EPSP evoked by the second pulse in relation to that evoked by the first

If the second stimulus is given during this early IPSP the resulting field EPSP has a smaller amplitude. Furthermore, we cannot exclude that other types of hyperpolarization caused by intrinsic membrane permeability changes may also contribute to this "late" positive component of the evoked potential.

We are now in the position to describe the changes occurring in field potentials during kindling. The main changes found were:

- The amplitude of the extracellular EPSP to the first stimulus increased as well as the duration, particularly the time constant of the decay.
- The "late" positive component tended to disappear and to be substituted by a potential with negative polarity.
- The ratio between the amplitude of the extracellular EPSP evoked by the first and by the second stimulus changed drastically in favor of the second response.

Discussion

The changes in field potentials evoked by Schaffer collaterals onto CA_1 pyramidal cells occurring during kindling may be interpreted as follows:

- The increase in amplitude of the initial extracellular EPSP may be due to an increase in the efficiency of excitatory synapses, for example, increased transmitter mobilization and/or a decrease in feed-forward inhibition.
- The conspicuous disappearance of the "late" positive component can be accounted for by a decrease in the inhibition responsible for the initial IPSP; in addition, the decrease in this "late" component might also result from a change in intrinsic membrane permeability for Ca^{2+} and/or K^+ ions. The first possibility is discussed below but there is, at present, no sufficient experimental data to distinguish between the possible contributions of these different processes.
- The change in the ratio between the first and second responses is a consequence of the changes indicated above; the decrease in inhibition will unmask the phenomenon of paired-pulse facilitation, which, as described above, can be found in control situations, when the stimulus intensity is low.

A central feature in all these changes is the disinhibition which appears to occur in the neuronal network at the kindling site. Therefore, we reasoned that if this is indeed a basic process responsible for the change in neuronal excitability during kindling, it should be possible to find a change in the transmitter system responsible for the inhibition. It is known that the initial IPSP is mediated by GABAergic synapses, i.e., synapses which have gamma-aminobutyric acid as transmitter (Knowles et al. 1984). Therefore, we investigated whether neurons could be visualized in the hippocampus of kindled rats using specific antibodies for GABA. Preliminary results (Kamphuis et al. 1986) indicate that a change in GABA immunoreactivity indeed takes place. This may therefore represent an important element in the development of kindling epileptogenesis. It has also been shown by Wadman et al. (1985) that in hippocampal slices of kindled rats there is a conspicuous change in the dynamics of extracellular Ca^{2+}. In response to a train of pulses or tetanus, slices of kindled animals show a significantly larger decrease in extracellular Ca^{2+} than those of controls. This suggests that the cellular elements within a kindled epileptogenic focus present an increased permeability for Ca^{2+}. It is, as yet, not clear how these changes in Ca^{2+} permeability may be related to changes in the field potentials. An answer to this question may indeed provide a cornerstone to our understanding of the fundamental cellular processes of kindling epileptogenesis.

Part B: Hippocampal Kindling Affects Spatial Memory Performance

There is ample evidence, most of it resulting from studies of bilateral lesions, that the hippocampus plays a role in memory tasks involving spatial cues as revealed in radial mazes (Olton et al. 1979; Jarrard 1983). We reasoned that if hippocampal kindling would produce long-lasting deficits in hippocampal function, these could be made apparent by testing rats in such a maze during the process of epileptogenesis. Preliminary results have been published elsewhere (Lopes da Silva et al. 1986).

Material and Methods

Fourteen male Wistar rats were successfully implanted as described in part A. The rats were maintained at an average weight of 300–350 g during the whole experimental period and in a reverse light-dark cycle of 12:12 h. The behavioral tests were performed in an eight-arm radial maze similar to the one described by Olton and Samuelson (1976); the behavioral procedures followed were as described by those authors. In our case five adjacent arms were baited with food at the beginning of each trial and the other three arms were always empty. The correct response of the rat was to enter each of the five baited arms only once. Deviations from this response were scored as errors and could be divided into two different classes: (1) if a rat entered an arm for the second time this was scored as a working memory (WM) error; (2) when it entered an unbaited arm for the first time this was scored as a reference memory (RM) error.

If a rat entered an unbaited arm more than once, this was scored as a WM error; since it seldom happened in the kindled rats (6 out of a total of 196 trials) and never in controls, scoring these errors as RM instead of WM errors did not affect the conclusions. Comparisons between kindled rats and controls were carried out by using the scores of rats from identical periods of observation. Since no significant differences were found for controls in the various periods, the mean score for controls over the whole time span could equally well have been used; doing so had no influence on the conclusions shown in Table 1. The statistical tests used are indicated in the heading of this table.

Rats were initially trained until a stable performance was obtained; each trial lasted until they had chosen all five baited arms or until a time limit of 15 min was reached, which however rarely happened. After implantation the rats were allowed 2 weeks for recovery, followed by 25 days of retraining until stable performance was regained. Rats were randomly assigned to a control and a kindling group of equal size, which were treated identically except for the tetanization. Each rat was daily submitted to a behavioral test 1-2 h after the beginning of the dark period. A train of pulses or tetanus was applied 1-2 h after this test; an average of 5 h, a second tetanus was administered to reduce the time needed for epileptogenesis. Each stimulus evoked an afterdischarge that lasted between 20 s in the beginning and 100 s at the end of the kindling process. Kindling was continued until the first generalized seizure was observed (stage 5 of Racine 1978). This period varied from rat to rat (range 12-23 days) and will be called period I. From then onward the tetanus was applied once a day until nine generalized seizures had been observed (period II). Thereafter the behavioral tests were continued for seven consecutive days without tetanization, to study possible recovery (period III).

The performance of the rats in WM as well as in RM are given as a function of time in Fig. 2. Statistical comparison between the two groups and between different periods are summarized in Table 1. Error score was calculated as the mean number of errors per daily trial. No significant difference in WM or in RM scores was found between the two groups in the period before kindling started (two-tailed Mann-Whitney U-test).

First, we analyzed the comparison between kindled animals and controls. For the RM task, there were significant differences during all periods; the difference increased slightly from period I to II and was still present after kindling was discontinued (III). For the WM task a significant difference between kindled animals and controls was also found during kindling (I, II), but when stimulation was stopped, performance improved and it was no longer different from controls in period III.

Fig. 2 a, b. Distribution along trials (days) of errors made in WM (**a**) and RM (**b**) by kindled rats and controls. The period during which the animals showed generalized seizures is indicated by the *hatched bar* ▶

Table 1. Mean error scores in WM and RM. Mean number of errors per trial in working and reference memory for rats ($n=7$) and controls ($n=7$) for the three experimental periods (I, II, III). For controls, no significant differences among periods were found

Mean error scores					Comparison among periods for kindled rats[b]		
Experimental period		Kindled rats	Controls	Comparison: Controls - Kindled[a]		WM	RM
I	WM	0.22	0.01	$P<0.01$	I vs II	$P<0.05$	ns
	RM	1.65	0.95	$P<0.05$			
II	WM	0.62	0.06	$P<0.001$	II vs III	$P<0.01$	ns
	RM	1.60	0.68	$P<0.001$			
III	WM	0.18	0.11	ns	I vs III	ns	ns
	RM	1.39	0.56	$P<0.001$			

[a] Mann-Whitney U-test (two-tailed)
[b] Wilcoxon matched pairs sign-ranked test (two-tailed)

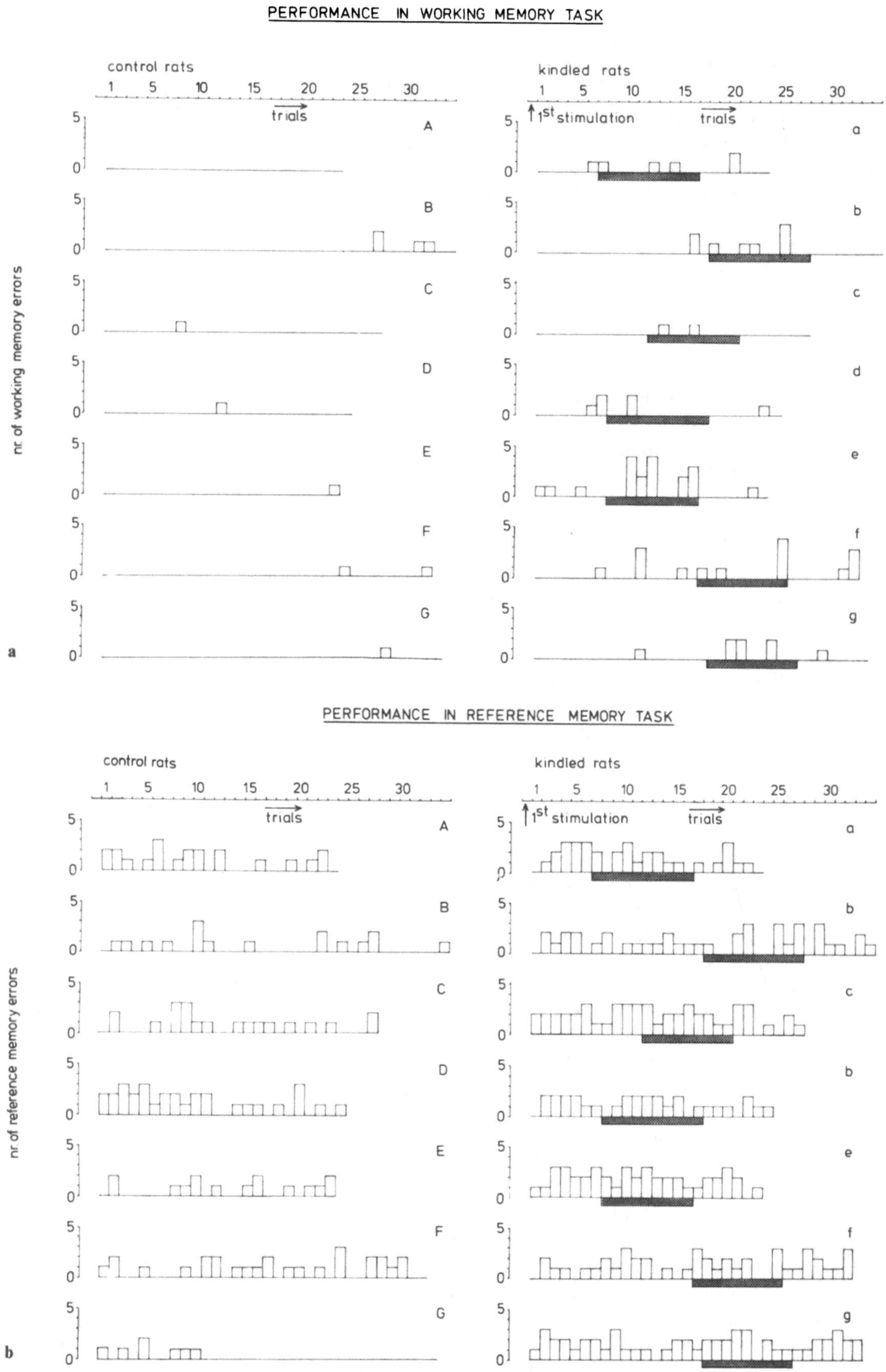
PERFORMANCE IN WORKING MEMORY TASK
control rats
kindled rats
trials
1st stimulation
nr. of working memory errors
A
B
C
D
E
F
G
a
b
c
d
e
f
g
PERFORMANCE IN REFERENCE MEMORY TASK
control rats
kindled rats
trials
1st stimulation
nr. of reference memory errors

Second, we compared the performance of the kindled rats among the three periods. Comparisons between any two different periods were made by Wilcoxon matched-pairs signed rank test (see Table 1). RM did not show significant changes between periods I and II, nor between periods II and III, which implies that there was hardly any recovery. However, WM performance significantly degraded from period I to II, but recovered from period II to III, even to such an extent that in period III it was not significantly different from in period I.

Discussion

It must be concluded that kindling of the left dorsal hippocampus is sufficient to disrupt the normal functioning of both hippocampi in such a way that a deficit in WM and RM is apparent. Furthermore, we were able to demonstrate that after stopping the kindling stimulation a certain degree of recovery is possible, which was at least significant for WM. In this respect, there appears to be a differentiation between the deficits of WM and RM, the latter lasting longer.

The behavioral deficits seen after left dorsal hippocampal kindling should be compared with behavioral changes produced in a similar or related way. Changes in working memory have only been reported after bilateral disconnection of the hippocampus (Olton et al. 1979), but a similar impairment in reference memory has been shown recently after kainic acid lesions of the hippocampus (Jarrard 1983). Ehlers and Koob (1985) showed that hippocampal kindling may cause a decrease in locomotor activity but this was limited to 2 h after a seizure; rats treated with injections of tetanus toxin in the hippocampus (George and Mellanby 1982) showed an amnestic syndrome in a discrimination task, but this treatment did not cause lasting impairment of short-term memory.

Each kindling stimulus may produce a certain degree of retrograde amnesia (Olton and Wolf 1981). The time between the test and the tetanus was, however, never shorter than 45 min and usually 1-2 h; the next behavioral test was 18-23 h later. Therefore, the findings presented here represent a long-lasting disruption of the processes involved in the spatial task that uses reference and working memory. We should, however, stress that the task relies on spatial cues (O'Keefe and Nadel 1978). This implies that we cannot generalize our findings without further explicit evidence to other memory tasks, which do not involve such cues. It seems of particular interest to relate these findings to the clinical observation that epileptic patients with limbic foci may have verbal memory deficits and that impairments in a number of cognitive functions correlate with focal limbic epileptiform activity (Wieser et al. 1985). In summary, hippocampus kindling provides not only a model of focal motor seizures but also of behavioral deficits correlated with epileptogenesis.

General Conclusions

The results obtained with the kindling model of epilepsy in the hippocampus of the rat described here enable the conclusion to be drawn that repeated electrical stimulation provokes a long-lasting impairment of hippocampal neural networks, which is revealed not only in a number of electrophysiological properties but also in specific behavioral deficits. The electrophysiological changes appear to consist mainly of a decrease of inhibitory processes, probably of the recurrent type. This will lead to a decrease in the stability of the network. In this connection, the hypothesis which we formulate here is that the kindling process will lead to a decrease in function of GABA-mediated synapses. This hypothesis is currently being verified using immunocytochemical methods. As mentioned in part A, one main question which remains unsolved is how such changes in GABA inhibition are related to the large increase in Ca^{2+} permeability which was shown to occur in kindled tissue by Wadman et al. (1985). This problem is a matter for future research.

Acknowledgments. Data reported here have been obtained in collaboration with W.J. Wadman, W. Kamphuis, and J. Gorter. Their criticisms and suggestions are gratefully acknowledged.

References

Andersen P, Eccles JC, Loyning Y (1964) Pathway of postsynaptic inhibition in the hippocampus. J Neurophysiol 27: 608-619

Andersen P, Dingledine R, Gjerstad L, Langmoen IA, Mosfeldt-Laursen A (1980) The different responses of hippocampal pyramidal cells to application of GABA. J Physiol 305: 279-296

Buszaki G, Eidelberg E (1982) Convergence of pyramidal cells of the rat hippocampus. Brain Res 237: 283-295

Creager R, Dunwiddie T, Lynch G (1980) Paired-pulse and frequency facilitation in the CA_1 region of the in vitro rat hippocampus. J Physiol 229: 409-424

Ehlers CL, Koob GF (1985) Locomotor behaviour following kindling in three different brain sites. Brain Res 326: 71-79

George G, Mellanby J (1982) Memory deficits in an experimental hippocampal epileptiform syndrome in rats. Exp Neurology 75: 678-689

Goddard GV (1967) Development of epileptic seizures through brain stimulation at low intensity. Nature 214: 1020-1021

Jarrard LE (1983) Selective hippocampal lesions and behaviour: effects of kainic acid lesions on performance of place and cue tasks. Behav Neurosci 97, no 6: 873-889

Kamphuis W, Wadman WJ, Buys RM, Lopes da Silva FH (1986) Decrease in number of gamma-aminobutyric acid (GABA) immunoreactive cells in the rat kindling model of epilepsy. Exp Brain Res 64(3) 491-495

Knowles WD, Schneiderman JH, Wheal HV, Stafstrom CE, Schwartzkroin PA (1984) Hyperpolarizing potentials in guinea pig hippocampal CA_3 neurons. Cell Mol Neurobiol 4(3): 207-230

Leung LS (1979) Orthodromic activation of hippocampal CA_1 region of the rat. Brain Res 176: 49-67

Leung LS, Lopes da Silva FH, Wadman WJ (1982) Spectral characteristics of the hippocampal EEG in the freely moving rat. Electroencephalogr Clin Neurophysiol 54: 203-219

Lopes da Silva FH, Wadman WJ, Arnolds DEAT, Veeken C, Holsheimer J (1984) Hippocampus: behaviour and modulation of local circuits. In: Reinoso-Suarez F, Ajmone-Marsan C (eds) Cortical integration. Raven, New York, pp 147-170

Lopes da Silva FH, Gorter J, Wadman WJ (1986) Kindling of the hippocampus induces spatial memory deficits in the rat. Neurosci Lett 63: 115-120

McNamara JO, Byrne MC, Dashief RM, Fitz JG (1980) The kindling model of epilepsy: a review. Prog Neurobiol 15: 139-159

O'Keefe J, Nadel L (1978) The hippocampus as a cognitive map. Clarendon, Oxford

Olton DS, Samuelson RJ (1976) Remembrance of places passed: spatial memory in rats. J Exp Psychol [Anim Behav] 2: 97-116

Olton DS, Wolf WA (1981) Hippocampal seizures produce retrograde amnesia without a temporal gradient when they reset working memory. Behav Neural Biol 33: 437-452

Olton DS, Becker JT, Handelman GE (1979) Hippocampus space and memory. Behav Brain Sci 2: 313-365

Racine R (1978) Kindling: the first decade. J Neurosurg 3: 234-252

Ribak CE (1985) Axon terminals of GABAergic chandelier cells are lost at epileptic foci. Brain Res 326: 251-260

Ribak CE, Anderson L (1980) Ultrastructure of the pyramidal basket cells in the dentate gyrus. J Comp Neurol 192: 903-916

Schwartzkroin PA (1975) Characteristics of CA1 neurons recorded intracellularly in the hippocampal in vitro slice preparation. Brain Res 85: 423-436

Storm-Mathisen J, Ottersen OP (1984) Neurotransmitters in the hippocampal formation. In: Reinoso-Suarez F, Ajmone-Marsan C (eds) Cortical integration. Raven, New York, pp 105-130

Wadman WJ, Lopes da Silva FH, Leung LS (1983) Two types of interictal transients of reversed polarity in rat hippocampus during kindling. Electroencephalogr Clin Neurophysiol 55: 314-319

Wadman WJ, Heinemann U, Konnerth A, Neuhaus S (1985) Hippocampal slices of kindled rats reveal calcium involvement in epileptogenesis. Exp Brain Res 57: 404-407

Wieser HG, Hailemariam S, Regard M, Landis T (1985) Unilateral limbic epileptic status activity: stereo EEG, behavioral, and cognitive data. Epilepsia 26(1): 19-29

Generation of Epileptiform Field Potentials in the Cerebral Cortex

E.-J. SPECKMANN and J. WALDEN

This chapter deals with the generation of potentials detectable in the space surrounding cellular elements of central nervous structures. Such field potentials have a prominent significance in the investigation of brain functions. Thus, the conventional electroencephalogram represents an essential tool in the diagnosis and classification of epileptic seizures as well as in the control of antiepileptic therapy. Also in experimental epileptology, field potentials have often been used in the functional analysis of neuronal populations. In the following, the mechanisms underlying the generation of seizure potentials in the cerebral cortex will be described. In this context, field potentials occurring during partial and generalized tonic-clonic seizures will be considered successively (Andersen and Andersson 1968; Caspers et al. 1984; Creutzfeldt and Houchin 1974; Goldensohn 1984; Klee et al. 1981; Lopes da Silva and van Rotterdam 1982; Prince 1974; Speckmann 1985; Speckmann and Caspers 1979a; Speckmann and Elger 1982, 1984; Speckmann et al. 1984).

Partial Seizure Activity

The findings described were obtained in animal experiments using model epilepsies. Such model epilepsies facilitate the analysis of field potential generation since, under these conditions, neuronal activity was found to be synchronized. Consequently, single elements can be taken as representative of the neuronal population concerned. Thus, recordings from single neurons allow a reliable interpretation of field potentials during seizure activity (Purpura et al. 1972).

Observations referred to in this section were made on the basis of partial seizure activity elicited by local application of penicillin. Typical potentials, led from the surface of the cortex and from a superficial cortical neuron after epicortical application of penicillin, are displayed in Fig. 1. In neurons contacted by the epileptogenic agent, at first flat depolarizations develop which trigger a group of action potentials and are terminated by steep repolarizations (Fig. 1 A, C). In the further course, these membrane potential (MP) changes are transformed into characteristic paroxysmal depolarization shifts. They consist of a steep depolarization giving rise to a burst of action potentials, of a long-lasting plateau-like diminution of the MP during which the somatic generator of action potentials is inactivated, and of a steep repolarization (Fig. 1 C). Concomitant with the appearance of the epileptic neuronal depolarizations, negative potential fluctuations develop in the space surrounding the neurons in question (Fig. 1 B). These relations between the changes of the intracellularly recorded MP and the extracellularly recorded field potential are strong. With an ongoing action of penicillin, the neuronal depolarizations and the negative field potentials increase in amplitude and duration. Finally, the extent of the two bioelectric events reaches a value which stays constant for some hours. These findings demonstrate that, in a fully established epileptic focus, paroxysmal neuronal depolarization shifts are associated with negative field potentials. Thus, the paroxysmal neuronal MP changes lead obviously to those currents in the extracellular space which are responsible for the generation of the typical negative seizure potentials (Creutzfeldt et al. 1966; Elger and Speckmann 1983a, b; Prince 1974; Speckmann and Caspers 1979b; Speckmann and Elger 1984).

With respect to clinical and experimental evaluations of field potentials, a prominant significance has to be attributed to the properties of the extracellular potential fields. In this context special attention has to be paid to the vertical extension of potential fields within the cortex, e.g., to the question which cortical layers can be explored reliably from the cortical surface with the aid of electrophysiological techniques. A first answer to this question can be de-

Department of Experimental Epileptology, Institute of Physiology - Neurophysiology, University of Münster, Robert-Koch-Str. 27 a, 4400 Münster, Federal Republic of Germany

Presurgical Evaluation of Epileptics
Edited by H. G. Wieser and C. E. Elger

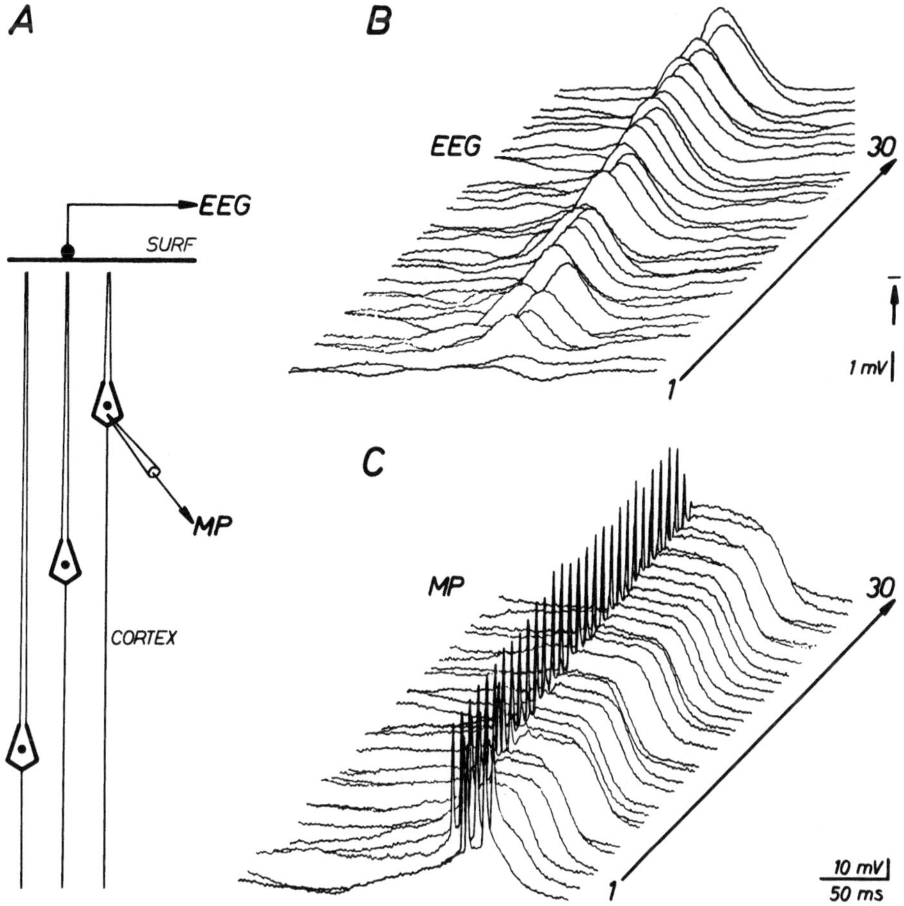

Fig. 1 A–C. Simultaneous appearance of neuronal paroxysmal depolarizations in superficial cortical layers and of seizure potentials in the EEG at the cortical surface during the development of focal epileptic activity. Seizure activity was elicited by local epicortical application of penicillin. **A** Schematic drawing of the electrode arrangement. *SURF*, cortical surface; *MP*, membrane potential. **B, C** Graphic superimposition of 30 successive potentials recorded simultaneously with the development of epileptic activity. (From Speckmann 1986; modified after Elger and Speckmann 1983b)

rived from the experiments displayed in Figs. 2 and 3, which demonstrate the development of an epileptic focus within the cerebral cortex. Parts A and B of Figs. 2 and 3 show that a strong correlation between the potential fluctuations at the cortical surface and the simultaneously appearing changes of field potentials and of MP in superficial cortical layers exist if penicillin is applied to the cortical surface. However, comparing the bioelectric events occurring in middle and deep cortical laminae with those at the surface, discrepancies in polarity and time course become apparent in various experimental situations (Fig. 2A1, 2 and 2B1; Fig. 3A). Such deviations from the epicortical EEG potentials can be detected even in superficial cortical laminae if penicillin is applied into the depth of the cortex (Fig. 2C). The differences in field potentials recorded epi- and intracortically coincide with differences in the cortical output. As shown in Fig. 2A and B, similar epicortical seizure potentials may (part B2) or may not (parts A1, 2 and B1) be associated with descending neuronal activity into the spinal cord. As a whole, the spatial extent of potential detection from the cortical surface is very restricted. Thus, epicortical EEG potentials are not in all cases representative for the neuronal activities in the different cortical layers (Elger and Speckmann 1980, 1983a, b; Elger et al. 1981, 1982; Gumnit et al. 1970; Petsche et al. 1981; Pockberger et al. 1983, 1984a, b; Speckmann and Caspers 1979b).

Generalized Seizure Activity

Basic mechanisms underlying the generation of field potentials during generalized seizures will be described, using tonic-clonic attacks elicited in animal experiments as a model. Field potentials appearing in the cerebral cortex with such tonic-clonic seizure activity can be divided into two types with respect to

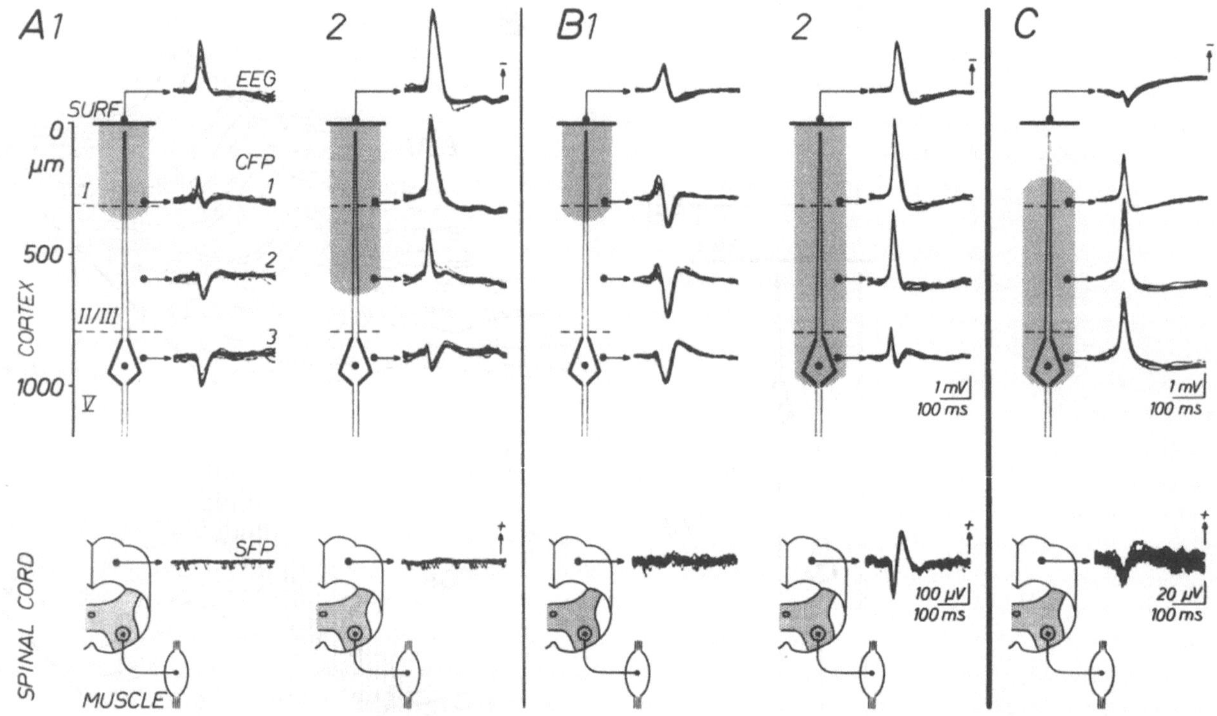

Fig. 2 A–C. Distribution of field potentials within the motor cortex during focal seizure discharges and descending neuronal activity to the spinal cord. Simultaneous recordings of epicortical *(EEG)*, intracortical *(CFP 1–3)*, and spinal *(SFP)* field potentials. Superimposition of 15 single potentials. *SURF*, cortical surface; *I*, *II/III*, *V*, cortical laminae. Seizure activity was elicited by local application of penicillin onto the cortical surface (**A, B**) and in deeper laminae of the cortex (**C**). **A** *1*, *2* and **B** *1*, *2*, recordings 5 min *(1)* and 15 min *(2)* after the epicortical penicillin application. Additional epicortical application of penicillinase before (**A**) and after (**B**) the penicillin administration. Vertical extension of seizure activity is indicated by *hatched columns*. (From Speckmann 1986; modified after Elger et al. 1981)

their time courses. On the one hand, slow shifts of the cortical DC potential develop which may last several minutes. On the other hand, series of fast potential fluctuations occur which are superimposed on the shifts of the DC potential. The relations between activity changes of single neurons and the shifts of epicortical and laminar DC potentials will be described in a first step. The mechanisms responsible for the generation of the fast potential fluctuations will be dealt with in a second step. In this context, the relations to the activity of pyramidal tract neurons will be specially pointed out because of their functional significance.

Tonic-clonic seizure activity was elicited in anesthetized and artificially ventilated rats and cats by repetitive systemic administrations of pentylenetetrazol (Caspers and Speckmann 1969, 1974; Purpura et al. 1972; Speckmann and Caspers 1974).

In the first step, the shifts of cortical DC potentials are related to MP changes of pyramidal tract cells. Typical recordings are displayed in Fig. 4. Part B of the Fig. shows that, with the commencement of a seizure, a nearly uninterrupted series of paroxysmal depolarization shifts appears. This neuronal activity is associated with a negative shift of the epicortical DC potential. Similar changes are also found in humans (Chatrian et al. 1964, 1968; Cohn 1954, 1963, 1964). The graphic superimposition of the changes of the MP and shifts of the DC potentials performed in part C demonstrate that the mean neuronal depolarization is associated with a negative displacement of the DC potential of the cortical surface. DC potential changes occurring with successive seizures and recorded from the surface, as well as from the depth of the cortex, are graphically superimposed in part D of Fig. 4. The tracings show that the DC potential, led from the layer of the pyramidal tract cells, was shifted also in the negative direction during the seizure activity. A comparison of all recordings displayed in Fig. 4 reveals that, together with the neuronal depolarization, the epicortical and the laminar DC potential are shifted into the negative direction *during* a fully established tonic-clonic seizure (Caspers and Speckmann 1969; Gumnit 1974; O'Leary and Goldring 1964; Speckmann and Caspers 1979 b; Speckmann and Elger 1982, 1984).

Besides the correspondence there are also discrepancies between the bioelectrical cortical events de-

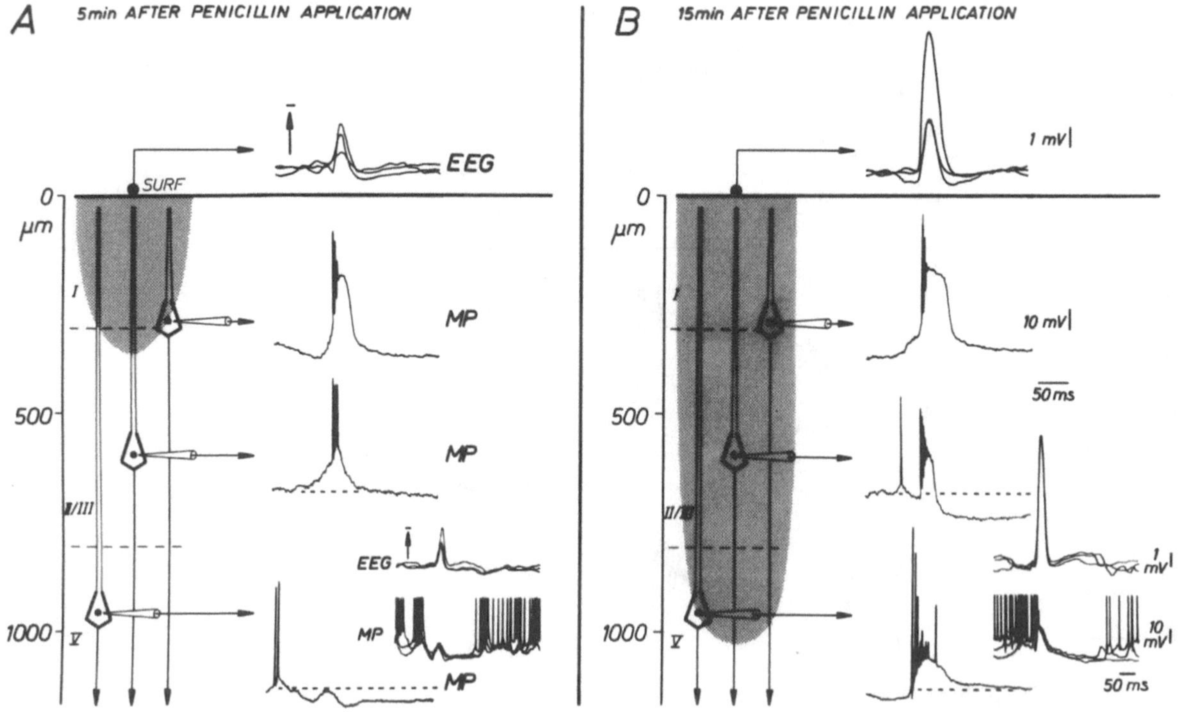

Fig. 3 A, B. Membrane potential changes *(MP)* of single neurons in different motor cortical laminae *(I, II/III, V)* during focal epileptic activity of various vertical extension. Seizure activity was elicited by local application of penicillin onto the cortical surface *(SURF)*. Recordings 5 min (**A**) and 15 min (**B**) after penicillin application. The extension of seizure activity is marked by *hatched columns*. The MP changes are related to each other in time by the epicortical seizure potentials *(EEG)*. *Insets* Recordings of the epicortical EEG and of the MP of a deep neuron with a reduced time scale. (From Speckmann 1986; modified after Elger and Speckmann 1983 a)

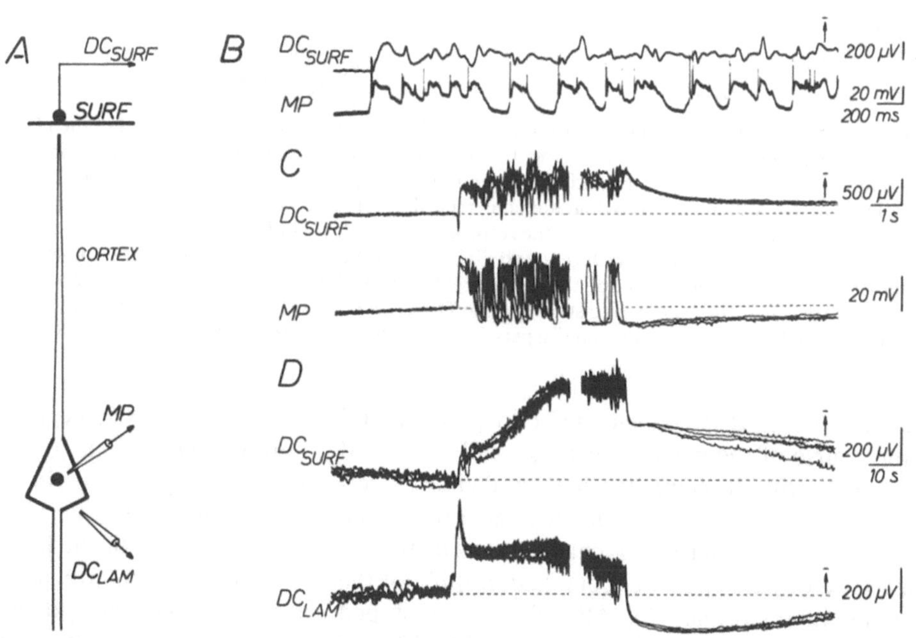

Fig. 4 A–D. Relationship between changes of epicortical and laminar DC potentials and of membrane potential of pyramidal cells *(MP)* during generalized tonic-clonic seizures. Epileptic activity was elicited by repetitive systemic administrations of pentylenetetrazol. **A** Schematic drawing of the electrode arrangement. *SURF*, cortical surface; DC_{SURF} and DC_{LAM}, epicortical and laminar DC potential. **B** Simultaneous recordings of DC_{SURF} and *MP* of a pyramidal tract cell by an oscilloscope. **C, D** Graphic superimposition of simultaneous recordings of DC_{SURF} and *MP*(**C**) and of DC_{SURF} and DC_{LAM} (**D**). Interruption of recordings: 30–60 s. (From Speckmann 1986; modified after Speckmann and Caspers 1979 b)

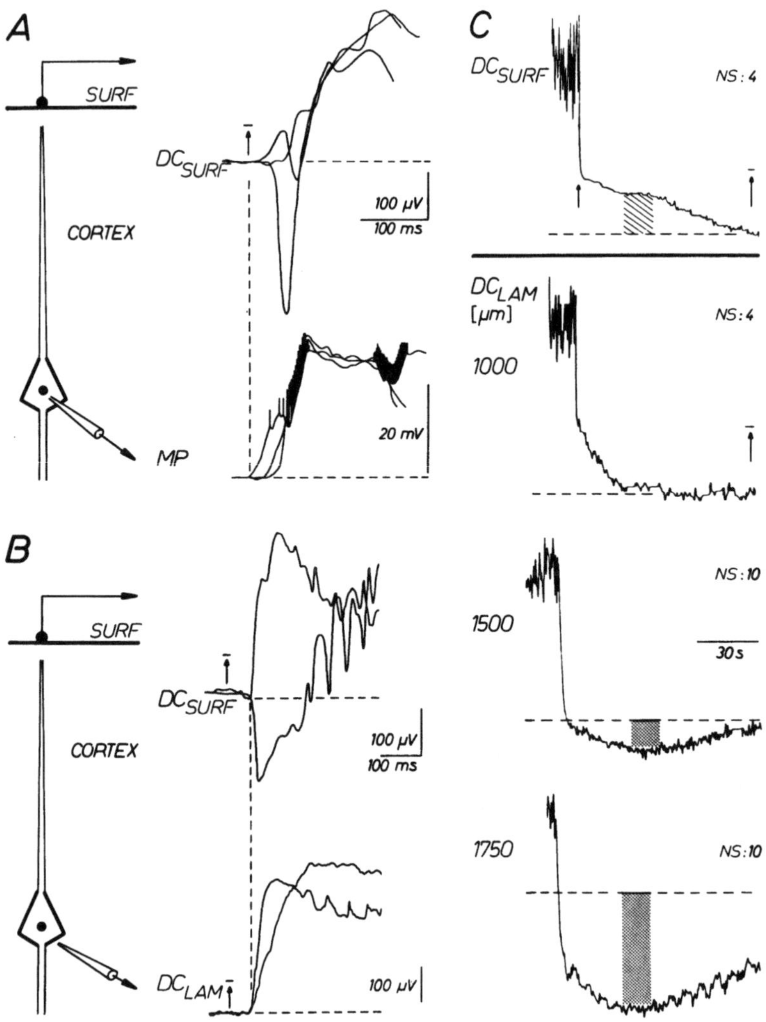

Fig. 5 A–C. Comparison of fluctuations of epicortical and laminar DC potentials and of membrane potential changes *(MP)* of a pyramidal tract cell with the commencement (**A, B**) and after the termination (**C**) of tonic-clonic seizures. The generalized epileptic activity was elicited by systemic administration of pentylenetetrazol. *SURF*, cortical surface. DC_{SURF} and DC_{LAM}, epicortical and laminar DC potential. **A, B** Simultaneous recording of DC_{SURF} and of the *MP* of a pyramidal tract cell (**A**) as well as of DC_{SURF} and DC_{LAm} (**B**). The position of the recording electrodes is indicated in the schematic drawing. Graphic superimposition of single recordings. **C** Shifts of DC_{SURF} and DC_{LAM} after the termination *(arrow)* of tonic-clonic seizures. Average curves. *NS*, number of sweeps. The depth of the laminar DC recordings is marked *on the left margin*. Pyramidal tract cell layer, 1500–1700 µm. *Hatched columns*, distance to the DC level before the onset of seizure activity *(broken line)*. [From Speckmann 1986; modified after Speckmann and Caspers 1979b (**A, B**) and after Speckmann et al. 1978 (**C**)]

scribed. These discrepancies refer to polarity and time course of the potential changes and are found especially with the commencement and after the termination of a seizure. As already mentioned, the *commencement* of seizures is characterized by the appearance of paroxysmal depolarization shifts of the pyramidal tract cells (Fig. 4 B). These stereotyped changes of the neuronal MP coincide with different fluctuations of the potentials at the cortical surface which precede the long-lasting shifts of the DC potential. The recordings in Part A of Fig. 5 demonstrate that the epicortical potential fluctuations in question consist in solitary negative and positive, as well as in diphasic negative-positive, waves. During comparable periods, uniform negative fluctuations of the DC potentials in the layer of the pyramidal tract cells appear. Part B of Fig. 5 shows that also these reactions of the laminar DC potentials are independent of polarity and time course of the simultaneously occurring potential fluctuations at the cortical surface. Combining the described observations (Fig. 5 A, B), it becomes apparent that, during the initial period of a tonic-clonic seizure, a strong relationship between the MP changes and the fluctuations of the laminar DC potential exists, whereas a correspondence between epicortical and laminar poten-

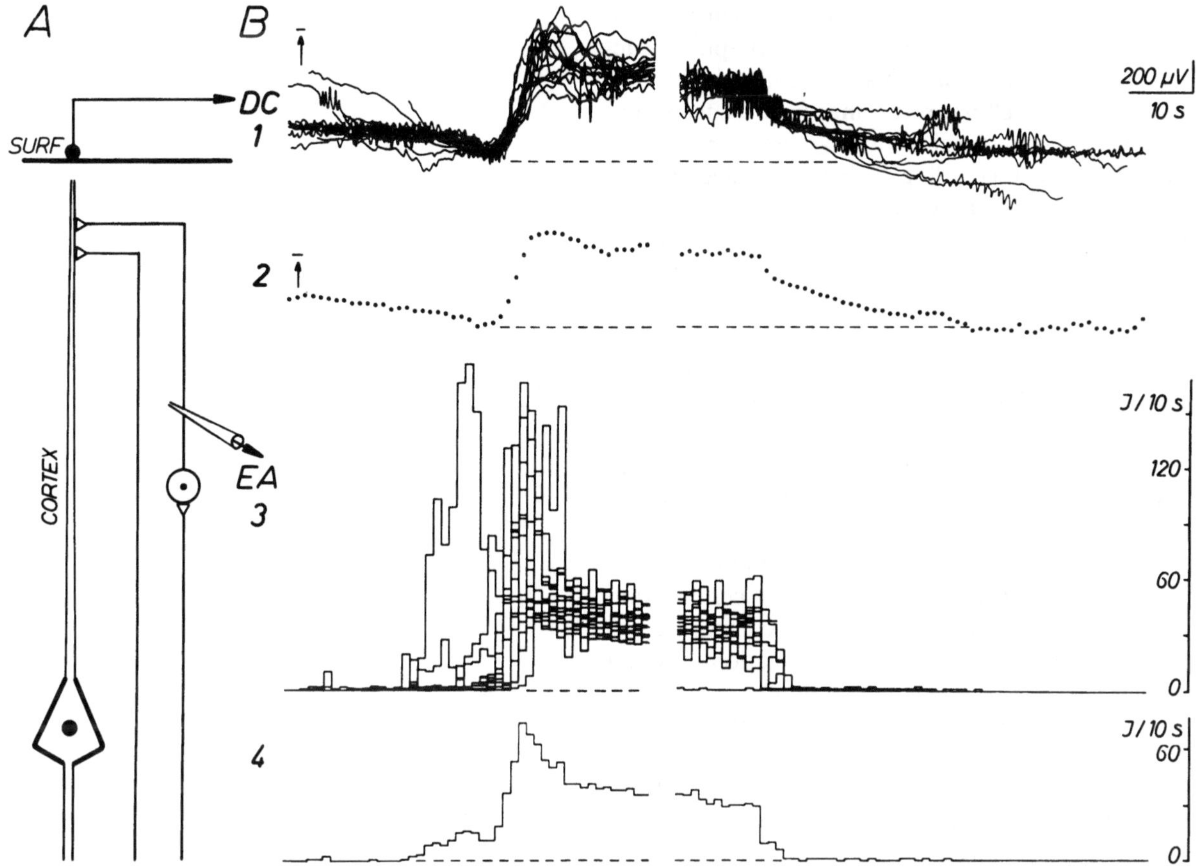

Fig. 6 A, B. Relationships between the shifts of the epicortical DC potential and changes of the discharge frequency of a neuronal element in a superficial cortical lamina during successive tonic-clonic seizures. The generalized epileptic activity was elicited by systemic application of pentylenetetrazol. **A** Schematic drawing of the electrode arrangement. *SURF,* cortical surface. *EA,* extracellularly recorded neuronal activity in a superficial cortical layer. **B** Single recordings (graphic superimposition, *1,3*) and their averaged values *(2,4).* Interruption of recordings, ca. 60 s. (From Speckmann 1986; modified after Speckmann et al. 1978)

tials is lacking. After the *termination* of a tonic-clonic seizure, a steep re- and hyperpolarization is found in all cortical neurons (Fig. 4 C). During these changes of MP, the DC potential at the cortical surface returns only slowly to the pre-attack level. Such a slow return of the negative shift of the DC potential is found to be typical of recordings from the cortical surface (Fig. 5 C). With the electrode progressively penetrating the cortex, the return of the negative shift of the DC potential becomes faster. As can be derived from Part C of Fig. 5, it has its maximal steepness, and, moreover, turns to a transient positive deviation in the layer of the pyramidal tract cells. Thus, the shifts of the DC potential recorded in this layer reflect the MP changes of pyramidal tract cells to a great extent (Fig. 4 C, D). A survey on the findings demonstrates finally that a correspondence between epicortical and laminar potentials is also lacking after the termination of seizures (Caspers et al. 1980; Speckmann and Caspers 1979 b; Speckmann et al. 1972, 1978).

The findings described demonstrate that the shifts of DC potentials in deep cortical layers and the changes of MP of pyramidal tract cells show a strong parallelism during generalized seizure activity. This observation allows the conclusion that the laminar DC potentials are generated primarily by the somata of pyramidal tract cells and adjacent neuronal structures. The discrepancies, found between the shifts of the epicortical DC potential and the changes of the DC potentials and the MP in deep cortical layers, led one to assume that the displacement of the superficial DC potential is at least not due solely to MP changes in deep laminae and that, therefore, superfical generator structures have to be taken into account (Speckmann and Caspers 1979 b; Speckmann and Elger 1984; Speckmann et al. 1972, 1978).

The partial correspondence of all the described bioelectric phenomena which were found to exist in the period between the commencement and termination of a fully developed tonic-clonic seizure (Fig. 4) led to the hypothesis that the superficial generator structures of the epicortical DC potential and the somata of the pyramidal tract cells are synaptically activated at the same time. This assumption is substantiated by extracellular recordings from neurons and fibers in the superficial cortical laminae. The evaluation of a typical recording during successive seizures is displayed in Fig. 6. As symbolized by the schematic drawing of the recording arrangement in Part A of this figure, a great part of the neuronal discharges picked up extracellularly represents obviously afferent activity to superficial structures. The recordings in Part B of Fig. 6 show that the negative displacements which develop during generalized tonic-clonic seizures are associated with a remarkable rise in the neuronal discharge frequency (Speckmann and Caspers 1979b; Speckmann and Elger 1984; Speckmann et al. 1972, 1978).

The question arises as to the contribution of glial cells to the cortical DC shifts. Such a contribution has proved to be uncertain because, comparing different experimental situations, no relation is found between glial activity and the behavior of the DC potential. Thus, monophasic depolarizations of glial cells are accompanied by negative DC fluctuations during electrical stimulation and seizure activity (Fig. 7 A, B), by positive DC shifts elicited by an isolated hypercapnia (Fig. 7 C), and by polyphasic DC displacements with a primary asphyxiation (Fig. 7 D; Caspers and Speckmann 1974; Caspers et al. 1979; De Robertis and Carrea 1965; Kuffler and Nicholls 1966; Kuffler et al. 1966; Orkand et al. 1966; Palay and Chan-Palay 1977; Somjen 1973, 1975; Somjen and Trachtenberg 1979).

In the second step, the fast potential flucutations which are superimposed on the epicortical negative DC shifts during seizures are compared with those appearing in deeper cortical layers and with MP changes of pyramidal tract cells. For this purpose three different patterns of potential fluctuations will be taken into account which, among others, are detectable during tonic-clonic seizures and are displayed in Fig. 8. They consist in solitary negative and positive waves (potentials 2 and 4 in Fig. 8 A–C) as well as in diphasic positive-negative fluctuations (potentials 3 in Fig. 8 A, B, and potential 1 in Fig. 8 C). The recordings in Fig. 8 demonstrate that each of the selected pattern of potential fluctuations at the cortical surface coincides with a paroxysmal depolarization shift in pyramidal tract cells (Fig. 8 A, C) and with a solitary negative fluctuation of the laminar DC potential (Fig. 8 B). Consequently the laminar potential fluctuations reflect the somatic depolarizations of pyramidal tract cells. The pattern of the epicortical potential fluctuations, however, is independent of the laminar potential changes (Kostopoulos et al. 1983; Speckmann and Caspers 1979b; Speckmann et al. 1972, 1978).

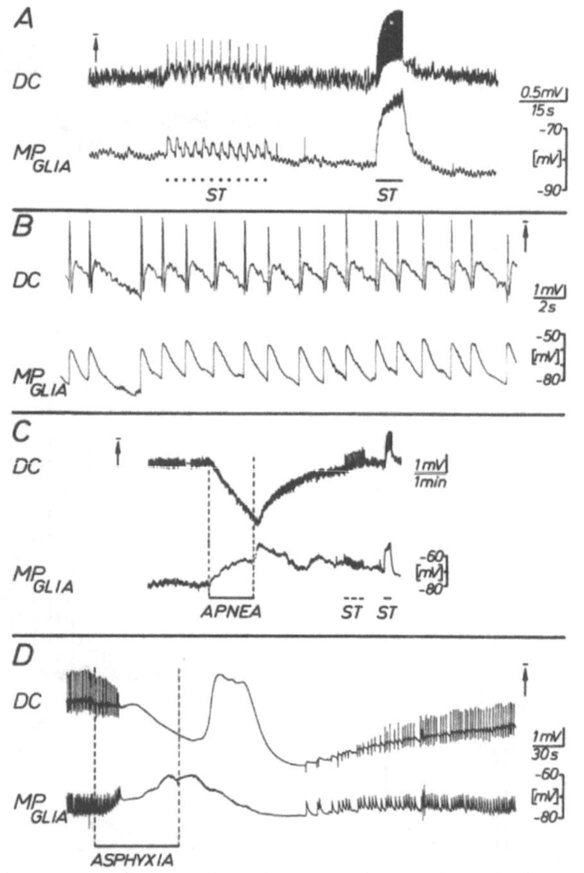

Fig. 7 A–D. Relationships between fluctuations and shifts of the epicortical DC potential and membrane potential changes *(MP)* of motor cortical glial cells in different experimental situations. **A** Single *(dots)* and high-frequency *(horizontal bar)* epicortical stimulation *(ST)*. See also Part **C**. **B** Focal epileptic discharges elicited by epicortical application of penicillin. **C** Apnea: interruption of artificial ventilation with pure oxygen. Elevation of the carbon dioxide pressure in tissue with the oxygen pressure above normal levels. **D** Asphyxia: interruption of artificial ventilation with room air. Simultaneous decrease and increase of the oxygen and carbon dioxide pressure, respectively. Focal epileptic discharges superimposed. (Experiments in collaboration with H. Caspers, C. E. Elger, and O. W. Witte)

The observation that with stereotyped changes of intra- and extracellular potentials in the pyramidal tract cell layer different patterns of epicortical potentials appear may be interpreted on the basis of the general principle showing a close relation between the pattern of fast potential fluctuations at the cortical surface and the level of the epicortical DC potential. Thus, a typical distribution becomes apparent if the different patterns of epicortical potential fluctuations are related to the amount of the negative dis-

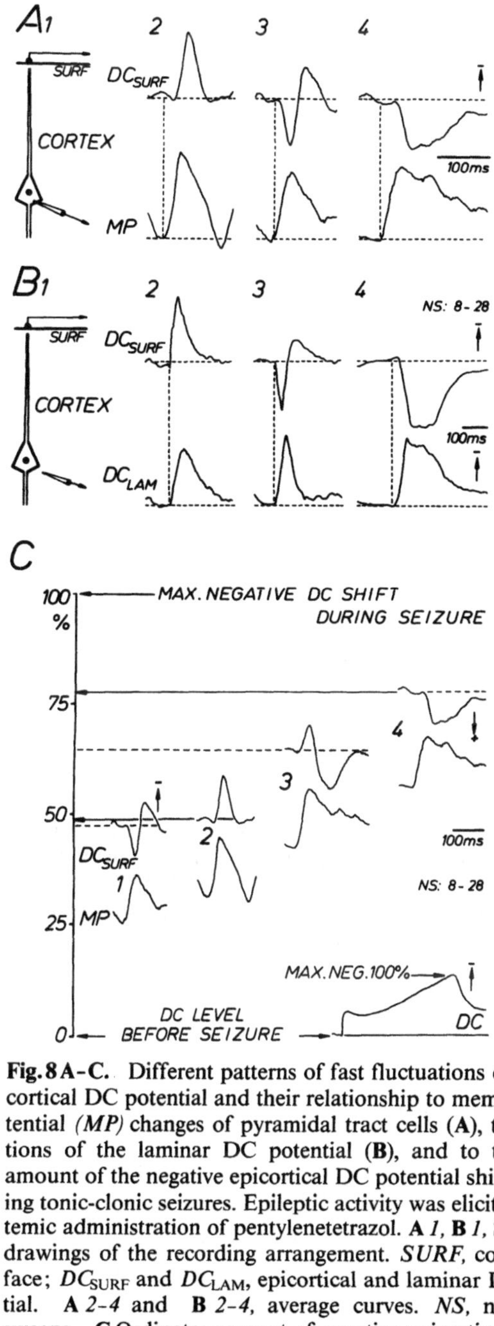

Fig. 8 A–C. Different patterns of fast fluctuations of the epicortical DC potential and their relationship to membrane potential *(MP)* changes of pyramidal tract cells (**A**), to fluctuations of the laminar DC potential (**B**), and to the actual amount of the negative epicortical DC potential shift (**C**) during tonic-clonic seizures. Epileptic activity was elicited by systemic administration of pentylenetetrazol. **A** *1*, **B** *1*, Schematic drawings of the recording arrangement. *SURF*, cortical surface; DC_{SURF} and DC_{LAM}, epicortical and laminar DC potential. **A** *2-4* and **B** *2-4*, average curves. *NS*, number of sweeps. **C** Ordinate: amount of negative epicortical DC potential shifts during seizures (see *inset*). *1-4*, different patterns of fast potential fluctuations of the epicortical DC potential. Average potentials. (From Speckmann 1986; modified after Speckmann and Caspers 1979b)

placement of the epicortical DC potential during tonic-clonic seizures. Such an analysis is given in Fig. 8C. It shows that potential patterns consisting mainly of negative components appear with a small negative shift of the DC potential, and that solitary positive potential fluctuations occur only if the negative displacement of the DC potential exceeds about 75% of its maximal value during a seizure. That holds true also for sequences of waves with the same polarity (Caspers 1959, 1963; Goldring and O'Leary 1951, 1954).

It has been described above (a) that the shape of the epicortical potential fluctuations is strongly related to the simultaneously appearing negative DC potential shifts (Fig. 8C) and (b) that the negative shift of the DC potential at the cortical surface is associated with a remarkable increase in discharge frequency of superficial neuronal structures (Fig. 6). Since the action potentials recorded extracellularly in upper cortical layers can be assumed to represent afferent activity, insights into the elementary mechanisms underlying the generation of the fast potential fluctuations at the cortical surface can be expected when the neuronal discharges in question are analyzed in more detail. Typical recordings and histographic evaluations are presented in Fig. 9. They show that with the beginning of a negative potential fluctuation, the discharge frequency increases, starting from a low preexisting level (Fig. 9 B1, C1); and that during a positive potential fluctuation, the discharge frequency decreases, starting from a high preexisting level (Fig. 9 B2, C3; see also Fig. 9 C2; Speckmann and Caspers 1979b; Speckmann et al. 1972, 1978).

The experimental observations described may be combined into a functional model which would explain the generation of different fast potential fluctuations at the cortical surface during tonic-clonic seizure activity. The model will be discussed using the flow chart in Fig. 10. The amount of the afferent input to superficial generator structures, which can be regarded as being predominantly asynchronous, is symbolized by the width of the hatched arrows. Correspondingly, the afferent input and the resulting negative DC shift at the cortical surface are smaller in Part A than in Part B. Depending on the different preexisting situations, a highly synchronized afferent input (symbolized by heavy lines in Fig. 10) leads to opposite neuronal reactions in upper cortical layers. In the case of the situation presented in Part A of Fig. 10, the synchronous afferent input reaches not only deep but, via intercalated neurons, also superficial cortical structures. This elicits simultaneously a paroxysmal depolarization shift in pyramidal tract

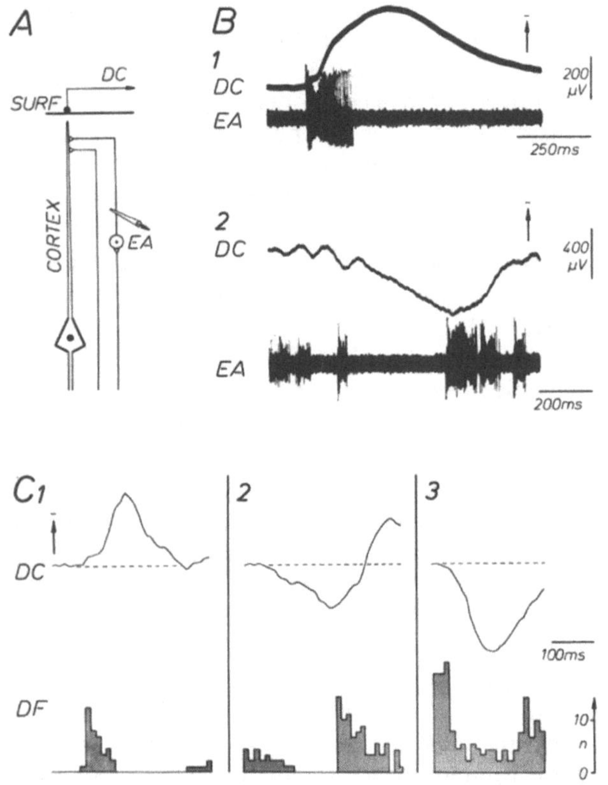

Fig. 9 A–C. Different patterns of fast fluctuations of the epicortical DC potential and their relationship to the discharge frequency *(DF)* of a neuronal element in superficial cortical laminae during tonic-clonic seizures. The generalized epileptic activity was elicited by systemic administration of pentylenetetrazol. **A** Schematic drawing of the recording arrangement. *SURF,* cortical surface; *EA,* extracellularly recorded neuronal activity. **B** Original recordings by an oscilloscope. **C** Histographic evaluations of up to 16 single events. [Modified after Speckmann et al. 1978 (**B**) and Speckmann and Caspers 1979b (**C**)]

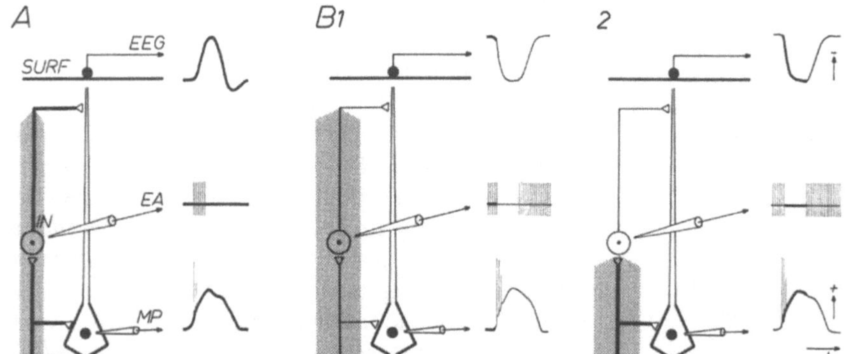

Fig. 10 A, B. Hypothetical flow chart for an interpretation of fast DC potential fluctuations of opposite polarity occurring during generalized tonic-clonic seizures. The *hatched arrows* symbolize the asynchronous input to the cerebral cortex and the *heavy lines* phasic volleys initiating single convulsive discharges. *PTN,* pyramidal tract neuron; *IN,* interneuron; *MP,* membrane potential; *EA,* extracellularly recorded neuronal activity. **A** With a moderate asynchronous input to the cortex *(small hatched arrow),* a burst of *EA* triggers a paroxysmal depolarization shift in a *PTN,* evokes a depolarization of superficial neuronal structures, and thus a negative fluctuation in the epicortical DC recording. **B** With an increased asynchronous input to the cortex *(widened hatched arrow,* **B** *1),* a phasic volley reaching the cortex triggers paroxysmal depolarization shifts in the *PTN* and interrupts the asynchronous input by an inactivation of *IN* (**B** *2*). The latter process results in a disfacilitation of superficial neuronal elements and thus in a positive fluctuation of the epicortical DC potential. (From Speckmann 1986; drawings after original tracings from Speckmann et al. 1978)

cells and depolarizations of neuronal elements close to the cortical surface. Thus, the paroxysmal depolarization, occurring in the soma of pyramidal tract cells, is associated with a negative potential fluctuation at the surface (Fig. 8 A2, C2). In the case of the situation presented in Part B of Fig. 10, the interneuronal activity is elevated due to the high preexisting afferent inflow. If, under these conditions, a synchronized afferent input appears, a paroxysmal depolarization is still triggered in the pyramidal tract cells as in Part A. The activity of interneurons, however, is transiently interrupted because of a critical synaptic depolarization. Thus, the excitatory synaptic input to superficial neuronal elements is discontinued. This disfacilitation leads to a positive field potential at the cortical surface, associated with a stereotyped paroxysmal depolarization and a monophasic negative field potential at depth (potentials 4 in Fig. 8 A-C; Speckmann and Caspers 1979b; Speckmann and Elger 1982, 1984; Speckmann et al. 1972, 1978, 1984).

As a whole, with respect to the origin of DC potential shifts and fast potential fluctations in the cerebral cortex, one can state that deep and superficial field potentials are generated at functionally different structures. Since epicortical field potentials are not constantly related to neuronal activity within the cortex, they can be interpreted in different ways.

References

Andersen P, Andersson SA (1968) Physiological basis of the alpha rhythm. Meredith, New York

Caspers H (1959) Über die Beziehungen zwischen Dendritenpotential und Gleichspannung an der Hirnrinde. Pflügers Arch 269: 157-181

Caspers H (1963) Relations of steady potential shifts in the cortex to the wakefulness-sleep spectrum. In: Brazier MAB (ed) Brain function. University of California Press, Berkeley, pp 177-213

Caspers H, Speckmann E-J (1969) DC potential shifts in paroxysmal states. In: Jasper HH, Ward AA Jr, Pope A (eds) Basic mechanisms of the epilepsies. Little, Brown, Boston, pp 375-388

Caspers H, Speckmann E-J (1974) Cortical DC shifts associated with changes of gas tension in blood and tissue. In: Rémond A (ed) Handbook of electroencephalography and clinical neurophysiology, vol 10, part A. Elsevier, Amsterdam, pp 41-65

Caspers H, Speckmann E-J, Lehmenkühler A (1979) Effects of CO_2 on cortical field potentials in relation to neuronal activity. In: Speckmann E-J, Caspers H (eds) Origin of cerebral field potentials. Thieme, Stuttgart, pp 151-163

Caspers H, Speckmann E-J, Lehmenkühler A (1980) Electrogenesis of cortical DC potentials. In: Kornhuber HH, Deecke L (eds) Motivation, motor and sensory processes of the brain: electrical potentials, behavior and clinical use. Prog Brain Res 54: 3-15

Caspers H, Speckmann E-J, Lehmenkühler A (1984) Electrogenesis of slow potentials of the brain. In: Elbert T, Rockstroh B, Lutzenberger W, Birbaumer N (eds) Self-regulation of the brain and behavior. Springer, Berlin Heidelberg New York Tokyo, pp 26-41

Chatrian GE, Shaw C-M, Plum F (1964) Focal periodic slow transients in epilepsia partialis continua: clinical and pathological correlations in two cases. Electroencephalogr Clin Neurophysiol 16: 387-393

Chatrian GE, Somasundaram M, Tassinari CA (1968) DC changes recorded transcranially during "typical" three per second spike and wave discharges in man. Epilepsia 9: 185-209

Cohn R (1954) DC recordings of paroxysmal activity from the intact human head. Electroencephalogr Clin Neurophysiol 6: 692

Cohn R (1963) DC potential variations in paroxysmal discharges. Electroencephalogr Clin Neurophysiol 15: 151

Cohn R (1964) DC recordings of paroxysmal disorders in man. Electroencephalogr Clin Neurophysiol 17: 17-24

Creutzfeldt OD, Houchin J (1974) Neuronal basis of EEG waves. In: Rémond A (ed) Handbook of electroencephalography and clinical neurophysiology, vol 2, part C. Elsevier, Amsterdam, pp 5-55

Creutzfeldt OD, Watanabe S, Lux HD (1966) Relations between EEG phenomena and potentials of single cortical cells. II. Spontaneous and convulsoid activity. Electroencephalogr Clin Neurophysiol 20: 19-37

De Robertis EDP, Carrea R (eds) (1965) Biology of neuroglia. Prog Brain Res, vol 15. Elsevier, Amsterdam

Elger CE, Speckmann E-J (1980) Focal interictal epileptiform discharges (FIED) in the epicortical EEG and their relations to spinal field potentials in the rat. Electroencephalogr Clin Neurophysiol 48: 447-460

Elger CE, Speckmann E-J (1983 a) Penicillin induced epileptic foci in the motor cortex: vertical inhibition. Electroencephalogr Clin Neurophysiol 56: 604-622

Elger CE, Speckmann E-J (1983 b) Vertical inhibition in motor cortical epileptic foci and its consequences for descending neuronal activity to the spinal cord. In: Speckmann E-J, Elger CE (eds) Epilepsy and motor system. Urban und Schwarzenberg, München, pp 152-160

Elger CE, Speckmann E-J, Prohaska O, Caspers H (1981) Pattern of intracortical potential distribution during focal interictal epileptiform discharges (FIED) and its relation to spinal field potentials in the rat. Electroencephalogr Clin Neurophysiol 51: 393-402

Elger CE, Speckmann E-J, Caspers H, Prohaska O (1982) Focal interictal epileptiform discharges in the cortex of the rat: laminar restriction and its consequences for activity descending to the spinal cord. In: Klee MR, Lux HD, Speckmann E-J (eds) Physiology and pharmacology of epileptogenic phenomena. Raven, New York, pp 13-19

Goldensohn ES (1984) Neurophysiologische Grundlagen der EEG-Aktivität. In: Klass DW, Daly DD (eds) Klinische Elektroenzephalographie. Fischer, Stuttgart, pp 379-395

Goldring S, O'Leary JL (1951) Experimentally derived correlates between ECG and steady cortical potential. J Neurophysiol 14: 275-288

Goldring S, O'Leary JL (1954) Correlation between steady transcortical potential and evoked response. II. Effect of veratrine and strychnine upon the responsiveness of visual cortex. Electroencephalogr Clin Neurophysiol 6: 201-212

Gumnit R (1974) DC shifts accompanying seizure activity. In: Rémond A (ed) Handbook of electroencephalography and clinical neurophysiology, vol 10, part A. Elsevier, Amsterdam, pp 66-77

Gumnit RJ, Matsumoto H, Vasconetto C (1970) DC activity in the depth of an experimental epileptic focus. Electroencephalogr Clin Neurophysiol 28: 333-339

Klee MR, Speckmann E-J, Lux HD (eds) (1981) Physiology and pharmacology of epileptogenic phenomena. Raven, New York

Kostopoulos G, Avoli M, Gloor P (1983) Participation of cortical recurrent inhibition in the genesis of spike and wave discharges in feline generalized penicillin epilepsy. Brain Res 267: 101-112

Kuffler SW, Nicholls JG (1966) The physiology of neuroglia cells. Erg Physiol 57: 1-90

Kuffler SW, Nicholls JG, Orkand RK (1966) Physiological properties of glial cells in the central nervous system of amphibia. J Neurophysiol 29: 768-787

Lopes da Silva F, van Rotterdam A (1982) Biophysical aspects of EEG and MEG generation. In: Niedermeyer E, Lopes da Silva F (eds) Electroencephalography. Urban and Schwarzenberg, Baltimore, pp 15-26

O'Leary JL, Goldring S (1964) DC potentials of the brain. Physiol Rev 44: 91-125

Orkand RK, Nicholls JG, Kuffler SW (1966) Effect of nerve impulses on the membrane potential of glial cells in the central nervous system of amphibia. J Neurophysiol 29: 788-806

Palay SL, Chan-Palay V (1977) General morphology of neurons and neuroglia. In: Kandel ER (ed) Handbook of physiology. The nervous system, vol 1, part 1. American Physiological Society, Bethesda, pp 5-37

Petsche H, Pockberger H, Rappelsberger P (1981) Current source density studies of epileptic phenomena and the morphology of the rabbit's striate cortex. In: Klee MR, Lux HD, Speckmann E-J (eds) Physiology and pharmacology of epileptogenic phenomena. Raven, New York, pp 53-63

Pockberger H, Petsche H, Rappelsberger P (1983) Intracortical aspects of penicillin-induced seizure patterns in the rabbit's motor cortex. In: Speckmann E-J, Elger CE (eds) Epilepsy and motor system. Urban und Schwarzenberg, München, pp 161-178

Pockberger H, Rappelsberger P, Petsche H (1984a) Penicillin-induced epileptic phenomena in the rabbit's neocortex. I. The development of interictal spikes after epicortical application of penicillin. Brain Res 309: 247-260

Pockberger H, Rappelsberger P, Petsche H (1984b) Penicillin-induced epileptic phenomena in the rabbit's neocortex. II. Laminar specific generation of interictal spikes after the application of penicillin to different cortical depths. Brain Res 309: 261-269

Prince DA (1974) Neuronal correlates of epileptiform discharges and cortical DC potentials. In: Rémond A (ed) Handbook of electroencephalography and clinical neurophysiology, vol 2, part C. Elsevier, Amsterdam, pp 56-70

Purpura DP, Penry JK, Tower DB, Woodbury DM, Walter RD (eds) (1972) Experimental models of epilepsy. Raven, New York

Somjen GG (1973) Electrogenesis of sustained potentials. Prog Neurobiol 1: 199-237

Somjen GG (1975) Electrophysiology of neuroglia. Ann Rev Physiol 37: 163-190

Somjen GG, Trachtenberg M (1979) Neuroglia as generator of extracellular current. In: Speckmann E-J, Caspers H (eds) Origin of cerebral field potentials. Thieme, Stuttgart, pp 21-32

Speckmann E-J (1986) Experimentelle Epilepsieforschung. Wissenschaftliche Buchgesellschaft, Darmstadt

Speckmann E-J, Caspers H (1974) The effect of O_2- and CO_2-tensions in the nervous tissue on neuronal activity and DC-potentials. In: Rémond A (ed) Handbook of electroencephalography and clinical neurophysiology, vol 2, part C. Elsevier, Amsterdam, pp 71-89

Speckmann E-J, Caspers H (eds) (1979a) Origin of cerebral field potentials. Thieme, Stuttgart

Speckmann E-J, Caspers H (1979b) Cortical field potentials in relation to neuronal activities in seizure conditions. In: Speckmann E-J, Caspers H (eds) Origin of cerebral field potentials. Thieme, Stuttgart, pp 205-213

Speckmann E-J, Elger CE (1982) Neurophysiological basis of the EEG and of DC potentials. In: Niedermeyer E, Lopes da Silva F (eds) Electroencephalography. Basic principles, clinical applications and related fields. Urban and Schwarzenberg, Baltimore, pp 1-13

Speckmann E-J, Elger CE (1984) The neurophysiological basis of epileptic activity: a condensed overview. In: Degen R, Niedermeyer E (eds) Epilepsy, sleep and sleep deprivation. Elsevier, Amsterdam, pp 23-34

Speckmann E-J, Caspers H, Janzen RWC (1972) Relations between cortical DC shifts and membrane potential changes of cortical neurons associated with seizure activity. In: Petsche H, Brazier MAB (eds) Synchronization of EEG activity in epilepsies. Springer, Vienna New York, pp 93-111

Speckmann E-J, Caspers H, Janzen RWC (1978) Laminar distribution of cortical field potentials in relation to neuronal activities during seizure discharges. In: Brazier MAB, Petsche H (eds) Architectonics of the cerebral cortex. IBRO monograph series, vol 3. Raven, New York, pp 191-209

Speckmann E-J, Caspers H, Elger CE (1984) Neuronal mechanisms underlying the generation of field potentials. In: Elbert T, Rockstroh B, Lutzenberger W, Birbaumer N (eds) Self-regulation of the brain and behavior. Springer, Berlin Heidelberg New York Tokyo, pp 9-25

Volume Conductor Principles: Their Application to the Surface and Depth Electroencephalogram

P. GLOOR

Introduction

The success of neurosurgical treatment of drug-resistant partial epilepsy depends upon the eradication by surgical excision of the area of gray matter responsible for initiating the patient's seizures. Accurate identification of this area is therefore a prerequisite for guiding the surgical act. Among the data contributing to such a localizing diagnosis, epileptiform abnormalities recorded in the EEG are of paramount importance, because they reflect most directly the presence of epileptogenic neuronal hyperexcitability in the form of interictal or ictal epileptiform discharges (Gloor 1975). In spite of this, EEG data present the interpreter with a number of problems: (a) extracranial recordings fail to provide access to some areas known to be involved in seizure precipitation; (b) epileptiform abnormalities, even if localized, may reflect abnormal activity originating from a distant site, not directly accessible to EEG recording, a factor which presents a problem in both extracranial and intracranial EEG recordings; and (c) EEG potentials are not narrowly confined to the area that generates them. They can be recorded over a considerable distance by virtue of volume conduction. The relationship between the topography of the potential distribution, either extra- or intracranially, and the location, extent, and configuration of the generator is anything but simple and straightforward (Nunez 1981; Gloor 1985). The importance of this aspect of the problem is often not sufficiently appreciated. This chapter will therefore focus on this particular issue.

Laboratory for EEG and Clinical Neurophysiology, Montreal Neurological Institute and Hospital Department of Neurology and Neurosurgery, McGill University, 3801 University Street, Montréal, Québec, H3A 2B4, Canada

Generators of the EEG

The unitary generator of EEG potentials is the pyramidal neuron of the cerebral cortex (Creutzfeldt and Houchin 1974; Gloor 1985). When excited (or inhibited) it behaves as a dipole (Fig. 1 A) and produces an "open field" with a significant spatial spread to areas outside its immediate confines. This is a consequence of its particular geometry and of the segregation of specific classes of synaptic inputs on the surface of its soma-dendritic membrane. This, in itself, however, is not sufficient for producing a recordable EEG signal at a distance. An additional requirement is that the activity of a large population of pyramidal neurons arranged in parallel to each other, as is the case in the cerebral cortex, is synchronized. If such synchronization takes place over a large enough area of cortex, electrical fields of sufficient strength are generated to make possible the recording of potentials at a considerable distance from the location of the generator. The reasons for this become evident when applying the solid angle theorem to volume conductor theory (Woodbury 1960; Gloor 1985).

When applied to a single excitable cell, this theorem postulates that in a referential recording the potential P recorded in a homogeneous conductor at some distance from the excitable element is proportional to the solid angle subtended at that point by the cross section of the cell which separates its excited from its non excited segment (Woodbury 1960) (Fig. 1 A):

$$P=\frac{\pm e}{4\pi}\Omega$$

This cross section – respresents the transition from the area of extracellular negativity (excited segment) to that of extracellular positivity (nonexcited segment) which, for simplicity's sake, is assumed to be stepwise (Fig. 1 A). The electrical sign of the potential is negative at points facing the "negative" side of the cross-sectional area and positive at those facing the "positive" side. In Fig. 1 A it is assumed that the pyramidal cell is excited by a set of synapses

Presurgical Evaluation of Epileptics
Edited by H. G. Wieser and C. E. Elger

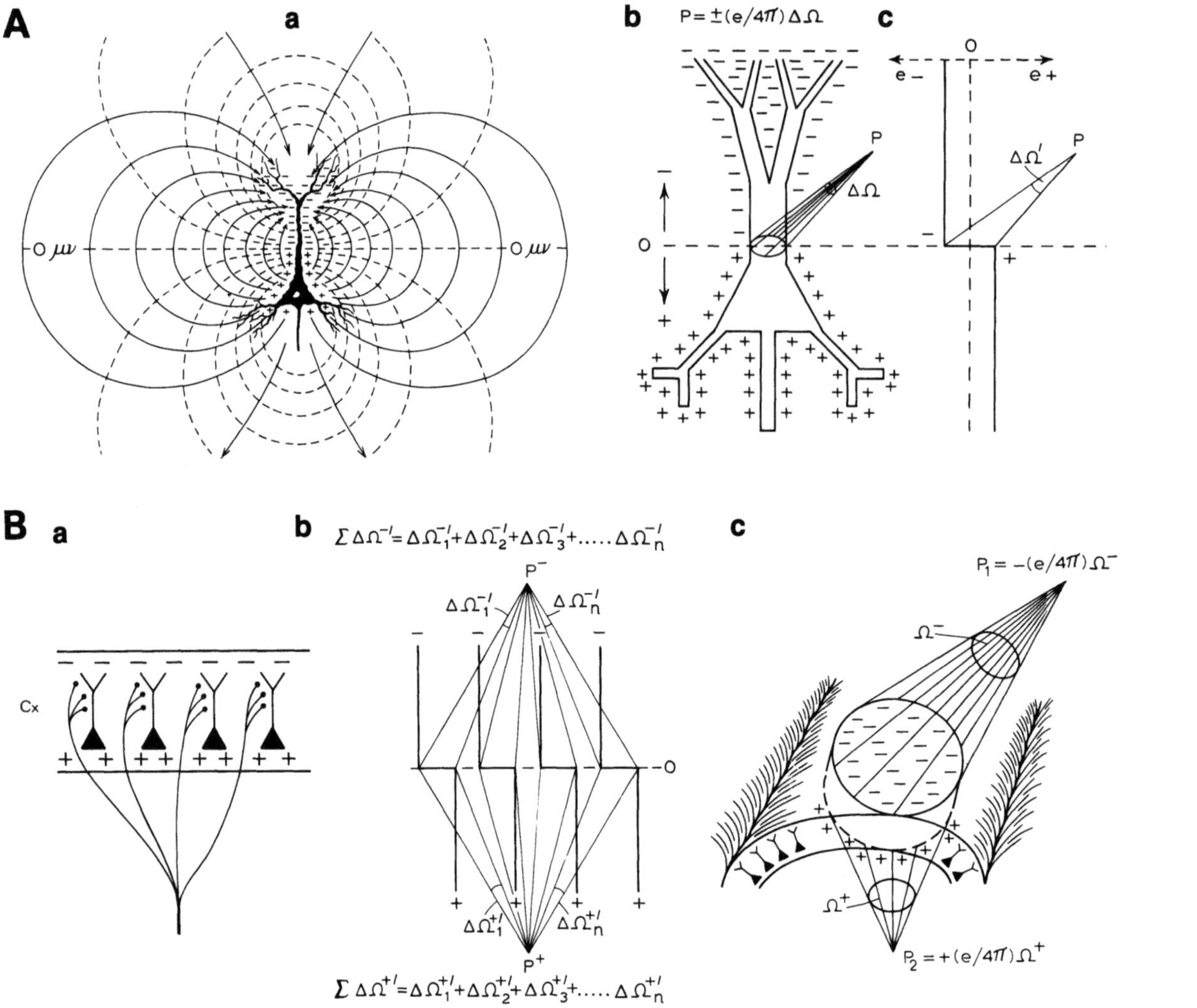

Fig. 1. **A** Electrical field in a volume conductor around a single pyramidal neuron excited by activation of a set of synapses on the apical dendrites. *a* Extracellular current flow *(solid lines with arrows)* and isopotential surfaces *(dashed lines)*. The electrical field resembles that of a dipole. *b* The potential at any point *P* around the excited pyramidal neuron is proportional to the solid angle $\Delta\Omega$ subtended at *P* by the cross-section of the area of transition between the excited and non-excited portion of the neuron. *c* Diagrammatic representation of the situation in *b*, the *horizontal line* representing the cross-section of the neuron at the transition between the excited and non-excited segment of the neuron subtended at *P* by $\Delta\Omega'$, the plane angle homologue of the solid angle $\Delta\Omega$. – **B** Electrical field generated by a population of synchronously excited pyramidal neurons. *a* Schematic diagram of electrical changes generated in a population of pyramidal neurons simultaneously excited through a synchronous synaptic volley arriving at their apical dendrites through the branches of a single afferent fiber. *b* Diagrammatic representation as in **A** *c* of this situation: the individual solid angles $\Delta\Omega'$ of each excited neuron add up to a large solid angle $\Sigma\Delta\Omega$. *c* Three - dimensional representation of the situation depicted in **B***a* and **B** *b*: the potential recorded at any point *P* depends on the size of the solid angle Ω subtended at *P* by the area of synchronously excited cortex. The electrical sign depends on that of the surface of the generator facing *P* (negative at P_1, positive at P_2). [**A***a*, **A***b*, **A***c*, and **B***c* from Gloor P (1985) Neuronal generators, the problem of localization and volume conductor theory in electroencephalography. J Clin Neurophysiol 2: 327–354, with permission of Raven Press, New York]

which make contact with its apical dendrites. The solid angle subtended at any point in the surrounding medium by the cross-section of a single excited pyramidal cell is small and therefore the extracellular potential recorded outside this cell is small, except for a microelectrode that can approach it closely without injuring it and thus becomes capable to "see" this cross-sectional area under a large solid angle. (The metaphoric term of "seeing" is quite appropriate and conceptually helpful, because the visual angle under which we see objects determines their apparent size; it is the one solid angle with which we are intuitively most familiar).

If a whole population of pyramidal cells lying in parallel alongside each other in the cerebral cortex receives synchronous excitatory synaptic inputs, as diagramatically depicted in Fig. 1 Ba, then the small individual solid angles of each excited pyramidal neuron add to form a large solid angle which represents the sum $\Sigma \Delta \Omega$ of all the small individual solid angles of each excited pyramidal neuron. (In Fig. 1 Ab, 1 Ba and b, as in all subsequent figures, solid angles are represented by corresponding plane angles.) A large area of synchronously activated cortex thus represents a dipole layer (Fig. 1 Bb, c). It can be subtended by a relatively large solid angle $\Omega = \Sigma \Delta \Omega$ even at a considerable distance. This is the reason why an extracranial EEG can be recorded.

Extracranial EEG Recordings

The generators of EEG signals recorded extracranially are thus dipole layers of macroscopic extent occupying a contiguous area of cerebral cortex. The simplest model of such a generator is that represented by a flat area of cortex as schematically depicted in Fig. 1 Bc. Generators occupying exclusively the top of a gyrus (Fig. 2 A) or the wall of a sulcus (Fig. 3 A) approximate this situation most closely. Most generators of EEG signals, however, are complexly shaped convoluted dipole layers. To conceptualize the relationship between size, orientation, and configuration of such a convoluted generator surface, and the size and polarity of the potential recorded at a given electrode location, the solid angle theorem is most useful. The potential measured at any electrode position in a referential (monopolar) recording is proportional to the apparent size of the generator surface as "seen" at the electrode position, i.e., it is proportional to the solid angle subtended by that surface at that position. Its electrical sign depends upon whether the electrode faces the negative or the positive side of the generator surface. Figures 2 and 3 illustrate how a variety of geometrical configurations of cortical generators would affect the recording of EEG signals from the overlying scalp. A generator localized on the crown of a gyrus produces a fairly localized signal of uniform polarity in the scalp EEG, with a relatively rapid falloff with increasing distances from the generator (Fig. 2 A). Only one side (of negative polarity in Fig. 2 A) of the generator surface is "seen" from the three scalp electrodes P_1, P_2, and P_3. The potential profile along these electrodes arranged in a straight line would thus be bell shaped, since at P_1 and P_3 the generator surface is subtended by smaller solid angles (Ω_1^- and Ω_3^-) than at P_2 (Ω_2^-). If the generator extends into the proximal walls of the two sulci delineating the gyrus, the potential profile is similar (Fig. 2 B), even though electrodes P_1 and P_3 also "see" the positive side of the dipole layer in the wall of the more distant sulcus. Thus at electrode P_1 the potential is proportional to the solid angle subtended at P_1 by the negative side of the generator occupying the crown of the gyrus and the wall of the sulcus proximal to the electrode (Ω_1^-), minus the solid angle subtended by the positive side of the generator occupying the wall of the sulcus distant to the electrode (Ω_1^+). The effective solid angle is thus $\Omega_{\text{eff}}^{(-)} = \Omega_1^- - \Omega_1^+$. The same would apply to electrode P_3. Both generators depicted in Fig. 2 produce a bell-shaped potential profile along a straight electrode row on the scalp. Such a field is similar to that produced by a single dipole oriented vertically to the scalp surface and can be modeled as such mathematically. Attractive as this may appear, it must be realized, however, that the corresponding current dipole and its computed location is fictitious and does not have any anatomical or physiological reality.

Figure 3 A depicts the situation resulting from a generator occupying the surface of one wall of a sulcus. A monopolar recording taken on the scalp shows a phase reversal between the two electrodes straddling the vertically oriented generator surface, since electrodes P_1 and P_2 "see" the positive, while electrode P_3 "sees" the negative side of the generator. There are thus two potential maxima in the scalp EEG, one positive, the other negative. Therefore there would be two phase reversals in a bipolar recording with the generator located roughly half-way between them. The field on the scalp produced by such a generator has a wide distribution and its profile along a straight electrode row conforms to that of a horizontal dipole. Again such a dipole is purely fictitious. Finally if a generator occupies both walls of a sulcus (Fig. 3 B), the differences between the solid angles subtended at the three electrode positions

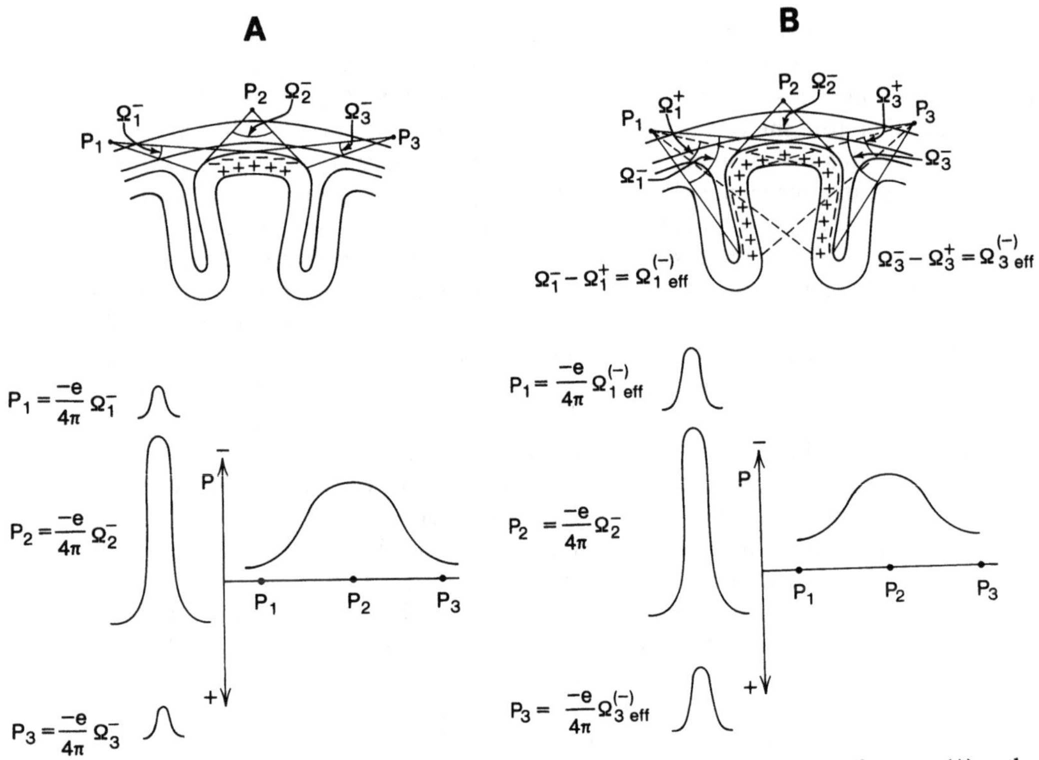

Fig. 2 A, B. Application of the solid angle principle to generators occupying the crown of a gyrus (**A**) or the whole surface of a gyrus including its crown (**B**) and the proximal walls of the two adjacent sulci delineating the gyrus. The *lower part* of the figure shows how the signals recorded at electrode positions P_1, P_2 and P_3 would appear in a referential (monopolar) recording. Also shown is the profile of the potential distribution along a straight line encompassing electrodes P_1, P_2, and P_3

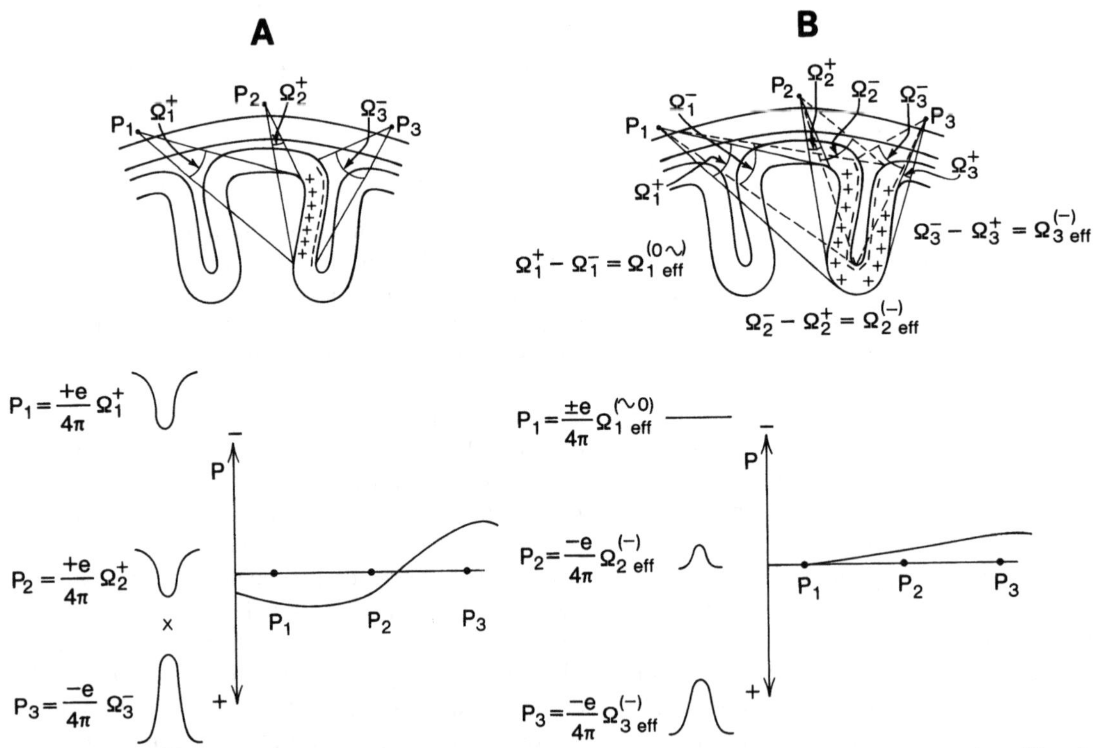

Fig. 3 A, B. Application of the solid angle principle to generators occupying cortex in one wall of a sulcus (**A**) or both walls (**B**). For further explanation see legend to Fig. 2

by the negative and positive sides of the generator tend to approach zero and any signal on the surface generated by such a generator would be very small and probably undetectable within the "noise" of the EEG background activity.

The localization, polarity, field configuration, and even recordability of a potential generated by a cortical generator is therefore dictated not only by its location in relation to the recording electrodes, but also by its size, its geometrical orientation, and its configuration. EEG generators are not single dipoles, but convoluted dipole layers. If the fields on the scalp or brain surface appear to conform to those generated by single dipole sources which can be classified as "vertical" or "horizontal," one must regard this terminology for what it is, a useful shorthand term which, however, in no way should be construed as indicating that an EEG signal is in reality generated by such a dipole. Any computation of the localization of the source of an EEG signal based on such an assumption inevitably leads to erroneous results, the error increasing with the size of the generator (Gloor 1985). This is as true for magnetoencephalography as for electroencephalography.

The electrical fields on the scalp are distorted by the electrical inhomogeneities of tissues placed between the cerebral generators and the scalp electrodes. These distortions do not seriously degrade the localizing information derived from scalp EEG recordings, except sometimes in the presence of skull defects (Nunez 1981).

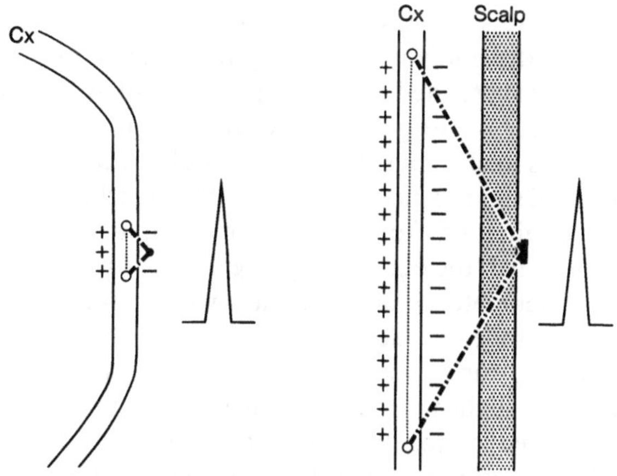

Fig. 4. *Left* A very small cortical generator (dipole layer) is subtended by a fairly large solid angle (depicted by a plane angle) at the site of an intracerebral depth electrode in close contact with the generator. *Right* To be subtended by a solid angle of the same size and thus produce a potential of the same amplitude in a scalp recording, a cortical generator must be several times larger. (Electrical inhomogeneities contributed by the skull and other tissues interposed between the brain and the recording electrode are not taken into account). *Cx*, cortex. [From Gloor P (1984) in: Porter RJ, Mattson R, Ward AA, Dam M (eds) Advances in epileptology: XVth epilepsy international symposium. Raven, New York, pp 433–437]

Intracranial EEG Recordings

In intracranial recordings, particularly those obtained with intracerebral depth electrodes, some biophysical conditions are encountered which never arise in extracranial recordings (Gloor 1984). In the latter the generators are always located at a distance of at least 1–2 cm from the recording electrodes. In intracerebral recordings, the electrodes are, however, in direct physical contact with some neuronal generators of EEG signals, while others are as distant as those generating recordable signals in extracranial recordings. This has important consequences: an electrode in direct contact with a generator of very small size can record a very large potential, because even though the generator is small, the solid angle subtended by the small dipole layer at the electrode in direct contact with it can be very large (Fig. 4), particularly if it is located at the concave side of such a generator, e.g., within the depth of a sulcus (Fig. 8). The signal derived from such an electrode may look

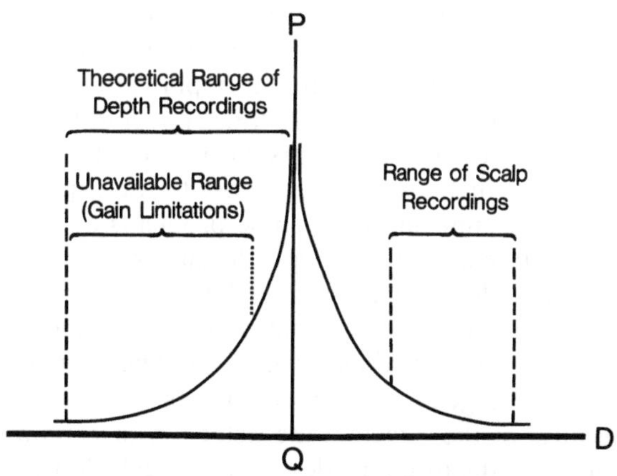

Fig. 5. The potential *(P)* decreases with the distance *(D)* from a charge *(Q)* according to a parabolic function. The theoretical range for depth electrode recordings is therefore wider, but in practice is limited to the upper, steep segment of the curve, because gains have to be kept low. Depth electrode recordings are therefore biased in favor of recording the activity of generators in close contact with the electrodes and may miss the contribution of somewhat more distant generators which is available in scalp recordings. [From Gloor P (1984) in: Porter RJ, Mattson R, Ward AA, Dam M (eds) Advances in epileptology: XVth epilepsy international symposium. Raven, New York, pp 433–437]

quite impressive. A signal of equal magnitude recorded extracranially could only be generated by a generator subtended at the scalp by a solid angle of equal size, i.e., the generator would have to be very large. It is likely that such a signal would be of greater clinical significance, considering that it reflects activity originating from a very large population of neurons, than the equally large signal recorded from a small generator in direct contact with an intracerebral electrode. In depth electrode recordings encountering such a small generator is a matter of chance. A slightly different electrode trajectory may have missed it altogether. The recording from such a small generator may, however, overshadow the contribution of much larger generators located at a distance comparable to that separating scalp electrodes from their generators. Because the amplification used in intracranial recordings must be kept lower than in scalp recordings, such potentials are likely to be missed. This is due to the fact that potentials fall off according to a parabolic function with increasing distance from their source (Fig. 5). The steepest potential decrease occurs near the source, while the slope at a larger distance is much gentler. In intracranial recordings, because of the extreme proximity of some of the generators, one is forced to operate in the steep, high-voltage part of the slope of many potential profiles. Hence the necessity to reduce the gains of the recording instrument. This severely limits the range within which one can record signals generated by neuronal populations which are still relatively close to the exploring electrode. Because of this unavoidable reduction in amplification, the lower segment of the potential slope shown in Fig. 5 is lost to recording. This renders the activity of many generators, which are relatively close to the exploring electrode, undetectable. These problems do not arise in extracranial recordings. Because a minimal distance of at least 1 cm separates the electrode from the generator, a recorded signal is always produced by a large generator which reflects the synchronous activity of a large population of neurons.

Because of these factors the detection of certain signals with depth electrodes is very much dependent upon the number of electrodes used, the localization of electrode contacts in relation to gyri and sulci, and the number of them incorporated in a montage with either the monopolar or bipolar technique. Figures 6-8 illustrate some of these problems. In these figures the trajectory of a depth electrode projected on a section of the Talairach Atlas (Talairach et al. 1967) is shown as it penetrates the brain through the second temporal convolution. It carries ten contacts 5 mm apart, with the deepest contact lying in the amygdala. Along its course, the electrode penetrates the walls of the superior temporal sulcus and the cortex at the bottom of the Sylvian fissure at the transition between the temporal and insular cortex. The position of two scalp electrodes approximately at positions C_3 and T_3 of the 10/20 International System are also shown. In Fig. 6 a generator occupying the crown of the second temporal convolution is portrayed. The moment in time when the signal produced by this generator is surface-negative was chosen for this diagram. At this moment the superficial cortex of the second temporal convolution forms a dipole layer with negativity at the cortical surface and positivity at its boundary with the underlying white matter (− and + signs in Fig. 6). The solid angles subtended by this generator surface at the various electrode contacts are depicted by corresponding plane angles, those subtending the negative side in red, those subtending the positive one in green. In the monopolar montage M shown in Fig. 6, the equivalent plane angles for each electrode position are shown in parentheses together with their appropriate electrical sign. The potentials recorded at each electrode in an ideal monopolar recording are drawn to scale in proportion to the sizes of the corresponding angles. These monopolar recordings show the potential to appear with the largest amplitude at contacts 8, 9, and 10. Contacts 9 and 10 straddle the generator and thus the electrical sign at 10 is negative and positive at 8 and 9 as well as at the deeper electrode contacts, where the signal rapidly becomes smaller. Because contacts 8, 9, and 10 are in direct contiguity with or at least very close to the generator, they "see" it under very large angles (equivalent to 205° at contact 9). At the other depth electrode contacts, the generator surface is not subtended by angles which are significantly different from those on the scalp, although the potential there would suffer some additional attenuation because of the interposition of electrical inhomogeneities, particularly the high-resistance skull bone. (This is not taken into account in Fig. 6.) In a bipolar recording utilizing all intracranial electrodes of the chain (column B1 in Fig. 6), the signal appears with a phase reversal at electrode 9. A very large amplitude is recorded by the electrode linkage 9-10, since the potential is proportional to the algebraic difference of the two angles at these two electrode positions: $+205°-(-95°)=+300°$. This reflects the fact that a potential difference between a large positive and a large negative potential is measured.

If bipolar montages using only some contacts of the whole chain are used (Fig. 6 B2, B3), the potential may be missed. Just substituting electrode 9 for electrode 10 (compare B2 and B3 of Fig. 6), a mere 5-mm shift renders the potential so small that it could easily be lost in the EEG background noise. Looking at column B1 it also becomes evident that if one had picked out just a few bipolar pairs by connecting a few electrode contacts but without recording the entire potential profile along an unbroken chain of contacts, the signal could have been missed unless one had picked out the pair 9-10, or perhaps 8-9.

In Fig. 7 the same depth electrode trajectory through the temporal lobe is shown, but the generator now occupies an area of basal temporal cortex at a distance of about 1-1.5 cm from the electrode tract which runs roughly parallel to the plane of the cortical generator. None of the electrodes are in direct contact with it. Only the positive side of the generator surface is subtended at all the contacts by solid angles which are of similar size. Hence, in a monopolar montage all electrodes pick up the signal with a similar amplitude (Fig. 7M), but the signal is virtually lost in bipolar recordings since the subtraction of the angles subtended at the electrodes used in the bipolar montages yields values approaching zero.

Figure 8 depicts the situation created by a generator occupying the fundus of the Sylvian fissure at the boundary between the temporal and insular cortex. The electrode trajectory is the same as that shown in Fig. 6 and 7. Contact 4 in the fundus of the fissure is located on the concave side of the curved generator which, at this position, is subtended by an exceedingly large solid angle (equivalent to $-250°$, encompassed by the "outer side" of the solid red lines in the insert in the top left corner of Fig. 8). Hence, the potential recorded at contact 4 in a monopolar montage is exceedingly large, but very small at all other electrode contacts, because the effective angles there are small or even virtually zero in spite of the fact that some of the contacts are close to the generator. This is so because at all electrode contacts other than 4 the generator surface is "seen" both from its negative and from its positive side (outlined by solid green and dashed red lines in Fig. 8) since they lie on the convex side of the generator. The differences resulting from the subtraction of these angles are very small, approaching zero. Thus, in the monopolar recording shown in Fig. 8, the potential is recordable virtually only at contact 4. Displacing contact 4 by only a few millimeters would cause the potential to virtually disappear from the recording. In a bipolar recording incorporating all electrode contacts of the chain (Fig. 8B1) the potential appears with a very large amplitude in two channels only, namely those connecting contacts 3-4 and 4-5, with a phase reversal at 4. When not all contacts are included in a bipolar recording, only those incorporating contact 4 show the signal at a very large amplitude, but when contact 3 only 5 mm away is substituted for contact 4, the signal becomes virtually undetectable (compare B2 and B3 of Fig. 8). Figure 2 in Gloor (1984) shows this situation in an actual depth electrode recording.

In depth electrode explorations recordings are sometimes taken from brain structures that do not produce "open fields" like the cerebral cortex, but "closed fields" (Lorente de Nó 1947; Rall 1962). Such structures, in contrast to cortex, are made up of multipolar neurons or neurons which, although they may have a pyramidal type of morphology, fail to show the orderly parallel arrangement of their main dendritic processes typical for cerebral cortex. A structure of this type which is frequently the target of stereotaxic depth electrode recordings is the amygdala. The generators of signals generated by amygdaloid neurons thus do not form dipole layers and the solid-angle theorem cannot be applied in this instance. "Closed fields" do not have any appreciable spatial spread beyond the confines of a neuronal population involved in generating it. An electrical signal produced by amygdaloid neurons could thus not be recorded outside this nuclear mass. Figure 9 shows a depth electrode penetrating the temporal lobe horizontally with its three most medial contacts located within the amygdala. Synchronous activation of its multipolar neurons produces a signal which, in monopolar recordings, can only be recorded from the three electrodes located within the amygdala itself (column M, Fig. 9). In a bipolar recording the signal is recorded when a contact within the amygdaloid gray matter is connected to an electrode outside it. The potential appears nowhere else.

The following conclusions can be drawn from these considerations:

1. Since the range of brain tissue that can be explored by a single intracranial electrode contact is limited, it is to be expected that many parts of the brain, even in relatively close proximity to electrode contacts, will remain unexplored. Target selection in electrode implantation is therefore important in order to include structures which in the light of preimplantation clinical and EEG data are likely sites of seizure onset. Intracranial electrode recordings should be devised in such a manner that likely alternative working hypotheses concerning seizure onset can be distinguished from each other by using proper targetting of electrodes and proper montage design. There always will remain a residual uncertainty about the precise area of seizure onset. Depth

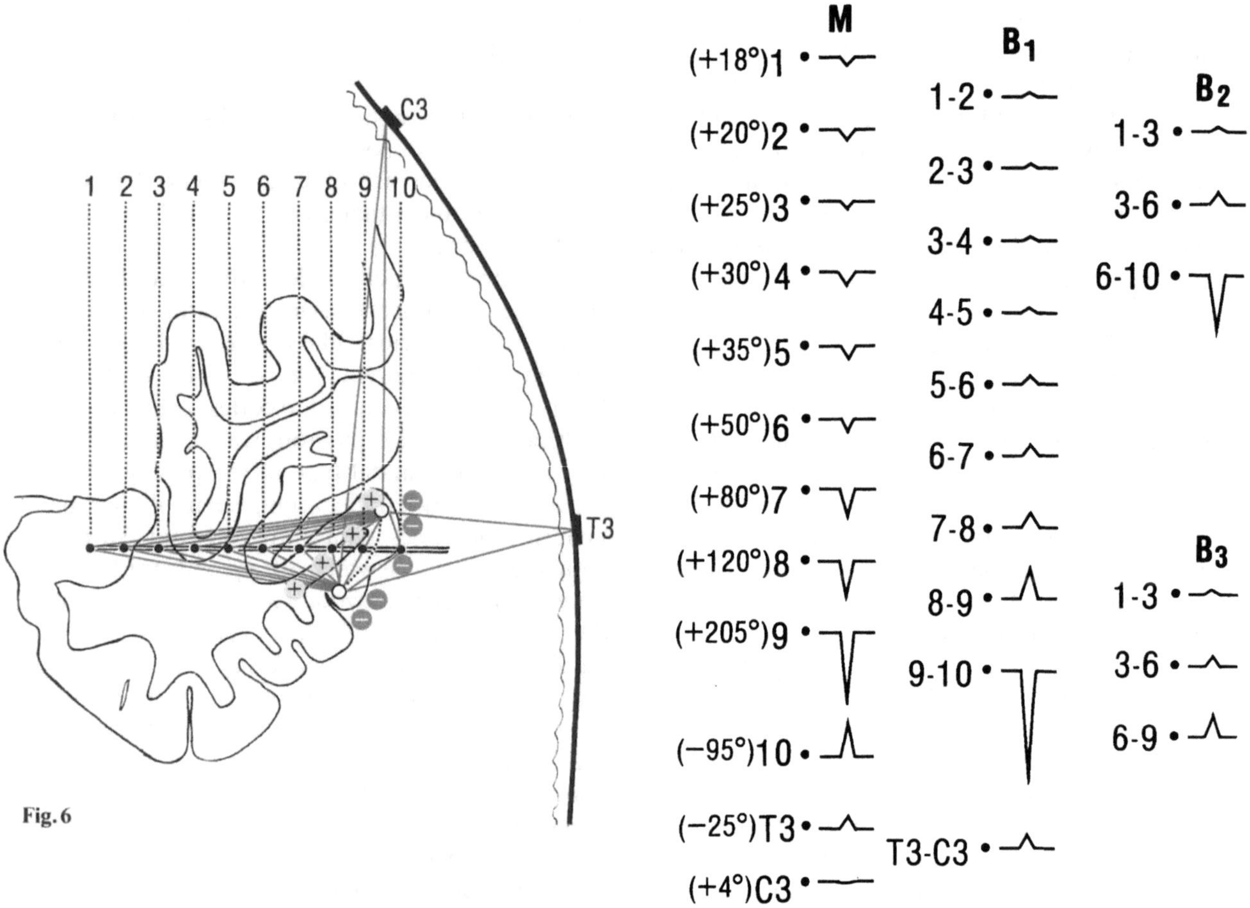

Fig. 6

Fig. 6. Generator occupying the crown of the second temporal convolution recorded with a depth electrode penetrating the brain horizontally, carrying ten contacts. *C3* and *T3*, scalp electrodes. The brain map in Figs. 6–8 portrays a frontal section drawn from the the Talairach Atlas (Talairach et al. 1967). The generator forms a relatively flat dipole layer with its negative side *(blue circles)* occupying the pial surface, and with its positive side *(yellow circles)* the white matter surface of the cortex. The solid angles subtended at the various electrode positions by the negative side of the generator surface are portrayed by *red*, those subtended by its positive side by *green* plane angles (the values of their plane angle homologues are listed on the left side of column M). *M*, monopolar, B_1 bipolar recordings from all electrode contacts; B_2 and B_3, bipolar recordings utilizing only some of the contacts

Fig. 7. Generator occupying basal temporal cortex. For further explanations see legend to Fig. 6

Fig. 8. Generator occupying the fundus of the Sylvian fissure. *Insert on top left* shows a magnification of this area. *Dashed red lines* indicate angles subtended by the positive surface of the generator "seen through" its convex side (as in Fig. 3 B). For other explanations see legend to Fig. 6

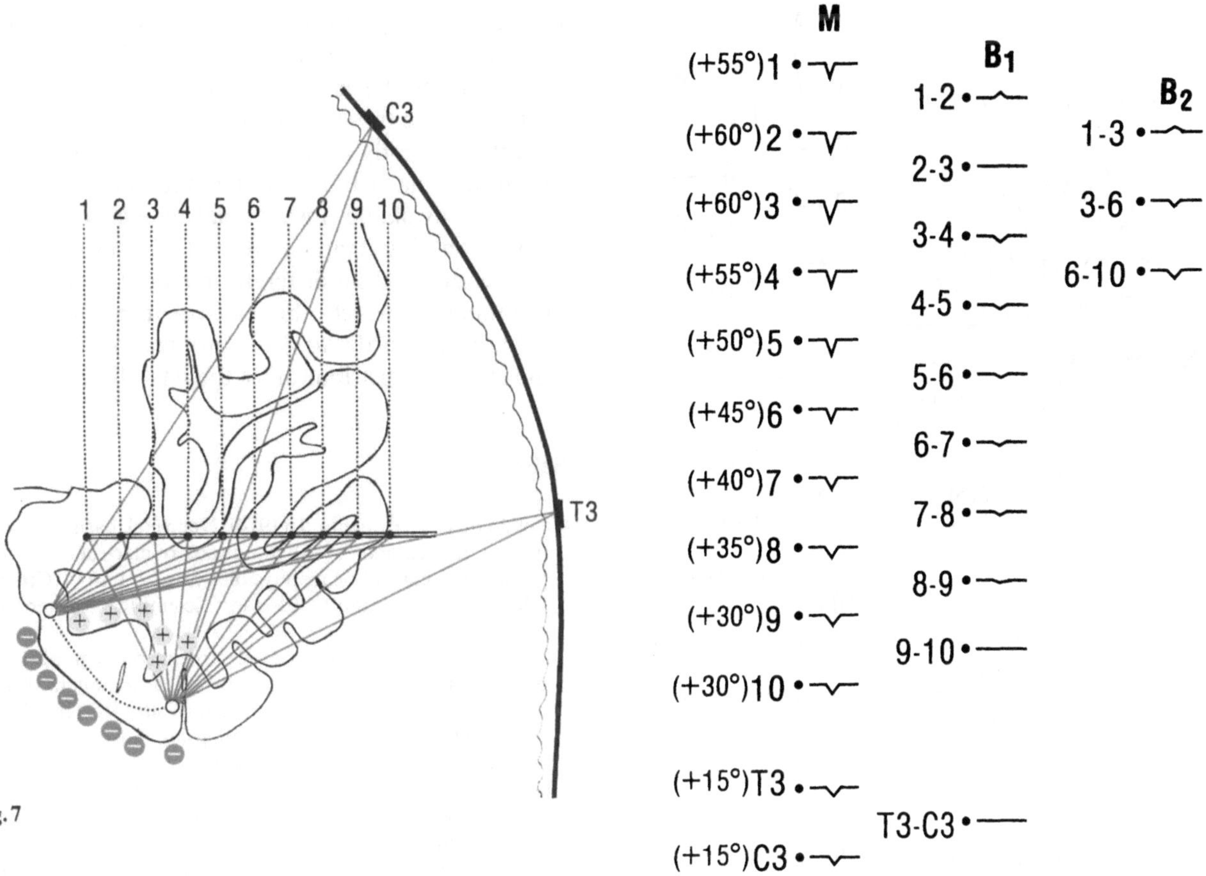
C3
T3
1 2 3 4 5 6 7 8 9 10
M
(+55°)1
(+60°)2
(+60°)3
(+55°)4
(+50°)5
(+45°)6
(+40°)7
(+35°)8
(+30°)9
(+30°)10
(+15°)T3
(+15°)C3
B1
1-2
2-3
3-4
4-5
5-6
6-7
7-8
8-9
9-10
T3-C3
B2
1-3
3-6
6-10

Fig. 7

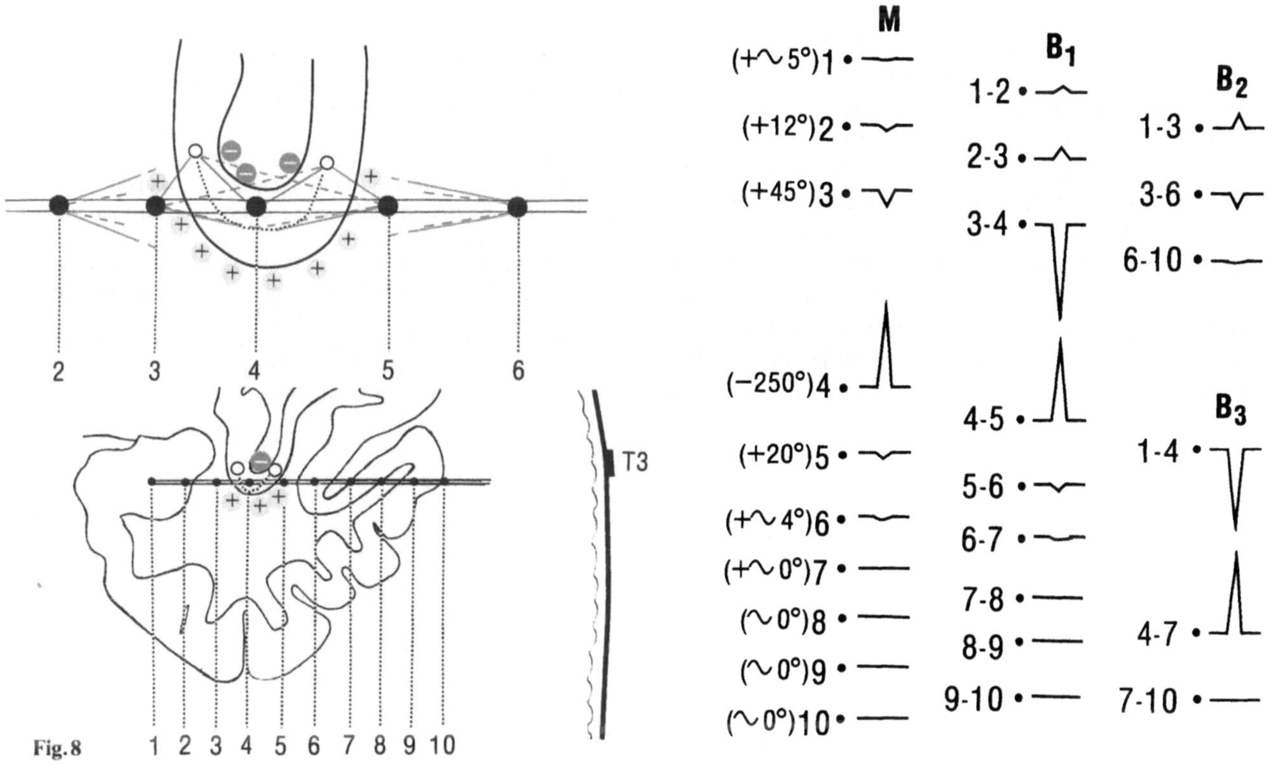
2 3 4 5 6
T3
1 2 3 4 5 6 7 8 9 10
M
(+∼5°)1
(+12°)2
(+45°)3
(−250°)4
(+20°)5
(+∼4°)6
(+∼0°)7
(∼0°)8
(∼0°)9
(∼0°)10
B1
1-2
2-3
3-4
4-5
5-6
6-7
7-8
8-9
9-10
B2
1-3
3-6
6-10
B3
1-4
4-7
7-10

Fig. 8

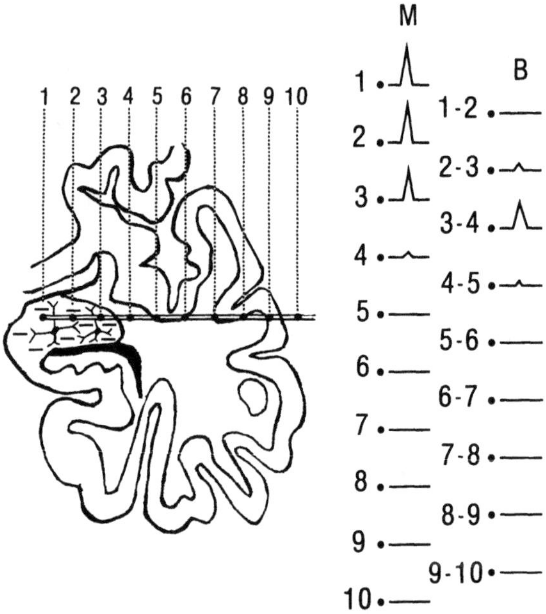

Fig. 9. Generator located in the amygdala consisting of neurons lacking the orderly geometrical arrangement characteristic of cortex and thus creating a "closed field." The brain map represents a horizontal section drawn from the Talairach Atlas (Talairach et al. 1967). For other explanations see legend to Fig. 6

electrode exploration if properly designed can delineate regions within which a seizure focus is localized, but it can never pinpoint it accurately to a very small volume of tissue.

2. Ideally one would want to explore selected target areas with as large a number of closely spaced electrode contacts as possible. When using bipolar recordings, preference should be given to unbroken chains incorporating as many contacts as possible, rather than to scattered closely spaced unconnected electrode pairs. The latter may by chance encounter a small insignificant generator producing a large signal or they may lie on an isopotential surface which could be quite close to a relevant generator. In practice, the number of channels available on an EEG machine imposes restrictions on the number of areas that ideally ought to be explored and on the montage design. Compromises must therefore be made. Changing montages between examinations can in part overcome these limitations. An attainable ideal with the presently available technology is to record the intracranial EEG with a monopolar technique from all available intracranial contacts simultaneously on digital tape. A record of a seizure could then be replayed on different monopolar and bipolar montages which the electroencephalographer could select in order to get an optimal picture of the topographical distribution of the ictal discharge.

References

Creutzfeldt OD, Houchin J (1974) Neuronal basis of EEG waves. In: Rémond A (ed) Handbook of electroencephalography and clinical neurophysiology, vol 2, part C (Creutzfeldt O, ed) section 1. Elsevier Scientific, Amsterdam, pp 5-54

Gloor P (1975) Contributions of electroencephalography and electrocorticography to the neurosurgical treatment of the epilepsies. In: Purpura DP, Penry JK, Walter RD (eds) Neurosurgical management of the epilepsies. Advances in neurology, vol 8. Raven, New York, pp 59-105

Gloor P (1984) Electroencephalography and the role of intracerebral depth electrode recordings in the selection of patients for surgical treatment of epilepsy. In: Porter RJ, Mattson R, Ward AA, Dam M (eds) Advances in epileptology: XVth epilepsy international symposium. Raven, New York, pp 433-437

Gloor P (1985) Neuronal generators, the problem of localization and volume conductor theory in electroencephalography. J Clin Neurophysiol 2: 327-354

Lorente de Nó R (1947) Action potentials of the motoneurons of the hypoglossus nucleus. J Cell Comp Physiol 29: 207-287

Nunez PL (1981) Electric fields of the brain. The neurophysics of EEG. Oxford University Press, New York, 484 pp

Rall W (1962) Electrophysiology of a dendritic neuron model. Biophys J 2: 145-167

Talairach J, Szikla G, Tournoux P, Prossalentis A, Bordas-Ferrer M, Covello L, Iacob M, Mempel E (1967) Atlas d'anatomie stéréotaxique du télencéphale. Masson, Paris, 323 pp

Woodbury JW (1960) Potentials in a volume conductor. In: Rush TC, Fulton JF (eds) Medical physiology and biophysics. Saunders, Philadelphia, pp 83-91

Editors' Commentary 2

Pathophysiology

The relevance of data derived from the huge number of experimental studies on epilepsy for its treatment in humans depends on how transferable these results are. One possible solution to this question is given by the experiments done on brain slices from human cortex resected for the purpose of epilepsy therapy. Concerning this aspect, the results published so far are inconclusive. Clearly, more studies are necessary because a better knowledge of the clinical relevance of experimental epilepsy models is essential to improve (1) development of new anticonvulsants, (2) understanding of clinical, electroencephalographic, and metabolic seizure manifestations, and (3) knowledge about the development of pathological alterations of brain tissue in relation to seizures and their consequences for treatment.

Summarizing the *discussion* on the topic 'pathophysiology,' two points have to be mentioned. One central question dealt with was whether or not a diminution of inhibition is of relevance for the development of seizure activity. There seemed to be general agreement that the reduction of inhibition influences the spread of epileptic activity. However, there was no clear evidence that the diminution of inhibition is a prerequisite for the generation of the endogenous epileptic discharge itself. In the kindling model of epilepsy, the reduction of the GABA-mediated inhibition in the hippocampus seemed to be decisive for initiation of the kindling process.

The second point dealt with the role of kindling phenomena in human epilepsies. This is of great importance for the surgical treatment of epilepsies for the following reasons. Firstly, the possibility of secondary epileptogenesis raises the questions: how can one test reliably whether 'secondary foci' are independent or dependent from the primary focus? Does the presence of dependent and/or independent foci have any influence on the results of surgical treatment?

Secondly, if secondary independent foci exist, what is the time course of their development?, and, does the duration of epilepsy before surgery influence the results?

Thirdly, what is the relation between the development of neuropsychological deficits in epilepsy patients and the duration and character of their epilepsy?

Since the recent studies of Morrell et al. (1987) provided clear evidence that secondary foci do develop in human epilepsies, it was agreed that solutions to the above-mentioned problems should be the aim of future studies.

Potential Generation

The purpose of these chapters was to deal with the generation of field potentials under seizure conditions especially from the viewpoint of recording them adequately. The experimental data clearly indicate that, in deep cortical layers, very restricted seizure foci may exist which do not show characteristic field potentials in the surface recording. The *discussion* revealed that similar phenomena were observed in clinical electrophysiology, e.g., in highly focalized motor seizures depth recordings from the corresponding motor cortex depicted narrowly confined epileptiform potentials, whereas simultaneous epicortical and/or scalp recordings did not show any conspicuous potentials.

Experimental studies indicate (1) that seizure activity may be restricted to either deep or superficial cortical layers depending on the experimental conditions, and (2) that the different participation of superficial and/or deep layers determines both the recordability of epileptiform field potentials from the surface of the cortex, and the presence or absence of clinical seizure accompaniments, such as tonic-clonic motor phenomena in the case of the motor cortex. Some participants, however, remarked that there is no clear evidence that clinical observations can be explained on the basis of these experimental data.

Gloor, by virtue of the volume conductor theory and the solid angle theorem, exemplified the most important biophysical factors which ought to be considered when inferences as to the localization of field potential generators (i.e., cortical dipole layer of macroscopic extent) will be made. Electroencephalographers, and in particular those involved in planning and designing intracranial recordings, were grateful for this contribution. They agreed with most of Gloor's conclusions. Concerning the limitations of depth EEG recordings it was emphasized that most European centers combine intracerebral depth recording with scalp EEG, and that only this combination (which of course requires a large number of recording channels) constitutes a proper 'stereoelectroencephalographic exploration' (SEEG).

In summarizing the *discussion* on the possibilities of recording interictal and ictal events with different types of electrodes, the following problem seems most relevant. In order to localize the origin of a seizure as precisely as possible, small electrodes have to be close to the region in question. This, however, results in recordings of very narrow regions of the brain. Therefore, proper target selection on the basis of preimplantation clinical and EEG data (which must be translated into alternative working hypotheses) remains crucial, i.e., the quality of preimplantation data, experience with various targets in different seizure types, and careful analyses of the clinical seizure phenomena are most important.

Reference

Morrell F, Wada J, Engel J Jr (1987) Potential relevance of kindling and secondary epileptogenesis to the consideration of surgical treatment for epilepsy. In: Engel J Jr (ed) Surgical treatment of the epilepsies. Raven, New York, pp 701-707

Pathological Findings Caused by Seizures, with Particular Reference to the Temporal Lobe

C. J. BRUTON

Epilepsy in humans may occur in association with a number of systemic disorders or with almost any local disease of the brain, and it is usual to separate this "secondary" or symptomatic form of the disorder from "primary" or cryptogenic epilepsy, in which the cause of the seizures is unknown (Meldrum and Corsellis 1984). The absence of an obvious cause, however, for the fits in patients with cryptogenic epilepsy does not necessarily imply the absence of brain damage; indeed Spielmeyer (1927) and Margerison and Corsellis (1966) found visible pathology either in the hippocampus or in the cerebellum in over 50% of their series of patients.

The pathological lesions of cryptogenic epilepsy usually take the form of a variable degree of nerve cell loss and gliosis which has a fairly characteristic pattern of distribution in the cerebral hemispheres and in the cerebellum; the effects of recent damage may be seen in the brains of some patients who die in status epilepticus but these lesions are less common than the more long-standing pathology, which is characterized by neuronal loss, fibrous gliosis, and atrophy.

The following review of epileptic pathology aims to identify the more common forms of brain damage; after this the pathogenesis of the lesions will be discussed.

Status Epilepticus

The most devastating and most acute neuropathological lesion associated with seizures may be seen in status epilepticus. The mortality of status is highest in infancy (Aicardi and Chevrie 1970) and for that reason the damage has been studied most widely in children (Ounsted et al. 1966). Nevertheless similar changes may be found in adults although the severity of the lesions is less severe (Corsellis and Bruton 1983). At postmortem the brain may be normal or simply congested. This is in sharp contrast, however, to the massive destruction which may be visible under the microscope (Fig. 1 a, b). The most commonly damaged areas are the hippocampus and the cerebellum, often with extensive involvement of the cerebral cortex, the thalamus, and the striatum. The affected nerve cells are unusually pale in Nissl preparations, whilst in haematoxylin and eosin stains the cell body appears eosinophilic with a triangular-shaped, shrunken, and hyperchromatic nucleus resembling the appearances of severe hypoxia or ischaemia. The cortical nerve cell damage frequently has a laminar distribution with destruction of the middle layers, particularly the third, and in addition there is nearly always an intense reactive gliosis.

Hemiatrophy

Occasionally the process of nerve cell loss and gliosis which characterizes "seizure pathology" may affect one hemisphere predominantly to produce a noticeable hemiatrophy which at times can be so severe that the damaged hemisphere may become little more than a fluid-filled remnant of scarred cortex and white matter (Figs. 2, 3). The selective vulnerability to epileptic insult of one hemisphere rather than the other is poorly understood. Taylor (1969, 1981), however, using temporal lobectomy material, has investigated the possible relationships between the side of the lesion, the sex of the patient, and the age at first fit, and these considerations may also apply in cases of hemiatrophy.

Cerebellar Lesions

The cerebellum is frequently involved in epileptic pathology, and Margerison and Corsellis (1966) found some degree of cerebellar damage in the brains of 45% of a series of patients with chronic

Department of Neuropathology, Runwell Hospital, Wickford, Essex SS11 7QE, and Division of Psychiatry, C.R.C. Northwick Park Hospital, Harrow, Middlesex, United Kingdom

Presurgical Evaluation of Epileptics
Edited by H. G. Wieser and C. E. Elger

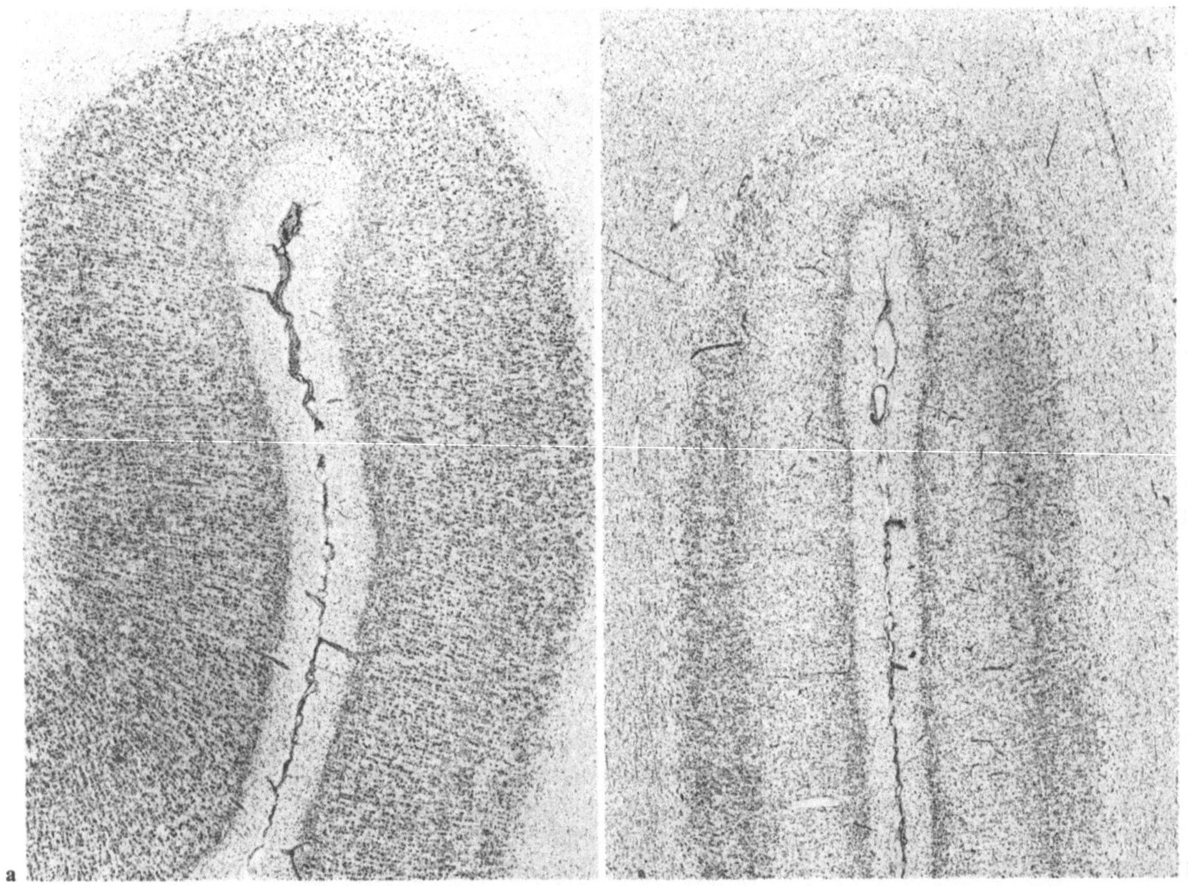

Fig. 1. **a** Low-power view of normal cerebral cortex. **b** Cerebral cortex of infant dying in status epilepticus. Note the extensive neuronal necrosis affecting the middle layers, particularly the third. (Nissl, ×12. By courtesy of Prof. J. A. N. Corsellis and the editors of *Greenfield's Neuropathology*)

idiopathic epilepsy. Their patients, however, were institutionalized epileptics, and in this respect probably represent the most severely afflicted end of the epileptic spectrum.

Histologically, the cerebellar Purkinje cells are the most vulnerable, with proliferation of the Bergmann glia, destruction of granule cells and fibrous gliosis of the folial white matter leading on occasions to considerable shrinkage and atrophy of the cerebellum as a whole (Figs. 4, 5).

Temporal Lobe Pathology

The most widely studied and most vulnerable site of epileptic damage is the hippocampus. In cases of fatal status epilepticus the hippocampus may show an acute, extensive, neuronal necrosis with a reactive astrocytosis resembling the acute lesions produced by artificially induced seizures in animals (Meldrum 1983). Nevertheless, the more common pathology is the final stage of scarring known variously as Ammon's horn sclerosis (Bouchet and Cazauvieilh 1825), incisural sclerosis (Earle et al. 1953), pararhinal sclerosis (Gastaut 1959), and mesial temporal sclerosis (Falconer et al. 1964). And although for historical reasons the term Ammon's horn sclerosis is most commonly used, the purely descriptive term of mesial temporal sclerosis is more accurate anatomically as the sclerotic process spreads frequently beyond the confines of the Ammon's horn to involve other "mesial" temporal structures such as the parahippocampal gyrus, the amygdaloid nucleus, the uncus, and the temporal lobe white matter (Falconer et al. 1964; Bruton 1984).

Microscopically, the lesion is a patterned loss of nerve cells accompanied by fibrous gliosis and eventually shrinkage and atrophy. The nerve cell loss varies from case to case, and three degrees of destruction can be identified. In the most common type the neuronal loss is most severe in two areas (the Sommer sector h1 and the end folium h3-5),

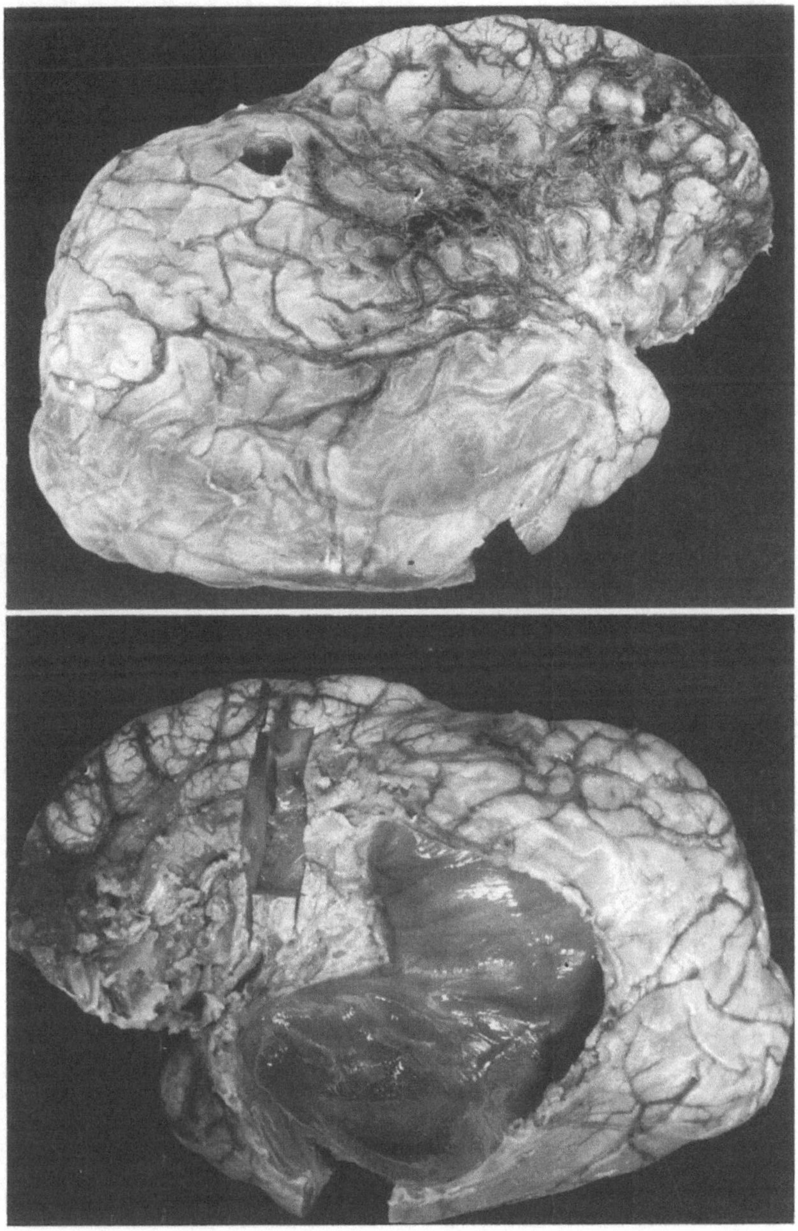

a

b

Fig. 2a, b. Lateral (**a**) and medial (**b**) aspects of a hemispherectomy specimen showing severe hemiatrophy. The hemisphere is represented by no more than a scarred remnant of cortex and subcortical white matter surrounding the greatly dilated ventricle

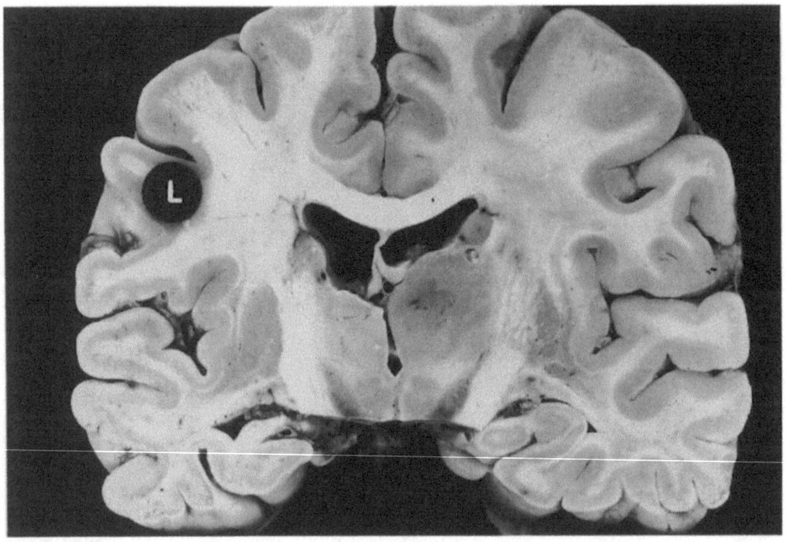

Fig. 3. Coronal slice of brain showing a mild hemiatrophy. The sclerotic hippocampus, shrunken thalamus, and dilated ventricular system of the left hemisphere are clearly seen. (× 0.5)

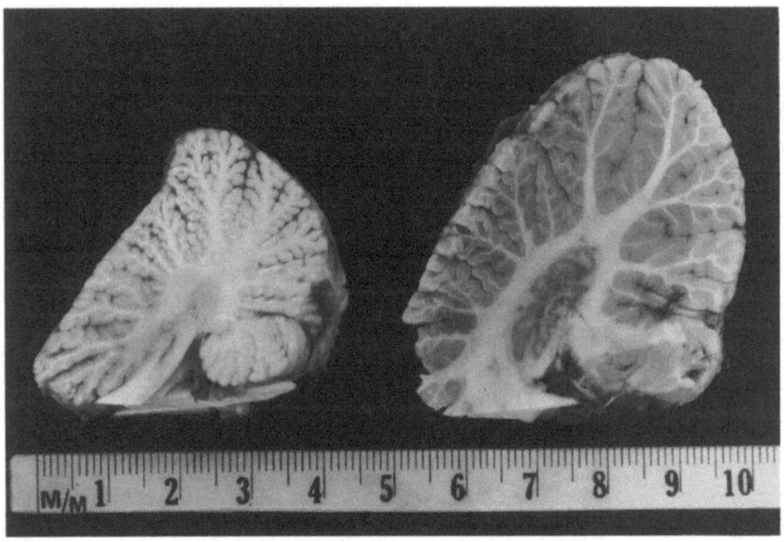

Fig. 4. A cut section through a normal cerebellum *(right)* compared with the cerebellum of a patient suffering from long-standing idiopathic epilepsy *(left)*. The folial atrophy of the latter is clearly visible. (Normal size)

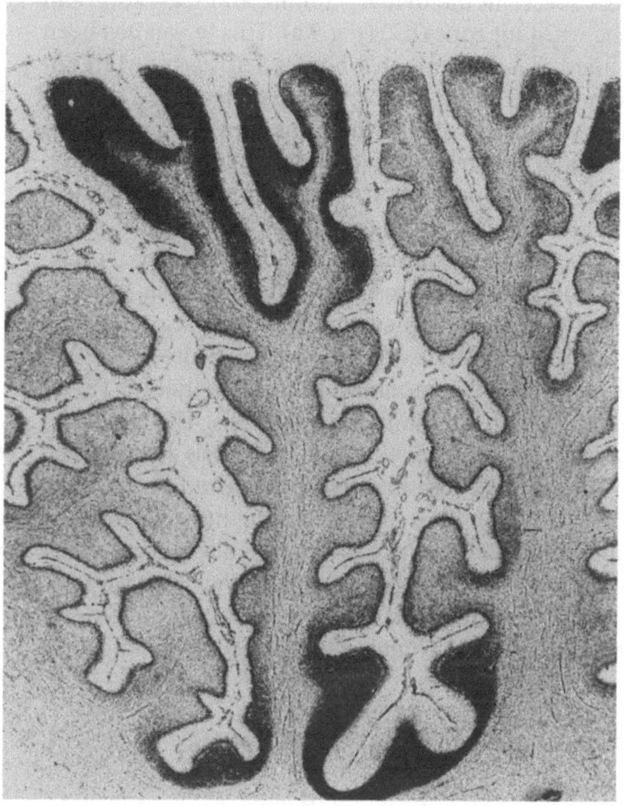

Fig. 5. Low-power view of cerebellar folia, with extensive loss of Purkinje cells and almost complete destruction of the granular layer. A few darkly stained patches of viable granule cells are seen. (Nissl, × 6)

Fig. 6. *(Top)* Coronal slice from a temporal lobectomy specimen shows a whitish, shrunken hippocampus with loss of the normal cortical ribbon. *(Bottom)* Normal temporal lobe and hippocampus for comparison. (× 0.5)

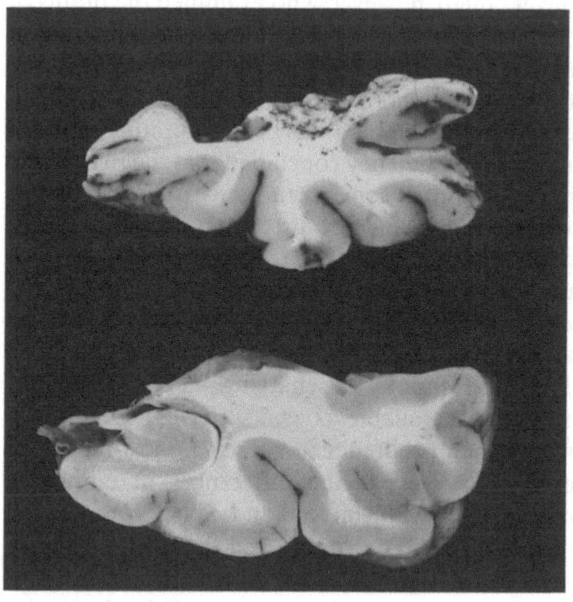

less severe in the dentate folium, and least severe in the resistant zone (h2) (Fig. 7a). This lesion has been called classic Ammon's horn sclerosis by Margerison and Corsellis (1966); they also described a much less common variety where the hippocampal nerve cell loss involved only the end folium (end folium sclerosis). Other cases, however, show an almost complete loss of hippocampal nerve cells in all areas, and this variation has been termed "total" Ammon's horn sclerosis (Bruton 1984) (Fig. 7b).

The relative frequency of the three sclerotic Ammon's horn lesions was reviewed by Bruton (1984) in a series of 249 temporal lobectomy patients; 107 of these had an Ammon's horn sclerosis. The classic lesion occurred in 57%, "total" sclerosis occurred in 40%, and end folium sclerosis occurred in 3%. Neither the type of the sclerotic lesion, however, nor the spread of pathology to other mesial temporal structures appeared to influence in any way the preoperative fit pattern or the beneficial postoperative prognosis which was found in the group as a whole.

Pathogenesis of Epileptic Brain Damage

Since the time of Sommer (1880), who produced the first histological description of hippocampal sclerosis, there has been controversy concerning the origin of epileptic brain damage. Sommer was convinced that the sclerotic lesions were the cause of seizures. Pfleger (1880), however, who examined the brains of epileptic patients who died during a fit, thought that the damage was of recent origin and a sequel to the fit. Spielmeyer (1927) and Scholz (1933) also believed that the sclerosis was the result rather than the cause of the fits. They concluded that the nerve cells died as a result of some disturbance of their blood or oxygen supply which occurred in association with a major epileptic convulsion. The precise nature of this disturbance was obscure but Spielmeyer suggested vascular spasm. Scholz emphasized that the damage was caused by a consumptive hypoxia in which the rapidly discharging nerve cells outstripped the available oxygen supply; the surrounding glia then proliferated as a reaction to the dying nerve cells. Despite the differences in emphasis, however, both Spielmeyer and Scholz considered the scar tissue to be "inert" and that it did not give rise to further epileptic fits.

The theory of the "inert" scar remained popular until the advent of EEG recordings on patients with temporal lobe seizures when it became clear that many epileptics with atrophy of the medial temporal region had frequent psychomotor fits and the characteristic electrical abnormalities associated with them. Earle et al. (1953) then proposed that Ammon's horn sclerosis was not the consequence of fits but that the initial damage occurred as a result of birth injury with herniation of the medial temporal gyri through the incisura of the tentorium cerebelli causing distortion of the hippocampal branches of the posterior cerebral artery and a consequent "incisural" sclerosis. The resultant scar then gradually "ripened" after many months or years into an active epileptogenic focus.

Although attractive, the hypothesis of birth injury as the sole cause of hippocampal sclerosis was soon questioned on anatomical grounds by Meyer et al. (1954), Falconer et al. (1964), Ounsted et al. (1966) and Veith (1970), who had also begun to find a frequent association of the Ammon's horn lesion with a clinical history of convulsions or of status in infancy. This latter observation suggested the hypothesis of "hypoxic" damage resulting from the episode of fits giving rise to nerve cell death, gliosis, and eventual scar formation.

The common thread of both theories, however, was the emphasis on the occurrence of a cerebral catastrophe either at birth or in infancy which gave rise to the initial neuronal damage in a particularly vulnerable area of the brain.

Bruton (1984) investigated the "possible precipitating factors" of psychomotor seizures in a series of temporal lobectomy patients. He found a highly significant association between Ammon's horn sclerosis and febrile convulsions in infancy ($P < 0.0001$) and a significant association of the lesion with status epilepticus ($P < 0.02$) and also with birth injury ($P < 0.02$), thus suggesting the possibility of several causative factors, although the significance of the association with febrile convulsions was by far the greatest.

In addition to the clinicopathological observations in man, however, there have been a number of notable experimental observations which have contributed to the understanding of seizure pathology (Meldrum, 1983, for detailed references). The earlier experiments involved the production of sustained generalized seizures in baboons using bicuculline, which gave a pattern of brain damage similar to that seen after status epilepticus. Later investigations, involving the induction of focal seizures with microinjections of kainic acid into the amygdala have sought to reproduce the pathological features found with the brains of patients with chronic idiopathic epilepsy. Meldrum (1983) summarized the current experimental evidence in the form of "a revised metabolic hypothesis for epileptic brain damage".

a

b

c

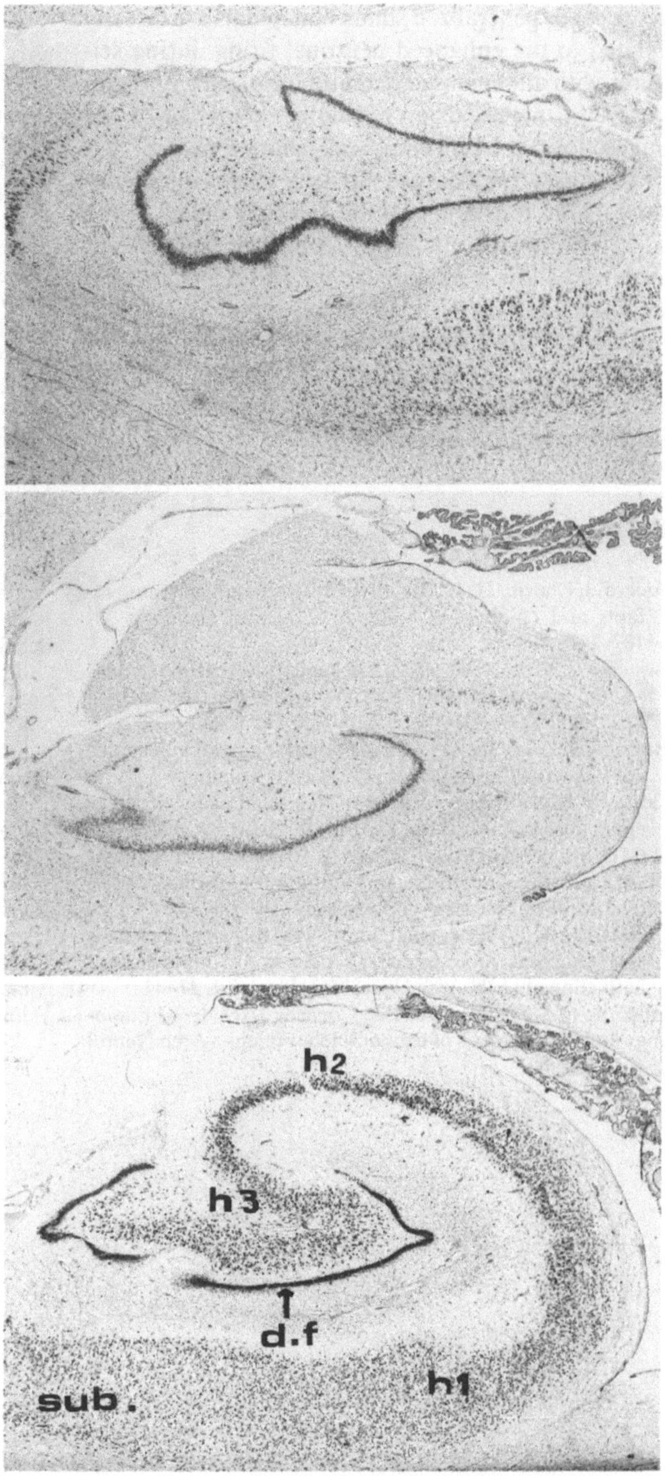

Fig. 7. **a** Classic Ammon's horn sclerosis with neuronal loss and gliosis in h 1 and to a lesser extent in h 3, the end folium. **b** "Total" Ammon's horn sclerosis, with almost complete loss of nerve cells in all sectors. **c** Normal labelled Ammon's horn for comparison. (Nissl, × 10)

He concluded that the selective neuronal changes after focal or generalized status epilepticus are directly related to the enhanced neuronal firing during seizures with an increased demand for oxygen and glucose and increased ion exchanges. This led to the suggestion that ischaemic cell change may be explained as a result of the toxicity of excess intracellular calcium ions and that the selective vulnerability of certain nerve cells arises from their greater tendency to show bursting activity and a paroxysmal depolarizing shift associated with enhanced calcium ion entry. When seizure activity is prolonged the capacity of the cell to extrude calcium ions is exceeded, producing the toxic changes which had led ultimately to neuronal death.

References

Aicardi J, Chevrie JJ (1970) Convulsive status epilepticus in infants and children. A study of 239 cases. Epilepsia 11: 187-197

Bouchet C, Cazauveilh (1825) De l'épilepsie considerée dans ses rapports avec l'alienation mentale. Arch Gén Méd 9: 510-542

Bruton CJ (1984) The neuropathology of temporal lobectomy, vol 1-2. MD Thesis, London University, London

Corsellis JAN, Bruton CJ (1983) Neuropathology of status epilepticus in humans. In: Delgado-Escueta AV, Wasterlain CG, Treiman D, Porter RJ (eds) Advances in neurology, vol 34. Status epilepticus: mechanisms of brain damage and treatment. Raven, New York, pp 129-139

Earle KM, Baldwin M, Penfield W (1953) Incisural sclerosis and temporal lobe seizures produced by hippocampal herniation at birth. Arch Neurol Psychiatry 69: 27-42

Falconer MA, Serafetinides EA, Corsellis JAN (1964) Etiology and pathogenesis of temporal lobe epilepsy. Arch Neurol 10: 233-248

Gastaut H (1959) Etiology, pathology and pathogenesis of temporal lobe epilepsy. Epilepsy News Letter no 15 (International League Against Epilepsy, Amsterdam) pp 15-24

Margerison JH, Corsellis JAN (1966) Epilepsy and the temporal lobes. A clinical, electroencephalographic and neuropathological study of the brain in epilepsy, with particular reference to the temporal lobes. Brain 89: 499-530

Meldrum BS (1983) Metabolic factors during prolonged seizures and their relation to nerve cell death. In: Delgado-Escueta AV, Wasterlain CG, Treiman D, Porter RJ (eds) Advances in neurology, vol 34. Status epilepticus: mechanisms of brain damage and treatment. Raven, New York, pp 261-275

Meldrum BS, Corsellis JAN (1984) Epilepsy. In: Adams JH, Corsellis JAN, Duchen LW (eds) Greenfield's neuropathology, 4th edn, chapter 19. Arnold, London, pp 921-950

Meyer A, Falconer MA, Beck E (1954) Pathological findings in temporal lobe epilepsy. J Neurol Psychiatry 17: 276-285

Ounsted C, Lindsay J, Norman R (1966) Biological factors in temporal lobe epilepsy. Clinics in developmental medicine 22. Heinemann, London

Pfleger L (1880) Beobachtungen über Schrumpfung und Sklerose des Ammonshorns bei Epilepsie. Allg Z Psychiatrie 36: 359-365

Scholz W (1933) Über die Entstehung des Hirnbefundes bei der Epilepsie. Z Gesamt Neurol Psychiatrie 145: 471

Sommer W (1880) Erkrankung des Ammonshornes als ätiologisches Moment der Epilepsie. Arch Psychiatrie Nervenkrankheiten 10: 631-675

Spielmeyer W (1927) Die Pathogenese des epileptischen Krampfes. Z Gesamt Neurol Psychiatrie 109: 501-520

Taylor DC (1969) Differential rates of cerebral maturation between sexes and hemispheres. Evidence from epilepsy. Lancet 2: 140-142

Taylor DC (1981) The influences of sexual deviation on growth development and disease. In: Davis J, Dobbin J (eds) Scientific foundations of paediatrics, 2nd edn. Heinemann, London

Veith G (1970) Anatomische Studie über die Ammonshornsklerose im Epileptikergehirn. Dtsch Z Nervenheilkd 197: 293-314

Pathological Changes Relevant for Seizure Generation

Y. ROBITAILLE

The type and extent of morphological changes which may be correlated with the electrophysiological phenomena characteristic of either the partial or generalized epilepsies cannot be dissociated from the molecular events underlying ictal activity. They must also reflect the abnormalities in neuronal-glial interrelationships which trigger or sustain neuronal seizure discharges. Neuronal pathological manifestations will first be explored, followed by relevant glial changes. The role of the blood brain barrier and other cell components will then be examined.

Morphological Substrates of Focal Epilepsy

Epileptogenic Neuronal Lesions

Hippocampal sclerosis (HS) is observed in 50%–83% of the surgical specimens from patients with temporal lobe epilepsy but similar abnormalities have also been reported in many postmortem series involving various other epilepsies. Since the works of Meynert (1868), Sommer (1880), Pfleger (1880), and Bratz (1899), it has been repeatedly described as a variable state of neuronal degeneration and gliosis. In the most severe cases the lesion encompasses Rose's fields H3 to H5, sector I, and the pro-subiculum, with various degrees of dentate gyrus involvement but a tendency to spare area H2. Although a particular relationship of H5 with partial complex seizures was recognized in a classical morphological study by Margerison and Corsellis (1966), it is not pathognomonic. Similar unilateral lesions are frequently found in severe generalized epilepsies (Sano and Malamud 1953) or other seizure types (Falconer 1974; Mouritzen-Dam 1982). Furthermore, postmortem studies involving both epileptic and nonepileptic subjects have all emphasized its nonspecificity. A wide variety of anoxic, ischemic, vascular, or inflammatory processes can give rise to similar morphological profiles (Morel and Wildi 1956; Haymaker et al. 1958). However, there are obvious limitations inherent in the morphological examination of remote brain injuries, and the concept of early perinatal or infantile lesions "ripening" to focal seizures after a latency of several years is strongly supported by clinical experience. Uncal herniation with compression of midline arteries, initially invoked by Earle et al. (1953) in the concept of "incisural sclerosis," has still not been fully established as a significant etiological factor in temporal lobe epilepsy. Falconer (1971) and Mathieson (1975) stress the high incidence of a familial history of epilepsy associated with mesial temporal sclerosis. The low seizure threshold in children between 6 months and 4 years of age was postulated to increase the vulnerability of the hippocampus to the effects of prolonged febrile convulsions, which had occurred in many of their patients undergoing temporal lobectomy for drug-resistant seizures.

Recent neuropathological studies emphasizing electro-morphometric correlations in patients with chronically implanted hippocampal depth electrodes have yielded much more specific data. These analyses were oriented along an anterior-posterior axis of the hippocampus and were restricted to patients without a significant history of previous generalized seizures. Babb et al. (1984b) found that the degree of neuronal loss was linked to the anatomical distribution of ictal activity. Patients who displayed pyramidal cell loss restricted to anterior hippocampi had a greater incidence of postoperative seizure control compared with those in whom both the anterior and posterior regions were affected. It appears that neurons adjacent to zones of hippocampal sclerosis were epileptogenic and that maximum surgical benefit was obtained by resecting the most damaged areas. In a further study of 43 temporal lobe specimens resected "en bloc," Babb et al. (1984a) reported that volumetric hippocampal neuronal densities could not be significantly correlated with age, side of operation, age at seizure onset, seizure frequency, or

Montreal Neurological Institute and Departments of Neuropathology, Neurology, and Neurosurgery, McGill University, 3801 University Street, Montréal, Québec H3A 2B4, Canada

Presurgical Evaluation of Epileptics
Edited by H. G. Wieser and C. E. Elger

total number of seizures. They postulated that the sclerotic hippocampal process occurred before seizure onset and that cell loss stimulated synaptic reorganization of the remaining and fringe neurons. These new networks gave rise to abnormal discharges that eventually propagated sufficiently to produce clinical seizures. The causes of the initial damage and the mechanisms of synaptic reorganization remain unknown in most cases. Hippocampal volumetric cell densities, however, were not significantly reduced when associated with lateral temporal pathology such as tumors, cysts, and hamartomas.

Deafferentation and "Ongoing" Degeneration

The concept of "ongoing" degeneration found support in ultrastructural studies of sclerotic human hippocampi derived from surgically treated epileptic patients (Brown 1973). Degenerative changes involving neuronal cell bodies, pre- and postsynaptic profiles, and myelinated axons admixed with numerous glial processes were demonstrated in the strata lacunosum and oriens, and near the alveus. The Golgi technique studies of the Scheibels (1973) revealed extensive dendritic spine loss and moniliform changes of pyramidal cell dendritic shafts, considered to reflect extensive neuronal deafferentation. The fascia dentata showed unusual "closed parasol" and "windswept" patterns involving the dendritic tree of granule cells. In a subsequent study swollen mossy tufts were described. Golgi studies of seizure-sensitive strains (SS) of Mongolian gerbils also revealed marked spine loss in the distal and proximal dendritic trees of CA_3 pyramidal cells. Swollen mossy tufts which harbored larger amounts of synaptic vesicles now occupied a larger proportion of the tuft area. Progressive deafferentation remains an attractive explanation of neuronal epileptogenicity through release of inhibitory input. Hippocampal or isocortical deafferentation experiments have effectively reproduced such dendritic spine loss (Colonnier 1964; Bernstein et al. 1975); however, the latter may also be secondary to seizure activity and excitotoxic damage. Such a view is supported by Golgi studies of CA_3 and CA_4 pyramidal cells of SS strains of gerbils at various ages before and after onset of seizure activity (Scheibel 1983). Furthermore, recent observations of isocortical neurons in tuberous sclerosis have revealed the presence of aberrant small multipolar stellate neurons endowed with poorly developed and beaded dendritic trees harboring few spines (Huttenlocher 1984). This suggests that abnormal spine development may mimic the process of "ongoing" degeneration.

Deafferentation and Loss of Inhibitory Input

Westrum et al. (1964) described dendritic spine loss and moniliform changes in the topical alumina-induced epileptic focus of primates. These were indistinguishable from those seen by the Scheibels in human temporal lobes. Using the reduced Silver methods of Nauta-Gygax and Fink-Heimer, "ongoing" cortical and white matter degeneration was also demonstrated in similar primate seizure models (Harris 1972). The role of GABAergic inhibitory input loss in seizure generation and maintenance was reinforced through morphometric evaluation of symmetrical type II synapses which reacted with immunocytochemically labeled glutamic acid decarboxylase (GAD); these type II synapses were markedly decreased in alumina-induced seizure foci (Ribak et al. 1979; Ribak et al. 1982). Isocortical pyramidal cells of primate sensorimotor epileptic cortex were found to have undergone extensive degeneration of axosomatic and proximal axo-axonic boutons originating from adjacent GABAergic basket and chandelier cells (Ribak 1985). This suggests that these cells may have become epileptogenic because of the loss of presynaptic inhibitory input. Other neurotransmitter systems may also contribute to such loss of inhibition. Although cobalt-induced seizure foci in various rodents have been reported to show transient decreases of GAD and GABA (Emson and Joseph 1975), intracortical noradrenergic and serotoninergic terminals have also been implicated. Trottier et al. (1983) reported reduced high-affinity noradrenaline and serotonin uptake in cobalt foci which peaked at the time of onset of maximal epileptic discharge. Similar changes were also observed in contralateral homotopic cortex but were less severe than in ipsilateral perifocal regions. Degeneration of noradrenergic terminals at days 8 and 14 around cobalt lesions was demonstrated with the glyoxylic acid method by Trottier et al. (1981), which were able to regenerate along with recovery of intracortical noradrenergic uptake to control values at day 40. Briere et al. (1986) noted a significant relative decrease in the density of alpha-1 adrenergic receptors in human epileptic spike foci compared with adjacent inactive areas of human temporal neocortex. The decreased adrenoceptor density was associated with a compensatory increase in the activity of tyrosine hydroxylase, the rate-controlling enzyme in catecholamine synthesis. In contrast, $[^3H]$flunitrazepam binding was remarkably similar in spiking and nonspiking areas. This is in keeping with the lack of persistent change in benzodiazepine receptors in various experimental models of focal epilepsy. Although isolated cortical slab experiments performed by the undercutting of afferent systems increase neuronal excitability (Halpern 1972), reports of secondary neurotransmitter abnormalities have been inconsistent. A significant decrease of cortical acetylcholinesterase and choline acetyl transferase and slight decrease of GAD were observed (Ulmar 1981) along with a marked loss of symmetrical synapses at the edges of the slabs (Ribak and Reiffenstein 1982). However, recent experimental evidence suggests that disinhibition could result from increased numbers of GABAergic terminals which synapse with adjacent GABAergic cells, which themselves inhibit hippocampal pyramidal cells of SS gerbils.

Lipid Peroxidation in Human and Experimental Epilepsy

Subpial iontophoresis of ferrous ($FeCl_2$) or ferric chloride ($FeCl_3$) produces chronic seizure foci with electrographic and behavioral seizures after a latency of 2–10 days. The morphology of these lesions resembles human epileptogenic scars resulting from trauma or remote intracortical hemorrhage (Willmore et al. 1978; Triggs and Willmore 1984). Transient lipid peroxidation manifested by early increased tissue levels of malonaldehyde occurs between 4 and 24 h postinjection. This process has been linked to the formation of superoxide radicals at the injection site. The extensive neuronal loss and gliosis and the epileptic activity triggered by these experiments can be largely prevented by pretreatment with the antioxidant α-tocopherol (Willmore and Rubin 1981). Reid and Sypert (1984), using rapid Golgi-Cox techniques, described severe neuronal loss, extensive dendritic spine loss, and dendritic shaft varicosities adjacent to the injection sites. Extracellular unit recording in acute iron-induced seizure models demonstrated the presence of numerous burst-firing neurons, mostly localized near the edges of the lesion. As deafferentation may cause dendritic spine loss, its presence in chronic $FeCl_3$-induced seizure models suggests that deafferented neurons may play a prominent role in the development and maintenance of some epileptogenic foci.

Epileptogenic Potential of Astroglial Changes

The view that neuronal loss is an important stimulus to proliferation and transformation of protoplasmic and fibrillary astrocytes into reactive forms capable of weaving dense feltworks of glial processes has been widely accepted by morphologists. Astrogliosis can develop from any type of injury and, although the precise mechanisms remain unknown, cerebral edema is always associated with damaged myelinated fibers and scattered reactive astrocytes. In the acute stage there is marked swelling of astrocytic cell bodies and perivascular end-feet. Experimental contusions of the cerebral cortex trigger intense endocytosis of horseradish peroxydase (HRP) between 4 and 72 h postinjury (Trachtenberg 1983). Normal astrocytes have the capacity to buffer excess extracellular (K^+) through Na^+-K^+ ATPase activity. In seizure foci there is evidence that the almost everpresent astrogliosis may afford some protection against extracellular water and electrolyte disturbances induced by seizure activity. Woodbury et al. (1984) reported that in cobalt foci of rats, acute or chronic administration of acetazolamide, an inhibitor of glial carbonic anhydrase, enhanced seizure activity in cobalt foci in rats.

However, astrogliosis may also be responsible for a series of detrimental effects which may overwhelm the normal buffering capacity of glial membranes. Harris and Jenkins (1975) infused HRP in the subarachnoid spaces overlying chronic epileptic foci in primates and observed a 67% decrease of HRP infiltration into the extracellular space. Such marked decrease of extracellular spaces hinders normal fluxes of anions, cations, or toxic substances which may not be cleared into the cerebrospinal fluid. Lewis et al. (1977) reported doubling of potassium clearance time following direct cortical stimulation in gliotic areas induced by alumina gel. There is also mounting evidence that glial Na^+-K^+ ATPase activity may be deficient in both experimental alumina foci (Harmony et al. 1968) and human chronic seizure foci (Grisar et al. 1983). Although abnormalities of (K^+) metabolism may enhance neuronal depolarization in epileptic foci, it does not appear to initiate the neuronal process of burst-firing since resting extracellular (K^+) was normal in gliotic seizure foci.

The GABA shunt is active within synaptosomes but not in glial cells because of the absence of glutamic acid decarboxylase (GAD). An alternative pathway for GABA formation could utilize glutamic acid derived from the glutamate-glutamine cycle catalyzed by the enzymes glutamine synthetase (GS) and glutaminase. In a series of patients submitted to neurosurgical treatment for intractable partial seizures, we have been able to compare the degree of astroglial change with glutamine synthetase activity as measured in homogenates prepared from immediately adjacent cortex. Glutamine synthetase has been shown to be an astroglial marker by means of

Fig. 1. Low-magnification coronal rat brain section autoradiograph prepared 30 min after i.p. injection of [^{14}C]-β-alanine (125 μCi/kg). There is uptake enhancement along the scar's cavity in the right motor cortex

immunohistochemical techniques. We found a marked increase in GS activity in homogenates prepared from all specimens of epileptogenic cortex in which there was a significant degree of astrocytic hyperplasia. This included several cases which involved dysgenetic processes (tuberous sclerosis or focal cortical dysplasia) (Sherwin et al. 1984).

Although GABA cannot be synthesized in astrocytes because of insufficient levels of GAD, it may still be significantly taken up by high-affinity systems (Hertz 1978; Giacobini and Guttierez 1983). We indirectly assessed the glial capacity to buffer-released GABA in $FeCl_3$-induced seizure foci in the rat by means of radioautography with [^{14}C]-β-alanine. This amino acid has the capacity to compete with GABA for a high-affinity glial uptake system. We found an apparent endothelial and astroglial-enhanced uptake in incubated scar edges 5, 30, and 120 days postinjection (Robitaille and Sherwin 1984). Results were tabulated by measuring autoradiographic optical densities within the inner 3.0-mm scar edges lining the cavities, which displayed the highest densities of reactive astrocytes (Robitaille et al., to be published). Statistically significant scar uptake was found in the models, compared with normal control brain (Fig. 1). The predominantly astrocytic uptake of β-alanine within scar edges was significantly suppressed within ipsilateral isocortex by the gliotoxic compound DL-α-amino adipic acid (Olney et al. 1971). By challenging the models with 3-mercaptopropionic acid, a strong inhibitor of GAD which selectively decreases cerebral GABA synthesis immediately prior to the onset of generalized seizures (Karlsson et al. 1974), GABA inhibition appeared to release β-alanine glial uptake which dramatically increased in scarred areas with significant recruitment of ipsilateral isocortical astrocytes (Figs. 2, 3). These data suggest that generalized seizures may be accompanied by substantial glial GABA uptake in preestablished $FeCl_3$-induced seizure foci.

Wyler (1977) and Ward (1983) pointed to the glial entrapment of neurons and dendrites in chronic alumina-induced epileptic foci which became more mechano-sensitive with respect to burst-firing and was more readily elicited upon deformation of neuronal membranes by micropipettes than normal cortex or penicillin-induced epileptic cortex. It thus appears that the astrogliosis of epileptogenic scars may provide protective as well as deleterious side effects on the epileptogenicity of adjacent pacemaker neurons. The combined effects of all seizure-protecting or -promoting factors vary according to the relative proportion of epileptogenic neurons and density of hypertrophic astrocytes and processes. Similar considerations may also be operative in the various pathological causes of human focal epilepsy.

Implications of Vascular Changes in Seizure Foci

Jasper (1970), noting the paucity of small vessels within human posttraumatic cerebral scars, specu-

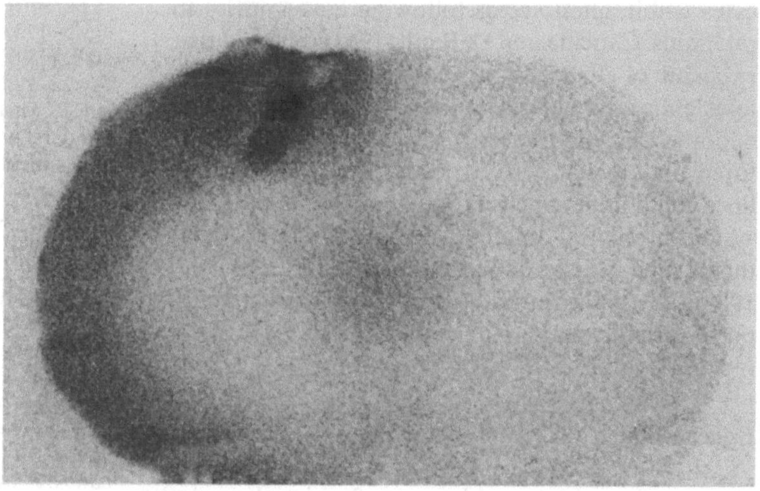

Fig. 2. Low-magnification coronal rat brain section autoradiograph prepared by i. p. [^{14}C]-β-alanine injection immediately following onset of generalized seizures provoked by 3-mercaptopropionic acid. There is marked enhancement of scar edge and ipsilateral isocortical uptake, which appears to be diffuse in the thalamus

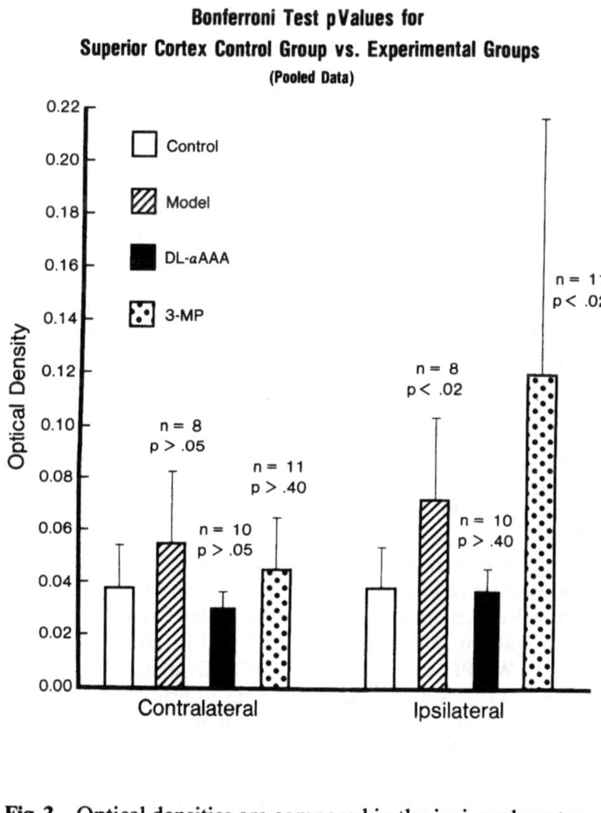

Fig. 3. Optical densities are compared in the ipsi- and contralateral superior cortex between the pooled data of control and experimental models. There is significant suppression of glial uptake bilaterally to control values through the effects of DL-α-AAA. DL-α-AAA, DL-α-amino adipic acid; 3-MP, 3-mercaptopropionic acid

lated on the role of the potentially ischemic effect of hypovascularization on the development of epileptogenicity. A detailed discussion of the blood brain barrier (BBB) is beyond the scope of this review. However, in vitro studies on $FeCl_3$-induced scar edges showed marked endothelial labeling of neocapillaries sprouting within the granulation tissue by [^{3}H]-β-alanine (Robitaille and Sherwin 1984), a low molecular weight neutral amino acid which normally displays a low brain uptake index (Oldendorf 1971). It is also well known that generalized seizures cause a transient opening of the BBB, which Petito et al. (1977) found to be related to abrupt increases of systemic blood pressure during seizures induced in rats. The regional distribution of BBB changes may also vary according to the type of chemical inductor (Suzuki et al. 1984). The ubiquitous finding of multiple clusters of hypertrophic astrocytes in a perivascular localization in the white matter immediately underlying human cortical seizure foci may possibly be related to bouts of vasogenic edema. It appears that repeated BBB disturbances, even transient, may enhance astroglial scarification, which could easily impinge on deeper cortical lamina.

Although the morphological substrate of the BBB resides in the tight junction network investing endothelial cells, perivascular glial end-feet are strategically localized to act as chemical buffers. Brightman et al. (1983) described a diffuse increase in the number of orthogonal arrays on the membranes of astrocytes within gliotic areas following cold injuries to rat brains. Cancilla and DeBault (1983) reported that exposure of glial cells to endothelial cell culture systems enhanced neutral amino acid transport across endothelial membranes. It also appeared that cultured endothelial cells exposed to C6 glioma cell lines could undergo changes in the polarity of the A-system of neutral amino acid transport from endothelial-glial to the opposite direction (Beck et al. 1984). The putative results of these in vitro changes on seizure foci remain to be investigated.

Conclusions

Human and experimental seizure foci involve relatively straightforward histopathological changes conventionally visualized as areas of neuronal loss, gliosis, and vascular anomalies. Other reported changes of a less specific nature include microglial activation, macrophage formation, and the accumulation of collagen at the junction between the lesion and adjacent pathological reactive changes. Oligodendrocytes and myelin breakdown also frequently occur and complicate the frequent "ongoing" degenerative process of active seizure foci. However, the extensive modifications of neurotransmitter pathways and water and electrolyte metabolism recorded in human and experimental seizure foci imply that extensive disturbances have taken place in the membranes of all cell types which may involve receptors, ionic channels, and neuronal-glial interrelationships. Their combined effects influence the degree of neuronal hyperexcitability, and their localization and extent determine the clinical seizure patterns. It appears that the epileptic focus cannot be defined by a single pathognomonic morphological change; it rather reflects the effects of various structural and molecular abnormalities on selectively vulnerable neuronal clusters and supporting cells.

References

Babb TL, Jann Brown W, Pretorius J, Davenport C, Lieb JP, Crandall PH (1984a) Temporal lobe volumetric cell densities in temporal lobe epilepsy. Epilepsia 25: 729-740

Babb TL, Lieb JP, Jann Brown W, Pretorius J, Crandall PH (1984b) Distribution of pyramidal cell density and hyperexcitability in the epileptic human hippocampal formation. Epilepsia 25: 721-728

Beck DW, Vinters HV, Hart MN, Cancilla PA (1984) Glial cells influence polarity of the blood brain barrier. J Neuropathol Exp Neurol 43: 219-224

Bernstein JJ, Wells MR, Bernstein ME (1975) Dendrites and neuroglia following hemisection of rat spinal cord: effects of puromycin. In: Kreutzberg GW (ed) Advances in neurology, vol 12. Raven, New York, pp 439-451

Bratz E (1899) Ammonshornbefunde bei Epileptischen. Arch Psychiatry Nervenkr 31: 820-836

Briere R, Sherwin AL, Robitaille Y, Olivier A, Quesney LF, Reader TA (1986) α-Adrenoreceptors are decreased in human epileptic foci. Ann Neurol 19: 26-30

Brightman MW, Zis K, Anders J (1983) Morphology of cerebral endothelium and astrocytes as determinants of the neuronal microenvironment. Acta Neuropathol [Suppl] VIII: 21-23

Brown WJ (1973) Structural substrates of seizure foci in the human temporal lobe. In: Brazier MAB (ed) Epilepsy: its phenomena in man. Academic, New York, pp 338-374

Cancilla PA, DeBault LE (1983) Neutral amino acid transport properties of cerebral endothelial cells in vitro. J Neuropathol Exp Neurol 42: 191-199

Colonnier M (1964) Experimental degeneration in the cerebral cortex. J Anatomy 98: 47-53

Earle KM, Baldwin M, Penfield W (1953) Incisural sclerosis and temporal lobe seizures produced by hippocampal herniation at birth. Arch Neurol Psychiatry 69: 27-42

Emson PC, Joseph MH (1975) Neurochemical and morphological changes during the development of cobalt-induced epilepsy in the rat. Brain Res 93: 91-110

Falconer MA (1971) Genetic and related aetiological factors in temporal lobe epilepsy: a review. Epilepsia 12: 13-31

Falconer MA (1974) Mesial temporal (Ammon's horn) sclerosis as a common cause of epilepsy: aetiology, treatment and prevention. Lancet 2: 767-770

Giacobini E, Guttierez MC (1983) GABA and pipecolic acid: a possible reciprocal modulation in the CNS, in glutamine, glutamate and GABA. In: Hertz L, Kvamme E, McGeer EG, Schousboe A (eds) The CNS. Liss, New York, pp 571-580

Grisar T, Franck G, Delgado-Escueta AV (1983) Glial contribution to seizure: K^+ activation of Na-K ATPase in bulk isolated glial cells and synaptosomes of epileptogenic cortex. Brain Res 261: 75-84

Halpern LM (1972) Chronically isolated aggregates of mammalian cerebral cortical neurons studied in situ. In: Purpura DP, Penry JK, Tower DB, Woodbury DM, Walter RD (eds) Experimental models of epilepsy. Raven, New York, pp 197

Harmony T, Urba-Holmgren R, Urbay CM, Szava S (1968) Na-K ATPase activity in experimental epileptogenic foci. Brain Res 11: 672-680

Harris AB (1972) Degeneration in experimental epileptic foci. Arch Neurol 26: 434-449

Harris AB, Jenkins DP (1975) Intercellular space in epileptic brain. Neurosci Abstr 1: 179

Haymaker W, Pentschew A, Margoles C, Bingham WG (1958) Occurrence of lesions in the temporal lobe in the absence of convulsive seizures. In: Baldwin M, Bailey P (eds) Temporal lobe epilepsy. Thomas, Springfield, pp 166-202

Hertz L, Wu PH, Schousboe AC (1978) Evidence for net uptake of GABA into mouse astrocytes in primary cultures - its sodium dependence and potassium independence. Neurochem Res 3: 313-323

Huttenlocher PR, Heydemann P (1984) Fine structure of cortical tubers in tuberous sclerosis: a Golgi study. Ann Neurol 16: 595-602

Jann Brown W (1973) Structural substrates of seizure foci in the human temporal lobe. In: Brazier MAB (ed) Epilepsy, its phenomena in man. Academic, New York, pp 339-374

Jasper HH (1970) Physiopathological mechanisms of post-traumatic epilepsy. Epilepsia 11: 73-80

Karlsson A, Fonnum F, Malthe-Sorenssen D, Storm-Mathisen J (1974) Effect of the convulsive agent 3-mercaptopropionic acid on the levels of GABA, other amino acids and glutamate decarboxylase in different regions of the rat brain. Biochem Pharmacol 23: 3053-3061

Lewis DV, Mutsuga N, Schuette WH, Van Buren J (1977) Potassium clearance and reactive gliosis in the alumina gel lesion. Epilepsia 18: 499-506

Margerison JH, Corsellis J (1966) Epilepsy and the temporal lobe: a clinical, electroencephalographic and neuropathological study of the brain in epilepsy, with particular reference to the temporal lobes. Brain 89: 499-530

Mathieson G (1975) Pathology of temporal lobe foci. In: Penry JK, Daly DD (eds) Advances in neurology, vol 11. Raven, New York, pp 163-185

Meynert T (1868) Der Bau der Großhirnrinde und seine örtlichen Verschiedenheiten, nebst einem pathologisch-anatomischen Corollarium. Engelmann, Leipzig

Morel F, Wildi E (1956) Sclérose ammonienne et epilepsies (étude anatomopathologique et statistique). Acta Neurol Belg 2: 61-74

Mouritzen-Dam A (1982) Hippocampal neuron loss in epilepsy and after experimental seizures. Acta Neurol Scand 66: 601-642

Oldendorf WH (1971) Brain uptake of radiolabeled amino acids, amines, and hexoses after arterial injection. Am J Physiol 221: 1629-1639

Olney JW, Ho LO, Rhee V (1971) Cytotoxic effects of acidic and sulphur containing amino acids on the infant mouse central nervous system. Exp Brain Res 14: 61-76

Petito CK, Schaefer JA, Plum F (1977) Ultrastructural characteristics of the brain and blood brain barrier in experimental seizures. Brain Res 127: 251-267

Pfleger L (1880) Beobachtungen über Schrumpfung und Sklerose des Ammonshorns bei Epilepsie. Allg Z Psychol 36: 359-365

Reid SA, Sypert GW (1984) Chronic models of epilepsy. In: Schwartzkroin PA, Wheal H (eds) Electrophysiology of epilepsy. Academic, New York, pp 137-151

Ribak CE (1985) Axon terminals of GABAergic chandelier cells are lost at epileptic foci. Brain Res 326: 251-260

Ribak CE, Reiffenstein RJ (1982) Selective inhibitory synapse loss in chronic cortical slabs: a morphological basis for epileptic susceptibility. Can J Physiol Pharmacol 60: 864-870

Ribak CE, Harris AB, Vaughn JE, Roberts E (1979) Inhibitory GABAergic nerve terminals decrease at sites of focal epilepsy. Science 205: 211-214

Ribak CE, Bradburne RM, Harris AB (1982) A preferential loss of GABAergic, symmetric synapses in epileptic foci: a quantitative ultrastructural analysis of monkey neocortex. J Neurosci 2: 1725-1735

Robitaille Y, Sherwin A (1984) High affinity {^{3}H}-β-alanine uptake by scar margins of ferric chloride-induced epileptogenic foci in rat isocortex. J Neuropathol Exp Neurol 43: 376-383

Sano K, Malamud NC (1953) Clinical significance of sclerosis of the cornu ammonis. Arch Neurol Psychiatry 70: 40-53

Scheibel ME, Scheibel AB (1973) Hippocampal pathology in temporal lobe epilepsy: a Golgi survey. In: Brazier MAB (ed) Epilepsy, its phenomena in man. Academic, New York, pp 311-337

Scheibel AB, Paul L, Fried I (1983) Some structural substrates of the epileptic state. In: Jasper HH (ed) Basic mechanisms of neuronal hyperexcitability. Liss, New York, pp 109-130

Sherwin A, Quesney F, Gauthier S, Olivier A, Robitaille Y et al. (1984) Enzyme changes in actively spiking areas of human epileptic cerebral cortex. Neurology 34: 927-933

Sommer W (1880) Erkrankung des Ammonshorns als aetiologisches Moment der Epilepsie. Arch Psychiatr Nervenkr 10: 631-675

Suzuki R, Nitsch C, Fujimara K, Klatzo I (1984) Regional changes in cerebral blood flow and blood brain barrier permeability during epileptiform seizures and in acute hypertension in rabbits. J Cereb Blood Flow Metab 4: 96-102

Trachtenberg MC (1983) Glial endocytosis of protein in the traumatized brain. J Neurosci Res 9: 413-423

Triggs WJ, Willmore LJ (1984) In vivo lipid peroxidation in rat brain following intracortical Fe^{2+} injection. J Neurochem 42: 976-980

Trottier S, Berger B, Chauvel P, Dedek J, Gay M (1981) Alterations of the cortical noradrenergic system in chronic cobalt epileptogenic foci in the rat: a histofluorescent and biochemical study. Neuroscience 6: 1069-1080

Trottier S, Claustre Y, Caboche J, Dedek J, Chauvel P, Nassif S, Scatton B (1983) Alterations of noradrenaline and serotonine uptake and metabolism in chronic cobalt-induced epilepsy in the rat. Brain Res 272: 255-262

Ulmar G (1981) Rat undercut cortex as a model for the study of central amino acid transmission. In: De Feudis G, Mandel P (eds) Amino acid neurotransmitters. Raven, New York, pp 105-113

Ward AA (1983) Physiological basis of chronic epilepsy and mechanisms of spread. In: Delgado-Escueta AV, Wasterlain CG, Treiman DM, Porter RJ (eds) Status epilepticus. Raven, New York, pp 189-197

Westrum LE, White LE, Ward AA Jr (1964) Morphology of the experimental focus. J Neurosurg 21: 1033-1046

Willmore LJ, Rubin JJ (1981) Antiperoxidant pretreatment and iron-induced epileptiform discharges in the rat: EEG and histopathologic studies. Neurology 31: 63-69

Willmore LJ, Sypert GW, Munson JB (1978) Recurrent seizures induced by cortical iron injection: a model of post-traumatic epilepsy. Ann Neurol 4: 329-336

Woodbury DM, Engstrom FL, White HS (1984) Ionic and acid-base regulation of neurons and glia during seizures. Ann Neurol [Suppl] 16: 135-144

Wyler AR (1977) Discrimination between epileptic and injury-induced repetitive firing in chronic epileptic cortex. Exp Neurol 55: 603-617

Chairman's Commentary

P. KLEIHUES

The session on morphological sequelae of chronic epilepsy was initiated by C.J. Bruton with an overview covering both historical and present views on the pathogenesis of epileptic brain damage. His own experience is based mainly on the unique collection of biopsies from 249 temporal lobectomy patients, who had been operated upon by the late Dr. Falconer. It was pointed out in the discussion that this series, although extremely valuable for pathologists, was collected before 1975, when CT, PET, and NMR were not yet available. Correlation with clinical data is, therefore, limited. This is, in part, also true for the morphological analyses. Modern techniques including immunohistochemistry and transmitter and receptor autoradiography were not performed, but Dr. Bruton announced that some immunohistochemical studies that can be carried out on paraffin-embedded material will be conducted in the near future.

The involvement of cerebral energy metabolism in epileptic seizures was recently reviewed by Chapman (1985). Metabolic activation accompanies seizures from the onset and includes efflux of potassium and influx of calcium, reduction of glucose and glycogen stores, and elevated levels of lactate, cyclic nucleotides, adenosine, free fatty acids, and prostaglandins. Regional inhibition of protein synthesis appears to be a very sensitive indicator for cells at risk of permanent damage. The selective vulnerability of certain nerve cell populations may be due to a relative energy deficit of maximally firing neurons, with available energy reserves being utilized for processes essential for cell survival (e.g., maintenance of ion gradients) rather than for macromolecular synthesis. Studies employing this technique have provided evidence that, in contrast to the hippocampus and cerebral cortex, Purkinje cell damage may not be due to seizure activity itself (Kiessling et al. 1984). In the cerebellum and brain stem amino acid incorporation was only inhibited in freely convulsing but not in anaesthetized, mechanically ventilated, and physiologically controlled rats, indicating that cerebellar nerve cell loss is a consequence of systemic seizure effects, e.g., asphyxia.

The contribution by Y. Robitaille concentrated on the role of altered transmitter metabolism, receptors for neurotransmitters, ionic channels, and amino acid uptake in the generation of epileptic seizures. He pointed out that glial-neuronal interrelationships are complex, and that microglial activation, macrophages, and connective tissue scars may all contribute to the formation of seizure foci. His review also expands on experiments with seizure induction by ferric chloride. In this model epileptogenesis can be prevented by the antioxidant, α-tocopherol. It remains to be shown whether prooxidant states with active oxygen species, lipid hydroperoxides and free radicals derived therefrom also contribute to other etiologies, e.g., epilepsy following subarachnoidal or intracerebral hemorrhage. Additional aspects of epileptic seizure generation may be found in the recent review by Benardo and Pedley (1985).

References

Benardo LS, Pedley TA (1985) Cellular mechanisms of focal epileptogenesis. In: Pedley TA, Meldrum BS (eds) Recent advances in epilepsy, vol 2. Churchill Livingstone, Edinburgh, pp 1-18

Chapman AG (1985) Cerebral energy metabolism and seizures. In: Pedley TA, Meldrum BS (eds) Recent advances in epilepsy, vol 2. Churchill Livingstone, Edinburgh, pp 19-64

Kiessling M, Xie Y, Kleihues P (1984) Regional impairment of protein synthesis in the rat brain during bicuculline-induced seizures. Brain Res 296: 1-13

Specific Detection of Epileptic Disorders in Clinical Neuropsychology*

E. Perret and M. Vetsch

Considering the ability of clinical neuropsychology to detect cerebral lesions and to provide clues about their localization, the absence of systematic studies about the role of the etiology of cerebral lesions is conspicuous. There are good reasons, however, to assume that an active epileptogenic focus modifies the cerebral functioning differently than an inert neoplastic lesion. Vascular malformations and vascular accidents must, still by their indirect action on distant brain areas, have other effects on cerebral functioning. Traumatic cerebral lesions, finally, are known to be multiple in many instances, and therefore do not allow the localization of the main origin of the cognitive deficits which have been assessed neuropsychologically.

The retrospective analysis of neuropsychological data presented here uses a wide variety of patients with cerebral lesions referred for clinical neuropsychological assessment. The reasons for referral vary from routine preoperative testing for later comparison with postoperative results, to questions of vocational guidance after a head injury. Every patient is examined with a large series of psychometric tests, each of which is known on the basis of previous experience to be sensitive to the primary function of one area of the brain, that is, laterally to the function of the left or right hemisphere and, within one hemisphere, to the function of the frontal, temporal, or parietal lobe. More recently, test procedures have proved sensitive to the functions of the brain stem, so that the clinical efforts are no longer concentrated on the cognitive performances of cortical structures. When a standard series of psychometric tests is applied systematically to every patient, a so-called *test profile* is established, in which each test result is represented on the same normalized scale. Basically, a unilateral lesion in the frontal, temporal, or parietal lobe yields a typical test profile, with lower values in the tests which assess the function of that lesioned area.

The inquiry into the specific detection of epileptic disorders can be made by comparing the test profiles of epileptic patients with those of patients in whom no epileptic disorders are reported. The comparisons are made in accordance with currently used localizational frames (left or right, and frontal, temporal, or parietal) and the combinations thereof (left or right hemisphere, independently of the affected lobe; and frontal, temporal, or parietal, independently of the affected hemisphere).

The test profiles can be compared either with regard to their distance from each other, independently of their shape, and, alternatively, with regard to their shape, independently of the distance separating them.

Methods

The test results of epileptic and nonepileptic patients were drawn from the data bank of the Neuropsychology Unit. All had been collected in the course of clinical-diagnostic neuropsychological assessments following referrals mostly from the Department of Neurosurgery and rarely from the Department of Neurology.

The *criteria for inclusion* in the *epileptic group* (EP) were a clear diagnosis of epilepsy in the patients' charts and a clear unilateral localization of the seizure origin in the frontal, temporal, or parietal lobe. No distinction was made, however, on the basis of different etiologies, which means that some patients with tumors were included. For the *nonepileptic group* (NE), the *absence* of mention of any epileptic disorders in the patients' charts was required, together with the clearly stated unilateral localization of the lesion in any of the frontal, temporal, or parietal

Neuropsychological Institute, Promenadengasse 18, 8001 Zürich, Switzerland

* The present work was supported in part by the Swiss National Science Foundation, Grants 3.771.80 and 3.816.79 to E. Perret.

Presurgical Evaluation of Epileptics
Edited by H. G. Wieser and C. E. Elger

Table 1. Descriptive data on the epileptic (EP) and nonepileptic (NE) patient subgroups

Hemisphere		Anteroposterior localization					
		Frontal		Temporal		Parietal	
		EP	NE	EP	NE	EP	NE
Left	*N*	5	55	34	104	5	50
	m/f (%)[a]	60/40	56/44	65/35	54/46	60/40	56/44
	Age (years)	33.0	46.9	27.0	44.2	26.8	44.8
Right	*N*	6	60	24	66	13	56
	m/f (%)[a]	67/33	62/38	58/42	53/47	54/46	59/41
	Age (years)	39.7	46.9	26.0	41.6	29.2	47.0

[a] Sex distribution, male vs. female as a percentage of *N*

lobes. Thus vast majority of the NE group was patients suffering from a tumor or a vascular accident. For *all patients*, the first neuropsychological assessment was considered, irrespective of whether or not a neurosurgical operation followed.

This selection yielded a group of 87 epileptics and another group of 391 nonepileptics. Both groups were subdivided according to the localization of the functional and/or structural lesion given in the charts, yielding left frontal, temporal, and parietal, and right frontal, temporal, and parietal EP and NE *subgroups*. Their size, the relative sex distribution, and the average age of the patients are shown in Table 1.

Forty-two *test variables* were selected from the body of test data collected in every patient. In brief, these included quantitative data from tests of verbal and spatial learning, verbal and figurative-spatial memory span and memory, perception of speech and space, and reasoning [for details see Perret (1973); similar test series have been described by Lezak (1983) and Taylor (1979)]. Each variable was normalized separately for each patient by subtracting it from the mean of a "norm group" of more than 3000 patients, including those considered here, and dividing the difference by the standard deviation of the norm group. For each subgroup of patients, finally, the mean of each test variable was computed, and the 42 averaged test variables made up the test profile of each subgroup. Results qualitatively below the norm (fewer correct responses, more errors, longer execution times) were given a minus sign.

The *distance* between the test profiles was estimated statistically by submitting the data of corresponding EP and NE test profiles to two-factorial analyses of variance, separately for the subgroups with left and right hemisphere lesions. The first factor was the presence or absence of epilepsy (EP vs NE); the second factor took into account the anteroposterior localization (frontal vs. temporal vs. parietal). The *shapes* of the test profiles were compared with each other on the basis of product-moment correlation coefficients between the corresponding test variables of the corresponding EP and NE subgroups. Correlation coefficients are frequently only apparently significant. For this reason, a *control* set of correlation coefficients was computed by *averaging* all correlation coefficients between the lobes of primary pathology and the neighboring lobes separately within the EP and within the NE groups. For instance, the correlation coefficients between the left frontal subgroup and the right frontal, left temporal, and left parietal subgroups of the EP and of the NE groups were averaged as a control for frontal lesions. This procedure was applied only for the control of the combined frontal, temporal, and parietal subgroups.

Results

Patient Selection

The selection of epileptic (EP) and nonepileptic (NE) patients has yielded heterogeneous groups. Not only is the size of the groups different, but there is also a significant difference in the respective *frequencies* of frontal, temporal, and parietal localizations of the lesions [chi-square(2) = 18.1, $P < 0.001$]. The analysis of variance based on *age* yields a significant difference between the (young) EP and (older) NE patients [$F(1,466) = 36.1$, $P < 0.001$], but the differences between the mean ages in the EP and NE subgroups are not significant [interaction between EP and NE and side of the lesions: $F(1,466) < 1$; interaction between EP and NE and anteroposterior localization of the lesions: $F(2,466) < 1$]. Only the relative *sex-distribution* does not vary conspicuously, either from the EP to the NE groups, or from one subgroup to another (the majority of men being usual in our experience).

Distance Between Test Profiles

The mean performance levels are presented in Table 2. With only two exceptions, the values are negative, i.e., the performances of the EP and NE patients in the psychometric tests were below those of the "norm group" composed of patients with cerebral lesions not fitting the inclusion criteria of this study.

The analysis of variance comparing the data of patients with lesions in the *left hemisphere* yields significant results throughout: between EP and NE [$F(1,295)=4.6$, $P<0.04$] and between the affected frontal, temporal, or parietal lobes irrespective of the EP vs. NE distinction [$F(2,295)=14.4$, $P<0.001$] as well as in interaction with it [$F(2,295)=3.7$, $P<0.03$]. The outcome of the analysis of variance comparing the data of patients with lesions in the *right hemisphere* is similar: significant difference between EP and NE [$F(1,299)=4.9$, $P<0.03$], between the frontal, temporal, and parietal subgroups [$F(2,299)=7.6$, $P=0.001$] and significant interaction effect of EP and NE combined with frontal, temporal, and parietal lobes [$F(2,299)=8.2$, $P<0.001$].

Looking at the actual averages in Table 2, no systematic or characteristic differences between the EP and NE data become apparent. As a matter of fact, the overall means of the two groups are almost identical (EP, −0.230; NE, −0.233), and the subgroup differences which entail the significant outcome of the analyses of variance are not distributed in any regular way.

Table 2. Performance level of the patient groups in normalized scale units (zero is set at the mean of the patient population; one scale unit equals the standard deviation of the patient population)

Hemisphere	Anteroposterior localization			
	Frontal	Temporal	Parietal	Combined[a]
Left				
Epileptics	−0.20	0.10	−0.49	−0.20
Nonepileptics	−0.23	−0.27	−0.48	−0.33
Right				
Epileptics	0.04	−0.26	−0.57	−0.26
Nonepileptics	−0.12	−0.16	−0.11	−0.13
Combined[b]				
Epileptics	−0.08	−0.08	−0.53	−0.23
Nonepileptics	−0.18	−0.22	−0.30	−0.23

[a] Average of frontal, temporal, and parietal means
[b] Average of left and right hemisphere means

Shapes of Test Profiles

The correlation coefficients between corresponding EP and NE subgroups are shown in Table 3.

All correlation coefficients are positive, indicating some parallelism of the test profiles, i.e., a similar shape. Only the coefficients of the right frontal and right temporal subgroups are not signficant. Those of the left frontal and combined (left and right) frontal subgroups are only moderately significant ($P<0.05$), and the remaining correlations are more highly so ($P<0.01$ or $P<0.001$).

The *control correlations* involving the neighboring structures within the EP and NE groups (last line of Table 3) are positive and reach significant values ($P<0.05$ to $P<0.01$).

Table 3. Product-moment correlation coefficients between the 42 corresponding psychometric test data of the epileptic and nonepileptic patient groups with similar localization of the cerebral lesions

Hemisphere	Anteroposterior localization			
	Frontal	Temporal	Parietal	Combined[a]
Left	0.31*	0.78***	0.44**	0.66***
Right	0.14	0.22	0.59***	0.49***
Combined[b]	0.36*	0.56***	0.71***	
Control,[c] average	0.38*	0.44**	0.47**	

* $P<0.05$; ** $P<0.01$; *** $P<0.001$
[a] Based on the mean of the corresponding test data of the frontal, temporal, and parietal groups
[b] Based on the mean of the corresponding test data of the left and right hemisphere groups
[c] See "Methods" for procedure

Discussion

It appears that the *distribution* of epileptic (EP) and nonepileptic (NE) patients referred for clinical neuropsychological assessment is neither random nor even. The localization of the lesions in the temporal lobes is the most frequent, especially in the EP group, which may be a universal fact corresponding to more frequent temporal than frontal and parietal epilepsies. The temporal dominance in the NE group, however, though less striking, remains to be explained. It is possible on the one hand that neoplastic and vascular lesions *do* involve the temporal lobes more frequently than the frontal and parietal lobes. On the other hand, there is a possibility that neuropsychological information about the patients and the possible extrapolations thereof to everyday life problems appears more useful to the referring physicians for patients with temporal lesions than in the presence of frontal or parietal lesions.

Concerning the *age differences*, the younger age of the EP than NE patients is a usual finding which does not require special comment here. Nevertheless, it is of certain interest to notice that the age difference between the two groups is not linked to any difference in performance level in the psychometric tests.

In general, the psychometric *performance level* of both EP and NE groups is lower than in other patients with cerebral lesions who did not correspond to the present criteria for inclusion. Comparing the results within the EP and NE groups, the performance of NE was inferior with lesions in the left hemisphere, and superior with lesions in the right hemisphere. In the absence of specific literature about basic variations in the cognitive test performances of patients with lesions of different *etiology*, this finding can only be recorded here - and perhaps be remembered when discrepancies of findings appear in studies carried out by other research groups and considering different etiologies of the cerebral lesions.

The *shapes of the test profiles*, as compared here on the basis of correlation coefficients, did not reveal significant differences between the EP and NE groups studied, at least in the global first approximation, in that all correlation coefficients relating EP to NE profiles are positive, and most of them in a significant manner. Two points are raised by these findings: (1) The correlations are not significant in two of the three right hemisphere lesions subgroups, and they are lower in the left, right, and combined frontal subgroups than in the corresponding temporal and parietal subgroups (the correlation coefficient is close to zero in the right frontal subgroup). There is no ready-made explanation for this finding. Concerning the frontal subgroups, however, a causal factor may have been the size of the lesions in the NE group. It is commonly accepted in clinical practice that tumors and vascular accidents must be larger in the frontal lobes, especially on the right side, to disturb the patients' adaptation to their social environment. To our knowledge this does not apply to epileptogenic foci in this area. The relatively low correlations in the frontal subgroups might therefore have proceeded from the presence of a larger dysfunctional volume in the NE than in the EP subgroups. The performance levels do not clearly support this hypothetical explanation, however, as they do not reach their lowest levels in the NE frontal subgroups, as would be expected of large lesions. Therefore, further factors must be involved to account for the present pattern of correlations. The large size of the human frontal lobes may be considered in this context: We cannot exclude the possibility that in a large proportion of EP patients the epileptogenic focus was localized in another part of the frontal lobes than the neoplastic or vascular lesion in the NE patients. While this aspect does not facilitate any statement as to the specificity of epileptic disorders in clinical neuropsychology, it will have to be taken into account in further work. (2) The averaged profile correlations *within* the EP and NE groups involving the neighboring localization subgroups (last line in Table 3) are equal to (frontal) or lower than (temporal and parietal) those *between* the EP and NE subgroups with the same localization. Thus it appears that the similarity of profile shapes is at least as marked between EP and NE as between lesions in neighboring cerebral areas - another indication of the nonspecificity of epileptic disorders in clinical neuropsychology.

The findings presented here may appear discouraging to those neuropsychologists who wish to reach the formulation of a differential diagnosis which goes beyond the mere localization of cerebral lesions. Such precision would overrate the present state of the art of clinical neuropsychology. The present results, however, preclude neither further developments nor the possibility of finding isolated differences between etiologically defined patient groups by the appropriate choice of single psychometric tests - whatever the validity and usefulness of the subsequent assertions may be.

References

Lezak MD (1983) Neuropsychological assessment, 2nd edn. Oxford University Press, New York

Perret E (1973) Gehirn und menschliches Verhalten: Neuropsychologie des Menschen. Huber, Bern

Taylor LB (1979) Psychological assessment of neurosurgical patients. In: Rasmussen T, Marino R (eds) Functional neurosurgery. Raven, New York, p 165-180

Problems of Learning and Memory: Comparison of Performances Before and After Surgical Therapy

T. Nadig[1] and H. G. Wieser[2]

Since 1981, our group has investigated drug-resistant patients who had frequent mesiobasal limbic seizures. These patients underwent a battery of neuropsychological tests after surgery for seizure control and, since 1983, both before and after the intervention. In this paper we compare the learning and memory performance of 17 patients before and after surgery.

The detrimental effect of lesions in hippocampus (H) or in amygdala and hippocampus (A+H) on the performance of learning and memory has been reported mainly by Milner (1974) in patients and Mishkin (1982) in animals. However, further research is required to study the structural-functional correlations between A+H and learning and memory. The unilateral surgical removal of A+H in Milner's patients (1980) always encompassed temporal neocortical structures. The role of A+H for learning and memory was primarily deduced from the differences in functional deficits between exclusively neocortical or neocortical *and* mesiobasal-limbic lesions.

Lateralized deficits in learning performance were found by Nadig et al. (1984) and Nadig (1986) after selective amygdalohippocampectomy (SAH).

While the neurosurgical removal of epileptogenic tissue dramatically improves the everyday life conditions of patients by decreasing the frequency of seizures, the effects on cognitive functions in general, and on learning and memory in particular, are still to be precisely defined.

Our investigation focused on the consequences of the surgical procedure in comparison with the detrimental functional effects of a long-standing epileptogenic focus. Learning and memory performance of epileptic patients was examined before and after a unilateral SAH.

[1] Department of Neurology, Neuropsychology Unit, University Hospital, Frauenklinikstr. 26, 8091 Zürich, Switzerland
[2] Department of Neurology, University Hospital, Frauenklinikstr. 26, 8091 Zürich, Switzerland

Methods

Patients

Seventeen patients were tested before and after operation. A specific test battery was used (Nadig 1986).

Twelve of these patients had the primary epileptogenic focus in the A+H region on one side. They were seizure free after surgical treatment (causal SAH).

The five remaining patients had their primary epileptogenic focus outside A+H, mainly in the opercular-insular or Wernicke's area. Such a localization typically precludes radical removal of the epileptogenic tissue. If secondary involvement of A+H had been proven, a "palliative" SAH was performed.

The mean age of both patient groups was 28.6 years. In all patients the primary epileptogenic focus had been localized by stereoelectroencephalography (SEEG; see Wieser and Yasargil 1982; Wieser 1983) with implanted depth electrodes. The postoperative examination took place about 15 weeks after SAH.

Twelve normal controls (mean age 33.2 years) were tested once to supply baseline data.

Materials and Procedure

Learning performance was tested with two series of 15 items each drawn or written in black ink on white cards (size, 15 × 10 cm). The items were presented to the subjects one by one. The items of the first series were nonsense designs (Rey 1966, 1968; Nadig 1986). The second series consisted of concrete words. The items were presented until the subjects were able to reproduce at least 12 items, but for a maximum of 5 trials.

Memory Performance

After a distraction interval of 30 min the subjects were required to reproduce as many words and designs as possible.

Presurgical Evaluation of Epileptics
Edited by H. G. Wieser and C. E. Elger

Parallel versions of these tests were devised for pre- and postoperative testing. Half of the patients were tested preoperatively with version A and the other half with version B.

In our study we used different materials corresponding to the specialization of each hemisphere, designs for the right, words for the left hemisphere. Both materials were tested in the visual modality.

Results

Causal Selective Amygdalohippocampectomy

Learning performance is represented by the number of trials to reach learning criterion (Table 1).

Neither of the hemispheres showed any decrease in performance after operation. In the contralateral (nonoperated) hemisphere we found an increase of efficiency, approaching a nearly normal score.

Memory performance is represented by the number of items recalled after the distraction interval (Table 2).

The preoperative score was clearly deficient. Postoperatively it improved strongly, almost reaching the level of the control group.

Table 1. Learning: mean number of trials to reach the criterion (max. 5) in pre- and postoperative testing in the causal SAH group (Wilcoxon test)

Material		Left $N=7$		Right $N=5$		Controls $N=12$
		Pre	Post	Pre	Post	
"Words"	Mean	4.2	4.2	4.6	3.3*	3.5
"Designs"	Mean	4.4	2.8**	5.0	4.7	2.9

* $P=0.05$; ** $P=0.01$

Table 2. Memory: mean number of items recalled (max. 15) in pre- and postoperative testing in the causal SAH group

Material		Left $N=7$		Right $N=5$		Controls $N=12$
		Pre	Post	Pre	Post	
"Words"	Mean	6.1	5.9	8.4	11.2	10.5
"Designs"	Mean	10.1	10.2	10.0	12.5	11.5

Palliative Selective Amygdalohippocampectomy

Learning performance is expressed by the number of trials necessary to reach the learning criterion (Table 3).

Table 3. Learning: mean number of trials to reach criterion (max. 5) in pre- and postoperative testing in the palliative SAH group

Material		Left $N=2$		Right $N=3$		Controls $N=12$
		Pre	Post	Pre	Post	
"Words"	Mean	4.5	5.0	3.0	4.0	3.5
"Designs"	Mean	3.5	5.0	3.3	4.0	2.9

Table 4. Memory: mean number of items recalled (max. 15) in pre- and postoperative testing in the palliative SAH group

Material		Left $N=2$		Right $N=3$		Controls $N=12$
		Pre	Post	Pre	Post	
"Words"	Mean	9.5	7.5	8.3	8.6	10.5
"Designs"	Mean	11.0	10.0	8.6	11.0	11.5

Before operation learning performance of the candidates for palliative *left* SAH is impaired in verbal learning, but not in the learning of designs; the deficit is material specific. Candidates for *right* SAH learn both materials as well as normal controls.

Postoperatively there occurs a material-specific deterioration on the operated side (words in the left, designs in the right hemisphere). Because of the small number of patients ($N=5$), these data must still be considered tentative.

Memory performance is represented by the number of items recalled after the distraction interval (Table 4).

The preoperative score was not deficient. The pre- to postoperative changes in memory performance are smaller in the palliative group than in the causal group.

Discussion

A unilateral primary epileptogenic focus in A+H inhibits not only ipsilateral but also contralateral learning and memory.

The outcome of a unilateral *causal* SAH on learning sheds some light on the nature of the performance deficiency produced by an epileptogenic focus. Causal removal of A+H structures does not increase preoperative deficits of hemisphere-specific information processing.

The effects of *palliative* SAH extend the preceding conclusions. Our results indicate that secondary epileptic involvement of A+H are of less influence in learning and memory.

Conclusion

Material-specific impairment of learning found after causal unilateral SAH is, according to our results, a consequence of two factors; first, the already preoperative bilaterally reduced performance and, second, postoperative improvement of the contralateral hemisphere performance. The latter may be impressive. In the absence of "noise" introduced by the active discharging focus, the contralateral side usually recovers and regains its functions.

The learning performance of the patients with secondary epileptic involvement of mesiobasal-limbic structures is less impaired than that with primary involvement.

The results of this study do not support the so-called hippocampal memory hypothesis, i.e., at least A+H cannot be considered as structures uniquely responsible for learning and memory. However, these structures seem to participate in the sense that they constitute elements in a complex learning and memory "circuit" or "system" (Mishkin 1982; Markowitsch 1984).

References

Markowitsch HJ (1984) Can amnesia be caused by damage of a single brain structure? Cortex 20 (1): 20-27

Milner B (1963) Cognitive defects in focal and nonfocal epilepsy. 71st Annual convention of the A. P. A., Philadelphia

Milner B (1974) Hemispheric specialization: scope and limits. In: Schmitt FO, Worden FG (eds) The neurosciences. Third study program. MIT Press, Cambridge, pp 75-89

Milner B (1980) Complementary functional specializations of the human cerebral hemispheres. In: Levi-Montalcini R (ed) Nerve cells, transmitters and behaviour. Pontificia Academia Scientiarum, Vatican City

Mishkin M (1982) A memory system in the monkey. Phil Trans R Soc Lond 298: 85

Nadig T (1986) Lern- und Neugedächtnisleistungen bei Hirngesunden und bei hirngeschädigten Patienten. Lang, Frankfurt

Nadig T, Perret E, Wieser HG (1984) Learning and memory performance after unilateral selective amygdalohippocampectomy. 7th European conference of the INS, Aachen

Rey A (1966) Les troubles de la memoire et leur examen psychometrique. Dessart, Bruxelles

Rey A (1986) Techniques inedites pour l'examen psychologique Fasc 3: epreuves mnesiques et d'apprentissage, Delachaux and Niestle, Neuchatel

Wieser HG (1983) Electroclinical features of the psychomotor seizure. Fischer/Butterworths, Stuttgart/London

Wieser HG, Yasargil MG (1982) Selective amygdalohippocampectomy as a surgical treatment of mesiobasal limbic epilepsy. Surg Neurol 17: 445-457

Intracarotid Amobarbital Testing and Assessment of Speech Laterality

H. Silfvenius

Intracarotid injection of sodium amobarbital (Amytal) transiently blocks the perfused cerebral hemisphere. It allows for investigations of various intra- and interhemispheric functions which must be clarified in the individual patient and which are of importance for the understanding of brain function in neurological disease. The bilateral intracarotid amobarbital test which Wada and Rasmussen (1960) used preoperatively on patients with intractable epilepsy has given otherwise unobtainable information on lateralization of speech and memory functions (Milner et al. 1962, 1964). The test may be used when there is uncertainty about speech lateralization and memory function, side of epileptic lesion, EEG, or neuropsychological findings.

Amobarbital dissolved in water or saline may be injected into cooperative adolescents and adults, and even also children after a short general anesthesia for catheterization. Threshold doses for EEG changes are 5-10 mg, for clinical 40-50 mg. Five to 10% solutions are generally used when injecting a bolus of 125-175 mg in 3-5 s. The transfemoral catheterization of the internal carotid artery is carried out on two consecutive daily sessions. Each test is usually preceded by an injection of contrast medium in order to visualize on frontal and lateral films the extent of ipsilateral and possible contralateral vascular fillings. These findings have to be considered when interpreting the test results.

The well-known clinical signs of hemisphere anesthetization appear instantly, last 3-8 min, and fade within 10-30 min. Speech is tested first, memory often 2-4 min after injection.

Speech

Bilateral Amytal tests on epilepsy patients - a population biased for non-right-handedness - have provided conclusive data on the relation between handedness and speech laterality/bilaterality, combined or not with early left hemisphere damage (Milner et al. 1964). Amytal induces a 3 to 8 min aphasia after injection into the speech-dominant side, a short anarthria when the other hemisphere is injected. If there is bilateral speech representation, aphasia may be absent or short with dissociated errors in naming and serial tasks. Some studies report remaining speech comprehension, others not (McGlone 1984; Milner et al. 1966; Oxbury and Oxbury 1984; Rausch and Walsh 1984). Amytal studies on epileptic patients have also provided the opportunity to study such problems as: writing hand posture vs. speech (Ajersch and Milner 1983; Volpe et al. 1981), hemisphere structural asymmetry vs. speech (Ratcliff et al. 1981; Wada and Rasmussen 1960), sex vs. speech asymmetry (McGlone and Fox 1982), and right-sided speech in righthanders (Rausch and Walsh 1984).

Intracarotid methohexital, a still shorter-acting barbiturate, has also been used for hemisphere speech studies (Willmore et al. 1978).

The present Amytal data include 41 cases, 26 males and 15 females, age 16-51 years (mean, 32 years). Their epilepsies started at age 0-38 years, (mean, 13 years). All had intractable simple or com-

Table 1. Handedness-speech relation in 41 epileptic patients given intracarotid Amytal (%)

Handedness			Hemisphere speech		
			Left	Bilateral	Right
Total $n=41$	Right-handed	31 (78)	29 (93)	2	–
	Left-handed	6 (16)	4	1	1
	Ambidexters	4 (10)	3	1	–
Females $n=15$	Right-handed	11 (73)	11 (100)	–	–
	Left-handed	3	1	1	1
	Ambidexters	1	–	1	–
Males $n=26$	Right-handed	20 (77)	18 (90)	2	–
	Left-handed	3	3	–	–
	Ambidexters	3	3	–	–

Percentages in parentheses

Department of Neurosurgery, University Hospital,
90185 Umeå, Sweden

Presurgical Evaluation of Epileptics
Edited by H. G. Wieser and C. E. Elger

plex partial seizures, with or without secondary generalization. Table 1 shows the handedness/speech laterality data.

The aphasia time range for 29 righthanders was 3–11.30 min (mean, 6 min 50 s), for 4 lefthanders 5.30–7.30 min (mean, 6 min 30 s). There was no sex difference in duration of aphasia. Speech comprehension was almost absent, and errors in naming and counting were typical during recovery. Four patients with bilateral speech had 3.30–7.25 min aphasia after left, and 4.15–7.0 min after right injections, which is longer than earlier reported (Milner et al. 1966). Two of those with bilateral speech had radiologically demonstrable left lesions. Of 29 with temporal lobe lesions, bilateral speech was seen only among the 13 with a left-sided lesion.

Memory

Milner et al. (1962) applied the Amytal technique for hemisphere memory studies. This ingenious method has greatly advanced the workup of epileptic patients and lessens the risks of postoperative complications. The test defines *retrograde* (interhemisphere) and *anterograde* (intrahemisphere) memory. Anterograde recognition results obtained on recovery, but not recall data, are relied upon for clinical decisions. Modifications of the original test technique have been published, which have facilitated the study of memory in various ways (Blume et al. 1973; Engel et al. 1981; Fedio and Weinberg 1971; Klöve et al. 1969; Milner 1976; Rausch et al. 1984; Risse and Gazzaniga 1978; Serafetinides 1966). Amytal results combined with data from visual half-field studies have not to our knowledge been published. Results from large series show reduced memory on the side with the epileptic lesion and usually a normal memory on the healthy side (Milner 1975).

We have used two test methods (Silfvenius et al. 1984; Christianson et al. 1985) with an early start of the presentation of postinjection tasks aimed at further quantifying hemisphere memory. Our initial 21 patients were studied with one technique, the following 20 with the other. Some results of the first study (Aasly and Silfvenius 1985) are presented first. The preinjection tasks were given 4 to 1 min before Amytal (color, number, sentence, nonverbal stereognostic). About 1 min after Amytal administration (175 mg, 10%), presentation of postinjection tasks began (three words, three pictures, nonverbal stereognostic, three pictures, three words) with the mean presentation time of the last task 4 min after injection.

The results show that the speech-dominant hemisphere had a mean 47% anterograde recognition of words and 62% of pictures. The nondominant side achieved 16% and 35% respectively. Figure 1 illustrates anterograde recognition of early- and late-presented words and pictures.

No statistically significant difference was seen for either hemisphere in recognition of early vs. late words or pictures ($P > 0.1$). An interhemispheric comparison, however, showed a highly significant difference ($P < 0.001$) for early/late words and pictures in favor of the speech-dominant side. Figure 2 shows recognition in relation to the site of epileptic lesion.

A comparison between two groups of patients with right-sided lesions showed that after left injec-

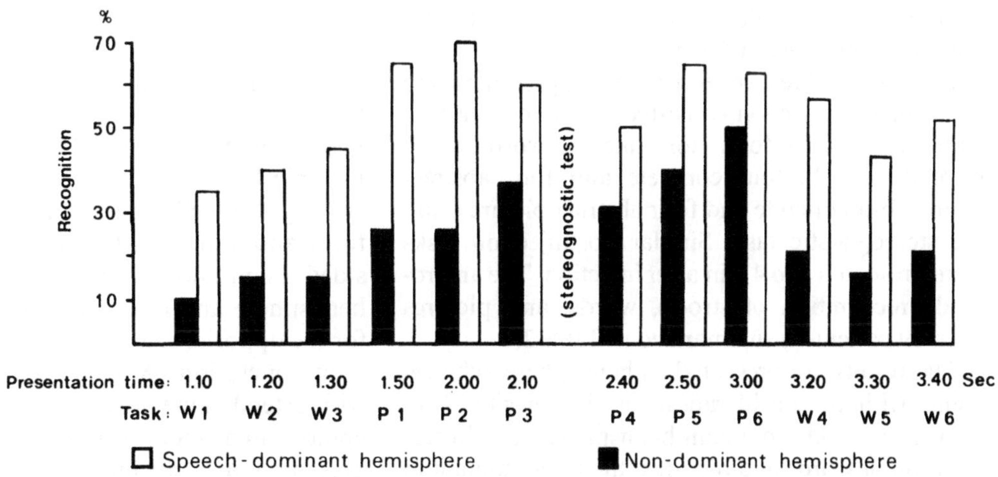

Fig. 1. Mean percentage anterograde recognition of postinjection word *(W)* and picture *(P)* tasks; *black bars*, nondominant hemisphere (opposite injected); *open bars*, speech-dominant hemisphere. The presentation time given is the mean of the testing times for both hemispheres. Between the early and the late tasks the stereognostic tests were performed

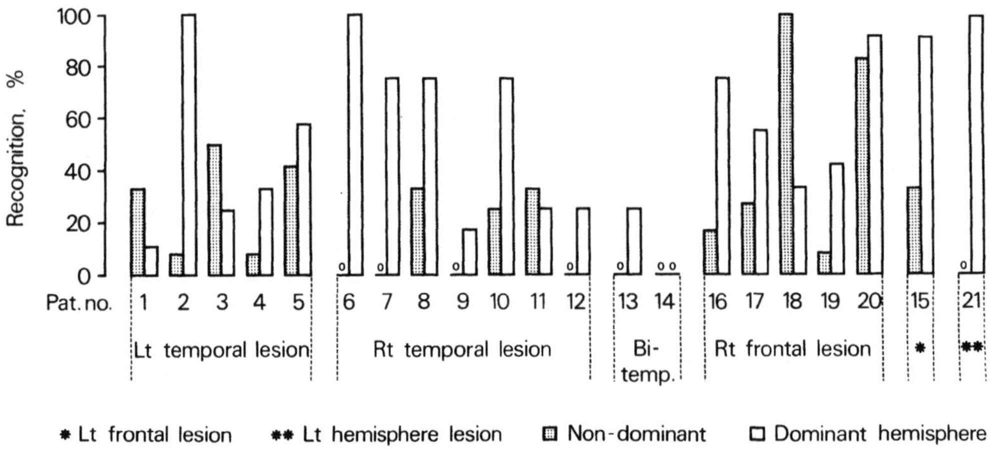

Fig. 2. Interhemispheric comparison of anterograde recognition related to localization of epileptic lesion in 21 epileptic patients. Each bar denotes summed percentages of word and picture recognition

tion, seven patients with right temporal lesions had poorer recognition for words and pictures than the five patients with right frontal lesions (P words <0.001; P pictures <0.01). No difference was seen when these patients performed with their "healthy" speech-dominant hemisphere. A corresponding comparison for left-sided lesions could not be carried out.

An interhemispheric comparison between five patients with left and seven with right temporal lesions showed that the former recognized 40% of the words and 42% of the pictures with their lesioned hemisphere, 17% and 40% with their "healthy" right side. Those with right-sided lesions recognized with that hemisphere 3% of the words and 24% of the pictures, their "healthy" left hemisphere achieving 39% and 69% respectively. There was a statistically significant difference in picture recognition ($P<0.05$) favoring the left "healthy" side.

In the second test version, the tasks presented 3 to 1 min before injection consisted of a stroop (a task with conflicting information, like the word "blue" printed in red), four concrete and four abstract words, four concrete and four abstract pictures, and the stereognostic task. Similar postinjection tasks were presented 1 to 4 min after injection. The anterograde recognition of stroop, words, and pictures from 19 patients is summarized in Table 2.

The results are presented as hemisphere achievements without consideration for side or site of lesion. The speech-dominant hemisphere leads in recognition of words and pictures; the nondominant in stroop recognition. The results on hemisphere nonverbal, nonvisual stereognosis in 25 patients are from Aasly and Silfvenius (1985). In 14 patients with temporal lobe lesions and unilateral speech, the relation between results from retrograde dichotic listening and anterograde word recognition are compared in Table 3.

Table 2. Summed results on anterograde correct recognition (%) of different tasks during bilateral Amytal tests in epileptic patients

	Tasks	Speech dominant side	Nondominant hemisphere
	Stroop	6	19
	Words,		
($n=19$)	concrete	50	31
	abstract	39	33
	Pictures,		
	concrete	58	22
	abstract	39	17
($n=25$)	Nonverbal Stereognostic	44	12

A right-sided lesion resulted in low scores for pictures and words of the same hemisphere. A left-sided lesion lowers word and picture scores for that hemisphere and also word, but not picture, scores for the opposite side.

In conclusion: Amytal speech testing combined with extended memory tests has defined speech laterality and memory in 41 epileptic patients. An early presentation of an increased number of postinjection tasks makes possible further characterization of hemisphere memory in patients considered for surgical treatment.

Table 3. Relationship between dichotic word listening, final cued recall, and anterograde recognition scores (%) in 14 epileptic patients

Lesion	Dichotic FCR		Anterograde recognition							
	Left	Right	Left				Right			
			cw	aw	cp	ap	cw	aw	cp	ap
Right TL ($n=8$)	14.8	10.4	56	34	66	41	31	31	19	9
Left TL ($n=6$)	20.1	20.5	21	33	29	17	13	25	42	33
Controls ($n=26$)	44.4	42.2								

cw, concrete word; aw, abstract word; cp, concrete picture; ap, abstract picture

References

Aasly I, Silfvenius H (1985) Methodological considerations on an extended intracarotid Amytal test for epileptic patients (unpublished)

Aasly I, Silfvenius H (1985) Stereognostic memory of cerebral hemispheres studied in epileptic patients during bilateral intracarotid Amytal testing. 13th World Congr Neurol, Hamburg

Ajersch MK, Milner B (1983) Handwriting posture as related to cerebral speech lateralization, sex, and writing hand. Human Neurobiol 2: 143-145

Blume WT, Grabow JD, Darley FL, Aronson AF (1973) Intracarotid amobarbital test of language and memory before temporal lobectomy for seizure control. Neurology 23: 812-819

Christianson S-Å, Silfvenius H, Nilsson L-G (1987) Hemisphere memory of concrete and abstract information determined with the intracarotid Sodium Amytal test Epilepsy Res 1: 185-193

Engel J Jr, Rausch R, Lieb JP, Kuhl DE, Crandall PH (1981) Correlation of criteria used for localizing epileptic foci in patients considered for surgical therapy of epilepsy. Ann Neurol 9: 215-224

Fedio B, Weinberg LK (1971) Dysnomia and impairment of verbal memory following intracarotid injection of sodium amytal. Brain Res 31: 159-168

Klöve H, Grabow JD, Trites RL (1969) Evaluation of memory functions with intracarotid sodium amytal. Trans Am Neurol Assoc 94: 76-80

McGlone J (1984) Speech comprehension after unilateral injection of sodium Amytal. Brain Lang 22: 150-157

McGlone J, Fox AJ (1982) Evidence from sodium amytal studies of greater asymmetry of verbal representation in men compared to women. In: Akimoto H, Kazamatsuri H, Seino M, Ward A (eds) Advances in epileptology. 13th epilepsy international symposium. Raven, New York, pp 389-391

Milner B (1975) Psychological aspects of focal epilepsy and its surgical management. In: Purpura DP, Penry JK, Walter RD (eds) Advances in neurology, vol 8. Raven, New York, pp 299-319

Milner B (1976) Hemispheric asymmetry in the control of gesture sequences. In: Kimura D (ed) Symposium on asymmetric processing in the cerebral hemispheres. Proceedings of the 21st international congress of psychology, Paris, p 149

Milner B, Branch, Rasmussen T (1962) Study of short-term memory after intracarotid injection of sodium amytal. Trans Am Neurol Assoc 87: 224-226

Milner B, Branch C, Rasmussen T (1964) Observations on cerebral dominance. In: de Reuck AVS, O'Connor M (eds) Ciba foundation symposium on disorders of language. Churchill, London, pp 200-214

Milner B, Branch C, Rasmussen T (1966) Evidence for bilateral speech representation in some non-right-handers. Trans Am Neurol Assoc 91: 306-308

Oxbury SM, Oxbury JM (1984) Intracarotid Amytal test in the assessment of language dominance. In: Rose FC (ed) Advances in neurology. Prog Aphasiol 42: 115-123

Ratcliff G, Dila D, Taylor L, Milner B (1981) The morphological asymmetry of the hemispheres and cerebral dominance for speech: a possible relationship. Brain Lang 11: 87-98

Rausch R, Walsh GO (1984) Right-hemisphere language dominance in righthanded epileptic patients. Arch Neurol 41: 1077-1080

Rausch R, Fedio P, Ary CM, Engel J, Crandall PH (1984) Resumption of behavior following intracarotid sodium amobarbital injection. Ann Neurol 15: 31-35

Risse GL, Gazzaniga MS (1978) Well-kept secrets of the right hemisphere: a carotid amytal study of restricted memory transfer. Neurology 28: 950-953

Serafetinides EA (1966) Auditory recall and visual recognition following intracarotid sodium amytal injections. Cortex 2: 367-372

Silfvenius H, Blom S, Nilsson L-G, Christianson S-Å (1984) Observations on verbal, pictorial and stereognostic memory in epileptic patients during intracarotid amytal testing. Acta Neurol Scand [Suppl] 99: 57-77

Volpe BT, Sidtis JJ, Gazzaniga MS (1981) Can left-handed writing posture predict cerebral language laterality? Arch Neurol 38: 637-638

Wada J, Rasmussen T (1960) Intracarotid injection of sodium amytal for the lateralization of cerebral speech dominance. Experimental and clinical observations. J Neurosurg 17: 266-282

Willmore LJ, Wilder BJ, Mayersdorf A, Ramsay RE, Sypert GW (1978) Identification of speech lateralization by intracarotid injection of methohexital. Ann Neurol 4: 86-88

Lateralization and Intrahemispheric Localization of Language and Memory During Surgical Therapy of Epilepsy*

G. A. Ojemann

The indications for resections for medically intractable epilepsy include not only the identification of an epileptic focus, but also that the focus be in what Penfield called "dispensible" brain (Penfield and Jasper 1954), that is brain the removal of which is unlikely to lead to a new functional deficit. In practice, this generally means that the focus is not in areas essential for language, that the homologous area of the opposite hemisphere is functionally intact, and if the focus is in Rolandic cortex, the patient is willing to accept a sensorimotor deficit, or already has one. For the most common of the resections for epilepsy, in temporal lobe, the functionally essential areas that must be avoided are those for language and recent verbal memory. Areas essential for language provide a posterior limit to temporal lobectomy in the dominant hemisphere. Substantial recent verbal memory deficits have been present in several series of dominant hemisphere temporal lobectomies (Novelly et al. 1984; Rausch and Crandall 1982); indeed it has been suggested that this operation should not be considered for patients whose occupations depend on facile memory (Delgado-Escueta et al. 1983). This chapter reviews the organization of language and memory in the human brain, with particular emphasis on organization in temporal lobe. The techniques used by the author's group for identifying areas essential for language and memory in individual patients are then discussed.

Language

In the great majority of patients, areas essential for language will be confined to the left cerebral hemisphere. This pattern is present in over 99% of right-handed patients and 60%-70% of left-handed patients (with intact right-hand function) based on changes in naming during left intracarotid amobarbital perfusion (Rasmussen and Milner 1977; Mateer and Dodrill 1983). That technique, often called the "Wada" test after Juhn Wada, who first described it (Wada and Rasmussen 1960), is the technique used in most centers when definitive information on language lateralization is needed. It is considered to be more reliable than other less invasive techniques for determining language lateralization, such as dichotic listening tasks (Kimura 1967), although there are few comparisons between the two techniques in the same patients.

Patients who do not have language lateralized to the left hemisphere include a group with language lateralized strongly to the right hemisphere, and a group with apparently bilateral representation of language. In a series of 99 patients evaluated for language lateralization with intracarotid amobarbital perfusion prior to epilepsy surgery at the author's institution, 9% demonstrated right dominance and 6% bilateral language representation (Mateer and Dodrill 1983). In most patients, essential areas for oral and written language output and understanding, control of sequential movements, and recent verbal memory are all lateralized to the left hemisphere. However, there is some evidence that occasionally language understanding can be processed independently by either hemisphere, although language output is confined to the left (Siditis et al. 1981) and areas essential for control of motor sequences may be contralateral to essential areas for language in some left-handed patients (Rasmussen and Milner 1977). Language lateralization can be altered by early life injury to the left hemisphere. Infants with removal of most of the left hemisphere cortex in the first few months of life, as treatment for unilateral Sturge-Weber disease, subsequently develop serviceable language presumably based only on the right hemisphere. However, this language is not quite normal, with less verbal facility and a reduced capacity for complex grammatical constructions (Dennis and Whitaker 1976). These defects are not seen in language of children subjected to the same

Department of Neurological Surgery, RI-20, University of Washington, School of Medicine, Seattle, WA 98118, USA

* This study was supported by NIH grants 21724, 17111, and 20482.

Presurgical Evaluation of Epileptics
Edited by H. G. Wieser and C. E. Elger

removal but in the right hemisphere. The exact age when this capacity for the right hemisphere to subsume language is lost is presently controversial, but may not extend much beyond 2 years of age (Woods 1983). Left frontal and parietal rather than temporal injury are more likely to be associated with a shift in language lateralization (Rasmussen and Milner 1977).

Within the language-dominant hemisphere, the areas classically considered essential for language include the posterior inferior frontal gyrus (Broca's area), posterior temporal-inferior parietal cortex (Wernicke's area) and the supplementary motor area in the medial posterior superior frontal gyrus (Penfield and Roberts 1959). This model of language localization was derived largely from investigation of patients with cerebral infarcts. More recent studies of the intrahemispheric organization of language, based on findings derived at operation using electrical stimulation-mapping techniques, suggest a somewhat different model. In these studies, areas essential for language in an individual patient are identified by the sites where stimulation evokes repeated errors in naming of common objects. For if a cortical resection encroaches on such sites, a postoperative aphasia is much more likely than when such sites are spared by the resection (Ojemann 1983).

Those sites identified as essential for language in an individual patient appear to be localized to "mosaics," of 1 or more cm^2 in extent, with sharp boundaries. Each patient has two or three (occasionally more) such mosaics, at least one in the posterior inferior frontal gyrus and one temporoparietally. Figure 1 illustrates this discrete localization of such sites in one patient. There is considerable variability in the exact location of these sites between patients. Figure 2 illustrates this variability for a series of cases, all left dominant for language. The only area where sites essential for language were identified in nearly all patients was the very posterior end of the inferior frontal gyrus, immediately in front of the face motor cortex (but distinguished from it by the absence of evoked face or tongue movements). However, elsewhere in lateral perisylvian cortex, no more than 50% of patients had essential sites for language in any one area, including the classical Wernicke's area. These include cases with essential areas for language as far forward in the temporal lobe as the level of the motor cortex, occasionally even in the middle temporal gyrus (Ojemann 1979). Those cases, which make up 17% of a 21-case series, have essential areas for language within 4.5 cm of the temporal tip, the usual landmark for "safe" dominant hemisphere temporal resections used by those who

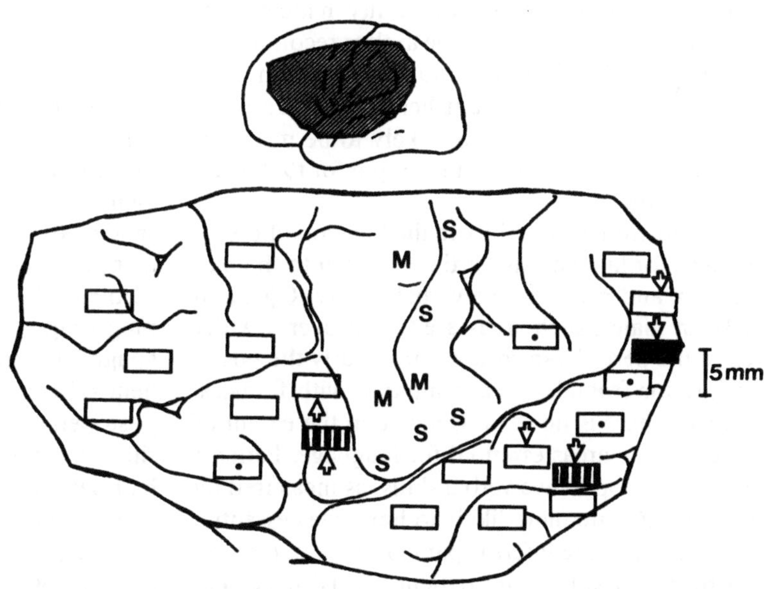

Fig. 1. Localization of stimulation-evoked changes in naming in lateral perisylvian cortex of the dominant hemisphere of a 24-year-old female. Drawing traced from photograph made at operation. Each site stimulated is represented by a *rectangle*. *Filled rectangle*, naming error evoked on all three samples with stimulation. *Vertical stripes*, naming errors evoked on two of three or three of four samples with stimulation. *Dot*, naming error evoked on one of three or four samples with stimulation. *Arrows* identify adjacent sites on continuous gyri separated by 5 mm or less, one site with repeated naming errors, the other with none at all. Stimulation with 4-s trains of 60 Hz 2.5-ms total duration biphasic pulses at 8 mA between pulse peaks through bipolar electrodes 5 mm apart. (From Ojemann 1983)

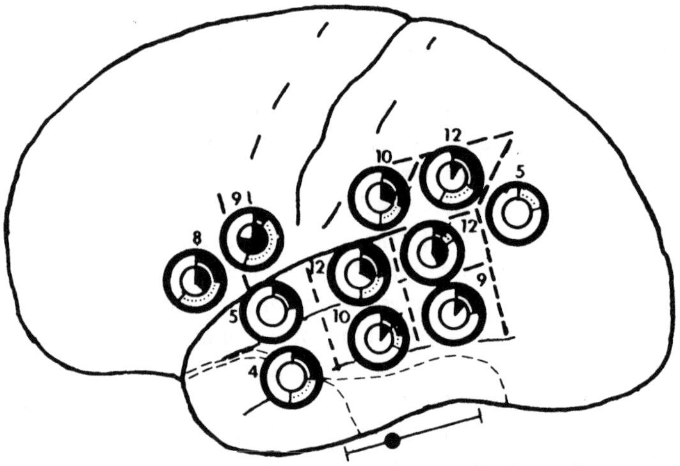

Fig. 2. Localization of changes in naming and short-term verbal memory evoked by intraoperative electrical stimulation mapping of lateral perisylvian cortex in the 14 patients of the study of Ojemann and Dodrill (1985 a). Stimulation sites in individual cases were aligned in relation to the motor cortex and the end of the Sylvian fissure, and assigned to one of the 11 zones indicated by the *dashed lines*. In each zone the number indicates cases with one or more samples there. The *inner circle graph* indicates the proportion of cases with naming errors evoked at sites within that zone. The *outer circle graph* shows the proportion of cases with short-term memory errors evoked at sites within that zone. The *filled outer portion* of these graphs indicates the proportion of cases with errors after stimulation during input or storage aspects of memory, the *dotted outer portion* cases with errors after stimulation at the time of retrieval. Stimulation at a mean current of 4.8 mA (range 2.5–8 mA) applied to an average of 8.4 sites per patient (range 4–13 sites)

do not individually map language localization (Falconer 1972). On the other hand, other patients have essential areas for language located quite far posteriorly in the temporal lobe so that very large resections can be accomplished without language deficits (Ojemann 1979). Because of this variability in location of essential areas for language, the author recommends assessing this localization intraoperatively in each patient in whom a dominant hemisphere temporal lobectomy (or other resection likely to be near language cortex) is planned. A technique for that is described below.

This individual variability in the location of essential areas for language is partially related to the patients' sex and preoperative verbal ability (Ojemann 1983). In males, such areas are found over a wider area of the perisylvian cortex, particularly the posterior inferior frontal gyrus, compared with females. However, there is no sex difference in the extent of temporal lobe areas essential for language. Patients with poor preoperative verbal skills, as measured by the verbal IQ, are more likely to have sites essential for language in the inferior parietal lobe, compared with more verbally facile patients. This individual variability in detailed functional localization of language corresponds to the detailed gyral anatomy and cytoarchitecture at the end of the Sylvian fissure (Rubens et al. 1976; Galaburda et al. 1978).

Assessment of the effects of stimulation on object naming is apparently adequate to identify most sites that must be spared in a resection if there are to be no postoperative deficits on standard aphasia batteries. However, there is some evidence that this assessment may not identify all areas essential for all language functions. In an individual patient, sites essential for naming the same object pictures in two different languages have been shown to be localized in separate though nearby areas (Ojemann and Whitaker 1978). Whether identifying sites essential for naming in a patient's primary language is adequate to preserve secondary languages in a multilingual population has not been determined. Sites essential for reading, as identified by stimulation mapping, are also sometimes separate from those for naming (Ojemann 1983). Indeed stimulation mapping suggests differential localization of many language functions in lateral cortex of the dominant hemisphere. The author has occasionally found postoperative reading deficits in patients in whom sites essential for language as measured by naming had been spared in the resection. These cases often had resections extending into the middle or posterior inferior temporal gyrus, although they represent a minority of cases with resections in that area in dominant hemisphere. However, those cases suggest that mapping of sites essential for sentence reading may also sometimes be necessary.

Memory

The observations of Penfield, Milner, Scoville, and their associates provided the first evidence that related the temporal lobe to memory (Penfield and Milner 1958; Scoville and Milner 1957). They observed global deficits in recent memory after resections involving the amygdala, uncus, and anterior several centimeters of hippocampus bilaterally, or a unilateral removal when the opposite hippocampus had major preexisting damage (Penfield and Mathieson 1974). Those deficits prevented the formation of new long-term memories, and extended in a retrograde fashion for several years, but did not affect earlier long-term memories. Immediate memory, for information within the digit span, was also intact in these patients (Drachman and Arbit 1966). Thus the deficit was predominately in postdistractional recent memory. Subsequently, Milner demonstrated more subtle defects in this type of memory after standard anterior temporal lobectomies, defects that were material specific according to the side of the resection (left, verbal; right, tone sequences and recognition of nonsense shapes). The severity of these deficits was related to the extent of the hippocampal removal (Milner 1967).

Disorders that lead to medically intractable epilepsy, or perhaps the repeated uncontrolled seizures themselves, can lead to bilateral hippocampal damage. However, in some cases one temporal lobe may both generate the seizures and contain neuronal circuits important to memory, while the other temporal lobe may not be functioning at all. This circumstance is said to set the stage for a global memory deficit after resection of the temporal lobe generating the seizures (Penfield and Mathieson 1974). The Montreal group reported that this circumstance could be identified preoperatively by testing memory during intracarotid amobarbital perfusion of the side of the proposed resection (Milner et al. 1962). A memory deficit for verbal material on that test suggested that the other hemisphere was damaged to the point where it could not sustain memory alone. The response to this finding varies somewhat between centers engaged in epilepsy surgery. Most will not undertake resections if there is a complete memory deficit after intracarotid amobarbital perfusion of the site of the planned resection. Centers using standardized resections in all cases will not consider cases with partial memory deficits above a certain level, while those surgeons who tailor resections usually spare the hippocampus in such cases (Blume et al. 1973).

Recent evidence suggests the need for revision in this approach. Substantial verbal memory deficits remain in cases with standard dominant hemisphere resections (Rausch and Crandall 1982; Novelly et al. 1984). However, little or no memory deficit seems to follow resections limited to medial temporal structures including anterior hippocampus (Wieser and Yaşargil 1982). Ojemann and Dodrill (1985a, b) reviewed the status of recent verbal memory in cases approached by tailoring the dominant hemisphere resections to minimize hippocampal removal in those patients with major memory deficit on the preoperative intracarotid amobarbital study. They found a persisting recent verbal memory deficit in about half the cases, at testing 1 month after operation. This deficit improved but did not entirely clear at testing 1 year after operation. The severity of the deficit correlated with the lateral, but not medial, extent of the resection. It was worse in patients with facile preoperative verbal memory performance. It was not predicted by memory performance during intracarotid amobarbital perfusion of that hemisphere, but was predicted by presence or absence of encroachment of the resection on lateral cortical sites identified by stimulation mapping as essential for language, or input or storage (but not retrieval) phases of recent verbal memory. In previous work, Ojemann (1978) described dominant hemisphere lateral cortical sites related to recent verbal memory. Those sites were adjacent to, but separate from, sites essential for language. Memory was measured with a technique (described below) that separates the effects of stimulation during the input to memory, when the memory was stored, or at the time of retrieval. Stimulation of temporal or parietal lateral cortical sites was more likely to alter memory when applied during the input, or storage phases, while stimulation of frontal sites was more likely to alter memory when applied during retrieval. There is also substantial individual variability in the location of these sites related to memory (Fig. 2). In some patients sites related to the input or storage phases of memory are located anteriorly to language areas in temporal lobe, and are thus more vulnerable to damage in an anterior temporal resection. It is damage to this part of the temporal lobe memory system that apparently accounts for the residual memory deficit in patients with modification of the hippocampal resection based on intracarotid amobarbital perfusion memory performance. Moreover, in two cases where there were 100% memory errors with amobarbital perfusion of the left hemisphere, no memory deficits followed left temporal resections that minimized both hippocampal removal and encroachment on lateral cortical sites related to language or input or storage aspects of memory (Ojemann and Dodrill 1985a). In this way temporal resections can be safely performed in such cases that otherwise would not be considered for operation for fear of a postoperative global memory deficit.

Although alterations in memory for visuospatial material are documented after nondominant anterior temporal lobectomies (Milner 1967), these deficits do not seem to be clinically significant. Few patients report them, in contrast to verbal memory deficits, which are regularly noticed. Recent visuospatial memory generally improves after dominant hemisphere anterior temporal lobectomy (Novelly et al. 1984; Ojemann and Dodrill 1985a). Most surgical centers respond to verbal memory deficits with intracarotid amobarbital perfusion of the *nondominant* hemisphere in the same way as they do to such deficits after dominant hemisphere perfusion, although the effectiveness of this approach has not been as extensively evaluated. Nondominant lateral temporal lobe sites have been related to visuospatial memory (Fried et al. 1982), but their role, if any, in memory changes after temporal lobectomy on that side has not been assessed.

Techniques for the Identification of Sites Essential for Language and Memory

As indicated previously, localization of sites essential for language and memory in an individual patient depends on two tests, one for lateralization and one for intrahemispheric localization. The author's group uses the intracarotid amobarbital perfusion technique for determining lateralization. This test, as previously described (Mateer and Dodrill 1983), is performed by retrograde femoral catheterization of the internal carotid artery, initially on the side of the proposed resection. The patient then engages in a task organized as a series of trials, each consisting of naming an object, reading a six to eight word sentence from a card, and then, in response to a verbal cue, repeating back the name of the object. The patient is given a maximum of 10 s to respond to each segment of a trial before proceeding to the next one. The patient is also instructed to keep his arms up. Following a 60-s control period, 75–125 mg (depending on the patient's weight) of sodium amobarbital is injected over a 4-s period. The duration of contralateral arm paresis is used as a measure of when the hemisphere is inactivated by the drug, and any changes in naming or reading, are related to this. Performance on memory retrieval is assessed for the 5 min after return of any deficits in naming. Injection of either hemisphere often produces a brief cessation of performance. In the language-dominant hemisphere, this is followed by naming and reading deficits in the presence of some speech output. A significant memory deficit is indicated by a substantial increase in the error rate over baseline and the duration of this deficit. Following the amobarbital perfusion, a carotid arteriogram is performed, to assess any potential for passage of the amobarbital to the opposite hemisphere which would compromise interpretation of the test. The catheter is then withdrawn into the aortic arch and, after about 30 min, inserted into the opposite internal carotid artery and the procedure repeated. Both sides are always done in each patient, for we have found that comparison of differences in performance is often necessary to be certain of language dominance, and of the ability of the hemisphere contralateral to the proposed resection to sustain recent verbal memory. The entire test is recorded on videotape as a check on the accuracy of the scoring of performance.

The intrahemispheric localization of areas essential for language and memory uses electrical stimulation-mapping techniques. Application of an electrical current to the cortical surface activates some neurons and fibers while blocking function of other neural elements. Both excitatory and inhibitory elements may be activated or blocked (Ranck 1975). Thus the behavioral response to application of the current cannot be predicted physiologically. However, empirically there are some cortical areas where activating effects seem to predominate, particularly primary motor and sensory areas. Application of a current to most association cortex evokes no behavioral responses in the quiet patient. But if the patient engages in an ongoing behavior, such as object naming, application of the current to some sites in the language-dominant hemisphere will block that function. Resections that encroach on sites where stimulation repeatedly blocks a language or memory function are likely to lead to a deficit in that function (Ojemann 1983; Ojemann and Dodrill 1985a), indicating that both stimulation and lesions denote the same cortical areas as essential to a function.

This technique of stimulation mapping to localize language in an individual patient was introduced by Penfield (Penfield and Roberts 1959). At some centers this mapping is done through chronically implanted subdural electrodes (Lesser et al. 1984). In the author's experience, that approach has not been entirely satisfactory, for it does not provide as detailed a localization as can be obtained with intraoperative mapping, and on occasion subdural electrode stimulation provided false-negative data, suggesting that a cortical area could be safely resected, when both intraoperative stimulation mapping and language testing during resection indicated that it could not.

Intraoperative stimulation mapping requires that the patient be awake and alert during at least a portion of the craniotomy. The author uses a mixture of ¼% bupivacane and ½% lidocaine as local anesthesia in a scalp field block around the incision, and as an intradural injection around the middle meningeal artery. This is supplemented by an intravenous neuroleptanalgesia mixture of droperidol (2.5 mg) and fentanyl (0.05 mg) given shortly before starting the operation. Subsequent sedating drugs are avoided until after any testing of language or memory. A wide cortical exposure is obtained, for the goal is to identify *both* sites essential for language, as well as those that are not. Only when both have been identified can the sites without evoked language changes be considered dispensible. A stimulation mapping study with no language changes at any site does not provide a guide to a safe resection.

The only stimulation parameter that is varied between patients is the current level. In all cases the author uses trains of biphasic square-wave pulses, 2.5 ms in total duration, at 60 Hz, delivered from a constant current stimulator through bipolar electrodes separated by 5 mm. The current level to be used with stimulation mapping in a particular patient is selected as the largest current that does not evoke afterdischarges in the area of cortex to be sampled. This current level is chosen as one likely to be large enough to alter local cortical function but not evoke seizures. In the author's experience, that current generally ranges between 2 and 8 mA, between peaks of the biphasic pulses, values that range between the threshold for motor effects from the Rolandic cortex and about four times that threshold.

The tasks used during stimulation mapping contain multiple standardized trials of uniform difficulty, with each segment of a trial presented for a predetermined uniform period. The general design of the mapping study is to intersperse trials with stimulation during some segment of the task with trials without stimulation, as a measure of control performance. At least three samples of stimulation effect during each trial segment are obtained for each site sampled. Thus, changes in performance evoked by stimulation at any site during any segment of the test can be compared statistically with control performance, using the single sample binomial test (Siegel 1956).

Object naming is used as a screening test for language localization. Objects to be named are presented on slides, usually at 4-s intervals. Slides with stimulation are pseudorandomly interspersed with those without stimulation. The current is applied to a particular site when the slide appears, and continued until the next slide. The choice of sites to be sampled is arbitrary, but should include both the area of proposed resection and its margin, as well as the likely location of areas essential for language in the posterior inferior frontal lobe and posterior superior temporal lobe. Fifteen or so sites can be mapped, with three samples of the effects of stimulation on naming at each site, in about 30 min. A site is identified as essential for language when stimulation there evokes repeated and statistically significant errors in naming, either arrests of speech or anomia, the inability to name correctly but with retained ability to speak. The particular type of naming error, omission of the name or misnaming, is of equal significance. Hesitations in naming may be clues as to the localization of language, but alone do not adequately identify the essential language areas.

When there are concerns about localization of sites related to reading, or to recent verbal memory, another task is then used. Each trial of this task includes three segments present on different slides: (1) a picture of an object to be named; (2) a simple sentence, which is read aloud; and (3) the word "recall," which acts as a cue to report the name of the object presented in this trial. Thus the object name is the input that must be stored during the distraction provided by the sentence reading, and then retrieved at the cue. The slide with the object picture and the "recall" slide are generally shown for 4 s, the reading slide for 8 or 10 s. On some trials stimulation is applied for the duration of the first segment, on others for the second, still others the third, and there must be interspersed trials without stimulation to measure control performance. The effects of stimulation of any one site are assessed for naming, reading, and three conditions of recent verbal memory: when stimulation occurred during the input of information, during the time the information had to be stored, or during retrieval of the information. Memory performance is analyzed only for trials where naming is intact, to ensure a correct input of information. As indicated above, lateral cortical sites related to the input or storage, but not retrieval aspects of memory seem to be the ones that must be spared to avoid postoperative memory deficits. This test of reading and memory imposes three stimulation conditions for each site. It is desirable to have at least three samples of stimulation effect for each condition at each site. And, interspersed trials without stimulation are essential, for there is usually a control error rate on the memory task. Thus this test can be quite time consuming. As a result its use is restricted to patients and sites where there is concern about the relation to reading or memory.

Stimulation mapping has generally been a reliable predictor of localization of areas essential for language. However, if there is an uncertainty about the mapping, or if the resection is very near to areas essential for language, the author also usually has the patient engage in naming, and occasionally also sentence reading, during removal of the areas of concern. Appearance of a deficit in these tasks is an indication to stop the resection. The patient will likely have a temporary language disturbance postoperatively, but not a permanent one. The author's experience suggests that the combination of stimulation mapping localization, and this additional check during resection, allows resection with a minimal risk of postoperative aphasia, of cortex such as that in posterior temporal lobe or inferior frontal lobe of the dominant hemisphere that classically is considered essential for language. Moreover, the use of stimulation mapping to localize lateral cortical sites related to memory, with the sparing of these sites along with hippocampus, allows temporal resections with minimal risk to memory, even when memory is totally dependent on that hemisphere.

References

Blume WT, Grabow JD, Darley FL, et al. (1973) Intracarotid amobarbital test of language and memory before temporal lobectomy for seizure control. Neurology 23: 812-819

Delgado-Escueta A, Treiman DM, Walsh GO (1983) The treatable epilepsies (second of two parts). N Engl J Med 308: 1576-1584

Dennis M, Whitaker H (1976) Language acquisition following hemidecortication. Brain Lang 3: 404-433

Drachman D, Arbit J (1966) Memory and the hippocampal complex: is memory a multiple process. Arch Neurol 15: 52-61

Falconer M (1972) Place of surgery for temporal lobe epilepsy during childhood. Br Med J 2: 631-635

Fried I, Mateer C, Ojemann G, Wohns R, Fedio P (1982) Organization of visuospatial functions in human cortex. Brain 105: 349-371

Galaburda A, Sanides F, Geschwind N (1978) Human brain: cytoarchitectonic left-right asymmetries in the temporal speech region. Arch Neurol 35: 812-817

Kimura D (1967) Functional asymmetry of the brain in dichotic listening. Cortex 3: 163-178

Lesser R, Lüders H, Dinner D, Hahn J, Cohen L (1984) The location of speech and writing functions in the frontal language area: results of extraoperative cortical stimulation. Brain 106: 275-291

Mateer C, Dodrill C (1983) Neuropsychological and linguistic correlates of a typical language lateralization: evidence from sodium amytal studies. Human Neurobiol 2: 135-142

Milner B (1967) Brain mechanisms suggested by studies of temporal lobes. In: Millikan CH, Darley FL (eds) Brain mechanisms underlying speech and language. Grune and Stratton, New York, pp 122-145

Milner B, Branch C, Rasmussen T (1962) Study of short-term memory after intracarotid injection of sodium amytal. Trans Am Neurol Assoc 87: 224-226

Novelly RA, Augustine EA, Mattson RH et al. (1984) Selective memory improvement and impairment in temporal lobectomy for epilepsy. Ann Neurol 15: 64-67

Ojemann GA (1978) Organization of short-term verbal memory in language areas of human cortex: evidence from electrical stimulation. Brain Lang 5: 331-340

Ojemann GA (1979) Individual variability in cortical localization of language. J Neurosurg 50: 164-169

Ojemann GA (1983) Brain organization for language from the perspective of electrical stimulation mapping. Behav Brain Sci 6: 189-230

Ojemann G, Dodrill C (1985a) Verbal memory deficits after left temporal lobectomy for epilepsy: mechanism and intraoperative prediction. J Neurosurg 62: 101-107

Ojemann G, Dodrill C (1985b) Intraoperative techniques for reducing language and memory deficits with left temporal lobectomy. Adv Epileptol 16: (to be published)

Ojemann G, Whitaker H (1978) The bilingual brain. Arch Neurol 35: 409-412

Penfield W, Jasper H (1954) Epilepsy and the functional anatomy of the human brain. Little, Brown, Boston, 896 pp

Penfield W, Mathieson G (1974) Memory, autopsy findings and comments on the role of hippocampus in experiential recall. Arch Neurol 31: 145-154

Penfield W, Milner B (1958) Memory deficit produced by bilateral lesions in the hippocampal zone. Arch Neurol Psychiatry 79: 475-497

Penfield W, Roberts L (1959) Speech and brain mechanisms. Princeton University Press, Princeton

Ranck J Jr (1975) Which elements are excited in electrical stimulation of mammalian central nervous system. A review. Brain Res 98: 417-40

Rausch R, Crandall PH (1982) Psychological status related to surgical control of temporal lobe seizures. Epilepsia 23: 191-202

Rasmussen T, Milner B (1977) The role of early left brain injury in determining lateralization of cerebral speech functions. Ann N Y Acad Sci 299: 355-369

Rubens A, Mahowald M, Hutton J (1976) Asymmetry of the lateral (Sylvian) fissures in man. Neurology 26: 620-624

Scoville WB, Milner B (1957) Loss of recent memory after bilateral hippocampal lesions. J Neurol Neurosurg Psychiatry 20: 11-21

Sidītis J, Volpe B, Wilson D, Rayport M, Gazzaniga M (1981) Variability in right hemisphere language function after callosal section: evidence for a continuum of generative capacity. J Neurosci 1: 323-331

Siegel S (1956) Nonparametric statistics for the behavioral sciences. McGraw-Hill, New York

Wada J, Rasmussen T (1960) Intracarotid injection of sodium amytal for the lateralization of cerebral speech dominance: experimental and clinical observations. J Neurosurg 17: 266-282

Wieser HG, Yaşargil MG (1982) Selective amygdalohippocampectomy as a surgical treatment of mesiobasal limbic epilepsy. Surg Neurol 17: 445-457

Woods BT (1983) Is the left hemisphere specialized for language at birth? Trends Neurosci 6: 115-117

Editors' Commentary 3

Within the context of surgical treatment of epilepsy careful and competent neuropsychological evaluation is of great importance with respect to:

1. Assessment and quantification of cognitive and affective deficits and release symptoms
2. Delineation of deficits associated with a particular brain locus (diagnostic purpose: lateralization/localization)
3. Providing guidelines for identification of those patients most likely to benefit from a planned operation - and also to identify those least likely to benefit from surgery or even with a high risk of worsening (predictive purpose in terms of: seizure control, cognitive changes, psychosocial outcome, abnormal brain organization)
4. Evaluation of direct effects of surgical treatment on cognitive and affective performance, comparing preoperative baseline with postoperative data. Moreover, if repeated and comparable testing is done, the evolution of brain function in a patient with a long-standing seizure career can be objectively documented, and effects of various drug regimes, as well as seizure frequency, etc., can be studied and compared with the effects of surgery

The presurgical neuropsychological evaluation assesses a wide array of brain functions. Interpretation of test performance takes into consideration subject variables such as age, socioeconomic status, profile analysis of scores in relation to normative data, and the qualitative nature of responses. A comprehensive summary of the most commonly applied selective neuropsychological measures used in the preoperative workup of epileptic patients has been presented at the recent Palm Desert Conference (Engel 1987) and is now available (Rausch 1987). In addition, Dodrill (1987) has summarized psychological predictors of surgical success, and Marilyn Jones-Gotman (1987) commented specifically on "testing hippocampal function" either by invasive (Wada Test) or by noninvasive tests. Furthermore, during the Palm Desert Conference an ad hoc workshop was formed to discuss pertinent issues involving the rationale and protocols for using the intracarotid sodium amytal (ISA) technique. A summary of this is being published by Rausch (1987).

The Zürich chapters on this subtopic address some important aspects: Perret and Vetsch show that only a comprehensive set of tests is effective in detecting organically impaired functioning and thus defining a "localized area" of dysfunction. The results shown by them, however, raise doubts on the probably too easily accepted assumption that tests valid with general neurology patients can be directly applied to the epileptic population. It is reasonable to assume that an epileptic focus without a radiologically demonstrable lesion may break down behavior in a different manner than a structural lesion. Clearly more research is needed which addresses these effects of the etiology of lesions. Even if the applied tests are not specific enough to allow etiological conclusions, their value for identifying brain dysfunction at an early stage has been convincingly demonstrated, with the implication that patients with more extensive defects, either bilateral or intrahemispheric, are less likely to benefit from restrictive resective surgery, such as temporal lobectomy (Bengzon et al. 1968).

Given the frequency of temporal lobe surgery and the body of evidence showing that a major function of the temporal lobe is the mediation of memory processes, memory evaluation becomes a prime goal in preoperative neuropsychological assessments. Nadig and Wieser's chapter addresses the role of the mesiobasal limbic structures in learning and memory by comparing preoperative with postoperative data. Their results raised several questions, the most important one centering around the discussion, which aspect of memory function is mediated by the hippocampal formations? An important point that emerged from the analysis of the Montreal data, which was gained with patients with varying extents of unilateral hippocampal excision (summarized in Jones-Gotman 1987), is that the hippocampus plays a role in some, but not all, memory tasks depending on the kind of stimuli and complexity of task. Even small differences in an experimental paradigm may be sufficient to make the performance defective on a task that is normally not sensitive to hippocampal function (Jones-Gotman 1987). The chapters by Silfvenius and Ojemann are complementary in many aspects. The intracarotid Sodium Amytal procedure is considered by many investigators to be a critical component of the presurgical workup of the epileptic patient who is a surgical candidate. As Silfvenius shows, application of this method provides diagnostic as well as prognostic information. By assessing language and memory functions the functional loss of the treated hemisphere can be determined and the functional integrity of the noninjected hemisphere can be evaluated. With respect to its predictive value, the ISA test continues to be used to identify those patients who are at risk for an amnestic syndrome if a temporal lobe is removed. Furthermore it gives the most definitive information as to which hemisphere is dominant for language. It is, however, not clear whether it can predict the degree of memory deficits seen after temporal lobe surgery, especially the extent of verbal memory problems encountered with dominant temporal lobe excisions. These seem to be more a reflection of intrahemispheric rather than interhemispheric mechanisms, as convincingly demonstrated in Ojemann's chapter. The location of areas within the dominant hemisphere essential for language and the identification of the lateral cortical component of short-term verbal memory by stimulation mapping is probably the most exact approach to localize these functions in an individual patient in order to avoid encroachment into "indispensible" brain areas. Because of its special requirements (patients have to be alert) and the fact that stimulation mapping during memory testing is time consuming, this testing is limited to patients considered at risk for postoperative verbal memory deficits, i.e., patients undergoing dominant hemisphere temporal resections with high scores on preoperative verbal memory measures or those with large decrements in verbal memory during ISA perfusion of the side of the proposed resection (Ojemann 1987).

References

Bengzon ARA, Rasmussen T, Gloor P, Dassault J, Stephens M (1968) Prognostic factors in the surgical treatment of temporal lobe epileptics. Neurology (Minn) 18: 717-731

Dodrill CB (1987) Comment: psychological evaluation. Predictors of success and the Washington Psychosocial Seizure Inventory. In: Engel J Jr (ed) Surgical treatment of the epilepsies. Raven, New York, pp 197-201

Engel J Jr (ed) (1987) Surgical treatment of the epilepsies. Raven, New York

Jones-Gotman M (1987) Comment: psychological evaluation. Testing hippocampal function. In: Engel J Jr (ed) Surgical treatment of the epilepsies. Raven, New York, pp 203-211

Ojemann GA (1987) Intraoperative functional mapping at the University of Washington, Seattle. In: Engel J Jr (ed) Surgical treatment of the epilepsies. Raven, New York, pp 696-697

Rausch R (1987) The neuropsychological evaluation. In: Engel J Jr (ed) Surgical treatment of the epilepsies. Raven, New York, pp 181-195

Electrophysiological and Other Diagnostic Procedures

Electrode Technology

P. Comte

Although the function of an electrode can be defined simply as the electrical link between the tissues and an electronic circuit, its construction involves the need for specialized design and a careful selection of materials.

A reliable electrode system must be able to resist the repeated mechanical stresses imposed by the environment, its contacting tip must be of nontoxic nature and allow signals to be transmitted with minimal electrical losses, and, finally, it must be able to withstand the chemical rigor of chronic implantation.

Choice of Biocompatible Materials

Biocompatibility can be described as the lack of interaction between the material and the tissues. In practice and even in the absence of imposed signals, metallic ions are released from the metal in different forms (oxides, hydroxides, complexes) through corrosion. These products are transported by diffusion into the tissues where reactions can occur. The type of tissue reaction is related to the nature of the material and its corrosion rate. Following Steinemann (1980), histological examination combined with in vivo corrosion tests have led to the results presented in Fig. 1.

In the case of highly corroding elements such as cobalt, copper, nickel, and vanadium, a severe tissue reaction is observed (toxic response) and is characterized by a sterile abscess. These materials must be avoided as implant materials. A vital reaction with loose and vascularized fibrous tissue is observed only for the best corrosion-resistant materials such as tantalum, niobium, zirconium, and titanium. Despite their relatively good corrosion resistance, stainless steel (type 316L) and cobalt alloys exhibit a capsule type of response due to the elements nickel and cobalt.

Avenue des Cerisiers, 1009 Pully, Switzerland

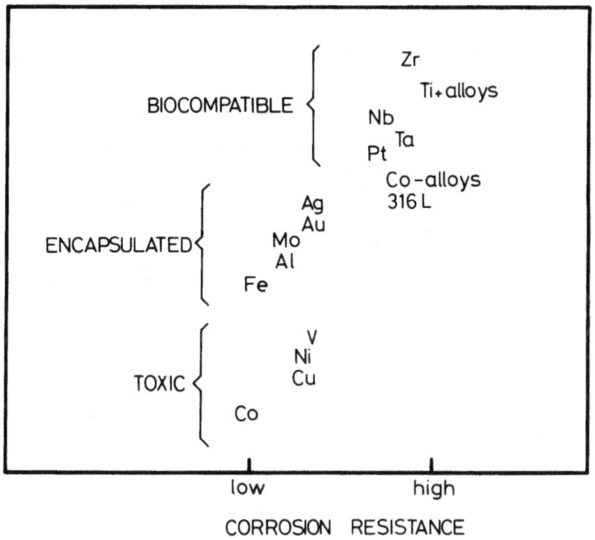

Fig. 1. Materials are grouped according to the type of tissue reaction and their resistance to corrosion

Special attention should be paid to the noble metals. Gold and silver can form complexes with chloride ions and involve a capsule type of response. They must be avoided as electrode materials. In contrast, platinum can be used safely as an electrode material when the area of the electrode is relatively large.

Lead Design

To withstand the large and repeated mechanical movements imposed by the body's environment, conductors have to be of high flexibility; however, they have to be made of very high strength and ductile material (Comte 1984). Figure 2 shows that good electrical conductors such as Ag, Cu, Au, and Al have poor mechanical strength. On the other hand, the presently used high-strength materials such as stainless steel 316L and cobalt alloys (Syntacoben, MP35N, Elgiloy) are bad conductors. Two metals, molybdenum and tungsten, are relative exceptions.

Presurgical Evaluation of Epileptics
Edited by H. G. Wieser and C. E. Elger

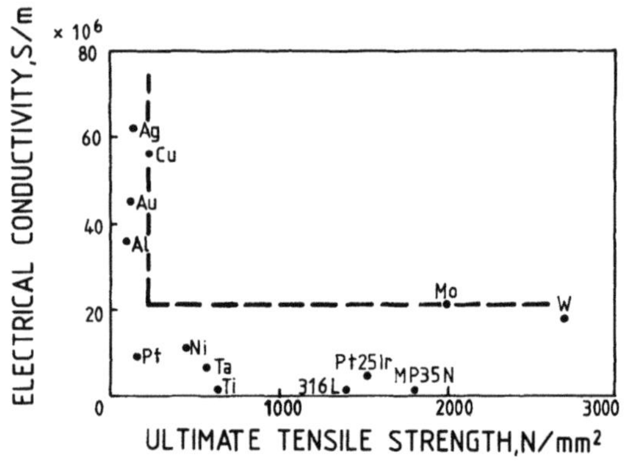

Fig. 2. Mechanical strength and electrical conductivity of metals and alloys

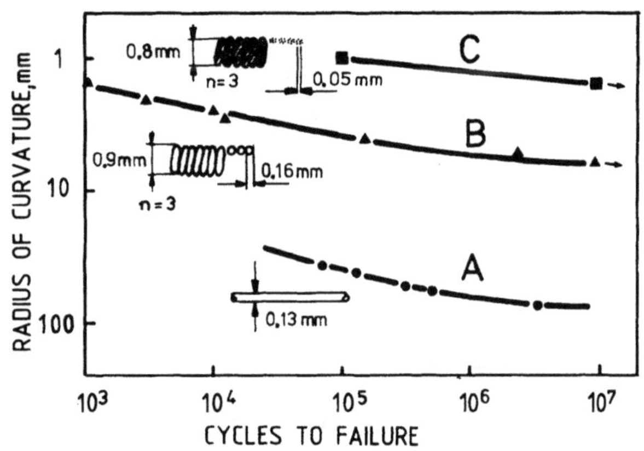

Fig. 3. Fatigue curves of cobalt base alloy (MP35N) with an ultimate tensile strength of 1800 N/mm². *A*, wire, *d* (wire diameter) = 0.13 mm; *B*, multiconductor coil, *d* = 0.16 mm; *C*, stranded multiconductor coil, *d* = 0.05 mm

They are of very high strength and have 18 times better conductivity than the cobalt alloys. Because of their very high ductilities they are therefore, with the cobalt alloys, candidates for coil design.

The requirement of high flexibility and fatigue resistance imposes on the designer the use of very small size conductors and special configurations such as coils - both conditions necessitated for reducing the mechanical stresses imposed at the outer surface of the wire which are responsible for fatigue failure.

Figure 3 shows the fatigue behavior of MP35 N wires in different configurations (straight, coiled, stranded and coiled) obtained in rotating bending tests. Straight wires of small dimensions display poor resistance to small fatigue curvatures, the multiconductor coils show a significant improvement, and the stranded coiled wires resist extremely small radii of curvatures (1.5 mm), closely approaching the limit required for the unbreakable conductor. From the metallurgical point of view one must realize that for the coiling operation of the wire a minimum amount of ductility is required in order to avoid surface cracking during the fabrication stage. Consequently, the amount of cold work needed for strengthening the wire must be carefully dosed in relation to the diameter of the coil.

Electrode Tip Design

The tip of the electrode must be able to exchange repeatedly high amounts of electrical charges (signals) with the tissue without altering the material itself (by corrosion) or the tissues. Materials can be selected for their capacity of exchanging high-density charges at low energy level involving only reversible reactions at the interface. In principle, the nonpolarizable electrodes of the type silver-silver chloride would be ideal but the release of silver chloride into the tissue makes them unsuitable for chronically implanted systems. Other metals such as platinum, stainless steel 316L, cobalt alloys, titanium, tantalum, niobium, and zirconium behave in physiological solutions as polarizable interfaces. Their ability to exchange charges is shown in Fig. 4, where their impedance (resistive and capacitive) has been reported.

Among the materials having the lowest polarization losses, platinum, carbon, and titanium nitride seem to offer about the same properties and should be preferred as electrode tip materials. Niobium, zirconium, and titanium dioxide (TiO_2) on the contrary exhibit very high polarizations.

Safe Charge Injection

It is essential that not only the materials but also the stimulus currents employed do not cause tissue damage. Mortimer et al. (1970) and Pudenz et al. (1977) have reported the critical amount of charges that have created tissue destruction based on the observation of the blood brain barrier breakdown and cell damages. Their results are represented graphically in Fig. 5, where the curves indicate the separation of the lesion/no lesion domains.

Because of the differences in the experimental conditions, these results must be interpreted with caution when choosing stimulation parameters. The production of tissue lesion is directly related to the electrochemical reactions involved at the electrode tip. Under certain conditions such as high-charge density, high frequencies, and certain types of sig-

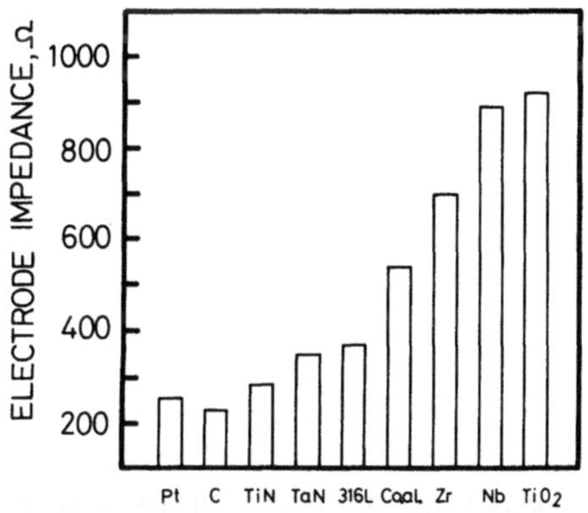

Fig. 4. Impedance of various electrode materials under 10 mA constant capacitively pulsed current in 0.9% NaCl (electrode area 10 mm^2)

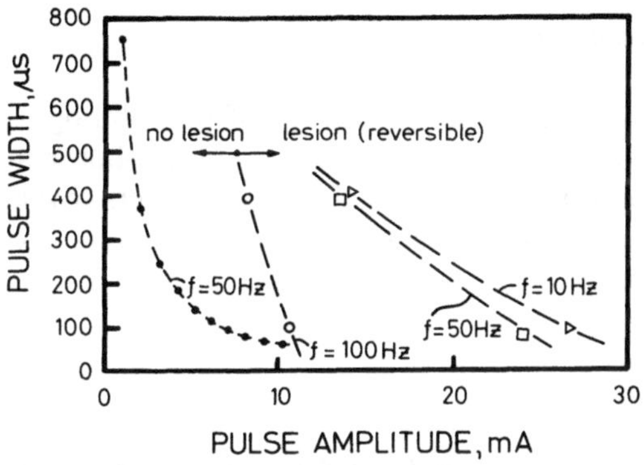

Fig. 5. Limits for safe electrical stimulation •, limit given by Pudenz et al. at 50 Hz and 1.5 mm^2 electrode surface area (cat brain). ○ □ △, limit given by Mortimer et al. at different frequencies and 5 mm^2 surface area of PtIr (cat brain)

nals, nonreversible reactions create local changes in the electrolyte composition. When these changes override the buffering capacity of the local system, permanent cell damage must be expected.

Causes of Electrode Failures

Electrode failures can be mechanical or electrical. The mechanical failures are usually breakages (by fatigue) at sharp bends, at the contact zone with hard bone or sharp edges, and at fixation points such as sutures. Electrical failures can be caused by too much tissue destruction at the vicinity of the electrode tip. In such cases, electrodes are practically insulated electrically with a thick capsule of connective tissue through which the transmission of signals is largely attenuated.

References

Comte P (1984) Metallurgical observations of biomaterials. In: Boretos JW, Eden M (eds) Contemporary biomaterials. Noyes, Park Ridge, pp 66-91

Mortimer JT, Sheally CN, Wheeler C (1970) Experimental nondestructive electrical stimulation of the brain and spinal cord. J Neurosurg 32: 553-559

Pudenz RH, Agnew WF, Yuen TGH, Bullara LA, Jaques S, Shelden CH (1977/78) Adverse effects of electrical energy applied to the nervous system. Appl Neurophysiol 40: 72-87

Steinemann SG (1980) Corrosion of surgical implants - in vivo and in vitro tests. In: Winter GB, Leray JL, de Groot K (eds) Evaluation of biomaterials. Wiley, New York, pp 1-34

Future Trends in Electrode Technology*

N. F. De Rooij

Introduction

The fabrication of microelectrodes with conventional techniques using glass micropipettes or thin metal wires to record the electrical activity of nerve cells is faced with difficulties in controlling the dimensions and impedance of such electrodes (Frank and Beeker 1964). The same situation is encountered in the conventional fabrication of microelectrodes and minielectrodes for ion-activity measurements.

Over the past 10 years research has been going on to develop minielectrodes and microelectrodes using so-called microfabrication techniques. These techniques have caused the revolution in solid-state electronics (Wise et al. 1970; Wise and Angell 1975; Bergveld 1972). In this article we will first briefly outline the various methods of microfabrication, followed by a discussion of the present state of electrodes produced using these methods. Finally we will indicate possible future trends in electrode technology using microfabrication techniques.

Microfabrication Methods and Techniques

The fabrication of minielectrodes and microelectrodes by microfabrication techniques is based on a series of sophisticated physical and chemical steps of processing the starting material, in most cases wafers of single-crystal silicon. By means of photolithography, oxidation, epitaxy, ion implantation, diffusion, micromachining, metalization, and bonding, the desired electrode structures are formed. Some simple devices like transistors or even complete integrated circuits may be integrated on the electrode structure. After slicing, bonding, and encapsulation, the electrodes are ready for use in their particular applications. The following process characteristics are of interest for the production of silicon-based mini- and microelectrodes (Middelhoek and Noorlag 1978).

Photolithography

This technique, based on the use of photolithographic masks, ultraviolet light, and layers of photoresistance produces patterns in the various layers covering the basic substrate such as insulating layers and metals.

Oxidation

Silicon wafers are oxidized for isolation purposes and for oxide masking by heating the wafer to a high temperature in a flow of oxygen with or without water vapor. Typical oxide thicknesses are between 0.5 and 1.0 μm, but much thinner layers can also be produced.

Impurity Doping

Diffusion and ion implantation are used to create n-type and p-type areas in the silicon substrate. Diffused layers have thicknesses between 1 and 20 μm depending on time and temperature. Surface impurity concentrations between 10^{14} and 10^{21} are possible. Ion implantation makes possible better control of the thickness and impurity profile.

Physical Vapor Deposition (PVD) and Chemical Vapor Deposition

These techniques allow the deposition of a great variety of materials varying from metals to insulators. These methods are especially important for the realization of electrodes.

Institute of Microtechnology, University of Neuchâtel, Rue A.-L. Breguet 2, 2000 Neuchâtel, Switzerland

* This study was supported by the Swiss National Science Foundation and Swiss Foundation for Microtechnology Research.

Presurgical Evaluation of Epileptics
Edited by H. G. Wieser and C. E. Elger

Micromachining

Using anisotropic as well as isotropic etching solutions, it is possible to obtain three-dimensional structures in silicon, such as needles and membranes. In combination with highly boron-doped regions, highly controlled etch stops can be obtained, resulting in a good control of, for instance, electrode dimensions, as we shall see later.

In the following section, some examples will be given of electrodes produced using the above-mentioned technological procedures.

Microelectrode Arrays for Potential Recording Purposes

Najafi et al. (1985) have recently reported the development of a multielectrode recording array for use in studies of the central nervous system and in the closed-loop control of neural prostheses. The electrode probe utilizes a silicon-supporting carrier, which is defined using a deep boron diffusion and an anisotropic etch stop. This carrier supports an array of polysilicon or tantalum thin-film conductors insulated above and below with silicon nitride and silicon dioxide. The probes have a shank length of 3 mm, a width of 50 µm, and a thickness of 15 µm. These probes are capable of simultaneous multichannel recording of neural activity in the cortex. One of the features of this probe is the compatibility of its fabrication process with the integration of n-chip MOS circuitry for signal amplification and multiplexing.

Another state-of-the-art probe is a multimicroelectrode for simultaneous recording of single action potentials, as reported by Takahashi and Matsuo (1984). They have fabricated a silicon probe, into which they have integrated the multimicroelectrode array as well as interface circuits. Their microprobe structure showed less parasitic capacitance from electrode to ground and less interelectrode coupling than conventional metal microelectrodes. The frequency bandwidth and noise characteristics were comparable to those of conventional microelectrodes.

Minielectrodes and Microelectrodes for Ion-Activity Recording Purposes

Another class of device under development deals with sensors for the measurement of activities of ions and dissolved gases in body fluids.

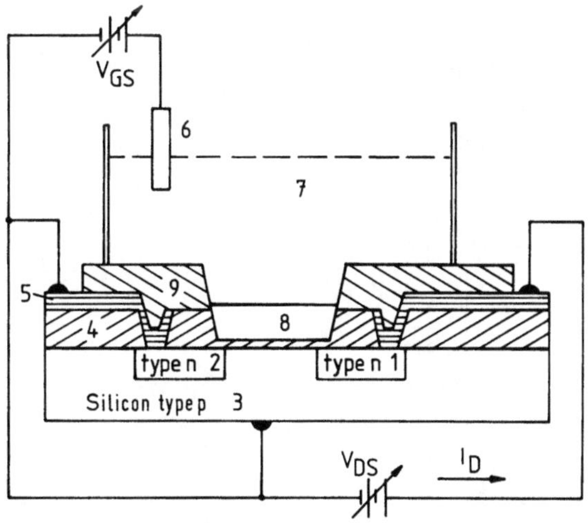

Fig. 1. Schematic diagram of an ISFET. *1*, drain; *2*, source; *3*, substrate; *4*, insulator; *5*, metal contacts; *6*, reference electrode; *7*, solution; *8*, membrane; *9*, encapsulant

A conventional approach for ion-activity measurement is the use of miniaturized ion-selective electrodes (ISEs). ISEs have been used for many years now. The theory of their behavior is relatively well known and recent advances in materials technology have led to the development of a wide range of novel ion-selective membranes, such as many new types of ion exchange glasses. Polymeric materials are developing fast, and techniques for depositing, for instance, PVC films loaded with liquid ion exchangers are becoming readily available (Meier et al. 1980). Mini- and micro-ISEs are characteristically high-impedance devices requiring very high input impedance amplifiers, which may present serious technical problems for small-sized devices.

Ion-selective electrodes also require several labor-intensive steps in their manufacture, which add to their cost. A novel approach for ion-activity measurements is the use of so-called ion-selective field effect transistors (ISFETs), reported for the first time by Bergveld (1972). These devices have been extensively studied and various review articles describe the state of the art (Janata and Huber 1980; Sibbald 1983). The ISFET is a hybrid between an ISE and a metal/oxide/semiconductor FET (MOSFET) and combines many of the advantages of both devices, such as the good ion-selectivity achieved with many ISEs and the high- to low-impedance transformation achieved using MOSFETs. The ISFET is basically a modified MOSFET and is similar in design and size, the main difference being the absence of the gate metal and its replacement by an ion-selective membrane. Figure 1 is a schematic diagram of these devices. Additional advantages are: small size, rugged solid-state structure, possibility of obtaining

Fig. 2. Photomicrograph of the five different designs of ISFETs

multi-ion sensors, and integration of signal-processing electronics.

Despite their promise, so far ISFETs have not appeared on the market, but many experimental data on their in vivo performance have been published, clearly indicating the potential of these devices for in vivo measurement and monitoring purposes (McKinley et al. 1984).

Certainly the most appealing feature of this new class of devices is the possibility of producing multi-ion sensors with integrated signal-processing circuitry. Recently published developments outline the first concrete multi-ion sensors, but with problems at the level of ion-selective membrane deposition and adhesion (Lauks et al. 1985; Brown et al. 1985).

We are currently developing a C-MOS-compatible pH-ISFET process (De Rooij et al. 1985). The first objective is to characterize these devices, especially with respect to their drift behavior and temperature sensitivity. Once these nonideal effects have been characterized, on-chip compensation may be possible using the C-MOS process for circuit implementation.

Five different designs of the C-MOS-compatible pH-ISFETs are shown in Fig. 2. Three of the designs have dimensions of 3.0×1.4 mm and contain two ISFETs and one MOSFET. The remaining two elements have dimensions of only 3.0×0.7 mm and contain only one ISFET. These latter designs are suitable for biomedical applications. As pH-selective material, Al_2O_3 has been used obtained by chemical vapour deposition (De Rooij et al. 1985). Typical characteristics of pH-sensitivity, temperature coefficient, and drift behavior are shown in Fig. 3a–c. These characteristics were obtained by connecting to a specially designed amplifier system, which operates the devices at constant drain current with constant drain to source voltage. The gate to source voltage V_{GS} is then the experimentally determined quantity and represents the electromotive force of an ISE (Arnoux et al. 1985).

When corrections are used for these effects, a very accurate pH-measurement can be made (De Rooij et al. 1985), where in vivo measurements of blood pH have been compared with blood-gas analyzer data. The data correspond with each other within ± 0.03 pH unit.

Another class of sensors is the so-called amperometric devices, which allow detection of partial oxygen pressure, but also form a base for more complex sensors. For instance, the reaction of glucose as catalyzed by the enzyme glucose oxidase (glucose $+ O_2 \rightarrow$ gluconic acid $+ H_2O_2$) will influence the measured current in the amperometric cell if the enzyme is situated near the reacting electrodes.

A recently developed amperometric sensor is shown in Fig. 4, consisting of a three-electrode system (working and auxiliary electrodes both of platinum, and the reference electrode consisting of silver/silver chloride). These devices are now under

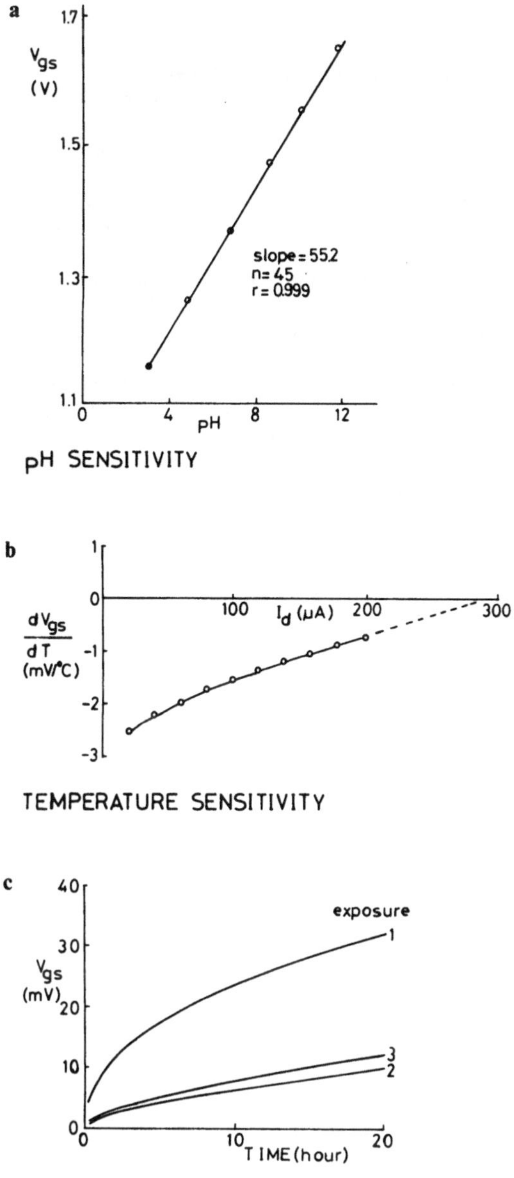

Fig. 3. **a** pH response of an ISFET at 25 °C. The pH was adjusted by adding a 0.5 *M* HCl or 0.5 *M* NaOH solution to a buffer containing 0.616 *M* Tris and 0.5 *M* HCl. **b** Temperature sensitivity of an ISFET as function of I_D in a 0.616 *M* Tris, 0.5 *M* HCl buffer (temperature range 15°-25 °C). **c** Drift of an ISFET at 25 °C in a 0.616 *M* Tris, 0.5 *M* HCl buffer. The time interval between exposures was 1 week, during which the ISFET was stored dry at room temperature

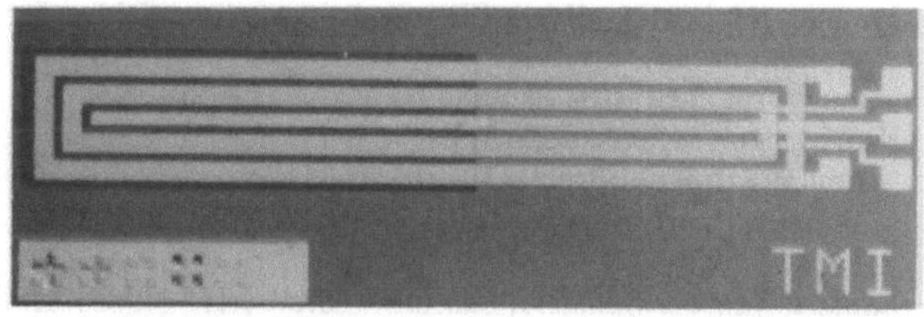

Fig. 4. Photomicrograph of a planar polarographic three-electrode system. The outer two electrodes are the auxiliary and working electrode (both of platinum); the inner electrode is the reference electrode of Ag/AgCl (here before the chloridation). Overall chip size, 0.5 × 3 mm

test (R. Buser and S. Gernet, 1985, personal communication).

Future Trends

The demands upon the future generation of silicon-based minielectrodes and microelectrodes will not only depend on the sensor itself but on the type and material of encapsulation, which are especially critical for applications in vivo. Much more effort will be needed to solve the problems related to biocompatibility and encapsulation.

What can be expected in the near future is the availability of catheter tip devices with sensors for a detection of a variety of chemical species. The sensors will be based on the FET principle, as well as on the polarographic one. Minielectrodes and microelectrodes will continue to be produced and perfected, for instance with backside contacts (Kim et al. 1985) and an integrated reference electrode (Smith and Scott 1984).

References

Arnoux C, Kloeck B, Van den Vlekkert H, De Rooij N (1985) Miniaturized silicon sensors for biomedical applications. Actes du 59e congrès SSC-ASMT, Interlaken, Switzerland. pp 37-43

Bergveld P (1972) Development, operation and application of the ion-sensitive field-effect transistor as a tool for electrophysiology. IEEE Trans Biomed Engin BME-19: 342-351

Brown R, Huber RJ, Petelenz D, Janata J (1985) An integrated multiple-sensor chemical transducer. Transducers 85. Digest of Technical Papers, pp 125-127

De Rooij NF, Haemmerli A, Grisel A (1985) C-MOS compatible pH ISFET with Al_2O_3 gate. In: Ko WH (ed) Implantable sensors for closed-loop prosthetic systems. Futura, Mount Risco, pp 139-151

Frank K, Beeker MC (1964) Microelectrodes for recording and stimulation. In: Nastuk WL (ed) Physical techniques in biological research, vol 5. Academic, New York

Janata J, Huber RJ (1980) Chemically sensitive field effect transistors. In: Freiser H (ed) Ion-selective electrodes in analytical chemistry. Plenum, New York, pp 107-175
Kim SJ, Kim M, Heetderks WJ (1985) Laser-induced fabrication of a transsubstrate microelectrode array and neurophysiological performance. IEEE Trans Biomed Engin BME-32: 497-502
Lauks I, Van der Spiegel J, Sansen W, Steyaert M (1985) Multispecies integrated electrochemical sensor with on-chip CMOS circuitry. Transducers' 85. Digest of Technical Papers, pp 122-124
McKinley BA, Houtchens BA, Janata J (1984) In-vivo application of ISFETs: summary of current laboratory research and probable future clinical detectors. Ion-Selective Electrode Rev 6: 173-208
Meier PC, Amman D, Morf WE, Simon W (1980) Liquid-membrane ion-selective electrodes and their biomedical applications. In: Konyta J (ed) Medical and biological applications of electrochemical devices. Wiley, Chichester, pp 13-91
Middelhoek S, Noorlag DJW (1978) Silicon micro-transducers: a new generation of measuring elements. In: Regtien PPL (ed) Modern electronic measuring systems. Delft University Press, Delft, pp 1-23
Najafi K, Wise KD, Mochitzuki T (1985) A high-yield IC-compatible multichannel recording array. IEEE Trans Electron Dev ED-32: 1206-1211
Sibbald A (1983) Chemical-sensitive field effect transistors. IEE Proc 130: 233-244
Smith RL, Scott DC (1984) A solid state miniature reference electrode. Proceedings of the IEEE symposium on biosensors, 61, Sept 15-17, Los Angeles
Takahashi K, Matsuo T (1984) Integration of multi-microelectrode and interface circuits by silicon planar and three-dimensional fabrication technology. Sensors Actuators 5: 89-99
Wise KD, Angell JB (1975) A low capacitance multielectrode probe for use in extracellular neurophysiology. IEEE Trans Biomed Engin BME-22: 212-219
Wise KD, Angell JB, Starr A (1970) An integrated-circuit approach to extracellular microelectrodes. IEEE Trans Biomed Engin BME-17: 238-246

Magnetoencephalography and Epilepsy*

J. VIETH

Introduction

In the exciting area of biomagnetism, magnetoencephalography is one of the new diagnostic tools usable for epilepsies, but the method is still in technical and theoretical development and thus much more benefit must be expected.

The main advantages of magnetoencephalography (MEG) over electroencephalography (EEG) are the almost undistorted and much more localized recording of fields of sources within or underneath the cortex, and the lack of the need to have a reference. Thus, after the development of large multi-channel systems in the near future, the main goal will be to provide a three-dimensional functional real-time image of the brain. This would be of great diagnostic help for patients with subcortical epileptic foci and may replace partly or fully the invasive diagnostic means. The three-dimensional localization of active focal epileptic sources in the brain represents one of the most promising clinical applications in biomagnetism (Ricci et al. 1985).

Recording Technique

Baule and McFee in 1963 first recorded, with large induction coils, the magnetic field of the human heart. And a few years later in 1970, the great breakthrough came in the development of the most sensitive device available, the so-called SQUID, the superconducting quantum interference device (Zimmerman 1971). The magnetic flux is registered by a superconducting detection coil and then coupled by another induction coil into a superconducting ring or cylinder at a weak junction in the form of magnetic quanta. This coupling functions with very weak magnetic fields. The result is a so-called tunneling current at the weak link, which is measured by appropriate means (Erné 1983; Williamson and Kaufman 1983).

The SQUID was first applied by Cohen et al. (1970) again in the heart and, shortly after that, in the brain as a one-channel system (Cohen 1972). Figure 1 is a schematic drawing of such a system.

The development of such a very sensitive device was necessary because biomagnetic fields are roughly one million to one billion times weaker than the earth's geomagnetic field or the environmental urban noise fields (Romani et al. 1982). Figure 2 shows the relationship of typical amplitudes and frequency ranges of biomagnetic fields. To date, the technique has advanced so far that biomagnetic fields can be recorded now without special, and of course expensive and space-consuming, means of shielding. And the multi-channel development has now reached the seven-channel state, producing much more information and a considerable saving in recording time (Romani 1984; Williamson et al. 1985; BTi Company, San Diego, personal communication).

Basic Principles of Magnetoencephalography

Biot and Savart first found that whenever an electric current flows, it is surrounded by a magnetic field. Since at all excitable membranes electric charges move, so currents flow and magnetic fields are generated (Williamson and Kaufman 1985; Fig. 3).

The source of the EEG, as well of the MEG, is almost exclusively the postsynaptic activity of neurons (Okada 1983a; Vieth 1970). The resulting flow of charges across the membrane gives rise to an extracellular distribution of currents registered by the EEG as potential differences across impedances, and gives rise to an intracellular current. Since the extracellular current, i.e., the volume current, has no appreciable contribution to the magnetic field, because of the widespread back flow with a relatively

---

Dept. of Experimental Neuropsychiatry, Neurological Clinic, University of Erlangen-Nürnberg, Schwabachanlage 6, 8520 Erlangen, Federal Republic of Germany

* Dedicated to Professor Dr. Hans Schaefer in honor of his 80th birthday.

Presurgical Evaluation of Epileptics
Edited by H. G. Wieser and C. E. Elger

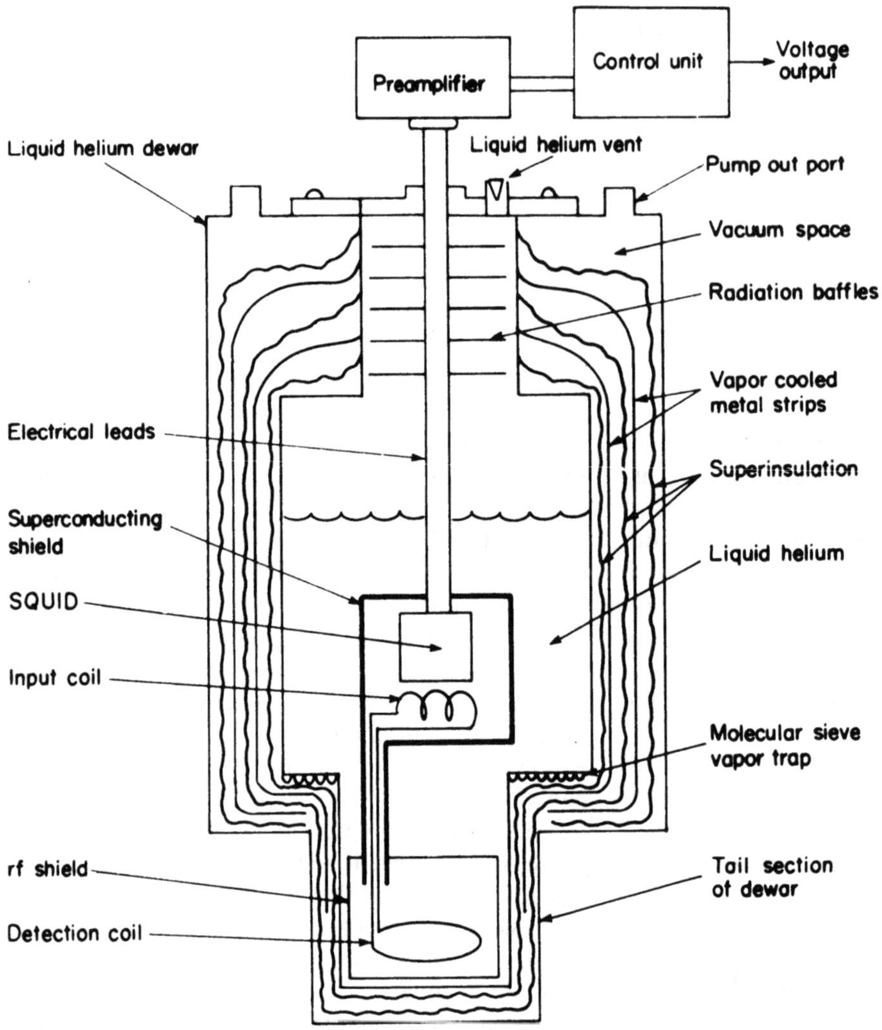

Fig. 1. Schematic drawing of a SQUID system mounted in an insulating Dewar vessel containing liquid helium. (From Zimmerman, 1983; by permission)

low current density along the external surface of the membrane, only the intracellular current may be considered as the primary source of the magnetic field outside the head (Okada 1983a; Wikswo 1983; Fig. 4).

With the assumption of a sphere (Cuffin and Cohen 1979) with concentric inhomogeneity in conductivity as the model of the head, primary current sources radially located to the surface do not contribute to the measurable radial magnetic field outside the head. That means only tangentially located current sources or the tangential component of tilted current sources contribute to the field (Geselowitz 1970; Grynszpan and Geselowitz 1973; Okada 1983a; Plonsey 1981; Tripp 1983; Williamson and Kaufman 1983; Fig. 5).

Cohen and Cuffin (1983) first tried to validate the sphere model. They compared the experimental data of a well-known neuronal source which appears to behave as a current dipole (i.e., the N20 component of the somatic evoked response) with the results of field calculations, and found an acceptable correspondence.

Furthermore, they made a comparison between MEG and EEG field maps due to the mentioned evoked current dipole. They stressed three differences seen in the predicted and the experimental data, showing that neuronal sources can be better differentiated and localized in some ways by the MEG than by the EEG: The first two differences are due only to the tangential component of the dipole. The MEG pattern of this component is rotated by 90° from the corresponding EEG map and the MEG is one-third tighter. The first difference enables the MEG to localize a tangential dipole better in a preferred direction, across the dipole, while the EEG

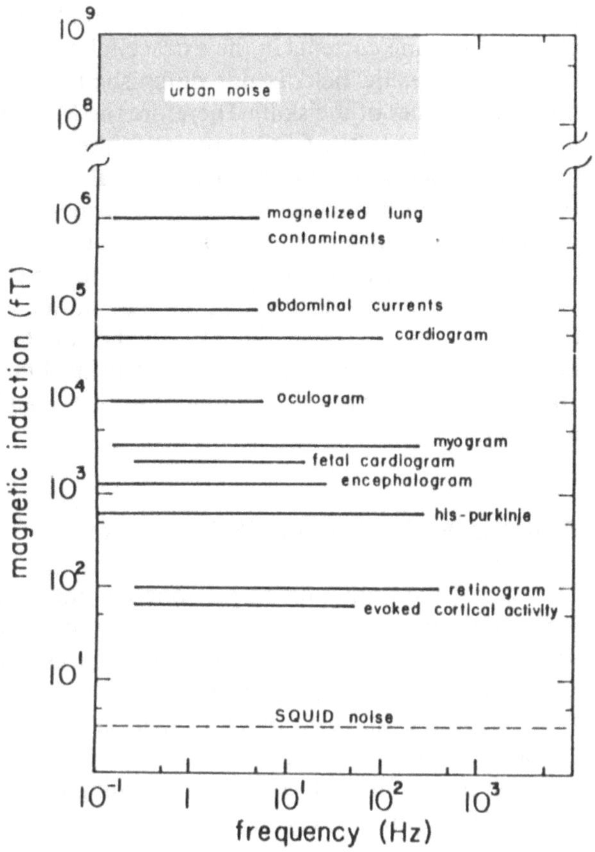

Fig. 2. Typical amplitudes and frequency ranges of different biomagnetic signals. Magnetic field strength is in femto-tesla (fT) or 10^{-15} tesla (T), that is 10^{-11} G (Gauss) in the cgs system. The *shaded area* represents the lower range of disturbing environmental fields amplitudes. The steady earth field is even stronger by 2^{10}. (From Romani 1984; by permission)

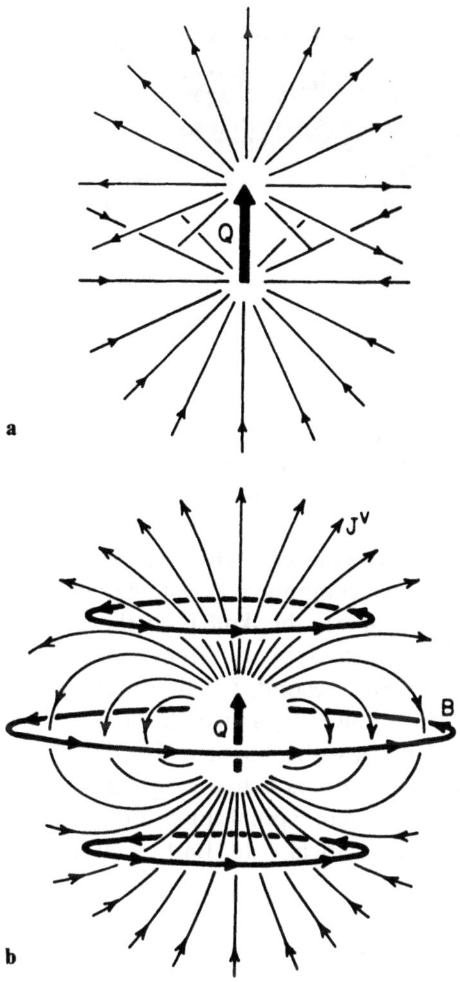

Fig. 3. **a** Diverging volume currents from the head of a current dipole Q and converging currents to its tail. **b** The dipolar current flow resulting from adding the two radial patterns and the encircling magnetic field arising from the current dipole only. (From Williamson and Kaufman 1981 b; by permission)

does so along the dipole. The second difference enables the MEG to localize better in its preferred direction than the EEG does in its preferred direction. The third difference is due only to the radial component of the dipole. While the MEG receives no contribution from this component, the EEG pattern is asymmetrical and is heavily weighted by the radial component. This is the reason the MEG is able to reveal tangential sources which are hidden by the radial sources in the EEG.

Many investigations of the human brain have demonstrated that the MEG is able to detect normal and pathological currents inside the skull three-dimensionally (Barth et al. 1984b; Beatty et al. 1983; Okada 1983a; Williamson and Kaufman 1981a). This is due to several properties of the MEG not present in the EEG (Barth et al. 1985a). First, the pattern of the component of the extracranial magnetic

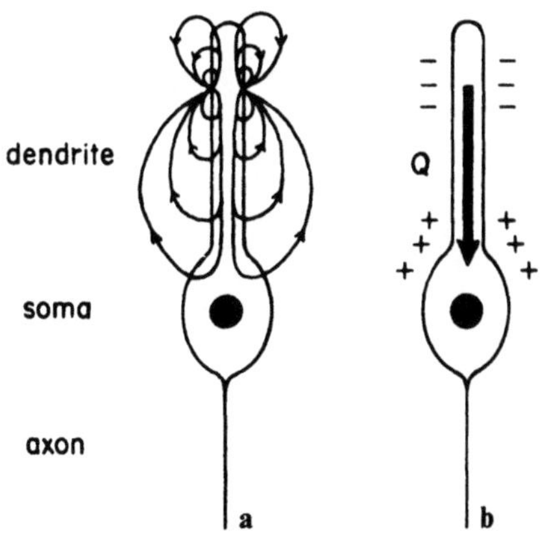

Fig. 4 a, b. Primary current source of a magnetic field of a neuron. **a** Intra- and extracellular current flow after an excitatory synaptic input on the dendrite. **b** Resulting intracellular current dipole Q, considered as the primary current source of the magnetic field of the neuron. (From Okada 1983 a; by permission)

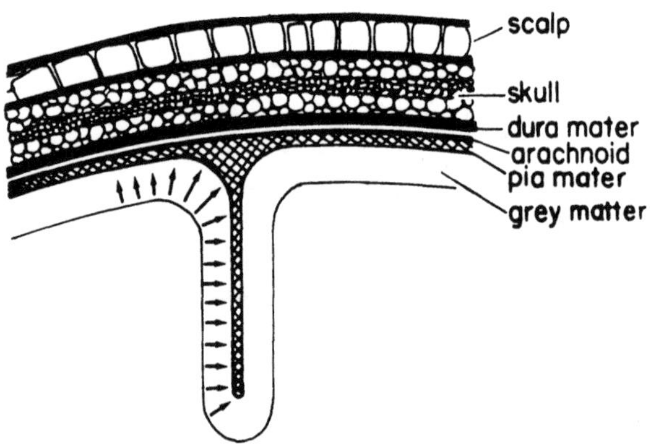

Fig. 5. Schematic representation of primary current sources *(arrows)* of cortical pyramidal cells with their different orientations to the surface. (From Okada 1983a; by permission)

field recorded by the present MEG sensors, that is, the radial or normal component, is not altered by the orientation of a source to the skull. Only the strength of the field is affected by different orientations (Barth et al. 1985a). Second, the interpretation of magnetic fields produced by a focal discharge is easier to interpret, because the magnetic field is not distorted by the uneven conductive properties of the boundaries at the cerebrospinal fluid, the bone and the skin, as is the electrical field (Barth et al. 1985a; Blum et al. 1984a; Ricci 1983). Third, unlike the electrical volume currents in the extracellular space, the neuromagnetic field is not diminished by the high impedances of the skull. Therefore the MEG is more sensitive than the EEG, especially in respect to the activity beneath the cortex (Barth et al. 1985a). Fourth, the electromagnetic theory indicates that predominantly only the intracellular current contributes to the magnetic field and not the complementary volume currents distributing throughout the extracellular space, whereas the EEG is the result of these volume currents (Cohen and Cuffin 1983; Okada 1983a; Tripp 1983; Williamson and Kaufman 1983).

Recording the Spontaneous MEG on the Scalp

Because of the limitation to just one channel, the first publications compared only the raw data of MEG and EEG and demonstrated differences and advantages of the MEG (Carelli et al. 1983; Cohen 1972; Hughes et al. 1976, 1977; Modena et al. 1982; Nicolas et al. 1983; Reite et al. 1976; Vieth 1984). An example of the first SQUID EEG recording of Cohen (1972) is shown in Fig. 6.

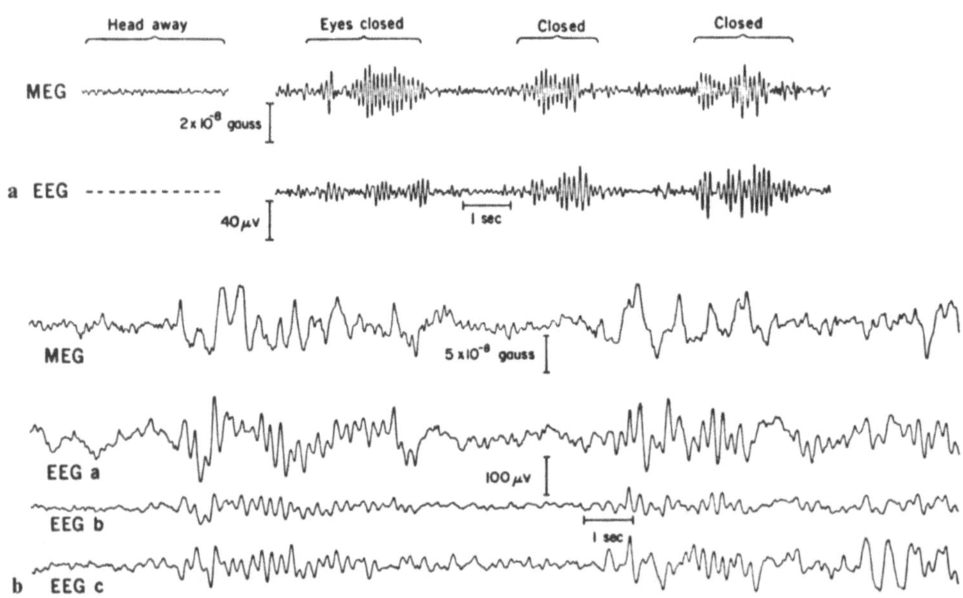

Fig. 6a, b. Simultaneous MEG and EEG recordings. **a** Alpha rhythm of a normal subject and the blockade by eyes opening. **b** Delta and theta activity of an epileptic subject in MEG and EEG. The difference is the lack of 5-Hz waves in MEG, present in all three EEG derivations. (From Cohen 1972; by permission of the author and publisher. Copyright by the American Association for the Advancement of Science, 1972)

Locating Current Sources

Due to the tighter and the less distorted recording of the MEG (Ricci et al. 1985b), the advantage of the MEG, the better localization of sources within the cortex of the brain and below it, even at the depth of several centimeters, became the goal of MEG research.

In spite of the encouraging results of all biomagnetic groups, we are still far from a unique solution to locate current sources exactly by the measured isofield maps on the surface of the skull, because of the complexity of the geometry of the skull and the brain, and because we have to make some fundamental assumptions, and because of technical limitations. But in clinical practice, especially when dealing with epileptic activity, it seems that only relatively minor difficulties may be expected, when trying to localize bioelectrical sources by magnetic measurements, as Romani (1984) has recently pointed out.

The relationship between an electrical source and the projected isofield lines on the surface of the skull allows two approaches (Tripp 1983; Williamson and Kaufman 1981a): The "forward problem" means the calculation of the distribution of magnetic fields and/or electric potentials at the surface of a volume conductor - in our case the head - as generated by a source of known strength and position. The "inverse problem" is the attempt to distinguish the sources within the volume conductor from the distribution of potentials and fields at the surface of the volume conductor as measured by the EEG and the MEG. And here there is no unique solution, because infinite source configurations may be responsible for the measured patterns. In locating current sources inside the skull, the method of Williamson and Kaufman (1981a) is mainly used.

But, as usual in the clinical field with the help of other information, the problems can be delimited and some specific hypotheses on the number and structure of the generators and on the physical properties of the volume conductor can be made, and thus an approximate and adequate solution can be achieved. The situation will be improved substantially when the large multi-channel system is available, hopefully in a few years. In the meantime, having only one MEG channel, the necessary field distributions can only be obtained by subsequent recordings of time-locked and iterative brain activity (Vieth 1984).

First, this can be achieved with stimulus-evoked responses, i.e., the evoked fields, similar to the evoked potentials of the EEG. And second, iterative focal epileptiform spike activity can be used for this purpose. Since in both situations the average technique has to be applied, the real-time will be lost.

Current Sources of the Evoked Magnetic Field

In all the investigations of stimulus and event-related responses, the accuracy of localization of the generators seems to be very good and higher than with the EEG. (For references see Kaufman and Williamson 1982; Weinberg et al. 1985; Williamson et al. 1983; and Vol. 2D, 1983, No. 2 of the Journal Il Nuovo Cimento). Williamson and Kaufman (1983) demonstrated very clearly the differentiation of current dipoles after the electrical stimulation of the right thumb, the index and the little finger and, further away, the right ankle (Fig. 7). In addition, it could be confirmed that the source of the recorded field is indeed the intracellular current.

Another example of good topographic differentiation is a tonotopic map by Romani et al. (1982a, 1982b) obtained from auditory evoked fields with different stimulation tone frequencies (Fig. 8).

Going more underneath the surface, Okada (1983c) and Okada et al. (1983) showed the generators of an event-related magnetic complex, corresponding to the N200 and the P300 of the event-related potential, deep in the brain, i.e., in the hippocampus. The localization of the current dipoles corresponded very well with the location found by stereoelectroencephalographic recordings

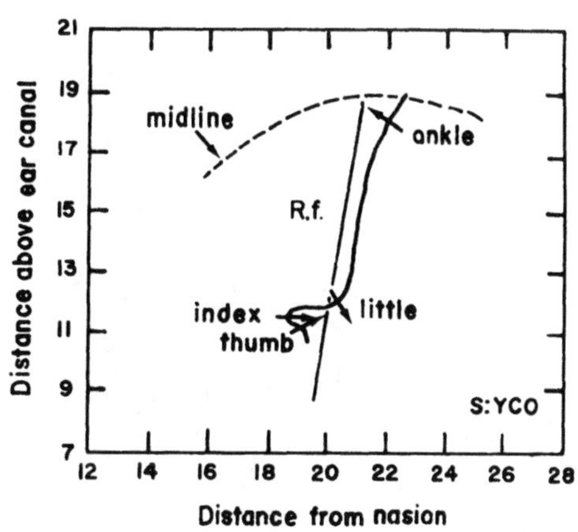

Fig. 7. Locations of the primary projection areas of the right thumb, index finger, little finger, and ankle. The estimated current dipoles *(arrows)* with direction and polarity of the measured field maps are shown with their projection on the scalp surface. The *smooth solid line* perpendicular to the dipoles represents the estimated location and shape of the Rolandic fissure. (From Okada 1983b; by permission)

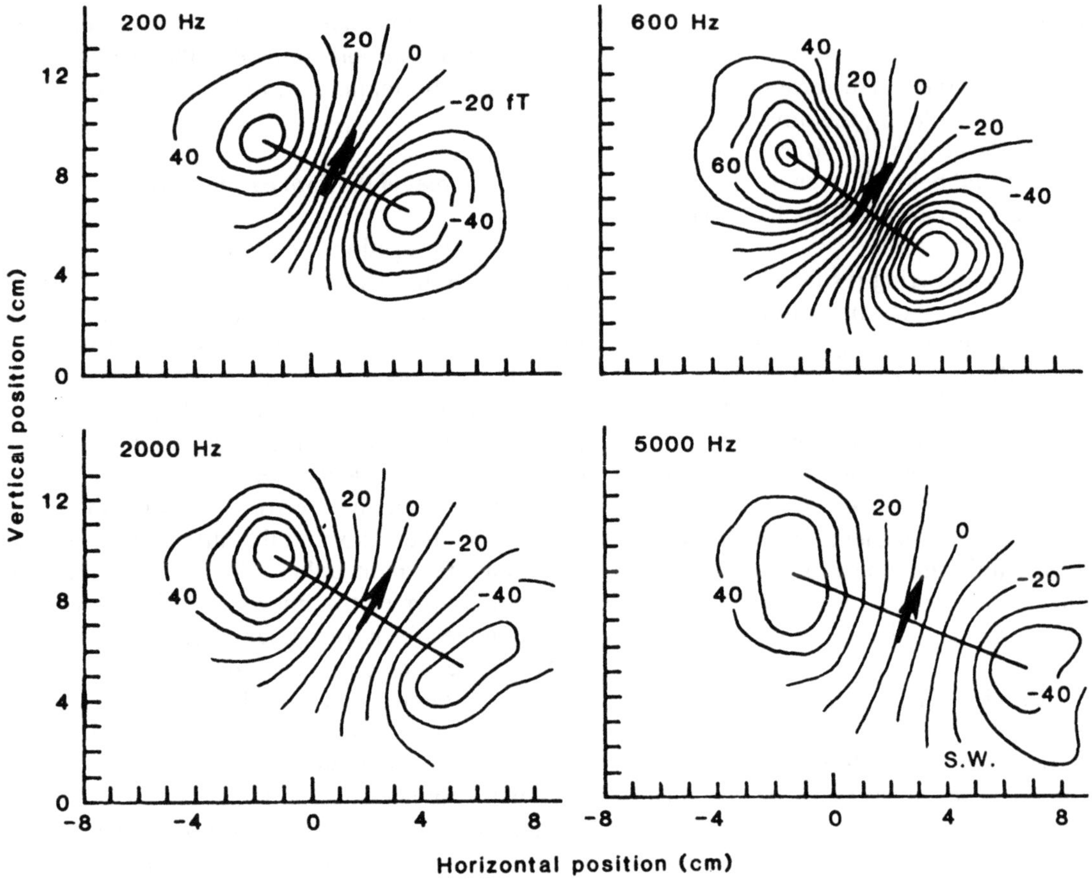

Fig. 8. Isofield maps for the component of the magnetic field normal to the scalp. *Arrows* show position and orientation of the equivalent current dipoles for 200-, 600-, 2000-, and 5000-Hz stimulation tones. The origin is at the ear canal; the corner of the eye lies at position '0.9'; the vertical axis points to the vertex. (From Romani et al. 1982a; by permission of the authors and publisher. Copyright by the American Association for the Advancement of Science, 1982)

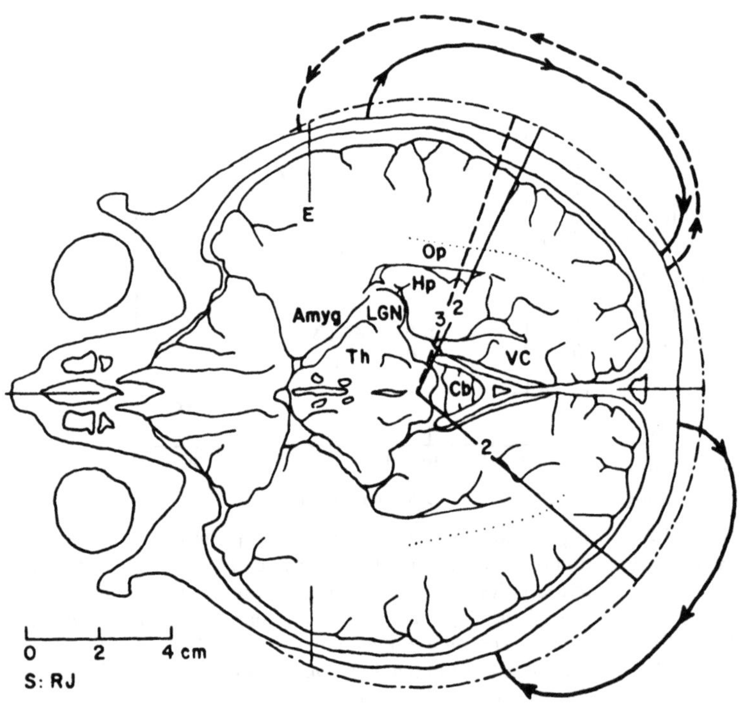

Fig. 9. Estimated locations of the equivalent current dipole sources for the 300-ms component "2" and 450-ms component "3" of the MEG projected onto the horizontal section 25 mm above the ear canals. The field extremes of the components projected onto the plane are shown along the perimeter of the skull. *Amyg*, amygdala; *Cb*, cerebellum; *Hp*, hippocampus; *LGN*, lateral geniculate body; *Op*, optic radiation; *Th*, thalamus; *VC*, visual cortex; *E*, location of the anterior position of the ear canal. (From Okada et al. 1983; by permission)

in epileptic patients (Halgren et al. 1980; Wood et al. 1980) (Fig. 9).

Another example of magnetic depth recording (Blum et al. 1984b) has recently been published. The authors registered the auditory evoked fields for the first time prenatally in the 35th fetal week, from a distance of 40-45 mm, and confirmed it postnatally in the newborn.

Current Sources of Epileptic Interictal Focal Spikes

The second method of obtaining a field distribution with a one-channel system is by using iterative focal epileptiform spike activity. Since there is no trigger coming from the equipment, the average process must be triggered by the spikes itself.

The Los Angeles biomagnetic group introduced this method (Barth et al. 1982) and were able to confirm a high localizing capability in their magnetic recordings when they interfaced their system for initial validation to X-ray or magnetic resonance imaging or positron emission computed tomography. Preliminary evidence from this group also indicates that the magnetic localization of the interictal epileptic spike focus in six patients correlates well with intracranial stereotaxic depth and with cortical EEG and with the other above-mentioned techniques (Barth et al. 1984b; Sutherling et al. 1985) (Fig. 10). This method even enables magnetic spatiotemporal propagation patterns to be detected between multiple adjacent epileptic foci (Barth et al. 1984a) (Fig. 11).

The Rome group has reported several times comparisons of MEG and EEG on patients with focal epilepsy (Modena et al. 1982; Ricci 1983). The authors stressed the high localizing power of the MEG compared with that of the EEG in respect to cortical epileptic foci. They also confirmed the finding of Hughes et al. (1977), who demonstrated a variable appearance of the spike-wave complex in the MEG compared with the EEG.

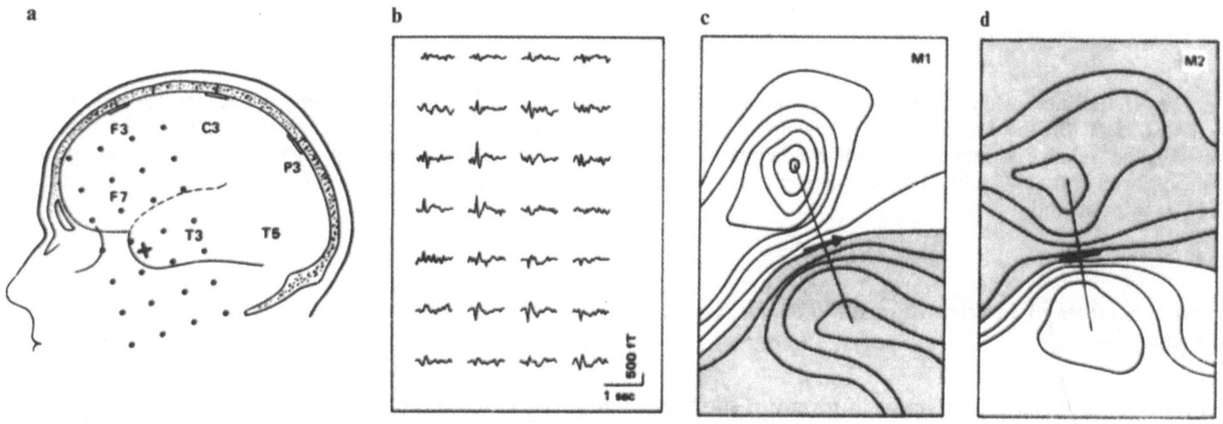

Fig. 10a-d. Magnetic fields of a focal interictal spike activity. **a** Rectangular MEG measurement matrix; the focus is cross-marked. **b** Average magnetic spike of each point of the matrix triggered by the EEG. **c, d** Isocontour maps showing the amplitude and the polarity distributions of two different generators M1 and M2. Calibration is 150 femto-tesla per line. (From Barth et al. 1982; by permission of the authors and the publisher. Copyright by the American Association for the Advancement of Science, 1982)

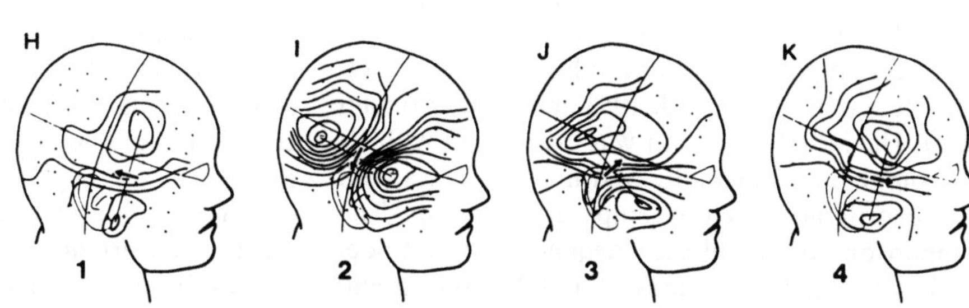

Fig. 11. Magnetic fields of two sequential equivalent magnetic sources of a focal temporal lobe spike. From *left to right* successive time intervals. *H*, first equivalent source with reversal of polarity at *K; I*, second source spatially discrete from the first with reversal in *J*. Calibration is 100 femto-tesla per line. (From Barth et al. 1984a; by permission of the authors and the publisher. Copyright by the American Association for the Advancement of Science, 1984)

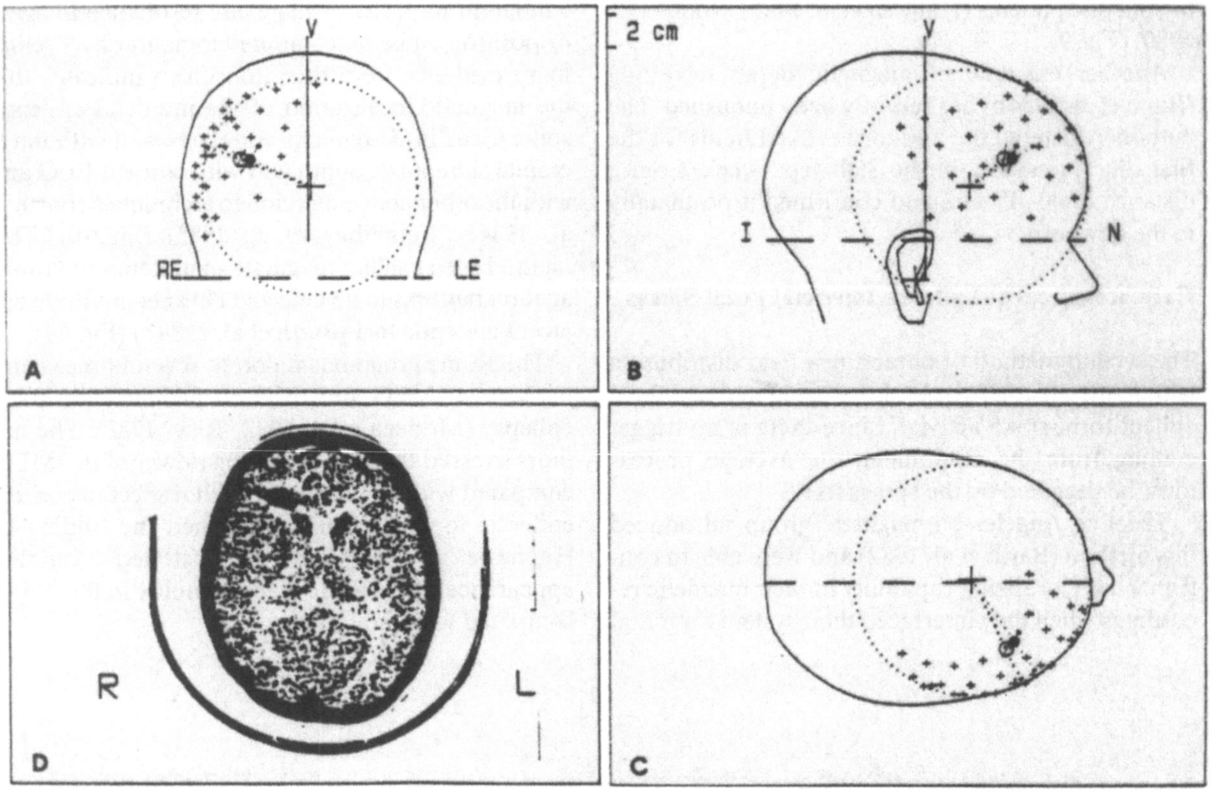

Fig. 12 A–D. Position of the current source of the interictal spike activity of a frontal lobe epileptic focus. Comparison of two methods: **A–C** The three-dimensional localization of the equivalent source is shown, an *open circle* representing the result obtained by the "relative covariance method" of Chapman et al. (1983, 1984), and a *closed circle* representing the spike average technique of Barth et al. (1984 a, b). **D** The equivalent pathological finding in a computerized X-ray tomography scan. (From Romani 1984; by permission)

Current Sources of Interictal Focal Nonspike Activity in Epileptics

When epileptic patients display high-amplitude interictal spikes, the above-mentioned spike average method of Barth et al. (1982) can be applied very well.

But for the numerous cases when interictally only a low-voltage and/or low-frequency focal activity can be found, Chapman and the Rome group proposed the so-called "relative covariance method" for the three-dimensional magnetic localization of epileptic foci (Chapman et al. 1983 and 1984; Romani 1984). The initial validation was carried out with X-ray computed tomography and in some cases with electrocortigraphy before temporal lobectomy. The comparison confirmed the adequacy of the model for foci located in the central, parietal, and occipital regions. The model seems partially sufficient for frontal areas and questionable for temporal lobe foci (Ricci et al. 1985 b). When interictal spike activity was investigated with the average method of Barth et al. (1982) and the "relative covariance method," the correspondence was high.

This method is used to locate the source of a chosen frequency band in the brain activity and obtain a temporal average of untriggerable activity. This will be reached by simultaneous recording of both the MEG and the EEG for a certain time. But the MEG is recorded in different positions sequentially, because of the channel limitation. Then the correlation between these signals will be studied. Variations in the covariance reflect the variation of the activity, provided that the amount of the activity remains constant. This will be accomplished by dividing the covariance of the signals of the magnetic and the electric channels by the variance of one chosen reference EEG signal. Negative or positive values of the "relative covariance coefficients" correspond to the out-of-phase or in-phase of magnetic and electrical signals, provided they come from the same source. When this procedure is repeated for all recording positions, with respect to the same EEG reference lead, a spatial distribution of the "relative covari-

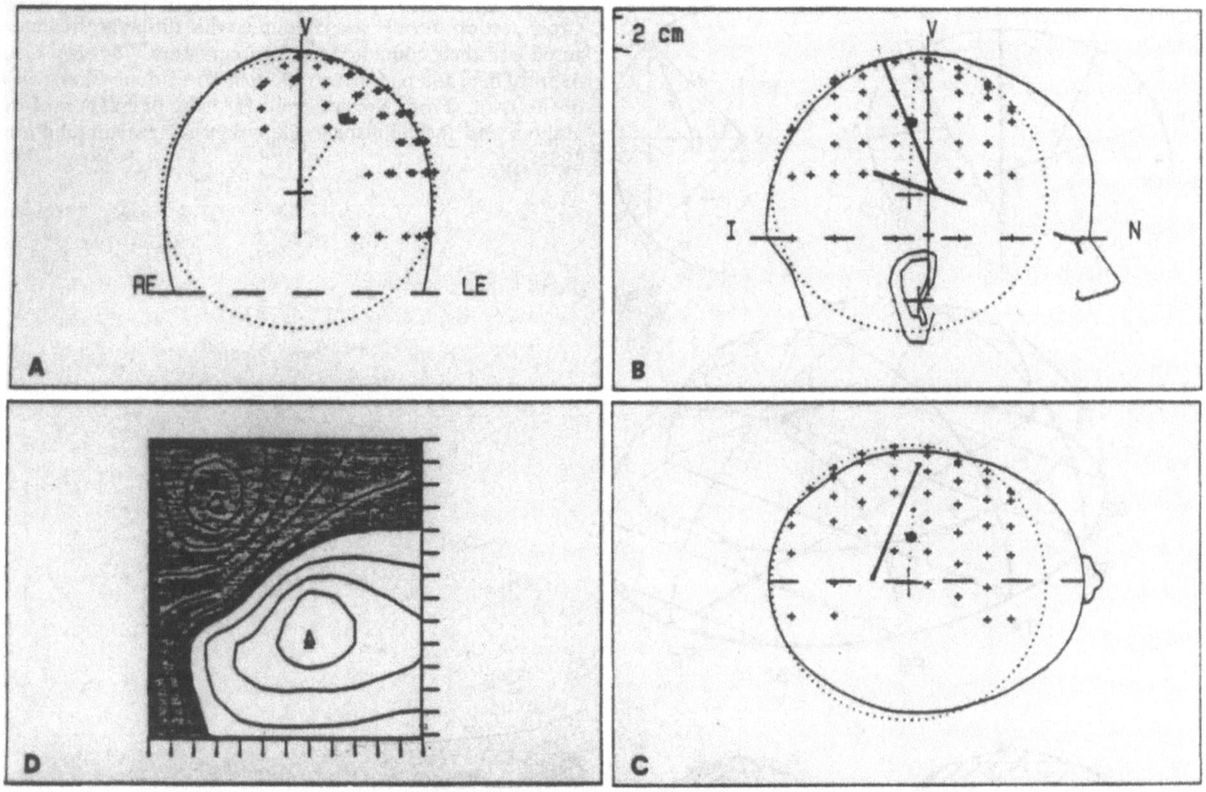

Fig. 13 A-D. Position of the current source of an interictal 2-Hz quasi-rhythmic activity of an epileptic focus. **A-C** The three-dimensional localization of the equivalent source shown with a *closed circle* obtained by the "relative covariance method" of Chapman et al. (1983, 1984). The approximate positions of the Sylvian and Rolandic fissures are represented by *solid lines*. **D** The spatial distribution of the relative covariance coefficient isolines over the Rolandic fissure (units in centimeters). (From Romani 1984; by permission)

ance coefficient" is provided, which reflects the spatial distribution of the magnetic field (Chapman et al. 1983, 1984; Romani 1984). Figure 12 shows a comparison between both methods when an interictal spike focus is available. A typical result of the "relative covariance method" alone is shown in Fig. 13.

Perspective

The technical development of the large multi-channel system still needs some time. But probably 40-50 channels will be ready for use in a couple of years. The recording time would thereby be reduced dramatically, because averaging and sequential recording is no longer needed for field determination. And this would be a benefit not only for the patient. In addition, simultaneous real-time recording is able to show the propagation of pathological activity, and probably much more real-time information, all very important especially for identifying the signs for single or multifocal activity. This has already been experienced with four- and five-channel systems (Romani 1984; Williamson et al. 1985).

More accuracy in localization of sources can be expected when the "inverse calculation" can be done on a more realistic model for the head than the usual assumed sphere. First attempts were made by Nicolas et al. (1985) with their piecewise circular model of the skull, as shown in Fig. 14. A further development was achieved by Meijs and Peters (1985) with a more realistically shaped multicompartment model of the head, calculated from nuclear magnetic resonance tomographs (Fig. 15).

Barth et al. (1985 b) have recently articulated the problems and the degree of error deriving from the nonsphericity of the MEG-measuring matrix about the human head. They compared the forward-calculated and the measured fields on the surface of a pure sphere and a human cranium coming from an artificial dipolar signal source in different positions and orientations. After using a more complete computer measuring procedure, adjusted for the non-

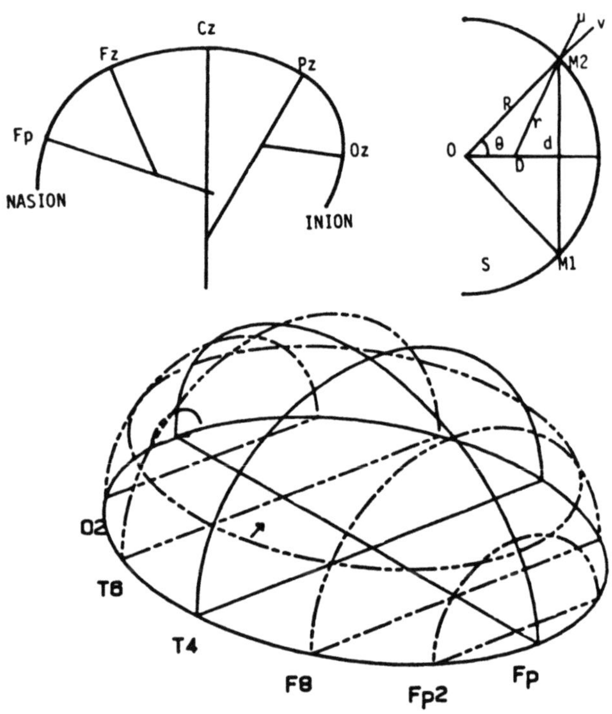

Fig. 14. Piecewise circular representation of the skull. *Top left* Cross section through the midline, with different arcs measured and their computed radii of curvature. *Top right* Geometrical data and parameters. *Bottom* Three-dimensional view of the skull. (From Nicolas et al. 1985; by permission of the authors and the publishers. Copyright by Pergamon Press, 1985)

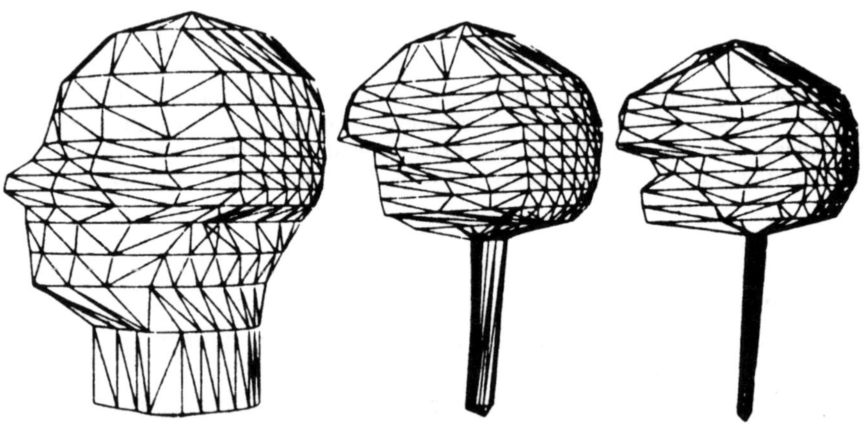

Fig. 15. Piecewise multicompartment model of the homogeneous compartments the scalp *(left)*, the skull *(middle)*, and the brain tissue *(right)*. Each surface is represented by *triangles*. Geometry of the compartments is based on the actual shape as found in nuclear magnetic resonance tomographic images. (From Meijs and Peters 1985; by permission)

spherical dimensions of the recording matrix, the accuracy of source localization was substantially improved.

Conclusions

We are just at the beginning of the development of an exciting new clinical diagnostic three-dimensional real-time tool to measure the electromagnetic activity of the brain. The localization of active focal epileptic sources in the brain is one of the most promising clinical applications. However, there are still some unsolved technical and theoretical problems.

References

Barth DS, Sutherling W, Engel J Jr, Beatty J (1982) Neuromagnetic localization of epileptiform spike activity in the human brain. Science 218: 891-894

Barth DS, Sutherling W, Engel J Jr, Beatty J (1984a) Neuromagnetic evidence of spatially distributed sources underlying epileptiform spikes in the human brain. Science 223: 293-296

Barth DS, Sutherling W, Engel J Jr, Beatty J (1984b) Neuromagnetic localization of single and multiple sources underlying epileptiform spikes in the human brain. In: Porter RJ, Mattson RH, Ward AA, Dam M (eds) Advances in epileptology: XVth epilepsy international symposium. Raven, New York, pp 379-384

Barth DS, Sutherling W, Beatty J (1985a) Animal neuromagnetometry and its specific application to the study of focal epileptic phenomena. In: Weinberg H, Stroink G, Katila T (eds) Biomagnetism: application and theory. Pergamon, New York, pp 237-248

Barth DS, Sutherling W, Broffman J, Beatty J (1985b) Magnetic localization of a dipolar current source implanted in a sphere and a human cranium. Electroencephalogr Clin Neurophysiol 63: 260-273

Baule G, McFee R (1963) Detection of the magnetic field of the heart. Am Heart J 66: 95-96

Beatty J, Richer F, Barth DS (1983) Magnetoencephalography. In: Coles MGH, Porges SW, Donchin E (eds) Psychophysiology: systems, processes and applications. Guilford, New York

Blum T, Bauer R, Kubicki St (1984a) Registrierung visuell evozierter neuromagnetischer Felder nach fovealen Partialfeldstimulationen. EEG-EMG 15: 27-33

Blum T, Saling E, Bauer R (1984b) Fetale Magnetoenzephalographie I: Erstmalige pränatale Registrierung eines auditorisch evozierten magnetischen Feldes. EEG-EMG 15: 34-37

Carelli P, Foglietti I, Modena I, Romani GL (1983) Magnetic study of the spontaneous brain activity of normal subjects. Il Nuovo Cimento 2D: 538-546

Chapman RM, Romani GL, Barbanera S, Leoni R, Modena I, Ricci GB, Campitelli F (1983) SQUID instrumentation and the relative covariance method for magnetic 3D localization of pathological cerebral sources. Lett Nuovo Cimento (2) 38: 549-554

Chapman RM, Ilmoniemi RJ, Barbanera S, Romani GL (1984) Selective localization of alpha brain activity with neuromagnetic measurements. Electroencephalogr Clin Neurophysiol 58: 560-572

Cohen D (1972) Magnetoencephalography: detection of the brain's electrical activity with a superconducting magnetometer. Science 175: 664-666

Cohen D, Cuffin BN (1983) Demonstration of useful differences between magnetoencephalogram and electroencephalogram. Electroencephalogr Clin Neurophysiol 56: 38-51

Cohen D, Edelsack EA, Zimmerman JE (1970) Magnetocardiograms taken inside a shielded room with a superconducting point-contact magnetometer. Appl Phys Lett 16: 278-280

Cuffin BN, Cohen D (1979) Comparison of the magnetoencephalogram and electroencephalogram. Electroencephalogr Clin Neurophysiol 47: 131-146

Erné SN (1983) SQUID sensors. In: Williamson SJ, Romani GL, Kaufman L, Modena I (eds) Biomagnetism, an interdisciplinary approach, vol 66. NATO advanced sciences institutes series. Plenum, New York, pp 69-84

Geselowitz DB (1970) On the magnetic field generated outside an inhomogenous volume conductor by internal current sources. IEEE Trans Mag MAG-6: 346-347

Grynszpan F, Geselowitz DB (1973) Model studies of the magnetocardiogram. Biophys J 13: 911-925

Halgren E, Squires NK, Wilson CS, Rohrbaugh JW, Babb TL, Crandall PH (1980) Endogenous potentials generated in the human hippocampal formation and amygdala by infrequent events. Science 210: 803-805

Hughes JR, Hendrix DE, Cohen J, Duffy FH, Mayman CI, Scholl ML, Cuffin BN (1976) Relationship of the magnetoencephalogram to the electroencephalogram. Normal wake and sleep activity. Electroencephalogr Clin Neurophysiol 40: 261-278

Hughes JR, Cohen J, Mayman CI, Scholl ML, Hendrix DE (1977) Relationship of the magnetoencephalogram to abnormal activity in the electroencephalogram. J Neurol 217: 93-97

Kaufman L, Williamson SJ (1982) Magnetic location of cortical activity. Ann N Y Acad Sci 388: 197-213

Meijs JWH, Peters MJ (1985) Computation of MEGs and EEGs using a realistic shaped multi-compartment model of the head. J Int Fed Med Biol Engineering, XIV ICMBE and VII ICMP, Espoo, Finland, pp 36-37

Modena I, Ricci GB, Barbanera S, Leoni R, Romani GL, Carelli P (1982) Biomagnetic measurements of spontaneous brain activity in epileptic patients. Electroencephalogr Clin Neurophysiol 54: 622-628

Nicolas P, Duret D, Teszner D, Tuomisto T (1983) Neuromagnetic measurements at hospital: instrumentation and preliminary tests. Il Nuovo Cimento 2D: 184-194

Nicolas P, Germain V, Thomas T, Duret D (1985) MEG representation based on piecewise circular models. In: Weinberg H, Stroink G, Katila T (eds) Biomagnetism: application and theory. Pergamon, New York, pp 294-298

Okada Y (1983a) Neurogenesis of evoked magnetic fields. In: Williamson SJ, Romani GL, Kaufman L, Modena I (eds) Biomagnetism, an interdisciplinary approach, vol 66. NATO advanced science institutes series. Plenum, New York, pp 399-408

Okada Y (1983b) Endogenous magnetic fields. In: Williamson SJ, Romani GL, Kaufman L, Modena I (eds) Biomagnetism, an interdisciplinary approach, vol 66. NATO advanced sciences institutes series. Plenum, New York, pp 460-468

Okada Y (1983c) Somatic evoked magnetic fields. In: Williamson SJ, Romani GL, Kaufman L, Modena I (eds) Biomagnetism, an interdisciplinary approach, vol 66. NATO advanced sciences institutes series. Plenum, New York, pp 409-421

Okada YC, Kaufman L, Williamson SJ (1983) The hippocampal formation as a source of the slow endogenous potential. Electroencephalogr Clin Neurophysiol 55: 417-426

Plonsey RW (1981) The magnetic field resulting from action currents on cylindrical fibers. Med Biol Eng Comp 19: 311-315

Reite M, Zimmerman JE, Edrich J, Zimerman J (1976) The human magnetoencephalogram: some EEG and related correlations. Electroencephalogr Clin Neurophysiol 40: 59-66

Ricci GB (1983) Clinical magnetoencephalography. Il Nuovo Cimento 2D: 517-537

Ricci GB, Buonomo S, Peresson M, Romani GL, Salustri C, Modena I (1985) Multichannel neuromagnetic investigation of focal epilepsy. J Int Fed Med Biol Engineering, XIV ICMBE and VII ICMP, Espoo, Finland. pp 42-43

Ricci GB, Leoni R, Romani GL, Campitelli F, Buonomo S, Modena I (1985b) 3-D neuromagnetic localization of sources of interictal activity in cases of focal epilepsy. In: Weinberg H, Stroink G, Katila T (eds) Biomagnetism: application and theory. Pergamon, New York, pp 304-310

Romani GL (1984) Biomagnetism: an application of SQUID sensors to medicine and physiology. Physica 126B: 70-81

Romani GL, Williamson SJ, Kaufman L (1982) Biomagnetic instrumentation. Rev Sci Instrum 53: 1815-1845

Romani GL, Williamson SJ, Kaufman L (1982a) Tonotopic organization of the human auditory cortex. Science 216: 1339-1340

Romani GL, Williamson SJ, Kaufman L, Brenner O (1982b) Characterization of the human auditory cortex by the neuromagnetic method. Exp Brain Res 47: 381-398

Sutherling W, Barth DS, Beatty J (1985) Magnetic fields of epileptic spike foci: equivalent localization and propagation. In: Weinberg H, Stroink G, Katila T (eds) Biomagnetism: application and theory. Pergamon, New York, pp 249-260

Tripp JH (1983) Physical concepts and sources. In: Williamson SJ, Romani GL, Kaufman L, Modena I (eds) Biomagnetism, an interdisciplinary approach, vol. 66. NATO advanced sciences institutes series. Plenum, New York, pp 101-139

Vieth J (1970) Biopotentiale. Elektromedizin 15: 99-111

Vieth J (1984) Die Magnetoenzephalographie, eine neue funktionsdiagnostische Methode. EEG - EMG 15: 111-118

Weinberg H, Stroink G, Katila T (1985) Biomagnetism: applications and theory. Pergamon, New York

Wikswo JP Jr (1983) Cellular action currents. In: Williamson SJ, Romani GL, Kaufman L, Modena I (eds) Biomagnetism, an interdisciplinary approach, vol 66. NATO advanced sciences institutes series. Plenum, New York, pp 173-207

Williamson SJ, Kaufman L (1981a) Magnetic fields of the cerebral cortex. In: Erne SN, Hahlbohm HD, Luebbig H (eds) Biomagnetism. de Gruyter, Berlin, pp 353-402

Williamson SJ, Kaufman L (1981b) Biomagnetism. J Magnetism Magn Mat 22: 129-201

Williamson SJ, Kaufman L (1983) Application of squid sensors to the investigation of neural activity in the human brain. IEEE Trans Mag MAG-19: 835-844

Williamson SJ, Kaufman L (1985) Frontiers in the new science of biomagnetism. In: Weinberg H, Stroink G, Katila T (eds) Biomagnetism: application and theory. Pergamon, New York, pp 471-490

Williamson SJ, Romani GL, Kaufman L, Modena I (eds) (1983) Biomagnetism, an interdisciplinary approach, vol 66. NATO advanced sciences institutes series. Plenum, New York

Williamson SJ, Pellinzone M, Okada Y, Kaufman L, Crum DB, Marsden JR (1985) Five channel SQUID installation for unshielded neuromagnetic measurements. In: Weinberg H, Stroink G, Katila T (eds) Biomagnetism: application and theory. Pergamon, New York, pp 46-51

Wood CC, Allison T, Goff WR, Williamson PD (1980) On the neural origin of P300 in man. In: Kornhuber HH, Deecke L (eds) Motivation, motor, and sensory processes of the brain. Elsevier, Amsterdam, pp 51-56

Zimmerman JE (1971) Sensitivity enhancement of superconducting quantum interference devices through the use of fractional-turn loops. J Appl Phys 42: 4483-4487

Zimmerman JE (1983) Cryogenics. In: Williamson SJ, Romani GL, Kaufman L, Modena I (eds) Biomagnetism, an interdisciplinary approach, vol 66. NATO advanced sciences institutes series. Plenum, New York, pp 43-67

Magnetic Resonance Imaging and the Epileptic Focus

A. Kertesz[1], R. S. McLachlan[2], S. E. Black[1], R. L. Nicholson[3], T. Carr[3], and W. T. Blume[2]

Nuclear magnetic resonance provides us with a new imaging modality which can be used not only to analyze the structure and structural pathology, but also to some extent the pathophysiology of the central nervous system. A large magnetic field and radio frequency pulse sequences are used to generate a signal consisting of several basic parameters of nuclear magnetic resonance: (1) proton density, (2) T_1 constant or spin-lattice (longitudinal) relaxation time, and (3) T_2 constant or spin-spin (horizontal) relaxation time. The resulting signal can be weighted for each of the parameters by altering the radio frequency pulse sequences. The most commonly used sequences are called inversion recovery (IR), which emphasizes T_1 constant and provides gray-white matter contrast, and spin echo (SE), which emphasizes T_2 constant and shows pathology. The signal is spatially encoded and computerized in a manner similar to computerized tomography (CT). For a detailed description of the technique, see Lauterbur (1973) and Pykett et al. (1982).

Magnetic resonance imaging (MRI) is a noninvasive technique which does not require the use of ionizing radiation. The ability to obtain high-resolution images in any plane, such as coronal and sagittal, allows better analysis of neuroanatomical relations than axial images alone. MRI is of particular interest to the study of neurological disease because of superior tissue contrast allowing better visualization of gray-white contrast and brain-CSF contrast. Variations in the proton MRI longitudinal and spin-spin relaxation times in biological tissue allows us to probe the nature of the molecular environment that is not seen in any other imaging modality. Good correlation between the relaxation times and water content has been demonstrated in tumor tissue and in stroke (Spetzler et al. 1983), and in demyelinating diseases of the brain (Lukes et al. 1983). Epilepsy, in contrast to these pathologically well defined conditions, is often a disorder of neuronal cell function in the absence of gross anatomical changes. However, seizures are accompanied by a number of morphological, physiological, and biochemical changes, as well as by several pathological features which could produce altered MRI properties.

In the investigation of focal epilepsies in consideration for surgical resection, pneumoencephalography and CT scanning show abnormalities in only about one-third of such cases (Engel et al. 1982; Wyler and Bolender 1983). Computer tomography with metrizamide enhancement increases the preoperative yield of temporal lobe abnormalities (Wyler and Bolender 1983) as does positron emission tomography (PET) (Engel et al. 1982). Most of these abnormalities are changes related to atrophy: enlarged lateral ventricle, enlarged sulci, and small temporal lobe. CT scanning is less able to detect small lesions in the temporal lobe than in other cortical regions due to the bone artifacts from middle cranial fossa edges. Unlike CT, such bone "artifact" does not appear on MRI. The former procedure requires intravenous contrast and only adds information about the edges of the temporal lobes without adding any direct information concerning temporal lobe tissue changes. The latter is only available at a few centers.

Our current results from MRI on 16 patients with uncontrolled temporal lobe epilepsy (TLE) and 10 normal controls suggest that this procedure is probably more effective than CT scanning in revealing atrophic changes in the temporal lobe in a selected population (McLachlan et al. 1984). Specifically, MRI reveals asymmetries in temporal lobe size, mesial temporal margins, temporal horns, and hemispheres as has also been shown by others (Sussman et al. 1984; Spencer et al. 1984, Abou Khalil et al. 1984; Riela et al. 1984; Schoerner et al. 1984). Theodore et al. (1985) compared CT, PET, and MRI.

[1] Department of Clinical Neurological Sciences, St. Joseph's Hospital, Research Institute, London, Ontario N6A 4V2, Canada

[2] Department of Clinical Neurological Sciences, Epilepsy Unit, University Hospital, London, Ontario N6A 5A5, Canada

[3] Department of Nuclear Medicine, St. Joseph's Hospital, London, Ontario N6A 4V2, Canada

Presurgical Evaluation of Epileptics
Edited by H. G. Wieser and C. E. Elger

Twenty-six patients had all three tests; ten had only CT and MRI. Seventeen patients had a localized EEG focus. PET was the most commonly positive modality (13 out of 14 cases). MRI was abnormal in nine and CT in four. In three out of eight patients with unilateral EEG focus, PET hypometabolism, and normal CT, the SE showed increased signal intensity; in one case, MRI had increased signal but normal PET; in another case, the signal was increased on the opposite side of the focus. Other reports of MRI in epilepsy are anecdotal or descriptive and none of them have as yet used a "blind" evaluative technique.

Subjects

We selected 16 patients, aged 17–54 years, who were surgical candidates for resection of a seizure focus on the basis of having uncontrolled, focally originating seizures from the temporal lobe, despite the use of adequate medications, as defined by McNaughton and Rasmussen (1975). Patients with gross lesions such as infarcts, haemorrhages, arteriovenous malformations, or tumors were also studied as a separate group.

Patients being considered for surgery undergo extensive investigation to verify the seizure focus, based on the recommendations of the Montreal group (McNaughton and Rasmussen 1975). Computer-controlled EEG telemetry provides us with both interictal and ictal data to localize the focus of electrical discharge, this being the primary standard. In some patients, subdural electrodes were inserted and left in place on the surface of the brain better to demonstrate the seizure focus. Neuropsychological assessment was also performed to localize cortical areas of functional deficit. Finally, skull X-rays and CT scans were obtained on all patients. Routine contrast enhancement was done in all cases and high-volume delayed contrast-enhanced scans are carried out in selected individuals with suspicious, but indefinite, changes using the standard procedures.

Imaging

Magnetic resonance imaging was carried out on each patient using a Technicare 0.15-tesla imager using a 256 × 128 matrix (pixel size 1.1 × 2.2 mm) displayed as a 256 × 256 image. Initially, a 28-cm-ID cylindrical receive/transmit radiofrequency coil was employed to obtain images approximately −15° to the orbitomeatal line according to the following protocol: the brain was surveyed using an anisotropic volume SE technique to obtain 32 T_2-weighted images in the axial plane [slice thickness (SLT) 1.7 cm; echo time (TE), 60; cycle repetition time (TR), 1000; all time measurements are given in milliseconds] and 20 T_1-weighted images in the coronal plane (SLT, 1.7 cm; TE, 30, TR, 250). Additional single axial slices through the middle of the temporal lobes were also collected by both SE (TE, 120; TR, 1000) and IR techniques (inversion time, 450; TE, 30; TR, 1500). Total imaging time was 1 h.

Installation of new hardware and software later on in this study, including a half-saddle (25-cm-ID) receive coil, allowed multiecho multislice collections which made possible a more comprehensive survey with better image quality in a shorter time. In the new protocol, the SE technique was used to yield 13 T_1- and T_2-weighted images in the coronal plane (SLT, 1 cm; TE, 30, 60; TR, 1060) and 15 T_2-weighted images in the horizontal plane (SLT, 0.75 cm; TE, 60, 120; TR, 2120) in approximately 30 min of imaging time. Half of the TLE patients and all control subjects were imaged using this protocol.

The MRI findings of the patients were compared with those of a normal control group consisting of 10 volunteers from hospital personnel (age range 20–40 years) and a control group of 18 multiple sclerosis (MS) patients. We used two independent observers who judged the scans for atrophy and for the presence or absence and the location of the T_1 or T_2 changes on the IR and SE images. These judges viewed all the scans in randomized order, in other words, they did not know if it was an epileptic or a control subject. They were also blind to the side of the seizure focus or the clinical type of the seizure. Disagreements were resolved by joint viewing. Judgments concerning anatomical symmetries in the size of the temporal lobes, the temporal horns, the mesial temporal borders, the Sylvian fissures, and the hemispheres were made on both coronal (TE, 30) and axial (TE, 60) slices. In addition, any increase in signal intensity on the T_2-weighted axial images was classified as localized or diffuse throughout the temporal lobe. These changes were further categorized as unilateral or bilateral. Comparisons of these findings were made in a nonblinded fashion to those obtained from 18 patients with MS. The results were then interpreted with respect to the electrographic and CT (GE 8800; SLT, 0.5 cm) findings in the TLE patients.

In the second phase of our study presently under way, we project these images by using a photoenlarger after randomly reversing the sides. The projected images are traced by two observers who are blind to the side of the image, the identity of the subject, whether it is a patient or a control, and the location of the seizure focus. After tracing the size of the cross-sectional area of the anterior and middle portions of the temporal lobes on the coronal cuts, the total hemispheric area on the axial cut above the third ventricles is determined and used to provide a temporal lobe-hemisphere ratio.

Results

The most convincing finding was a smaller temporal lobe in 12 out of 16 of the TLE cases, but only 1 of the 10 controls; and 4 out of 16 mesial temporal atrophy TLE cases and none of the controls (Table 1). An example of temporal lobe atrophy is illustrated in Fig. 1. In addition, four epileptics and no controls had mild hemispheric asymmetry (Fig. 2). Four TLE patients (25%) had no asymmetry of any kind. A larger temporal horn on one or the other side was found in 11 out of 16 TLE patients (Fig. 1) but also in 9 out of 18 MS patients and a surprising 6 out of 9 controls. In nine patients, the appearance of the temporal horns was in agreement with the MRI findings while in two cases they were seen only with MRI. The large temporal horn was often bilateral (6/11) and therefore less specific (Table 1).

Seizures were localized by ictal EEG recordings (scalp and sphenoidal electrodes in ten and subdural electrodes in two patients) to either right or left temporal lobes in eight patients and to both temporal lobes independently in four patients. In four other

Table 1. MRI and CT findings in temporal lobe epilepsy and controls

	TLE (16)	Controls (10)	TLE-CT (16)	MS (18)
Smaller temporal lobe	12	1	7	0
Mesial lobe atrophy	4	0	1	1
Smaller hemisphere	4	0	0	0
Large temporal horn	11[a]	6	9	7
Increased T_2 signal	8[b]	0	-	8[c]

[a] Bilateral in six
[b] Bilateral in three
[c] See text for description

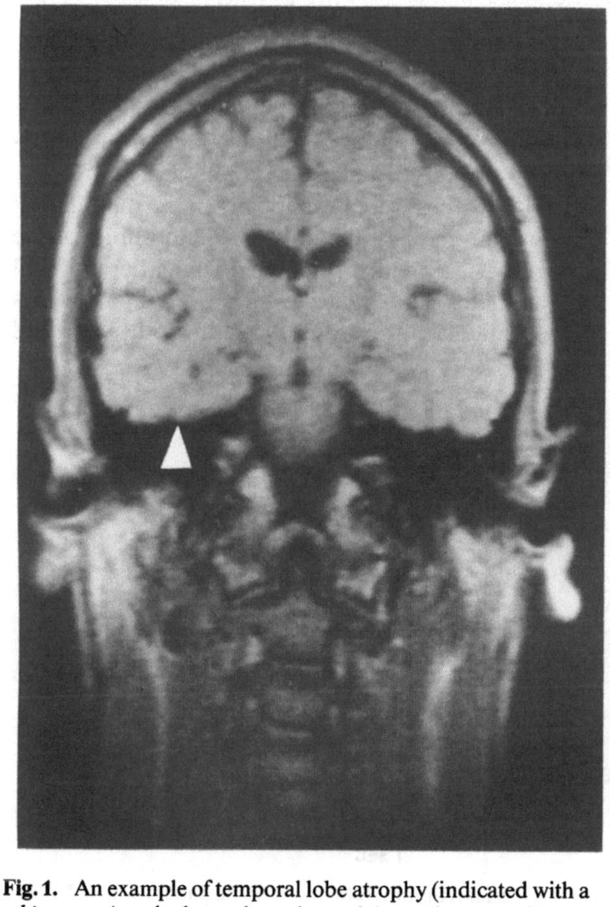

Fig. 1. An example of temporal lobe atrophy (indicated with a *white arrow*) and a larger lateral ventricle on the same side

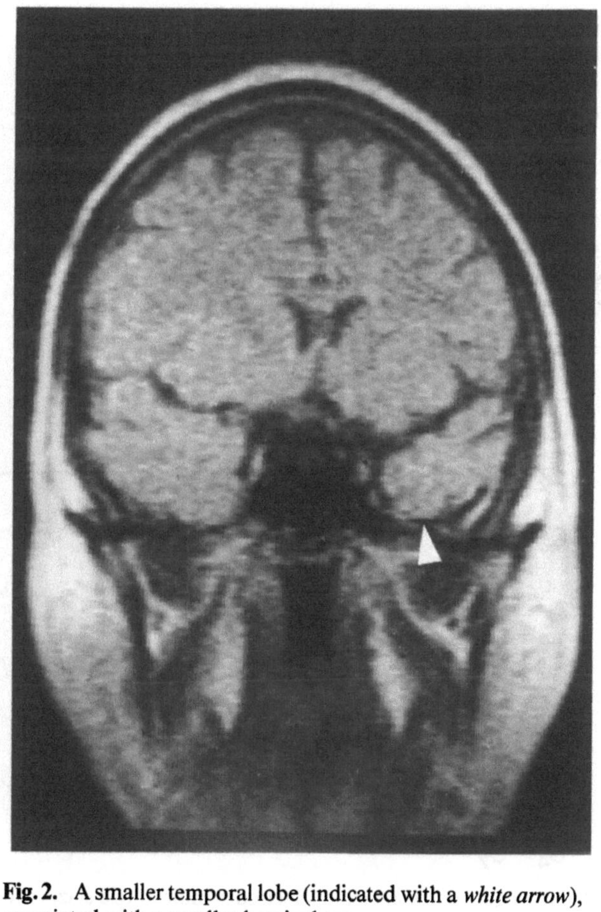

Fig. 2. A smaller temporal lobe (indicated with a *white arrow*), associated with a smaller hemisphere

Table 2. MRI findings and seizure focus

MRI findings	Unilateral focus		Bilateral focus	
	Same side	Opposite	Predominant	Opposite
Smaller temporal lobe	7	0	3	2
Mesial lobe atrophy	2	0	2	0
Smaller hemisphere	2	0	2	0
Larger temporal horn	2	1	1	1
Increased T_2 signal	3	1	4	0

patients, seizures were not recorded but interictal data strongly supported a right temporal origin in one and independent bitemporal origin in three. In the seven patients with independent bitemporal foci, ictal and interictal activity predominated in the right temporal lobe in four and was bitemporally, equal, in three. The smaller temporal lobe corresponded well with the location of the seizure focus on EEG, and with a higher positivity rate on the MRI. The smaller mesial temporal lobe was especially better seen on the MRI with high correspondence with the seizure focus.

Table 2 shows the distribution of the location and the correspondence of MRI findings with the seizure. In the unilateral instances of atrophy the correspondence was good with both CT and MRI, but the CTs were interpreted retrospectively (nonblindly). However, the asymmetries were better delineated by MRI primarily because the absence of bone artifacts allowed a clearer definition of the temporal lobe outlines. Asymmetries shown by MRI were not apparent in the CT scans of five patients. CT was also less likely to reveal asymmetries in mesial temporal structures or hemisphere size.

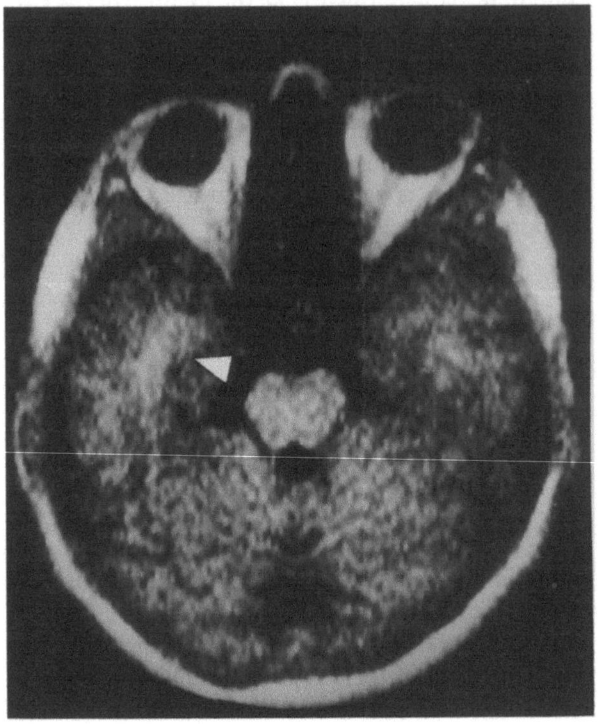

Fig. 3. Increased signal on T_2-weighted spin echo image (indicated with a *white arrow*) corresponding to the mesial temporal focus in Fig. 4

Among the focal MRI signal changes, the T_2-weighted SE techniques showed the most consistent abnormality. Increased signals indicating prolonged T_2 values were seen focally in the mesial temporal region in eight patients (50%) and none of the controls. These focal T_2 changes were either unifocal (five patients) or bilateral in homologous areas (three patients). They were seen in both axial and

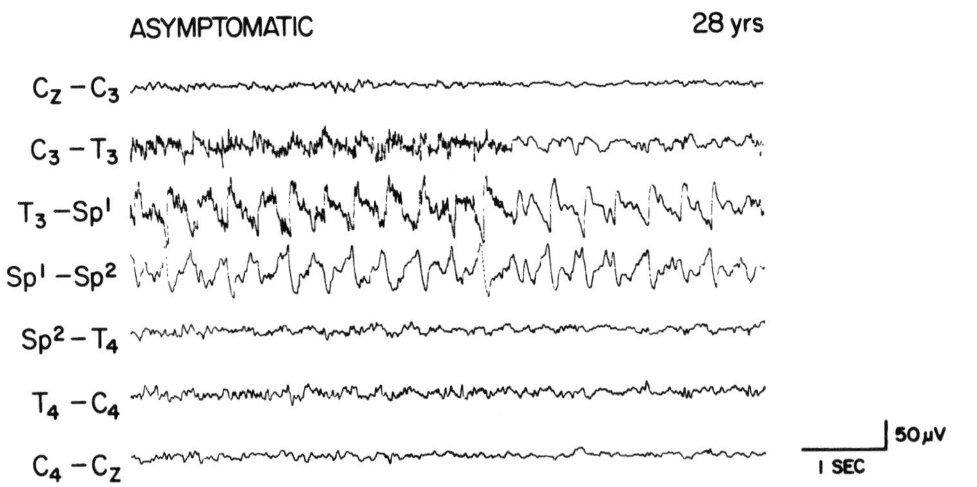

Fig. 4. EEG with sphenoidal electrodes indicating a seizure focus in the mesial temporal lobe in the same patient whose MRI is illustrated in Fig. 3

coronal projections. Only in one of the patients was the increased T_2 signal opposite to the unilateral seizure focus.

Two patients with continuous interictal spiking (30–60/min), at one sphenoidal electrode on the same day as the MRI study, had the most prominent focal increases in T_2 corresponding to the active focus. Figure 3 represents a typical increase in signal caused by prolonged T_2 relaxation time in an epileptic and corresponds to the mesial temporal focus found on the EEG (Fig. 4).

In an additional control group, nonblinded interpretations were made from the MRI of 18 MS patients. Neither temporal lobe nor hemisphere asymmetries were seen and only one patient had asymmetry in the size of the mesial temporal structures (Table 1). Temporal horns were visible in 12 patients and were thought to be "enlarged" in seven of these. In evaluating T_2-weighted signal changes, the MS patients represented an interesting control group because a significant number had some T_2 alteration on one side or the other. These, however, tended to be multifocal white matter and central periventricular alterations in contrast to the more diffuse and cortical changes in the TLE group.

Sufficient T_1 weighting of images was obtained in only seven patients. Because of lack of data in the other patients and controls, T_1 was not assessed in the blinded study. However, three of the seven showed a unilateral increase in T_1 in the temporal lobe white matter. In two of these three, T_2 was increased on the same side but in a different location involving the mesial temporal region. Figure 5 illustrates a patient who had a well-circumscribed left temporal T_1 change unilaterally confirmed by CT and EEG. The decreased signal in this IR sequence signifies prolonged T_1 constant at the area of involvement.

There are many instances where MRI shows an actual lesion as well or even better than CT. At times a temporal lobe seizure discharge will be produced by a more distant lesion, in this instance, an occipital lobe tumor (Fig. 6). The advantage of MRI, even when the CT is positive, is seen in this instance of a low-grade temporal lobe astrocytoma, which was shown on the CT but not as well as on the MRI, both on IR and the SE (Fig. 7).

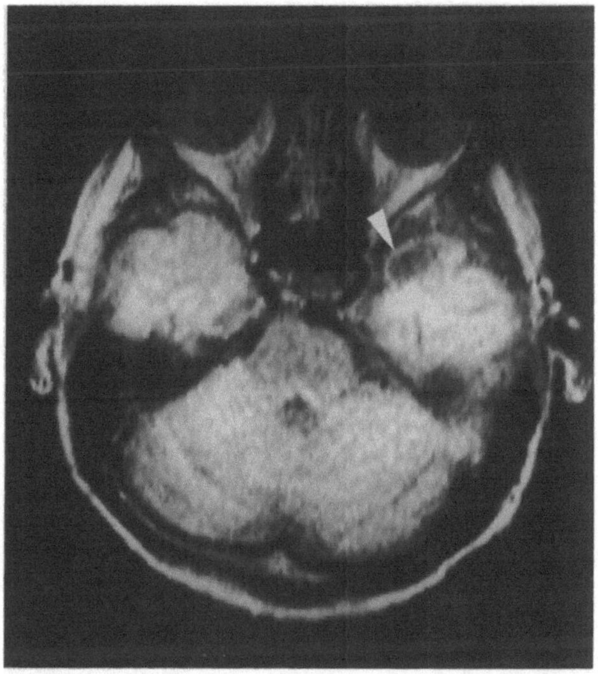

Fig. 5. The *white arrow* indicates decreased signal in a T_1-weighted inversion recovery image at the mesial temporal lobe

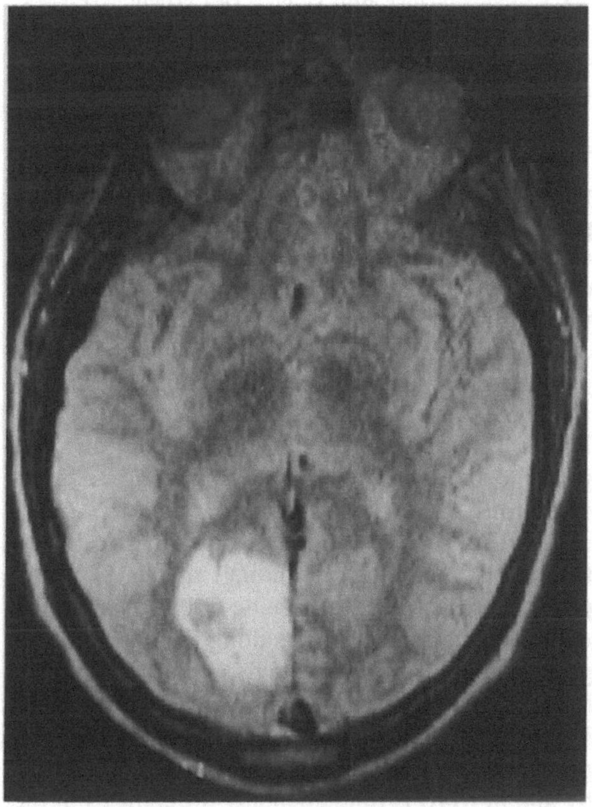

Fig. 6. Magnetic resonance imaging of occipital lobe tumor with temporal lobe seizure discharge

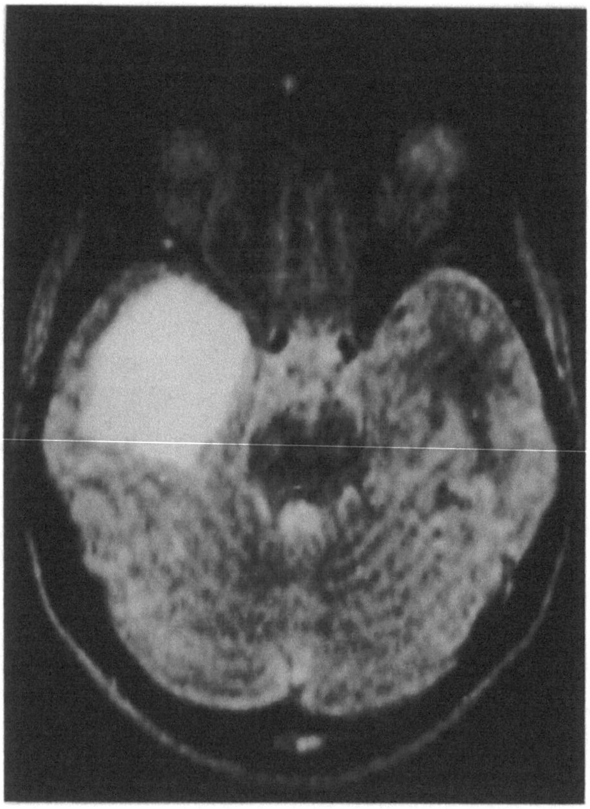

Fig. 7. Temporal lobe astrocytoma is clearly outlined on MRI

Discussion

These results suggest that the most reliable accompaniment of a temporal lobe focus on MRI is temporal lobe atrophy on coronal cuts. These are changes that may not be seen on CT. The reliability of this finding needs to be confirmed by quantification of a larger number of cases that includes tracing of temporal lobe areas blind to the side of the subject. Temporal horn asymmetry did not appear to be a reliable feature. There are also the areas of prolonged T_1 and T_2 signals that are related to the tissue proton content, pH, and the molecular environment of protons. We can only speculate about the nature of this alteration that appears on MRI. We know from clinicopathological findings that mesial temporal sclerosis (MTS) can be described as a combination of neuronal loss, resulting in induration of gray matter and gliosis due to hyperplasia of interstitial glial connective tissue. This typically involves the hippocampus but similar changes occur in the temporal neocortex as well. The replacement of neurons by glia in MTS is fundamentally different from the gliosis observed in MS plaques. In MS, the glia replaces myelin in the white matter rather than neurons as in MTS. The MRI features of demyelination are well known. The increased T_2 and T_1 signals appear mostly periventricular and focal, not specifically cortical as in MTS.

Apart from the morphological changes representing gliosis, there are other possible explanations of the increased signal on SE and the decreased signal on the IR images of epileptic brains. Experimental seizures result in increased permeability of the blood brain barrier at the seizure focus (Bakay 1956; Lorenzo et al. 1972). Prensky et al. (1973) found positive isotope brain scans in the area of repeated focal seizures, presumably reflecting a transient increase in blood brain barrier permeability. Transient focal lesions characteristic of edema have also been demonstrated by CT following focal seizures, and interictal seizure foci have been found to enhance slightly more than surrounding tissue (Heinz et al. 1983). There are complex alterations in water flow from the extracellular to the intracellular space as a result of ionic movement, and thus a reduction in the extracellular space in the acute experimental seizure focus (Lorenzo et al. 1972). In vivo MRI can detect changes in brain water content as early as 2 h after its induction in experimental animals. Relaxation times (T_1 and T_2) of the brain edema usually exceed that of gray and white matter. Our clinical studies of brain edema in stroke indicate better MRI sensitivity and much better contrast resolution than that in CT (Kertesz et al. 1985). Since MRI can give significant information about changes in brain water content, this procedure may actually allow imaging of the active seizure focus itself.

When both the anatomical asymmetries and the alterations in MRI signal are considered, 14 of the 16 TLE cases showed abnormalities relevant to the side of the seizure focus. This combination of features is clinically more productive than the individual features considered alone.

Temporal lobectomies were performed on eight of these TLE patients by Dr. J. Girvin. The atrophic changes on the MRI were confirmed by the surgical findings. The microscopic pathology, however, was not specific for the altered MRI signal. Although gliosis was present in the area of increased T_2 signal in four patients, it was not different than that seen in other areas in the same patients and in patients without T_2 changes (McLachlan et al. 1984).

In conclusion, our studies of focal temporal lobe epilepsy with MRI indicate that the most consistent finding is lobar atrophy, more reliably and obviously seen than on CT. Less frequently increased signal in the temporal lobe on SE technique, and at times decreased signal with IR pulse sequences, are seen, but the combination of atrophy and the signal changes appears diagnostic in a significant portion of cases. MRI is also quite effective in detecting glial tumors and at times shows more extensive or better-outlined lesions than CT. Presently, MRI shows more abnormality than CT and somewhat less than PET scans, but direct comparisons and more quantification are still needed. Future use of MRI in epilepsy may incorporate other creative pulse sequences, actual volume measurements, and the calculation of regional T_1, T_2, and proton density ratios. Spectroscopy will allow further probing of the biochemical changes intra- and interictally, and paramagnetic contrast agents may further help to outline some pathological alterations.

References

Abou Khalil BW, Sackellares JC, Latack JT, Vanderzant CN (1984) Magnetic resonance imaging in refractory partial epilepsy. Epilepsia 25: 650

Bakay L (1956) The blood brain barrier. Thomas, Springfield, 154 pp

Engel J, Brown WJ, Kuhl DE, Phelps ME, Mazziotta JC, Crandall PH (1982) Pathological findings underlying focal temporal lobe hypometabolism in partial epilepsy. Ann Neurol 12: 518-528

Heinz ER, Jabaily J, Drayer BP, Osbourne D (1983) Transient hyperdense focal cortial abnormalities on unenhanced and enhanced CT scans: beware. J Comp Assist Tomogr 7: 195

Kertesz A, Black SE, Cooper PE, Stewart SE, Nicholson LR, Shewan C (1985) The use of MRI in stroke. Ann Neurol 18 (11): 121

Lauterbur PC (1973) Image formation by induced local interactions: examples employing NMR. Nature 242: 190-191

Lorenzo AV, Shirahige IU, Liang M, Barlow CF (1972) Temporary alteration of cerebrovascular permeability of plasma protein during drug-induced seizures. Am J Physiol 223: 168-177

Lukes SA, Crooks LE, Aminoff MS, Kaufman L, Panitch HS, Mills C, Norman D (1983) Nuclear magnetic resonance imaging in multiple sclerosis. Ann Neurol 13: 592-601

McLachlan RS, Black SE, Nicholson RL, Blume WT (1984) Nuclear magnetic resonance (NMR), a new approach to imaging in temporal lobe epilepsy. Annual meeting of American epilepsy society, San Francisco, November 6-11

McNaughton FL, Rasmussen T (1975) Criteria for selection of patients for neurosurgical treatment. In: Purpura DP, Penry JK, Walter RD (eds) Neurosurgical management of the epilepsies. Raven, New York, pp 37-48

Prensky AL, Swisher CN, DeVivo DC (1973) Positive brain scans in children with idiopathic focal epileptic seizures. Neurology 23: 798-807

Pykett IL, Newhouse JH, Buonanno FS, Brady TJ, Goldman MR, Kistler JP, Pohost GM (1982) Principles of nuclear magnetic resonance imaging. Radiology 143: 157-168

Riela AV, Penry JK, Laster DW, Schartze M (1984) Magnetic resonance imaging versus computerized cranial tomography in complex partial seizures. Epilepsia 25: 650

Schoerner W, Felix R, Claussen C, Meencke D, Janz D (1984) Comparative evaluation of MR and CT in patients with temporal lobe epilepsy. Radiology 153: 142

Spencer DD, Spencer SS, Williamson PD, Sostman HD, Gore JC, Mattson RH (1984) Nuclear magnetic resonance imaging in refractory epilepsy patients. Epilepsia 25: 650

Spetzler RF, Zabramski JM, Kaufman B, Yeung HN (1983) Acute NMR changes during MCA occlusion: a preliminary study in primates. Stroke 14: 185-190

Sussman MN, Scanlon M, Garfinkle W, Callahan M, Barry E, Katz RI, Harner N (1984) Magnetic resonance imaging in temporal lobe epilepsy: comparison with EEG and computerized tomography. Epilepsia 25: 650-651

Theodore WH, Dorwart R, Holmes M, Porter RJ, DiChiro G (1985) Neuroimaging in epilepsy: comparison of PET, CT and MRI. Neurology 25 (S): 135

Wyler AR, Bolender N (1983) Preoperative CT diagnosis of mesial temporal sclerosis for surgical treatment of epilepsy. Ann Neurol 13: 59-64

The Use of Positron Emission Tomography in the Surgical Treatment of Epilepsy*

J. ENGEL, Jr.,[1, 2, 4, 5] L. D. CAHAN,[6] W. W. SUTHERLING,[1] P. H. CRANDALL,[3, 5] and M. E. PHELPS[4]

Electrophysiological Studies of Epileptiform Excitability May Be Misleading

The presurgical evaluation of epileptic patients who are candidates for resection of an epileptogenic lesion is aimed at localizing an abnormal area of the brain that is most responsible for generating the patient's habitual seizures. Traditional methods of identifying this region have depended heavily on interictal and ictal EEG recordings obtained using a variety of techniques (Ajmone-Marsan 1980; Engel et al. 1983; Gloor 1975). Whether EEG investigations concentrate on interictal spike activity, usually recorded chronically with extracranial electrodes, and during intraoperative electrocorticography; or whether localization is derived from ictal EEG recordings using extracranial electrodes, or invasive chronic procedures that insert depth electrodes directly into the brain (SEEG), or over the cortical surface, conclusions must be drawn from identification of epileptiform events that propagate widely. Consequently, it is impossible to be certain that the site at which interictal or ictal events occur reflects the actual epileptogenic lesion, or propagated activity from a distant lesion that is beyond the recording range of available electrodes. False localization occurs in 10%-20% of patients based on scalp and sphenoidal interictal and ictal EEG recordings, as well as interictal SEEG recordings (Engel et al. 1981; Engel et al. 1984; Lieb et al. 1981 a, b). It is difficult to determine the incidence of false localization from depth electrode recordings of ictal onset since, when found, this is usually the single most important criterion for resective surgery, and rarely would a surgical resection be undertaken at a distance from the site of depth-recorded ictal onset (Engel and Crandall 1983). Nevertheless, the 10%-20% of patients in most series who do not benefit from surgical resection based on this criterion may be taken as evidence for false localization with this technique.

Confirmatory Tests of Focal Functional Deficit Are Useful

In order to increase confidence in localization of epileptiform excitability based on EEG evaluation of interictal and ictal discharges, it is necessary to use independent confirmatory tests that identify areas of focal functional deficit (Engel et al. 1981; 1983). In most patients with partial epilepsy, the brain area responsible for generating habitual epileptic seizures is functionally abnormal during the interictal state, and this abnormality can often be identified by a variety of procedures: barbiturate activation of EEG fast activity, intracarotid amobarbital (Amytal) assessment of hemispheric language and memory function, neuropsychological testing, and functional imaging techniques that measure local changes in blood flow and metabolism.

Positron Emission Tomography is a Highly Reliable Confirmatory Test

Positron emission tomography (PET), using ^{18}F-fluorodeoxyglucose (FDG), can be used to image local cerebral metabolic rate for glucose (LCMRGlc) (Phelps et al. 1982). Patients with partial epilepsy frequently demonstrate zones of cerebral hypometabolism (Engel et al. 1982 c) that correspond to the site of the epileptogenic focus demon-

[1] Department of Neurology, Reed Neurological Research Center, UCLA School of Medicine, Los Angeles, CA 90024, USA
[2] Department of Anatomy, UCLA School of Medicine, Los Angeles, CA 90024, USA
[3] Department of Surgery, UCLA School of Medicine, Los Angeles, CA 90024, USA
[4] Laboratory of Nuclear Medicine and Biophysics, UCLA School of Medicine, Los Angeles, CA 90024, USA
[5] The Brain Research Institute, UCLA School of Medicine, Los Angeles, CA 90024, USA
[6] Division of Neurosurgery, University of California, Irvine Medical Center, Orange, CA 92668, USA

* This work was supported by grants NS-02808, NS-15654, and NS-20867 from the National Institutes of Health, and contract DE-AC03-76-SF00012 from the Department of Energy.

Presurgical Evaluation of Epileptics
Edited by H. G. Wieser and C. E. Elger

strated electrophysiologically (Engel et al. 1982b) and the epileptogenic lesion demonstrated by pathological analysis of resected brain tissue (Engel et al. 1982a, 1984). PET has been used in the preoperative evaluation of epileptic patients for resective surgery at UCLA since 1977, and this test became an important part of a protocol established in 1979 to identify patients who could undergo anterior temporal lobectomy without depth electrode recording. Anterior temporal lobectomy is now performed without invasive diagnostic procedures when: (1) an EEG onset is consistently observed from the temporal lobe which is also the site of the most prominent interictal EEG spike focus; (2) this temporal lobe is hypometabolic on an interictal PET scan with FDG; and (3) focal functional deficit in the same region is validated by two or more of the confirmatory tests noted previously, or there is evidence of nonspecific structural change, such as a dilated ventricle or calcification (Engel et al. 1983). Since most of the confirmatory tests for focal functional deficit appear to be more sensitive to mesial temporal lesions than neocortical lesions, this protocol has not yet proved useful for planning extratemporal resections, in the absence of an obvious structural abnormality.

For patients who do not meet the criteria for resective surgery without depth electrode recording, the preoperative PET scan has been useful for determining the target sites for depth electrode exploration. A zone of hypometabolism has also increased the confidence with which surgery could be performed following depth electrode evaluation in patients who have demonstrated regional ictal onsets (Walter 1973) that may indicate propagated activity (Engel et al. 1983), or occasional ictal onsets contralateral to the site of the presumed epileptogenic lesion (Engel and Crandall 1983).

The UCLA Experience with PET

Adequate pathological analysis has been carried out on tissue removed from 65 patients who underwent resective surgery based on this protocol between June 1978 and February 1985. Eighteen of these had anterior temporal lobectomy without depth electrode evaluation. Forty-three underwent anterior temporal lobectomy, partial frontal resection was carried out on two patients, and partial occipital resection was carried out on two others, all following depth electrode evaluation. Thirty-eight percent of patients, who received anterior temporal lobectomy since the new criteria were established in October 1979, have done so without invasive studies.

A zone of hypometabolism on PET was seen contralateral to the presumed epileptogenic lesion in two patients, early in this series, and this metabolic defect appeared to reflect changes induced by depth electrode implantation (Engel et al. 1982a, c). Subsequently, PET scans have not been performed on patients while depth electrodes were in place, and false lateralization has not occurred since.

Table 1 shows the correlation between positive PET scan results and positive pathological findings in the resected tissue of the 61 patients who underwent anterior temporal lobectomy. PET and pathology were both positive for 36 patients, and both negative for 9. There were five patients, however, who had zones of hypometabolism on their PET scans but a normal pathological evaluation. All of these patients continue to have seizures, but are improved, suggesting that at least part of the epileptogenic lesion was not removed. Eleven patients had pathological abnormalities that were not identified as a zone of hypometabolism on PET scan. All of these patients had hippocampal sclerosis and all have benefited from surgery. Three of the four patients who underwent extratemporal resections had hypometabolic zones in the area of the presumed epileptogenic focus. Two of the three with positive PET scans had structural lesions, one a glioma, and the other calcification. The one patient with hypometabolism and no structural lesion has been seizure free for 2 years, suggesting that brain tissue responsible for epileptic seizures was removed.

Table 2 demonstrates the yield of PET with specific pathological lesions. A zone of hypometabolism was seen in 74% of patients who had hippocampal sclerosis alone, in 75% of patients who had hippo-

Table 1. Correlation of PET and pathological findings

	Path +	Path −	Total
PET +	36	5	41
PET −	11	9	20
Total	47	14	61

Table 2. Positron emission tomography results for various pathological findings

	N	PET +
HS*	39	29 (74%)
HS plus small lesions	4	3 (75%)
Small lesions only	4	4 (100%)
No pathology	14	5 (36%)
Total	61	41 (67%)

* hippocampal sclerosis

campal sclerosis in addition to other small lesions such as tumors, heterotopia, and cysts that did not show up on X-ray CT scan, and in 100% of patients with only these small lesions that were not demonstrated by X-ray CT scan. More recently, small tumors have been found in three of our patients following anterior temporal lobectomy that were not identified with either X-ray CT or PET, but were seen with magnetic resonance imaging (MRI) (Sperling et al. 1986). Thirty-six percent of patients without evidence of pathological abnormality in their resected temporal lobe demonstrated a zone of hypometabolism on PET.

There is no difference in the outcome with respect to seizures following surgery between the 44 patients who had positive findings on PET scan and the 21 patients who had normal PET scans. When structural lesions were present, patients without hypometabolic zones did just as well as patients with hypometabolic zones. Interestingly, however, when no structural lesions were identified in the resected specimen, patients without hypometabolic zones did better than those with hypometabolic zones. Since the number of patients in the latter group was small, the meaning of this finding is not clear. Nevertheless, it does not appear that the PET scan, by itself, has prognostic value.

Positron emission tomography is also being used at UCLA to identify subtypes of the Lennox-Gastaut syndrome (Chugani et al. 1984). These children can demonstrate focal, unilateral, or diffuse hypometabolism, or PET scans can be normal. It may eventually become appropriate to evaluate such patients with focal hypometabolism for possible resective surgery, while patients with unilateral hypometabolism may be good candidates for corpus callosum section. However, surgery based on these findings at this time is not justified, and the value of PET for this purpose will be determined only by follow-up of patients who eventually undergo surgical treatment based on other criteria.

The Use of PET at Other Centers, and Other New Imaging Procedures

The results of PET studies with FDG at UCLA are being reproduced at other centers (Theodore et al. 1984; Yamamoto et al. 1983), and zones of decreased perfusion and oxygen metabolism also appear to be present in patients with temporal lobe epilepsy who are studied by PET with ^{15}O-oxygen (Bernardi et al. 1983). MRI does not appear to be able to identify hippocampal sclerosis, the most common lesion in temporal lobe epilepsy (Margerison and Corsellis 1966), but may reveal small focal lesions such as tumors and hamartomas that are missed by PET scan (Sperling et al. 1986). Single photon emission computed tomography (SPECT) is also capable of identifying zones of decreased cerebral blood flow that correspond to the epileptogenic lesion in patients with temporal lobe epilepsy (Sanabria et al. 1983; Bonte et al. 1983), and this procedure may prove useful in selecting patients for resective surgery.

Conclusion

Positron emission tomography studies with FDG have been extremely valuable at UCLA for localizing a zone of focal functional deficit and confirming the site of the epileptogenic lesion determined by electrophysiological means. PET has proven useful for identifying patients who can undergo surgery without invasive studies, for planning invasive procedures, and for providing supplemental information in addition to that derived from invasive studies. False lateralization appears to be rare with PET, in contrast to the 10%-20% risk of misidentifying propagated epileptiform activity with EEG studies. However, PET was positive in a few patients with negative pathological findings who, although improved, have not been cured of their epilepsy. This may indicate false localization to some extent with this technique. The failure of PET to identify hippocampal sclerosis in 12 of 43 patients represents an insensitivity that might be reduced by utilizing continuous performance tasks that selectively activate mesial temporal structures. Although PET is highly accurate for localizing an epileptogenic region that can be resected with benefit to the patient, a positive PET study per se does not appear to indicate a specific pathological process or have prognostic value. The role of PET in identifying patients who might be candidates for extratemporal resection, as well as corpus callosum section, has not yet been determined.

References

Ajmone-Marsan C (1980) Depth electrography and electrocorticography. In: Aminoff MJ (ed) Electrodiagnosis in clinical neurology. Churchill Livingstone, New York, pp 167-196

Bernardi S, Trimble MR, Frackowiak RSJ, Wise RJS, Jones T (1983) An interictal study of partial epilepsy using positron emission tomography and the oxygen-15 inhalation technique. J Neurol Neurosurg Psychiatry 46: 473-477

Bonte FJ, Stokely EM, Devous Sr MD, Homan RW (1983) Single photon tomographic study of regional cerebral blood flow in epilepsy. Arch Neurol 40: 267-270

Chugani H, Engel J Jr, Mazziotta J, Phelps M (1984) ^{18}F-2-fluorodeoxyglucose positron emission tomography in medically refractory childhood epilepsy. Neurology 34 (Suppl 1): 107

Engel J Jr, Crandall PH (1983) Falsely localizing ictal onsets with depth EEG telemetry during anticonvulsant withdrawal. Epilepsia 24: 344-355

Engel J Jr, Rausch R, Lieb JP, Kuhl DE, Crandall PH (1981) Correlation of criteria used for localizing epileptic foci in patients considered for surgical therapy of epilepsy. Ann Neurol 9: 215-224

Engel J Jr, Brown WJ, Kuhl DE, Phelps ME, Mazziotta JC, Crandall PH (1982a) Pathological findings underlying focal temporal lobe hypometabolism in partial epilepsy. Ann Neurol 12: 518-528

Engel J Jr, Kuhl DE, Phelps ME, Crandall PH (1982b) Comparative localization of epileptic foci in partial epilepsy by PET and EEG. Ann Neurol 12: 529-537

Engel J Jr, Kuhl DE, Phelps ME, Mazziotta JC (1982c) Interictal cerebral glucose metabolism in partial epilepsy and its relation to EEG changes. Ann Neurol 12: 510-517

Engel J Jr, Crandall PH, Rausch R (1983) The partial epilepsies. In: Rosenberg RN, Grossman RG, Schochet S, Heinz ER, Willis WD (eds) The clinical neurosciences, vol 2. Livingstone, New York, pp 1349-1380

Engel J Jr, Sutherling WW, Cahan L, Crandall PH, Kuhl DE, Phelps ME (1984) The role of positron emission tomography in the surgical therapy of epilepsy. In: Porter RJ, Mattson RH, Ward AA, Dam M (eds) Advances in epileptology: epilepsy international symposium. Raven, New York, pp 427-432

Gloor P (1975) Contributions of electroencephalography and electrocorticography to the neurosurgical treatment of the epilepsies. In: Purpura DP, Penry JK, Walter RD (eds) Advances in neurology, vol 8. Neurosurgical management of the epilepsies. Raven, New York, pp 59-105

Lieb JP, Engel J Jr, Brown WJ, Gevins AS, Crandall PH (1981a) Neuropathological findings following temporal lobectomy related to surface and deep EEG patterns. Epilepsia 22: 539-549

Lieb JP, Engel J Jr, Gevins AS, Crandall PH (1981b) Surface and deep EEG correlates of surgical outcome in temporal lobe epilepsy. Epilepsia 22: 515-538

Margerison JH, Corsellis JAN (1966) Epilepsy and the temporal lobes: a clinical, electroencephalographic and neuropathological study of the brain in epilepsy, with particular reference to the temporal lobes. Brain 89: 499-530

Phelps ME, Mazziotta JC, Huang SC (1982) Study of cerebral function with positron computed tomography. J Cereb Blood Flow Metab 2: 113-162

Sanabria E, Chauvel P, Askienazy S, Vignal JP, Trottier S, Chodkiewicz JP, Bancaud J (1983) Single photon emission computed tomography (SPECT) using ^{123}I-isopropyl-iodoamphetamine (IAMP) in partial epilepsy. In: Baldy-Moulinier, Ingvar D-H, Meldrum BS (eds) Current problems in epilepsy: cerebral blood flow, metabolism and epilepsy. Libbey, London, pp 82-87

Sperling MR, Wilson G, Engel J Jr, Babb TL, Phelps M, Bradley W (1986) Magnetic resonance imaging in intractable partial epilepsy: correlative studies. Ann Neurol 20: 57-62

Theodore WH, Newmark ME, Sato S, De LaPaz R, DiChiro G, Brooks R, Patronas N, Kessler RM, Manning R, Margolin R, Channing M, Porter RJ (1984) ^{18}F-fluorodeoxyglucose positron emission computed tomography in refractory complex partial seizures. Ann Neurol 14: 429-437

Walter RD (1973) Tactical considerations leading to surgical treatment of epilepsy. In: Brazier MAB (ed) Epilepsy: its phenomena in man, UCLA forum in medical sciences no. 17. Academic, New York, pp 99-119

Yamamoto YL, Ochs R, Gloor P, Ammann W, Meyer E, Evans AC, Cooke B, Sako K, Gotman J, Feindel WH, Diksic M, Thompson CJ, von Robitaille Y (1983) Patterns of rCBF and focal energy metabolic changes in relation to electroencephalographic abnormality in the inter-ictal phase of partial epilepsy. In: Baldy-Moulinier M, Ingvar D-H, Meldrum BS (eds) Current problems in epilepsy: cerebral blood flow, metabolism and epilepsy. Libbey, London, pp 51-62

Editors' Commentary 4

The *discussion* on the first two contributions centered around practical questions such as the biocompatibility of recording devices and how such information can be best achieved. The announcement that mini- and microelectrodes for ion-activity recording purposes might be available in the near future (which can be combined with some conventional electrodes used in SEEG) raised great expectations with a view toward future research in human epilepsy. Expectations were also directed towards MR spectroscopy, which may be used to study energy states by measuring ^{31}P and pH, despite poor spatial resolution at present. The UCLA experience with PET, which has been used to assess local changes in cerebral blood flow, blood volume, oxygen extraction, pH, protein synthesis, receptor binding, and oxygen and glucose metabolic rates (Mazziota and Engel 1985), demonstrated the value of this new technique both for further research into basic mechanisms of epilepsy and for the noninvasive evaluation of candidates for surgical treatment. In epilepsy most PET studies have measured glucose metabolism. Although a high correlation between the site of the interictal hypometabolic region and the site of ictal onset, recorded with scalp or depth EEG, was noted (Mazziota and Engel 1985), the use of PET in the ictal state for the purpose of localization is limited (Sperling et al. 1987). Current PET techniques require a nearby cyclotron, which is inordinate for many institutions. Although less expensive stand-alone medical cyclotrons may soon be available, SPECT is considerably less costly than PET and has therefore the potential for more widespread use. At the Zürich symposium SPECT was not specially represented. This was thought to be justified by the fact that only blood flow can be measured with the tracers now available. The discussion, however, revealed a strong interest in this technique. In particular, the advantages and disadvantages of the two detector types (rotating gamma camera vs that utilizing four sets of sodium iodide crystal scintillation detectors in a square arrangement that rotate about the subject's head) and the various tracers including ^{123}I-N-isopropyl-iodoamphetamine (IMP), ^{123}I-hydroxyiodopropyl-dianine (HIPDM), and ^{133}Xenon were discussed.

With regard to magneto-encephalography (MEG) most participants agreed that, although in some cases the magnetic field patterns may be simpler than that of EEG thus making localization easier, at this time MEG should be used together with EEG rather than replace it. The development of multichannel MEG recording systems, however, has the potential to unlock more powerful, noninvasive, clinical neurophysiology.

The potential of MRI in the surgical assessment of epileptic patients was fully acknowledged. All participants agreed that MRI is able to detect small nonsclerotic epileptogenic lesions frequently missed by CT and emphasized the advantages in detecting abnormalities, especially in the middle fossa, near bony structures. However, despite general enthusiasm it became evident that current MRI techniques are probably inadequate for detecting mesial temporal sclerosis.

When writing this comments we were in the comfortable situation of knowing that a competent summary on the basic principles of these new techniques as well as on the relevant patient studies is now available (Sperling et al. 1987). In addition this chapter of Sperling and collegues gives a valuable outlook for the future and is therefore recommended to the interested reader who wishes to delve further into these matters.

References

Mazziota JC, Engel J Jr (1985) Advanced neuro-imaging techniques in the study of human epilepsy: PET, SPECT and NMR-CT. In: Pedley TA, Meldrum BS (eds) Recent advances in epilepsy 2. Churchill Livingstone, New York, pp 65-99

Sperling MR, Sutherling WW, Nuwer MR (1987) New techniques for evaluating patients for epilepsy surgery. In: Engel J Jr (ed) Surgical treatment of the epilepsies. Raven, New York, pp 235-257

Specific Neuroradiological Examinations Necessary for Stereotactic Procedures

C. MUNARI[1, 2], A. T. GIALLONARDO[3], A. MUSOLINO[2], P. BRUNET[1], J. P. CHODKIEWICZ[2], J. BANCAUD[1, 2], and J. TALAIRACH[1, 2]

Introduction

Neuroradiological study of epileptic patients may be carried out in different ways according to the methodology employed and to the therapeutic goals of the clinicians. Our task here is confined to discussing the neuroradiological examinations which should be performed before exploring epileptic patients with intracerebral stereotactically implanted electrodes (stereo-EEG). According to our methodology, as often previously described (Bancaud 1959; Bancaud and Talairach 1970, 1975; Talairach and Bancaud 1963, 1973), the neuroradiological study is always carried out in stereotactic conditions. Such a procedure enables us (a) to precisely identify the different cerebral structures (cortical and subcortical), and (b) to explore these structures while avoiding, as well as possible, any dangerous approach.

In this paper we will, firstly, study the kinds of information that the different neuroradiological stereotactic investigations provide us with. Then we will examine if and how, in our experience, the introduction of the CT scan can improve the chances of discovering anatomical lesions. And, finally, we will try to compare the different radiological findings with the data obtained by the stereo-EEG.

Patients

In the period 1981-1984, we carried out 126 stereo-EEG explorations. For this study we excluded 13 patients having a cerebral epileptogenic tumor with evident mass effect and 9 patients in whom the stereo-EEG data were unclear. Also excluding another nine patients who had two Stereo-EEG investigations, this study concerns 95 patients.

There were 60 males and 35 females aged, at the time of the Stereo-EEG, from 9 to 49 years (mean, 24.5 years). The age at onset of the seizures varied from 1 day to 28 years (mean, 10.5 years). The duration of the epilepsy varied from 2 to 35 years (mean, 14 years).

The etiology, unknown in 41 cases, was perinatal in 19, traumatic in 16, infectious in 15, and surgical in 4. In seven cases, the epilepsy started with an epileptic status. The neurological examinations showed a motor deficit in 28 patients (severe in 9, slight in 19), sensitive disturbances in 10, various limitations of the visual field in 7, and other disturbances in 3. The neurological interictal examination was normal in 62 patients. Three patients had an intellectual impairment.

Seventy patients were right handed and four left handed. The WADA test (14 times unilateral, 7 times bilateral) was carried out in the 21 patients in whom the speech dominance was unclear; the dominant hemisphere was the right in 7, the left in 13. In one patient, we were unable to precisely define the cerebral dominance.

Methods

Neuroradiological Stereotactic Investigations

The radiological stereotactic investigations were carried out only when the preliminary investigations permitted the formulation of an apparently valuable hypothesis regarding the localization of the epileptogenic area(s) (Bancaud 1968; Bancaud et al. 1965, 1984; Munari and Bancaud 1985).

1. The head of the patient was fixed in the Talairach frame and then orthogonal teleradiographs (4.3 m) were taken on different projections, permitting a stereoscopic view.

2. Carotid angiography, always performed with stereoscopy, was unilateral in 61 patients (right 42, left 19) and bilateral in 34 patients. The stereoscopic study of such angiograms (Szikla et al. 1977; Szikla

[1] INSERM U 97, 2 ter rue d'Alésia, 75014 Paris, France
[2] Hôpital Sainte Anne, Service de Neurochirurgie, 1 rue Cabanis, 75014 Paris, France
[3] Va Clinica Neurologica Università degli Studi "La Sapienza", Viale dell' Università 30, 00100 Rome, Italy

Presurgical Evaluation of Epileptics
Edited by H. G. Wieser and C. E. Elger

1979; Talairach and Szikla 1979) enables us directly to identify sulci and gyri in any individual patient. It is clear that the correct identification of the sulci needs the simultaneous use of anteroposterior and lateral angiograms. Such a study should permit the observation of even small distortions of the vessels, provoked either by space-occupying lesions or by atrophic processes.

3. Pneumoencephalography was performed in only three patients.

4. Contrast ventriculography, also with stereoscopy, was carried out with a frontal approach in all patients and obtained a correct visualization of the entire ventricular system in almost all cases.

5. A CT scan was performed in almost every patient. However, because of the poor quality of several pictures, we will consider this examination in only 82/95 patients.

6. In 12 patients the CT scan data showed an important and extended hypodensity, raising the hypothesis of the existence of a (pseudo)cyst. In these patients we did a stereotactic exploration and, then, air-contrast cystography.

Stereo-EEG Investigations

The methodological details have been presented elsewhere (Bancaud et al. 1973; Bonis et al. 1981; Munari et al. 1983; Talairach and Bancaud 1971; Talairach et al. 1974). We would just like to stress that, in our methodology, the decision as to number and position of the intracerebral electrodes is based on previously collected anatomo-electroclinical data. This may explain why we only rarely explore only one lobe and why we do not systematically perform bilateral temporal exploration. In fact, the electrodes (6–11) explored only the right side in 50 patients, only the left side in 20, and both hemispheres in the remaining 25.

On the basis of the stereo-EEG recordings, we considered those regions to be a "lesional area" (LA) in which the exploring electrodes recorded permanent, high-voltage, slow waves or low-voltage activity ($<10\,\mu V$).

The complex concept of "epileptogenic area" (EA) should be discussed in detail. In this paper, we consider as "epileptogenic" those areas where the ictal discharges started and those involved during the first 10 s.

Results

Plain Skull Radiograms

In 43 patients this examination was normal. A cranial hemiatrophy was evident in 20. Abnormal calcifications were noted in 11. A local abnormality existed in 11 patients. The frontal sinuses were very asymmetrical in 16 (8 without global hemiatrophy) and agenesic in 2 patients.

Carotid Angiography

A displacement of the anterior cerebral artery was observed in 14 of the 20 patients having a global hemiatrophy (Table 1). It was always homolateral to the side of the atrophy, as well as to the displacement of the internal cerebral vein (three patients).

The stereoscopic study of the angiograms enabled us to appreciate various alterations of the central sulcus and/or of the insula in 24 patients. An avascular area was observed in 13 patients. In this series, we observed no vascular malformations. In 60 patients (53%) we were unable to identify any angiographic abnormality.

Table 1. Carotid Angiography

	Patients (N = 114)
Anterior cerebral artery displacement	14
Internal cerebral vein displacement	3
Avascular areas	13
Insular modifications	15
Central sulcus modifications	9
No abnormalities	60

Ventriculography

In 55 of 95 patients (58%; Table 2) we observed relatively localized alterations of the lateral ventricles, above all, temporal, and frontal. The ventriculography was normal in 28 patients. Modifications of the third and fourth ventricles were observed in some patients with a perinatal etiology.

CT Scan

In this investigation (Table 3) no lesions were identified in 31 patients (38% of the 82 correctly examined with this method). Very localized alterations (hypo-

Table 2. Ventriculography in 95 patients

	Patients
Localized abnormalities	55
Frontal	12
Temporal	16
Occipital	1
Frontotemporal	3
Frontoparietal	10
Temporoparietal	1
Unilateral ventricular dilatation	9
Bilateral ventricular dilatation	9
Septum deviations	8
Third and/or fourth ventricle modifications	5
No abnormalities	28

Table 3. CT scan in 82 patients

	Patients
Localized hypodensity	24
Localized calcification	5
Local atrophy	5
Localized dilatation of ventricles	2
Unilateral dilatation of ventricles	8
Bilateral dilatation of ventricles	3
Unilateral dilatation of sulci	2
Bilateral dilatation of sulci	1
Third and/or fourth ventricle modifications	3
No abnormalities	31

density, calcification, atrophy, localized dilatation of ventricles) were evident in 44 patients (54%). Bilateral alterations existed in 4 patients.

Comparative Study of Different Neuroradiological Investigations

All the radiological investigations (Table 4) were in accordance in 31 patients; 20 of them (18 with CT scan) presented no abnormalities. Excluding the skull radiograms, homogeneous pathological findings characterized all the radiological investigations in 22 patients. The ventriculography appears to be the most useful investigation in our patients; its rate of "nonpathological finding" is the lowest. The (pseudo)cysts are the type of pathology in which the most useful examination is the CT scan. Among the 12 patients having a (pseudo)cyst the CT scan (performed in 10) showed in all a very marked, well-localized hypodensity. The carotid angiography showed an avascular space in five, a mass effect in two, and seemed to be normal in the other five patients. The ventriculography made possible the direct visualization of the cyst in only two patients. In one patient it seemed normal. In the others we observed a mass effect (two) or a ventricular dilatation (localized, one; unilateral diffuse, six).

Table 4. Degree of accordance between data of different neuroradiological investigations (and CT scan)

	Patients
No abnormalities	
Rx+A+V+CT[a]	20
Pathological findings	
Rx+A+V+CT	11
A+V+CT	11
V+CT	5
Rx +CT	4
Rx+ V+CT	1
Rx+ A+CT	1
A+CT	1
Rx+ V	4
Rx+A+V	3
A+V	2

Rx, plain skull radiography; *A*, carotid angiography; *V*, ventriculography; CT, computerized tomography; [a] only 18/20 patients with CT

Comparison Between Neuroradiological and Stereo-EEG Data

The 20 patients not having any neuroradiological evidence of a brain lesion had a "lesional area" (LA) (see above) defined by the Stereo-EEG, which was localized in the frontal lobe (5), in the temporal lobe (4), in the homolateral frontotemporal region (4), temporoparietally (3), parietally (1), and frontocentrally (1). The extension of the radiological lesions and of the "lesional area" was the same in 37 patients. In nine patients no topographic relationships were observed between the site of the neuroradiological alterations and the LA. The LA was much more extended than the lesion shown by neuroradiology in 26 patients. The radiological lesions were more extended than the LA in three patients.

Comparison Between the Lesional and the Epileptogenic Areas

The LAs and the epileptogenic areas (EAs) seemed to us perfectly coincident in 44 patients. They were completely different in 13 patients. The LA was larger than the EA in 24 patients. The EA was more extended than the LA in 14 patients.

Discussion

In our experience, the neuroradiological examinations conducted with a surgical viewpoint should always be carried out in stereotactic conditions (Talairach et al. 1974). If one considers only the concrete chance of identifying a lesion responsible for epilepsy, the results may appear to be poor. In our patients with severe, partial, and drug-resistant epilepsy, skull radiograms were normal in 45% of the cases, carotid angiography was normal in 65%, and ventriculography was normal in 29%. On the other hand, this type of finding, especially that of skull radiography, does not seem to have any specificity for epileptic patients (McRae 1974). What may be specific are, probably, the insular and central modifications shown by the stereoscopic, stereotactic, carotid angiography (Szikla 1979).

The contribution provided by the CT scan, considered very important by many epileptologists, appears in our cases moderate, since very localized alterations (excluding calcifications) are evident in only 38% of the cases, exactly the same rate as nonaltered examinations.

In fact, neuroradiological examinations, carried out in stereotactic conditions, make possible the identification of the different cerebral structures and, of course, are the indispensable basis for the introduction of the intracerebral electrodes (Talairach and Szikla 1979; Munari and Bancaud 1985).

The existence of a lesion is actually considered the necessary anatomopathological basis of the severe partial epilepsies. However, in our opinion it is clear that a lesion may be demonstrated by several different diagnostic approaches: the clinical approach (neurological, neuropsychological examinations), neuroradiological investigations, Stereo-EEG findings, surgical macroscopic examination, and histopathological findings. According to the specific characteristics, we have different criteria to evaluate not only the existence of an epileptogenic lesion, but also its extension and its limits. In any case, even if an anatomical lesion may be demonstrated, this step cannot be considered as sufficient to define the limits of a therapeutic surgical treatment. Our data seem to show that:

1. The EAs and the LAs are exactly the same in less than one-half of patients (46%)
2. It may not be necessary to remove all the LA (in 25% of patients the LA is larger than the EA)
3. In 15% of the patients the removal of the LA would not be enough to cure the epilepsy (EA > LA); and in more than 13% the EA and the LA are completely different.

Such statements may be correct if one assumes that our criteria for defining a LA and an EA are correct. Our proposal is to verify, in the near future, whether or not the surgical results are in agreement with these criteria, especially considering those patients in whom the surgery of the EA was not as complete as it should be, because of functional risks.

As a concluding remark, we would like to stress that the existence of a CT scan hypodensity in patients with a severe partial epilepsy may indicate the existence of a lesion, but this fact itself does not allow us to identify this CT scan finding with the real extension of a lesion, to confuse this lesion with the EA, or to consider that only a CT scan alteration proves the existence of an anatomical lesion. In other words, a surgical perspective should not be considered only in patients with CT scan alterations and, moreover, the existence of a CT scan alteration does not in itself make possible the decision about a possible surgical intervention.

References

Bancaud J (1959) Apport de l'exploration fonctionnelle par voie stéréotaxique à la chirurgie de l'épilepsie. Neurochirurgie 5: 55-112

Bancaud J (1968) Contribution de la stéréo-électro-encéphalographie à l'étude physiopathologique et clinique des épilepsies (dans une perspective chirurgicale). Gaggi, Bologna pp 135-155

Bancaud J, Talairach J (1970) L'électroencéphalographie de profondeur (SEEG dans l'épilepsie). In: Modern problems of pharmaco-psychiatry, vol 4. Epilepsy. Karger, Basel, pp 29-41

Bancaud J, Talairach J (1975) Macro-stereo-electroencephalography in epilepsy. In: Bancaud J (ed) Handbook of electroencephalography and clinical neurophysiology, vol 10 part B. Elsevier, Amsterdam, pp 3-33

Bancaud J, Talairach J, Bonis A, et al. (1965) La stéréo-électroencéphalographie dans l'épilepsie. Masson, Paris, 321 pp

Bancaud J, Talairach J, Geier S, Scarabin JM (1973) EEG et SEEG dans les tumeurs cérébrales et l'épilepsie. Edifor, Paris, 351 pp

Bancaud J, Bonis A, Munari C, Szikla G, Chodkiewicz JP, Talairach J (1984) Localizing value of the clinical manifestations of the partial seizures. Acta Neurochir (Suppl) 33: 7-15

Bonis A, Chodkiewicz JP, Munari C, Szikla G, Bancaud J, Talairach J (1981) La chirurgie de l'épilepsie. Méthodologie stéréotaxique, résultats et indications. Gaz Méd France 88: 4125-4130

McRae DL (1974) Radiology in epilepsy. In: Magnus O, Lorente de Haas AM (eds) Handbook of clinical neurology, vol 15. The epilepsies. Elsevier, Amsterdam, pp 583-558

Munari C, Bancaud J (1985) The role of the stereo-electroencephalography (SEEG) in the evaluation of partial epileptic seizures. In: Porter RJ, Morselli PL (eds) The epilepsies. Butterworths, London, pp 267-306

Munari C, Talairach J, Musolino A, Szikla G, Bancaud J, Chodkiewicz JP (1983) Stereotaxic methodology of "functional" neurosurgery in tumoral epileptic patients. Ital J Neurol Sci (Suppl) 2: 69-82

Szikla G (1979) Stereotactic neuroradiology and functional neurosurgery: localisation of cortical structures by three-dimensional angiography. In: Rasmussen T, Marino R (eds) Functional neurosurgery. Raven, New York, pp 197-217

Szikla G, Bouvier G, Hori T, Petrov V (1977) Angiography of the human brain cortex. Atlas of vascular patterns and stereotactic cortical localization. Springer, Berlin Heidelberg New York

Talairach J, Bancaud J (1963) Processus de l'investigation fonctionnelle dans la neurochirurgie stéréotaxique. Neurophysiologie et neuropsychophysiopathologie. Actual Neurophysiol 6: 137-167

Talairach J, Bancaud J (1971) Traitement chirurgical des épilepsies. Encyclopédie Méd Chir 10 (9): 1-8

Talairach J, Bancaud J (1973) Stereotaxic approach to epilepsy. Methodology of anatomofunctional stereotaxic investigations. Prog Neurol Surg 5: 294-354

Talairach J, Szikla G (1979) Stereotactic neuroradiological concepts applied to surgical removal of cortical epileptogenic areas. In: Rasmussen T, Marino R (eds) Functional neurosurgery. Raven, New York, pp 219-242

Talairach J, Szikla G, Tournoux P, et al. (1967) Atlas d'anatomie stéréotaxique du télencéphale. Masson, Paris

Talairach J, Bancaud J, Szikla G, et al. (1974) Approche nouvelle de la neurochirurgie de l'épilepsie. Méthodologie stéréotaxique et résultats thérapeutiques. Neurochirurgie (suppl 1) 20: 1-240

Presurgical Epileptological Intensive Evaluation

H. STEFAN[1] and H. G. WIESER[2]

In this paper, different steps in the course of intensive presurgical evaluation are described. The program of evaluation we are practicing or intending to practice is outlined with special attention to the documentation of symptoms and findings. Presurgical evaluation serves to answer the following questions:

1. Where and how precisely can the epileptogenic process be located?
2. Does there exist in addition a morphological and/or functional lesion?
3. Do the findings give hints that surgery may lead to intolerable neurological or psychological effects?
4. Is surgery indicated and which surgical procedure should be selected?

To answer these questions, different investigation techniques of neuroradiology, neuropsychology and other special examinations, such as intensive monitoring, are applied. The epileptological examination has been supported in the past decade by the increased application of intensive monitoring.

The synchronized video-EEG documentation of behavior and biosignals is a very important part of intensive monitoring. This synchronization is performed in different technical ways, for instance in the form of the split-screen technique (Penin 1968) or by means of simultaneous multiplexed EEG and video recording. Here the multiplexed channels of the telemetry are recorded on an audio track of a video-recorder (Kaiser 1976), or the video-EEG synchronization is done by means of a digital clock (Saint-Hilaire et al. 1976). According to the mobility of the patient and the recording system, one can differentiate stationary, semi-mobile, or mobile intensive CCTV monitoring systems.

The transmission of biosignals can be performed by means of cable telemetry, radio telemetry, or with cassette recorders (Porter et al. 1971; Stefan and Burr 1982; Binnie et al 1981). Preamplifiers on the head near the electrode make artifact-reduced recordings possible. Sixteen-channel (simultaneous 2 × 8) cassette recording allows mobile registrations with or without video documentation in addition to stationary investigations (Stefan and Burr 1985).

Depending on the special aim of the planned investigation, different monitoring systems and modes of biosignal transmission can be combined (Stefan and Burr 1982; Binnie et al 1981). Cable telemetry can be performed with a high number of channels, quite good technical quality, and instantaneous artifact recognition and elimination. The mobility of the patient is restricted. In the case of radiotelemetry, this restriction is less, but limitations of the number of channels, disturbances of transmission in buildings, necessity of continuous registration of the technical quality and high costs have restricted its use up to now. Cassette recordings allow the highest mobility of the patient and comfortable long-term recordings, but limitation of the number of channels and band width, as well as fixed montage programs, are disadvantages. New technical developments in this field seem to have succeeded in incorporating the essential advantages of the different systems mentioned above into a single mobile recording unit (DR 16 EEG-Videometry System), as our first experiences suggest.

For the exact comparison of behavioral symptoms and polygraphic findings, an electronic digital clock is an essential element of video-EEG recording (Wieser 1982). For the scientific analysis of video documentations we use 1 inch size with variable slow-motion replay, fast picture search, programmed operation or at least ¾-inch Umatic recordings. Time lapse documentations are not suited for detailed analysis of seizure symptoms. The technique of video documentation should be standardized. The patients should wear only shirts and shorts so that the motor activity of all extremities can be better examined. The camera should take a total view and the room should be adequately illuminated. Recordings in darkened rooms require special

[1] Neurological Clinic, University Erlangen-Nürnberg, Schwabachanlage 6, 8520 Erlangen, Federal Republic of Germany

[2] Department of Neurology, University Hospital, Frauenklinikstr. 26, 8091 Zürich, Switzerland

Presurgical Evaluation of Epileptics
Edited by H. G. Wieser and C. E. Elger

cameras (for instance, infrared cameras). These simple criteria are often disregarded. The weakest link in this chain between recording and replay limits the quality of the whole documentation.

Long-term EEG recordings during presurgical intensive monitoring are usually performed with silver-silver-chloride collodium-fixed electrodes using the 10/20-system or modifications. At least 16 channels including EOG and ECG should be recorded. To answer specific questions, the polygraphic recording may be extended by using actograms, measurement of respiration, skin resistance, or reaction time, etc. The storage of the findings can be performed by paper chart, magnetic tape, or conversion to the videotape and can finally be put into the computer. Storage on paper allows continuous on-line observation to achieve a good quality, but no computer-assisted analysis is possible. It requires storage on magnetic tape. The computer assistance of visual analysis provides data reduction and more objective findings for statistical analysis. Selective data reduction in video-EEG monitoring can be performed in the following different ways:

1. A human observer indicates the time of occurrence of a seizure and the EEG is analyzed at this time.
2. The patient presses the time press button.
3. The automatic EEG analysis is used for seizure detection.

It is not intended to present a complete scheme covering the various computerized analysis approaches. To this end, reference is given to a survey by Burr (1982). In addition to the recognition of the spike wave (Quy et al. 1980) and interictal spikes (Carrie 1976; Gotman and Gloor 1976), the seizure detection was developed by Ives et al. (1976), Babb et al. (1974) and Gotman (1982). The buffering and selective editing of interesting EEG epochs (Gotman et al. 1979, 1982) put the data reduction into practice and combines this advantage with a chance of visual artifact recognition.

The documentation of electrophysiological findings has been intensively refined and cultivated in the past. The documentation of ictal symptoms is probably the least-developed area and because of its importance must be improved in the future. A few approaches with computer-assisted documentation of the seizure symptoms (Stefan et al. 1981; Wieser 1983) warrant further development. Within the schedule of our analytical approach, an important step is to gather observed data into a form suitable for computerized handling. The computer program for coding and storing the data contains a list of all relevant questions (Stefan et al. 1981). Events before, during, and after the polygraphic findings can be investigated exactly, including their temporal relationship, emphasizing ictal patterns of underlying stereotypes.

The use of refined investigation techniques makes possible the recognition of subclinical and subtle initial focal events which may give hints of regional predominance of epileptic activity also during generalized seizures.

The recognition of subclinical and subtle initial focal ictal events is shown in the following example. In this case, the visual observation of the seizure showed bilateral clonic activity in both arms. Analysis of the video-polygraphic documentation indicated that, before the onset of generalized spike wave paroxysms, focal myoclonic EMG activity started in the left musculus extensor digitorum communis. This subtle myoclonic activity was repeated with increasing intensity and finally propagated to the right arm (Fig. 1). Thus, this bilateral myoclonus was not synchronous. The left side always led. The initial focal character of the ictal symptomatology in this case was not recognizable if one looked only at the ictal surface EEG which showed general spike wave paroxysms. Video-polygraphic recording of high quality and videometric slow motion analysis was necessary to detect initial focal and secondary generalized seizure activity (Stefan et al. 1983). In other cases, before generalized spike wave paroxysms during an absence, subclinical autonomic changes of respiration and heart beat (pulse rigidity) were recognized. If such subclinical or subtle phenomena are not recognized, and therefore hints for initial focal epileptic activity are not noticed, the impulse for further intensive clarification is often lacking. With respect to surgery the differentiation of seizure types is also of great importance, as practiced by Delgado-Escueta et al. (1982) and Walsh et al. (1984), who differentiated types 1-3 of complex partial seizures.

Intensive monitoring can be divided into a first and second level (Fig. 2). *Level 1* includes long-term videomonitoring with basal extradural - such as sphenoidal - EEG recordings. In addition to routine activation procedures, methohexital (Musella et al. 1971; Gloor 1975) may activate focal epileptic activity and thiopental (Lombroso and Erba 1970; Engel et al. 1981) may show focal reduction of fast activities. *Level 2* comprises intracranial electrophysiological recordings, for instance epi- or subdural recordings (Goldring 1978; Goldring and Gregorie 1984; Lüders et al. 1982), foramen ovale recordings (Wieser et al. 1985), and depth recordings (Dodge 1953; Brazier 1956; Talairach et al. 1958; Bancaud 1959; Crandall et al. 1963; Gloor 1984; Wieser 1981, 1986).

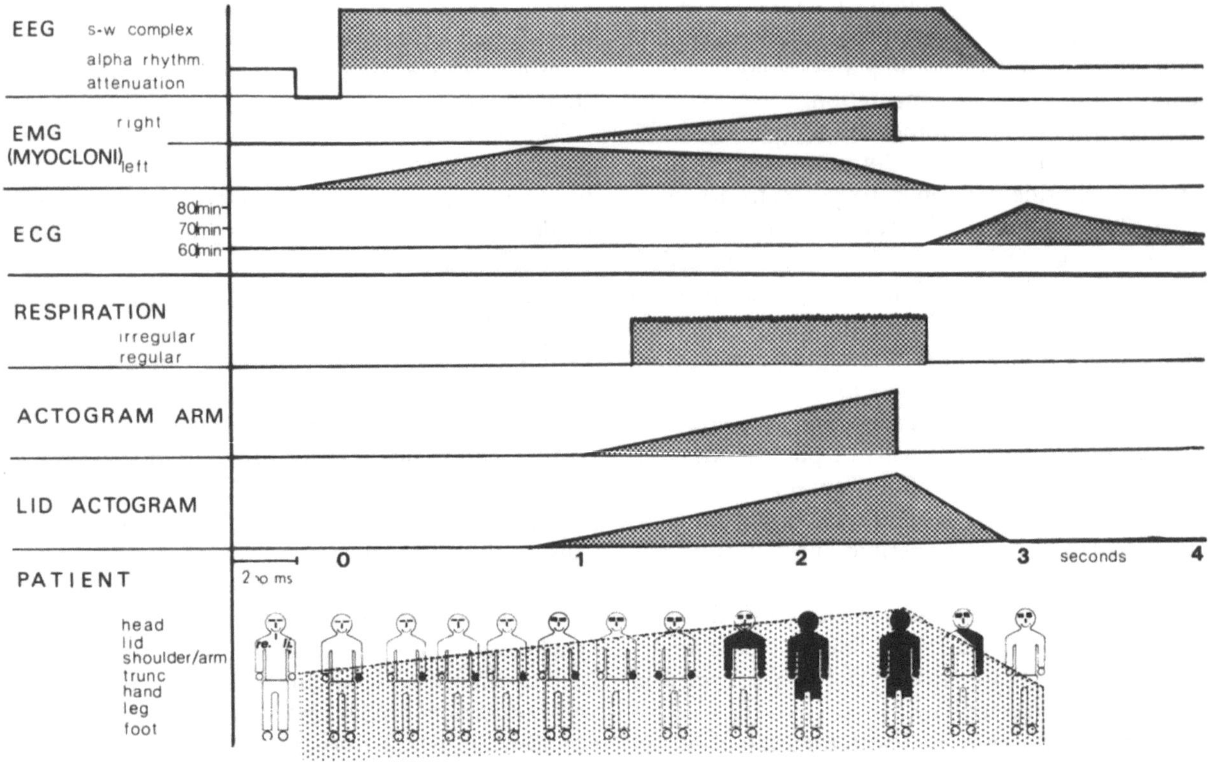

Fig. 1. Focal onset and secondary generalized motor activity during a seizure with generalized spike wave paroxysms in the EEG. The time course of the ictal motor symptoms and polygraphic findings shows the following succession: myoclonus (left hand) before occurrence of spike wave paroxysm and later propagation to bilateral asynchronous myoclonus

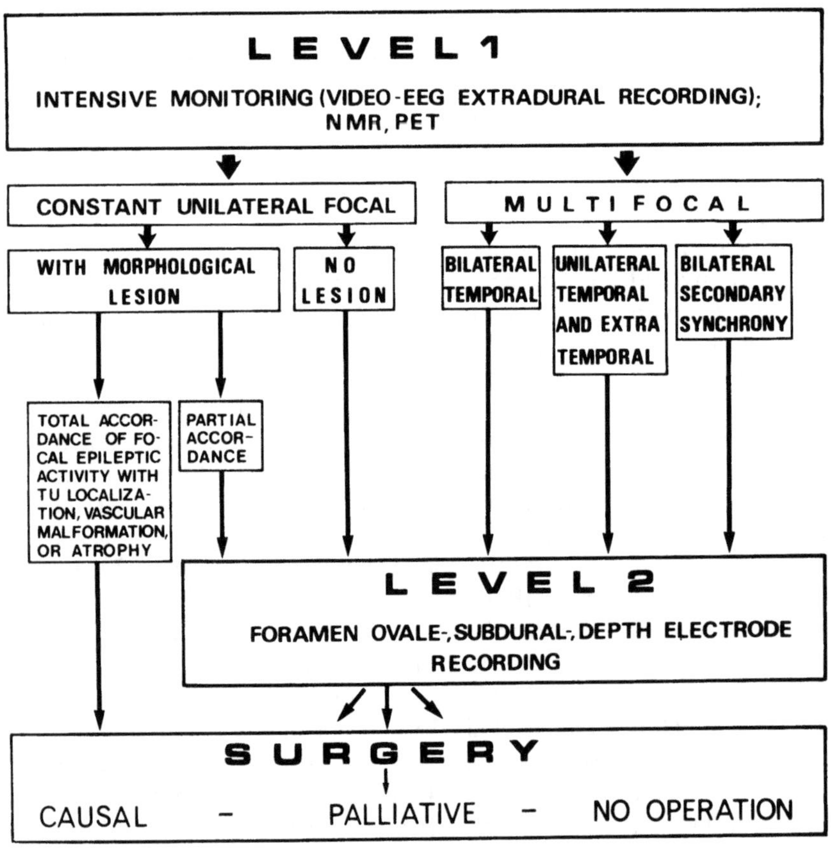

Fig. 2. Evaluation in levels 1 and 2 with respect to surgery

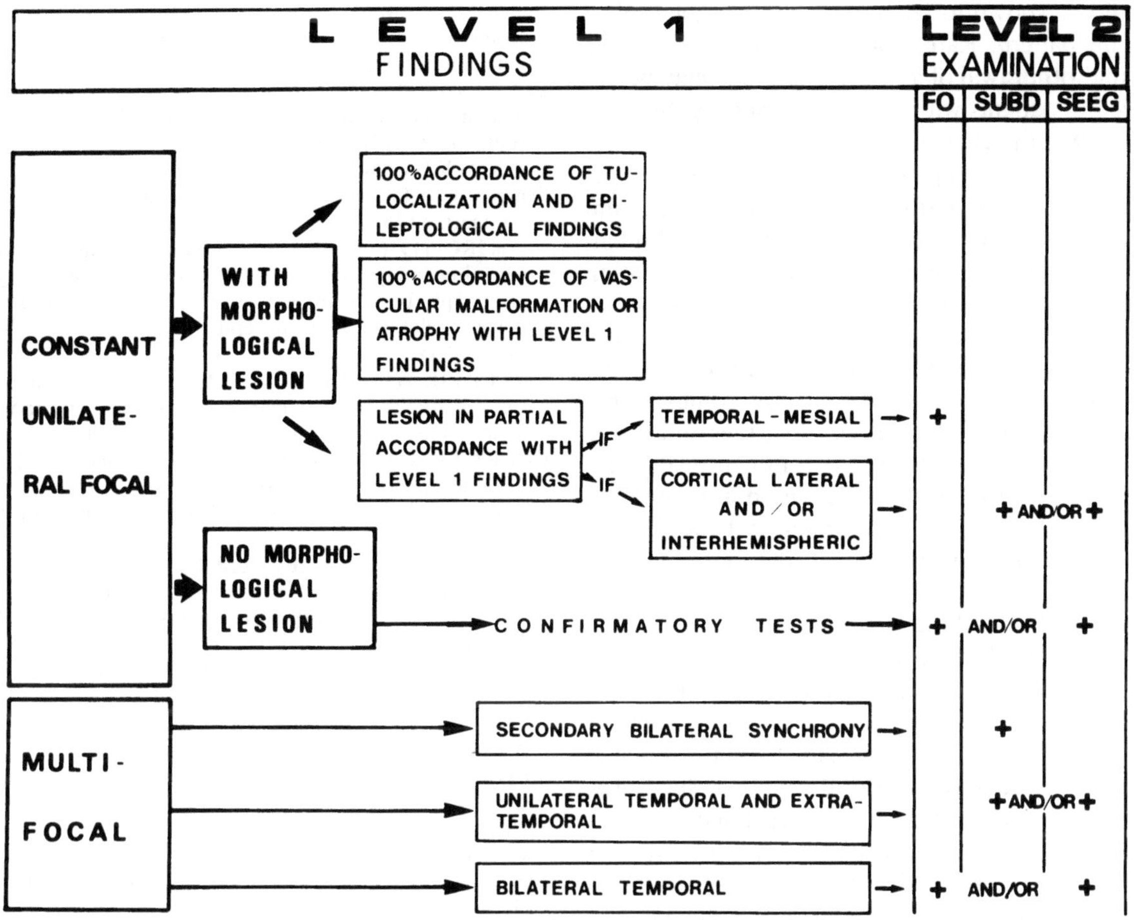

Fig. 3. Methods of different electrophysiological investigation techniques in levels 1 and 2

Other diagnostic *Level 2* procedures are the corticographic and cortical or intracerebral stimulations. As interictal epileptic activity provides only limited information (Spencer et al. 1981) and the side of seizure onset correlates with interictal activity only in approximately two-thirds of the cases (Wieser et al. 1979), the ictal documentation is of great importance. The recording of spontaneous seizures should be preferred to chemically induced or electrically induced seizures. If the localization of stimulation reproducing the patient's habitual seizure coincides with the area of onset observed in spontaneous attacks, the likelihood is good to having identified the patient's principal epileptogenic region (Gloor 1975). In critical cases epi- or subdural stimulation allows a presurgical mapping for localization of speech and somatomotor reaction, including somatotopic presentation (Goldring 1978; Goldring and Gregorie 1984). This information is of value for determining the extent of surgery.

Different steps of presurgical evaluation are shown as a flow chart in Fig. 2. In the first instance, the investigation has to identify one of several foci as the starting point of the patient's seizures. Important information is obtained during *Level 1* evaluation with interictal and ictal surface recordings. If a constant unilateral focus correlates with the region of seizure onset, the next step is to investigate whether this epileptic focal activity correlates with a morphological lesion or neuropsychological functional deficiency. In the case of complete coincidence of focal epileptic activity, for instance with vascular malformation or atrophy, the patient can be suggested for surgery. The success of surgery depends on an extremely exact localization of the epileptogenic region. Also, if *Level 2* diagnosis is necessary, the hypotheses of seizure origin must already have been put forward in *Level 1*. Here the basic strategic considerations for depth exploration are made, which will confirm the initial hypotheses of the origin of habitual seizures. Therefore during *Level 1* evaluations, already top-quality seizure recording is necessary. In our opinion, at least two seizures should be documented. The video-EEG documentation should be shown to relatives to confirm the typical seizure pattern. Rarely occurring atypical seizures during withdrawal of antiepileptic drugs may be observed in patients with multifocal epilepsy (Spencer et al. 1981; Engel et al. 1983).

If there is no complete coincidence between morphological lesion and functional deficit, important information is obtained by means of *Level 2* intensive monitoring. The use of different investigation techniques in *Levels 1* and *2* is schematically shown in Fig. 3. At the transition of *Level 1* with surface and extradural recording to *Level 2* with intracranial (epi-, subdural or intracerebral) recordings, sphenoidal recordings (Rovit et al. 1960; Ives and Gloor 1977) and registrations via the foramen ovale electrodes from mesial parts of the temporal lobe (Wieser et al. 1985) are of special interest, because with these techniques epileptic activity of infero-mesiobasal temporal structures can be recorded. In patients with partial accordance of lesions with *Level 1* electrophysiological findings, foramen ovale recordings can be used to good effect. Multifocal epilepsies, for instance with temporal and extratemporal activity or secondary bilateral synchrony, are to be investigated in *Level 2* with epi-or subdural or depth electrodes, before the principal epileptogenic region can be definitely determined. The succession and duration of the different phases during presurgical evaluation as practiced in Zürich are demonstrated in the chapter "Stereoelectroencephalography" (this volume). After localization of epileptic activity, careful studies of the risks of possible functional deficits, which might accompany resection of the focus, are performed. These studies often include the Wada test (Wada 1949). The final decision for causal, palliative, or no surgery is arrived at in discussion between neurosurgeon, epileptologist, and patient.

Considering the complexity of presurgical evaluation, with numerous anamnestic data and findings of intensive monitoring and other investigations, the problem of simplification and rationalization of data storage is not a trivial one. The many findings of the evaluation process are to be pooled for final hypothesis. To facilitate this, a protocol notebook has been developed in Zürich and appears in the Appendix.

References

Babb TL, Mariani E, Crandall PH (1974) An electronic circuit for detection of EEG seizures recorded with implanted electrodes. Electroencephalogr Clin Neurophysiol 37: 305-308

Bancaud J (1959) Apport de l'exploration fonctionelle par voie stéréotaxique à la chirurgie de l'épilepsie. Neuro-chirurgie 5: 55-112

Binnie CD, Rowan AJ, Overweg J, Meinardi H, Wisman T, Kamp A, Lopes da Silva F (1981) Telemetric EEG and video monitoring in epilepsy. Neurology 31: 298-303

Brazier MA (1956) Depth recordings from the amygdaloid region in patients with temporal lobe epilepsy. Electroencephalogr Clin Neurophysiol 8: 532-533

Burr W (1982) Computerized pattern analysis of MLE recordings. In: Stefan H, Burr W (eds) EEG monitoring. Fischer, Stuttgart, pp 257-287

Carrie JRG (1976) Computer-assisted EEG sharp-transient detection and quantification during overnight recordings in an epileptic patient. In: Kellaway P, Petersén I (eds) Quantitative analytic studies in epilepsy. Raven, New York, pp 225-235

Crandall PH, Walter RD, Ravel RW (1963) Clinical applications of studies on stereotactically implanted electrodes in temporal lobe epilepsy. J Neurosurg 20: 822-840

Delgado-Escueta AV, Enrile Bacsal Fe, Treiman DM (1982) Complex partial seizures on closed-circuit television and EEG: a study of 691 attacks in 79 patients. Ann Neurol 11: 292-300

Dodge HW, Bailey AA, Bickford RG, Petersen MC, Sem-Jacobsen CW, Miller RH (1953) Neurosurgical and neurologic applications of depth electrography. Proc Mayo Clin 29: 188-191

Engel J Jr, Rausch R, Lieb JP, Kuhl DE, Crandall PH (1981) Correlation of criteria used for localizing epileptic foci in patients considered for surgical therapy of epilepsy. Ann Neurol 9: 215-224

Engel J Jr, Crandall PH, Rausch R (1983) The partial epilepsies. In: Rosenberg RN, Grossman RG, Schochet S, Heinz ER, Willis WD (eds) The clinical neurosciences, vol 2. Livingstone, New York pp 1349-1380

Gloor P (1975) Contributions of electroencephalography and electrocorticography to the neurosurgical treatment of the epilepsies. In: Purpura DB, Penry JK, Walter RD (eds) Advances in neurology, vol 8. Neurosurgical management of the epilepsies. Raven, New York, pp 59-105

Gloor P (1984) Electroencephalography and the role of intracerebral depth electrode recordings in the selection of patients for surgical treatment of epilepsy. In: Porter RJ, Mattson R, Ward AA, Dam M (eds) Advances in epileptology. XVth epilepsy international symposium. Raven, New York, pp 433-437

Goldring S (1978) A method for surgical management of focal epilepsy, especially as it relates to children. J Neurosurg 49: 344-356

Goldring S, Gregorie EM (1984) Surgical management of epilepsy using epidural recordings to localize the seizure focus. J Neurosurg 60: 457-466

Gotman J (1982) Automatic recognition of epileptic seizures in the EEG. Electroencephalogr Clin Neurophysiol 54: 530-540

Gotman J, Gloor P (1976) Automatic recognition and quantification of interictal epileptic activity in the human scalp EEG. Electroencephalogr Clin Neurophysiol 41: 513-529

Gotman J, Ives JR, Gloor P (1979) Automatic recognition of interictal epileptic activity in prolonged EEG recordings. Electroencephalogr Clin Neurophysiol 46: 510-520

Gotman J, Ives JR, Gloor P (1982) A computer system for prolonged EEG recordings of ictal and interictal activity in epilepsy. In: Akimoto H, Kazamatsuri H, Seino M, Ward A (eds) Advances in epileptology, XIII epilepsy international symposium. Raven, New York, pp 167-172

Ives JR, Gloor P (1977) New sphenoidal electrode assembly to permit long-term monitoring of the patients ictal and interictal EEG. Electroencephalogr Clin Neurophysiol 42: 575-580

Ives JR, Thompson CJ, Gloor P (1976) Seizure monitoring: a new tool in electroencephalography. Electroencephalogr Clin Neurophysiol 41: 422-427

Kaiser E (1976) Telemetry and video recording on magnetic tape cassettes in long-term EEG. In: Kellaway P, Petersén J (eds) Quantitative analytic studies in epilepsy. Raven, New York, pp 279-288

Lombroso CT, Erba G (1970) Primary and secondary bilateral synchronic epilepsy. A clinical and electroencephalographic study. Arch Neurol 22: 321-334

Lüders H, Hahn J, Lesser RP, Dinner DS, Rothner D, Erenberg G (1982) Localization of epileptogenic spike foci: comparative study of closely spaced scalp electrodes, nasopharyngeal, sphenoidal, subdural and depth electrodes. In: Akimoto H, Kazamatsuri H, Seino M, Ward AA (eds) Advances in epileptology: XIIIth epilepsy international symposium. Raven, New York, pp 185-189

Musella L, Wilder BJ, Schmidt RP (1971) Electroencephalographic activation with intravenous methohexital in psychomotor epilepsy. Neurology 21: 594-602

Penin H (1968) Neuartige Diagnostik- und Forschungsanlagen in der Universitäts-Nervenklinik Bonn. Acta medicotechn. 16: 76-78

Porter RJ, Wolf AA Jr, Penry JK (1971) Human electroencephalographic telemetry: a review of systems and their applications and a new receiving system. Am J Electroencephalogr Technol 11: 145-159

Quy RJ, Fitch P, Willison RH (1980) High speed automatic analysis of EEG spike and wave activity using an analogue detection and microcomputer plotting system. Electroencephalogr Clin Neurophysiol 49: 187-189

Rovit RL, Gloor P, Henderson LR (1960) Temporal lobe epilepsy - a study using multiple basal electrodes. I. Description of methods. Neurochirurgia 3: 5-18

Saint-Hilaire JM, Bouvier G, Lymburger J, Picard J, Mercier R (1976) La stéréoelectroencephalographie synchronisée avec l'enregistrement visuel et sonore dans l'exploration chronique de l'epilepsie. Union Med Can 105: 1583-1541

Spencer SS, Spencer DD, Williamson PD, Mattson RH (1981) Ictal effects of anticonvulsant medication withdrawal in epileptic patients. Epilepsia 22: 297-307

Stefan H (1982) Epileptische Absencen. Studien zur Anfallsstruktur, Pathophysiologie und zum klinischen Verlauf. Thieme, Stuttgart

Stefan H, Burr W (eds) (1982) Mobile long-term EEG monitoring. Proceedings of the MLE symposium, Bonn. Fischer, Stuttgart

Stefan H, Burr W (1985) Atlas of mobile long-term EEG recording. Fischer, Stuttgart

Stefan H, Burr W, Hildenbrand K, Penin H (1981) Computer-supported documentation in the video analysis of absences: preictal-ictal phenomena: polygraphic findings. In: Dam M, Gram L, Penry JK (eds) Avances in epileptology: XIIth epilepsy international symposium. Raven, New York, pp 365-373

Stefan H, Burr W, Penin H (1983) Time structure analysis of motor phenomena in absence epilepsies. In: Speckmann EJ, Elger CE (eds) Epilepsy and motor system. Urban and Schwarzenberg, Munich, pp 310-324

Talairach J, David M, Tournoux P (1958) L'exploration chirurgicale stéréotaxique du lobe temporal dans l'épilepsie temporale. Repérage anatomique stéréotaxique et technique chirurgicale. Masson, Paris 136 pp

Wada J (1949) A new method for the determination of the side of cerebral speech dominance. A preliminary report on the intracarotid injection of sodium amytal in man. Igaku to Seibutssugaku (Jpn) 14: 221-222

Walsh GO, Delgado-Escueta AV (1984) Type II complex partial seizures: poor results of anterior temporal lobectomy. Neurology 34: 1-13

Wieser HG (1981) Stereoelectroencephalographic correlates of focal motor seizures. In: Speckmann EJ, Elger CE (eds) Epilepsy and motor system. Urban and Schwarzenberg, Munich, pp 287-309

Wieser HG (1982) Value of long-term EEG- and stereo-EEG in candidates of surgical epilepsy therapy. In: Stefan H, Burr W (eds) Mobile long-term EEG monitoring. Fischer, Stuttgart, pp 121-136

Wieser HG (1983) Electroclinical features of the psychomotor seizure. Fischer-Butterworth, Stuttgart-London, 242 pp

Wieser HG (1986) Psychomotor seizures of hippocampal-amygdalar origin. In: Pedley TA, Meldrum BS (eds) Recent advances in epilepsy 3. Livingstone, Edinburgh, pp 57-79

Wieser HG, Bancaud J, Talairach J, Bonis A, Szikla G (1979) Comparative value of spontaneous and chemically and electrically induced seizures in establishing the lateralization of temporal lobe seizures. Epilepsia 20: 47-59

Wieser HG, Elger CE, Stodieck SRG (1985) The 'foramen ovale electrode': a new recording method for the preoperative evaluation of patients suffering from mesiobasal temporal lobe epilepsy. Electroencephalogr Clin Neurophysiol 61: 314-322

Editors' Commentary 5

The preoperative workup of patients "has to make use of all evidence available, has to weigh the significance of each piece of evidence against others, and thus to arrive at the most logically consistent concept of what constitutes the nature of the underlying pathology, whether it is defined structurally, functionally, or both" (Gloor 1987a).

Besides the clinical history of a patient (with special emphasis on the patient's seizure pattern(s) and possible etiological factors) and a careful neurological examination, quite a number of tests both noninvasive and invasive have to be considered. For several reasons it is appropriate to differentiate these into noninvasive and invasive categories, and to classify the diagnostic armamentarium into (a) procedures which test for epileptic excitability (interictal epileptiform EEG activity, ictal EEG), (b) functional deficits (interictal and ictal behavior, neuropsychological assessment, "functional imaging techniques" such as PET, SPECT, and functional mapping), and (c) procedures which test for structural abnormalities (radiological tests, MRI, "lesional activity" in the EEG).

There was general agreement that restricting surgical treatment only to patients with gross structural lesions is too narrow. But even with a view towards such a limited approach, it is indispensable to demonstrate – prior to surgery – the relevant epileptogenicity of a radiologically detectable structural lesion. Although in general a structural lesion often is closely linked with the site of seizure origin, its precise anatomical site and extent are not necessarily the same.

In this context the connotation of the "epileptogenic lesion" had to be defined, i.e., the neuroradiological evidence of a brain lesion per se, and differentiated from the "lesional zone" with electrophysiological evidence of "lesional activity" (such as slow-wave activity in the EEG, attenuation of normal background rhythms, signs of hypoexcitability, and the evidence derived from pharmacological tests measuring, for example, focal attenuation of barbiturate-induced, fast EEG activity).

The definition of the "epileptogenic area," which is thus not synonymous or coexistent with the "lesional area," as "that of a positively defined area of seizure onset" (Gloor 1987a) remains the crucial step in the preoperative evaluation. However, at present this involves an individually flexible and by necessity highly interactive, step-wise, and goal-directed diagnostic process rather than the application of a rigid protocol. Whereas the goal (i.e., to arrive at the highest possible convergence of localizational data derived by independent methods) was generally accepted, quite a lot of discussion dealt with how it should best be achieved. In particular, questions were raised whether or not ictal data were always necessary, that is, whether interictal EEG data were sometimes sufficient or even preferable, and when nonivasive tests were sufficient. Two main factors were shown to critically influence this discussion (see also Gloor 1987a, b). They are referral patterns and the existing facilities, "philosophies" (i.e., strategies), and "familiar" approaches of an epilepsy surgical center (see also Engel 1987). For example, one center may advocate a rather restrictive approach by limiting surgical treatment to cases with unequivocal evidence of a radiologically demonstrable unilateral lesion (Ammon's horn sclerosis or indolent tumors on the side of the "EEG focus"), while another may offer surgical treatment to a wide range of patients including cases without demonstrable lesion and with less convincing evidence of a well-localized unifocal seizure disorder.

Furthermore, the surgical approach per se is influential insofar as the aim might be to pursue a highly "tailored" resection closely fitted to the "boundaries" of the epileptogenic area, or – conversely – a standard operation (such as temporal lobectomy or amygdalohippocampectomy) which might take care of the epileptogenic lesion. Finally there are centers where the feeling prevails that it is necessary to concentrate all efforts on patients who appear most likely to benefit from surgical treatment following preliminary findings, and to reject "difficult cases" at an early stage. Other centers hold that "difficult cases" should be just as carefully evaluated, if only with the limited scope of an amelioration ("palliative" surgical treatment) instead of a cure. There was a tacit agreement that contradicting or nonconvergent sets of noninvasive data require further, properly designed, intracranial recordings and that a center ought to have the potential to choose between or combine stereotactic depth and subdural or epidural recording techniques. If stereotactic depth recording is to solve the discrepancies or to provide the best answer for a particular question, specific neuroradiologic studies under stereotactic conditions are required, with a threefold aim: to select precisely the most appropriate targets, confirming the actual recording sites, and to avoid critical vascular and cerebral structures in order to minimize the risks of this procedere. The incorporation of recent brain imaging techniques and their use within a stereotactic reference system is highly desirable. It will enhance both potential and precision, and at the same time minimize the invasive nature of this part of the examinations. Olivier and colleagues (1987) have described an integrative approach for the stereotactic utilization of these modern techniques, such as CT, MRI, PET, and digital subtraction angiography (DSA).

References

Engel J Jr (1987) Approaches to localization of the epileptogenic lesion. In: Engel J Jr (ed) Surgical treatment of the epilepsies. Raven, New York, pp 75-95

Gloor P (1987a) Commentary: approaches to localization of the epileptogenic lesion. In: Engel J Jr (ed) Surgical treatment of the epilepsies. Raven, New York, pp 97-100

Gloor P (1987b) Postscript: when are noninvasive tests enough? In: Engel J Jr (ed) Surgical treatment of the epilepsies. Raven, New York, pp 259-261

Olivier A, Marchand E, Peters T, Tyler J (1987) Depth electrode implantation at the Montreal Neurological Institute and Hospital. In: Engel J Jr (ed) Surgical treatment of the epilepsies. Raven, New York, pp 595-601

Conventional Electroencephalography

R. M. Hess

Once upon a time, the EEG was, for the evaluation of epilepsy, the predominant method of investigation. Within the span of one generation, a host of shortcomings and limitations relating to the method's precision and reliability came to light, while new developments and alternative techniques progressed continuously. Many feared that electroencephalography might lapse into insignificance. There is therefore good reason to evaluate, within the context of the variety of methods of examination now available, the present place of the conventional EEG.

Electroencephalography also has developed and improved. Looking for a way to record a faithful image of the electrical activity under each electrode, truly indifferent reference points were proposed (Stephenson and Gibbs 1951; Sorel 1969). Enhanced discrimination of focal abnormalities was obtained by the "source derivation" (Hijort 1975), which may gain importance for localization of foci. Automatic frequency analysis (Grass and Gibbs 1938; Walter 1943a, b) and a variety of methods for quantifying the EEG by computer techniques (cf. Brazier 1961; Walter and Brazier 1968; Cobb and Van Duijn 1978) may become relevant to our subject, once multichannel on-line procedures are routine. Telemetric recording over longer periods is already an excellent diagnostic tool, particularly in combination with automatic pattern recognition, for detecting concealed focal epilepsies; if the computer is still less than perfect in this respect as compared with the eye of an experienced electroencephalographer, it has the advantage of not getting tired when looking through a 24-h record. Once still more parameters are used for programming, it is to be expected that the computer will perform better than man, lacking as it does the disadvantage of subjective prejudice.

All the technical improvements cannot circumvent one *serious failing,* namely, that one unavoidably deals with the potential changes of the scalp electrodes, and we know from the experience with depth recording how incomplete and distorted an image of the real electrical activity of the brain is picked up by them.

Since the early days, researchers have attempted to record from placements closer to the source of potential changes of diagnostic significance by means of pharyngeal, sphenoidal, and tympanic electrodes (MacLean 1949; Grinker and Serota 1938) or to reach cortical areas remote from the convexity by needle electrodes sunk into the brain. Such techniques will be reported separately in the following chapters. It should only be mentioned here that they have already contributed to a better understanding of some aspects of the scalp EEG and, by comparing simultaneous surface and depth records, there is hope for more such positive feedback.

The time-consuming and expensive procedure of stereoelectroencephalography (SEEG) can be used on a restricted number of patients only. The task of preselecting such candidates for possible surgery falls to the conventional EEG. The selection begins with the exclusion of *nonepileptic* seizure disorders. We know that even a normal EEG during the patient's habitual attacks is no absolute evidence against its being of an epileptic nature; it is, nevertheless, suggestive of some other etiology. An unspecific abnormality, on the other hand, is in favor of organic attacks of cerebral origin. A clear-cut slow-wave focus in the presence of seizures is strong evidence of focal epilepsy, and transient trains of focal slow rhythms of a few seconds duration are highly suggestive of such (Fig. 1); it may be surmised that at some distant place the slow waves are locked to sharp waves, as can be demonstrated occasionally in the scalp EEG, and more readily with depth electrodes (Fig. 2).

It is notorious that the *paroxysmal potential changes* are the actual hallmark of epilepsy, yet they are not invariably of absolute diagnostic value. Simple sharp waves in the midtemporal region in elderly people, or Rolandic sharp waves in children of school age, are not necessarily associated with clini-

Rietholzstr. 34, 8125 Zollikerberg, Switzerland

Presurgical Evaluation of Epileptics
Edited by H. G. Wieser and C. E. Elger

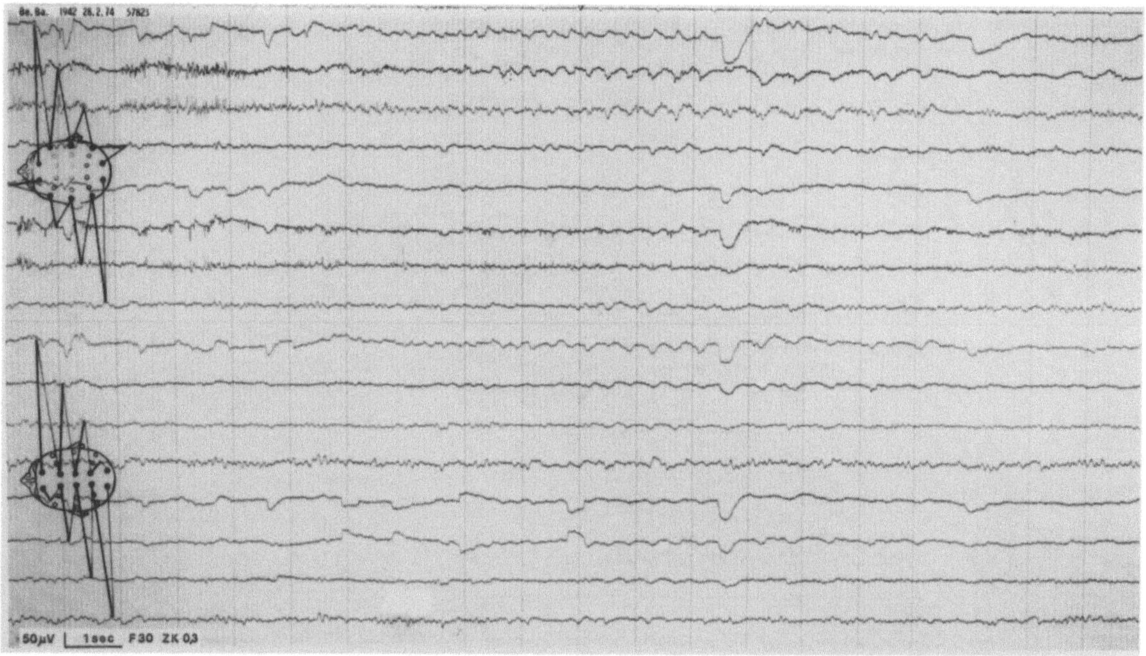

Fig. 1. Gradual buildup of a slow rhythm with right frontotemporal maximum amplitude and wide spread, ending with a few barely recognizable sharp waves. There is no sign or subjective experience of an attack in this instance. This patient complains of occasional short episodes of arrest of thought, unnoticed by witnesses. Years ago, he had suffered a few grand mal seizures of probable traumatic origin

cal epilepsy. Such a connection is, on the contrary, highly probable when 3-Hz spikes and waves or unequivocal spiking over one temporal pole are found. Diagnostic EEG patterns are, however, not always recorded at the first attempt. Repeated attempts are often required, if possible at the time or in the situation suspected to promote the natural seizures, or else right after an attack. An EEG after sleep deprivation or during a short nap is always desirable, whereas the tedious all-night sleep record does not usually provide much more information. Pharmacological activation is rarely appropriate at this stage: it is unsuitable for distinguishing epileptic seizures from others, mainly because of too many false-positive results. It still has its merits for activating a silent focus in a case of partial epilepsy as suspected from description or confirmed by clinical observation of the ictal event. (Inasmuch as the evidence so obtained is of practical importance almost exclusively for patients in whom surgical therapy is considered, the pharmacological activation has its real place during SEEG, when it is liable to yield much more information).

As a second step of the screening, we have to eliminate the primary generalized epilepsies, i.e., those seizure disorders in which a local origin cannot be determined. I do not mean to imply that such cases are totally inaccessible to surgical therapy. Successful interventions were made in many instances, and they may become current in the future, but the eventual role of the EEG in this context has yet to be defined.

Even now the occurrence of *bilateral discharges,* synchronous at visual inspection, is no unequivocal evidence against a hidden lesion as their primal cause. The best-known example is the "secondary bilateral synchrony" of Tükel and Jasper (1952), who described mainly mesial frontal or frontobasal sites of origin, but focal spiking at other locations may also trigger bilateral discharges (Jasper 1949; Stevens 1970). Whenever clinical or radiological data raise suspicions of a localized condition, the EEG has to be studied with particular care for additional focal abnormalities. In order to detect them, such above-mentioned methods as source derivation and long-term monitoring may be appropriate, and a sleep record is mandatory. Pharmacological activation is unlikely to be helpful, since bilateral discharges are, as a rule, more readily enhanced than the focal ones we are looking for. It needs hardly to be stressed that patients with bilateral paroxysms will be candidates for surgery on very rare occasions only. The great majority of such cases, at the present state of the art, have to be dismissed. This decision is

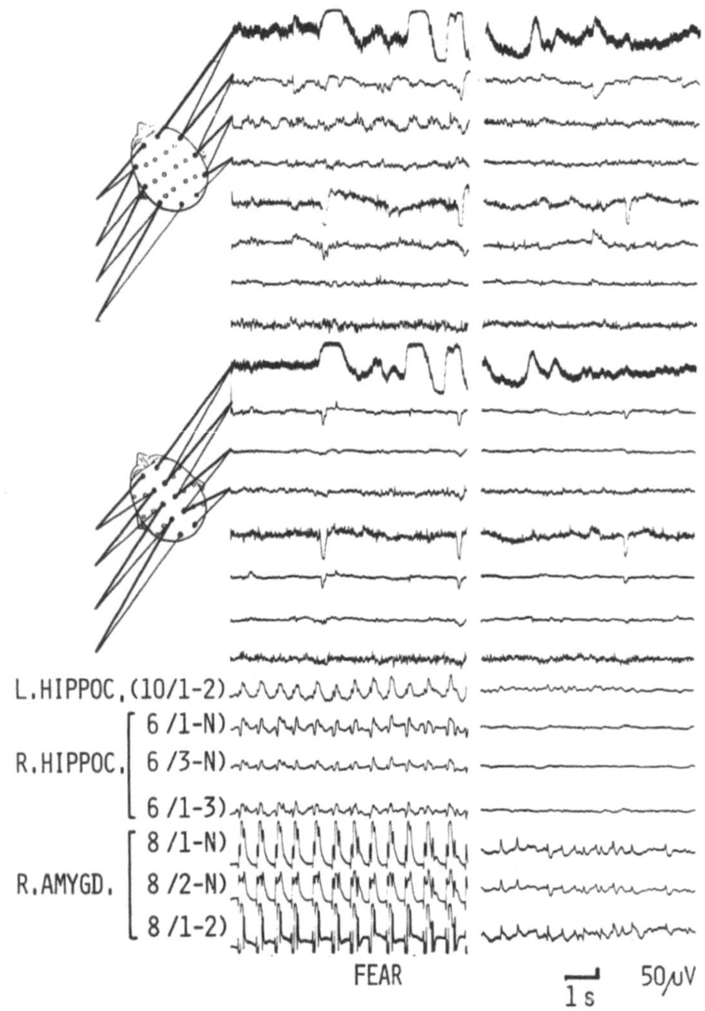

Fig. 2. Epileptic discharge in the right nucleus amygdalae, transmitted to the right and left hippocampus. At the surface it shows over the right temporal lobe in the form of rhythmic slow waves, synchronous with the spike complexes in the depth. The rest of the scalp EEG appears normal and desynchronized: the patient experiences fear during the ictal event. *Right side of figure,* essentially normal, with the exception of afterdischarge-like activity in right amygdala. *First 16 channels,* conventional EEG as shown in the schematic drawing. *Bottom 7 channels,* derived from innermost contacts (1, 2) of three needle electrodes (10, 6, 8). *N,* nose reference. (Courtesy of Dr. H. G. Wieser)

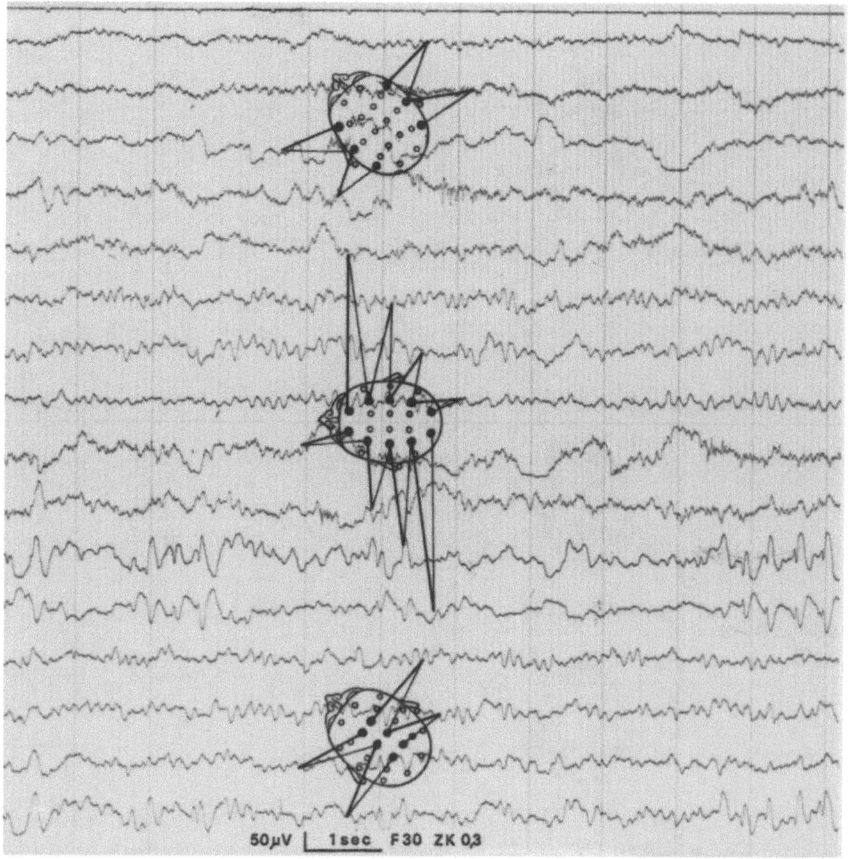

Fig. 3. Focal spike and sharp wave complexes in the left parietal region, where slow waves and marked reduction of alpha rhythm also suggest a gross cortical lesion. Five-year-old girl with atrophy of the left hemisphere and epileptic attacks after early brain damage of an unknown origin

arrived at with more confidence, the closer the EEG patterns resemble the "classical" spikes and waves (cf. Jasper and Droogleever-Fortuyn 1947), which are one of the most reliable findings in the EEG screening, although from the surgeon's point of view, it is a negative one.

Positive findings give more satisfaction, not least to the patient, who craves for relief from his intractable seizures. A clear-cut *epileptogenic focus* is also most rewarding for the electroencephalographer, but it does not dispense him from scrutinizing the records with all the auxiliary techniques available for additional abnormalities, such as occasional bilateral discharges or a possible secondary focus, which might bear an influence on subsequent investigations. Bancaud and coworkers (1965) produced ample evidence for the fact that the most obvious EEG focus is not always the one which is responsible for the clinical seizures. The real epileptogenic area may be situated at a distance from the interictal spiking, even at a quite different place and often hidden in the depth. Indirect signs, sometimes in the form of an unobtrusive transient unspecific abnormality, are then what we can pick up at best, on the surface (Fig. 2).

In the case of *neocortical focal epilepsy,* the situation is not too complicated as a rule. Gross lesions or superficial scars are often responsible, and the epileptogenic focus is commonly located in its vicinity. An unspecific abnormality, particularly an area of depressed background activity around the spike focus, adds to its significance (Fig. 3). Contralateral paroxysmal discharges may be secondary and do not necessarily preclude surgery, particularly if ("functional") hemispherectomy is considered. Apart from this, the well-known rules may be applied, namely, that frequent, stable sharp spikes are more likely to originate close by, whereas sporadic blunt triangular transients, especially if locked to slow waves, may be propagated from some distant point. It is, furthermore, my experience that paroxysmal complexes with positive main deflection sug-

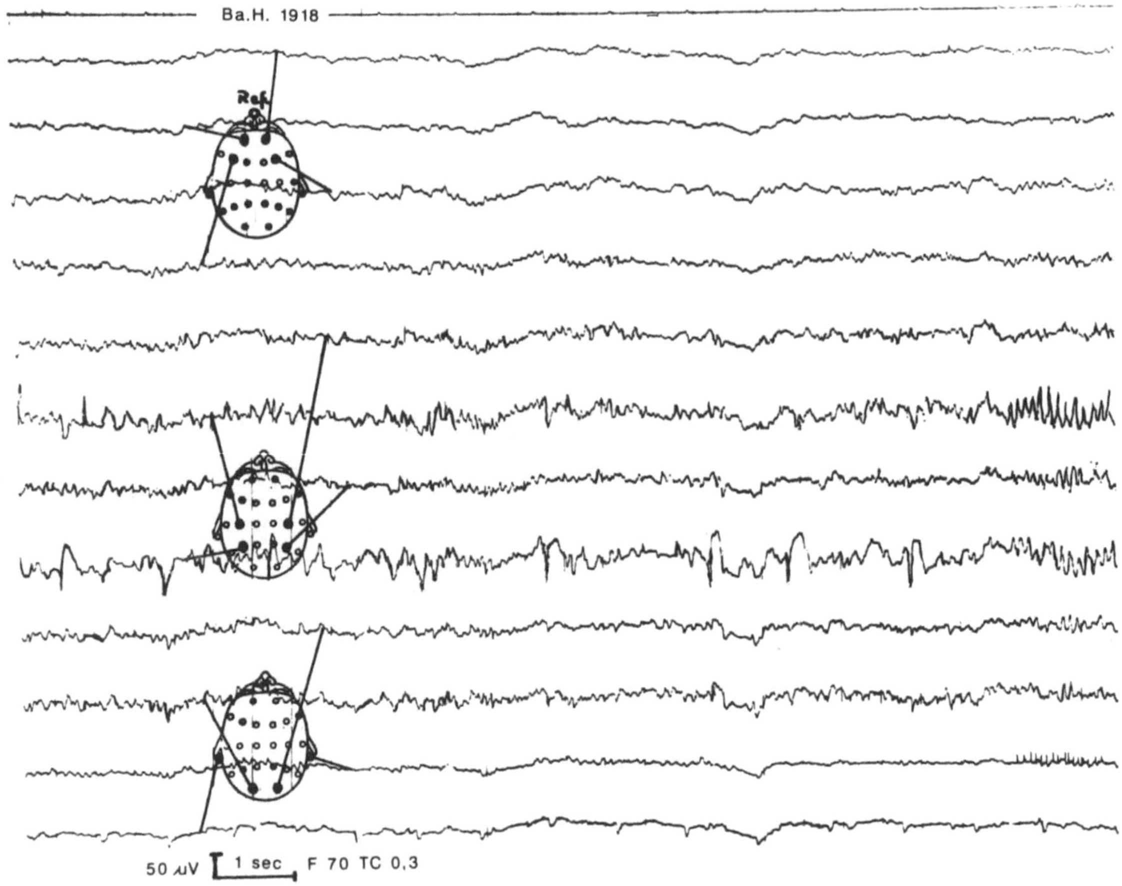

Fig. 4. Spike and sharp wave complexes with positive main deflection in the left parietal area, where a large intracerebral and subarachnoidal cyst had been operated on and a subcalibre projectile was extracted 25 years ago. No overt seizures were reported since the operation. Nose reference

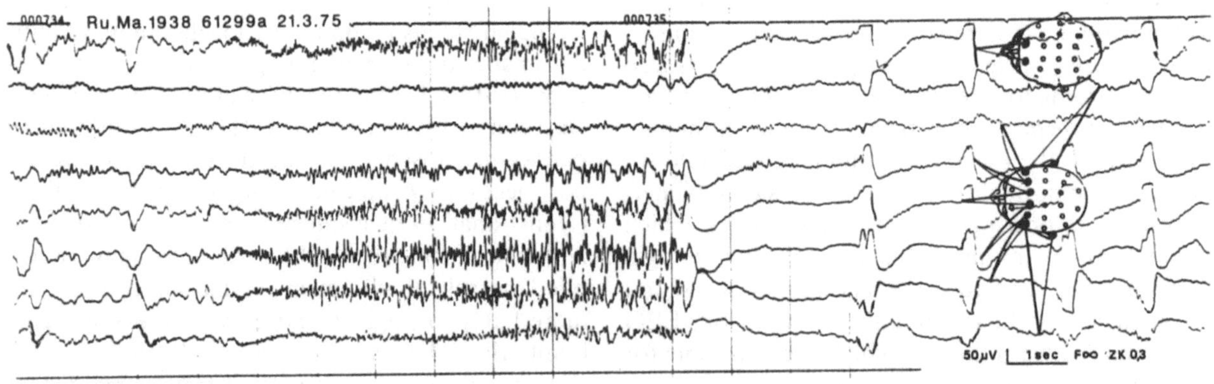

Fig. 5. Aborted focal seizure discharge in the left frontal region, with onset close to the frontal pole. No overt clinical signs. An epileptogenic lesion in the immediate neighborhood must be assumed

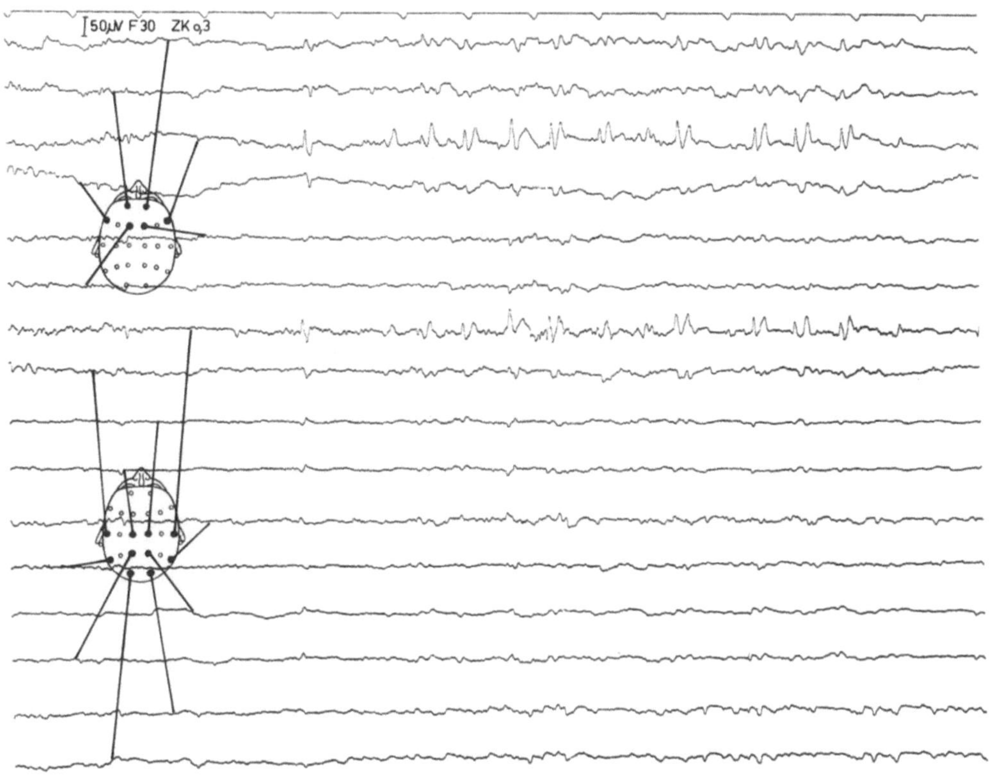

Fig. 6. Serial sharp waves over the right temporal pole in a patient with complex partial seizures. This "semi-ictal" discharge takes place during the lightest sleep: Note the slow swings in channels 3 and 4, due to lateral eye movements

gest the presence of a large cortical defect (Fig. 4). If by chance an ictal discharge can be recorded - with or without observable signs of a clinical attack - the epileptic potency of the focal abnormality is evident (Fig. 5).

In irritable lesions of the convexity, mirror foci are not uncommon, especially in children, and they are easy to recognize as a rule. Even if the relevance of a superficial focus seems undisputed, one has to make sure that a less obtrusive secondary focus has not been missed. At the base of the temporal lobe, damage is not infrequently caused by the same noxious event from which the superficial lesion resulted, or (still somewhat controversial) by local metabolic changes due to severe seizures.

Although the pioneers of surgical treatment of epilepsy operated successfully on a large variety of epileptogenic lesions all over the brain (Penfield and Erickson 1941; Penfield and Jasper 1954), the bulk of today's candidates for surgery are the so-called *"temporal epilepsies,"* i. e., cases where the limbic system is involved, the deep-lying site of which implies substantial limitations for the scalp EEG. The most common finding in between seizures is a frontotemporal sharp wave focus (Fig. 6). It is a reliable sign of limbic epilepsy, but it does not permit further differentiation, since the temporal pole seems to participate readily in the exaggerated activity of any part of the limbic system. Additional abnormalities must therefore be looked for. They will but rarely allow attribution of the case to any of the subgroups, which can be distinguished with the SEEG, but they may give an indication for placing the depth electrodes.

A contralateral focus is a common finding, and it must be determined whether it is a mere mirror focus propagated from the other side or whether it is entirely independent. Although the latter situation does not altogether preclude surgical therapy, it is certainly an argument against it. There are also intermediate situations, which must be evaluated according to circumstances. One obviously tries to record seizure discharges, which are often elicited by light sleep and sleep deprivation. Pharmacological activation cannot be relied upon for producing the habitual seizure form (Bancaud et al. 1968; Wieser et al. 1979). An ictal discharge recorded from the onset should contain all the information available at the surface of the head. Unfortunately, it is often obscured by heavy muscle artifact just in the so-important temporal regions, except when occurring during

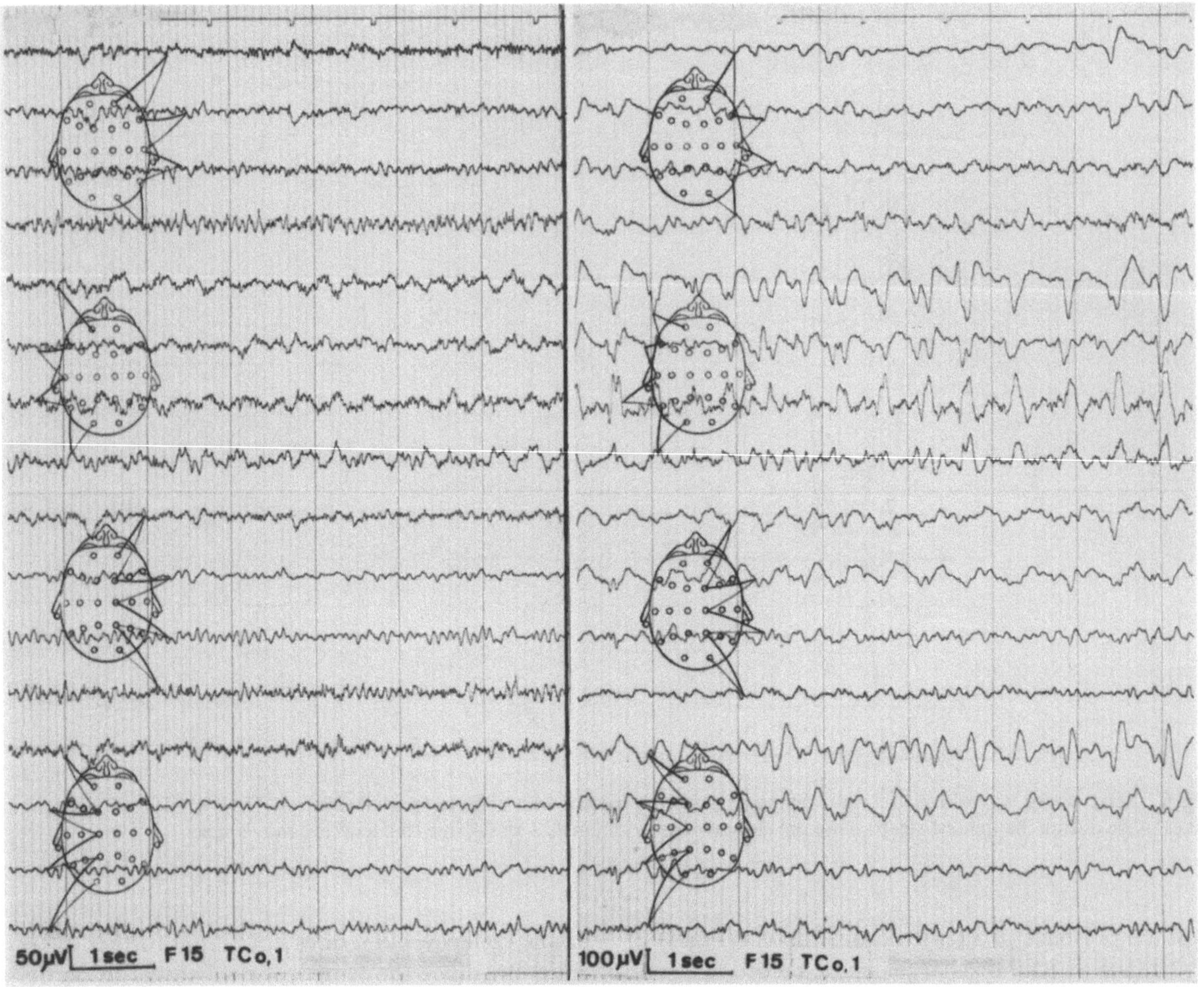

Fig. 7. Two sections of a recorded psychomotor (complex partial) attack. *Left* Fifteen minutes after onset, the patient is conscious and reports feeling strange. The EEG over the right hemisphere is virtually unchanged, in the left parasagittal regions there is some slowing, but the alpha activity persists. The left temporal lobe is dominated by widespread activity around 3 c/s. *Right* One minute later, the patient's consciousness appears impaired. The alpha rhythm is broken up and diminished, slowing is generalized, and frontal delta waves are linked in time with pseudorhythmical sharp waves in the left anterior temporal region (note decreased gain)

sleep. Even then, the conventional EEG misses the very beginning of the discharges, which take place in deep structures, and it reflects secondary projection phenomena only, and eventually the late spread of seizure activity (Fig. 7).

It is no wonder that the complex and multifarious EEG patterns have so far defied any reliable correlation with clinical seizure forms and that they are difficult to classify, as Gastaut and Vigouroux (1958) testified, after analyzing over 500 psychomotor seizures of their patient series. The starting point of the discharges can rarely be identified with confidence. Quite often the initial change consists in a general flattening of the record, which may be taken as an arousal reaction to a subjective aura. In other cases the flattening is local, probably as the concomitant of circumfocal inhibition, but often widely spread, which limits its localizing value. It is not unreasonable to hope that further experience with depth electrodes will permit the pointing out of structures which are more likely than others to cause, when excessively active, such a local depression: this would greatly increase this sign's significance. In the meantime, a record of an abortive seizure, with or without unobtrusive clinical symptoms or signs, or a prolonged onset (not marred by artifact) of a full seizure, may be more informative, as it allows the tentative conclusion that the limbic structures are involved in the excessive discharges.

This is then *the whole bioelectrical evidence* - obviously complemented by clinical and radiological data - which the pioneers of epilepsy surgery had at hand for their interventions. Their success cannot be disputed. Yet the worldwide tendency toward ever larger excisions and, in the case of limbic epilepsy, to remove the whole temporal lobe as far back as possible without risk, does not convey the impression of adequate precision of the presurgical evaluation. I remember well my frustration when, during such operations, after each partial excision new spike foci kept popping up in the border areas, and above all in the remaining limbic structures. All too often, furthermore, the abnormally functioning cortex seemed to extend beyond the field made accessible by the craniotomy. The shortcomings of the EEG for delimiting the epileptogenic area were manifest.

The conventional EEG represents then, in the presurgical evaluation of epilepsies, mainly the method of a first broad screening. It serves to eliminate clearly unsuitable cases, indicates the gross localization of irritative foci and can possibly suggest ways of propagation of seizure discharges. It thus provides basic data for planning the subsequent, more intricate stages of investigation.

References

Bancaud J, Talairach J, Bonis A, Schaub C, Szikla G, Morel P, Bordas-Ferer M (1965) La stéréo-électroencéphalographie dans l'épilepsie. Masson, Paris, 321 pp

Bancaud J, Talairach J, Waltregny P, Bresson M, Morel P (1968) L'activation par la mégimide dans le diagnostic topographique des épilepsies corticales focales. Rev Neurol 119: 320-325

Brazier MAB (ed) (1961) Computer techniques in EEG analysis. Electroencephalogr Clin Neurophysiol [Suppl 20]

Cobb WA, Van Duijn H (eds) (1978) Contemporary neurophysiology. Electroencephalogr Clin Neurophysiol [Suppl 34]

Gastaut H, Vigouroux M (1958) Electro-clinical correlations in 500 cases of psychomotor seizures. In: Baldwin M, Bailey P (eds) Temporal lobe epilepsy. Thomas, Springfield, pp 118-128

Grass AM, Gibbs FA (1938) A fourier transform of the electroencephalogram. J Neurophysiol 1: 521-526

Grinker RR, Serota HM (1938) Studies on corticohypothalamic relations in the cat and man. J Neurophysiol 1: 573-589

Hijort B (1975) An on-line transformation of EEG scalp potentials into orthogonal source derivations. Electroencephalogr Clin Neurophysiol 39: 526-530

Jasper HH (1949) The electroencephalography in neurosurgery. Proceed. IVe Congrès Neurologique International I: 21-26

Jasper HH, Droogleever-Fortuyn J (1947) Experimental studies on the functional anatomy of petit mal epilepsy. Res Publ Assoc Res Nerv Ment Dis 26: 272-298

MacLean PD (1949) A new nasopharyngeal lead. Electroencephalogr Clin Neurophysiol 1: 110-112

Penfield W, Erickson TC (1941) Epilepsy and cerebral localization. Thomas, Springfield, 623 pp

Penfield W, Jasper HH (1954) Epilepsy and the functional anatomy of the brain. Little, Brown, Boston, 896 pp

Sorel L (1969) Les montages verticaux en électroencéphalographie. Bull Acad r Méd Belg 9: 587-632

Stephenson WA, Gibbs FA (1951) A balanced non-cephalic reference electrode. Electroencephalogr Clin Neurophysiol 3: 237-240

Stevens JR (1970) Focal abnormality in petit mal epilepsy. Neurology 20: 1069-1076

Tükel K, Jasper HH (1952) The electroencephalogram in parasagittal lesions. Electroencephalogr Clin Neurophysiol 4: 481-494

Walter DO, Brazier MAB (eds) (1968) Advances in EEG analysis. Electroencephalogr Clin Neurophysiol [Suppl 27]

Walter GW (1943a) An automatic low frequency analyser. Electron Engin 16: 9-13

Walter GW (1943b) An improved low frequency analyser. Electron Engin 16: 236-240

Wieser HG, Bancaud J, Talairach J, Bonis A, Szikla G (1979) Comparative value of spontaneous and chemically and electrically induced seizures in establishing the lateralization of temporal lobe seizures. Epilepsia 20: 47-59

Special Extracranial Electrodes

L. F. Quesney and P. Gloor

Introduction

The main objective of the preoperative EEG investigation in patients with partial or complex partial seizures resides in providing reliable localization of their seizure problem. This task is best attained by means of prolonged EEG monitoring which is primarily aimed at the recording of the patient's habitual attacks (Gotman et al. 1985 for review). This method can supply localizing evidence deriving from several sources: documentation of the clinical seizure pattern; recording of the ictal electrographic features, ideally including its correlation with the clinical behavioral manifestations, and recording of the interictal EEG abnormalities.

Clinical localization in partial seizures is unreliable, except in patients with simple partial motor or sensory seizures. Complex partial seizures originating in the temporal lobe normally do not provide clinical lateralizing clues. Somatomotor ictal manifestations frequently observed in complex partial seizures may result from spread of seizure discharge from temporal to extratemporal regions (Bossi et al. 1984) and they do not provide reliable lateralization or localization of the seizure onset (Ochs et al. 1984). Complex partial seizures of extratemporal origin sometimes produce little or no distinctive ictal behavioral manifestations while the seizure discharge is confined to its extratemporal site of origin. Subsequent seizure spread to temporal lobe frequently leads to ictal automatic behavior, wrongly suggesting a temporal lobe seizure onset (Livingston and Escobar 1971; Quesney et al. 1984; Quesney and Gloor 1985). Ictal automatisms may, however, be observed during seizures which remain confined to the frontal lobe, particularly in the orbitofrontal region (Geier et al. 1975, 1976, 1977; Ludwig et al. 1975; Munari et al. 1981; Quesney et al. 1984; Quesney and Gloor 1985).

Department of Neurology and Neurosurgery, McGill University, Montreal Neurological Institute and Hospital, 3801 University Street, Montreal, Quebec H3A 2B4, Canada

Presurgical Evaluation of Epileptics
Edited by H. G. Wieser and C. E. Elger

Despite the recent development of sophisticated neuro-imaging diagnostic techniques, the role of electroencephalography in the localization of epileptic foci remains undisputed. The preoperative EEG localization of seizure onset should coincide with the localizing evidence deriving from neurological examination, neuroradiological abnormality, and neuropsychological dysfunction. These ancillary methods provide helpful localizing evidence (Penfield and Jasper 1954; Rasmussen and Jasper 1958; Rasmussen and Branch 1962; Milner 1975; Rasmussen 1982), but they mainly indicate brain regions with predominant damage, which do not always correspond to the site of origin of the patient's seizures (Gloor 1975; Engel et al. 1981; Quesney and Gloor 1985). Focal interictal epileptiform abnormality, which is commonly and easily recorded in patients with partial seizures, may also provide useful localizing evidence particularly when its anatomical distribution agrees with the site of origin of the spontaneous seizures (Jasper and Kershman 1941; Penfield and Jasper 1954; Zivin and Ajmone-Marsan 1968; Gloor 1975; Quesney and Gloor 1985; Rovit et al. 1961).

This chapter is devoted to the contribution of EEG monitoring with special extracranial electrodes towards the preoperative localization of epileptic foci. Substantial evidence in the literature supports the view that reliable electrographic localization of epileptic foci is attainable by means of EEG recording using extracranial electrodes (Bloom et al. 1959/60; Engel et al. 1975, 1981, 1983; Gloor 1975; Laws et al. 1970; Lieb et al. 1976, 1981; Lüders et al. 1982; Penfield and Erikson 1941; Penfield and Jasper 1954; Rasmussen 1963, 1975, 1979, 1982; Rasmussen and Jasper 1958; Quesney and Gloor 1985; Rovit et al. 1961).

The localizing evidence provided by ictal and interictal epileptiform abnormalities suffers, however, from limitations which must receive proper consideration before an indication for surgical therapy is decided.

Limitations of Extracranial EEG Recordings in the Localization of Epileptic Foci

Perhaps one of the most significant and frequently underestimated limitations of extracranial EEG recordings derives from the fact that the voltage of electrical potentials generated by the brain drops off significantly in an inversely proportional ratio to the square of the distance (Gloor 1975; Morris and Lüders 1985). Scalp electrodes are invariably located at a distance from the generator, even if this is located in the brain's surface. Generators located in the mesial and basal surfaces of the brain and generators buried in sulci or fissures are farther away and, therefore, focal epileptiform discharges originating in these areas may go undetected. In some of these cases the EEG discloses widespread interictal epileptiform potentials in areas of cerebral cortex receiving projections from the inaccessible structures where the primary epileptic focus is located. Anatomical pathways which may mediate the propagation of electrophysiological signals between temporal and frontal lobes include cingulum and the uncinate fasciculus (Buser et al. 1973). It is therefore possible that an EEG tracing obtained with scalp electrodes may miss the primary epileptic focus located in the orbitofrontal region, amygdala or hippocampus, only to disclose more widespread interictal epileptic abnormality recorded from the frontotemporal convexity (Figs. 2, 5, 14, 16). Similar localizing problems may arise in electrocorticography (Figs. 15, 17).

Bitemporal independent epileptiform potentials with consistent unilateral predominance have been reported in 25%-28% of patients with complex partial seizures of temporal lobe origin submitted to surgical therapy (Engel et al. 1975; Van Buren et al. 1975; Rasmussen 1982). Such bilateral discharges may at least partially reflect a process of secondary epileptogenesis, also referred to as "mirror focus" (Morell et al. 1959/1960; Bancaud et al. 1969). Background activity abnormality is often recorded from the primary epileptic focus, whereas spiking emerging from areas of secondary epileptogenesis is commonly associated with a normal background activity.

Bifrontal interictal or ictal epileptic abnormalities are commonly recorded in patients with unilateral frontal lobe seizures (Table 1). This phenomenon is more likely to occur in epileptic foci located in the orbitofrontal, cingulate, or mesial parasagittal regions (Tharp 1972; Ludwig et al. 1975; Goldring 1972; Tükel and Jasper 1952; Mazars 1970).

The localizing accuracy of interictal and ictal epileptiform abnormality can be increased by using special extracranial electrodes which render basal regions of the frontal and temporal lobes more accessible to EEG recording.

Table 1. Localizing effectiveness of seizure monitoring. [From Quesney and Gloor (1985) with permission]

	Group I	Group II
Seizures recorded	311	302
Number of patients	30	16
Localized seizure onset	127 (40.4%)	66 (22%)
Lateralized seizure onset	29 (9.4%)	34 (11%)
Bilateral or generalized seizure onset	71 (22.8%)	111 (37%)
Uninterpretable	84 (27.4%)	57 (19%)
Absence of ictal EEG changes	0	34 (11%)

Contribution of Special Extracranial Electrodes to the Localization of Epileptic Foci

Several special extracranial electrodes have been developed in order to improve the EEG diagnostic effectiveness in the inferomesial temporal lobe structures and in the orbitofrontal region. Currently, the most commonly employed special extracranial electrodes in the investigation of patients with temporal or frontal lobe epilepsy include nasopharyngeal (Pg1-Pg2), sphenoidal (Sp1-Sp2), nasoethmoidal (NE1-NE2), and supraorbital (SO1-SO2) electrodes.

Nasopharyngeal Electrodes

Following their introduction in the late forties, nasopharyngeal leads were widely accepted as an effective and simple method to obtain EEG recording from the mesial temporal lobe structures (Gastaut 1948; Roubicek and Hill 1948; MacLean 1949). They are semirigid coated electrodes introduced through the nostrils into the posterior-lateral nasopharynx, with their tips located immediately under the anterior portion of the mesial temporal lobe surface (Fig. 1). Much of the popularity of these electrodes derived from the lack of other practical and noninvasive means to record from this region, as well as from the simplicity of their application, which can be performed by EEG technologists. These considerations outweighed some of the known disadvantages of these electrodes, such as discomfort to patients due to local irritation of the nasal mucosa and frequent recording of breathing or pulse artifacts, which often render the tracing unin-

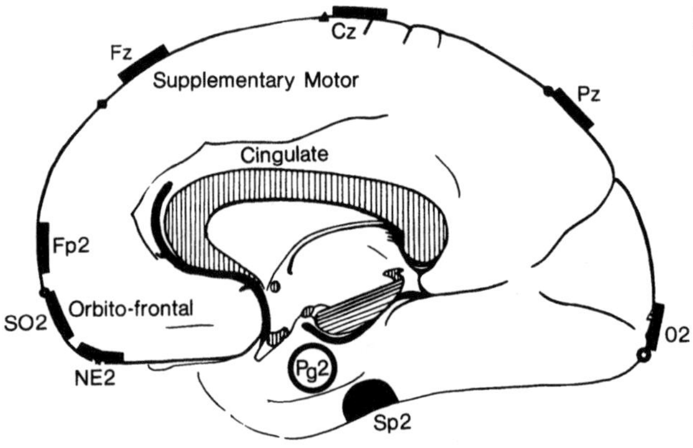

Fig. 1. Recording sites of commonly employed special extracranial electrodes

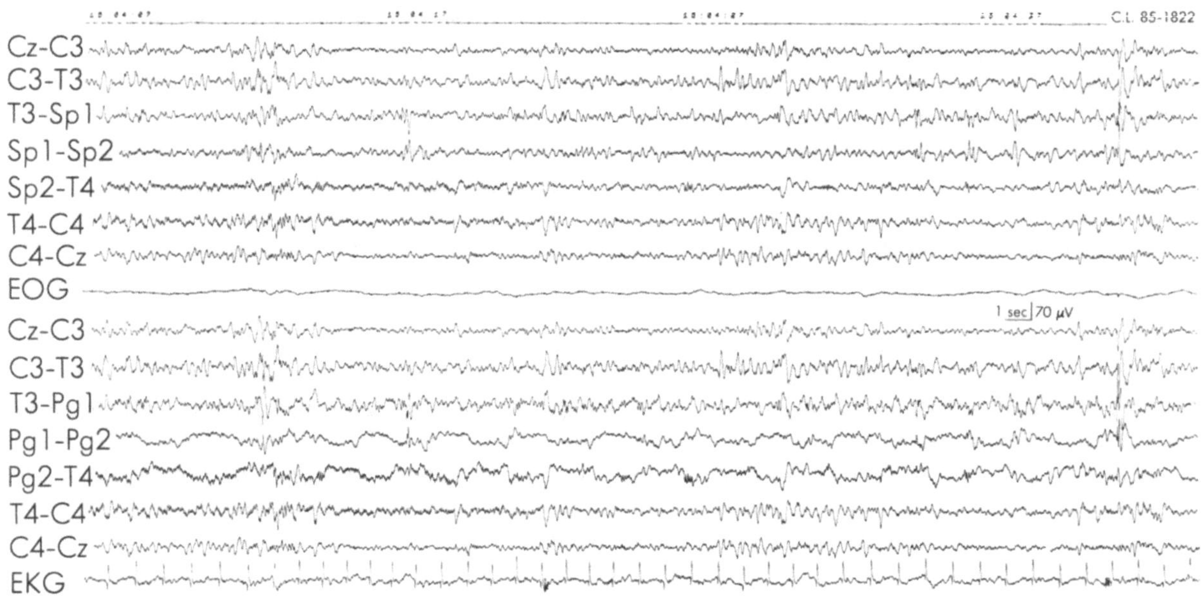

Fig. 2. Simultaneous recording with nasopharyngeal and sphenoidal electrodes in a patient with complex partial seizures. Notice the breathing artifact recorded from the right nasopharyngeal lead *(Pg2)*. Frequent spiking was recorded from the left sphenoidal electrode with only occasional reflection in the left nasopharyngeal lead *(Pg1)*

terpretable (Fig. 2). Despite these limitations, a vast and encouraging experience accumulated over the years regarding the usefulness of these electrodes in the investigation of patients with complex partial seizures of suspected temporal lobe origin (Mavor and Hellen 1964). Reevaluation of the efficacy of nasopharyngeal recordings was prompted in the early sixties by the introduction of anterior temporal (Silverman 1960) and sphenoidal electrodes (Jones 1951; Pertuiset and Capdevielle-Arfel 1951), techniques which offered a novel alternative to obtain reliable recording of temporal lobe epileptogenic foci. More recently, the relative effectiveness of nasopharyngeal and anterior temporal electrodes in terms of recording epileptiform potentials has been reevaluated by two different centers, yielding similar conclusions. A UCLA study (Sperling and Engel 1985) comprising 103 EEGs obtained in 98 patients suspected of suffering from epilepsy showed that epileptiform discharges recorded by nasopharyngeal leads were invariably detected by anterior temporal electrodes. That observation led this group to conclude that the use of nasopharyngeal electrodes offers no additional benefit in the recording of temporal lobe epileptiform discharges. The second study (Morris and Lüders 1985) comprised 26 patients with complex partial seizures. Their results clearly indicate that anterior temporal electrodes (T1, T2) record temporal lobe epileptiform discharges more effectively than nasopharyngeal, anterior Sylvian (F7, F8), and midtemporal (T3-T4) electrodes (Fig. 3).

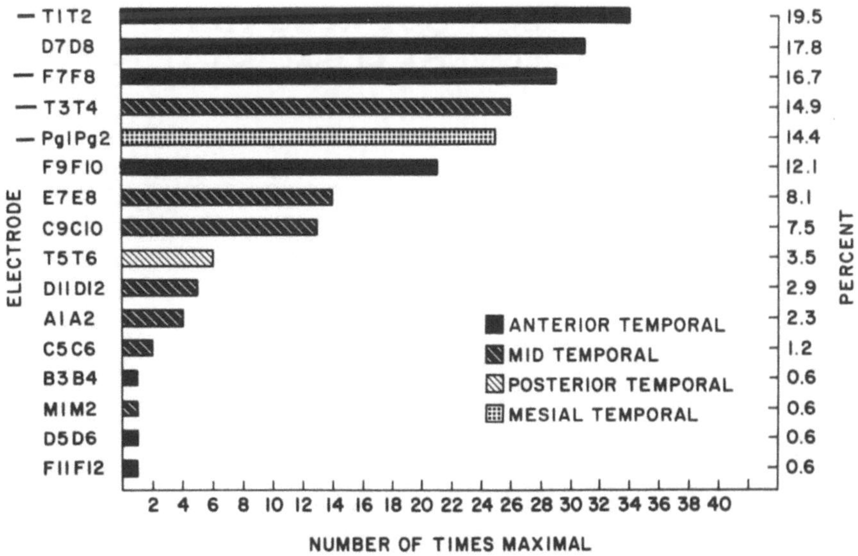

Fig. 3. Bar graphs representing the effectiveness of different extracranial electrodes in the recording of temporal lobe epileptic abnormality. [From Morris and Lüders (1985) with permission]

Sphenoidal Electrodes

The undisputed role of mesiobasal temporal lobe structures in the pathogenesis of temporal lobe epilepsy prompted the development of special extracranial electrodes to record from this brain region, which is otherwise inaccessible to scalp electrodes (Fig. 1). Sphenoidal leads were introduced shortly after the description of nasopharyngeal electrodes (Jones 1951; Pertuiset and Capdevielle-Arfel 1951). The original technique proposed needle electrodes inserted transcutaneously below the zygomatic arch with their tips lying in the vicinity of the foramen ovale. Several reports confirmed the usefulness of this technique in the recording of focal epileptiform discharges involving the mesiobasal temporal lobe structures (Penfield and Jasper 1954; Pampiglione and Kerridge 1956). A modification of the original technique proposing stainless steel wires instead of needles enabled safer recording of ictal events to be made (Rovit et al. 1960). Ives and Gloor (1977, 1978) introduced the technique of chronic sphenoidal teflon-coated silver wires, which can be left in situ for several days up to 2 weeks without significant morbidity or unpleasant side effects, thus permitting a safe and effective preoperative seizure monitoring. The insertion of sphenoidal electrodes should be carried out by a physician.

Sphenoidal electrodes provide high-quality recordings with very little artifactual activity (Figs. 4-6). A recent study demonstrated that chronic silver wire sphenoidal electrodes do displace outwardly along their track of insertion during the course of prolonged EEG monitoring (Wilkus et al. 1985). This electrode displacement does not, however, affect the recording quality of ictal or interictal epileptiform abnormalities in patients with temporal lobe seizures, a finding which suggests that deep insertion of sphenoidal electrodes is less critical than originally assumed. This view is also shared by those postulating satisfactory recording of mesiobasal temporal lobe epileptiform abnormalities using minisphenoidal needle electrodes (Laxer 1984), which are inserted only 1 cm deep below the zygomatic arch or by means of surface zygomatic electrodes (Sindrup et al. 1981). Although it is generally accepted that the usefulness of sphenoidal electrodes resides mainly in providing reliable recording of epileptic foci involving the uncinate temporal lobe region, it is possible that epileptiform activation of other basal temporal lobe structures, such as parahippocampal and fusiform gyri as well as the third temporal gyrus, might add a significant contribution to the electrical field of the epileptic potentials recorded by sphenoidal electrodes.

The use of sphenoidal electrodes in the investigation of patients with complex partial seizures of temporal lobe origin offers several undisputed advantages:

1. It may disclose unequivocal epileptiform disturbance which is not properly recorded by standard scalp or nasopharyngeal electrodes (Fig. 4). Using sphenoidal leads, Kristensen and Sindrup (1978) found definite epileptic abnormality in 16% of

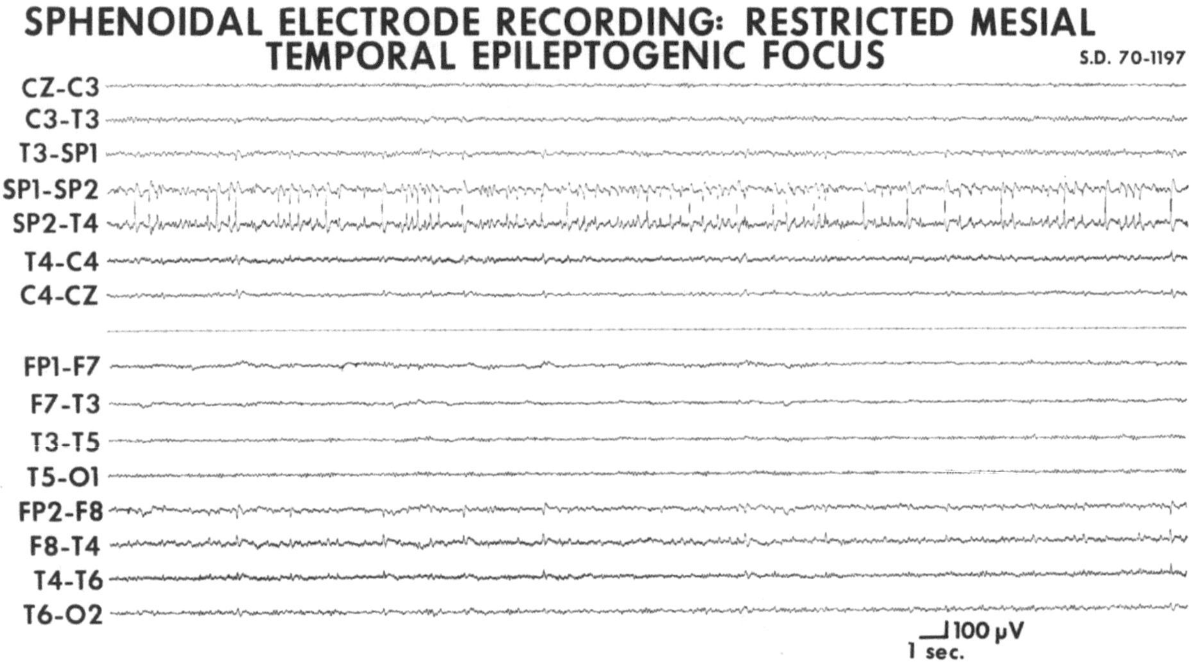

Fig. 4. Electroencephalogram obtained in a patient with complex partial seizures disclosing active spiking restricted to the right mesial temporal region. Notice phase reversal of the epileptiform discharge at the right sphenoidal electrode *(SP2)*. [From Gloor (1975) with permission]

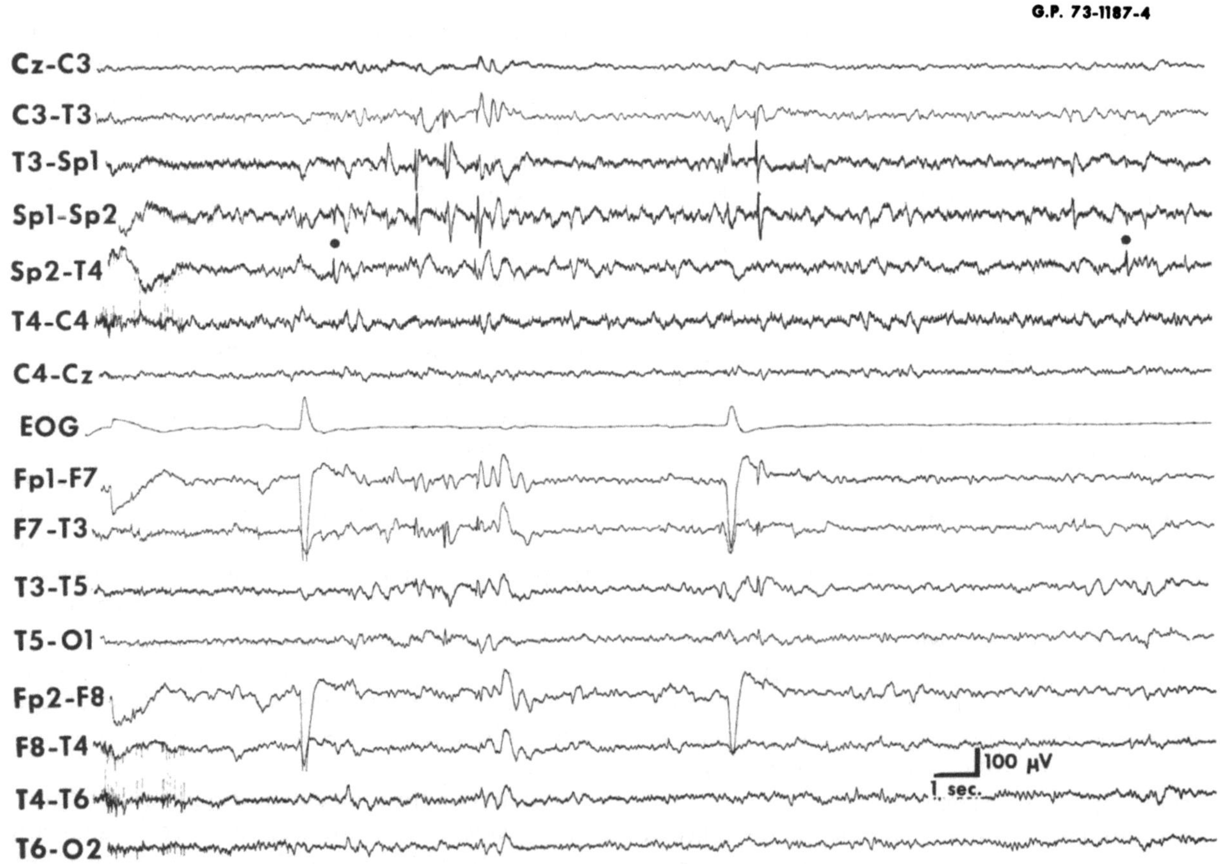

Fig. 5. Bitemporal independent spiking recorded with sphenoidal electrodes in a patient with complex partial seizures. The left temporal lobe epileptiform abnormality is more frequent and it often extends into the anterior and mid-temporal regions

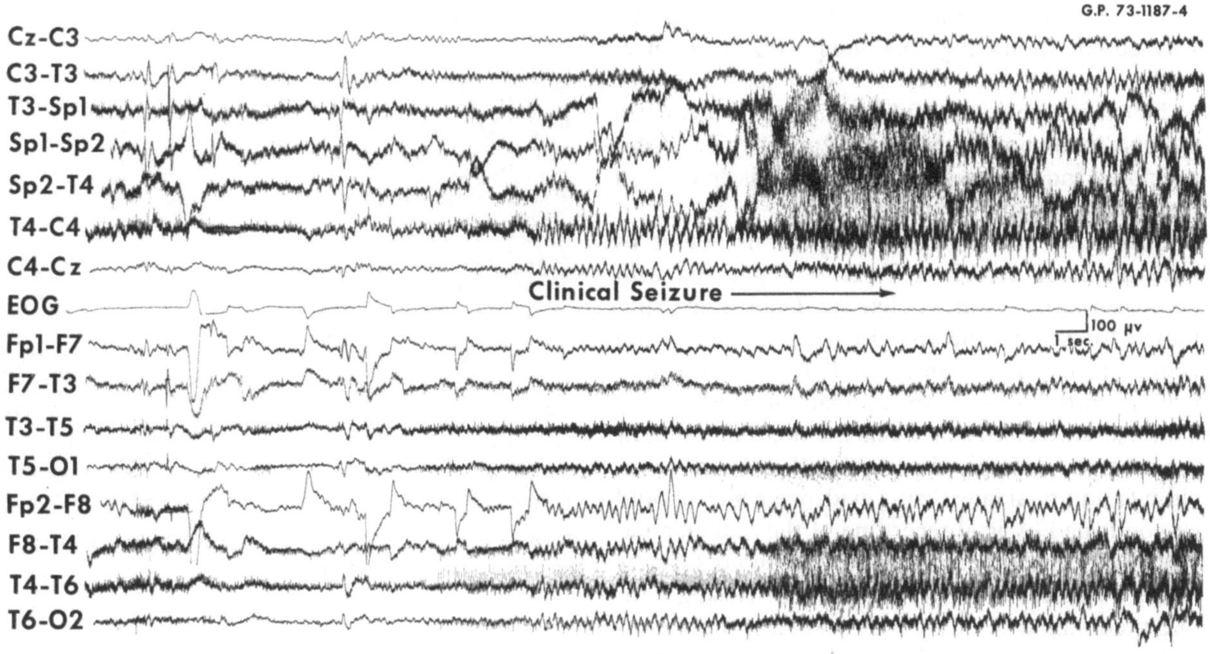

Fig. 6. Clinical seizure recorded in the same patient as in Fig. 5. Ictal electrographic changes consisted of rhythmic 4- to 5-Hz sharp wave activity lateralized to the right temporal lobe

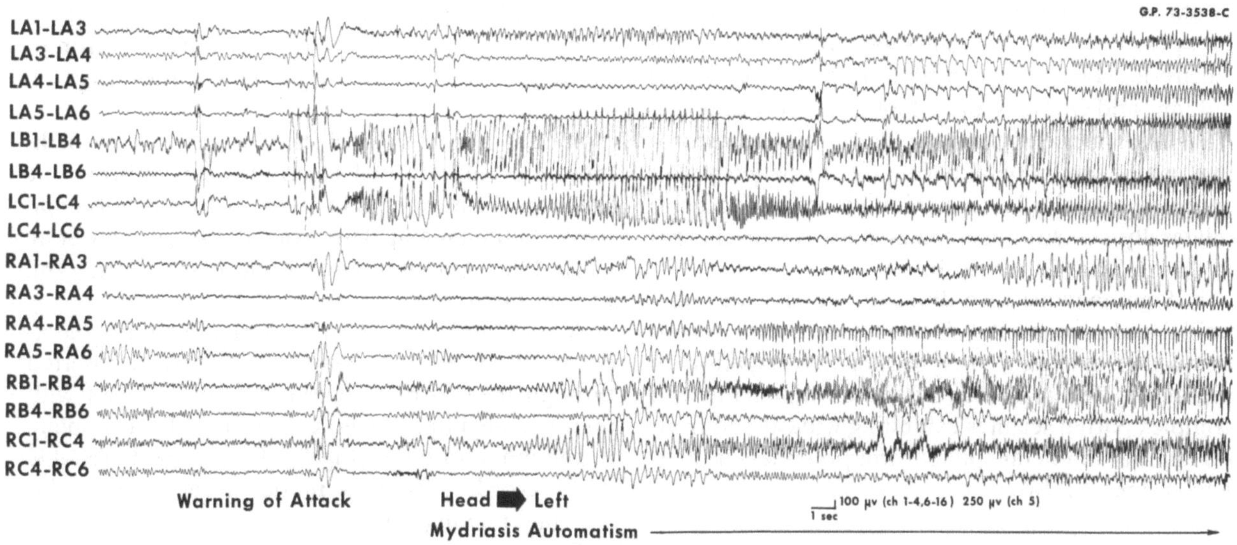

Fig. 7. Ictal recording obtained with chronically implanted depth electrodes in same patient as in Figs. 5 and 6. Depth electrodes were stereotactically implanted in both amygdalae *(LA* and *RA)*, anterior hippocampi *(LB* and *RB)*, and posterior hippocampi *(LC* and *RC)*. The patient's habitual attacks exhibited a regional seizure onset involving the amygdaloid *(LA1-LA3)*, anterior hippocampal *(LB1-LB4)*, and posterior hippocampal structures *(LC1-LC4)* of the left temporal lobe

294 patients with complex partial seizures, which was absent in tracings obtained with scalp electrodes.

2. It may provide useful lateralizing evidence in patients with complex partial seizures exhibiting bitemporal independent epileptiform discharges in their EEGs (Fig. 5).

Bitemporal independent epileptiform potentials have been reported in 25%-28% of patients with complex partial seizures of temporal lobe origin submitted to surgical therapy (Engel et al. 1975; Van Buren et al. 1975; Rasmussen 1982). In these patients recordings with sphenoidal electrodes may demonstrate a consistent unilateral predominance of the interictal epileptic abnormality (Fig. 5), thus providing additional lateralizing evidence which should be confirmed by ictal recording of the patient's habitual attacks. Commonly, the spontaneous seizures in these patients arise from the side disclosing definite predominance of the interictal epileptic abnormality recorded with sphenoidal electrodes (Gloor et al. 1980). There are, however, cases in which the lateralization of interictal and ictal abnormalities yields conflicting results (Fig. 6) (Gentilomo et al. 1975; Ajmone-Marsan 1978; Wieser et al. 1979; Gloor et al. 1980; Engel et al. 1981), thus requiring preoperative EEG investigation with intracranial electrodes (Fig. 7; see contribution by Wieser in this volume, and Quesney and Gloor 1985 for review).

3. Less frequently, sphenoidal electrode recordings may reveal a focal pacemaker in patients with complex partial seizures exhibiting generalized and bilaterally synchronous spike-and-wave discharges in their EEGs (Gloor 1975).

4. Simultaneous recordings obtained with sphenoidal and supraorbital electrodes are useful in confirming or excluding a significant participation of mesiobasal temporal lobe structures in patients with complex partial seizures exhibiting widespread frontotemporal interictal epileptiform potentials in their EEGs (Figs. 12, 16).

Nasoethmoidal and Supraorbital Electrodes

Several frontal lobe structures which are capable of generating distinctive ictal behavioral manifestations during spontaneous or electrically induced seizures are largely inaccessible to recordings with scalp electrodes (Fig. 1). They include: the orbital-frontal region; the mesial interhemispheric convexity, particularly the supplementary motor area; the cingulate gyrus; and, to a similar extent, the depth of cerebral sulci within the frontal lobe (Ludwig et al. 1975; Gloor 1975; Geier et al. 1977; Quesney et al. 1981, 1984; Mazars 1970). Epileptic foci arising from these regions can be easily missed in tracings obtained with scalp electrodes. Frequently, however, small epileptogenic foci located in some of these frontal lobe structures are reflected over the fronto-parasagittal or frontotemporal convexities (Figs. 14, 16).

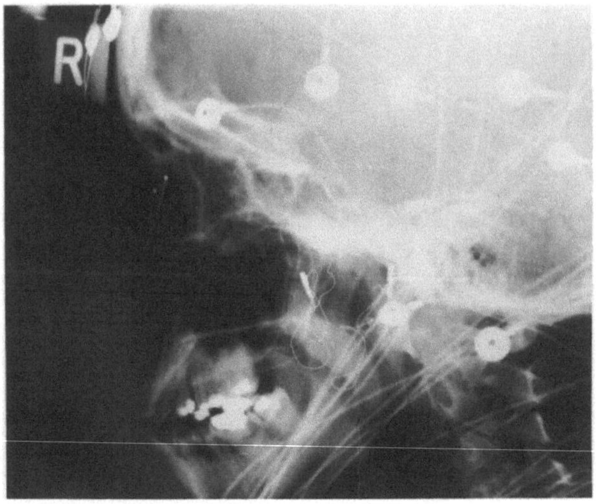

Fig. 8. Lateral skull X-ray showing nasoethmoidal electrode tips under the cribriform plate

Special extracranial electrodes such as nasoethmoidal or supraorbital electrodes can improve the EEG localization cffectiveness in frontal lobe epilepsy (Lehtinen and Bergstrom 1970; Quesney et al. 1981, 1984; Quesney and Gloor 1985; Morris and Lüders 1985).

Nasoethmoidal electrodes are flexible teflon-coated silver wires (1.5-2 mm in diameter) inserted through the nostrils into the ethmoid sinuses with their rounded tips placed under the cribriform plate of the ethmoid bone (Fig. 8). They provide reliable EEG recording from the mesial surface of the orbitofrontal region (Fig. 1).

The insertion of nasoethmoidal electrodes should preferably be performed by an ENT specialist and it is advisable to check the position of the electrodes with ap and lateral skull X-rays.

Of the patients submitted to nasoethmoidal electrode insertion, 30%-40% may report unpleasant side effects such as sneezing and eye tearing probably due to local irritation of the nasal and conjunctival mucosa. This symptomatology is, however, transient and chronic nasoethmoidal electrode recording (up to 7 days) has been reported to be free of complication (Morris and Lüders 1985).

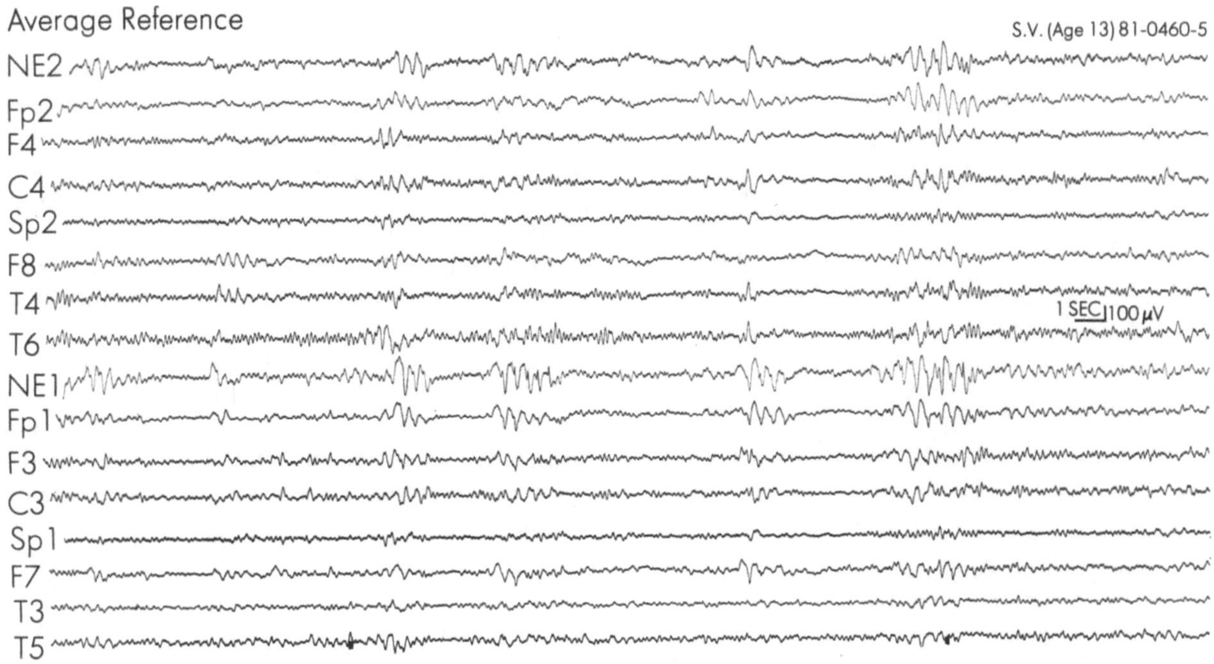

Fig. 9. Referential recording showing bifrontal epileptiform activity with amplitude predominance on the left side

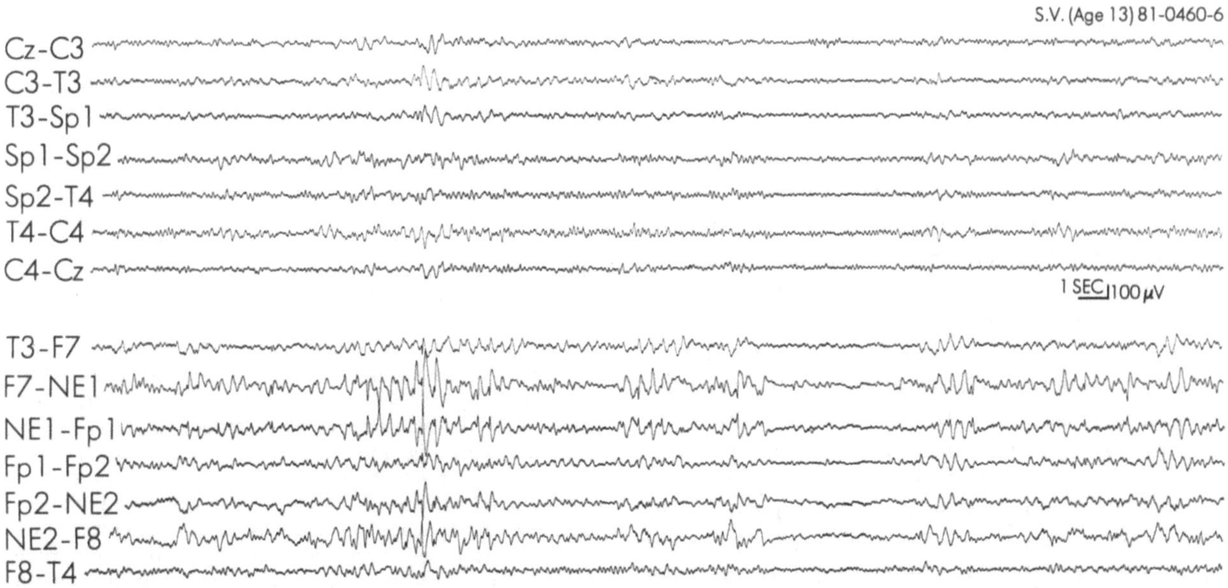

Fig. 10. Bipolar EEG recording obtained with sphenoidal and nasoethmoidal electrodes in the same patient as in Fig. 9. Notice focal independent spikes recorded at the left nasoethmoidal electrode *(NE1)* as well as a bifrontal epileptiform disturbance predominating at *NE1*

If properly inserted, nasoethmoidal electrodes provide reliable and high-quality EEG recordings. Eye movement is a common source of artifactual activity during wakefulness, but this problem is significantly reduced during sleep tracings. Breathing artifact is frequently recorded and its occurrence is rather unpredictable.

Nasoethmoidal electrodes are particularly helpful in the localization of epileptic foci in patients with complex partial seizures in whom the interictal epi-

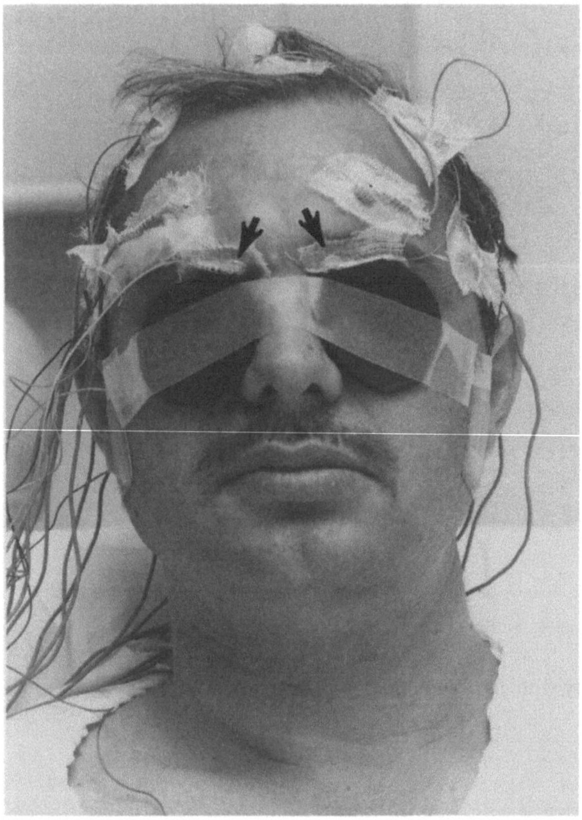

Fig. 11. Supraorbital electrode placement

leptic abnormality is widespread and lateralized to the frontocentrotemporal convexity (Figs. 9, 10, 13). They may also provide lateralizing evidence in cases exhibiting bifrontal epileptiform abnormalities. The latter was the situation in a 13-year-old right-handed female patient (SV) with a history of recurrent complex partial seizures starting at age 9 years. Her habitual seizures were characterized by an ill-defined epigastric sensation followed shortly afterwards by loss of consciousness, oral automatism, and clonic jerking of the right upper extremity. Several preoperative EEG tracings revealed bifrontal sharp-and-slow wave activity at 2.5-3 Hz with some left-sided predominance (Fig. 9). EEG recordings obtained with nasoethmoidal electrodes disclosed independent spiking recorded from the left orbitofrontal region (Fig. 10). Electrocorticography confirmed the preoperative localization.

Supraorbital electrodes, which also record from the orbitofrontal region (Fig. 1), are routine scalp electrodes placed on the supraorbital ridge, 1.5-2 cm from the nasal bone (Fig. 11). The application of these electrodes is performed by EEG technologists and no unpleasant side effects are reported by patients. Eye movement artifactual activity is common during wakefulness but it reduces significantly during sleep tracings.

In our experience supraorbital electrodes have gradually replaced nasoethmoidal leads on account of the simplicity of their application, lack of unpleasant side effects, and optimal recording quality (Fig. 12), which is only comparable to that achieved by nasoethmoidal electrodes (Fig. 13).

Recording with supraorbital and/or nasoethmoidal electrodes may be indicated in the following conditions:

1. Clinical suspicion of orbitofrontal seizures.

2. Localization of epileptic foci in patients with complex partial seizures in whom the EEGs exhibit widespread interictal epileptic abnormality in the frontocentrotemporal convexity. Two pertinent clinical cases will illustrate the usefulness of recordings obtained with supraorbital electrodes in this situation:

D.F., a 10-year-old male patient, suffered from poorly controlled complex partial seizures starting at 18 months of age. The patient's habitual seizure pattern was characterized by initial vocalization and loss of consciousness, followed by urination and automatic behavior during which the patient would pat his chest with his right hand, occasionally turning his head and body toward the right side. Neurological examination and CT were unremarkable. Several preoperative EEG tracings revealed frequent spike-and-sharp waves recorded from the right frontocentral region (Fig. 14). Recording with supraorbital electrodes showed equipotentiality of the abnormal potentials between the right supraorbital (SO2) and frontotemporal (F8) electrodes (see inserts of Fig. 14), thus indicating definite accentuation of the epileptogenic abnormality in the orbitofrontal region. At electrocorticography, continuous spiking was recorded from the inferomesial frontal lobe structures, which were densely gliotic (Fig. 15).

P.F., a 20-year-old patient, presented to us with poorly controlled complex partial seizures characterized by tonic head turning toward the right side, associated with tonic extension of the right upper extremity and loss of consciousness. A left frontal pole hypodensity was found on a brain CT scan and neuropsychological testing revealed a reduced IQ level. Preoperative EEG investigation with sphenoidal electrodes revealed widespread spike activity recorded from the left frontocentral and temporal convexities. During tracings obtained with supraorbital and sphenoidal electrodes, focal spiking was independently recorded from the left supraorbital electrode (Fig. 16). Electrocorticography revealed frequent high-amplitude slow sharp wave activity recorded from the orbitofrontal structures (Fig. 17). Less abundant sharp wave activity was recorded from the first, second, and third frontal gyri anteriorly.

3. Identification of a focal pacemaker within the frontal lobe in patients with complex partial seizures exhibiting bifrontal or generalized epileptiform discharges in their EEGs (see Figs. 9, 10).

4. Diagnostic EEG confirmations in patients with suspected frontal lobe seizures exhibiting no interictal epileptic abnormality in tracings obtained with standard scalp electrodes.

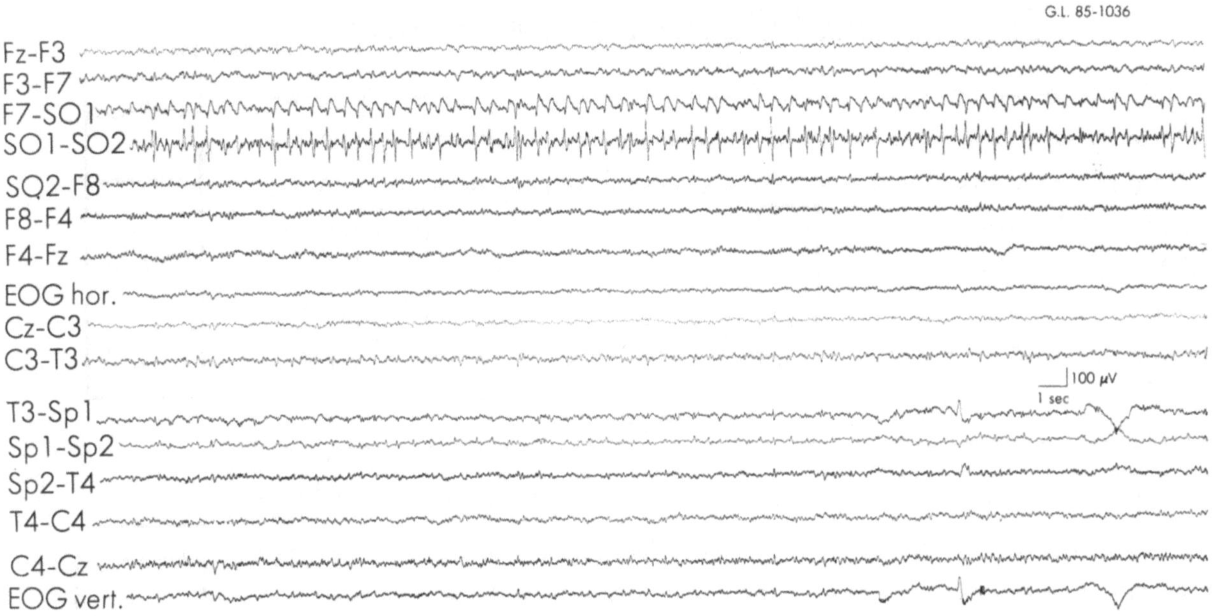

Fig. 12. Electroencephalogram in a patient with complex partial seizures disclosing continuous spiking restricted to the left orbitofrontal region. Notice phase reversal of the abnormal potentials at the left supraorbital electrode *(SO1)*

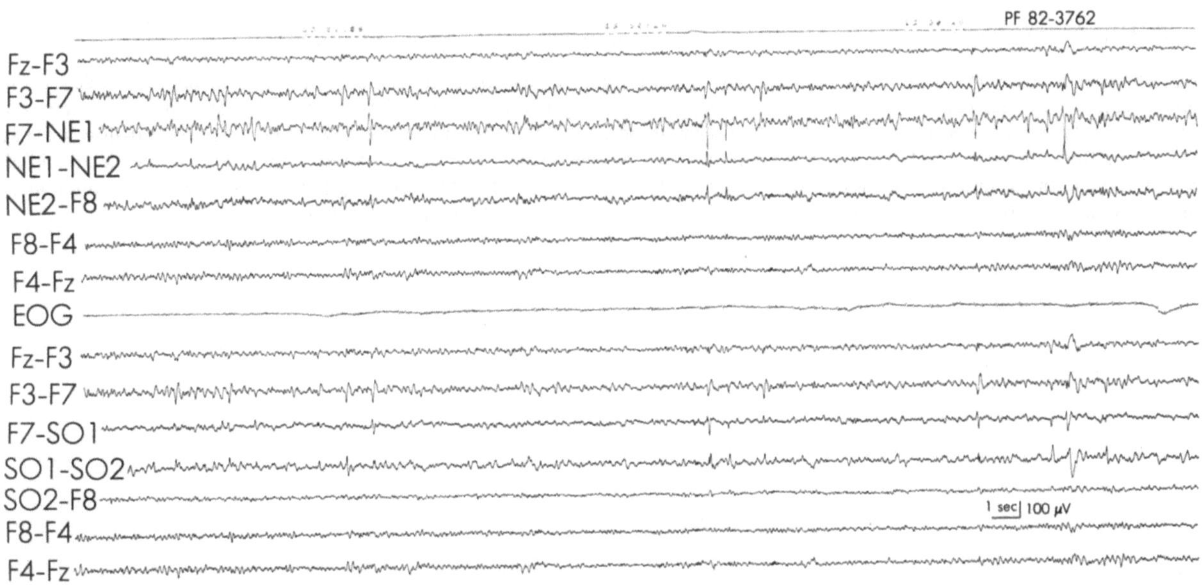

Fig. 13. Simultaneous nasoethmoidal and supraorbital electrode recording in the same patient as in Figs. 16 and 17. Supraorbital electrodes are as effective as nasoethmoidal leads in terms of recording epileptic discharges arising from the orbitofrontal cortex

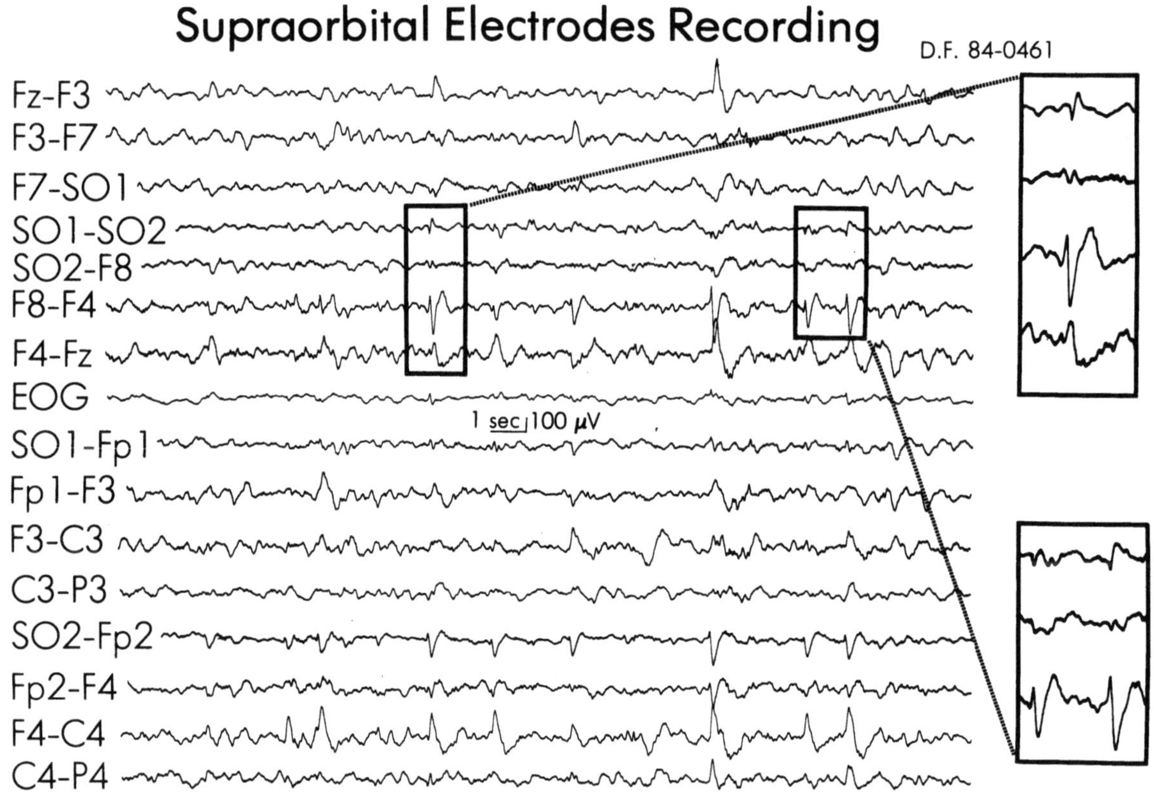

Fig. 14. Electroencephalogram obtained with scalp and supraorbital electrodes in a patient with right frontal lobe epilepsy. Frequent widespread sharp-and-slow waves are recorded from the right frontal region. Phase reversals occur at *F4* and across an equipotential zone between *Fp2* and *F4* in the longitudinal montage. Note the simultaneously recorded sharp-and-slow wave activity from the right orbitofrontal and frontal opercular regions (see *inserts*), as indicated by equipotentiality of the electrical field between the right supraorbital electrode *(SO2)* and *F4* with phase reveral at *SO2*, indicating that the discharge is maximal in the orbitofrontal and frontal opercular regions. [From Quesney and Gloor (1985) with permission]

Fig. 16. Preoperative EEG tracing obtained with sphenoidal and supraorbital electrodes. Notice active spiking with phase reversals at the left supraorbital electrode *(SO1)* and occasional involvement of the left centrotemporal convexity ►

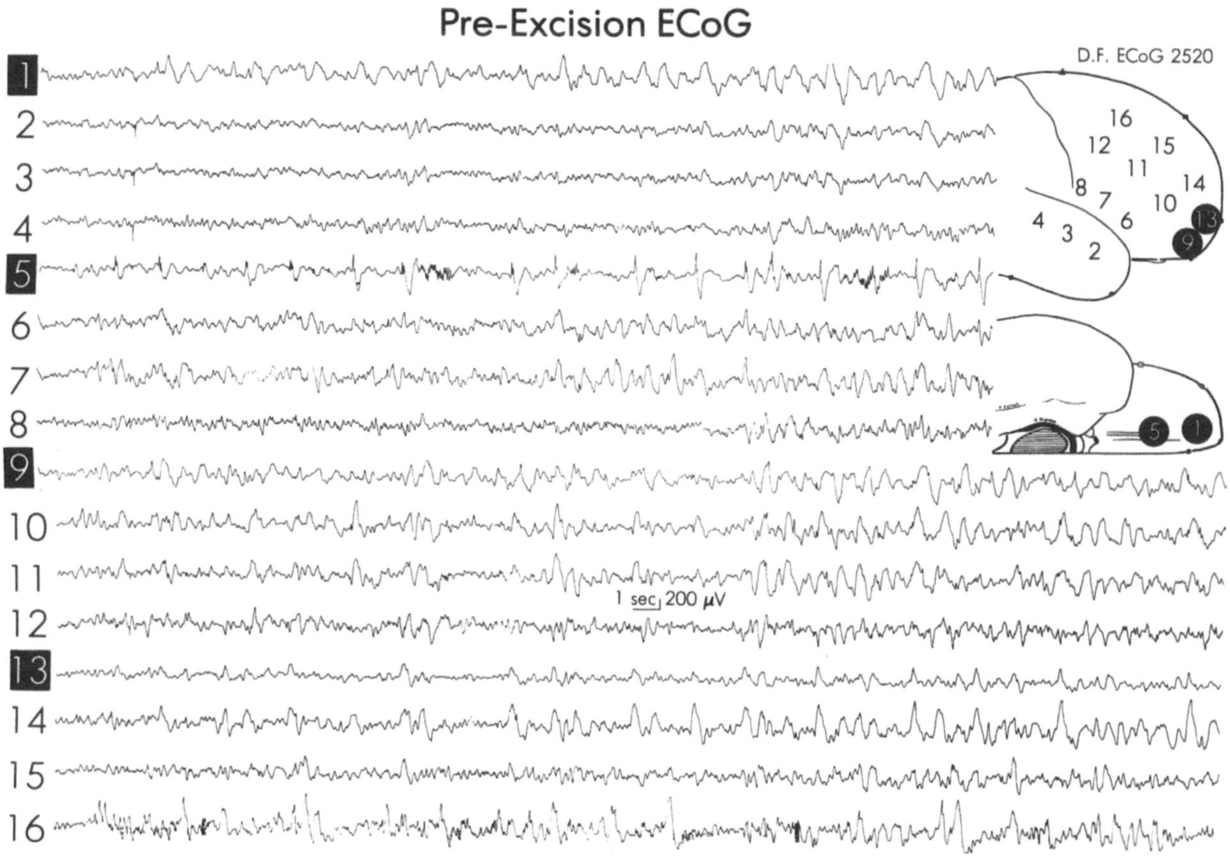

Fig. 15. Pre-excision electrocorticogram. Cortical electrode placement is indicated by *numbers* on the brain diagram. Electrodes *1* and *5* are placed on the orbitofrontal and electrodes *9* and *13* on the frontopolar cortex *(white numbers on black circles).* The most active epileptiform spike discharge is recorded from the orbitofrontal cortex at electrode *5,* with some spread of sharp wave activity to the cortical surface (electrodes *14, 13,* and *16*). The orbitofrontal cortex was densely gliotic. Referential recording with a reference electrode on the edge of the bone flap. [From Quesney and Gloor (1985) with permission]

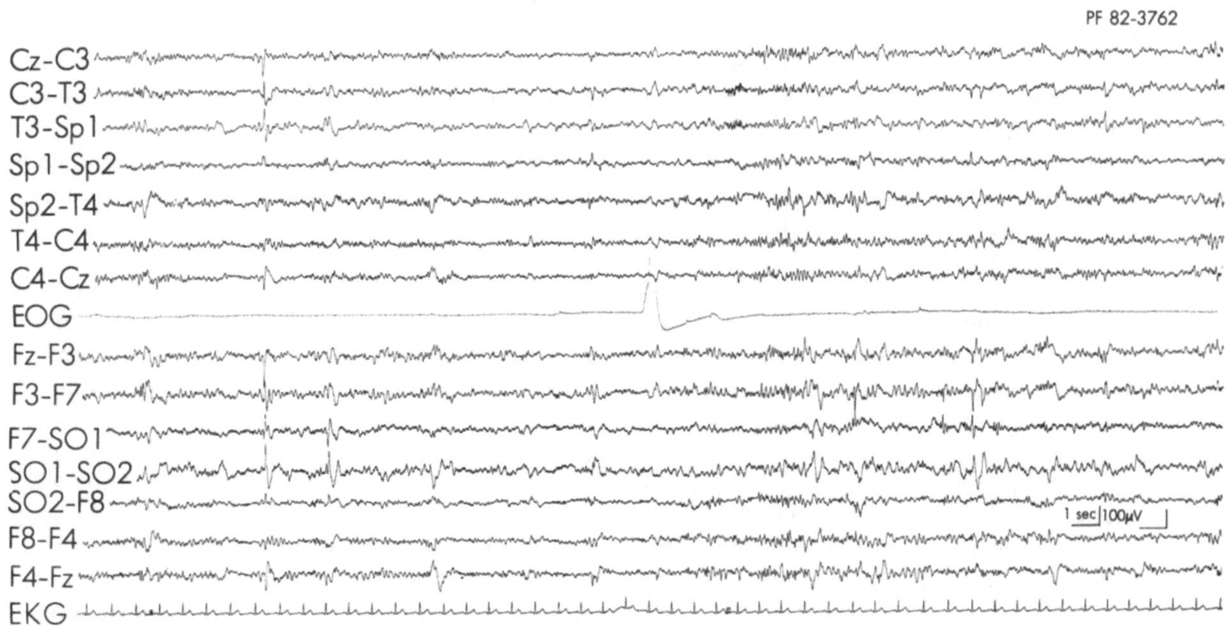

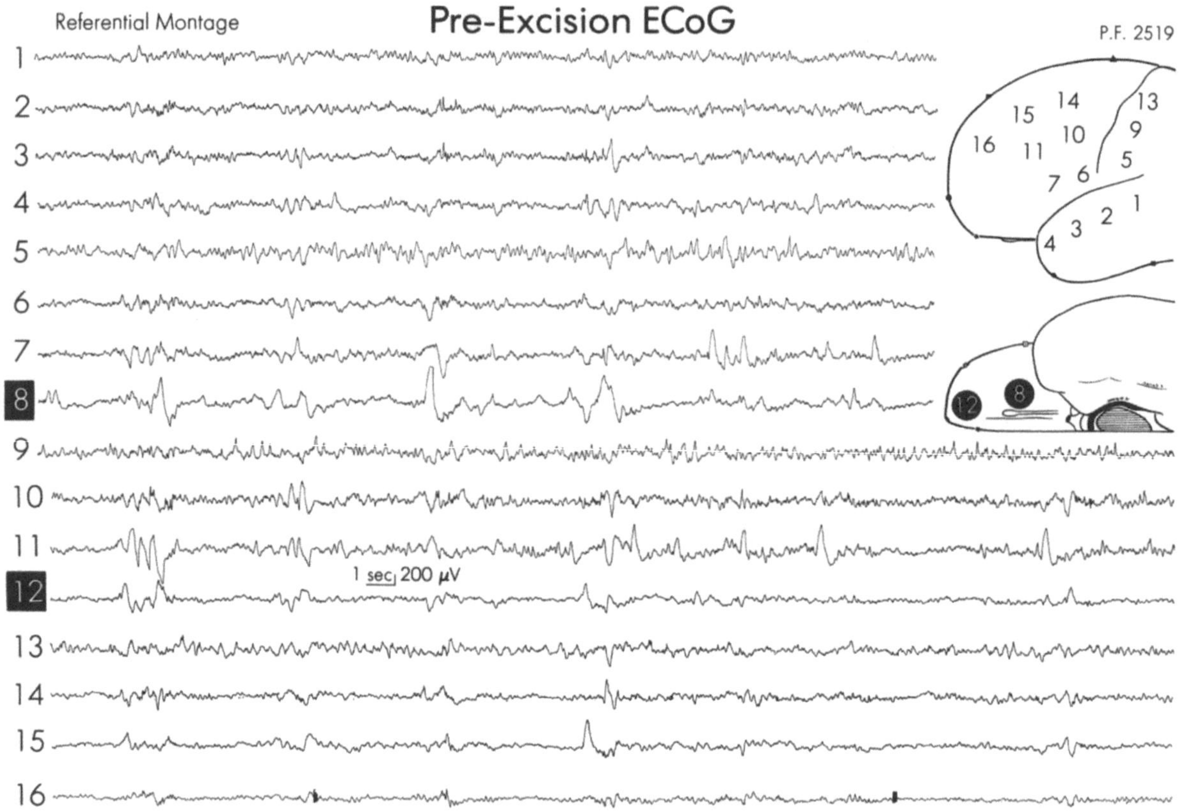

Fig. 17. Pre-excision electrocorticogram. Cortical electrode placement is indicated by *numbers* on the brain diagram. Notice high-amplitude sharp wave activity recorded from the orbitofrontal and mesial frontal regions (electrodes *12* and *8*) with widespread involvement of the frontal convexity (electrodes *7, 11,* and *15*)

Effectiveness of Ictal Recordings Obtained with Extracranial Electrodes in the Localization of Seizure Onset

The localizing effectiveness of preoperative seizure monitoring has been recently reported in a retrospective study performed in 50 randomly selected patients with complex partial seizures predominantly of temporal lobe origin (Table 1, group I) and in 22 patients with frontal lobe epilepsy (Table 1, group II), who underwent surgical therapy at the Montreal Neurological Hospital (Quesney and Gloor 1985). Patients who required depth electrode investigation were not included in this study and therefore the results apply only to the localizing effectiveness of seizures recorded with scalp, sphenoidal, and nasoethmoidal or supraorbital electrodes.

A total of 311 spontaneous seizures were available for analysis in 30 patients of group I, comprising mostly temporal lobe epileptics. A localized seizure onset in the EEG, commonly in the form of a regional onset, was observed in 127 of the recorded attacks (40.4%). Lateralized seizure onset within one cerebral hemisphere, without further localization, was observed in 29 attacks (9.4%). A bilateral or generalized seizure onset with or without unilateral predominance of the ictal discharge was recorded in 71 seizures (22.8%). Eighty-four ictal events (27.4%) were uninterpretable on account of excessive muscle or movement artifact.

Three hundred and two seizures were recorded in 16 patients of the frontal lobe group (Table 1, group II). Reliable localization of seizure onset within one frontal lobe was obtained in 66 seizures (22%). A lateralized seizure onset involving the frontal and temporal regions of one cerebral hemisphere was recorded in 34 ictal events (11%). Bilateral or generalized seizure onset was the most common finding occurring in 111 seizures (37%). In 57 seizures the ictal tracings were uninterpretable (19%) and absence of the ictal electrographic changes occurred in 34 clinical seizures (11%). In these patients localizing evidence pointing to a frontal lobe seizure problem was based on interictal EEG data as well as on localizing evidence deriving from neuroradiological and neuropsychological studies. The localizing effectiveness of seizure monitoring in frontal lobe epileptic patients can be increased up to 30% using nasoethmoidal or supraorbital electrodes.

References

Ajmone-Marsan C (1978) Clinical-electrographic correlations of partial seizures. In: Wada JA (ed) Modern perspectives in epilepsy. Eden, Montreal, pp 76-98

Bancaud J (1969) Physiopathogenesis of generalized epilepsies of organic nature (stereo-electroencephalographic study). In: Gastaut H, Jasper H, Bancaud J, Waltregny A (eds) The physiopathogenesis of the epilepsies. Thomas, Springfield, pp 158-185

Bloom D, Jasper H, Rasmussen T (1959/60) Surgical therapy in patients with temporal lobe seizures and bilateral EEG abnormality. Epilepsia 1: 351-365

Bossi L, Munari C, Stoffels C, Bonis A, Bacia T, Talairach J, Bancaud J (1984) Somatomotor manifestations in temporal lobe seizures. Epilepsia 25: 70-76

Buser P, Bancaud J, Talairach J (1973) Depth recordings in man in temporal lobe epilepsy. In: Brazier MAB (ed) Epilepsy, its phenomena in man. Academic, New York, pp 67-97

Engel J Jr, Crandall PH (1983) Falsely localizing ictal onsets with depth EEG telemetry during anticonvulsant withdrawal. Epilepsia 24: 344-355

Engel J Jr, Driver MV, Falconer MA (1975) Electrophysiological correlates of pathology and surgical results in temporal lobe epilepsy. Brain 98: 129-156

Engel J Jr, Rausch R, Lieb JP, Kuhl DE, Crandall PH (1981) Correlation of criteria used for localizing epileptic foci in patients considered for surgical therapy of epilepsy. Ann Neurol 9: 215-224

Falconer MA, Serafetinides EA (1963) A follow-up study of surgery in temporal lobe epilepsy. J Neurol Neurosurg Psychiatry 26: 154-165

Gastaut H (1948) Présentation d'une électrode pharyngée bipolaire. Rev Neurol (Paris) 80: 623-624

Geier S, Bancaud J, Talairach J, Bonis A, Szikla G, Enjelvin M (1975) Clinical note: clinical and telestereo-EEG findings in a patient with psychomotor seizures. Epilepsia 16: 119-125

Geier S, Bancaud J, Talairach J, Bonis A, Enjelvin M, Hossard-Bouchard H (1976) Automatisms during frontal lobe epileptic seizures. Brain 99: 447-458

Geier S, Bancaud J, Talairach J, Bonis A, Szikla G, Enjelvin M (1977) The seizures of frontal lobe epilepsy. A study of clinical manifestations. Neurology 27: 951-958

Gentilomo A, Colicchio G, Pola P, Rossi GF, Scerrati M (1975) Brain depth recording of interictal epileptic potentials during sleep in man. In: Levin P, Koella WP (eds) Sleep. Karger, Basel, pp 444-446

Gloor P (1975) Contributions of electroencephalography and electrocorticography to the neurosurgical treatment of the epilepsies. In: Purpura DP, Penry JK, Walter RD (eds) Neurosurgical management of the epilepsies. Advances in neurology, vol 8. Raven, New York, pp 59-105

Gloor P, Olivier A, Ives J (1980) Prolonged seizure monitoring with stereotaxically implanted depth electrodes in patients with bilateral interictal temporal epileptic foci: how bilateral is bitemporal epilepsy? In: Wada JA, Penry JK (eds) Advances in epileptology. Xth epilepsy international symposium. Raven, New York, pp 83-88

Goldring S (1972) The role of prefrontal cortex in grand mal convulsion. Arch Neurol 26: 109-119

Gotman J, Ives JR, Gloor P, Quesney LF, Bergsma P (1985) Long-term monitoring at the Montreal Neurological Institute. In: Gotman J, Ives JR, Gloor P (eds) Long-term monitoring in epilepsy. Electroencephalogr Clin Neurophysiol [Suppl 37]: 327-340

Ives JR, Gloor P (1977) New sphenoidal electrode assembly to permit long-term monitoring of the patient's ictal or interictal EEG. Electroencephalogr Clin Neurophysiol 42: 575-580

Ives JR, Gloor P (1978) Update: chronic sphenoidal electrodes. Electroencephalogr Clin Neurophysiol 44: 789-790

Jasper H, Kershman J (1941) Electroencephalographic classification of seizures. Arch Neurol Psychiatry 45: 903-943

Jones DP (1951) Recording of the basal electroencephalogram with sphenoidal needle electrodes. Electroencephalogr Clin Neurophysiol 3: 100

Kristensen O, Sindrup EH (1978) Sphenoidal electrodes, their use in the electroencephalographic investigation of complex partial epilepsy. Acta Neurol Scand 58: 157-166

Laws ER Jr, Niedermeyer E, Walker AE (1970) Diagnostic significance of scalp and depth EEG findings in patients with temporal and frontal lobe epilepsy. Johns Hopkins Med J 126: 146-153

Laxer KD (1984) Mini-sphenoidal electrodes in the investigation of seizures. Electroencephalogr Clin Neurophysiol 58: 127-129

Lehtinen LOJ, Bergstrom L (1970) Naso-ethmoidal electrode for recording the electrical activity of the inferior surface of the frontal lobe. Electroencephalogr Clin Neurophysiol 29: 303-305

Lieb JP, Walsh GO, Babb TL, Walter RD, Crandall PH (1976) A comparison of EEG seizure patterns recorded with surface and depth electrodes in patients with temporal lobe epilepsy. Epilepsia 12: 137-160

Lieb JP, Engel J Jr, Gevins A, Crandall PH (1981) Surface and deep EEG correlates of surgical outcome in temporal lobe epilepsy. Epilepsia 22: 515-538

Livingston KE, Escobar A (1971) Anatomical bias of the limbic system concept. A proposed reorientation. Arch Neurol 24: 17-21

Lüders H, Hahn J, Lesser RP, Dinner DS, Rothner D, Erenberg G (1982) Localization of epileptogenic spike foci: comparative study of closely spaced scalp electrodes, nasopharyngeal, sphenoidal, subdural and depth electrodes. In: Akimoto H, Kazamatsuri H, Seino M, Ward AA (eds) Advances in epileptology. XIIIth Epilepsy international symposium. Raven, New York, pp 185-189

Ludwig BI, Ajmone-Marsan C, Van Buren J (1975) Cerebral seizures of probable orbitofrontal origin. Epilepsia 16: 141-158

MacLean P (1949) A new nasopharyngeal lead. Electroencephalogr Clin Neurophysiol 1: 110-112

Mavor H, Hellen MK (1964) Nasopharyngeal electrode recording. Am J EEG Technol 4: 43-50

Mazars G (1970) Criteria for identifying cingulate epilepsies. Epilepsia 11: 41-47

Milner B (1975) Psychological aspects of focal epilepsy and its neurosurgical management. In: Purpura DP, Penry JK, Walter RD (eds) Neurosurgical management of the epilepsies. Advances in neurology, vol 8. Raven, New York, pp 299-322

Morrell F (1959/60) Secondary epileptogenic lesions. Epilepsia 1: 538-560

Morris HH III, Lüders H (1985) Electrodes. In: Gotman J,

Ives JR, Gloor P (eds) Long-term monitoring in epilepsy. Electroencephalogr Clin Neurophysiol [Suppl 37] : 3-26
Munari C, Stoffels C, Bossi L, Bonis A, Talairach J, Bancaud J (1981) Automatic activities during frontal and temporal lobe seizures: are they the same? In: Dam M, Gram L, Penry JK (eds) Advances in epileptology. XIIth Epilepsy international symposium. Raven, New York, pp 287-291
Ochs R, Gloor P, Quesney LF, Ives J, Olivier A (1984) Does head-turning during a seizure have lateralizing or localizing significance? Neurology 34: 884-890
Pampiglione G, Kerridge JC (1956) EEG abnormalities from the temporal lobe studied with sphenoidal electrodes. J Neurol Neurosurg Psychiatry 19: 117-129
Penfield W, Erickson T (1941) Epilepsy and cerebral localization. Thomas, Springfield, 623 pp
Penfield W, Jasper HH (1954) Epilepsy and the functional anatomy of the human brain. Little, Brown, Boston, 896 pp
Pertuiset B, Capdevielle-Arfel G (1951) Deux techniques particulières d'exploration basale. Les électrodes sphéno-ptérygoidiennes et orbitaires. Rev Neurol 84: 606-612
Quesney LF, Gloor P (1985) Localization of epileptic foci. In: Gotman J, Ives JR, Gloor P (eds) Long-term monitoring in epilepsy. Electroencephalogr Clin Neurophysiol [Suppl 37]: 165-200
Quesney LF, Katsarkas A, Gloor P, Andermann F (1981) Contribution of the naso-ethmoidal electrode recording in the electrographic exploration of frontal and temporal lobe epilepsy. In: Dam M, Gram L, Penry JK (eds) Advances in epileptology. XIIth Epilepsy international symposium. Raven, New York, pp 293-304
Quesney LF, Krieger C, Leitner C, Gloor P, Olivier A (1984) Frontal lobe epilepsy: clinical and electrographic presentation. In: Porter RJ, Mattson RH, Ward AA, Dam M (eds) Advances in epileptology, XVth epilepsy international symposium. Raven, New York, pp 503-508
Rasmussen T (1963) Surgical therapy of frontal lobe epilepsy. Epilepsia 4: 181-198
Rasmussen T (1975) Surgical treatment of patients with complex partial seizures. In: Penry JK, Daly DD (eds) Complex partial seizures and their treatment. Advances in neurology, vol. 11. Raven, New York, pp 415-449
Rasmussen T (1979) Cortical resection for medically refractory focal epilepsy: results, lessons and questions. In: Rasmussen T, Marino R (eds) Functional neurosurgery. Raven, New York, pp 253-269
Rasmussen T (1982) Localizational aspects of epileptic seizure phenomena. In: Thompson RA, Green JR (eds) New perspectives in cerebral localization. Raven, New York, pp 177-203
Rasmussen T, Branch C (1962) Temporal lobe epilepsy. Indications for and results of surgical therapy. Postgrad Med 31: 9-14
Rasmussen T, Jasper HH (1958) Temporal lobe epilepsy: indications for operation and surgical technique. In: Baldwin M, Bailey P (eds) Temporal lobe epilepsy. Thomas, Springfield, pp 440-460
Roubicek J, Hill D (1948) Electroencephalography with pharyngeal electrodes. Brain 71: 77-87
Rovit RR, Gloor P, Henderson I (1960) Temporal lobe epilepsy - a study using multiple basal electrodes. I. Description of method. Neurochirurgia 3: 6-19
Rovit RR, Gloor P, Rasmussen T (1961) Spenoidal electrodes in the electroencephalographic study of patients with temporal lobe epilepsy. J Neurosurg 18: 151-158
Silverman D (1960) The anterior temporal electrode and the ten-twenty system. Electroencephalogr Clin Neurophysiol 12: 735-737
Sindrup E, Thygesen N, Kristensen O, Alving J (1981) Zygomatic electrodes: their use and value in complex partial epilepsy. In: Dam H, Gram L, Penry JK (eds) Advances in epileptology. XIIth epilepsy international symposium. Raven, New York, pp 313-318
Sperling MR, Engel J Jr (1985) Electroencephalographic recording from the temporal lobes: a comparison of ear, anterior temporal, and nasopharyngeal electrodes. Ann Neurol 17: 510-513
Tharp BR (1972) Orbital frontal seizures. An unique electroencephalographic and clinical syndrome. Epilepsia 13: 627-642
Tükel K, Jasper H (1952) The electroencephalogram in parasagittal lesions. Electroencephalogr Clin Neurophysiol 4: 481-494
Van Buren JM, Ajmone-Marsan C, Matsuga N, Sadowsky D (1975) Surgery of temporal lobe epilepsy. In: Purpura DP, Penry JK, Walter RD (eds) Neurosurgical management of the epilepsies. Advances in neurology, vol 8. Raven, New York, pp 155-196
Wieser HG, Bancaud J, Talairach J, Bonis A, Szikla G (1979) Comparative value of spontaneous and chemically and electrically induced seizures in establishing the lateralization of temporal lobe seizures. Epilepsia 20: 47-59
Wilkus RJ, Thompson PM (1985) Sphenoidal electrode positions and basal EEG during long term monitoring. Epilepsia 26: 137-142
Zivin C, Ajmone-Marsan C (1968) Incidence and prognostic significance of 'epileptiform' activity in the EEG of non-epileptic subjects. Brain 91: 751-778

Foramen Ovale Electrode Recordings

C. E. ELGER[1] and H. G. WIESER[2]

The success of surgical treatment of patients suffering from drug-resistant focal epilepsies depends on the precise localization of the epileptic focus. As has been shown already by early attempts of epilepsy surgery, routine scalp-EEG recordings are, as a rule, not sufficient to determine exactly the location of the initiating structure(s) of a seizure event. Therefore, several special *extracranial* recording techniques were developed in order to overcome this problem. In case of temporal lobe epilepsy this, however, proved to be difficult for several reasons. Firstly, temporal lobe activity, especially in its basal and medial parts, can be recorded only in part even by sphenoidal and nasopharyngeal electrodes. Secondly, temporal lobe seizures often run in parallel with oral automatisms, which result in severe artifacts in recordings from the types of extracranial electrodes mentioned above. In addition, mediobasal temporal lobe epilepsy, the most difficult to record by extracranial electrodes, has proved to be the most common type of temporal lobe epilepsy (Wieser 1983).

In order to avoid unnecessary large resections and, especially with the view towards microsurgical selective operations, the precise localization of the epileptic focus can only be achieved by *intracranial* recordings with epidural or intracerebral electrodes. Both methods require a considerable amount of technique, and especially in Stereo-EEG (SEEG) exploration, the knowledge of stereotactical procedures. Furthermore, there is an additional risk for the patient due to intracerebral implantation and anesthesia (see chapter on Stereo-EEG by Wieser, this volume).

Based on experience with more than 100 patients, using SEEG exploration and treating them successfully by a selective amygdalohippocampectomy, confirmatory evidence in special cases by less invasive ictal and interictal EEG data collection has been found to be sufficient as a basis for a decision to operate; if the lateralization to one mediobasal temporal lobe is reliable, even a selective resection might be indicated (Wieser 1986).

Due to the anatomical situation, an electrode placed through the foramen ovale is situated *extracerebrally* and adjacent to the limbic structures (Fig. 1 A). Animal experiments and recordings from 20 patients, mostly in parallel with SEEG recordings, showed that epileptic activity from the limbic system can be picked up with great reliability if the amount of nervous tissue involved in the generation of seizure activity reaches a certain quantity. Details of these investigations have already been published elsewhere (Wieser and Elger 1985; Wieser et al. 1985).

In the following, details of the electrode and its placement will be described. Examples of recordings via electrodes inserted through the foramen ovale (FO) will be presented. Also, a suggestion will be made concerning which questions in epilepsy surgery can possibly be solved by using 'FO electrode' recordings.

Electrode Data and Placement

The foramen ovale electrode used in our recordings consisted of a cylindrical Ag-AgCl pellet 0.4 mm in diameter and 4 mm in length. It was sintered around the tip of a silver wire of 0.25 mm, which was insulated by teflon. Ag-AgCl was used in order to permit DC recordings as well as AC. DC recordings proved to be very useful. For this reason in most cases a disc-shaped Ag-AgCl electrode of 12.5 mm diameter subgaleally implanted at C_z, served as a reference.

The electrode impedance (AC 100 Hz and 200 Hz; DC) ranged from less than 100 Ω and 210 Ω, respectively, while the DC offset potential was less than 2 mV. The EEG signals were amplified using conventional techniques with a time constant of 0 s (DC) and 0.3 s (AC), and an upper frequency limit of 1 Hz (DC) and 70 Hz (AC). Parallel to this, scalp EEG re-

[1] Neurological Clinic, University of Münster, Albert-Schweitzer-Str. 33, 4400 Münster, Federal Republic of Germany
[2] Department of Neurology, University Hospital, Frauenklinikstr. 26, 8091 Zürich, Switzerland

Presurgical Evaluation of Epileptics
Edited by H. G. Wieser and C. E. Elger

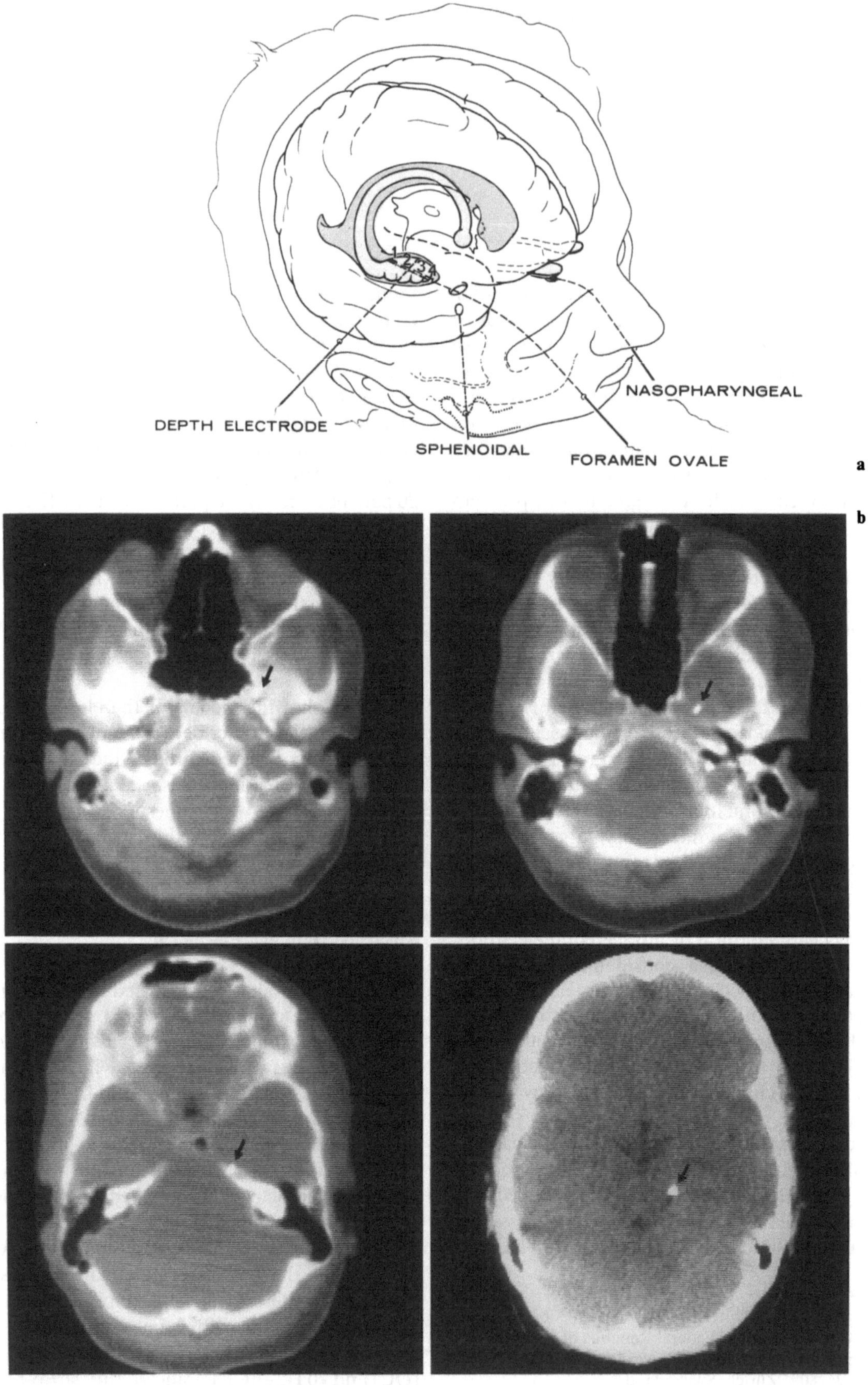
NASOPHARYNGEAL
DEPTH ELECTRODE
SPHENOIDAL
FORAMEN OVALE
a
b

◀ **Fig. 1.** **a** Schematic drawing indicating the position of the FO electrode in relation to a multicontact intracerebral depth electrode, sphenoidal electrode, and nasopharyngeal electrode. **b** CT scan of a patient with a right FO electrode (from Wieser and Elger 1985)

cordings and in most cases SEEG recordings were performed.

Kirschner (1932) introduced a technique which was subsequently used by many neurosurgeons for the so-called percutaneous trigeminal rhizotomy. The cannula used for the application of the coagulation device proved to fit well for the FO electrode application. After the patient has been anesthetized with an intravenous application of 30-50 mg of methohexital, a 19-gauge spinal needle is placed in a stereotactic fashion by freehand manipulation with the use of percutaneous landmarks (Härtel 1914; for details see Tew et al. 1979). As soon as the needle passes the entrance of the foramen ovale, signalled by a wince and a brief contraction of the masseter muscle, the stylet is withdrawn in order to collect CSF as an indicator of correct placement. The recording electrode is then introduced under cineradiographic control until the placement adjacent to the medial aspect of the temporal lobe is satisfying. The cannula is then withdrawn over the electrode wire, which finally is fixed to the buccal skin. A control of the placement of the electrode is easy in the case of a simultaneous implantation of depth electrodes, because stereotactic target points are available. However, one should be able to use FO electrodes without SEEG. In these cases a CT scan can be done in addition to conventional X-rays of lateral and submental views. In Fig. 1 a CT with a right-sided FO electrode is shown, together with a schematic drawing of the location of several electrode types in relation to the mediobasal limbic structures. The FO electrodes were left in place under antibiotic protection for up to 10 days.

To date, a total number of 20 patients have been subjected to FO electrode recordings. In 11 the electrode was inserted unilaterally and in 9, bilaterally. All patients tolerated the insertion procedure, the recording, and the withdrawal (without anesthesia) without major difficulties. Minor troubles during eating which did not result in a premature end of the recording were reported by two patients. Clinical tests of trigeminal function, such as the corneal reflex or examination of the touch perception, exhibited no disturbance of trigeminal function. Two patients reported a tingling sensation during withdrawal of the electrode. As a whole, the experience up to now has shown that electrode placement is relatively easy to perform, well tolerated by the patient, and safe.

Recordings

The amount of tissue involved in the epileptic activity necessary for a successful recording with the FO electrode can be derived from the interictal and ictal recordings shown in Fig. 2. Together with surface

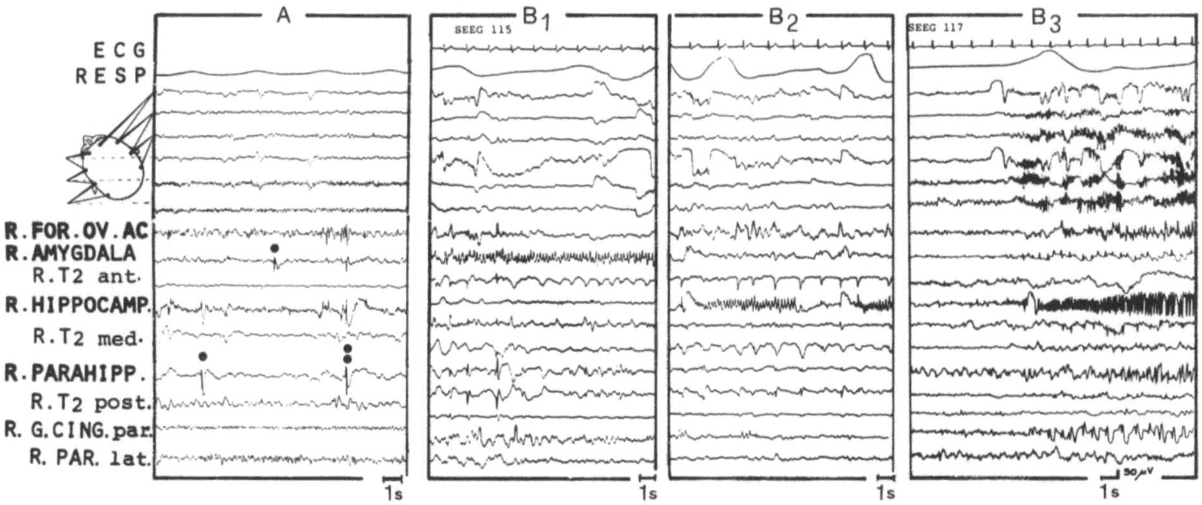

Fig. 2. Surface EEG, SEEG of limbic structures, and FO-recordings *(FOR. OV. AC)* during interictal *(A)* and ictal *(B)* states. The interictal events involving different areas of the limbic structures are marked by *dots* in part A. The locations of the SEEG electrodes are described at the left margin (modified from Wieser et al. 1985)

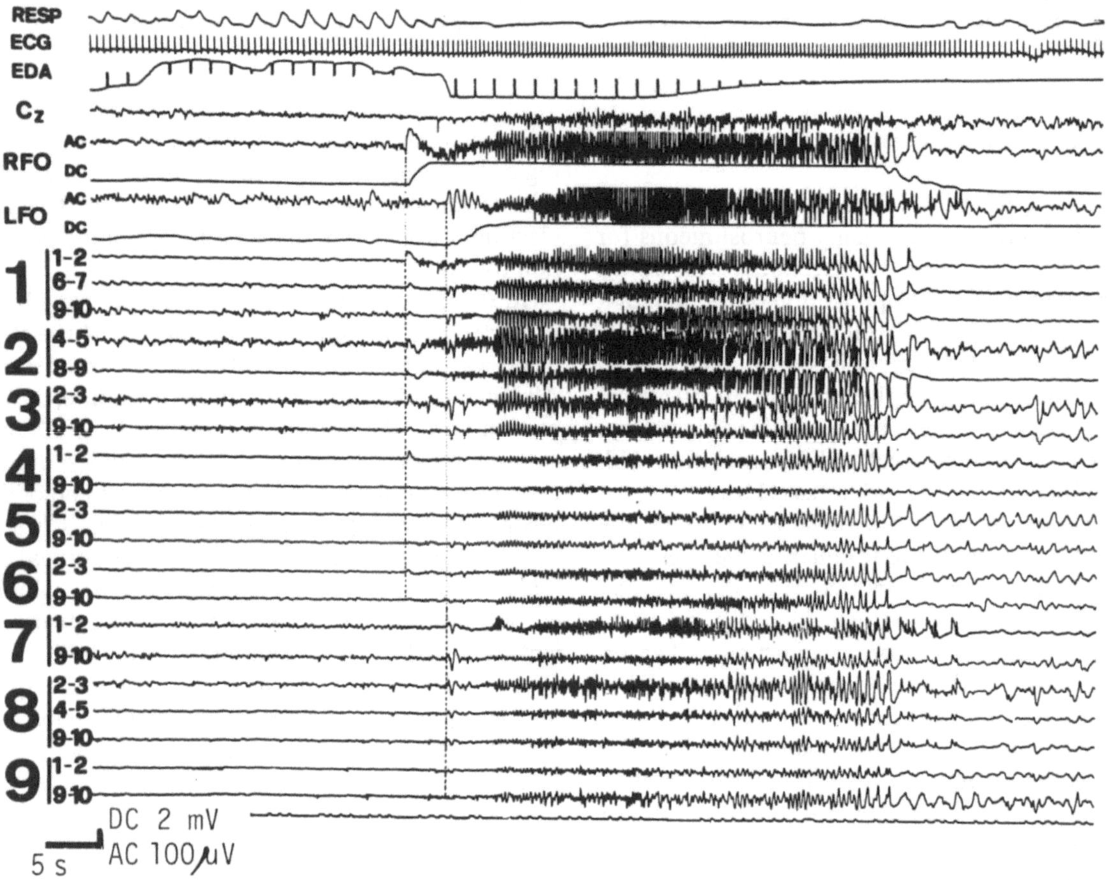

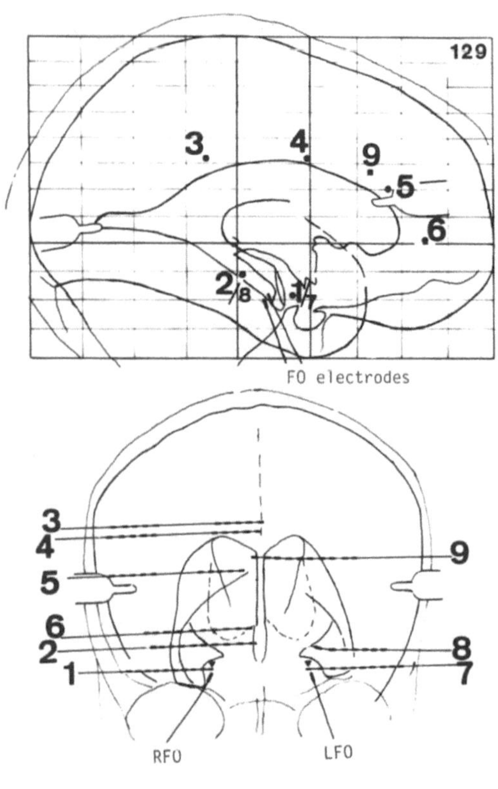

Fig. 3. AC and DC recordings with FO electrodes implanted bilaterally *(RFO, LFO)* in relation to depth recordings. The right and left seizure onset is marked by a *dashed line*. The location of the depth electrodes is indicated in the *schematic brainmap*. The contacts are numbered from medial to lateral with 1-10. *RESP*, respiration; *ECG*, electrocardiogram; C_Z, *surface EEG of the reference electrodes used for the DC recordings; EDA*, electrodermal activity

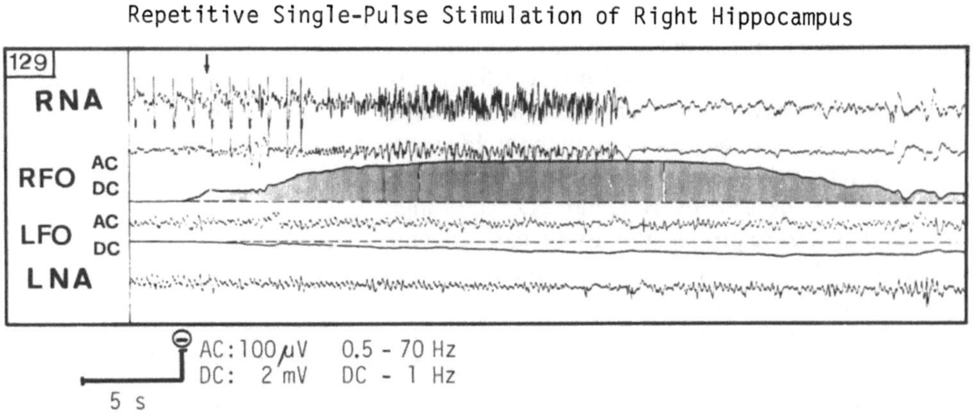

Fig. 4. AC and DC recordings with the right *(RFO)* and left *(LFO)* FO electrode together with SEEG from the right *(RNA)* and left *(LNA)* amygdala during a seizure induced by repetitive, single-pulse electrical stimulation of the right hippocampus. The *arrow* indicates the onset of seizure characterized by an olfactory aura

recordings, the signal picked up by the FO electrode is compared with signals from various intracerebral electrodes. Part A of the figure shows the situation under *interictal* conditions. As can be seen, small subcompartments of the limbic system generating an interictal spike do not lead to a specific signal in the FO electrode recordings (1 dot). If, however, a larger portion of the limbic structure is synchronously active (2 dots), an easily detectable spike can be seen deriving from the FO electrode.

In a similar fashion to the interictal activity, *ictal* activity restricted to the amygdala (Fig. 2, B1) or to the hippocampus (Fig. 2, B2) is not clearly reflected in the FO recordings. A seizure, however, originating in the mediobasal limbic structures and spreading to other limbic sites (Fig. 2, B3) is picked up by the FO electrode without artifacts, which dominate the picture in the surface leads.

One major problem in determining the lateralization of seizure origin within the mediobasal limbic structures is the frequently occurring seizure spread to the contralateral temporal limbic structures within seconds (Wieser 1983). Such a case is presented in Fig. 3. As can be derived from the picture, the start of the seizure in the right mediobasal temporal lobe is exactly documented in the right FO recording. The fast spread of the seizure to the contralateral side is clearly seen in the recording obtained from the left FO electrode. The time delay seen in the FO recordings corresponds to the one in the SEEG recordings.

It is well-known from animal experiments that slow extracellular potential fluctuations recorded by DC-coupled amplifiers accompany seizure activity in the CNS (Speckmann and Caspers 1979). Since these so-called DC potentials are thought to reflect slow membrane potential fluctuations of the adjacent neurones, seizure activity runs in parallel with a negative shift of the DC potential. As can be seen in Fig. 3, the onset of the seizure activity in the SEEG and FO recordings (AC mode) runs in parallel with a steep negative shift of the DC potential in the recording of the right FO electrode, which was connected to a DC-coupled amplifier. Several seconds later the DC signal picked up by the left FO electrode becomes negative too. This corresponds to the spread of the seizure to the contralateral side. Due to the less synchronized activity, this negative deflection is not so steep.

The experience obtained up to now with DC recordings indicates that the DC signal might be slightly more sensitive in showing interictal and ictal activity (c.f. Wieser et al. 1985). A finding pointing in this direction is shown in Fig. 4. Electrical stimulation of the right hippocampus led to a seizure event with characteristic activity in the SEEG of the right amygdala (RNA, arrow). The onset of the seizure activity is not clearly reflected by the AC recording with the right FO electrode (RFOAC). A negative DC shift, however, runs in parallel with the discharges in the amygdala (RFODC). The positive DC shift in the contralateral recording (LFODC) is probably due to a disfacilitation.

Conclusions

The FO electrode recording technique - unilateral or bilateral - has proved to be a useful and safe tool for evaluating certain patients suffering from drug-resistant temporal lobe seizures. The placement of the FO electrode can be done easily by freehand manipulation in centers which have experience with the technique of percutaneous thermocoagulation of the Gasserian ganglion. The FO electrode is well tolerated with no relevant side-effects. The recordings performed to date have shown that mediobasal-limbic epileptic activity can be picked up with the FO electrode if a relatively large volume of these structures is involved in the seizure. Bilateral FO recordings showed that seizure propagation to the contralateral side can be detected and that a differentiation between primarily and secondarily involved structures is possible. Recording of the signals via a DC-coupled amplifier seems to be slightly superior to AC recordings. Since mediobasal-limbic epilepsy is the most common type of the drug-resistant temporal lobe epilepsies, the FO electrodes can be used for exploration in a large proportion of candidates for epilepsy surgery.

Several authors have reported satisfactory results with sphenoidal electrodes in the evaluation of patients with temporal lobe seizures (Pampiglione and Kerridge 1956; Rovit and Gloor 1960; Rovit et al. 1960, 1961; Laxer 1984). In our experience, however, the data obtained by the use of sphenoidal electrodes are not sufficient to provide the necessary information for a highly tailored neurosurgical intervention in the case of mediobasal-limbic epilepsy. Subdural strip electrodes might reveal data comparable to the FO electrode (Wyler et al. 1984); however, the risks with subdural strip electrode placement and removal are greater. Although the information gained with the FO electrode is limited compared with SEEG recordings, the FO electrode technique has the main advantage of not penetrating the brain and therefore reducing the risk for the patient.

The proper use of the FO electrode requires a careful selection of the patients to be evaluated. As a rule, patients with seizures highly suspected to be of mediobasal-temporal origin from their clinical semiology are good candidates. In these cases the lateralization of the seizure onset can be determined. Especially in cases where a lesion close to the mediobasal-limbic structures has to be removed, recordings via the FO electrode help to decide whether or not the seizure-prone mediobasal-limbic structures should be removed in addition to the lesion itself. Further studies using the FO electrode simultaneously with intracerebral depth electrodes are necessary to determine the value of the information gained by the FO electrode recordings in the differentiation of mediobasal vs lateral neocortical temporal lobe seizures and in the differentiation between complex partial seizures of frontal and temporal lobe origin (Delgado-Escueta et al. 1982; Wieser 1983).

References

Delgado-Escueta AV, Enrile-Bacsal F, Treiman DM (1982) Complex partial seizures on closed-circuit television and EEG: a study of 691 attacks in 79 patients. Ann Neurol 11: 292-300

Härtel F (1914) Über die intracranielle Injectionsbehandlung der Trigeminusneuralgie. Med Klin (Barc) 10: 582-584

Kirschner M (1932) Elektrocoagulation des Ganglion Gasseri. Zentralbl Chir 47: 2841-2843

Laxer KD (1984) Mini-sphenoidal electrodes in the investigation of seizures. Electroencephalogr Clin Neurophysiol 58: 127-129

Pampiglione G, Kerridge J (1956) EEG abnormalities from the temporal lobe studied with sphenoidal electrodes. J Neurol Neurosurg Psychiatry 19: 117-129

Rovit RR, Gloor P (1960) Temporal lobe epilepsy - a study using multiple basal electrodes: I Description of method. Neurochirurgia (Stuttg) 3: 6-19

Rovit RR, Gloor P, Henderson LR (1960) Temporal lobe epilepsy - a study using multiple basal electrodes: II Clinical EEG findings. Neurochirurgia (Stuttg) 3: 19-34

Rovit RR, Gloor P, Rasmussen T (1961) Sphenoidal electrodes in the electrographic study of patients with temporal lobe epilepsy: an evaluation. J Neurosurg 18: 151-158

Speckmann E-J, Caspers H (1979) Cortical field potentials in relation to neuronal activities in seizure conditions. In: Speckmann E-J, Caspers H (eds) Origin of cortical field potentials. Thieme, Stuttgart, pp 205-213

Tew JM Jr, Keller JT, Williams DS (1979) Functional surgery of the trigeminal nerve: treatment of trigeminal neuralgia. In: Rasmussen T, Marino R (eds) Functional neurosurgery. Raven, New York, pp 129-141

Wieser HG (1983) Electroclinical features of the psychomotor seizure. Fischer/Butterworths, Stuttgart/London

Wieser HG (1986) Selective amygdalohippocampectomy: indications, investigative technique and results. In: Symon L, Brihaye J, Guidetti B, Loew F, Miller JD, Nornes H, Pásztor E, Pertuiset B, Yaşargil MG (eds) Advances and technical standards in neurosurgery, vol 13. Springer, Vienna pp 39-133

Wieser HG, Elger CE (1985) Der Einsatz der Foramen-Ovale-Elektrode im Rahmen der Abklärung von Kandidaten einer chirurgischen Epilepsietherapie. In: Kruse R (ed) Epilepsie 84. Einhorn, Hamburg, pp 256-263

Wieser HG, Elger CE, Stodieck SRG (1985) The 'foramen ovale electrode': a new recording method for the preoperative evaluation of patients suffering from mediobasal temporal lobe epilepsy. Electroencephalogr Clin Neurophysiol 61: 314-322

Wyler AR, Ojemann GA, Lettich E, Ward AA Jr (1984) Subdural strip electrodes for localizing epileptogenic foci. J Neurosurg 60: 1195-1200

Electrocorticography

A. R. Wyler

The object of this paper is to review several approaches to electrocorticography that are presently used for the surgery of epilepsy. In this discussion, the functional definition of electrocorticography (ECoG) will be the recording of cortical (and some subcortical structures) electrical activity for the purpose of delineating the extent and location of an epileptogenic focus. Although immaculate surgical technique is important in obtaining an excellent postoperative result, even more crucial to the success of epilepsy surgery is the ability to localize single epileptogenic foci or to recognize the presence of multiple foci that should not be resected. Therefore, ECoG is the single most important factor in the success of surgery for epilepsy. Although the past decade has seen the introduction of highly sophisticated neuroimaging techniques like CT, MRI, and PET scanning, the cornerstone of successful epilepsy surgery will remain ECoG. This is because the imaging techniques can only show obvious structural and some gross metabolic abnormalities, and not necessarily the extent or location of the epileptogenic focus. Presently, the most accurate signature of the epileptic process is still the epileptiform spike, and hence ECoG is the most accurate method of localizing the focus.

Preoperative Localization of the Focus

Once a patient's seizures have been found to be refractory to standard anticonvulsant medication and the question of surgery has been seriously considered, the next step is to localize the epileptogenic focus sufficiently well to optimize the success of the surgery. This requires that all the patient's previous standard EEGs be evaluated to determine the stability of interictal discharges. If, for example, a patient has had very focal epileptic discharges from the anterior temporal region of one side without any spread to the contralateral side, and if the clinical seizure is compatible with the location of the EEG focus, and the discharges have been stable over multiple interictal recordings, then prolonged EEG/video monitoring is not necessary before proceeding with surgery, if acute ECoG will be done during the surgical resection. However, because of the fact that the basic underlying mechanism responsible for the epileptiform EEG spike is not known, and because the epileptiform spike is a propagated electrical event, it is felt by many epileptologists that the presence of interictal spikes alone is insufficient evidence for localizing an epileptogenic focus. Therefore, many believe that foci may only be localized with any degree of confidence by recording the onset of ictal events. The factual evidence for that opinion is lacking, however. It should be pointed out though that successful epilepsy surgery was carried out for years by Penfield prior to the advent of prolonged monitoring, and Penfield's decisions were primarily based on the results of interictal recordings and intraoperative ECoG (Penfield and Jasper 1954). However, there is logical appeal to the concept that ictal recordings will have more accurate localizing value than only interictal recordings, and thus this principle is adhered to whenever possible, since long-term video/EEG monitoring is now available in any major epilepsy center.

If bilateral or diffuse discharges are present, or if the clinical seizure does not coincide with the EEG abnormalities, then prolonged scalp EEG/video monitoring is mandatory and should include several clearly recorded spontaneous seizures. The focal onset of the seizures should be unequivocal and reproducible. Moreover, the seizures must be typical clinically to the patient's habitual seizures. (This can be determined by having a family member who is familiar with the patient's seizure view the videotape.) Finally, the EEG onset of the seizure must either precede or coincide with the onset of the clinical seizure. If any of these criteria are not met, then the use of preresection intracranial recordings should be

Department of Neurosurgery, University of Tennessee, Memphis and Semmes-Murphy-Clinic, Coleman Building A202, 956 Court Avenue, Memphis, TN 38163, USA

Presurgical Evaluation of Epileptics
Edited by H. G. Wieser and C. E. Elger

strongly considered. It should be emphasized that not all surgeons share the view that intracranial recordings need or should be done. Walker (1975) has criticized the use of such recordings, and others have reasoned that if scalp recordings can be falsely localizing, then so could depth recordings. There are several series of epilepsy surgery performed in which no preoperative intracranial recordings were done, and there are also several series in which they are routinely done. Although it would be tempting to compare the results of both types of approaches in order to determine if such recordings are justified by generating a superior surgical outcome, the series are most likely not comparable because of patient selection differences. However, a commonsense approach seems to be to implant electrodes in patients that do not clearly have a well-localized focus by scalp recordings. It is certainly not justified to take the stance of (1) never doing chronic intracranial recordings or (2) submitting every patient to depth electrodes.

The technique for intracranial depth electrode recordings was popularized by Bancaud and Talairach (Bancaud et al. 1965; Talairach et al. 1962, 1974). Since then, various surgeons have devised their own electrodes and methods of insertion depending upon the clinical problem dealt with (Brazier 1956; Crandall 1973). The techniques and results of depth electrode studies have been discussed fully in the chapter by Wieser in this volume and will not be described further.

Strip Electrodes

Although depth electrodes have been successfully used for localizing epileptogenic foci, the objective may be equally well realized in many cases from the somewhat simpler technique of subdural strip electrodes (Wyler et al. 1984). Each electrode consists of four stainless steel disks residing in a flexible silastic strip. The strips are placed through a burr hole to lay over the cortex. Because of the thinness and flexibility of the strips, they can be easily slid to otherwise inaccessible cortex, for example, under the temporal lobe or along the mesial side of the frontal lobe. They are routinely used on all my seizure surgery as part of the routine ECoG at the time of craniotomy for cortical resection (Fig. 1). For example, in cases of temporal lobectomy, the strips are commonly used to help determine the need for more extensive posterior hippocampal resection in some patients because, as pointed out by Spencer et al. (1984), approximately 20% of temporal lobe foci extend more posteriorly along the hippocampus than are commonly removed during standard temporal lobectomy.

Although the strip electrodes are now routinely used acutely during epilepsy surgery, they were originally intended for the preresection evaluation of patients with foci that were difficult to localize using standard scalp and sphenoidal recordings, i. e., those patients who would otherwise be candidates for placement of depth electrodes. Most commonly, these were patients who showed either (1) bitemporal discharges without clear lateralization at seizure onset, (2) bifrontal discharges without clear lateralization, or (3) to determine if a focus resided in a temporal or frontal lobe when scalp recordings were ambiguous but lateralized to one hemisphere. For these patients, from four to eight electrodes were implanted intracranially for periods up to 3 weeks. The recordings are commonly of excellent quality with a minimum of artifacts. An example is shown in Fig. 2. These electrodes provide excellent-quality recordings for prolonged periods, and of special importance is that they generate little artifact when spontaneous ictal events are recorded.

The usefulness of these electrodes compares favorably with results from using depth electrodes: in a series of 28 patients who had undergone 30 implantations (Wyler et al. 1984), the recordings localized an EEG epileptogenic focus in 20 (71%) patients who have subsequently undergone surgery. In one patient, the recordings were inconclusive and no surgical decision could be made. Seven patients (25%) were found to have either primary generalized epilepsy or bilateral foci (either frontal or temporal), and they were thought not to be appropriate candidates for a cortical resection. Thus, the data obtained from the implantation of 27 of 28 patients (96%) was critical in the decision to perform or not to perform surgery. Of the 20 patients who did undergo surgery, 18 have been followed for over 2 years, and these can be used to evaluate the outcome of the surgery: five (28%) are seizure free, six (33%) are significantly improved, and seven (39%) were not significantly improved. Thus, if both the seizure-free and significantly improved patients are combined, 61% of those who underwent chronic subdural strip electrode monitoring and who subsequently had surgery obtained beneficial results. Although these figures may not seem impressive, it should be remembered that subdural strip electrode recordings were reserved for only those patients who were the most difficult diagnostic problems, and many of these patients would not be considered surgical candidates at all if intracranial recordings were not available.

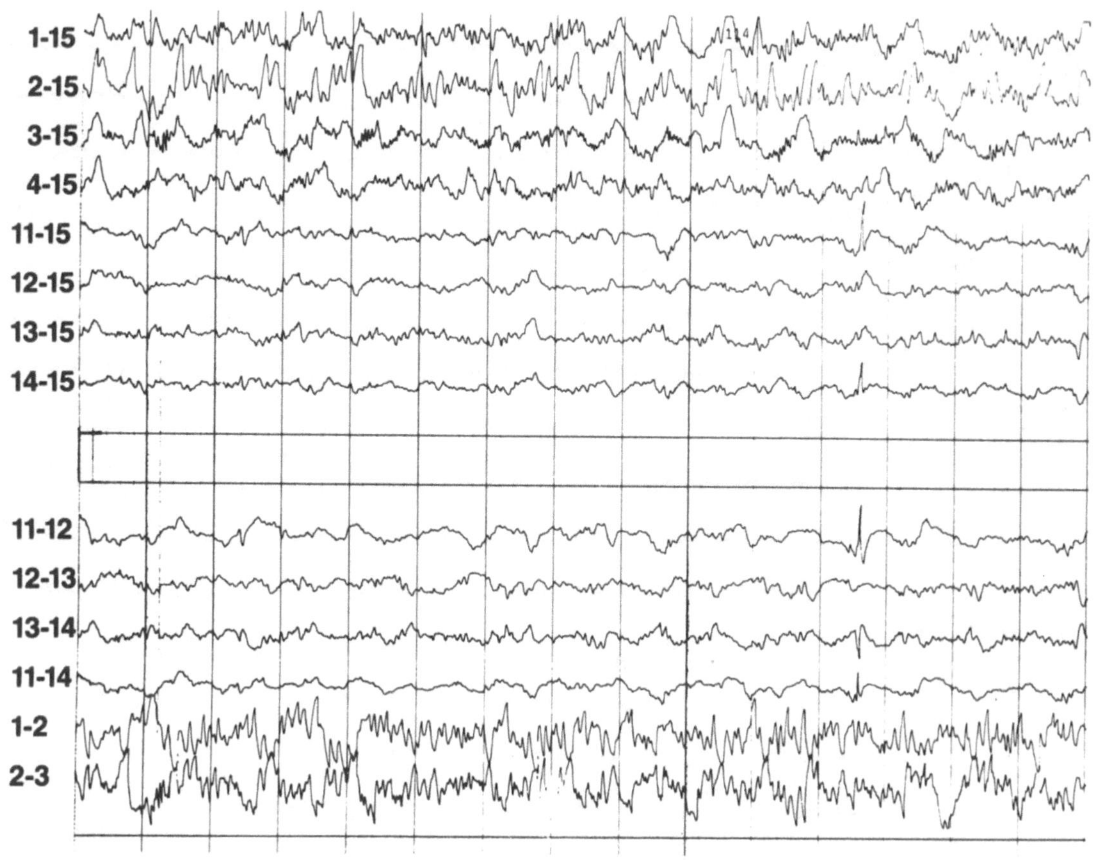

Fig. 1. Electrocorticographic recordings from the temporal lobe during craniotomy. Electrode contacts *11-14* were made from a subdural "strip electrode" which was placed under the midtemporal lobe in a lateral to medial direction. Electrodes *1-4* were placed over the lateral temporal lobe and electrode *15* was placed over the occiput. Note the epileptiform spikes that were recorded from the "strip electrode" but were not recorded from the lateral temporal cortex

From the 30 strip electrode implantation procedures, two complications occurred for a morbidity rate of 8%. One patient developed a small brain abscess which was treated successfully with twist drill aspiration and antibiotic therapy. The second patient developed a small cortical contusion which resolved without neurologic deficit. These two complications occurred earlier in the series of patients when the techniques was being developed, and future use may well have a lower morbidity rate. It is difficult to compare this complication rate with that of depth electrodes, since the latter is not known. However, some of the complications that have occurred with depth electrodes are (1) intracerebral hemorrhage with residual neurological deficits or death, (2) brain abscess, and (3) Creutzfeldt-Jakob syndrome has been transmitted by this route (Bernoulli et al. 1977).

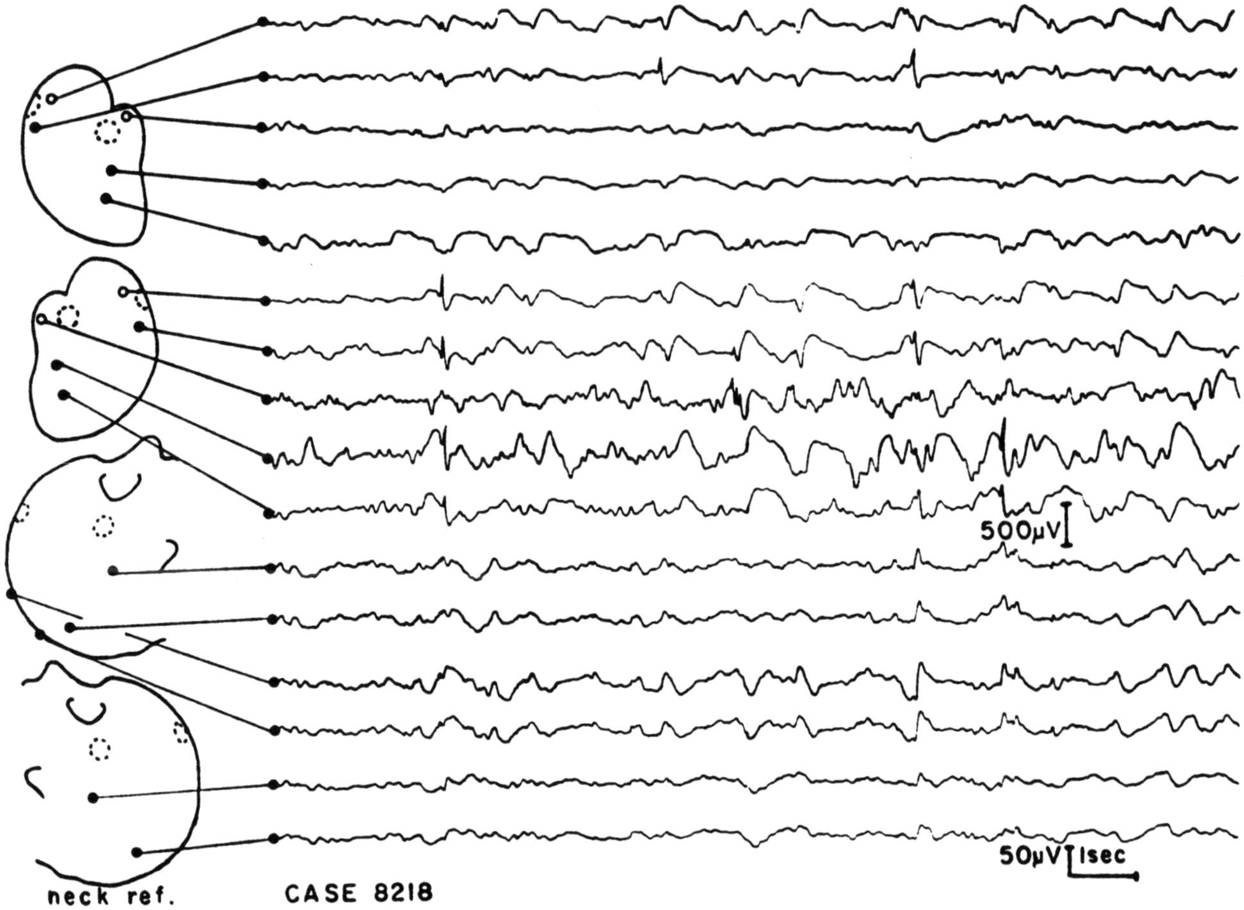

Fig. 2. Recording of ECoG from subdural "strip electrodes" placed over the frontal and temporal cortex bilaterally and some posterior scalp EEG electrodes. (All recordings were referenced to a neck electrode.) Notice the active spikes recorded from the left frontal and posterior temporal regions (strip electrode) that were not well demonstrated in the posterior temporal scalp recordings

Comparison of Strip Electrodes with Depth Electrodes

It is accepted that epileptogenic foci reside in gray matter rather than white matter, since the pathophysiology arises from the neuronal somata and dendrites, and not their axons. Thus, recordings to localize the focus should provide as much information regarding the ECoG as possible, while causing minimal risk to the patient. Regarding the first goal, the recordings should not be too discrete, because the goal is to scan as large an expanse of cortex as possible to localize a focus not otherwise known. Thus, the electrodes should be larger rather than smaller. By design, many depth electrodes have electrode contacts along the electrode's shaft. These electrode contacts are quite small, usually being the diameter of the wire. As a result, the solid angle of the individual contacts is relatively small, so that the volume of brain that is recorded is close to the contact. In contrast, each contact along the strip electrode is 0.5 cm in diameter. Many of the contacts of a depth electrode will come to lie within the white matter rather than gray matter. Up to four strip electrodes can be placed easily through one burr hole, thus providing a large expanse of cortical recordings. For example, since each electrode has four cortical contacts each separated by 1 cm, four electrodes in one burr hole will result in the sampling of 16 cortical sites. If significant sampling of different areas is to be obtained with depth electrodes, multiple electrodes must be inserted into the brain. Although clinical sequelae are rare, the biological facts are clear that many neurons are destroyed with each depth electrode insertion. Rasmussen (1976) has pointed out that the "ultimate risks of inserting electrodes into the good hemisphere of a brain that has already demonstrated its ability to generate a clinically significant seizure tendency still await assessment." The subdural strip electrode is applied *over* the cortical surface, not through it. For the above-stated reasons, this author prefers to use the strip electrode technique rather than depth electrodes when intracranial chronic recordings are indicated.

On the other hand, there are occasions for which stereotaxically implanted electrodes may be preferable (Wieser 1983), i. e., to confirm the presence of a clinically suspected focus in such difficult to reach regions of the brain as the amygdala. In these situations, depth electrodes provide localized recordings, and the recording sites can be precisely placed by stereotaxic implantation at the site desired (Crandall et al. 1963). For example, the new surgical procedure of selective amygdalohippocampectomy devised by Wieser and Yasargil (1982) requires that the presence of a mesial temporal focus be confirmed prior to surgery since intraoperative ECoG cannot be done with this approach. In their series, 93% of patients who have had this procedure have had a beneficial outcome (Wieser 1986).

Justification for Using Intracranial Recordings

The justification for using intracranial recordings is poorly documented statistically, although quite well documented anecdotally. In a recent review of the literature, Spencer (1980) discussed the results of using depth electrode recordings in the selection of patients for epilepsy surgery. She estimated that such recordings allowed 36% of patients to be operated upon who otherwise would have been diagnosed as having nonfocal epilepsy. In addition, she stated that such recordings could have prevented surgery in another 18% of cases by demonstrating that multiple independent foci were present. She concluded that depth electrode recording has an important role in evaluating surgical candidates since it may have the potential to alter the surgical decision in more than 50% of patients. However, she also pointed out that only large epilepsy centers are performing these recordings and, by the nature of their clinical practices, they attract more difficult cases for surgical consideration. Her conclusions fit well with our own experience since the strip electrode recordings have excluded 25% of patients from having *focal resections* (some of these went on to have corpus callosum sectioning). As noted in a previous section, of those patients in whom a focus was ultimately found, because of the strip electrode recordings, 61% have had a beneficial outcome. These patients would probably not have had surgery if long-term intracranial monitoring had not been done. Thus, the use of these recordings has benefited patients in two ways: (1) they have helped the surgeons select the most appropriate surgery, i. e., focal resection, corpus callosum section, or selective amygdalohippocampectomy, and (2) they have helped identify patients in whom focal resections are not indicated.

Approaches to Cortical Resection

After the epileptogenic focus has been localized, a craniotomy is performed for its resection. There are essentially two approaches to focal cortical resections: (1) the en bloc resection and the (2) electrocortigraphic guided resection. The en bloc resection assumes the surgeon will do a standard lobectomy, i. e., if the left (speech dominant) temporal lobe is removed, then a resection measuring 4½ cm from the tip is done. If the resection involves the nondominant temporal lobe, 5½ cm of lobe are removed. This type of resection may be done after only scalp localization, as was the case with the patients of Falconer (1971), or it may follow the implantation of depth electrodes as is done by the UCLA and Yale groups (Crandall et al. 1963; Spencer 1980). A very special type of resection that may be considered within the en bloc category is the selective amygdalohippocampectomy described by Wieser and Yasargil (1982). In this approach, the patients' focus is determined by extensive depth electrode monitoring to be within the mesial structures of one temporal lobe. The surgeon then does a selective mesial temporal resection under general anesthesia without any additional electrocorticography.

The second approach is similar to what was popularized by Penfield, and will be termed the "tailored" approach (Penfield and Jasper 1954; Rasmussen 1975). The cortical regions thought to be the epileptogenic focus and its surround are exposed during craniotomy under local anesthesia. A rigid electrode holder (Grass Instrument Company, Quincy, Ma.) is secured to the skull margin, and multiple electrodes are placed over the cortex to be investigated. The author has found that graphite tip electrodes provide recording characteristics superior to the older "cotton wick" electrodes. Unless dealing with a well-defined central focus, I routinely use at least one four-contact strip electrode to investigate regions of unexposed cortex, for example, the undersurface of the temporal lobe including the region of hippocampus (as was shown in Fig. 1). Initially, the montage is comprised of three to four linear arrays across the exposed cortex. For the temporal lobe, it would be comprised of a linear array of four electrodes (from anterior to posterior) along the presumed suprasylvi-

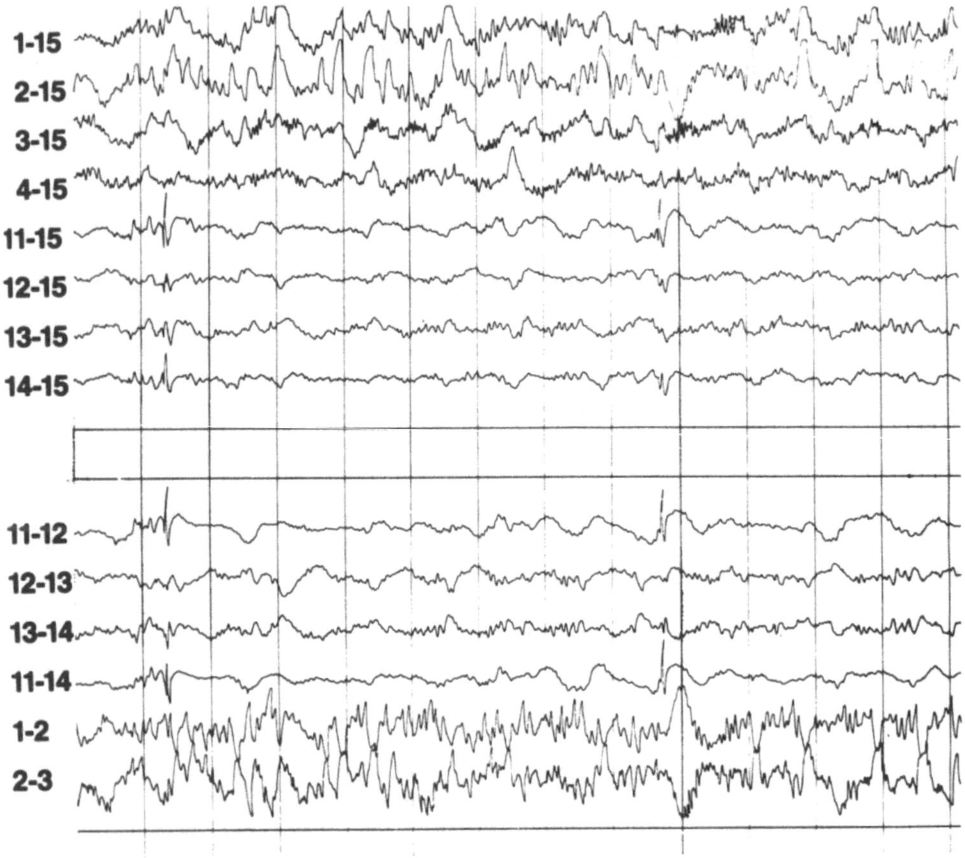

Fig. 3. Comparison of referential and bipolar recordings from the same electrodes. As in Fig. 1, the upper four traces are from the lateral temporal cortex; leads *11–14* were slipped under the temporal lobe to record from the hippocampus, uncus, and mesial inferior temporal gyrus. Electrode *15* was applied to the occipital neck region prior to the patient entering the operating room and was used as a reference for the upper eight traces. The lower six traces are made with a bipolar montage using the same electrodes. Active spiking is recorded from either electrode montage and can be demonstrated equally both ways

an cortex, and three parallel rows of five electrodes, one each along the superior, middle, and inferior temporal gyri. The recordings may be made in a bipolar or referential configuration, depending upon the preference of the electroencephalographer. Our preference is to apply reference electrodes to the base of the occiput in the midline just before the patient is taken to the operating room. This serves as an excellent reference electrode and has not resulted in any cardiac or muscle artifact in our experience. An alternative is to pick one cortical area devoid of spiking and use this as a reference electrode. Although there may be controversy over the superiority of bipolar to referential recording, we have found both to yield excellent results, and will use either, depending upon the specific needs during the corticography. An example of a recording from the temporal lobe is given in Fig. 3. The top eight traces were referenced to electrode 15 (neck reference), and the lower six traces are bipolar to adjacent electrodes and include all but two of the electrode contacts in the upper eight traces. As in Fig. 1, electrode numbers 11–14 were from a strip electrode placed to record from the undersurface of the temporal lobe.

The above electrode array can be manipulated better to define the limits of active spiking as needed by the surgeon. It is usual that between 30 min and 1 h of recording is needed fully to evaluate the focus. During this time electrical stimulation of the cortex can be undertaken to identify regions of cortical function (Ojemann and Dodrill 1985). For example, when exploring a temporal lobe in the dominant (for speech) hemisphere, it is routine to test for speech-related cortex in or near the margins of the presumed focus. This approach is also useful for identifying the pre- and postcentral gyri.

After the cortex has been "mapped" to determine the margin of cortex that can be removed safely (without incurring an unacceptable neurologic deficit) and the extent of the focus is approximated by ECoG, the cortical resection is done. Following the resection, the ECoG should be repeated. This will often yield additional regions, most often in continuity with the previous resection site that will continue to harbor active spiking. A strip electrode is also slipped under the more posterior hippocampus to determine if this remains active and, if so, it is resected. However, it has been known for many years that postresection spikes may occur over the insular cortex following temporal lobe resection, and this phenomenon has no prognostic significance. In contrast, if spiking remains in regions other than the insula, the outcome is generally poorer than if the postresection recording is completely devoid of epileptiform spikes.

This ECoG approach described above allows a precise delineation of the epileptogenic focus while, at the same time, it allows the surgeon to determine if the cortex under consideration contains essential neurologic function, such as speech. However, the strategy does not work well if patients are not completely cooperative and, therefore, cannot be used in many situations. Moreover, the length of time for ECoG recordings may be considered too limited by some surgeons. One alternative to these criticisms has been developed by Goldring and Gregorie (1984) and Gregorie and Goldring (1984). Two to three days prior to ablative surgery, a craniotomy is done over the proposed resection site. A large grid of electrodes is placed over the exposed dura and the wound closed. This grid can be used for both stimulation of the underlying brain and ECoG recordings. This allows the cortical region of interest to be mapped for important functions (i. e., language) and the regions of active spiking to be identified. During this testing, the patient is within the familiar environment of the laboratory rather than the operating room. After the patient has been studied in this manner, he or she is returned to the operating room where the cortical resection can be done with the patient under general anesthesia. Thus, this technique is especially attractive for use with patients who, for one reason or another, would not tolerate operation under local anesthesia. The obvious disadvantages to this approach are the need for two separate operations and the potential for an increased risk of infection.

Recordings with General Anesthetics

In certain cases, there will be no need for extensive cortical mapping or long periods of ECoG recording with the patient awake. In these uncomplicated situations, all that may be needed is some ECoG verification of the limits of the cortical focus, and this may be done quite well with the patient under general anesthesia. The management of anesthesia in these situations is straightforward. Premedicant and anesthetic drugs are used that do not suppress the EEG. Demerol and scopolamine are given preoperatively using the standard-dose schedule. The patient is brought to the operating room where an i.v. infusion of dextrose in lactated Ringer's solution is started and an arterial line is inserted to monitor blood pressure and arterial blood gases. A warming mattress is placed under the patient and core temperature is monitored.

General anesthesia is induced with i. v. thiopental, 1-3 mg/kg, and i. v. fentanyl, 50-150 µg, in an average adult. A nondepolarizing muscle relaxant, pancuronium bromide (Pavulon), 0.1-0.15 mg/kg, is given i. v. to facilitate endotracheal intubation. Anesthesia is maintained with nitrous oxide and oxygen supplemented with isoflurane in low concentrations, 0.5%-1.0% inspired. The proposed line of incision is infiltrated with a mixture of (equal volumes) lidocaine, 1% (with epinephrine 1:100000), and 0.5% marcaine. This mixture provides a rapid onset of local anesthesia that will last throughout the entire case if prolonged recordings are needed with relatively light anesthesia. Arterial blood pressure and body temperature are maintained at normal levels; arterial pCO_2 is maintained near 40 mm. Ten to fifteen minutes before cortical mapping, isoflurane is discontinued. The patient is maintained on nitrous oxide and oxygen with supplemental fentanyl, 50-150 µg, and Pavulon as necessary. If it is anticipated that electrical stimulation of the cortex will be needed to identify motor cortex, then the Pavulon must not be given. At this time the EEG is usually very active. Following cortical mapping, administration of isoflurane is resumed and is continued as in any other surgical patient. Another combination which has also worked is 50% N_2O and O_2 and halothane up to 2%. If this is used, then during cortical stimulation, halothane concentration is reduced to 0.5%.

If there is need to define the limits of pre- and postcentral gyrus while the patient is under general anesthesia, the method described by Gregorie and Goldring (1984) is quite useful. They have a silastic template that holds three rows of eight linearly oriented electrodes. The interelectrode distance is 1 cm, and each row is separated by 1.5 cm. The template is then placed over the cortex, and simultaneous records are made from each adjacent pair of electrodes while the contralateral median nerve is electrically stimulated transcutaneously at 1/s at the wrist. The strength of the stimulus is adjusted to be just above threshold for eliciting movement of the thumb. A typical somatosensory evoked response is looked for and, if not found, the recordings are made from the next row of electrodes. They have found this a reliable method for accurately finding the pre- and postcentral gyri, because somatosensory evoked responses occur only in the primary somatosensory area, and variably in the motor cortex. They have been able to confirm precentral cortex by evoking peripheral movement from the electrical stimulation of cortex.

Brevital Activation

Many is the time that the ECoG is not very active and the presence of the focus cannot easily be confirmed. In these situations, an intravenous methohexital sodium (Brevital) activation is often useful. This procedure can prove effective for patients under either local or general anesthesia. While recording the ECoG, the anesthesiologist administers 50-100 mg Brevital (depending upon the patient's weight) intravenously over a few seconds. Within 90 s after injection, the ECoG should show a profound activation with an increase in epileptiform spiking. We have found this to be helpful in augmenting spiking in 85% of our cases when this has been attempted and, therefore, have come to rely on this procedure almost routinely during corticography.

Summary

The factor most critical to the success of epilepsy surgery is the ability to identify accurately the location and limits of the epileptogenic focus to be resected. In some simple cases, this can be adequately done by reviewing multiple routine scalp EEGs collected over time. If the focus is clearly very focal and has been demonstrated to give rise to the onset of typical clinical seizures by prolonged EEG/video monitoring, then surgery can be done successfully without the placement of intracranial electrodes. However, if the seizure focus is not clearly identified from prolonged scalp EEG monitoring, then intracranial recordings are highly desirable, since they can localize single foci and, in some cases, identify the presence of multiple foci, and both situations influence highly the decision for or against surgery. The choice of intracortical recording techniques is debatable, with the most popular methods being the placement of intracortical depth electrodes. A more recent method is the use of subdural strip electrodes. Classically, surgery has been done under local anesthesia so that the cortex could be explored by electrical stimulation as well as by ECoG. This strategy is especially important when resecting cortex nearby critical areas, such as speech cortex. However, that approach is often not reasonable for many clinical situations. The alternative is to proceed with surgery with general anesthesia, since this also can yield excellent ECoG localization of foci. During general anesthesia, evoked potentials can accurately localize the postcentral gyrus, and cortical electrical stimulation can be used to elicit peripheral movements for the identification of precentral cortex. The use of strip electrodes has also been advantageous during routine ECoG for defining the extent of foci in regions of cortex not easily assessable with standard cortical electrode arrays. All of the techniques mentioned can be used fruitfully by the surgical epileptologist, and should be given thoughtful consideration in evaluating each individual case considered for surgery.

References

Bancaud J, Talairach J, Bonis A, Schaub C, Szikla G, Morel P, Bordas-Ferer M (1965) La stéréo-électroencéphalographie dans l'épilepsie. Masson, Paris, 321 pp

Bernoulli C, Siegfried J, Baumgartner G, et al. (1977) Danger of accidental person-to-person transmission by Creutzfeld-Jakob disease by surgery. Lancet 1: 478-479

Brazier MAB (1956) Depth recordings from the amygdaloid region in patients with temporal lobe epilepsy. Electroencephalogr Clin Neurophysiol 8: 532-533

Crandall PH (1973) Developments in direct recordings from epileptogenic regions in the surgical treatment of partial epilepsies. In: Brazier MAB (ed) Epilepsy: its phenomena in man. Academic, New York, pp 288-310

Crandall PH, Walter RD, Rand RW (1963) Clinical applications of studies on stereotactically implanted electrodes in temporal lobe epilepsy. J Neurosurg 20: 827-840

Falconer MA (1971) Anterior temporal lobectomy for epilepsy. In: Logue V (ed) Operative surgery, vol 14, neurosurgery. Butterworths, London, pp 142-149

Goldring S, Gregorie EM (1984) Surgical management of epilepsy using epidural electrodes to localize the seizure focus. Review of 100 cases. J Neurosurg 60 (3): 457-66

Gregorie EM, Goldring S (1984) Localization of function in the excision of lesions from the sensorimotor region. J Neurosurg 61: 1047-54

Ojemann GA, Dodrill CG (1985) Verbal memory deficits after left temporal lobectomy for epilepsy. Mechanism and intraoperative prediction. J Neurosurg 62: 101-107

Penfield W, Jasper H (1954) Epilepsy and the functional anatomy of the human brain. Little, Brown, Boston, pp 748-785

Rapport RL, Ojemann GA, Wyler AR, et al. (1977) Surgical management of epilepsy. West J Med 127: 185-189

Rasmussen T (1975) Cortical resection in the treatment of focal epilepsy. In: Purpura DP, Penry JK, Walter RD (eds)

Neurosurgical management of the epilepsy. Advances in neurology, vol. 8. Raven, New York, pp 139-154
Rasmussen T (1976) The place of surgery in the treatment of epilepsy. In: Morely TP (ed) Current controversies in neurosurgery. Saunders, Philadelphia, pp 463-477
Spencer SS (1980) Depth electroencephalography in selection of refractory epilepsy for surgery. Trans Am Neurol Assoc 104: 241-243
Spencer DD, Spencer SS, et al. (1984) Access to the posterior temporal lobe structures in the surgical treatment of temporal lobe epilepsy. Neurosurgery 15: 667-671
Talairach J, Bancaud J, Bonis A, et al. (1962) Functional stereotaxic exploration of epilepsy. Confin Neurol 22: 328-330
Talairach J, Bancaud J, Szikla G, et al. (1974) Approche nouvelle de la neurochirurgie de l'épilepsie. Méthodologie stéréotaxique et résultats thérapeutiques. Neurochirurgie 20 [Suppl 1]: 1-240
Walker AE (1975) Critique and perspectives. In: Purpura DP, Penry JK, Walter RD (eds) Neurosurgical management of the epilepsies. Advances in neurology, vol 8. Raven, New York, pp 155-196
Wieser HG (1983) Electroclinical features of the psychomotor seizure. A stereoelectroencephalographic study of ictal symptoms and chronotopographical seizure patterns including clinical effects of intracerebral stimulation. Fischer/Butterworths, Stuttgart/London
Wieser HG (1986) Selective amygdalohippocampectomy: indications, investigative technique and results. In: Symon L, Brihaye J, Guidetti B, Loew F, Miller JD, Nornes H, Pásztor E, Pertuiset B, Yaşargil MG (eds) Advances and technical standards in neurosurgery, vol 13. Springer, Vienna, pp 39-133
Wieser HG, Yasargil MG (1982) Selective amygdalohippocampectomy as a surgical treatment of mesiobasal limbic epilepsy. Surg Neurol 17: 445-457
Wyler AR, Ojemann GA, Lettich E, Ward AA Jr (1984) Subdural strip electrodes for localizing epileptogenic foci. J Neurosurg 60: 1195-1200

Stereo-Electroencephalography

H. G. Wieser

Introduction

The rationale for resective surgical therapy is based on the concept of the epileptic focus. Therefore, successful surgery for epilepsy depends on finding, as precisely as possible, an answer to two questions.

The first pertains to the location of the neuronal population which triggers the seizure. Nearly inextricably related to this is the determination of the location and extent of brain substance that is essential for recruitment into the abnormal activity in order to produce the patient's habitual seizures. With a view towards therapeutic consequences, the second question refers to the location and extent of the focal epileptogenic area that must be removed in order to produce satisfactory control of the patient's seizure tendency.

At present, there is no single method which can simply provide unequivocal answers to these questions. Consequently, presurgical evaluation of a patient consists of a multi-directional, step-by-step, goal-directed decision process, which is methodologically highly interactive. Furthermore, it is a dynamic process which makes use of several feedback loops and involves the formulation of hypotheses which have subsequently to be confirmed or disproved. It is in this context that stereo-electroencephalography has to be seen.

The term stereo-electroencephalography (stereo-EEG, SEEG) was coined by the Paris school (Talairach et al. 1952; Bancaud 1959) in order to emphasize first, that this method relies on stereotactic principles and second, that this method aims to measure the electrical activity of the brain in a three-dimensional way.

The first intracerebral recordings from man were made by Berger (1931) and by Foerster and Altenburger (1935) using insulated needles. Multicontact electrodes were first developed by Hayne and colleagues (1949).

The fabrication and application of human stereotactic instruments (Leksell 1949; Hayne and Meyers 1950; Spiegel and Wycis 1952; Talairach et al. 1958; Riechert and Mundinger 1955) provided the means for accurate electrode placements.

Department of Neurology, University Hospital, Frauenklinikstr. 26, 8091 Zürich, Switzerland

Presurgical Evaluation of Epileptics
Edited by H. G. Wieser and C. E. Elger

In epilepsy, stereotactic methods were first applied by Spiegel and Wycis (1950, 1951) in functionally investigating thalamus and hypothalamus in petit mal epilepsy.

Together with similar techniques, employed during this time by Williams and Parsons Smith (1949), Jung et al. (1951) and Kirikae and Wada (1951), such functional investigations were directed to the treatment of seizures within the concept of "centrencephalic" seizure origin. Determination of focal cortical epilepsies remained the domain of electrocorticography as systematized by Jasper et al. (1951).

Multi-target depth electrode studies in humans date back to 1953 (Ribstein, 1960). At this time, however, placement of several flexible electrodes, for instance, in the rhinencephalic structures, was rather imprecise and did not depend on stereotactic principles.

The use of simultaneous functional stereotactic exploration of many cortical and sub-cortical structures had to await the publication of atlases providing their spatial coordinates (Talairach et al. 1957, 1967).

The Role of Stereo-EEG in the Presurgical Evaluation Protocol

The use of stereo-EEG, because of its invasive character, requires rigorous criteria. In our understanding, it is only a method for presurgically investigating potential candidates for surgical epilepsy therapy. This implies that the generally accepted criteria for judging a patient as a candidate for epilepsy surgery must be fulfilled. Furthermore, depth recordings should be restricted to those patients in whom SEEG alone can provide essential information towards solution of pending diagnostic problems.

Studies of intracerebral electrical activity in the epileptic patient have contributed enormously to our understanding of the origin and spread of seizure discharges (Ajmone-Marsan 1980; Angeleri et al. 1964; Babb and Crandall 1976; Bancaud et al. 1965; Bickford et al. 1953; Bickford 1956; Brazier 1956, 1973; Brazier et al. 1954; Buser et al. 1973; Crandall et al. 1963; Crandall 1973; Delgado 1956; Delgado-Escueta and Walsh 1983; Engel and Crandall 1983; Gloor 1975; Lieb et al. 1976; Ludwig et al. 1975; Quesney and Gloor 1985; Rossi 1973; Sem-Jacob-

Table 1. Current Presurgical Evaluation Protocol at Zürich

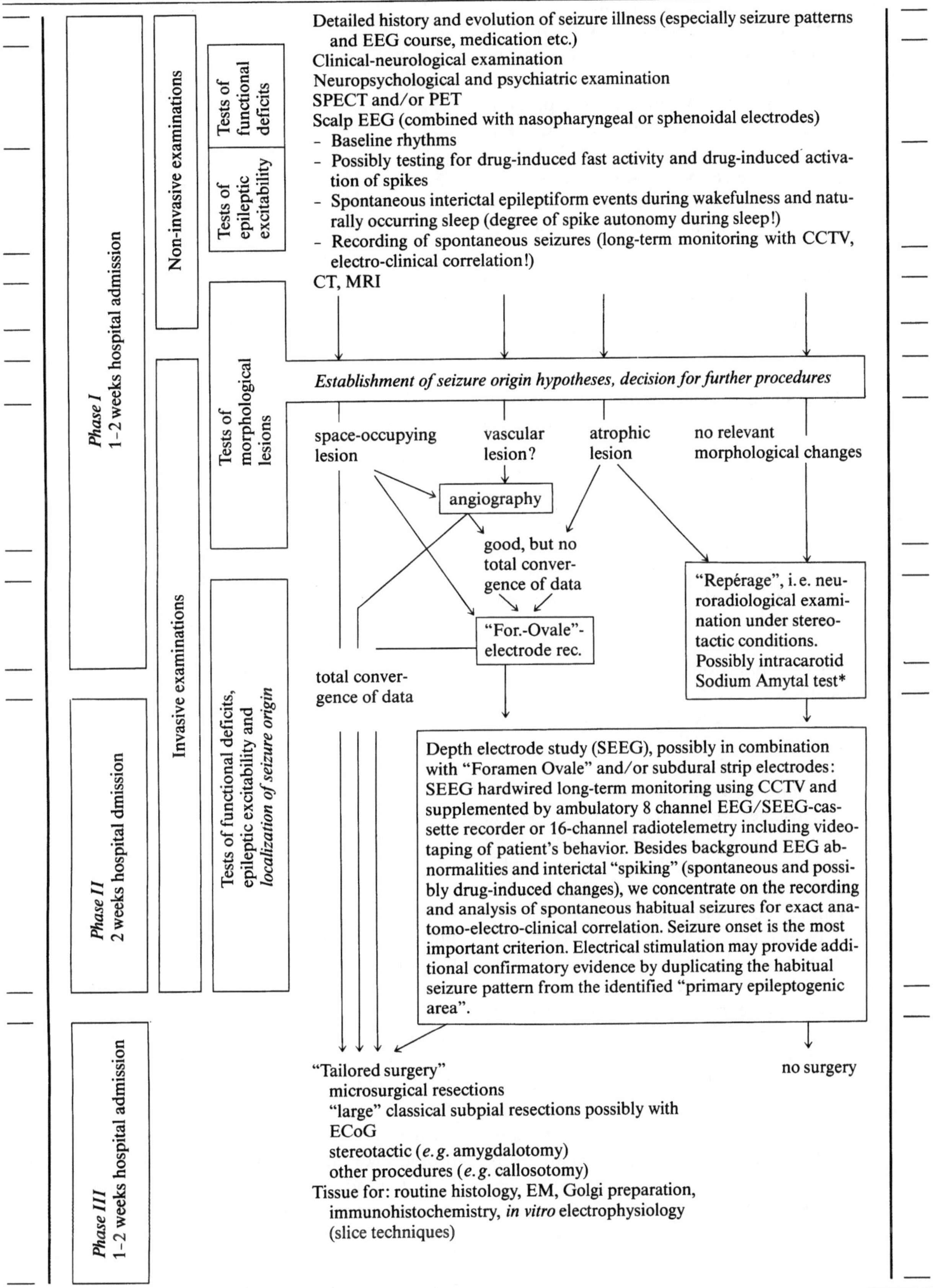

* A "Selective Amytal Temporal Lobe Memory Test" has been developed with temporary balloon occlusion distal to the origin of the anterior choroidal artery and with subsequent intracarotid Amytal inactivation of the temporal lobe only (Wieser et al. 1987)

sen 1968; Spiegel and Wycis 1952, 1961; Talairach et al. 1958, 1974; Talairach and Szikla 1980; Van Buren et al. 1975; Walker 1962; Wieser 1983; Williamson et al. 1980). Gloor (1975, and this volume) has reviewed the electrophysiology of the epilepsies, emphasizing the types of information that may properly be sought by depth electrode implantation. Certainly it should not be a "fishing expedition". Only a limited number of electrodes may justifiably be placed, and these must be positioned in the most highly suspect structures based on clinical and non-invasive studies.

Stereo-EEG is therefore a *diagnostic phase II procedure* (see Table 1) with the explicit aim of (a) improving inconclusive or insufficient data of the phase I diagnosis to such an extent that sound judgements with regard to the possibility of surgical treatment can be made and/or (b) resolving discrepancies with regard to previously obtained non-invasive investigations, which in themselves would preclude a therapeutic operation. It is obvious that the types of surgery which can be offered by a given center determine to a large extent whether stereo-EEG will be necessary or not, insofar as certain selective and highly tailored operations necessitate a much greater precision of presurgical data than more crude and standardized resective procedures.

It is platitudinous to state that presurgical evaluation is under the obligation to ensure that every effort is made and directed to provide the most effective patient care. As recurrent seizures define the very nature of epilepsy, it is obvious that the study of the patient's habitual seizures constitutes an important, if not the most important, aspect.

Seizures themselves are characterized by a specific neuronal dysfunction which at present can best be measured by electrophysiological means in combination with careful observation of the subjective and objective clinical concomitants. The generally unpredictable occurrence of seizures and their time course demand methods which are able to provide both long-term monitoring and high time resolution.

It is important to emphasize that stereo-EEG is not equivalent to depth recording only, but is the simultaneous registration of surface *and* depth EEG. In this combined procedure, the shortcomings of each are compensated by the other, i.e. the extremely sharp but relatively "tunnelled vision" (Quesney and Gloor 1985) of depth recordings is balanced by the distorted and "blurred vision" of the scalp EEG which guarantees that a larger potential generator at some distance will not be missed. The sampling problem inherent in depth EEG is thus optimally minimized.

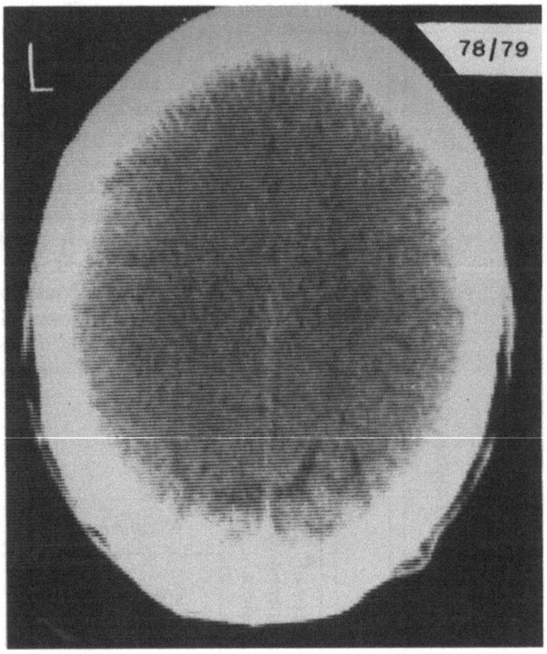

Fig. 1. Computer tomogram of a 30-year-old patient (78/79) at the time of stereo-EEG, showing marked bilateral occipital cortical atrophy. At age 8, following splenectomy because of panmyelopathy, the seizure illness started with a series of grand mal convulsions. In the following years, under anticonvulsive medication, grand mal seizures were controlled, but adversive and later on drug-resistant psychomotor seizures persisted. Initially elementary and subsequently complex visual auras were frequently reported; during the last years before stereo-EEG however, the patient's typical aura was an epigastric one

The documentation of a *structural lesion* is very important, but it is essential to remember that it need not correspond to the site of the epileptogenic focus. From this, it follows that a mandatory step before surgery is to demonstrate relevant epileptogenicity of a documented morphological and/or functional lesion. In principle, this requirement also applies in the case of tumors and vascular abnormalities which demand surgical treatment in their own right, because if seizures are associated, surgical intervention should take into account the "epileptogenic zone". In fact, although rare, the morphological lesion and even the electrophysiologically defined "lesional activity" might be located at a considerable distance from the actual seizure onset area, and might thus be misleading. Figures 1 and 2 illustrate this.

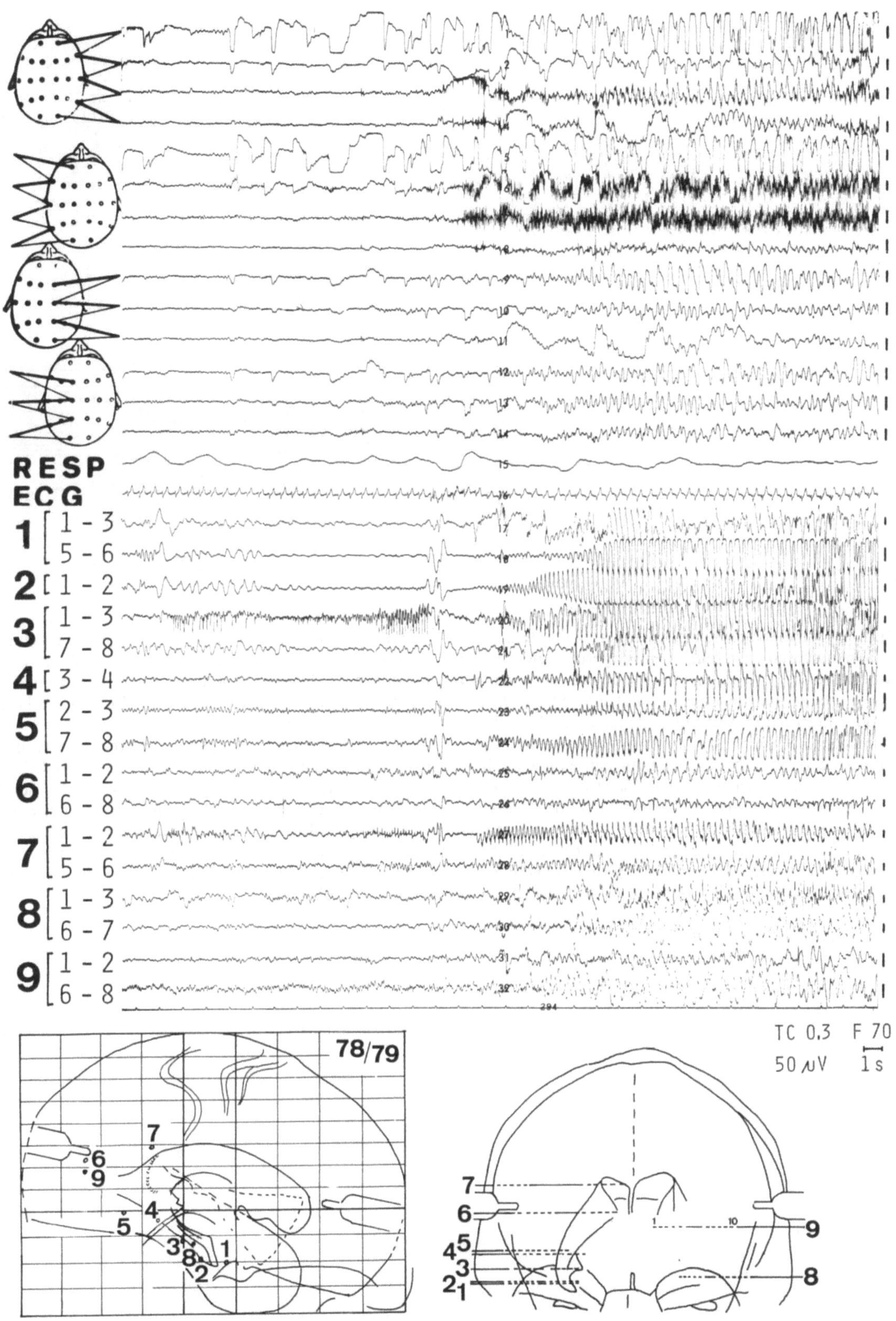

Fig. 2. Seizure onset of a typical attack of patient 78/79, showing a focal discharge at the right hippocampus (3/1-3) and the right posterior cingulate gyrus (7/1-2). Based on this and other seizures with well-localized focal seizure onset, a right selective amygdalohippocampectomy was performed which left the patient seizure free. Follow-up is 6 years

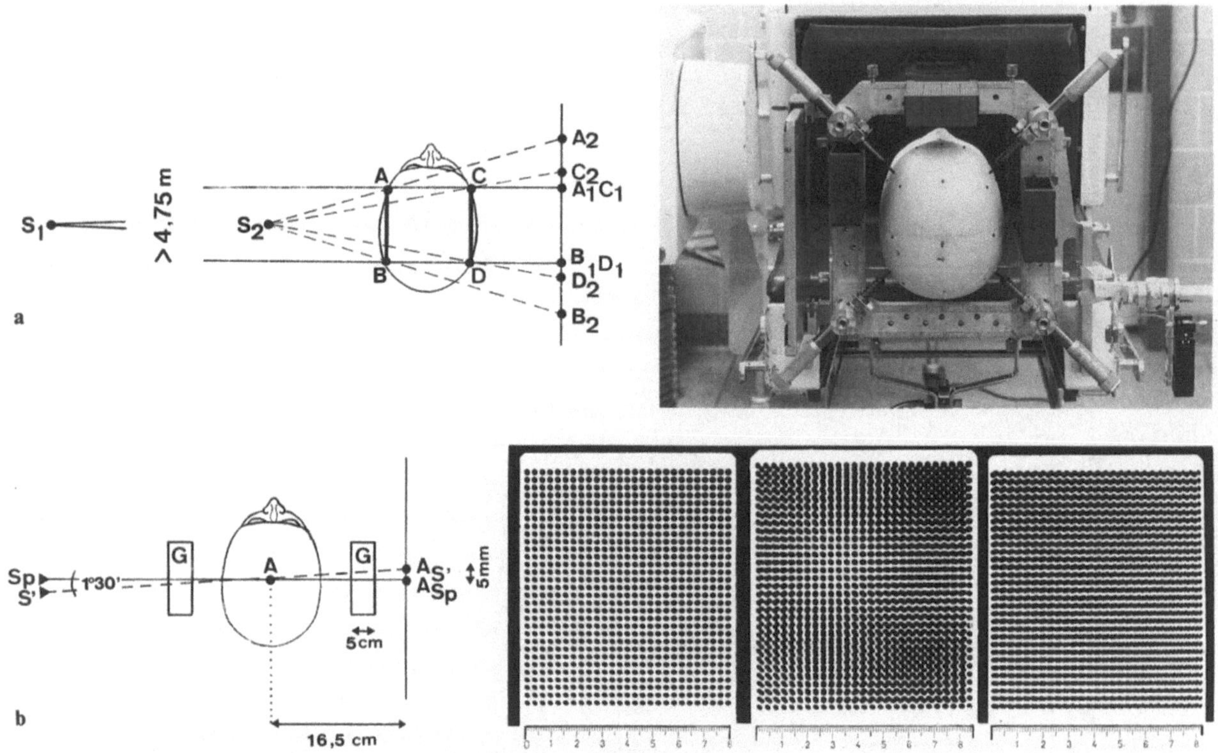

Fig. 3. **a** Demonstration of the advantages of teleradiography. In order to prevent distortion of the object due to magnification, the X-rays should be parallel. This is approximately achieved if the source-object distance is >4.75 m (S_1). A source at short distance to the object (S_2) magnifies the object. As the magnification due to the divergence of the X-rays is a function of the distance between the source, the object (characterized by *A–B, C–D*) and the plate, structures in the plane *A–B* undergo a larger magnification than structures situated in the plane *C–D*. **b** The second important precondition of this technique is accurate perpendicular incidence of the X-rays, which has to be assured and verified for each picture. It is illustrated by the fact that if the incident X-ray were to differ by 1° 30′ from the perpendicular, the image of a point situated in the middle of the skull would be 5 mm away from its normal projection. Control is provided by the double grids *(G)*. These have the appearance of a single grid when the X-rays are correctly aligned. **c** Stereotactic frame of Talairach with its double grid control and sighting apparatus for anteroposterior and lateral teleradiography. **d** *Left*, Correct teleradiographic sighting of the double grid. *Middle*, Incorrect sighting with enlargement of the grid image due to short source-object distance and displacement of the orifices, which is greatest at the periphery of the grid. *Right*, Incorrect sighting with enlargement of grid and displacement of all orifices due to "bad centring", i.e. voluntary displacement of the X-ray tube which, however, is at 4.75 m distance

Prerequisites for, and Techniques of, Depth Electrode Implantation

Before implanting depth electrodes, the anatomy of the brain and in particular its blood vessels must be known in terms of a stereotactic reference system, in order to (a) precisely identify possible lesions, the target areas and electrode trajectories and (b) minimize the risk of injuring vessels.

Direct and Indirect Stereotactic Localization

Neuroradiological examination under stereotactic conditions is done under general anaesthesia at the end of phase I (see Table 1). The head of the patient is maintained by the stereotactic frame fastened to the cranial vault by four metal pins so as to avoid any displacement (see Fig. 3C). The apparatus is placed in such a manner that the midsagittal plane is parallel to the long axis and perpendicular to the plane of the frame; thus the radiologic plate fixed to the apparatus is also parallel (lateral view) or perpendicular (anteroposterior view) to the median plane. Asymmetries of the skull, especially of the external auditory meatus, may cause slight rotation of the head in relation to the frame. In order to prevent distortions due to magnification, teleradiography - i.e., a source-object distance of at least 4.75 m - is employed. This permits the use of teleradiographic documents without calculating the corrections, and thus gives a true 1:1 picture of the anatomical situation (see Fig. 3).

Indirect localization of the targets and electrode trajectories can be attained by the use of stereotactic atlases.

Spiegel and Wycis (1952) published the first stereotactic atlas of the human brain, with subsequent publications by Talairach et al. (1957, 1967) and Schaltenbrand and Bailey (1960). In each atlas, careful measurements of the model brains were made in relation to the three planes of Cartesian coordinates constructed over a base line. Spiegel and Wycis used the foramen of Monro-pineal gland line. Talairach and Schaltenbrand used, with minor variations between them, the anterior-posterior bicommissural line (AC-PC). While Talairach employed a tangent to the upper limit of the anterior commissure and inferior limit of the posterior commissure, Schaltenbrand used the center of both commissures.

At the end of the neuroradiological examination a series of pictures has been produced which - because of the possibility of their direct superimposition - allow for precise and uncomplicated calculation of the electrode trajectories and targets, thus avoiding any injury of vessels. By applying the AC-PC reference system, the individual brain structures can be compared with the most appropriate model brain of the Talairach atlas and also with the statistically derived brain in the standardized proportional system (see Fig. 4).

It is obvious that the incorporation of computed tomography (CT), magnetic resonance imaging (MRI) and positron emission tomography (PET) into stereotactic examination could substantially improve the diagnostic value of this procedure and replace pneumoencephalography and ventriculography. Moreover, the application of these modern techniques, together with high resolution digital subtraction angiography under stereotactic conditions, would reduce its invasive nature to a minimum.

Implantation of the Recording Electrodes

The implantation of the depth electrodes is performed at the beginning of phase II, i.e. about 4 weeks following phase I. This has the advantage that the position of the electrodes can be exactly calculated outside the operating theatre, i.e. no time is lost with the patient under general anaesthesia. After repositioning of the stereotactic frame, X-rays are performed to check for its correct alignment.

In the next step, small transcutaneous-osseous trepanations are made at the calculated points with a 2.45-mm diameter twist drill. After perforation of the skin and the bone, the dura is coagulated and perforated with a pointed electrode of the same diameter. Before the introduction of the depth electrodes, the exact position of the trepanation orifices is confirmed once again by X-ray (Fig. 5).

The electrodes used currently are implanted as follows: the flexible hollow-core electrode is guided with a central stylet which is introduced 9 cm from the tip of the electrode into its central lumen. This offers the advantage that the connector can be mounted at the factory to ensure the best electrode reliability. The electrode, with the stylet inside, is guided through the Talairach double grid by using a special guide inserted into the bony trepanation orifice (Fig. 6A). In order to have enough space between the skull and the grid, a metallic extension to the Talairach frame is used, as illustrated in Fig. 6.

The electrode contacts consist of 10 teflon-insulated, helically wound stainless steel (type 316) wires arranged in two groups of five poles. The external diameter of the coil is 1 mm, with a distance between contacts of 4 mm (center to center). Each pole is cylindrical, with a tissue contact length of 0.8 mm and a real surface area of 3.8 mm^2 (Fig. 6C). To make its surface very smooth, the electrode is covered with a

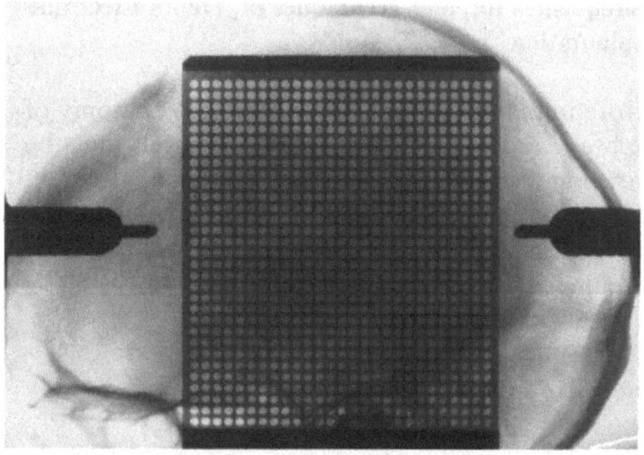

a

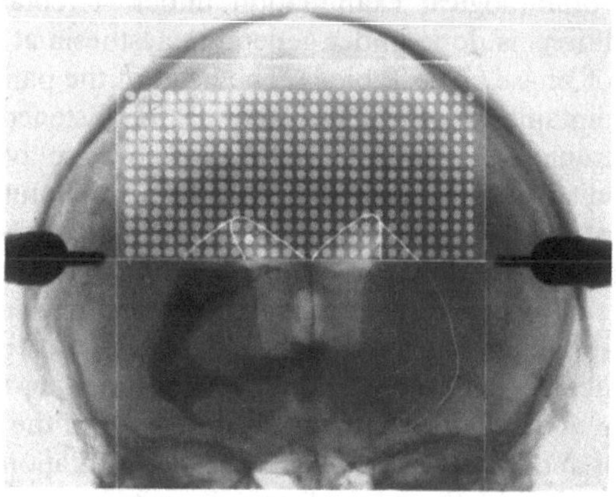

d

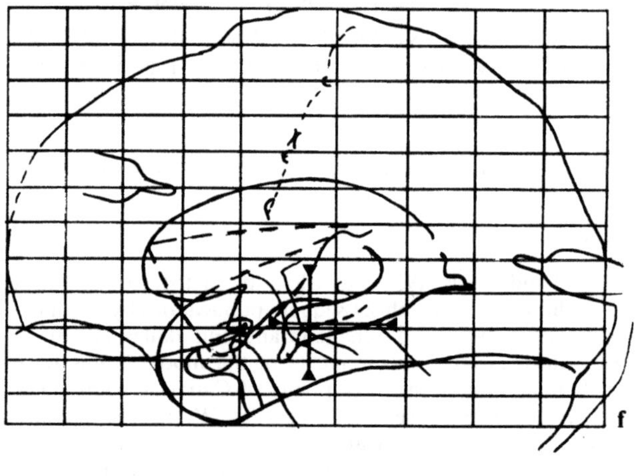

f

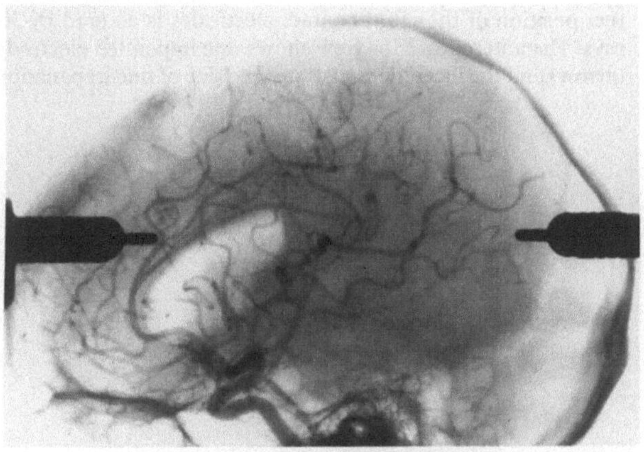

b

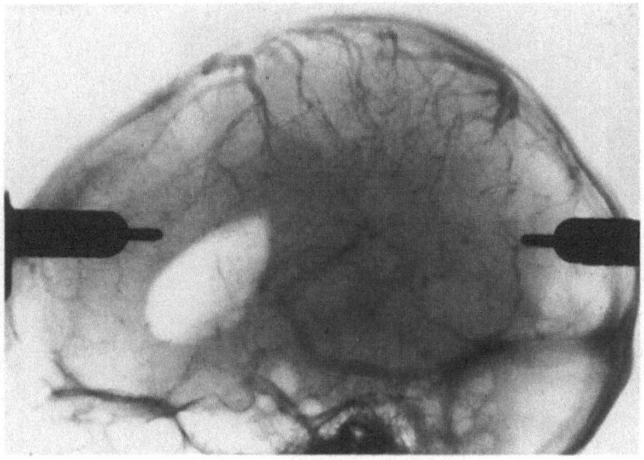

c

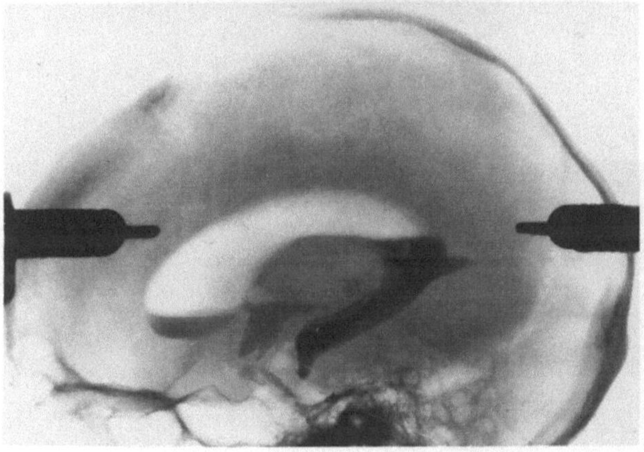

e

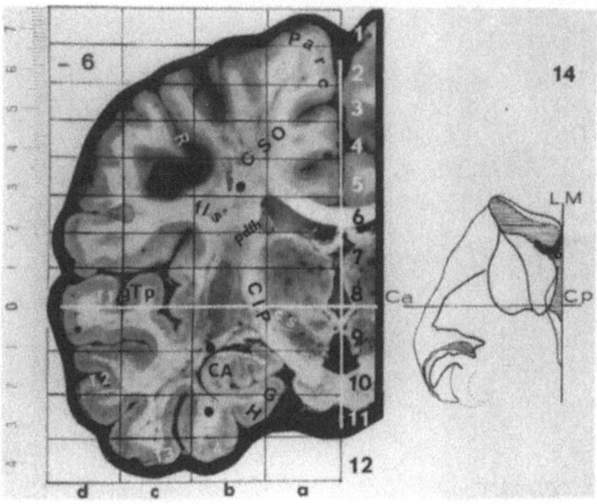

g

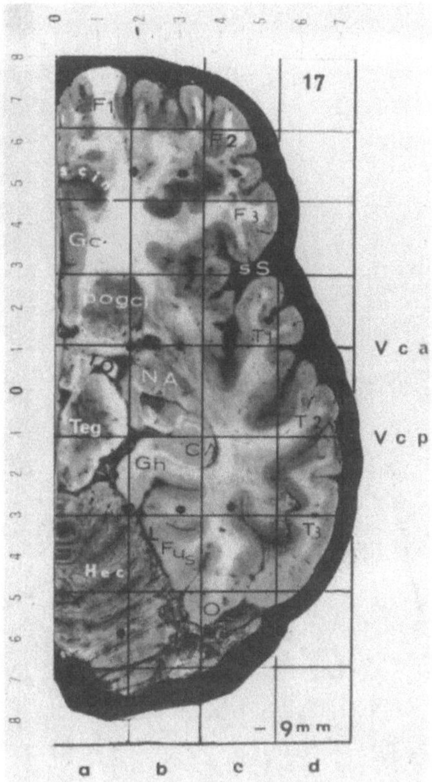

h

Fig. 4 a–e. Teleradiographic examination under stereotactic conditions with direct delineation of the most important brain structures and the vessels. The electrode trajectories have to avoid the vessels, and therefore compromises with regard to the envisaged targets must, if necessary, be made. **f-h** illustrate the indirect localization of an electrode trajectory using the appropriate pictures in the Talairach Atlas (Talairach et al. 1967) (**g**, vertico-frontal section of Brain VF48, 6 mm rear of the midcommissural line, with projection of the ventricular system, p. 101; and, **h** horizontal section of Brain HD6, 9 mm below the AC–PC line, p. 137) and, **f** the drawing of the patient's brain landmarks in the standardized proportional system. Atlas sections are indicated by ┼. Abbreviations: *parc,* lobulus paracentralis; *CSO,* centrum semiovale; *fls,* fasciculus longitudinalis superior; *pdth,* pedunculus dorsalis thalami; *CIP,* capsula interna, crus posterior; *cs,* corpus subthalamicum; *sS,* sulcus lateralis cerebri (Sylvii); *s cing,* sulcus cinguli; *Gc,* gyrus cinguli; *pogcl,* pontes grisei caudatolenticulares; *TO,* tractus opticus; *Teg,* Tegmentum mesencephali; *Gh, GH,* gyrus parahippocampalis; *L Fus,* gyrus fusiformis; *Hec,* hemisphaerium cerebelli; *O,* gyri occipitales; *NA,* nuclei amygdalae; *CA,* cornu Ammonis; *T1,* gyrus temporalis superior; *T2,* gyrus temporalis medius; *T3,* gyrus temporalis inferior; *F1,* gyrus frontalis superior; *F2,* gyrus frontalis medius; *F3,* gyrus frontalis inferior

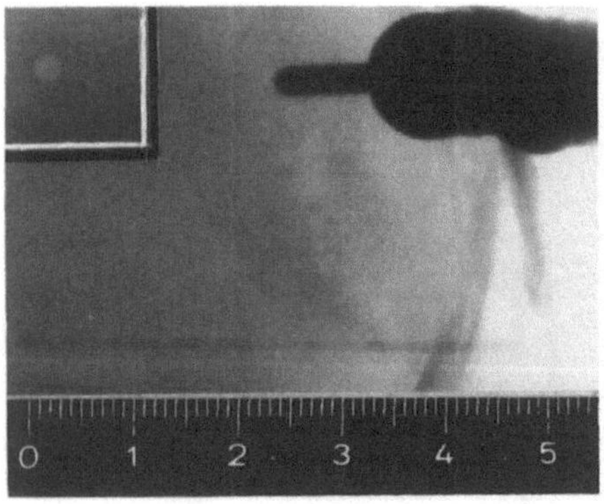

Fig. 5. The correct position of the trepanations and the correct position of the multi-contact electrodes is assured by X-rays. The anteroposterior view shows one implanted electrode (arrow) and the insert demonstrates the size of one trepanation

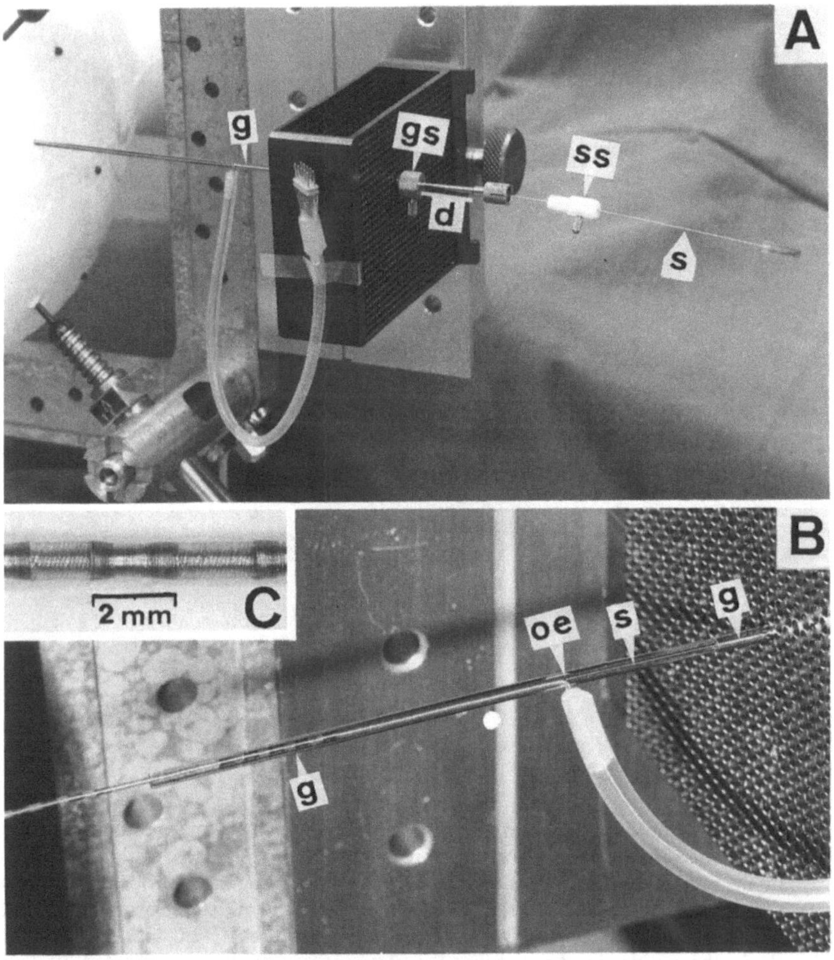

Fig. 6A-C. Electrode and guiding device mounted on a Talairach frame. The guiding cannula *(g)* enters the bony orificium. Then the distance between the guide stopper *(gs)* and the end of the guide for the stylet is measured *(d)* and added to the calculated position of the stylet stopper *(ss)*. To allow placement of the hollow-core electrode with mounted connector on the rigid stylet *(s)* between the grid and the patient's head, the double grid has had to be moved back by 6 cm. **A** shows an electrode with its 10-pin connector and how the electrode is guided through the Talairach grid; **B** shows how the guiding stylet is introduced into the lumen of the electrode *(oe)*. **C** visualizes details of the slightly recessed contacts

thin layer of polyurethane elastomer. Electrodes are available in six different types, i.e. the distance from contact 5 to 6 varies between 4 mm and 29 mm (center to center). The appropriate electrode type is chosen according to the distance between the target and the dura.

Recording and Monitoring

Proper stereo-electroencephalographic examination of a potential candidate for epilepsy surgery should include the simultaneous recordings of both the combined scalp and depth EEG and the clinical behavior. Since the ictal events provide the most valuable information, long-term monitoring facilities are required which guarantee a maximum of freedom for the patient, at least most of the time.

At present, probably most surgically active centers agree that alternation between daily scheduled direct laboratory recording and integrated telemetry-computer-video system monitoring during, for example, evening, night and weekend, might provide the most advantageous solution. One advantage of a hard-wired laboratory recording is that, without special equipment, a large number of channels (32 or more) can be recorded simultaneously. Out of the usually large number of intracerebral contacts available, the most informative must be chosen and appropriate montages should be used, which either consist of unbroken chains or utilize a single common reference. The latter permits the replay of data recorded by a fixed montage from a given set of electrodes in any desired montage. A more comprehensive discussion of the principle considerations important for reliable recordings is given by Gloor elsewhere in this volume.

Today, outside the laboratory, some kind of integrated telemetry-computer-video system should be an integral part of an advanced monitoring system. Gotman et al. (1985) have recently updated the "state of the art", and elsewhere we have tried to review some of the most important aspects of data acquisition and analysis (Stefan and Wieser 1986; Wieser 1987). Computer-aided data reduction is needed and increasingly carried out at the most advanced centers, which use the computer as a kind of "intelligent editor" of the EEG, retaining only valuable and specific information (Gotman et al. 1985).

Specific Information Obtained by Stereo-EEG

Reliable recording and subsequent analysis of the electro-clinical features of the patient's typical ictal events, in terms of localization of the *seizure onset,* are considered by us to constitute the most important aspect of the assessment of a patient. The study of the *propagation* of the ictal discharges and the definition of the *seizure sustaining substratum* are closely related. *Interictal epileptiform events* ("spiking") and *abnormalities of the background activity,* as well as such techniques as pharmacological tests (drug-induced fast activity) and peripheral and direct electrical stimulation of the brain, provide further important information. The relative value of the information obtained by these different methods with regard to prediction of surgical outcome has been recently discussed (Wieser 1987). Here it may suffice to mention that the localizationally invariable, high-frequency, low-voltage, focal seizure onset, in conjunction with certain other factors, is the best predictive sign and thus the most reliable criterion for definition of the "primary epileptogenic zone" in terms of surgical consequences. Figure 7 illustrates one such focal seizure onset, and aims at the same time to demonstrate the different information content of interictal and ictal events. Furthermore, it is an example of how montages can be specifically applied in order to provide reliable answers to particular questions, e.g. whether, in the given patient at this stage of the examination, an exclusively mesiobasal limbic seizure onset (amygdala and/or hippocampus) without participation of the lateral neocortex could be ascertained.

Concluding Remarks

The importance of stereo-EEG for the progress to date in epilepsy surgery is unquestionable. However, besides its invasive nature, this method has undisputed drawbacks and limitations. Because the number of brain sites that can be monitored is limited, the accuracy of the technique depends on previous identification of the most likely sites of seizure generation prior to electrode implantation; i.e. coherent hypotheses with regard to possible seizure origin have to be elaborated in advance. Therefore with the advent of new technologies, such as MRI, SPECT and PET, the hope has arisen during the last years that these techniques could replace the invasive stereo-EEG.

Indeed several surgically active centers, which have at their disposal the whole range of modern

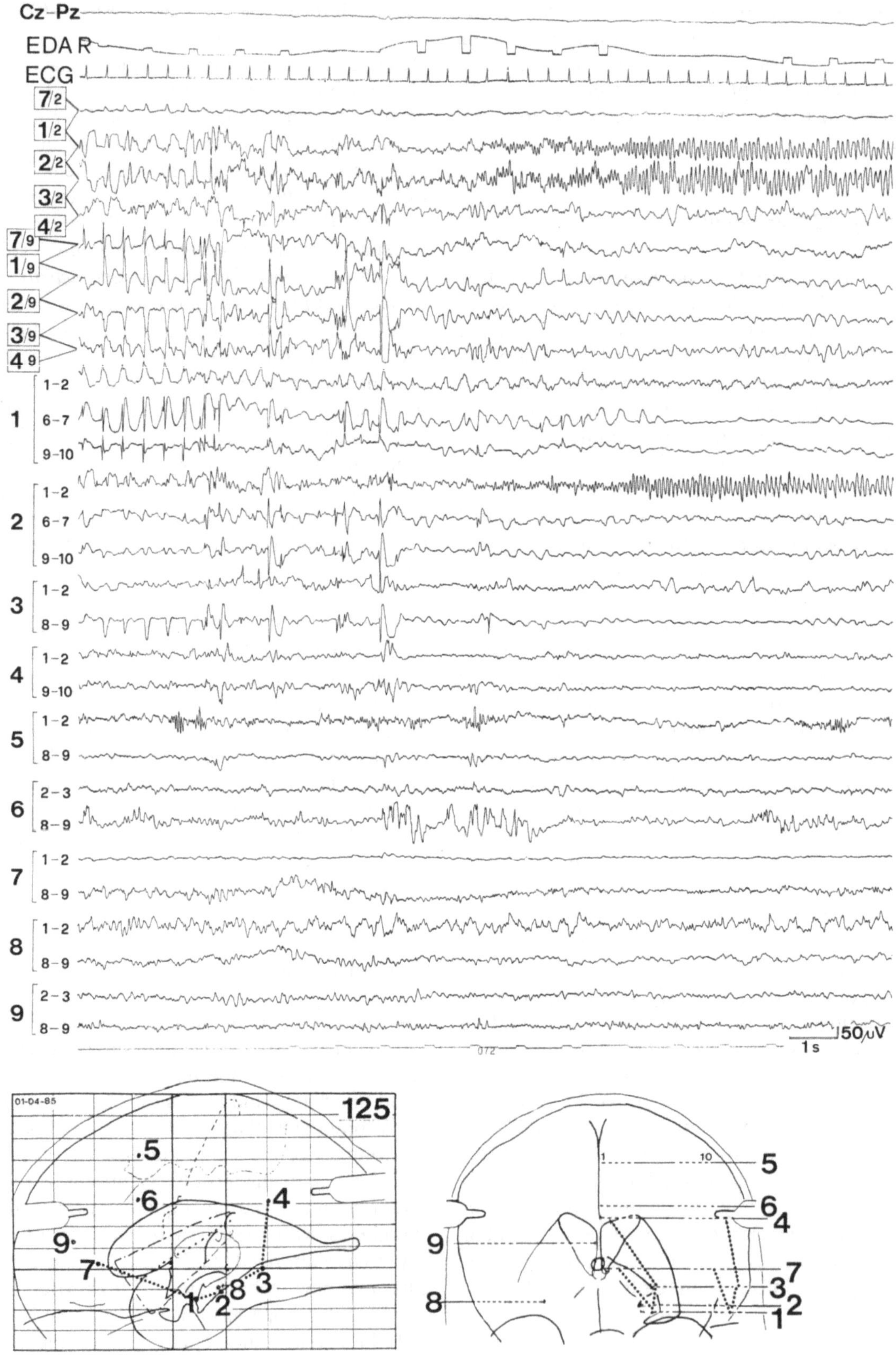
Cz–Pz
EDA R
ECG
7/2
1/2
2/2
3/2
4/2
7/9
1/9
2/9
3/9
4/9
1 1-2 6-7 9-10
2 1-2 6-7 9-10
3 1-2 8-9
4 1-2 9-10
5 1-2 8-9
6 2-3 8-9
7 1-2 8-9
8 1-2 8-9
9 2-3 8-9
50 μV
1 s
01-04-85
125
.5
·6
.4
9·
7
1
2
8
3
5
6
4
9
7
3
2
1
2
8
1
10

◀ **Fig. 7.** Example of depth recordings at a seizure onset, characterized by an aura of detachment. Note the special montage, which consists of uninterrupted bipolar long-distance derivations in the parasagittal plane (... in the brain map), connecting the inner (channels 4-7) and the outer contacts (channels 8-11) of electrodes 7, 1, 2, 3, 4, and of interrupted short-distance derivations along electrodes 1-9 (channels 12-31). Whereas the pre-ictal epileptiform activity consists of fairly regular 2/sec spike-slow-wave complexes with a clear predominance in the anterior lateral temporal neocortex *(left)*, the low-voltage high-frequency ictal discharge *(right)* is clearly localized to the anterior hippocampus (electrode 2, contact 2). The position of the electrodes is indicated on the brain map. The ten contacts of an electrode are numbered from inside outwards (small numbers). In addition to the depth EEG, one scalp derivation *(Cz-Pz)*, the electrocardiogram *(ECG)* and the electrodermal activity of right fingers *(EDA R)* are shown. Calibration refers to surface and depth EEG only

diagnostic tools, reported that the incorporation of them into the presurgical evaluation protocol allowed a certain percentage of patients to be operated on without stereo-EEG examination. These patients would probably have undergone depth recording without the confirmatory evidence obtained with these new methods. However, it seems that although the modern morphological and functional neuroimaging techniques possess an indisputably great diagnostic potential, they do not provide the same kind of information as electrophysiological procedures and so far cannot substitute for direct intracranial EEG recording, at least in the more complicated cases.

At present, chronic stereo-EEG remains most useful in identifying patients for temporal lobe surgery, especially with a view towards selective resections of the mesiobasal limbic structures. In combination with other intracranial recording techniques, such as subdural strips, stereo-EEG also helps to identify and differentiate neocortical epilepsies of other origin, for example, those with seizure onset in mesial or basal parts of the frontal or occipital lobes.

References

Ajmone-Marsan C (1980) Depth electrography and electrocorticography. In: Aminoff MJ (ed) Electrodiagnosis in clinical neurology. Churchill Livingstone, Edinburgh, pp 167-196

Angeleri F, Ferro-Milone F, Parigi S (1964) Electrical activity and reactivity of the rhinencephalic, pararhinencephalic, and thalamic structures: prolonged implantation of electrodes in man. Electroencephalogr Clin Neurophysiol 16: 100-129

Babb TL, Crandall PH (1976) Epileptogenesis of human limbic neurons in psychomotor epileptics. Electroencephalogr Clin Neurophysiol 40: 225-243

Bancaud J (1959) Apport de l'exploration fonctionelle par voie stéréotaxique à la chirurgie de l'épilepsie. Neurochirurgie 5: 55-112

Bancaud J, Talairach J, Bonis A, Schaub C, Szikla G, Morel P, Bordas-Ferer M (1965) La stéréo-électroencéphalographie dans l'épilepsie. Masson, Paris, 321 pp

Berger H (1931) Über das Elektroenkephalogramm des Menschen III. Arch Psychiatr Nervenheilk 94: 16-60

Bickford RG (1956) The application of depth electrography in some varieties of epilepsy. Electroencephalogr Clin Neurophysiol 8: 526-527

Bickford RG, Dodge HW Jr, Peterson MC, Sem-Jacobsen CW (1953) A new method of recording from subcortical regions of the human brain. Electroencephalogr Clin Neurophysiol 5: 464

Brazier MAB (1956) Depth recording from the amygdaloid region in patients with temporal lobe seizures. Electroencephalogr Clin Neurophysiol 8: 532-533

Brazier MAB (1973) Electrical seizure discharges within the human brain: the problem of spread. In: Brazier MAB (ed) Epilepsy: its phenomena in man. Academic, New York, pp 153-170

Brazier MAB, Schroeder H, Chapman WP, Geyer C, Fager C, Poppen JL, Solomon HC, Vakovley PI (1954) Electroencephalographic recordings from depth electrodes implanted in the amygdaloid regions in man. Electroencephalogr Clin Neurophysiol 6: 702

Buser P, Bancaud J, Talairach J (1973) Depth recordings in man in temporal lobe epilepsy. In: Brazier MAB (ed) Epilepsy: its phenomena in man. Academic, New York, pp 67-97

Crandall PH (1973) Developments in direct recordings from the epileptogenic regions in the surgical treatment of partial epilepsies. In: Brazier MAB (ed) Epilepsy: its phenomena in man. Academic, New York, pp 288-310

Crandall PH, Walter RD, Rand RW (1963) Clinical applications of studies on stereotactically implanted electrodes in temporal lobe epilepsy. J Neurosurg 20: 827-840

Delgado JMR (1956) Use of intracerebral electrodes in human patients. Electroencephalogr Clin Neurophysiol 8: 528-530

Delgado-Escueta AV, Walsh GD (1983) The selection process for surgery of intractable complex partial seizures: surface EEG and depth electrography. Ass Res Nerv Ment Dis 61: 295-326

Engel J Jr, Crandall PH (1983) Falsely localizing onsets with depth EEG telemetry during anticonvulsant withdrawal. Epilepsia 24: 344-355

Foerster O, Altenburger H (1935) Elektrobiologische Vorgänge an der menschlichen Hirnrinde. Verhandlungen der Gesellschaft Deutscher Nervenärzte (22. Jahresversammlung) Von Vogel FCW, Berlin, pp 93-102

Gloor P (1975) Contributions of electroencephalography and electrocorticography to the neurosurgical treatment of the epilepsies. In: Purpura DP, Penry JK, Walter RD (eds) Advances in neurology, vol 8. Raven, New York, pp 59-105

Gotman J, Ives JR, Gloor P (eds) (1985) Long-term monitoring in epilepsy. Elsevier, Amsterdam (EEG suppl. no. 37)

Hayne R, Meyers R (1950) An improved model of a human stereotaxic instrument. J Neurosurg 7: 463-475

Hayne RA, Belinson L, Gibbs FA (1949) Electrical activity of subcortical areas in epilepsy. Electroencephalogr Clin Neurophysiol 1: 437-445

Jasper H, Pertuiset B, Flanigin H (1951) EEG and cortical electrograms in patients with temporal lobe seizures. Arch Neurol Psychiat (Chic) 65: 272-290

Jung R, Riechert T, Heines KD (1951) Zur Technik und Bedeutung der operativen Elektrocorticographie und subcorticalen Hirnpotentialableitung. Nervenarzt 22: 433-436

Kirikae T, Wada J (1951) Electrothalamogram of petit mal seizures. Med Biol 20: 253

Leksell L (1949) A stereotaxic apparatus for intracranial neurosurgery. Acta Chir Scand 99: 229-253

Lieb JP, Walsh GO, Babb TL, Walter RD, Crandall PH (1976) A comparison of EEG seizure patterns recorded with surface and depth electrodes in patients with temporal lobe epilepsy. Epilepsia 17: 137-160

Ludwig BI, Ajmone-Marsan C, Van Buren JM (1975) Depth and direct cortical recording in seizure disorders of extratemporal origin. Neurology 26: 1085-1099

Quesney LF, Gloor P (1985) Localization of epileptic foci. In: Gotman J, Ives JR, Gloor P (eds) Long-term monitoring in epilepsy. Elsevier, Amsterdam, (EEG suppl no 37), pp 132-156

Ribstein M (1960) Exploration du cerveau humain par électrodes profundes. Electroencephalogr Clin Neurophysiol [Suppl] 16

Riechert T, Mundinger F (1955) Beschreibung und Anwendung eines Zielgerätes für stereotaktische Hirnoperationen: II Modell. Acta Neurochir [Suppl] 3: 308-337

Rossi GF (1973) Problems of analysis and interpretation of electrocerebral signals in human epilepsy: a neurosurgeon's view. In: Brazier MAB (ed) Epilepsy: its phenomena in man. Academic, New York, pp 259-285

Schaltenbrand G, Bailey P (1960) Introduction to stereotaxis with an atlas of the human brain. Thieme, Stuttgart

Sem-Jacobsen CW (1968) Depth electrographic stimulation of the human brain and behavior. Thomas, Springfield

Spiegel EA, Wycis HT (1950) Thalamic recordings in man, with special reference to seizure discharges. Electroencephalogr Clin Neurophysiol 2: 23-29

Spiegel EA, Wycis HT (1951) Diencephalic mechanisms in petit-mal epilepsy. Electroencephalogr Clin Neurophysiol 3: 473-475

Spiegel EA, Wycis HT (1952) Stereoencephalotomy Part I. Grune and Stratton, New York

Spiegel EA, Wycis HT (1961) Chronic implantation of intracerebral electrodes. In: Sheer DE (ed) Electrical stimulation of the brain. University of Texas Press, Austin, pp 37-44

Talairach J, Szikla G (1980) Application of stereotactic concepts to the surgery of epilepsy. Acta Neurochir [Suppl 1] 30: 35-54

Talairach J, De Ajuriaguerra J, David M (1952) Études stéréotaxique et structures encéphaliques profundes chez l'homme. Presse Méd 28: 605-609

Talairach J, David M, Tournoux P, Corredor H, Kvasina T (1957) Atlas d'anatomie stéréotaxique des noyaux gris centraux. Masson, Paris, 294 pp

Talairach J, David M, Tournoux P (1958) L'exploration chirurgicale stéréotaxique du lobe temporal dans l'épilepsie temporale: repérage anatomique stéréotaxique et technique chirurgicale. Masson, Paris, 136 pp

Talairach J, Szikla G, Tournoux P, Prossalentis A, Bordas-Ferrer M, Covello L, Iacob M, Mempel E, Buser P, Bancaud J (1967) Atlas d'anatomie stéréotaxique du télencéphale. Masson, Paris, 326 pp

Talairach J, Bancaud J, Szikla G, Bonis A, Geier S, Vedrenne C, Fontaine C, Hémon E, Hori H, Recoules-Arche R, Scarabin J-M, Serrano E, Trottier S (1974) Approche nouvelle de la neurochirurgie de l'épilepsie. Méthodologie stéréotaxique et résultats thérapeutiques. Neurochirurgie 20 [Suppl 1]: 1-240

Van Buren JM, Ajmone-Marsan C, Mutsaga N, Sadowsky R (1975) Temporal lobe seizures with additional foci treated by resection. J Neurosurg 43: 596-607

Walker AE (1962) Stereotaxic methods for the study of subcortical activity in epilepsy. Confinia Neurologica 22: 217-222

Wieser HG (1983) Electroclinical features of the psychomotor seizure. Fischer, Stuttgart and Butterworths, London

Wieser HG (1987) Data analysis. In: Engel J Jr (ed) Surgical treatment of the epilepsies. Raven, New York pp 335-360

Wieser HG, Valavanis A, Regard M, Landis T, Yasargil G (1987) "Selective" Amytal temporal lobe memory test. (Abstr) 17th Epilepsy Internat Congress Jerusalem

Williams D, Parsons Smith G (1949) The spontaneous electrical activity of the human thalamus. Brain 72: 450-482

Williamson PD, Spencer DD, Spencer SS, Mattson RH (1980) Presurgical intensive monitoring using depth electroencephalography in temporal lobe epilepsy. In: Wada JA, Penry JK (eds) Advances in epileptology: the 10th Epilepsy International Symposium. Raven, New York, pp 73-81

Comparison Between Electrophysiological Approaches for Presurgical Evaluation*

J. ENGEL, Jr.

Electrophysiological evaluation of patients with epilepsy who are candidates for surgical resection is performed for two specific purposes. The most important is precise localization of the epileptogenic lesion responsible for most or all of the patient's habitual seizures. The second involves functional mapping that may be required to identify primary motor and sensory cortical areas that cannot be removed without producing unacceptable neurological deficit. Considerable controversy exists concerning the necessary and sufficient electrophysiological studies for both localization and functional mapping, and a variety of approaches are currently utilized.

Localization of the Epileptogenic Lesion

A most basic difference of opinion concerns the localizing value of interictal EEG spike activity, as opposed to identification of the site of ictal onset. Both indicators of epileptiform excitability are subject to false localization due to propagation of electrical discharges since interictal spikes, as well as ictal onset, may be recorded at a distance from the actual epileptogenic lesion (Ajmone-Marsan 1980; Engel et al. 1983; Gloor 1975; Lieb et al. 1981 a, b). For this reason, tests of focal functional deficit are also important to confirm a localized abnormality at the site of the identified epileptic focus (Engel et al. 1983).

The most common approach which depends on interictal spike localization utilizes scalp EEG recording, usually with sphenoidal electrodes as well, to localize the site of predominant interictal spike activity. Based on this information, a craniotomy is then performed and electrocorticography (ECoG) is carried out to determine the extent of the resection (Gloor 1975). This presurgical evaluation approach is relatively inexpensive, compared with more sophisticated long-term monitoring procedures that are required for capturing ictal events (see below). False localization based on surface recording of interictal spikes, however, occurs in 10%-20% of patients (Engel et al. 1981; Lieb et al. 1981 a, b), and since craniotomy is performed on one side only, intraoperative ECoG is incapable of correcting an error in lateralization.

Identification of the epileptogenic region on the basis of interictal spikes recorded from the surface of the brain during surgery suffers from the same false localization problems as interictal spikes recorded with EEG. Interictal spikes may be recorded from wide areas of the brain, bilaterally, and there is no definitive way to determine which interictal spikes actually reflect the epileptogenic lesion responsible for generating habitual spontaneous seizures. This is particularly true when the epileptogenic lesion is in the mesial temporal lobe (Ajmone-Marsan 1980; Engel et al. 1983; Gloor 1975). Furthermore, there is considerable disagreement concerning the prognostic value of postresection interictal spikes recorded intraoperatively (Bengzon et al. 1968; Engel et al. 1975), and no definitive criteria for using ECoG-recorded spikes to determine the extent of the lesion have been developed.

Localization of the epileptogenic lesion on the basis of ictal recordings was first carried out using stereotactically implanted depth electrodes (SEEG) (Talairach et al. 1974; Rand et al. 1964). Epileptic seizures can be induced pharmacologically, or with electrical stimulation, but such seizures are not as useful for reliably identifying the site of habitual seizure origin as are spontaneous seizures (Wieser et al. 1979). Recording spontaneous seizures requires long-term monitoring, usually in combination with video recording so that EEG and behavioral ictal events can be correlated (Gotman et al. 1985; Gumnit 1987; Engel et al. 1985). Depth electrodes are usually implanted stereotactically into mesial tem-

Department of Neurology and Anatomy, UCLA School of Medicine, Reed Neurological Research Center, University of California, 710 Westwood Plaza, Los Angeles, CA 90024, USA

* This work was supported by grants NS-02808, NS-15654, and NS-20867 from the National Institutes of Health and contract DE-AC03-76-SF00012 from the Department of Energy.

Presurgical Evaluation of Epileptics
Edited by H. G. Wieser and C. E. Elger

poral structures, and other brain areas implicated on the basis of additional clinical data. The number of intracerebral sites that can be monitored is limited, however, and the accuracy of this technique depends upon the ability to identify the most likely sites of epileptic seizure generation prior to electrode implantation.

Chronic SEEG recordings are most useful for identifying epileptogenic lesions in mesial temporal structures, particularly when bilateral epileptiform abnormalities exist and it is necessary to determine which side gives rise to the habitual seizures. Chronic SEEG electrode recording is also indicated for differentiating between mesial temporal and extratemporal lesions, particularly when suspected areas for the extratemporal lesion are mesial or basal aspects of frontal or occipital lobes.

When chronic SEEG recording has identified with a high degree of reliability an epileptogenic lesion in one anterior temporal lobe, many workers choose to perform a standard anterior temporal lobectomy without intraoperative ECoG. This is based on the view that the best results are obtained when the most tissue is removed, and lobectomy should therefore consist of the largest standard resection possible based on known functional anatomy (Falconer 1976). A recent operative approach, when the epileptogenic lesion can be localized to one hippocampus or amygdala, is to perform a selective amygdalohippocampectomy, using microsurgical technique (Yaşargil et al. 1985). It has been suggested that postoperative neurological deficits can be reduced if lateral temporal cortex is not removed. This procedure also does not require intraoperative corticography. ECoG is necessary, however, if the results of the preoperative evaluation raise questions concerning the possibility of a more posterior, or extratemporal, epileptogenic lesion. Others prefer routinely to perform intraoperative ECoG recordings following chronic SEEG evaluation and use interictal spikes to determine the extent of the resection.

A third approach utilizes long-term scalp and sphenoidal EEG telemetry and video monitoring to determine the site of ictal onset. When ictal onset can be consistently localized to one temporal lobe, and tests of focal functional deficit confirm dysfunction in the same temporal lobe, anterior temporal lobectomy is recommended without requiring invasive recordings (Engel et al. 1983). This approach avoids the expense and risk of chronic SEEG recording, but with present techniques only 30%–40% of patients who undergo anterior temporal lobectomy have their epileptogenic lesion localized noninvasively with sufficient confidence to allow them to avoid the depth electrode procedure (Engel et al. 1984). When the epileptogenic lesion can be localized to one anterior temporal lobe by noninvasive means, a standard anterior temporal lobectomy is usually performed without ECoG. If questions arise during the initial noninvasive evaluation procedure, the approach is usually to proceed with invasive recordings, and resort to ECoG identification of the site or extent of the epileptogenic lesion only if questions remain following chronic invasive evaluation.

A fourth approach utilizes another chronic invasive recording technique to identify the site of seizure onset. Electrode strips, or tresses, are inserted through burr holes into the subdural space to record chronically over the cortical surface (Wyler et al. 1984). This approach may be considered less invasive than depth electrodes, since there is no penetration of brain parenchyma. Subdural electrodes of this type identify epileptogenic lesions located on the neocortical convexity better than stereotactically implanted depth electrodes, but are less useful when the epileptogenic lesion is in mesial limbic structures. The number of electrode strips or tresses that can be inserted are also limited with this technique, and prior evaluation is necessary to determine the cerebral areas most likely to be generating habitual seizures.

A fifth approach, most recently devised, utilizes a cortical grid or array of electrodes imbedded in Silastic and placed over a large area of cortical surface either epidurally (Goldring 1978) or subdurally (Lesser et al. 1984). This technique is more invasive than the other two chronic intracranial recording techniques since a craniotomy is required and the chances for infection are increased. Ideally this procedure should be used unilaterally when lateralization is not an issue, but bilateral craniotomies can be performed if necessary. The cortical grid is best suited for suspected neocortical lesions located over the convexity and the site and extent of the epileptogenic lesion can be accurately mapped on the basis of ictal and interictal recordings. Smaller grids can also be placed under the temporal lobe and may record from inferior temporal regions, although the accuracy of this technique, compared with SEEG recording, for identifying mesial temporal abnormalities remains to be demonstrated. Since the epileptogenic zone can be chronically mapped using both interictal and ictal discharges, there is no need for intraoperative ECoG for this purpose in most cases.

Functional Mapping

There is debate concerning the reliability of utilizing observable anatomical landmarks for planning the extent of a safe resection, and the degree of variability from one patient to another in cerebral localization of specific functions. Functional mapping is usually carried out intraoperatively using evoked potential techniques and cortical stimulation (Gloor 1975; Ojemann 1979). Functional mapping can also be done chronically with cortical grids (Lesser et al. 1984; Goldring 1978). Intraoperative functional mapping prolongs the surgical procedure and increases patient discomfort when this must be done with the patient under local anesthesia. On the other hand, intraoperative monitoring offers the flexibility of moving electrodes, if necessary, while chronically inserted cortical grid electrodes cannot be moved. The chronic electrodes are sufficiently close together, however, that this may not introduce a serious problem, while offering the advantage of repeated studies on separate days, with a comfortable and cooperative patient.

When extratemporal resections are performed near primary sensory and motor areas, it is generally agreed that anatomical landmarks are not sufficiently reliable to allow safe excision of brain tissue without first mapping cortical function using sensory evoked potentials to identify primary visual, somatosensory, and motor cortex. Once the central sulcus has been identified, it is usually not necessary to carry out stimulation to map the extent of the precentral gyrus. Stimulation is more useful, however, for identifying language areas and is important when dominant hemispheric resections are carried out in the inferior frontal and posterior temporal regions (Ojemann 1979).

There is controversy concerning the need for mapping of language function when performing a standard anterior temporal lobectomy in the dominant hemisphere. Those who advocate use of this procedure in all cases argue that interpatient variability may cause important language areas to be damaged by a standard "safe" resection, while a more generous resection might be possible in other patients (Ojemann 1979). The latter point is particularly valid for those who believe that chances for surgical benefit are increased when more tissue is removed. The former point is not clearly substantiated by postoperative results, however. Although those who routinely tailor the anterior temporal lobectomy according to mapping of language function do not produce postoperative aphasia in their patients, the postoperative aphasia that occasionally occurs following a standard anterior temporal lobectomy is almost always transient. Consequently, demonstration that stimulation of an area of brain causes aphasia does not necessarily mean that this area is essential for normal language function.

Conclusions

Interictal EEG spike activity will provide adequate information for surgical resection in many, if not most, patients. There is a group of patients, however, in whom the epileptogenic lesion cannot be identified by this means, and they should be referred for more sophisticated evaluation. More importantly, there is another group of patients in whom false localization will occur, and there is a danger that normal tissue will be removed. How often such errors are made is not possible to determine. However, since patients with medically refractory partial seizures most likely need all the normal brain they have, there is a serious possibility that removal of normal tissue will impair cognitive function. Whether the opportunity to offer improvement to many patients with minimum discomfort and expense justifies the risk of error in a few that could be avoided can be debated.

The epileptogenic lesion can be identified noninvasively with a higher degree of confidence by using scalp and sphenoidal telemetry and confirmatory tests of focal functional deficit. However, the most liberal criteria that will allow effective resection without a serious risk of error remains to be determined. When invasive procedures are necessary, SEEG remains the most reliable approach for mesial temporal lesions, but is not accurate for lateral neocortical lesions. Subdural strip or tress electrodes are more accurate than depth electrodes for identifying neocortical epileptogenic lesions, but mesial and basal brain areas are not accessible to this technique. Cortical grids are more invasive than these other two chronic techniques, but can be used for mapping epileptic activity, as well as functional regions over large areas of neocortex, and may also provide adequate recording from the inferior temporal lobe. It would seem that when mesial temporal lesions are suspected, SEEG recording remains the procedure of choice, while subdural strips or tresses, or cortical grids are the procedures of choice for neocortical lesions. The problem arises when it is necessary to differentiate between a mesial temporal and lateral temporal or extratemporal lesion. Approaches for the use of SEEG in combination with these other techniques have not yet been developed.

Functional mapping, whether intraoperative or with chronic cortical grids, is necessary when planned resection is adjacent to the primary motor or sensory cortex. The need for mapping of language function for standard anterior temporal lobectomy carried out in the dominant hemisphere remains controversial. This may be useful when it allows a larger resection to be performed, but smaller resections based on such mapping techniques may not be necessary.

References

Ajmone-Marsan C (1980) Depth electrography and electrocorticography. In: Aminoff MJ (ed) Electrodiagnosis in clinical neurology. Livingstone, New York, pp 167-196

Bengzon ARA, Rasmussen T, Gloor P, Dussault J, Stephens M (1968) Prognostic factors in the surgical treatment of temporal lobe epileptics. Neurology 18: 717-731

Engel J Jr, Driver MV, Falconer MA (1975) Electrophysiological correlates of pathology and surgical results in temporal lobe epilepsy. Brain 98: 129-156

Engel J Jr, Rausch R, Lieb JP, Kuhl DE, Crandall PH (1981) Correlation of criteria used for localizing epileptic foci in patients considered for surgical therapy of epilepsy. Ann Neurol 9: 215-224

Engel J Jr, Crandall PH, Rausch R (1983) The partial epilepsies. In: Rosenberg RN, Grossman RG, Schochet S, Heinz ER, Willis WD (eds) The clinical neurosciences, vol 2. Livingstone, New York, pp 1349-1380

Engel J Jr, Sutherling WW, Cahan L, Crandall PH, Kuhl DE, Phelps ME (1984) The role of positron emission tomography in the surgical therapy of epilepsy. In: Porter RJ, Mattson RH, Ward AA, Dam M (eds) Advances in epileptology: XVth epilepsy international symposium. Raven, New York, pp 427-432

Engel J Jr, Ebersole JS, Burchfiel JL, Gates JR, Gotman J, Homan RW, Ives JR, King DW, Sato S, Wilkus RJ (1985) Guidelines for long-term neurodiagnostic monitoring in epilepsy. J Clin Neurophysiol 2: 419-452

Falconer MA (1976) Anterior temporal lobectomy for epilepsy. In: Rob C, Smith R (eds) Operative surgery, vol 14 2nd ed. Butterworths, London, pp 142-149

Gloor P (1975) Contributions of electroencephalography and electrocorticography to the neurosurgical treatment of the epilepsies. In: Purpura DP, Penry JK, Walter RD (eds) Advances in neurology, vol 8. Neurosurgical management of the epilepsies. Raven, New York, pp 59-105

Goldring S (1978) A method for surgical management of focal epilepsy, especially as it relates to children. J Neurosurg 49: 344-356

Gotman J, Ives JR, Gloor P (eds) (1985) Long-term monitoring in epilepsy. Electroencephalogy Clin Neurophysiol [Suppl 37], 444 pp

Gumnit R (ed) (1987) Intensive neurodiagnostic monitoring. Advances in neurology, vol. 46, Raven, New York, 336 pp

Lesser RP, Dinner DS, Lüders H, Morris HH III (1984) Differential diagnosis and treatment of intractable seizures. Cleve Clin Q 51: 227-240

Lieb JP, Engel J Jr, Brown WJ, Gevins AS, Crandall PH (1981a) Neuropathological findings following temporal lobectomy related to surface and deep EEG patterns. Epilepsia 22: 539-549

Lieb JP, Engel J Jr, Gevins AS, Crandall PH (1981b) Surface and deep EEG correlates of surgical outcome in temporal lobe epilepsy. Epilepsia 22: 515-538

Ojemann GA (1979) Individual variability in cortical localization of language. J Neurosurg 50: 164-169

Rand RW, Crandall PH, Walter RW (1964) Chronic stereotactic implantation of depth electrodes for psychomotor epilepsy. Acta Neurochir 11: 609-630

Talairach J, Bancaud J, Szikla G, et al. (1974) Approche nouvelle de la neurochirurgie de l'épilepsie. Méthodologie stéréotaxique et résultats thérapeutiques. Neurochirurgie 20 [Suppl 1]: 1-240

Wieser HG, Bancaud J, Talairach J, Bonis A, Szikla G (1979) Comparative value of spontaneous and chemically and electrically induced seizures in establishing the lateralization of temporal lobe seizures. Epilepsia 20: 47-59

Wyler AR, Ojemann GA, Lettich E, Ward AA Jr (1984) Subdural strip electrodes for localizing epileptogenic foci. J Neurosurg 60: 1195-1200

Yaşargil MG, Teddy PJ, Roth P (1985) Selective amygdalohippocampectomy: operative anatomy and surgical technique. In: Symon L, Brihaye J, Guidetti B, Loew F, Miller JD, Nornes H, Pásztor E, Pertuiset B, Yaşargil MG (eds) Advances and technical standards in neurosurgery, vol 12. Springer, Vienna, pp 93-123

Spread of Epileptic Seizure Activity in Experimental and Clinical Epilepsy: The Use of Mutual Information Analysis*

F. H. Lopes da Silva[1] and N. J. I. Mars[2]

The mechanism of spread of epileptic seizures remains an important unsolved question in the process of epileptogenesis. This question is not only of fundamental importance for understanding how focal epileptic activity propagates to neighboring and/or distant brain areas but also of practical interest. This last aspect is particularly relevant where the primary site of an epileptogenic focus must be determined, namely in those patients in whom the surgical removal of the focus is being considered. The propagation of focal epileptic activity is not easy to assess by way of visual examination of a large number of EEG records even if recorded from subdural or intracerebral electrodes. It is generally assumed that the EEG signal recorded from the focal area will present a time lead over signals recorded from distant areas. However, a main difficulty is that it is not possible to assess visually small time differences in the order of a few milliseconds which may exist between such EEG signals. Therefore, computer-based analysis methods have been devised to obtain estimates of time delays between such EEG signals. Brazier (1972, 1973) carried out the first type of such an analysis. Based on the phase angle (φ_o in degrees) at a particular frequency (f_o in Hz) for which coherence was maximal, a time delay (Δt) was calculated according to: $\varphi_o/360 f_o$. This method, however, may give ambiguous results since the phase angle at one particular frequency can be transformed into a time delay only if the phase as function of frequency fits a straight line which intersects the axes at the origin. However, this does not always occur in practice. Sometimes phase functions are encountered between EEG signals which at zero frequency intersect the phase axis at $\pm 180°$. This depends on the fact that EEG signals of one given structure may be recorded from either side of an equivalent current dipole which represents the neural activity of, for example, a cortical gyrus. Such signals will have a phase difference of 180°. This type of problem has been avoided by Gotman (1983) using an extension of the method employed by Brazier (1973) for the measurement of time differences between two EEG channels. According to this method, a certain frequency range (Δf) is first defined for which coherence is significant; second, the linearity of the phase spectrum is assessed, and, third, the corresponding slope ($\Delta\varphi$) is converted to a time delay ($\Delta t = \Delta\varphi/\Delta f \cdot 360°$). The same approach has been used earlier by Mars et al. (1977) with regard to EEG signals recorded during kindled seizures in dogs, but these authors noted that time delays could be estimated in this way in some cases but not in all. A possible interpretation of such findings was that the relationship between the EEG signals may not be linear, whereas measurements based on coherence and phase spectra assume a linear relation between the signals.

This led to a novel approach, based on the average amount of mutual information (AAMI) method of analysis. Mutual information (MI) has been defined in the area of information theory by Shannon and Weaver (1949), elaborated by Gel'fand and Yaglom (1959), implemented in a computer program by Mars and Van Arragon (1982), and applied to EEGs recorded during epileptic seizures induced by kindling in dogs (Mars and Lopes da Silva 1983) and in patients (Mars et al. 1985). A few of the corresponding results are reviewed here. In part 1, the experimental results are presented; in part 2, the clinical results. The basis of the AAMI method of analysis is presented first.

Analysis Method and Estimation of Time Delays

The definition of mutual information (MI) for continuous stochastic signals x and y can be given as fol-

[1] Neurophysiology Group, Department of General Zoology Biological Centre, University of Amsterdam, Kruislaan 320, 1098 SM Amsterdam, The Netherlands
[2] Department of Computer Science, University of Twente, P.O. Box 217, 7500 AE Enschede, The Netherlands

* This work was partly supported by a grant from the CLEO-TNO Committee for Epilepsy Research of the Netherlands.

Presurgical Evaluation of Epileptics
Edited by H. G. Wieser and C. E. Elger

lows, assuming that the signals have marginal probability densities $f_X(x)$ and $f_Y(y)$ and a joint probability density $f_{XY}(x,y)$:

$$MI_{xy} = \iint f_{XY}(x,y) \log \frac{f_{XY}(x,y)}{f_X(x) \cdot f_Y(y)} dxdy$$

MI can be interpreted as how well one can predict, on average, the values of one signal given the values of the other. MI is equal to zero if the signals are independent and greater than zero if the signals are interdependent. In order to estimate MI, it is necessary to estimate the joint and marginal probability density functions from a limited number of observations. The density functions were estimated using the Parzen estimator (1962). Recently, Moddemeijer (1985) has developed a simpler and quicker algorithm for estimating MI using a histogram method to make discrete the joint probability density of x and y. Using the MI method, no restrictions have to be assumed on the underlying model of the relation between EEG channels. In this way, time delays may be estimated by computing the average amount of mutual information for different time shifts. The shift for which AAMI reaches the largest value is taken to indicate the delay between the two signals.

Part 1: Studies in Kindled Dogs

Material and Methods

Electrodes were implanted for stimulation and recording in several brain areas in dogs (beagles) as described before (Mars and Lopes da Silva 1983), namely in the olfactory cortex or prepyriform cortex (PPC) and amygdala (AM). Kindling was elicited by daily delivering a tetanus consisting of a 1-s train of biphasic pulses (60/s, 1 ms duration, 100 μA) to two electrodes placed in the PPC.

The EEG signals recorded from the above-mentioned and other areas were filtered with antialiasing filters, digitized off-line at a rate of 102.4 or 204.8 Hz and stored on magnetic disk.

Results

A visual inspection of the EEG records during a kindled seizure shows that the characteristics of the signals change rapidly in the course of a few seconds as illustrated in Fig. 1. By trial and error, it was found that the best compromise between a short analysis

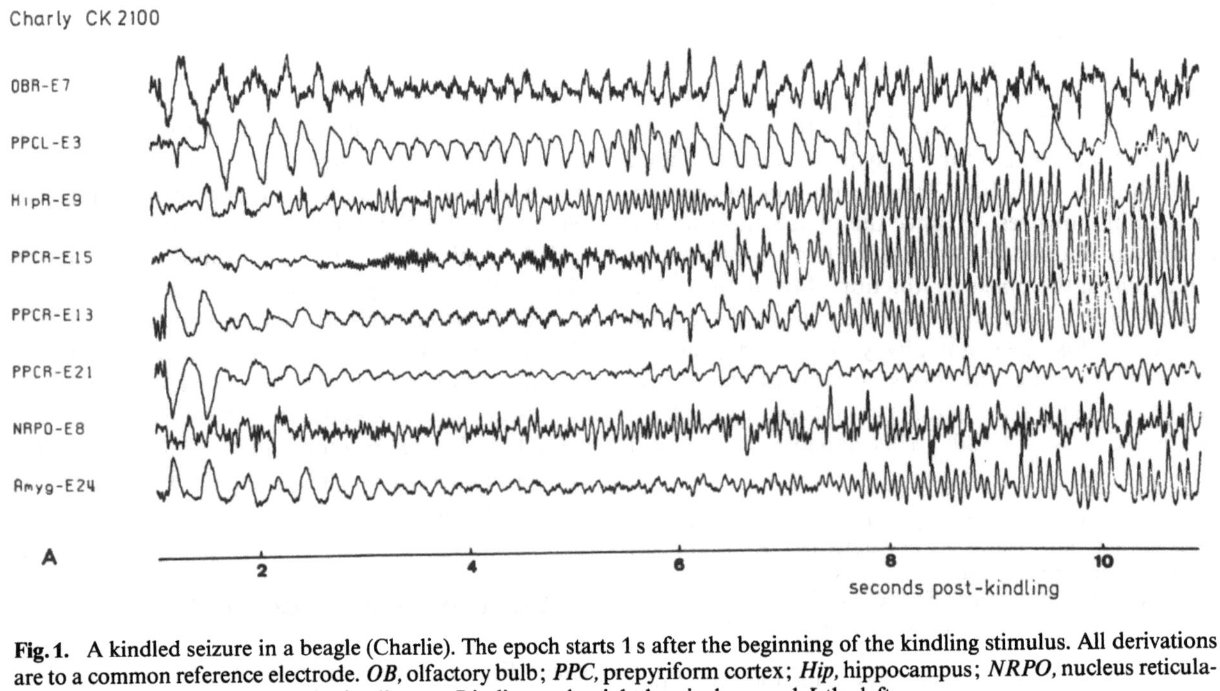

Fig. 1. A kindled seizure in a beagle (Charlie). The epoch starts 1 s after the beginning of the kindling stimulus. All derivations are to a common reference electrode. *OB*, olfactory bulb; *PPC*, prepyriform cortex; *Hip*, hippocampus; *NRPO*, nucleus reticularis pontis oralis; *Amyg*, amygdala; in all cases *R* indicates the right hemisphere and *L* the left

epoch, which should be chosen in view of the rapid changes in signal characteristics, and an epoch sufficiently long in order to obtain a reliable estimate of AAMI, was to choose epochs of 2.5 s. The values of AAMI between PPC and AM signals were computed for consecutive epochs in three dogs.

The main results can be summarized as follows:

- The development of the pattern of kindled seizures shows a large variability among animals and, in some cases, also within the same animal among different seizures.
- Nevertheless, a characteristic pattern which appeared regularly was that, in a kindled seizure, three periods may be distinguished as shown in Fig. 2.
  1. A first period lasting about 5 s where AAMI values were low (<0.3)
  2. A second period lasting about 8 s where AAMI values were larger than 0.3. The corresponding time delays ranged in one dog from 19.1 to 23.0 ms at the beginning and from 11.3 to 16.0 ms at the end of this period
  3. A third period in which AAMI values decreased to around 0.3 and estimated time delays were inconsistent.

A comparison was made between the time delays computed using the AAMI method and the method based on coherence and phase spectrum. The latter was carried out following the procedure introduced by Carter (1976). It was found that, in general, the results obtained using either method were comparable (Mars and Lopes da Silva 1983), although the AAMI method yields results with a relative smaller variance. It is of interest to note that in the third period mentioned above ranging from 20 to 28 s after the beginning of the seizure, two clusters of time delays could be found (Fig. 2): one with positive values ranging between 5 and 25 ms and another one of time delays within the same range of magnitudes but with negative values. There were also two cases of time delays very close to zero in the time window 28-29 s; the latter may be due to the fact that if a mixture of positive and negative time delays occurred in the same epoch the average would tend to zero. It is of importance to note that the values of time delays clearly clustered around a positive and a negative value and were not randomly distributed around zero. The fact that two clusters of time delays tend to occur is suggestive that in this phase of the seizure one area, either PPC or amygdala, may lead the other.

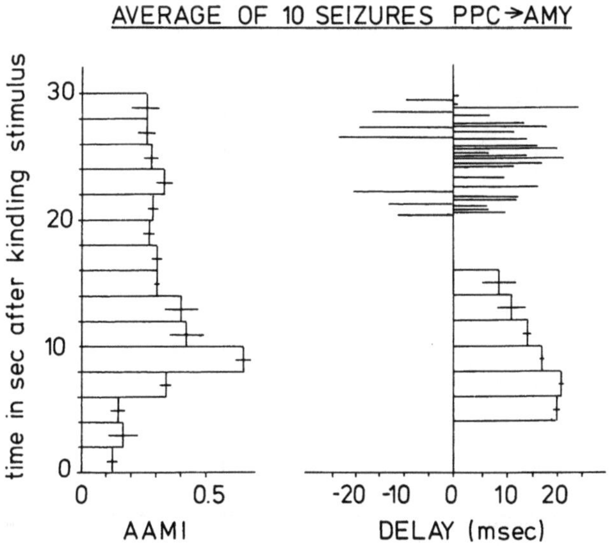

Fig. 2. Running means of the average amount of mutual information (AAMI) and time delay between the PPC and AMY for ten seizures in dog Charlie. Means and standard errors of the mean are indicated. Time runs vertically. The time after the kindling stimulus has been divided into time windows of 2 s starting from below. Means were computed from AAMI and delay values corresponding to equivalent epochs after the kindling. The means were allocated to the time window at which the analyzed epoch started. Note that three periods can be distinguished which are characterized, respectively, by low *(1)*, large *(2)*, and intermediate *(3)* values of AAMI. In period *1* no time delay could be estimated; in period *2* consistent positive time delays were found which decreased as a function of time; in period *3* time delays tended to cluster around a positive and a negative value so that the individual values are plotted. There are two time delays which are approximately zero

Part 2: Studies in Epileptic Patients

Materials and Methods

As described in detail by Mars et al. (1985), AAMI analysis was carried out on three epileptic patients who were evaluated for surgical treatment at the University of Washington Epilepsy Center at Seattle. The patients were provided with subdurally placed strip electrodes (1 cm apart) as described by Wyler et al. (1984). Intracranial recordings were made using Pz (10-20 system) as extracranial reference. The EEG signals were low-pass filtered (50 Hz) and digitized at 100 Hz for analysis using a PDP11-34 computer. AAMI analyses were carried out for EEG signals recorded from electrodes on the same strip and between strips overlying homologous areas on the two hemispheres. In this case, AAMI values were computed for epochs lasting for 5.1 s, which in this case was found to be a good compromise.

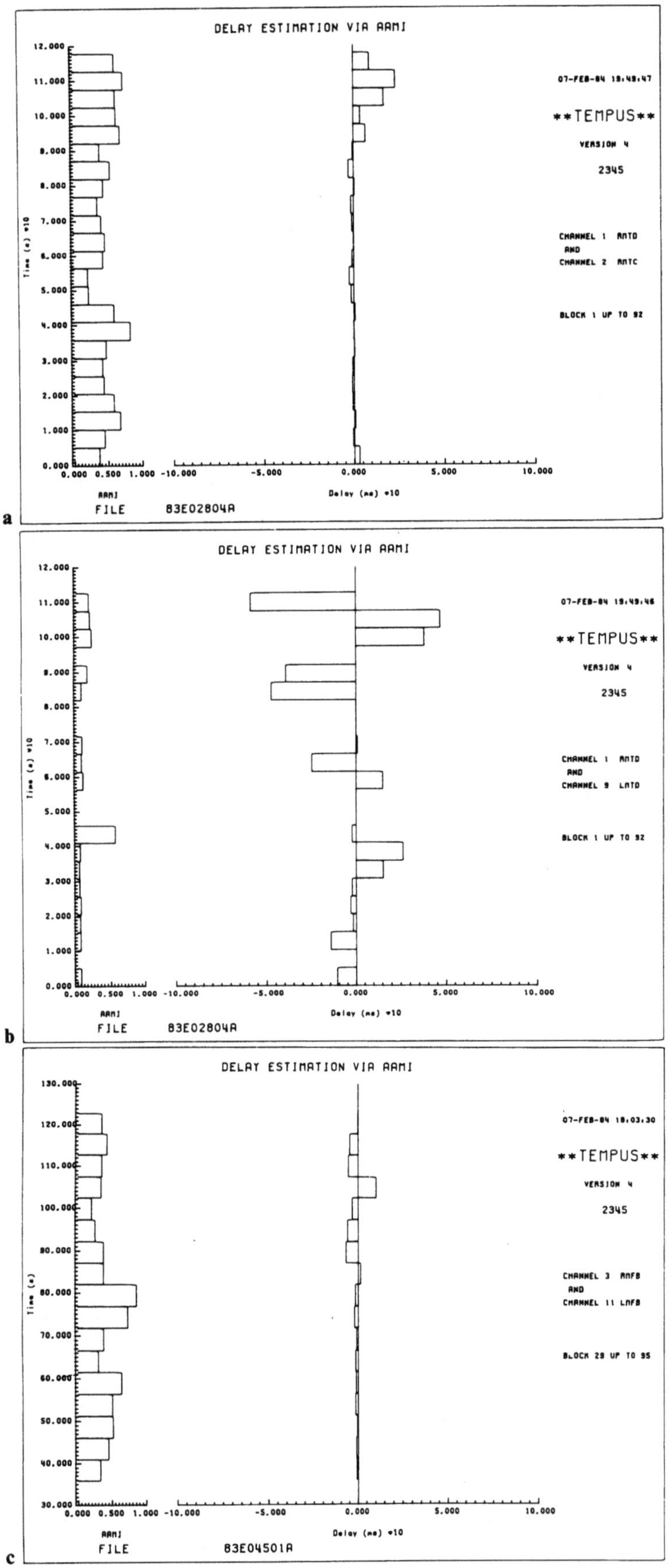

Fig. 3. **a** AAMI *(left)* and time delay *(right)* for contiguous 5.1-s epochs calculated for an ictal EEG recorded from two adjacent electrodes near the seizure focus (case 1). Positive time delays indicate those epochs in which the first channel *(RMTD)* leads the second *(RMTC)*. **b** Comparable analysis for an EEG recorded from two adjacent electrodes contralateral to the seizure focus in case 1. Note that AAMI values are larger for *RMTD-RMTC* than for *LMTD-LMTC*. **c** Interhemispheric AAMI of an EEG recorded from the right *(RMFB)* and from the left *(LMFB)* medial frontal areas in case 3. Note that for a long sequence of epochs *RMFB* lags (negative time delay) relative to *LMFB*

Results

For each patient, three seizures were analyzed. The main results can be summarized for the three patients as follows:

1. Out of the nine seizures analyzed in five patients (three in case 1 and two in case 2 according to Mars et al. 1985) no consistent lead, or lag, was encountered for one side (temporal or frontal) relative to the contralateral side throughout the course of a seizure.

2. In the other four seizures, there were consistent time delays between signals recorded from homologous areas.

In one of them the EEG from the right lateral temporal surface led the left side significantly more often throughout the ictus ($P<0.025$); in this case (case 2) the patient had a herniation of the right medial temporal area revealed by a CT scan: a right temporal lobectomy was performed and the patient was seizure free 1 year after surgery. In another patient (case 3), all three seizures showed the left medial frontal EEG signals leading the contralateral signals ($P<0.025$) (see Fig. 3c). In this case, the CT scan indicated only a right frontal lobe low-density area; after a right frontal cortical resection with division of the anterior half of the corpus callosum, the patient still had seizures although less frequently.

3. The AAMI values for electrode pairs within the strip overlying the focus both in case 1 (area of decreased density revealed by CT in the right medial temporal region) and in case 3 showed significantly larger values ($P<0.025$) than at all other sites (see Fig. 3). In case 3, in only one seizure, significantly larger AAMI values were found on the right inferior and medial frontal areas.

Discussion

The results of both studies, in kindled dogs and in epileptic patients, demonstrate that the estimation of time delays between EEG signals recorded during an epileptic seizure, although possible, remains a difficult problem. On the one hand even during one particular seizure the estimated time delays may change in the course of a seizure; on the other, there is a considerable variability between different seizures in the same subject. The first aspect appears to reflect a basic (patho)physiological phenomenon since in ten kindled seizures it was possible to find a consistent pattern of evolution of AAMI values and time delays in the course of seizures. Moreover, the same type of pattern was confirmed in another dog and more recently in our laboratory in rats which were kindled in the entorhinal cortex (J. P. Pijn, unpublished observations). This finding is of theoretical and practical importance.

Theoretically, it allows the conclusion to be drawn that during a kindled seizure after a latent period of variable length, anatomically related brain areas are strongly coupled for a period of seconds and thereafter tend to become independent although both continue to produce epileptiform activity. Practically, it implies that the analysis of coupling between EEG signals recorded from different areas during epileptic seizures has to be carried out by dividing the EEGs in short segments. The optimal length of the segments has to be determined experimentally in each case but, in our experience, it should lie between 1 and 5 s. In a number of seizures in patients, a consistent time lead of the EEG signals recorded from one side over those of the contralateral side was found. Whether this finding may have clinical relevance for the assessment of the primary epileptogenic area has to be studied in a larger collection of data.

A significant finding was that in two patients with focal seizure onset larger values of AAMI were consistently encountered over the brain region (subdural strip electrodes) where the focus was apparently present relative to other areas. This was not the case for the patient with generalized seizures (case 3). This may be interpreted as an indication that none of the electrodes used in this case was placed near the epileptogenic focus.

We should note that the results presented here were based on the computation of AAMI values and time delays over a large number of EEG segments. Other investigators have inferred the spread of seizures from a few time intervals (Gotman 1983). We chose to use the entire period in order to try to find statistically consistent results and to identify the possible existence of patterns of change. This last aspect has to be emphasized since our studies in kindled animals have shown that the period, during which an accurate prediction of the direction of spread of epileptic activity can be made, is brief. This was also found by Gotman (1983).

An important aspect is that during some seizures or parts thereof, only low AAMI values could be obtained. This holds again both for kindled animals and for patients. From this, we can infer that in such cases, the EEG signals recorded during these seizures are statistically independent. In patients, this may be due to a less fortunate placement of the electrodes. But it is difficult to account for the same effect in kindled dogs in the same way. In this respect, a possible explanation may be that during seizures different brain areas produce simultaneously complex bursts of electrical activity which will spread to a large number of anatomically interrelated areas possibly not only along orthodromic pathways but also antidromically (Schwartzkroin et al. 1975). Moreover, the possibility of ephaptic mechanisms of spread (Taylor et al. 1983) and/or of spread through gap junctions (MacVicar and Dudek 1981) has to be taken into account. In this way a complex pattern of feedback loops will develop. This may preclude the emergence of clear-cut time delays. In this sense, it may be useful to investigate this type of multichannel EEG seizure activity using a more general method of mutual information analysis (Geweke 1982; Gersch 1972) which may allow the detection of causal relationships between complex signals in a more explicit way.

Acknowledgments. The advice and criticisms of Drs. W.J. Wadman and J.P. Pijn are gratefully acknowledged, as well as the secretarial help of J.C. Cabi and O. van Vliet and the collaboration of Simon van Mechelen with the photography.
This work was partly subsidized by the Division of Health Research TNO (Grant CLEO A 49).

References

Brazier MAB (1972) Spread of seizure discharges in epilepsy: anatomical and electrophysiological considerations. Exp Neurol 36: 263-272

Brazier MAB (1973) Electrical seizure discharges within the human brain: the problem of spread. In: Brazier MAB (ed) Epilepsy: its phenomena in man. Academic, New York, pp 153-170

Carter GC (1976) Time delay estimation. Naval Underwater Systems Center, New London, CO Report TR-5335

Gel'fand JM, Yaglom AM (1959) Calculation of the amount of information about a random function contained in another such function. American Mathematical Society Translations 12: 199-246

Gersch W (1972) Causality or driving in electrophysiological signal analysis method (with G.V. Goddard). Science 150: 701-702

Geweke J (1982) Measurement of linear dependence and feedback between multiple time series. J Am Statistical Assoc 77: 304-314

Gotman J (1983) Measurement of small time differences between EEG channels: method and application to epileptic seizure propagation. Electroencephalogr Clin Neurophysiol 56: 501-514

MacVicar B, Dudek FE (1981) Electrotonic coupling between pyramidal cells: a direct demonstration in rat hippocampal slices. Science 213: 782-785

Mars NJI, Lopes da Silva FH (1983) Propagation of seizure activity in kindled dogs. Electroencephalogr Clin Neurophysiol 56: 194-209

Mars NJI, Van Arragon GW (1982) Time delay estimation in non-linear systems using average amount of mutual information analysis. Signal Processing 4: 139-153

Mars NJI, Lopes da Silva FH, Van Hulten K, Lommen JG (1977) Computer assisted analysis of EEGs during seizures; localisation of an epileptogenic area. Electroencephalogr Clin Neurophysiol 43: 575

Mars NJI, Thompson PM, Wilkus RJ (1985) The spread of epileptic seizure activity in humans. Epilepsia 26: 85-94

Moddemeijer R (1985) Estimation of entropy and mutual information of continuous distribution. In: Vinck AJ (ed) Proceeding 6th symposium on information theory in the Benelux. Enschede

Parzen E (1962) On estimation of a probability density function and mode. Annals of Mathematical Statistics 35: 1065-1076

Schwartzkroin PA, Mutani R, Prince DA (1975) Orthodromic and antidromic effects of a cortical epileptiform focus on ventrolateral nucleus of the cat. J Neurophysiol 38: 795-811

Shannon CE, Weaver W (1949) The mathematical theory of communication. University of Illinois Press, Urbana

Taylor C, Krnjević K, Ropert N (1983) Field interactions in CA_3 hippocampal pyramidal layer in rats in vivo. Can J Pharmacol 61: A32-A33

Wyler AR, Ojemann GA, Lettich E, Ward AA Jr (1984) Subdural strip electrodes for localizing epileptogenic foci. J Neurosurg 60: 1195-1200

Evaluation of Relationships Between Seizure Potentials

P. RAPPELSBERGER, H. POCKBERGER, and H. PETSCHE

Introduction

For the evaluation of relationships between seizure potentials different methods can be applied. The choice of method largely depends on the kind of signals to be analyzed and on the question which is to be answered. An essential problem in epileptology concerns the development of a seizure initiated by a focus and the mechanisms leading to spreading and generalization. However, since in most cases the site of the focus is unknown the reverse problem arises, namely, the localization of the focus. Therefore, examinations and analyses have to be aimed at this localization problem.

The involvement of cortical or subcortical areas into a seizure is time delayed with respect to the initiating discharges, according to basic mechanisms of propagation of action potentials in nerve fibers and hitherto largely unknown mechanisms leading to the buildup of feedback loops between and within neuronal masses necessary for self-sustained seizure discharges. Therefore in the evaluation of relationships between seizure potentials, most emphasis has to be laid on the time relationships between seizure potentials recorded at different sites and their measurement.

This paper is aimed at describing different methods for the evaluation of relationships between seizure potentials. Examples of a study of thalamocortical relationships during experimental seizure activities are to demonstrate the different methods.

Methods

Preparation and Recording

Experiments are made in rabbits. The methods of preparation are described extensively in Pockberger et al. (1984a). For intracortical recordings multielectrodes produced in thin-film technology are used (Prohaska et al. 1979). These electrodes consist of a glass carrier needle with 16 Ag/AgCl electrode contacts of 25 μm in diameter and 125 μm apart. The multi-electrode is introduced into the rabbit's cortex under microscopic control until the uppermost contact just disappears under the cortical surface. With these electrodes, simultaneous recordings of field potentials (EEG) from all cortical layers can be made.

For the thalamus recordings two platin semi-microelectrodes are used with a tip diameter of about 20 μm. According to the recordings from the visual cortex the thalamus electrodes are placed either in the geniculate body or in the pulvinar.

Seizures are elicited by local epicortical or intracortical application of penicillin (Pockberger et al. 1984a, b).

The recorded signals are stored on analogue tape and processed offline. Three different methods are applied to estimate relationships between cortical and thalamic signals.

Spectral Analysis

The first method consists of the estimation of coherence and phase spectra. Data are sampled at 256/s and epochs of 2 s (=4 blocks of 128 data) are Fourier transformed. To reduce the considerable variance of the raw spectral estimates, smoothing windows are applied (Jenkins and Watts 1968). Better estimates can also be obtained by averaging two or more short time spectra of the 2-s epochs. This, however, requires stationary signals in the interval of interest, e.g., 10 s. Stationarity means that the statistical properties of the EEG signals do not change within the analysis interval (Jenkins and Watts 1968). This, however, is hardly the case with seizure potentials.

The coherence spectrum is the normalized cross-power spectrum. The interpretation is as follows: a

Institute of Neurophysiology, University of Vienna, and Brain Research Institute, Austrian Academy of Sciences, Währinger Str. 17, 1090 Vienna, Austria

Presurgical Evaluation of Epileptics
Edited by H. G. Wieser and C. E. Elger

coherence value of 1 means that the two signals are identical and a coherence value of 0 shows that there is no linear relationship between the two EEG signals. Each coherence value relates to an individual frequency component and may therefore also be interpreted as correlation per frequency. The entity of all coherence values in the obtained frequency range is called coherence function, or coherence spectrum.

For the description of the common properties of two EEG signals besides the coherence spectrum, the phase-spectrum must also be considered. The phase spectrum shows whether a certain frequency component of the first EEG signal is leading or lagging the same frequency component of the second EEG signal. From the phase shift, the phase delay of this frequency component between the two EEG signals can be calculated. The phase delay time then indicates the time difference of the occurrence of this frequency component at the two recording sites.

In most cases, seizure potentials do not consist of only a dominant frequency but of various frequency components. In this case an individual frequency component does not describe the signal properties appropriately and, for the evaluation of time delays, the group delay has to be considered. The group delay may be interpreted as the delay of the envelope of a signal. Phase delay and group delay do not have to be identical. This property is called dispersion. The group delay time is computed from the slope of the phase spectrum in the frequency range of interest.

Correlation Analysis

The second method for the evaluation of relationships between the EEG signals is the cross-correlation analysis (Jenkins and Watts 1968). In this case, correlations between two EEG signals are calculated, whereby one signal is shifted with respect to the other signal, yielding a function of correlation values versus time shift. The maximum value of the cross-correlation function is a measure of the relationship between the two EEGs, and the site of the maximum with respect to the delay scale indicates which EEG signal is leading the other. As mentioned above, coherence analysis and cross-correlation analysis are closely related to each other and both require linear relationships between the two EEG signals.

Average Amount of Mutual Information (AAMI) Analysis

The term linear means that the common portions of both EEG signals differ only in amplitude; however, a time delay may occur. Since it can be expected that the relationships between seizure potentials are not only linear but also that nonlinearities probably exist, e.g., rectifying or squaring properties can sometimes be observed, a method has been developed (Mars and Van Arragon 1982) to account also for existing nonlinearities. This method is called the average amount of mutual information method (AAMI) and is also described by Lopes da Silva and Mars in this volume.

Besides the possibility of applying the AAMI technique to nonlinear systems, this method has a further advantage with respect to correlation analysis. The AAMI method estimates amplitude probability density functions using a special procedure which reduces the variance of the estimates considerably. Therefore this method can be applied to relatively short epochs of data with great reliability. In contrast to this, correlation functions of short epochs show in most cases considerable variance, which makes the interpretation difficult.

Examples

Figure 1 shows a section of 12-s seizure activity immediately after the onset of the seizure. In the upper part the traces C1, C4, C7, C10, and C13 relate to the electrode contacts of the intracortical multi-electrode. Electrode contact C1 was close to the cortical surface and contact C13 in about layers V–VI of the visual area. In the lower part of the figure simultaneously recorded seizure activity of the geniculate body is presented.

The seizure activity consists of tonic discharges (negativity upwards) with slowly decreasing frequency from about 13/s at the beginning of the record to about 8/s 12 s later.

Spectral Analysis

Figure 2 shows the results of spectral analysis of the first 4-s seizure activity of Fig. 1 (block 31–38). The spectra were computed by averaging two short-time spectra of consecutive 2-s epochs and additionally smoothed by a spectral window (Jenkins and Watts 1968). In the left and right part, the power spectra of cortex channel C7 and the thalamus electrode are

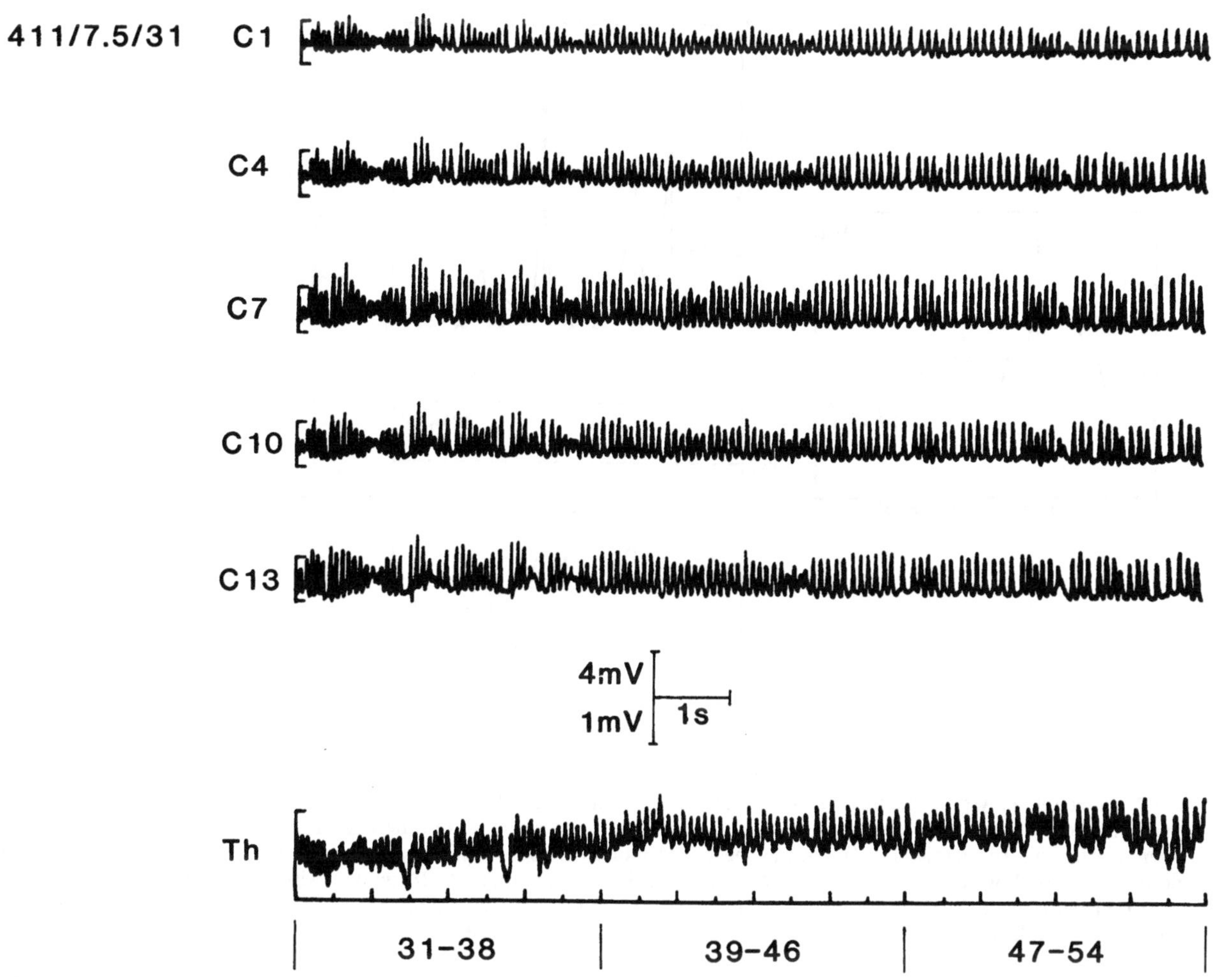

Fig. 1. Section of 12-s seizure EEG immediately after its beginning. The seizure was elicited by application of 500 IU penicillin onto the rabbit's visual cortex. Channels C1, C4, C7, C10, and C13 relate to the contacts of the intracortical multi-electrode. The lowermost trace shows simultaneously recorded seizure activity of the geniculate body. The *numbers* on the time axis relate to the block numbers of the stored EEG (1 block = 0.5 s). The 12-s interval is sectioned into three epochs of 4 s, the analysis intervals for the following analyses. The seizure activity consists of tonic discharges with decreasing frequency from about 13/s at the beginning of the seizure to about 8/s during the last second of the record

presented on a logarithmic scale. The cortical power spectrum shows a peak at about 12–13 Hz. This peak also occurs in the power spectrum of the thalamus EEG but is less dominant (arrows).

In the middle part of Fig. 2 the squared coherence and the phase spectra between the cortical and the thalamic EEGs are shown. In the coherence spectrum there are three peaks at 8, 12 (arrow), and 22.5 Hz. But only at 12 Hz is there a considerable amount of power, as can be seen from the power spectra. The amount of squared coherence at 12 Hz is 0.35. However, the computation of 95% confidence intervals (Benignus 1969a, b; Rappelsberger 1977) shows that this value is statistically not significantly different from zero. Therefore it must be stated, with an error probability of 5%, that the 12-Hz components in both EEGs are not linearly dependent. The confidence intervals are indicated in the

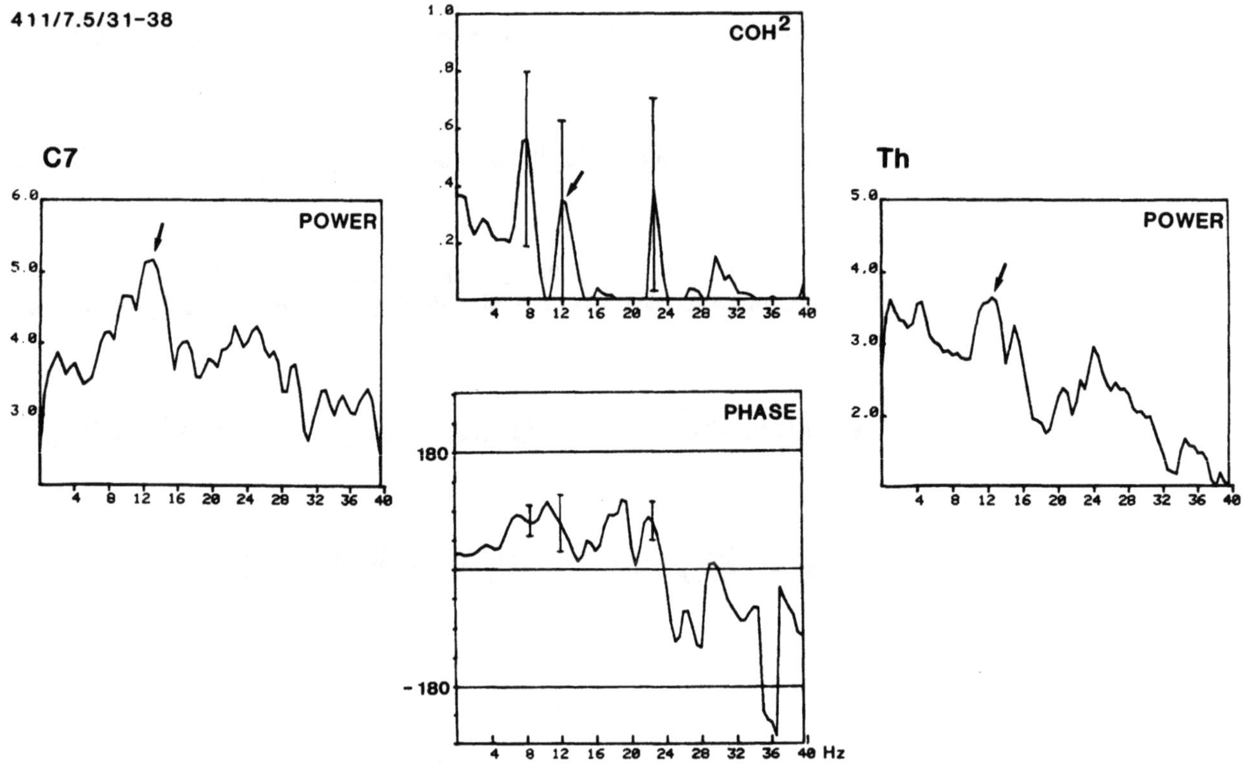

Fig. 2. Spectral analyses of the first 4-s interval of Fig. 1. To the left and to the right the power spectra of cortex channel C7 and thalamus are shown, respectively. In the middle of the picture the corresponding coherence and phase spectra are presented. Frequency scales are from 0 to 40 Hz. The power spectra are plotted on logarithmic scales; the phase scale is in degrees. The cortical power spectrum shows a clear peak at 12 Hz *(arrow)* according to the dominant 12/s activity within this analysis interval. A 12-Hz peak is also present in the thalamic power spectrum but with much less power than in the cortical power spectrum (note the different ranges of the logarithmic power scales). Three peaks can be distinguished in the coherence spectrum. The coherence value at 12 Hz *(arrow)* is not significantly different from zero as indicated by the confidence interval *(bar)*. Therefore, a significant linear relationship between the cortical and thalamic 12/s component cannot be assumed. The coherence values at 8 Hz and 22.5 Hz are significantly different from zero; however, at these frequencies no significant amount of power can be observed. An interpretation of phase is only sensible if a certain amount of coherence is present. This is not the case for the 12-Hz component. An interpretation of the phase of the 8-Hz and 22.5-Hz components is difficult because of the missing amount of power within these frequency bands

coherence spectrum of Fig. 2 by the bars. The broad confidence intervals reflect the high variance of the coherence estimates due to the short analysis interval of only 4 s.

In the lower middle part of Fig. 2 the corresponding phase spectrum is shown. Since there is only a small amount of coherence the interpretation of phases is not sensible. This is also reflected by the large confidence region of the phase which is dependent on the amount of coherence (Jenkins and Watts 1968).

From the phase spectrum a phase of 71° is obtained for the 12-Hz component. The corresponding confidence interval (Rappelsberger 1977) ranges from 37° to 105° and according to the formula $\tau(\text{ms})=\frac{\varphi(\text{degree})\ 1000}{360\ f(\text{Hz})}$ the region for the phase delay is $9\ \text{ms} \leq 16\ \text{ms} \leq 24\ \text{ms}$. The phase delay is positive, hinting at the cortex leading the thalamus. However, as already mentioned, this result is not undisputed since the corresponding coherence value is not significantly different from zero.

Figure 3 shows the results of spectral analysis of the second 4-s interval of Fig. 1 (block 39–46). In the power spectra significant peaks at about 9.5 Hz can be observed according to the 9.5/s rhythm of the seizure activity. There is significantly high coherence within this frequency band of about 0.90 and from the phase spectrum a phase delay for the 9.5-Hz component of 37 ms can be calculated. According to the high coherence, the confidence region is very small and ranges from 35 to 39 ms.

In the phase spectrum there is also a clear slope between 8 and 20 Hz visible (broken line). From this positive slope a group delay of 37 ms can be computed. These results mean that the cortical seizure activity is leading the thalamic seizure activity by about 37 ms.

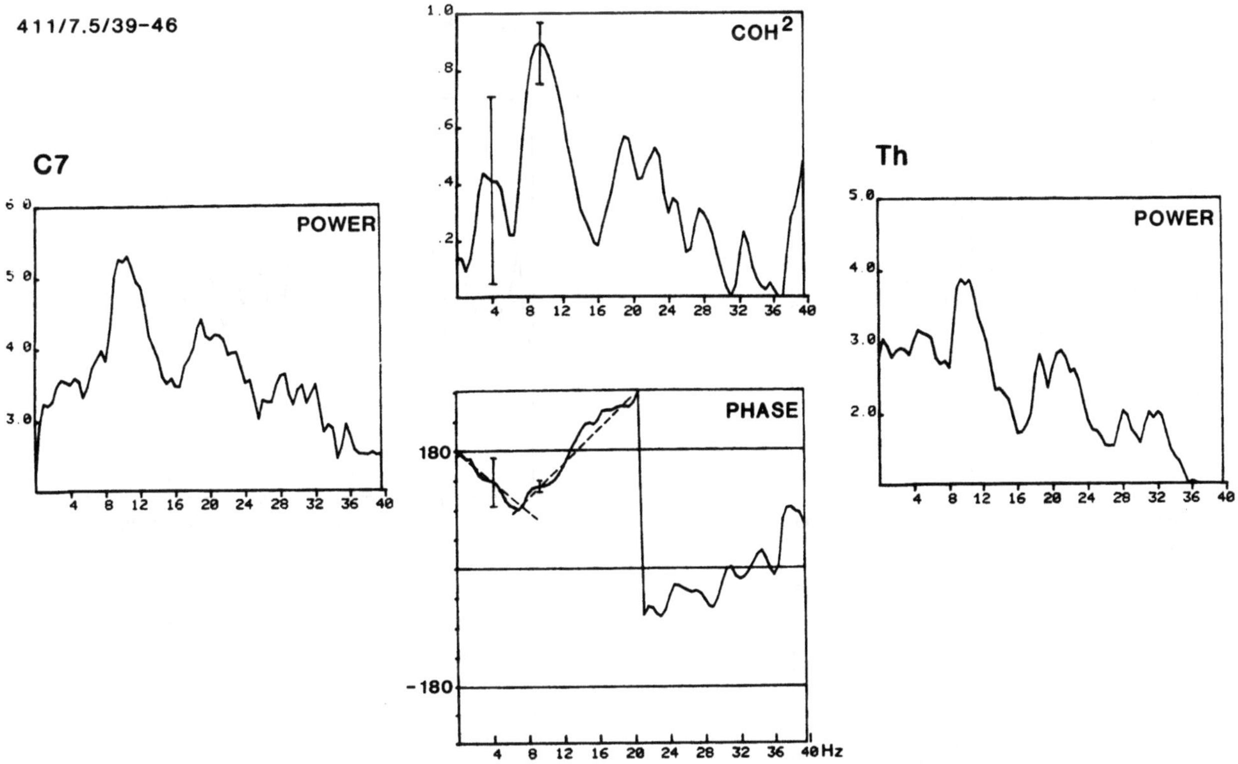

Fig. 3. Spectral analysis of the second 4-s interval of Fig. 1. The arrangement of the spectra is the same as in Fig. 2. The power spectra show clear peaks at about 9.5 Hz according to the rhythmic seizure activity. There is also a clear peak in the coherence spectrum within this frequency band, indicating a strong linear relationship of the 9.5-s component between the cortex and thalamus. The positive slope in the phase spectrum between 8 and 20 Hz *(broken line)* indicates a group delay between the cortex and thalamus with the cortex leading. The negative slope between 0.5 and 8 Hz indicates a group delay within this frequency band with the thalamus leading. However, this finding must be interpreted very carefully because coherence is low within this frequency band

However, there is a second slope in the phase spectrum of Fig. 3 between 0.5 and 8 Hz. This slope is negative, reflecting a negative group delay of about 31 ms. This means that the thalamic EEG leads the cortical EEG in the frequency range of 0.5–8 Hz. However, coherence within this frequency band is relatively small and also clear peaks are missing in the power spectra. Therefore, this result should be interpreted very carefully.

For the third 4-s interval of Fig. 1 (block 47–54) about the same results were obtained as for the second 4-s interval. For comparison the essential results are listed in Table 1 and may be summarized as follows. During the first 4 s after the onset of the seizure no significant linear relationship between the cortical and thalamic seizure activity is observed. A coupling of the cortical and thalamic 12-Hz component is indicated; however, coherence is not significantly different from zero. During the next seconds, characterized by a slowing down of the dominant frequency, a strong linear relationship between the cortex and thalamus is found with the cortex leading.

Theoretical considerations and practical applications of the spectral analytical methods to human seizure EEGs can be found in Gotman (1983).

Correlation Analysis

Figure 4 shows two cross-correlation functions. The time scale is from −500 ms to +500 ms. In the upper part, the correlation function between cortical seizure activity C7 and thalamic seizure activity of the first 4 s (block 31-38) interval of Fig. 1 is presented. The maximum is situated to the right of the zero delay line (arrow), and this means that the cortical EEG is leading the thalamic EEG with about 13 ms. However, the amount of correlation is low.

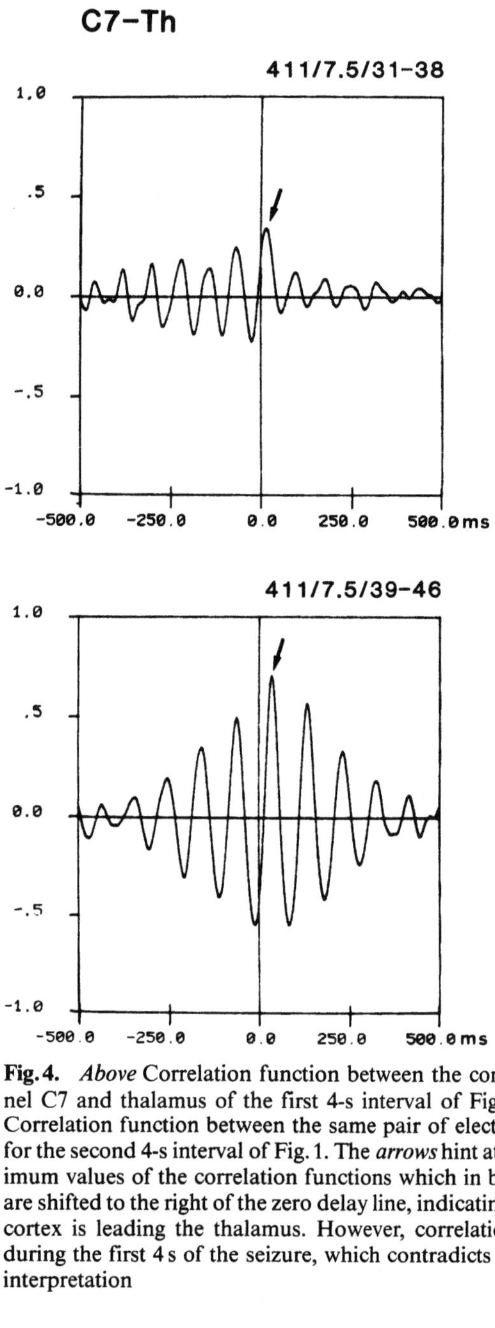

Fig. 4. *Above* Correlation function between the cortex channel C7 and thalamus of the first 4-s interval of Fig. 1. *Below* Correlation function between the same pair of electrodes but for the second 4-s interval of Fig. 1. The *arrows* hint at the maximum values of the correlation functions which in both cases are shifted to the right of the zero delay line, indicating that the cortex is leading the thalamus. However, correlation is low during the first 4 s of the seizure, which contradicts a reliable interpretation

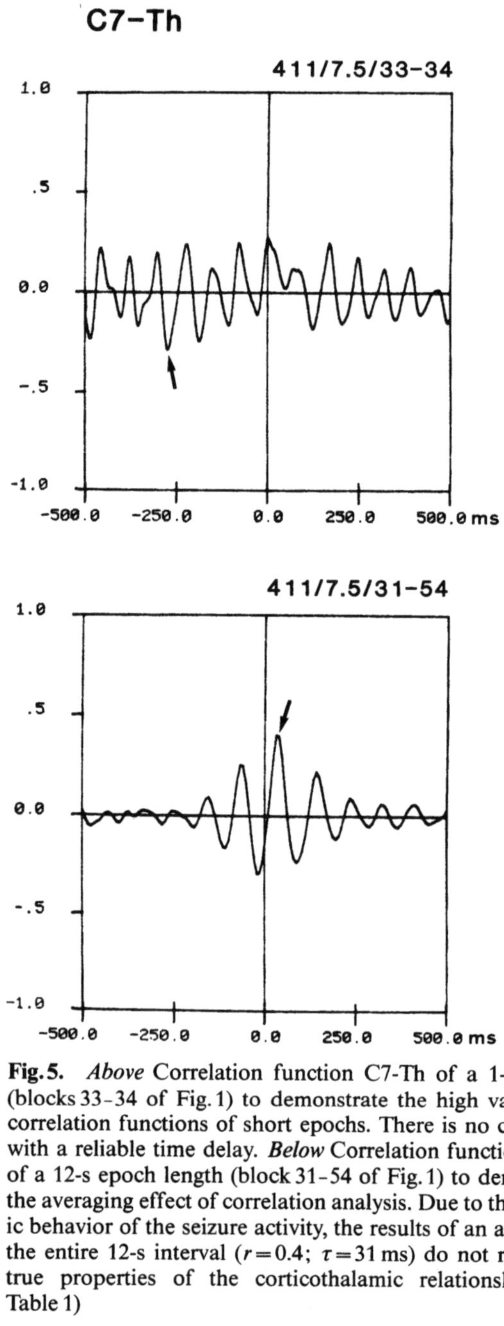

Fig. 5. *Above* Correlation function C7-Th of a 1-s interval (blocks 33-34 of Fig. 1) to demonstrate the high variance of correlation functions of short epochs. There is no clear peak with a reliable time delay. *Below* Correlation function C7-Th of a 12-s epoch length (block 31-54 of Fig. 1) to demonstrate the averaging effect of correlation analysis. Due to the dynamic behavior of the seizure activity, the results of an analysis of the entire 12-s interval ($r=0.4$; $\tau=31$ ms) do not reflect the true properties of the corticothalamic relationships (see Table 1)

The oscillations of the correlation function are due to the 12/s activity during the analyzed time interval.

In the lower part of Fig. 4 the correlation function between C7 and thalamus for the second 4-s (block 39-46) interval of Fig. 1 is shown. This correlation function shows its maximum at about +36-ms time delay with a correlation value of about 0.71. The oscillations of the correlation function in this case correspond to the approximate 10/s activity. Similar results were obtained for the third 4-s interval of Fig. 1. These are listed in Table 1.

If one compares the results of correlation analysis with those of spectral analysis, no significant differences can be observed with respect to the measured time delays.

A method for the estimation of the variance of cross-correlation functions and the computation of confidence intervals is described in Box and Jenkins (1970). However, the practical application of those theoretical formulas is very complicated and confidence intervals of cross-correlations are usually not used in EEG analysis. Frequently significance levels are assumed by experience but are highly dependent on the length of the analysis interval and the kind of activity to be analyzed.

The choice of too short epochs can result in correlation functions with high variance which can lead to erroneous results. An example is shown in Fig. 5, upper part. In this graph the analysis interval was 1 s. Due to the high variance, the maximum correlation (negative maximum) appears at time delay −273 ms (arrow), which is uninterpretable in physiological terms.

On the other hand, too long an analysis interval may cover essential properties due to the averaging effect of correlation analysis. An example is shown in Fig. 5, lower part. This graph shows the correlation function between C7 and thalamus of the entire 12-s interval of Fig. 1. The maximum peak is situated at +31 ms and the corresponding correlation value is 0.4. If one compares these results with those listed in Table 1 for the three 4-s epochs, it can be seen that a delay of 31 ms does not agree with the results of one of the 4-s segments. This is due to the dynamic behavior of the seizure activity showing no significant linear relation between the cortex and thalamus within the first 4 s and a strong linear relation during the following 4-s intervals. Therefore, to find out such dynamics, an appropriate segmentation has to be carried out.

Average Amount of Mutual Information

Examples of the AAMI method, which is extensively described in Mars and Van Arragon (1982), are shown in Fig. 6. In Fig. 6a, AAMI values and corresponding time delays are presented for the three 4-s intervals of Fig. 1. The arrow in the AAMI graph denotes the significance level (0.3) as proposed by Mars and Lopes da Silva (1983). During the first 4-s epoch, the AAMI value is below this level of significance. Therefore an interpretation of the corresponding delay is not sensible. The second epoch yields an AAMI value of 0.37 with a corresponding time delay of +38 ms. A similar result is shown for the third epoch. For comparison the results are listed in Table 1.

As mentioned in "Methods", AAMI analysis can be applied to nonlinear systems as they may be expected in the brain. In contrast to this, coherence and correlation analyses yield reliable results only in linear systems. The agreement of the results obtained with coherence, correlation, and AAMI analysis, as shown in Table 1, permits the conclusion that no essential nonlinearities exist in the examples

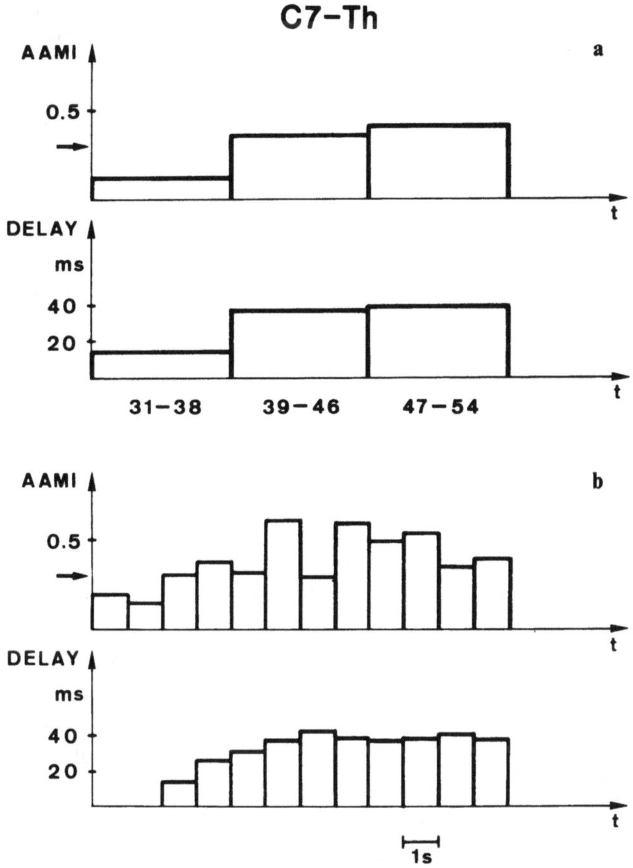

Fig. 6. **a** Results of an average amount of mutual information analysis (C7-Th) of the three sections indicated in Fig. 1. In the first interval, mutual information is below the level of significance *(arrow)* and the interpretation of the positive time delay is therefore not sensible. In the second and third interval, there is a significant amount of mutual information and the positive time delays of about 40 ms indicate that the cortex is leading the thalamus. **b** Results of AAMI analyses of 12 consecutive 1-s epochs. The segmentation was done to account for the dynamic behavior of the seizure activity and the results show that significant AAMI values are obtained for the 3rd and 4th second of the seizure and that there is a slight increase of the positive time delays *(lowermost plot)*. Only the 1st and 2nd epochs show no significant average amount of mutual information and no measurable time delays

demonstrated. Indeed, our experience with the AAMI method hitherto shows that there are only a few cases where the AAMI technique is superior to coherence or correlation analysis. However, in epileptology those few cases may be essential.

Undoubtedly the AAMI technique has an advantage concerning the segmentation of the data. Whereas correlation functions and coherence spectra of only short epochs have the property of high variances and the frequency resolution of coherence spectra additionally is considerably reduced, the

Table 1. Comparison of the results obtained by spectral analysis, correlation analysis and average amount of mutual information analysis of cortex channel C7 and thalamus for the three sections indicated in Fig. 1

	C7-Th	Blocks 31-38	Blocks 39-46	Blocks 47-54
Spectral analysis	Phase delay time	9 ≤ *16* ≤ 24 ms	35 ≤ *37* ≤ 39 ms	37 ≤ *40* ≤ 43 ms
	Coherence	0 ≤ *0.35* ≤ 0.63 (12 Hz) N.S.	0.76 ≤ *0.90* ≤ 0.96 (9.5 Hz)	0.73 ≤ *0.88* ≤ 0.95 (9 Hz)
	Group delay time	-	37 ms (8-20 Hz)	40 ms (8-20 Hz)
Correlation analysis	Delay	13 ms	36 ms	38 ms
	Correlation	0.35	0.71	0.69
AAMI analysis	Delay	14 ms	38 ms	39 ms
	AAMI	0.12 N.S.	0.37	0.42

AAMI method also yields reliable results if only short epochs are analyzed. This essential property allows the study of dynamic, nonstationary processes as they can be observed at the beginning of numerous seizures.

Figure 6b shows the results of AAMI analyses with segments of 1 s. From the upper graph of Fig. 6b, it can be seen that a significant relationship between the cortex and thalamus starts to develop during the 3rd second which cannot be detected by analyzing 4-s epochs (Fig. 6a). The time delay graph of Fig. 6b shows the development of positive time relationships, which are about 14 ms during the 3rd second, 26 ms during the 4th second, 32 ms during the 5th second, and so on. This means that time relationships during seizures may change within short intervals and that such a dynamic behavior can only be detected by analyzing short epochs.

Practical applications of the AAMI method to seizure activities in kindled dogs are found in Mars and Lopes da Silva (1983). Applications to human seizures are described in Mars et al. (1985). Some results of correlation analyses and AAMI analyses in experimental seizures are described in Rappelsberger et al. (1983).

References

Benignus VVA (1969a) Estimation of the coherence spectrum and its confidence interval using fast Fourier transform. IEEE Trans. on Audio and Electroacoustics, vol AU-17, no 2

Benignus VAA (1969b) Estimation of the coherence spectrum of non-Gaussian time series populations. IEEE Trans. on Audio and Electroacoustics, vol AU-17, no 3

Box GEP, Jenkins GM (1970) Time series analysis. Holden Day, San Francisco

Gotman J (1983) Measurement of small time differences between EEG channels: method and application to epileptic seizure propagation. Electroencephalogr Clin Neurophysiol 56: 501-514

Jenkins GM, Watts DG (1968) Spectral analysis and its application. Holden Day, San Francisco

Mars NJI, Lopes da Silva FH (1983) Propagation of seizure activity in kindled dogs. Electroencephalogr Clin Neurophysiol 56: 194-209

Mars NJI, Van Arragon GW (1982) Time delay estimation in non-linear systems using average amount of mutual information analysis. Signal Processing 4: 139-153

Mars NJI, Thompson PM, Wilkus RJ (1985) Spread of epileptic seizure activity in humans. Epilepsia 26 (1): 85-94

Pockberger H, Rappelsberger P, Petsche H (1984a) Penicillin induced epileptic phenomena in the rabbit's neocortex. I. The development of interictal spikes after epicortical application of penicillin. Brain Res 309: 247-260

Pockberger H, Rappelsberger P, Petsche H (1984b) Penicillin induced epileptic phenomena in the rabbit's neocortex. II. Laminar specific generation of interictal spikes after the application of penicillin to different cortical depths. Brain Res 309: 261-269

Prohaska O, Pacha F, Pfundner P, Petsche H (1979) A 16-fold semi-microelectrode for intracortical recordings of potentials. Electroencephalogr Clin Neurophysiol 47: 629-631

Rappelsberger P (1977) Spektrale Kurzzeitanalyse stochastischer Signale zur Auswertung von Elektroenzephalogrammen. Thesis, Technische Universität, Wien

Rappelsberger P, Pockberger H, Petsche H (1983) Messung von Zeitbeziehungen zwischen neokortikalen und intrazerebralen EEG Signalen. In: Bergmann H, Gilly H, Kenner Th, Schuy St, Steinbereithner K (eds) Monitoring in der Anästhesiologie und Intensivmedizin. Maudrich, Vienna, pp 299-303

Editors' Commentary 6

There was general agreement that, at the present time, the precise localization of the epileptogenic area requires the analysis of field potentials recorded by various electrophysiological techniques. However, concerning the different techniques, various experiences, interpretations, and approaches became apparent.

The discussion centered on the following aspects:

1. What are the criteria for using EEG, ECoG, and SEEG in a given patient for presurgical evaluation?
2. What is the value of interictal data vs ictal data?
3. Are chemical activation techniques of any value?
4. How reliable are seizure events induced by drug reduction?
5. Are highly sophisticated mathematical analyses of EEG data of any value for the presurgical evaluation?

First, it became obvious that different experiences with the various techniques have accumulated in the groups. Concerning non-invasive[1] techniques in temporal lobe epilepsy, there was general agreement on the point that the data derived from sphenoidal electrodes are superior to those from the surface EEG, even when additional and closely spaced electrodes are applied.

The use of the different invasive techniques depends on what data have already been accumulated and on the type of operation planned. At the moment selective operations involving deep brain structures require SEEG or in certain cases of TL epilepsy foramen ovale electrode recordings. The latter technique especially provoked discussion. The promising results obtained with this new technique, enabling reliable recording of mesiobasal-limbic epileptic events, should lead to its wider use.

The classical resective surgery shows excellent results with acute and chronic electrocorticography. It is still unknown whether ECoG and SEEG can be used interchangeably with equally good results or whether the technique has to be selected according to the particular case, and they also may have to be combined in certain cases. There are indications that ECoG is more appropriate for lateral activity whereas SEEG is necessary for exploring mesial aspects of the brain, especially of the temporal lobe. No presentation was specifically devoted to the subdural or epidural "grid technique." The huge amount of literature which appeared in the months following the Zürich symposium shows that these recording and mapping possibilities will extend to a great extent the potentials available to surgery, particularly in extratemporal regions close to functionally important areas.

Second, compared with ictal events there is a tremendous amount of interictal epileptogenic activity in any kind of recording. The discussion of its interpretation was controversial. Though the location of the epileptogenic area, determined by ictal events, is often confirmed using interictal data, unexpected differences in a number of cases were reported, which obviously limit the value of interictal activity. However, it seems that a constant "interictal focus" is a good predictor for a successful surgical result and therefore the value of interictal data should not be underestimated.

Third, a number of participants reported their experiences with pharmacological tests and drug activation of epileptic discharges using barbiturates, benzodiazepines, and convulsants such as bemegride, pentylenetetrazole (cardiazole), and penicillin. In this field computerized evaluation has started, and the results as interpreted are controversial. It seems that at present the use of pharmacological activation in routine presurgical evaluation provides only confirmation and that further scientific studies are necessary to define its future role.

Fourth, the facilities for presurgical evaluation are still small compared with the number of patients referred. Thus, the recording time for a single patient is limited. In order to achieve ictal information within

[1] Here the term "non-invasive" is used for techniques that do not violate the calvarium.

a reasonable amount of time, many centers reduce anticonvulsant medications in order to induce seizures. Recent evaluations of the Montreal group indicate that the seizures recorded under drug-reduced conditions are similar to spontaneous events. Though there were reports of occasional exceptions, this point was generally agreed on.

Fifth, the two presentations dealing with very sophisticated computerized evaluations of seizure activity under experimental conditions, as well as from patients, received much attention in the discussion. It became obvious that there is a great need for computerized analyses. However, whether the results derived from the average amount of mutual information (AAMI) technique can be used for clinical decisions depends on further studies. The promising data should encourage many groups to collaborate with those who already have experience. All participants agreed that considerable data reduction with the help of computers is possible and practically feasible, and that some form of seizure detection as pioneered in Montreal and UCLA is essential in order to enhance the efficiency of long-term monitoring.

Summarizing the impression of the discussion on electrophysiological techniques used in focus localization, one must admit that there are obviously different ways leading to a similar benefit for patients. In order to optimize this costly and time-consuming part of presurgical evaluation, standardized approaches should be developed.

Electroclinical Semiology of the Epilepsies

Limbic Seizures

H. G. WIESER and W. KAUSEL

Introduction

Before beginning a discussion of limbic seizures, it seems appropriate to define briefly the connotation of the term "limbic system." In 1878, Broca described this phylogenetically old, trilaminar part around the brain stem as "le grande lobe limbique." This apparatus became very popular after Papez (1937) had proposed his famous "circuit" and hypothetically linked it, because of its capability to reverberate, with the emotional-affective sphere. In 1952, Paul MacLean suggested the term "limbic system" as a designation for the limbic cortex and structures of the brain stem with which it has primary connections. As a result of more recent intensive anatomical studies, the borders of the limbic system concept have been expanded (Nauta 1979). This has led to a certain vagueness in its anatomical connotation. Obviously it is a matter of preference whether these diencephalic and mesencephalic structures, with which the core structures of the limbic region of the cerebral hemisphere (i. e., amygdala and hippocampal formation) are known to be associated by largely reciprocal fiber connections, should be incorporated. Thus the term limbic system, as currently used, is also at least partly based on physiological grounds. Morphologists, for this reason, sometimes find that the term "system" is not an adequate one insofar as different brain areas are constituents of it.

According to MacLean (1973) in the most simple terms, "the limbic system comprises three major constellations of neurons, each centered around a particular group of nuclei. The limbic cortex of the frontotemporal region is largely organized around the nuclear components of the amygdala, through which impulses arrive from and return to the brainstem. The nuclei of the septum serve in a similar capacity for a second constellation, including much of the hippocampal and parahippocampal cortex. Finally, the anterior thalamus serves as a gravitational center for the large constellation of neurons of the cingulate cortex." These main structures and their most important connections are schematically illustrated in Fig. 1. (For more precise information, see the chapter by Pandya and Yeterian, this volume.)

Koella (this volume) has extensively reviewed the functional role of the limbic structures in the making of behavior as experimentally studied in animals. Experimental and clinical findings attribute to the limbic subdivision associated with the amygdala mainly somatovisceral and endocrine functions which serve the organism in a global sense as self-protection, i. e., feeding, fighting, etc. At least parts of the septal subdivision seem to be occupied predominantly with the promotion of species procreation, i. e., they are involved in expressive and feeling states (MacLean 1970).

The involvement of the cingulate gyrus in visceral functions seems to be evident both from microelectrode studies (MacLean 1973) and from effects of stimulation, with a variety of autonomic manifestations including cardiovascular and respiratory changes, gastrointestinal effects, and genital tumescence (Smith 1945; Ward 1948). MacLean (1973) strongly ascribed the functions involved with sexual and especially maternal behavior to the mamillothalamocingulate subdivision. As this subsystem articulates with the nucleus dorsomedialis thalami which projects to the prefrontal cortex, he furthermore hypothesized that "This spearhead of the neocortex performs the transformation of insight into foresight that underlies human concern in planning and working for the welfare and preservation of the species."

These introductory remarks concerning the function of the limbic subdivisions represent an outline which is generally correct but which carries within itself certain serious limitations. It must be asked what actual evidence there is in humans for such assumptions, and here for obvious reasons the evidence is in fact relatively scanty. There is one exception, namely the insights gained from psychomotor epilepsies which either primarily or secondarily involve the

Department of Neurology, University Hospital, Frauenklinikstr. 26, 8091 Zürich, Switzerland

Presurgical Evaluation of Epileptics
Edited by H. G. Wieser and C. E. Elger

A

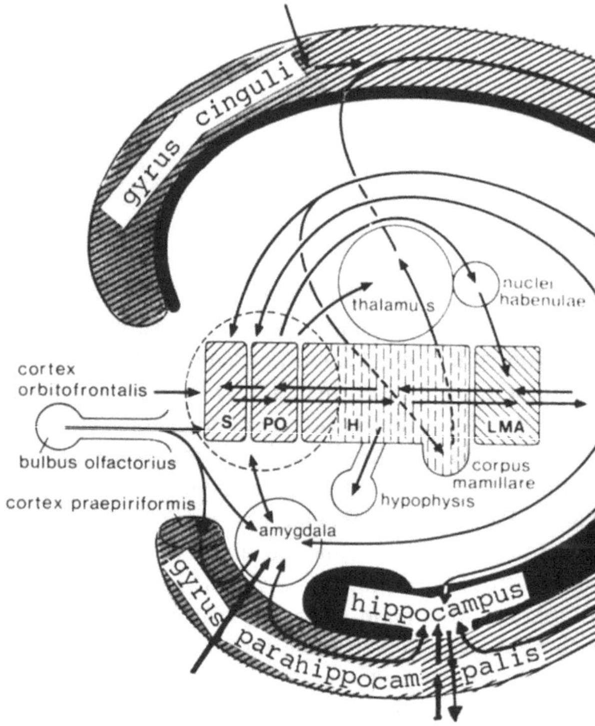
gyrus cinguli
nuclei habenulae
thalamus
cortex orbitofrontalis
S
PO
H
LMA
bulbus olfactorius
corpus mamillare
hypophysis
cortex praepiriformis
amygdala
gyrus parahippocampalis
hippocampus

B

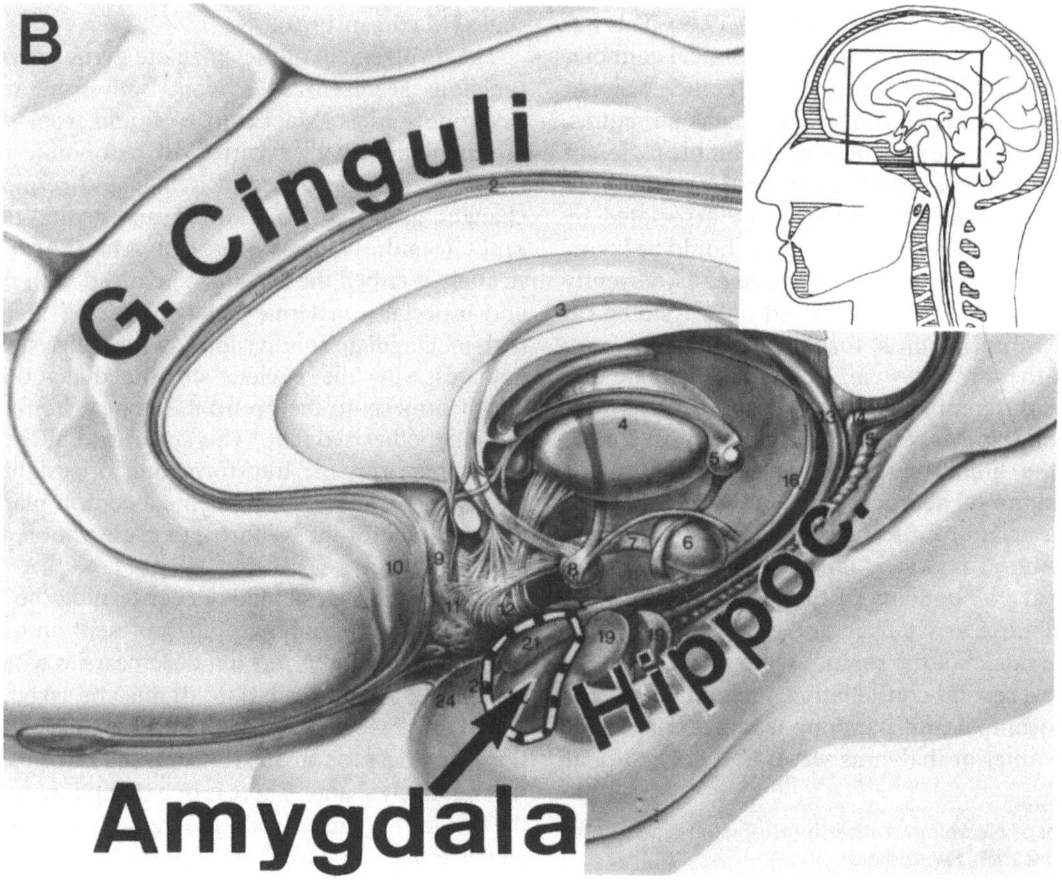
G. Cinguli
Hippoc.
Amygdala

◀ **Fig. 1.** **A** Structures of the limbic and olfactory systems with some afferent and efferent connections. *S*, septum; *PO*, preoptic area; *H*, hypothalamus; *LMA*, limbic midbrain area (from Nieuweenhuys et al. 1980, p 199, with slight modification). **B** Anatomical situation of the core structures of the limbic system, i.e., amygdala, hippocampal formation, and cingulate gyrus, together with some of the most important circuitry (from Nieuwenhuys et al. 1980, p 202, with slight modification). *2*, indusium griseum; *3*, stria terminalis; *4*, nucleus medialis thalami; *5*, nuclei habenulae; *6*, nucleus ruber; *7*, fasciculus telencephalicus medialis; *8*, corpus mamillare; *9*, septum verum; *10*, area subcallosa; *11*, gyrus diagonalis; *12*, fibrae amygdalofugales ventrales; *13*, crus fornicis; *14*, gyrus fasciolaris; *15*, fasciola cinerea; *16*, fissura choroidea; *17*, gyrus dentatus; *18*, subiculum; *19*, cornu Ammonis; *21*, nucleus corticalis amygdalae; *22*, nucleus anterior amygdalae; *23*, nuclei basalis + lateralis amygdalae; *24*, cortex praepiriformis

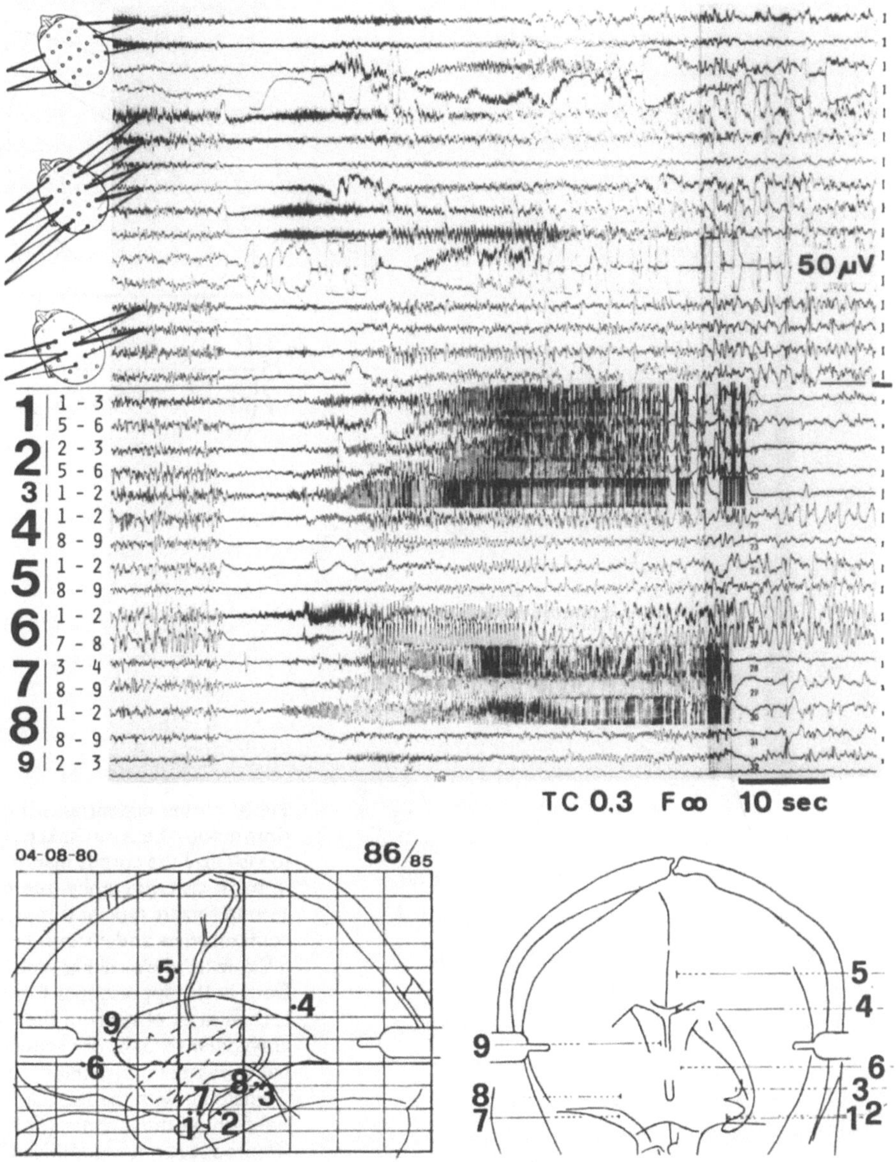

Fig. 2. Example of a "systemic limbic seizure" with orbitofrontal seizure onset occurring while the patient was sleepy. Note the generalized flattening which marks the beginning of the low-voltage, high-frequency discharge in 6/1-2. Later on both hippocampal formations (8/1-2 and 3/1-2) start discharging. Finally, a bitemporal seizure with participation of the lateral temporal cortex results. Clinically, adversive movement of the head to the right was initially observed, followed by loss of consciousness. After recovery, the patient did not remember the seizure. (Reproduced with slight modification from Wieser 1983 a)

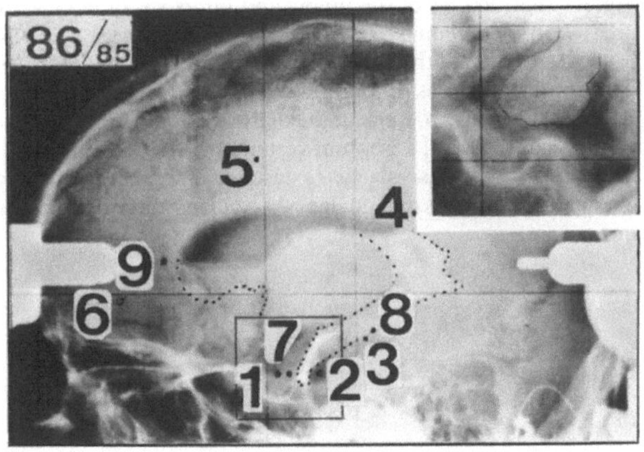

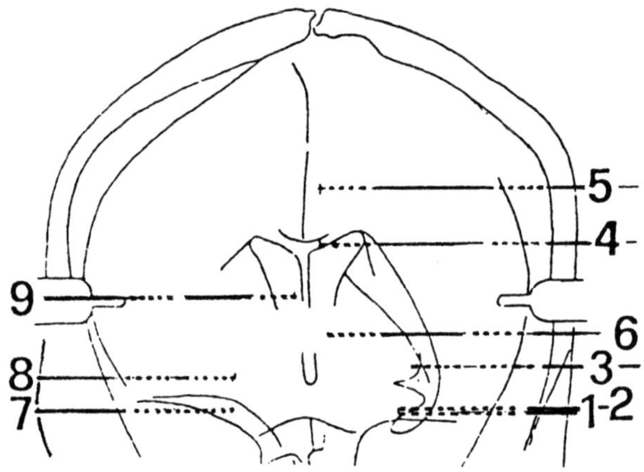

Fig. 3. Same patient as in Fig. 2. The electrode positions are drawn onto the X-ray taken during ventriculography in order to visualize the corpus mamillare glioma, which was best seen in basal cisternography (see *insert*). This figure illustrates the second seizure type of this patient with clouded but no loss of consciousness and no amnesia. At the beginning this unilateral left seizure shows slight modification of the activity recorded from both hippocampal formations. The epileptic discharge is then most obvious in the left amygdalar (1/1-3) and peri-amygdalar (1/5-6) derivations before the left parahippocampal gyrus (3/1-2) exhibits the maximum of the discharge. Coincident with the left parahippocampal gyrus, the left mesial fronto-orbital cortex (6/1-2) participates in the discharge. (Reproduced with modifications from Wieser 1983a)

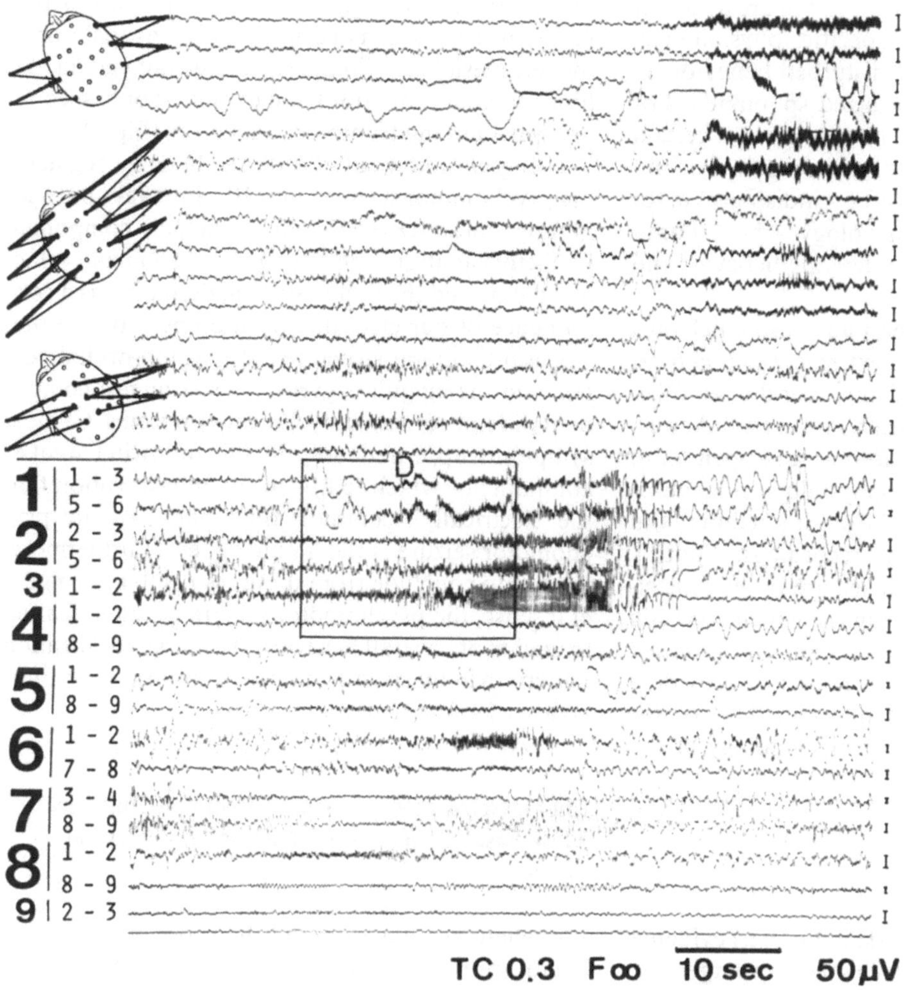
D
1 1 - 3
5 - 6
2 2 - 3
5 - 6
3 1 - 2
4 1 - 2
8 - 9
5 1 - 2
8 - 9
6 1 - 2
7 - 8
7 3 - 4
8 - 9
8 1 - 2
8 - 9
9 2 - 3
TC 0.3 F∞ 10 sec 50μV

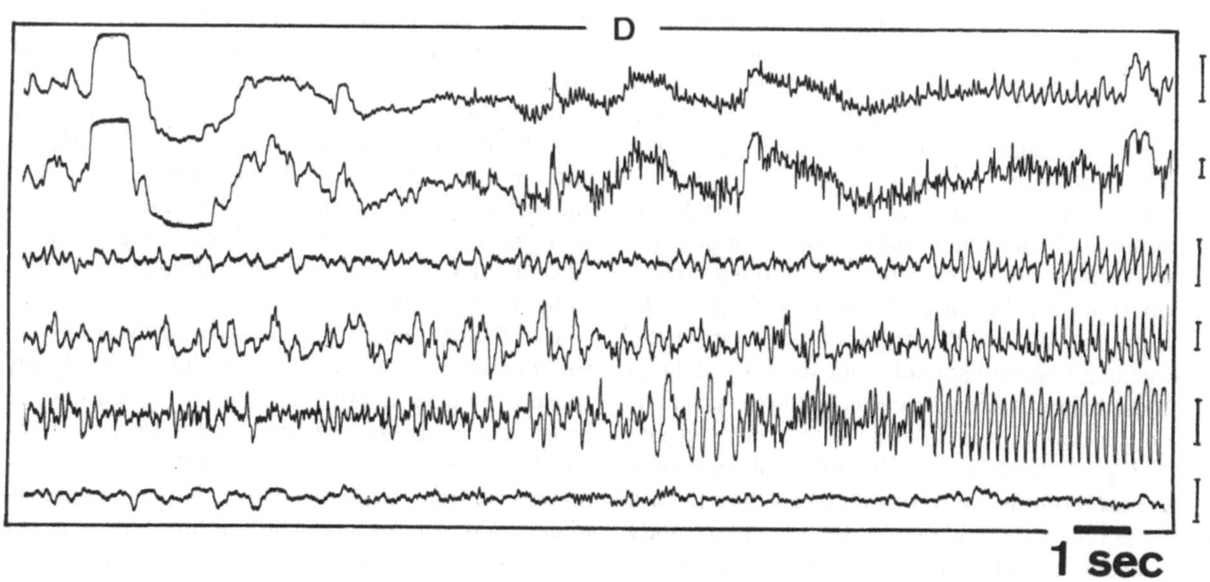
D
1 sec

limbic system, or at least parts of it. We are aware of the problems of inferring specific physiological functions from watching the disturbed behavior resulting from more or less localized spontaneous or induced epileptic discharges. Nevertheless, if certain preconditions are met, careful anatomoelectroclinical correlation studies of seizures - indispensible for understanding their phenomenology - best illuminate the brain mechanisms underlying behavior and mental functions in humans.

In this paper we will concentrate not so much on a systematic description of the extremely rich clinical phenomenology of limbic seizures (for this see Wieser 1987a; and also the "Zürich Presurgical Admission Questionnaire," reproduced as an appendix to this volume) as on the electrophysiological localization of seizure origin inferred from intracranial and especially stereotactic depth recordings. Starting from certain relatively typical seizure onset constellations, we shall then try to examine whether certain characteristic clinical features can be grouped around them. In this way, it is hoped to provide a bridge between the other chapters in this volume concerned with the electroclinical semiology of seizures.

Systemic Limbic Seizures

Unlike with experimental work, in humans evidence for discharges affecting simultaneously large portions of the "limbic system" in the broader sense is meager. By far the most often encountered seizure discharge onset within the limbic system is a focal one and, at least in the temporal lobe, unilateral. It must be said parenthetically that in this context the epileptic discharge consists of high-frequency, low-amplitude activity.

Propagation of discharges originating focally in the limbic system usually follows preferential pathways. Thus, most of the above-mentioned limbic subdivisions can be traced if the recording conditions permit. We had the opportunity, for instance, to document seizure spread from the amygdala to the hypothalamus and to record discharges originating in the midbrain and, along with a more global flattening of the EEG, the homolateral amygdala only. Furthermore, we have published data from a rather unique case of glioma of the mamillary body with severe drug-resistant seizures of different clinical phenomenology (Wieser 1983a). In this patient we recorded several seizures which originated in the frontobasal-cingulate region and then involved both temporal lobes with emphasis on the mesiobasal-limbic structures (Fig. 2). Other seizures of this patient, however, originated in the left amygdala or in the left mesiobasal-limbic structures (amygdala and hippocampal formation) and exhibited only secondary spread to the fronto-orbital cortex (Fig. 3).

Using an admittedly crude method, by segmenting psychomotor seizures into 10-s intervals and submitting these ($n=563$) to principal component (=factor) analysis, the strongest component (=factor) delineated the limbic system, defined - as a consequence of our electrode positioning - by the mesiobasal-temporal structures, the fronto-orbital cortex, and the cingulate gyrus. This strongest component did not contain any clinical symptoms. Clinical symptoms appeared only in the following weaker components, which led us to the interpretation that the electrically "common final path" of the psychomotor seizure is the limbic system. Furthermore, we had to conclude from these results that, in a general sense, the psychomotor seizure is better characterized by its overall electrical discharge localization pattern than by specific clinical signs and symptoms (Wieser 1983b).

Limbic Seizures with Focal or Regional Unilateral Mesiobasal Onset and Consistent Contralateral Spread

Elsewhere, we have studied the problem of contralateral spread statistically and have come to the conclusion that 44% of 213 analyzed psychomotor seizures showed contralateral propagation, 60% of these within the first 10 s, another 24% in the following 10 s; 7.5% affected contralateral structures in the third 10-s interval, and 8.5% in the fourth. Comparison of the seizure-discharging brain sites just before and immediately after the contralateral spread showed that the contralateral hippocampal formation most often received the discharge (Wieser 1983b). Figure 13 illustrates this contralateral hippocampal propagation.

The factors which are decisive with regard to whether seizure spread - as monitored by closely spaced multiple depth electrodes - appears to be constantly progressive from station to station, or whether it seemingly abruptly jumps without showing any buildup in the respective channel are poorly understood. The progressive buildup, however, can very often be observed when repetitive single-pulse stimulation is applied to a given brain structure. As shown in Fig. 4B, this buildup, ending either in stimulus-dependent, so-called paroxysmal responses (Buser et al. 1968), or even in self-sustained afterdischarges, may be observed transhemispherically. In the example illustrated we are inclined to regard this phenomenon of focal contralateral paroxysmal responses (in the right amygdala following stimulation of the left hippocampus) as an indication of the primary epileptogenic zone. One reason for this is that in this patient spontaneous seizures originated in the identical brain structure, i.e., the right amygdala (as illustrated in Fig. 5), and right-sided stereotactic amygdalotomy left the patient seizure-free with a follow-up of 10 years and a completely normalized scalp EEG (Fig. 6). This is remarkable because, as illustrated in Fig. 4C, at least in nasopharyngeal recordings, this patient had independent bilateral epileptiform potentials. Numerically, however, they predominated on the right.

Limbic Seizures with Focal or Regional Unilateral Mesiobasal Onset and Consistent Homolateral Spread

Spread to the Cingulate Gyrus

Focally originating, spontaneous seizure discharges in the hippocampus or in the amygdala with secondary spread to the hippocampus may remain confined to these structures. More often, however, they show a relatively typical propagation towards the homolateral cingulate cortex, especially into its posterior parts. This type of spread is so characteristic, that - if doubt exists regarding the location of the "driving site" because of inadequate electrode positioning - the hippocampal participation can be inferred from it. On several occasions we have illustrated this connection. Here it may suffice to demonstrate it with an electrically induced afterdischarge following stimulation of the amygdala and primarily affecting both the stimulated amygdala and the homolateral hippocampal formation (Fig. 7).

Spread to the Orbital and Lateral Frontal Cortex

Seizure discharges originating in the amygdala very often propagate to the orbitofrontal and lateral frontal cortex. This may be mediated through the uncinate fasciculus, as Brodmann's area 24 need not be involved. In such cases adversive movements of the head and eyes and de novo automatisms (see Delgado-Escueta et al., this volume) are frequently observed. If the supplementary motor area is also involved, the typical M2e posture (as illustrated in the "Zürich Presurgical Admission Questionnaire," this volume) may result. Strong participation of the anterior cingulate gyrus is heralded by the appearance of marked autonomous-visceral symptoms. Figure 8 illustrates this propagation to the frontal cortex and at the same time the secondary involvement of the lateral temporal cortex.

Involvement of Thalamic Nuclei

As little is known about the participation of the thalamic nuclei during seizure spread in humans, we would like to illustrate briefly an electrically induced afterdischarge following stimulation of the pulvinar and propagating to the homolateral amygdala (Fig. 9). This outlines the experimentally deduced connections of this thalamic nucleus to the inferotemporal cortex, which finally feeds into the amygdala (Fig. 10). This example of a successful operation for left occipital epilepsy at the same time serves to illustrate the secondary involvement of the temporal lobe and especially its amygdalar subdivision in primary occipital epilepsies. It is noteworthy that this patient suffered from complex partial seizures with complex visual auras and at the same time from strong visceromotor and viscerosensitive symptoms. The depth-EEG study with documentation of the constantly occurring secondary involvement of the homolateral amygdala (1) offered a satisfactory explanation of the observed symptomatology and (2) directed the operation towards an occipital lobe resection instead of the temporal lobe resection proposed initially at another institution. The postoperative outcome with full seizure control without drugs (follow-up 15 years) is conclusive that this was the correct interpretation.

Electrodes were implanted in the nucleus dorsomedialis thalami in several patients with suspected frontal seizure origin. In no instance could we record high-frequency seizure discharges in this structure, either in spontaneous seizures or following electrical stimulation. In one patient with a considerably abnormal scalp EEG, nearly continuous, rhythmic, spike-wave discharges were recorded in the left interhemispheric orbital frontal cortex, with synchronous slow-wave activity in the homolateral nucleus dorsomedialis (Fig. 11). The scalp EEG consisted of more or less regular, 2/s, slow-wave rhythms with left frontal predominance and was associated with the clinical picture of a semistuporous vigilance and mental slowing as well as frequently intermingled, secondarily generalized tonic-clonic convulsions.

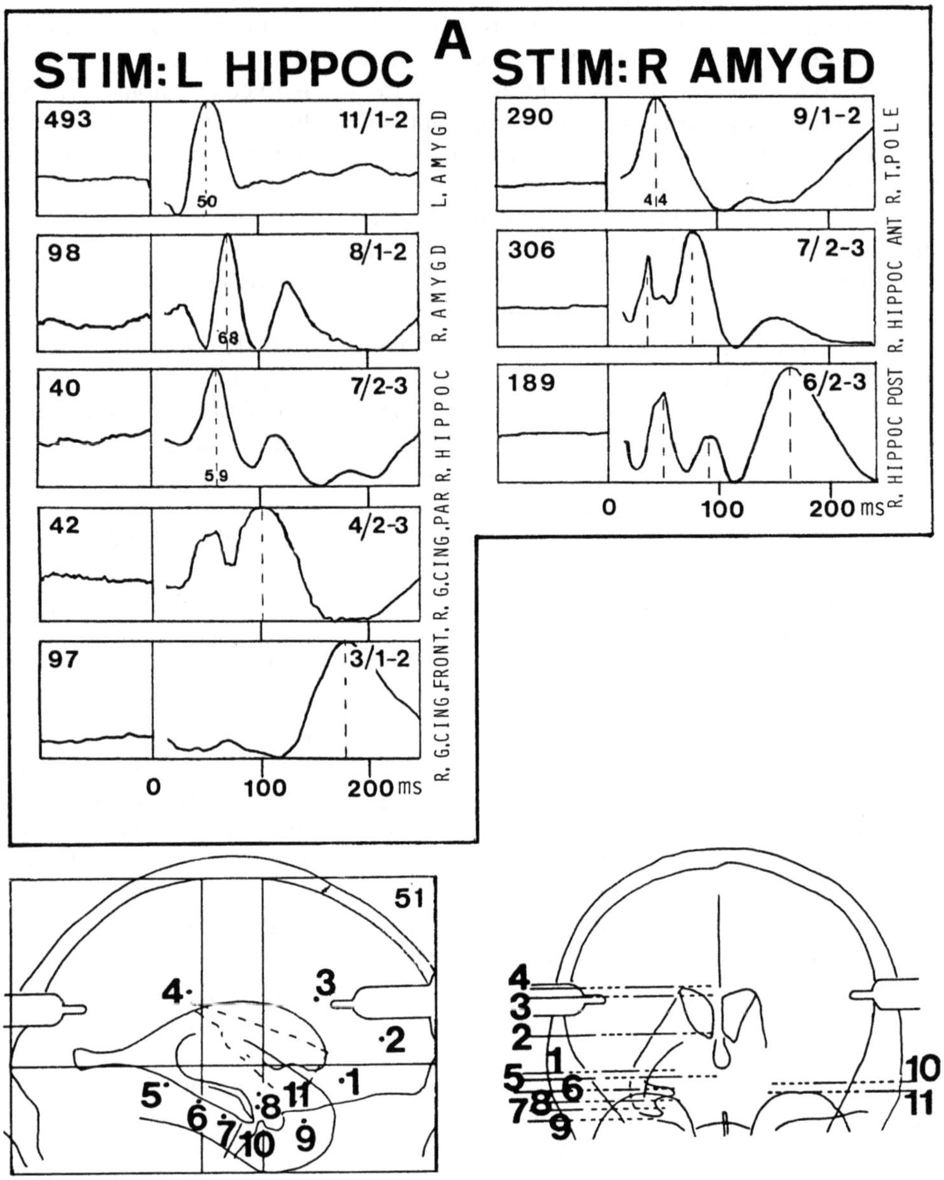
A
STIM: L HIPPOC
493
11/1-2
50
L. AMYGD
98
8/1-2
68
R. AMYGD
40
7/2-3
59
R. HIPPOC
42
4/2-3
R. G.CING.PAR
97
3/1-2
R. G.CING.FRONT.
0
100
200 ms
STIM: R AMYGD
290
9/1-2
44
R. T.POLE
306
7/2-3
R. HIPPOC ANT
189
6/2-3
R. HIPPOC POST
0
100
200 ms
51
1
2
3
4
5
6
7
8
9
10
11

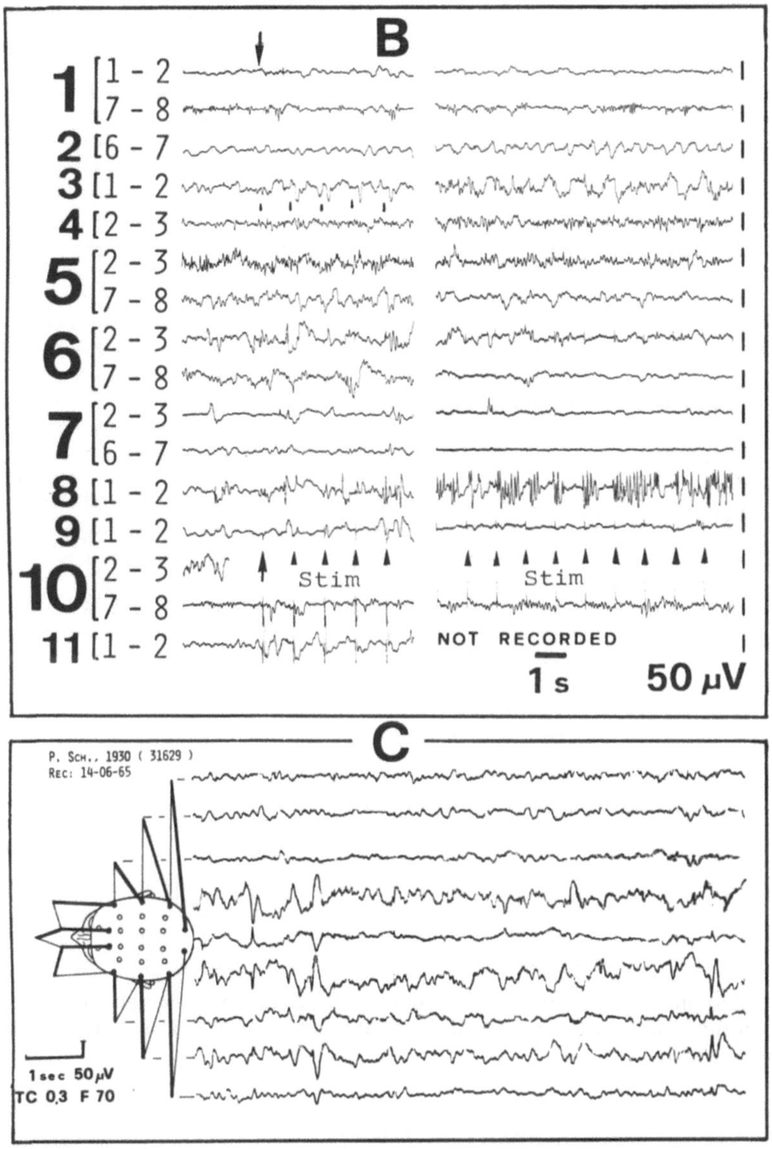

Fig. 4. **A** Intracerebral evoked potentials (EPs) (averaged responses, $n = 112$) following stimulation of the left hippocampal formation (10/2-3) and the right amygdala (8/1-2) simultaneously recorded at the sites indicated at the *right margin* of the EP graphs (electrode contacts and the respective anatomical location). The repetitive, single-pulse stimuli were delivered at time 0 and plotted for 236 ms following the stimulus. Prestimulus period is 100 ms. The amplitudes are normalized. The highest peak-to-peak amplitudes within the plots are indicated by the *numbers* (relative measures) at the left corner of the EP graphs. Note the different latencies as well as shapes of the EPs and, in particular, the transhemispherical EPs. **B** Repetitive single pulse stimulation of the left hippocampus (10/2-3) showing the build-up of "paroxysmal responses" at the contralateral, i.e., right, amygdala (8/1-2) which in this patient was the locus of spontaneous seizure origin (see Fig. 5). **C** Nasopharyngeal and surface EEG recordings showing bilateral and independent anterior temporobasal epileptiform potentials

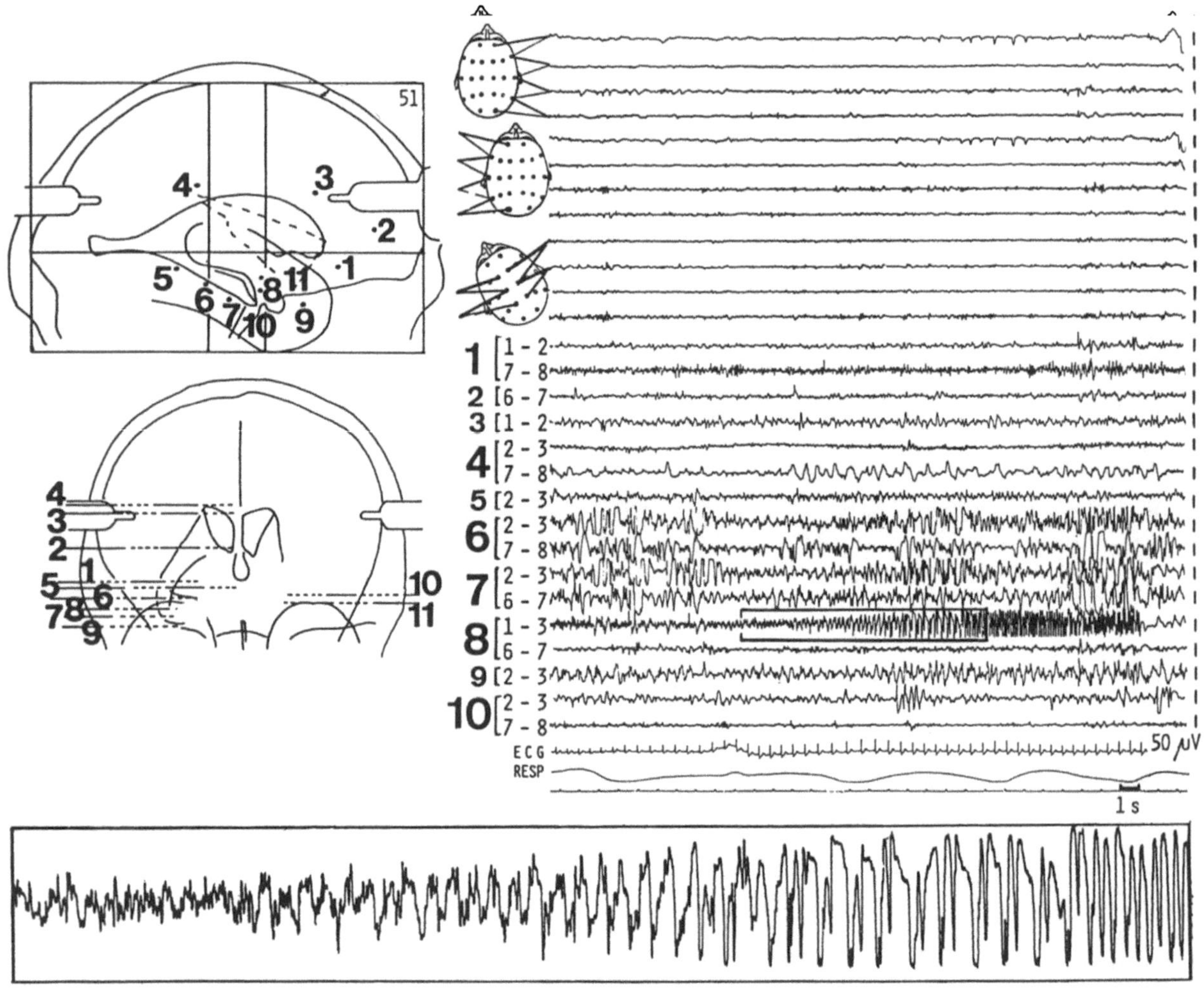

Fig. 5. Combined scalp and depth EEG-recorded, spontaneous, short-lived seizure discharge confined to the right amygdala (8/1-3) and accompanied by a fearful "strange" feeling with gastric sensation and vegetative signs, such as flush, bradycardia, and change of respiration (same patient as in Figs. 4 and 6)

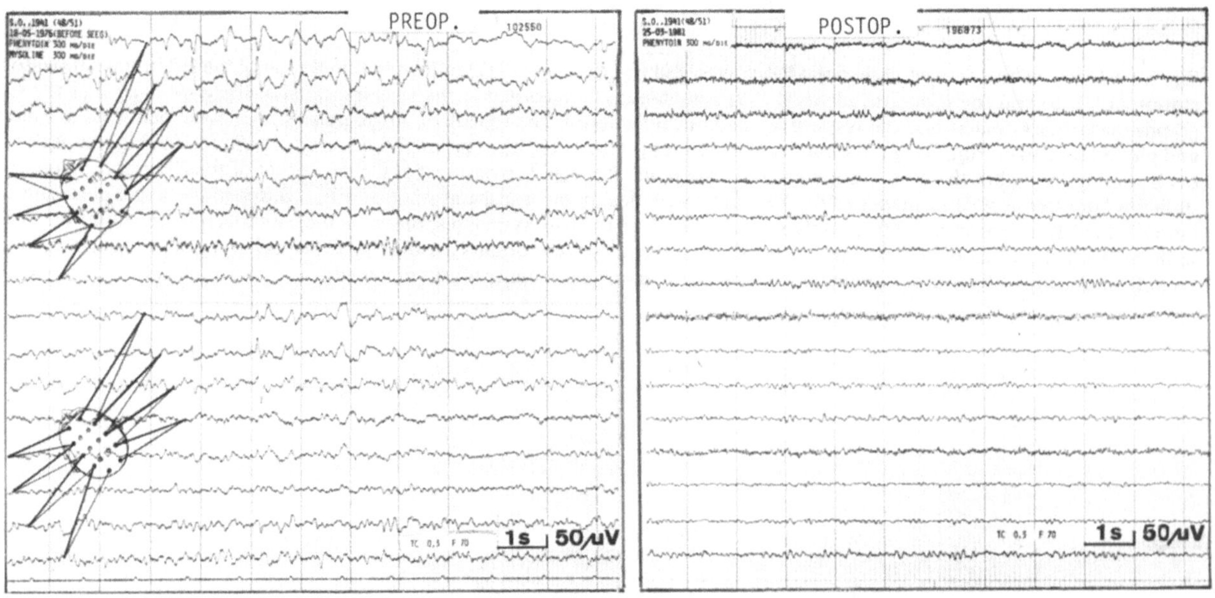

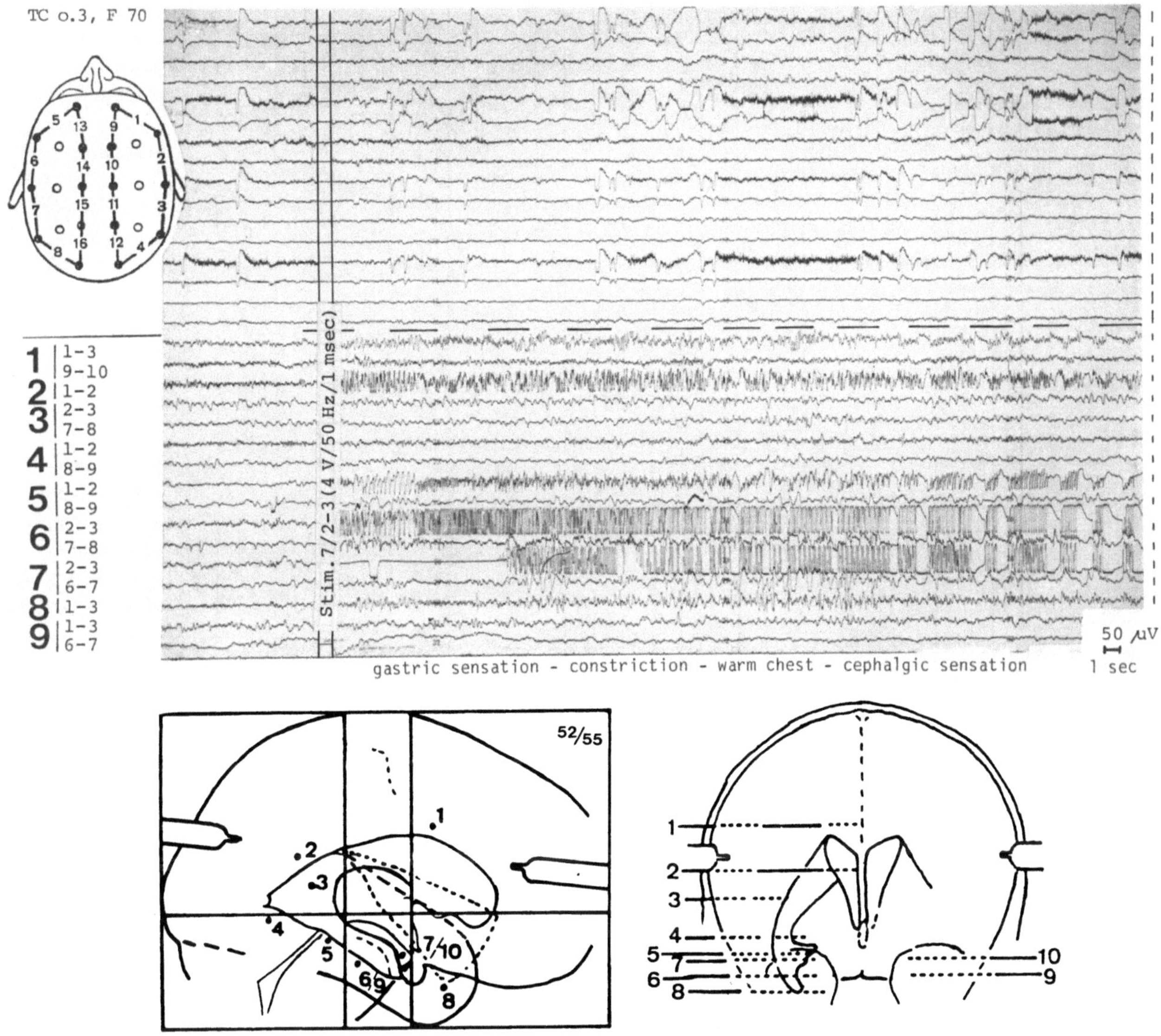

Fig. 7. Combined scalp and depth EEG-recorded, unilateral limbic afterdischarge. Following electrical stimulation of the right amygdala (7/2-3), a seizure discharge was recorded at the stimulated site and the homolateral hippocampal formation (6/2-3 and 5/1-2). Propagation can be traced to the posterior (2/1-2) as well as to the anterior cingulate gyrus (1/1-3). A typical rising gastric aura accompanied this limbic discharge

◂ **Fig. 6.** Two scalp EEG sections illustrating preoperatively right frontotemporal repetitive sharp waves of about 1.5/s. Five years after right-sided stereotactic amygdalotomy the EEG is normal *(right section)* (same patient as in Fig. 4 and 5)

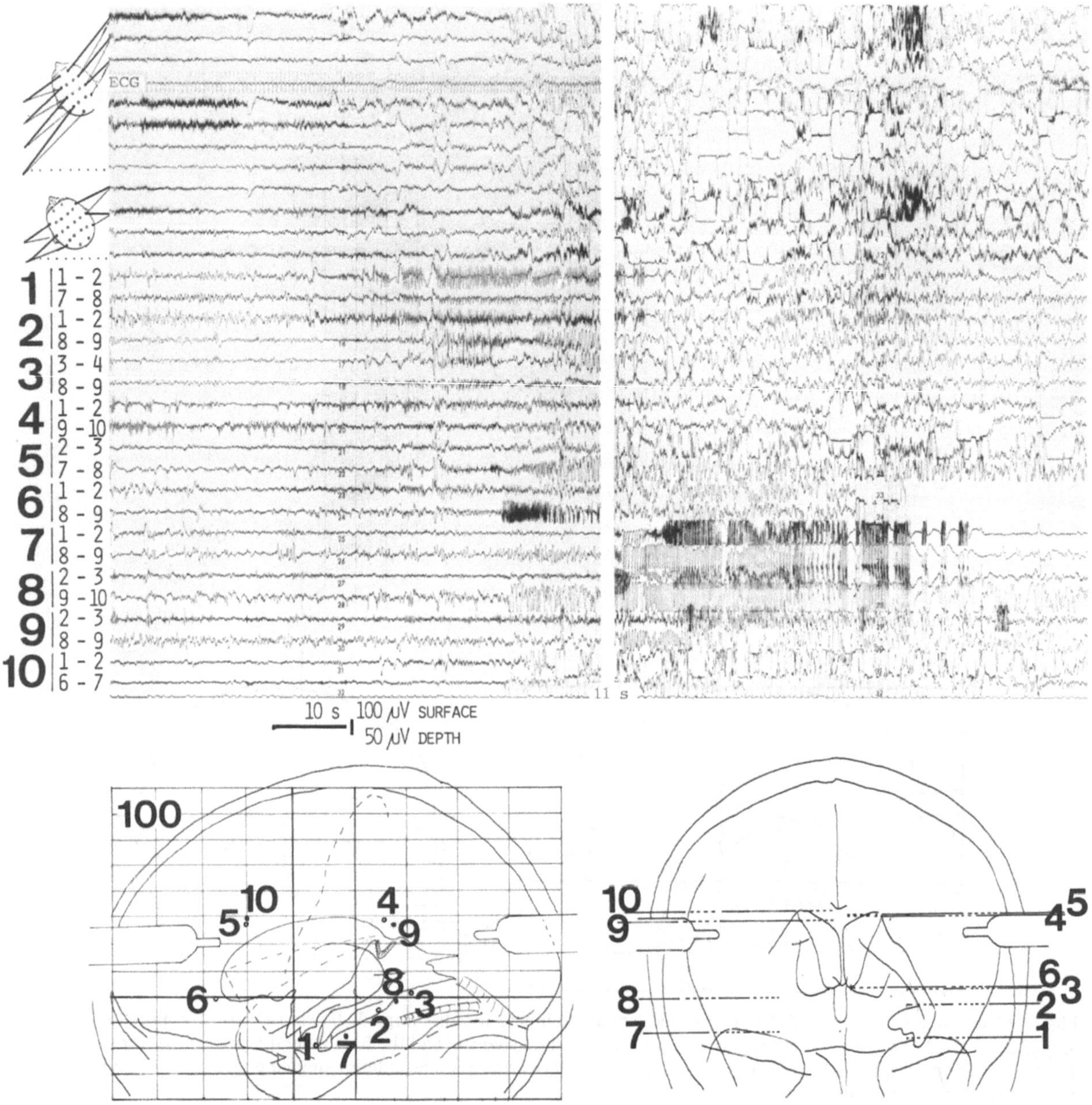

Fig. 8. Combined scalp and depth EEG-recorded, spontaneous seizure. It originated in the left mesiobasal-limbic structures (2/1-2, left hippocampus; 1/1-2, left amygdala). About 27 s later it spread to the frontobasal lateral cortex of the same side (6/8-9), before exhibiting its maximum at the right temporal lobe (electrodes 7 and 8)

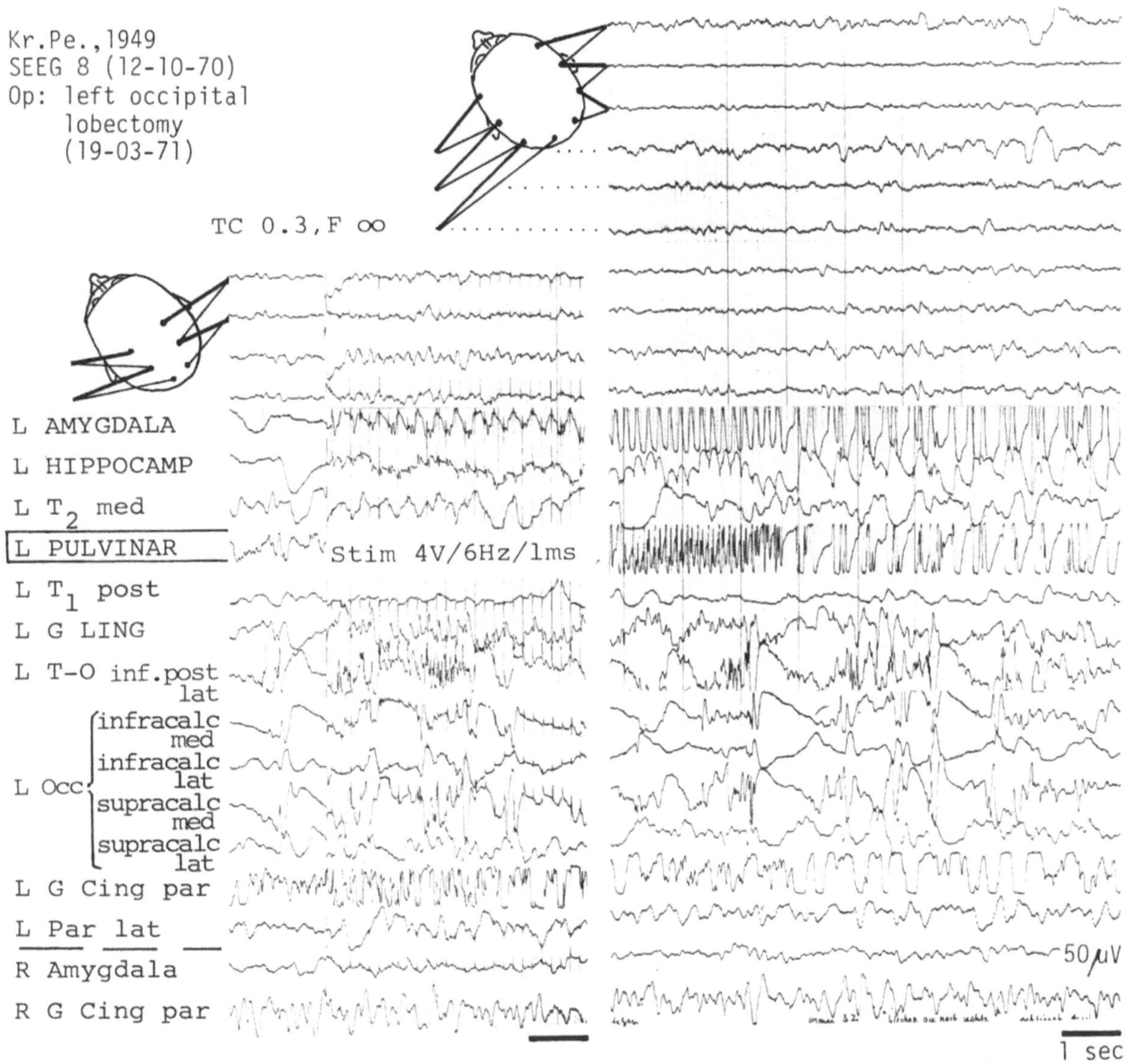

Fig. 9. Afterdischarge following stimulation of the left pulvinar thalami and spreading to the left amygdala. During repetitive electrical stimulation of the pulvinar (*left*), evoked potentials are seen in the left posterior cingulate gyrus *(L G Cing par)*, the left hippocampus, and other structures, such as the lingual gyrus *(LG LING)*

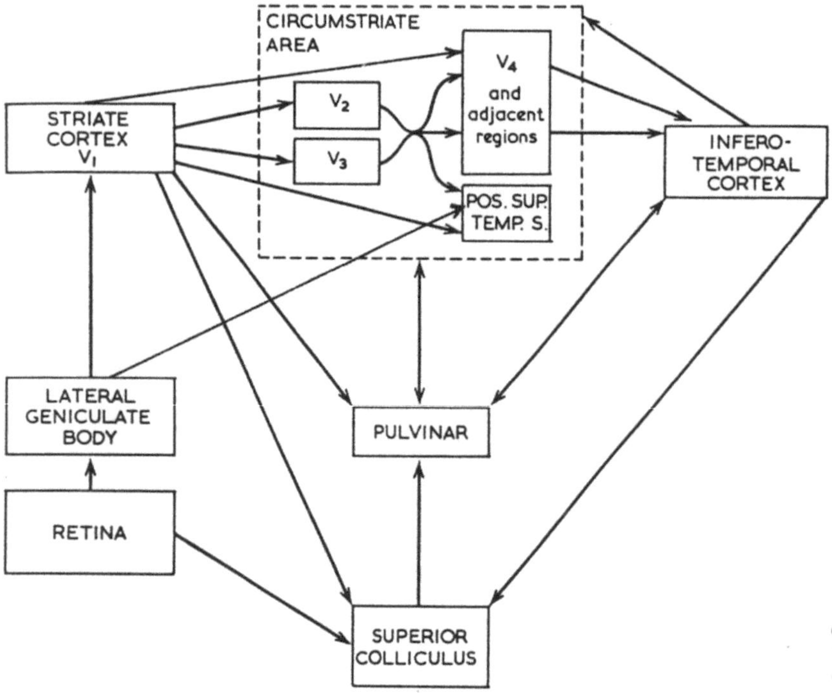

Fig. 10. The pulvinar connections (modified from Bindman and Lippold, 1981, p 191, with permission). Compare with Fig. 9

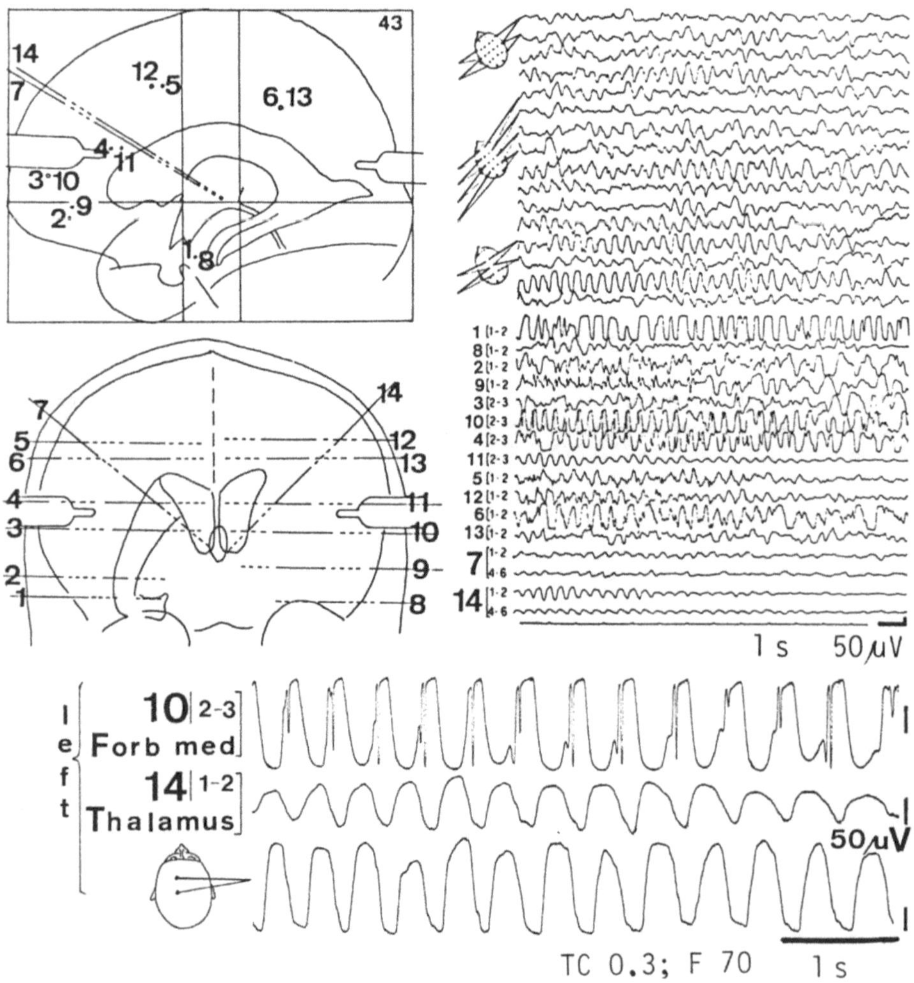

Secondary Involvement of Mesiobasal-Limbic Structures in Seizures Originating Elsewhere

This constellation is frequently encountered and is dealt with more comprehensively in the chapter by Wieser and Müller (this volume). Here it may suffice to underline the important implications we have deduced from it. As discussed elsewhere in more detail (Wieser 1986a, b), the fact that epileptogenic foci in surgically inaccessible brain areas, such as the posterior lateral temporal cortex close to Wernicke's speach area, frequently spread to the homolateral mesiobasal-limbic stations which can then act as a secondary "pacemaker zone" with further distribution and maintenance of the ictal event led us to perform so-called palliative selective amygdalohippocampectomies in such a constellation. The results obtained so far are sufficiently favorable to encourage use of this "palliative" approach in otherwise intractable patients. As may be expected, many of these patients continue to have seizures, although significantly less frequently, of shorter duration, and - even more important - often with only a rudimentary ictal clinical symptomatology.

Specific Diagnostic Problems for Surgical Therapy Encountered with Limbic Seizures

Although a certain amount of knowledge has now accumulated, which in general allows safe and successful judgements for tailored surgical therapy of at least certain subtypes of limbic seizures, there remain quite a lot of poorly understood findings.

It is probable that in most patients who do not profit from carefully planned and well performed operations (although presurgically and even after retrospective re-evaluation of the data they are judged as ideal candidates with an anticipated successful surgical outcome), data has been misinterpreted according to a false rationale.

Although at a first glance it might seem strange, probably the most frequently encountered diagnostic problem in mesiobasal-limbic seizures remains the lateralization of the primary focus. If seizure onset is unequivocally confined to one temporal lobe and no, or only late, contralateral propagation occurs, and if the recorded seizures are absolutely typical for the habitual seizures of the patient and also encompass all observed and experienced signs and symptoms, one can usually be absolutely safe. The greatest care, however, must be exercised with patients who report a change in their "usual aura" and, even more importantly, if a "new aura" additionally appears. This should not be considered simply an extension of the customary aura as it may express the establishment of a second focus. Such a finding is exemplified by Figs. 12 and 13, which show the "new aura" to be associated with restricted, contralateral, spontaneous seizure discharges. In this instance, the patient underwent a right-sided selective amygdalohippocampectomy, remaining free of seizures with 5 years follow-up. The consequences of a false lateralization leading to operation on the wrong side would probably have been catastrophic, with little or no beneficial effect on seizures, and with additional marked neuropsychological deficits. These patients have therefore usually been considered inoperable. Today, with further special examinations such as the bilateral intracarotid or "selective" TL Amytal test (Wada 1949; Wieser et al. 1987), the methohexital suppression test (Morrel and Whisler 1980) and other neuropsychological and pharmacological tests, which allow quantification of the amount of drug-induced fast beta activity (Engel et al. 1981), operations can be performed with good, even excellent, seizure relief and without neuropsychological deficits.

DC recordings with so-called foramen ovale electrodes (Wieser et al. 1985) turned out to be especially helpful for the lateralization of the origin of mesiobasal-limbic seizures with rapid contralateral spread. This is dealt with elsewhere in this volume (see chapter by Elger and Wieser) and therefore need not be further elaborated here.

From the careful follow-up of patients with such a bilateral constellation who had been operated upon, we learned that a "running down" of the contralateral focus can be expected, provided that the operation is performed while the secondary focus is still in what Morrel (1985) calls the (1) "dependent" or (2) "intermediate" stage.

A second problem frequently encountered is the "initial flattening." This phenomenon might present itself locally or regionally or even be generalized. In Fig. 14 all three of the following important aspects can be exemplified: (1) the local flattening preceding a high-frequency ictal discharge, (2) the regionally

◄ **Fig. 11.** Combined scalp and depth EEG from a patient with an atypical petit mal stupor. As better visualized by the enlarged detail, the spike/slow-wave discharges of about 2/s recorded at the left medial fronto-orbital cortex (10/2-3, *Forb med*) were coincident with the slow-wave rhythms recorded from the homolateral nucleus dorsomedialis thalami (14/1-2) and from the right fronto-precentral scalp EEG

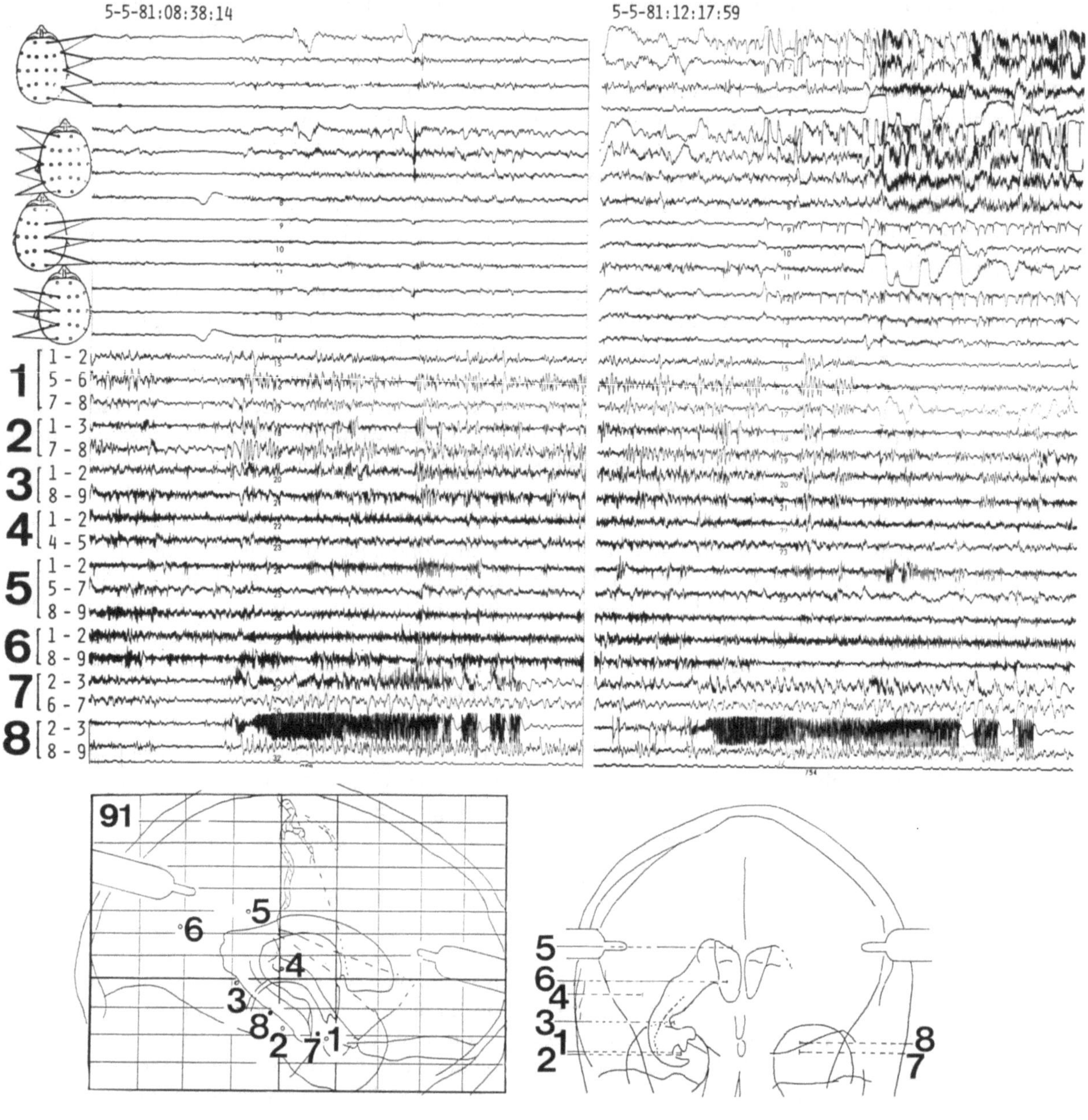
5-5-81:08:38:14
5-5-81:12:17:59
1 1-2 5-6 7-8
2 1-3 7-8
3 1-2 8-9
4 1-2 4-5
5 1-2 5-7 8-9
6 1-2 8-9
7 2-3 6-7
8 2-3 8-9
91

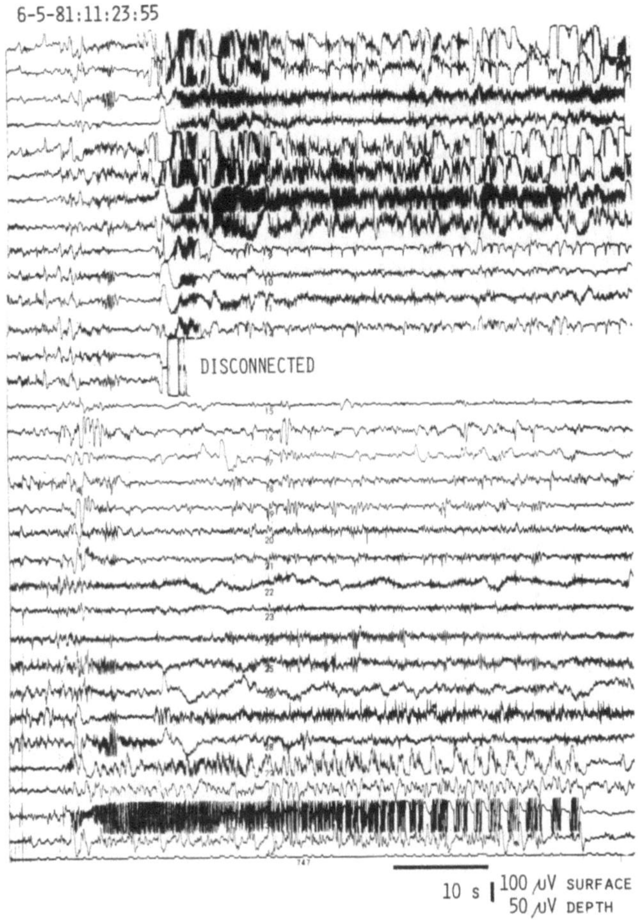

Fig. 12. Three left hippocampal spontaneous seizure discharges. All three illustrated were accompanied by an aura, which was judged by the patient to represent the "new" aura type. This, a feeling of uncommon familiarity, has been experienced only during the last 8 months in addition to the persisting "usual aura." The usual aura consisted of micropsia and vegetative symptoms and has persisted since early childhood; the patient clearly distinguished it from the "new aura" (compare with Fig. 13)

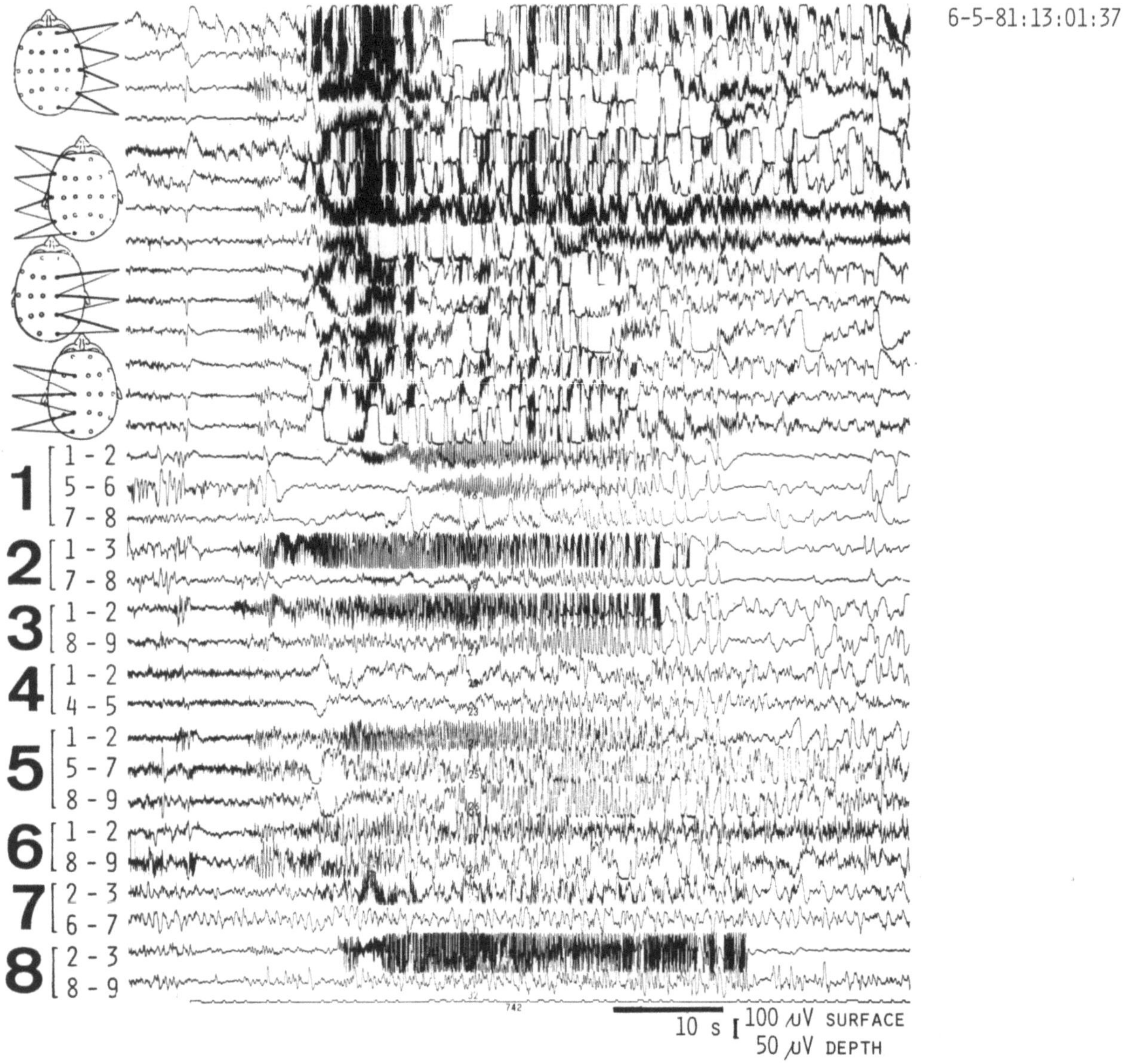

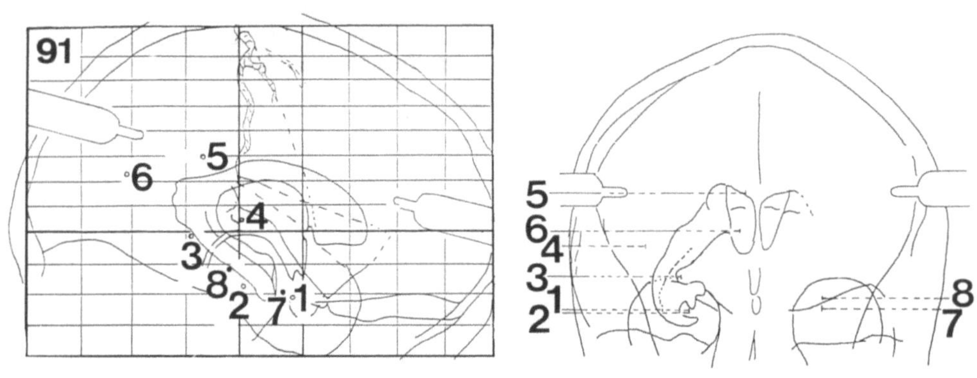

Fig. 13. Same patient as in Fig. 12. This seizure accompanied by the "usual aura" and originating in the right hippocampal formation (2/1-2 and 3/1-2) represents the patient's typical habitual seizure. Note the contralateral, i.e., left, hippocampal discharge (8/2-3) which appears 10 s after right hippocampal seizure onset

and neocortically accentuated flattening associated with high-frequency mesiobasal-limbic ictal discharges, and (3) the low-voltage surface EEG, which can sometimes be linked to the phenomenon first described by Landolt (1953) as "forced normalization." This change of the EEG, with the disappearance of the otherwise present epileptiform potentials and the replacement of the background rhythms by a low-voltage "desynchronized" EEG activity, can be associated with a worsening of the patient's mental status. This condition is also known as "alternative psychosis" (Tellenbach 1965). While such an antithetical relationship was indeed observed in the patient illustrated in Fig. 14, the forced normalization remains a controversial issue, insofar as it is not present in many epileptics entering psychotic episodes (Wolf 1986).

To explain the local flattening preceding the high-frequency ictal discharge, several possibilities have to be considered. First, in the given example it might be that the recording conditions (interrupted bipolar depth derivations) account for it. This would indeed be the case if the electrical field - produced by an epileptiform localized generator at a distance - affected both electrode contacts equally. With enlargement of the epileptically discharging brain volume, more and more neurons adjacent to the recording contacts would be recruited. Thus, after a certain time the discharge would become recordable, because the locally generated epileptic fields would affect the electrodes differently. Therefore, as Gloor (this volume) rightly points out, a montage using uninterrupted long chains should be used. However, as similar initial local flattening was also observed with recordings against a common average reference, this explanation alone cannot account for all observations of this type. It would be tempting to link the flattening with the inhibitory surround. It seems that simultaneous AC and DC recordings, possible with foramen ovale electrodes, might help clarify this question.

Clinical Concomitants of Mesiobasal-Limbic Seizures

As should be obvious from the foregoing, because of the secondary propagation of seizures involving different brain sites at a distance from that of the seizure onset, the clinical manifestations vary considerably. This protean symptomatology precludes forming a reliable inference as to the exact seizure onset area from the observed or experienced clinical signs. Although in a statistical manner some ictal symptoms show a higher correlative adherence to certain discharging brain sites than others, a single sign or symptom remains of questionable localizational value, especially with a view towards the precision necessary for surgical decisions. Reliable correlation studies have to include *multidimensional* analysis, i.e., of electroclinical and chronotopographical aspects.

The overall gestalt of a seizure starting in the mesiobasal-temporal lobe and successively progressing to functionally linked structures is probably best summarized with the old term "psychomotor seizure." This is because initially the patient usually reports what the Montreal group call an "experiential" aura often associated with a strange and/or strong emotional-affective tone. Objectively, not infrequently after a warning of the imminent seizure, the patient may exhibit a motionless stare. Then, with progressive clouding of consciousness, automatisms may occur. The "arrest reaction," however, is neither an obligatory symptom in mesiobasal-temporal lobe seizures nor localizationally confined to the temporal lobe. Strong vegetative signs and oroalimentary automatisms are likewise not exclusively bound to amygdala seizures, although in a statistical sense their correlation with the amygdalar limbic subdivision is relatively strong.

Concluding Remarks

The experience accumulated by localization of resectable epileptic foci has revealed the paramount importance of the mesiobasal limbic structures in both originating and maintaining the most frequently encountered subtype of psychomotor seizures. The characterization of limbic seizures at the present, at least for surgical purposes, depends largely on intracranial and especially stereotactically implanted intracerebral recordings. The most important consequence has been the inauguration of the selective amygdalohippocampectomy, which translated this advance in diagnostic precision into the actuality of a highly tailored therapeutic operation. This has proven to be beneficial for patients suffering from this frequent seizure type.

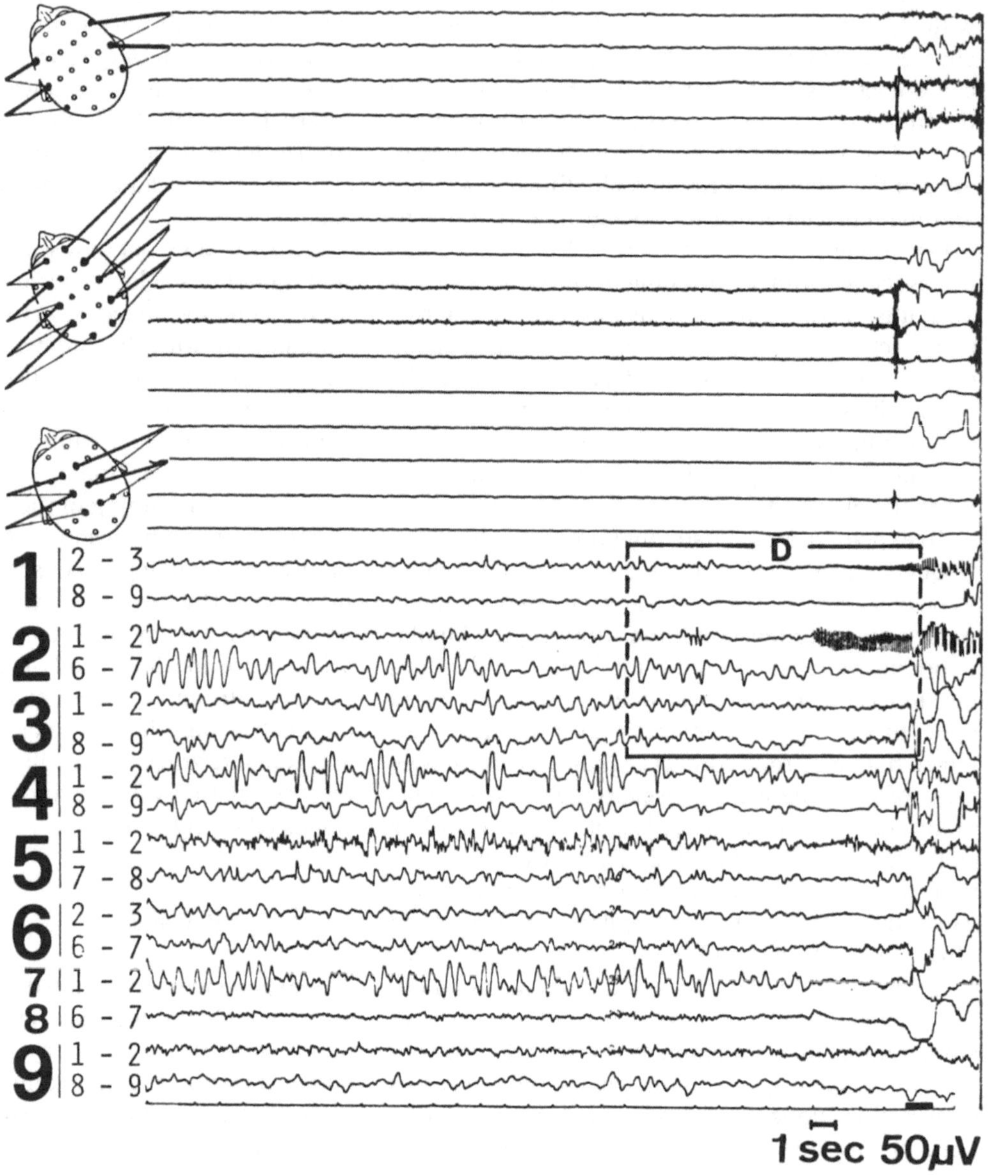
D
1
2 - 3
8 - 9
2
1 - 2
6 - 7
3
1 - 2
8 - 9
4
1 - 2
8 - 9
5
1 - 2
7 - 8
6
2 - 3
6 - 7
7
1 - 2
8
6 - 7
9
1 - 2
8 - 9
1 sec 50µV

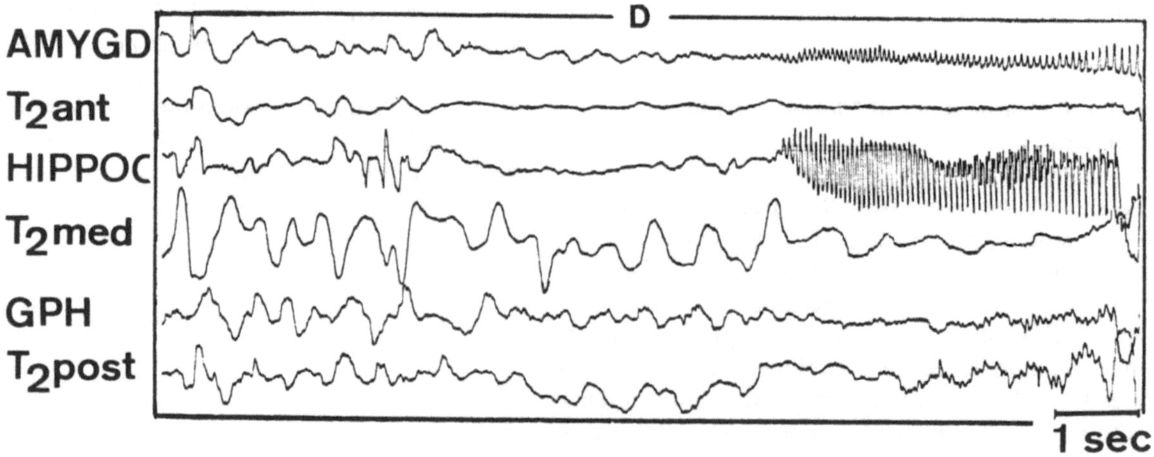
D
AMYGD
T2ant
HIPPOC
T2med
GPH
T2post
1 sec

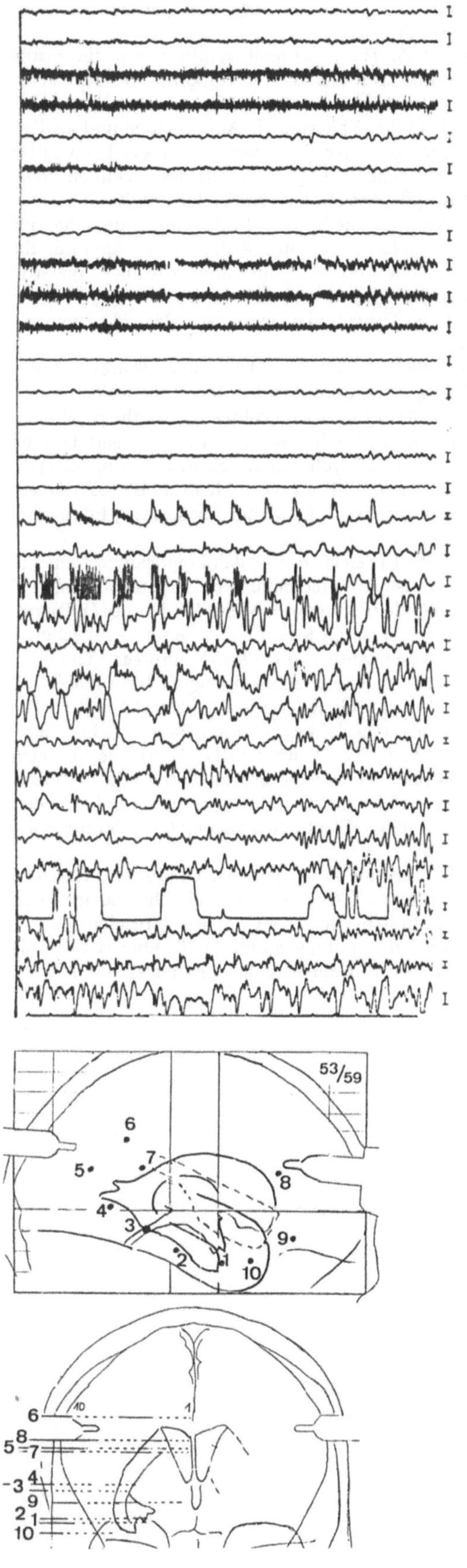

Fig. 14. Combined scalp and depth EEG-recorded, spontaneous, right mesiobasal-limbic seizure (1/2-3, amygdala; 2/1-2, hippocampus). Note the local flattening preceding the high-frequency ictal discharge, the regionally and neocortically accentuated flattening associated with high-frequency, mesiobasal-limbic ictal discharges, and the low-voltage scalp EEG

References

Bindman L, Lippold O (1981) The neurophysiology of the cerebral cortex. Edward Arnold, London, 495 pp

Broca P (1878) Anatomie comparée des circonvolutions cérébrales. Le grand lobe limbique et la scissure limbique dans la série des mammifères. Rev Anthrop, Seri 2, 1: 385-498

Buser PA, Talairach J, Bancaud J (1968) Potentiels évoqués, réponses paroxystiques aux stimulations périphériques et zone épileptogène. Rev Neurol 118: 514

Engel J Jr, Rausch R, Lieb JP, Kuhl DE, Crandall PH (1981) Correlation of criteria for localizing epileptic foci in patients considered for surgical therapy of epilepsy. Ann Neurol 9: 215-224

Landolt H (1953) Some clinical-electroencephalographical correlations in epileptic psychoses (twilight states). Electroencephalogr Clin Neurophysiol 5: 121

MacLean PD (1952) Some psychiatric implications of physiological studies on fronto-temporal portion of limbic system (visceral brain). Electroencephalogr Clin Neurophysiol 4: 407-418

MacLean PD (1970) The triune brain, emotion, and scientific bias. In: Schmitt FO (ed) The neurosciences: second study program. The Rockefeller University Press, New York, pp 336-349

MacLean PD (1973) The limbic galaxy. In: Heppner F (ed) Limbisches System und Epilepsie. Hans Huber, Berne, pp 52-61

Morrel F (1985) Secondary epileptogenesis in man. Arch Neurol 42: 318-335

Morrel F, Whisler WW (1980) Secondary epileptogenic lesions in man: prediction of the results of surgical excision of the primary focus. In: Canger R, Angeleri F, Penry JK (eds) Advances in epileptology: XIth epilepsy international symposium. Raven, New York, pp 123-128

Nauta WJH (1979) Expanding borders of the limbic system concept. In: Rasmussen T, Marino R (eds) Functional neurosurgery. Raven, New York, pp 7-23

Nieuwenhuys R, Voogd J, van Huijzen C (1980) Das Zentralnervensystem des Menschen (Atlas mit Begleittext). Springer, Berlin Heidelberg New York

Papez JW (1937) A proposed mechanisms of emotion. Arch Neurol 38: 725-743

Smith WK (1945) The functional significance of the rostral cingular cortex as revealed by its responses to electrical excitation. J Neurophysiol 8: 241-255

Tellenbach H (1965) Epilepsie als Anfallsleiden und als Psychose. Über alternative Psychosen paranoider Prägung bei 'forcierter Normalisierung' (Landolt) des Elektroencephalogramms Epileptischer. Nervenarzt 36: 190-202

Wada J (1949) A new method for the determination of the side of cerebral speech dominance. A preliminary report on the intracarotid injection of sodium amytal in man. Igaku to Seibutssugaku (Jpn) 14: 221-222

Ward AA (1948) The cingulate gyrus: area 24. J Neurophysiol 11: 13-23

Wieser HG (1983a) Depth recorded limbic seizures and psychopathology. Neurosci Biobehav Rev 7: 427-440

Wieser HG (1983b) Electroclinical features of the psychomotor seizure. Gustav Fischer, Butterworths/Stuttgart, London

Wieser HG (1986a) Psychomotor seizures of hippocampal-amygdalar origin. In: Pedley TA, Meldrum BS (eds) Recent advances in epilepsy 3. Churchill-Livingstone, Edinburgh, pp 57-79

Wieser HG (1986b) Selective amygdalohippocampectomy: indications, investigative techniques and results. In: Symon L, Brihaye J, Guidetti B, Loew F, Miller JD, Nornes H, Pásztor E, Pertuiset B, Yasargil MG (eds) Advances and technical standards in neurosurgery 13: 39-133, Springer, Vienna

Wieser HG (1987a) The phenomenology of focal limbic seizures. In: Wieser HG, Speckmann EJ, Engel J Jr (eds) The epileptic focus. John Libbey, London, Paris, pp 113-136

Wieser HG, Elger CE, Stodieck SRG (1985) The 'foramen ovale electrode': a new recording method for the preoperative evaluation of patients suffering from mesio-basal temporal lobe epilepsy. Electroencephalogr Clin Neurophysiol 61: 314-322

Wieser HG, Valavanis A, Regard M, Landis T, Yaşargil G (1987b) "Selective" Amytal temporal lobe memory test. (Abstr) 17th Epilepsy Internat Congress, Jerusalem

Wolf P (1986) Forced normalization. In: Trimble MR, Bolwig TG (eds) Aspects of epilepsy and psychiatry. John Wiley, Chichester, pp 101-115

The Significance of Phase Structure in Mesiotemporal Seizures in Relation to Clinical Manifestations

Y. Mayanagi

Depth EEG recordings of mesiotemporal structures have been carried out with a view towards a final decision for surgical intervention in cases of uncontrollable complex partial seizures with temporal spiking on EEG. Flexible wire electrodes with four contacts were placed stereotactically in the amygdala and the hippocampus through the temporal bone on both sides, according to Crandall's technique (Crandall et al. 1963; Engel et al. 1983). When satisfactory EEG recordings were obtained, usually in the second and third week, clinical manifestations of seizures were videotaped through a videomonitoring system for analysis.

In 20 patients, 70 seizures of mesiotemporal origin were recorded: 58 originated in the hippocampus and 12 simultaneously in the amygdala and hippocampus. Based upon the patterns of scalp and depth EEG, the course of seizures can be divided into three phases:

Phase A: Localized mesiotemporal discharge (Fig. 1)

Phase B: Limited propagation to the temporal or other cortex

Phase C: Secondary generalization

Among 70 seizures recorded, 26 remained in Phase A (mainly in the hippocampus), 24 developed to Phase B, and 20 went into Phase C. Figure 2 shows the variety of phase development of 15 representative seizures, associated with clinical symptoms. In Phase A, patients usually remain asymptomatic, subjectively and objectively. However, sometimes subjective sensory or autonomic symptoms are experienced. In a few seizures, oral automatism or slightly delayed response to external stimuli can also be observed. The typical symptoms of complex partial seizures, such as unresponsiveness, staring, oral, verbal or short gestural automatisms, respiratory suppression and others are associated with Phase B (Jasper 1964). Twitching of face muscles and adversion are the signs of impending Phase C.

Tokyo Metropolitan Police Hospital, and Department of Neurosurgery, University of Tokyo, 2-10-41, Fujimi, Chiyodaku, Tokyo, Japan 102

The pattern of development from Phase A to B seems variable, depending upon differences in intensity of initial firings. Thus, a relatively long period of aura without impairment of consciousness may be observed even when seizure discharge of mesiotemporal region is associated with unilateral temporal rhythmic waves. In these cases, impairment of consciousness may occur relatively slowly with growing extension of rhythmic slow waves in wide cortical areas. Abrupt loss of consciousness is often related to the onset of bilateral seizure involvement, which is indicated by a run of rhythmic spikes in both hippocampi or diffuse theta rhythm in bilateral cortical leads.

I would like to emphasize the significance of the concept of "phase structure" (Fukuzawa et al. 1971) in the course of temporal lobe seizures for a better understanding of the complexity of clinicoencephalographic correlation in complex partial seizures. Based upon the scalp EEG and polygraphic recordings and clinical observations of many seizures, Fukuzawa et al. (1971) classified four phases in the course of psychomotor seizures: Phase 0: aura; Phase 1: psychomotor lapse; Phase 2: oral automatisms; and Phase 3: behavioral automatisms. They concluded that the phases of psychomotor lapse and oral automatism, which are associated with series of diffuse high voltage and rhythmical slow waves, initially in 4-6 Hz and later in 2-3 Hz, comprise the main portions of a psychomotor seizure and that behavioral automatisms are the postictal aspect.

The results of this study on 70 mesiotemporal seizures by means of depth EEG and videomonitoring seem to support fairly well this theory of phase structure of temporal lobe seizures. It is emphasized that Phases A, B and C in mesiotemporal seizures were analyzed by means of depth EEG.

The results of experimental temporal lobe epilepsy in the monkey with chronic alum focus showed that seizure activities spread along certain preferential pathways systematically into the other parts of the brain (Mayanagi and Walker 1974). The most important pathways are those connecting temporal

Presurgical Evaluation of Epileptics
Edited by H. G. Wieser and C. E. Elger

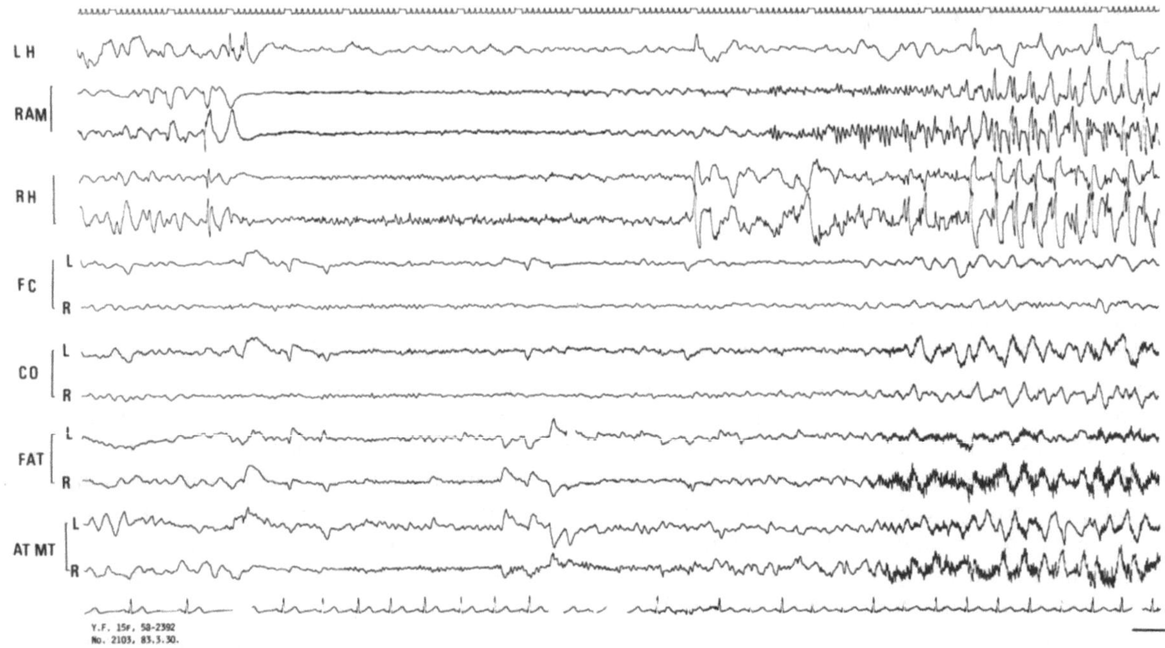

Fig. 1. Depth and scalp EEG recordings of a mesiotemporal seizure to show a phase development. *Upper five channels* are depth EEG, derived bipolarly from a depth electrode with four contacts, with 5-mm distance each. *LH*, left hippocampus; *RAM*, right amygdala; *upper trace* was derived from the deepest and the second contacts, *lower trace* from the second and the third contacts; *RH*, right hippocampus, in the same way. *Lower eight channels* are scalp EEG. *FC*, frontal-central; *CO*, central-occipital; *FAT*, frontal-anterior temporal; *ATMT*, anterior temporal-mid-temporal; *L*, left; *R*, right. Calibration: 1 s, 100 μV for the depth and 50 μV for the scalp EEG

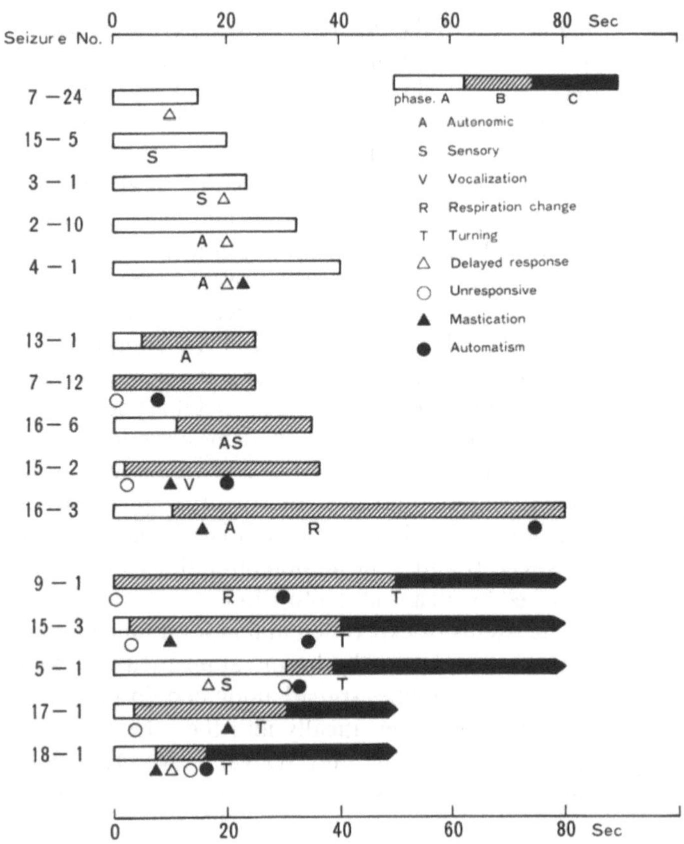

Fig. 2. Phase structures of 15 seizures associated with clinical manifestations. Numbers at the left indicate patient and seizure. For example, *7-24* refers to the 24th seizure recorded from patient No. 7. Square endings of bars indicate cessation of seizures, and pyramid endings termination of EEG recordings of ongoing seizures

lobe structures with each other, the hippocampus, amygdala, and lateral neocortex. In the stage of full development of a seizure, these three latter structures fire together so intensely that the location of original firing cannot be identified any more. Other important pathways of spread seem to be the fornix from the hippocampus, and the stria terminalis and amygdalofugal ventral fibers from the amygdala. By means of these pathways, the septal region, hypothalamus, anterior and medial thalamus, cingulate gyrus, and other diencephalic structures are functionally connected. Commissural fibers, the anterior commissure, and the psalterium convey impulses to the contralateral temporal structures. The pathways of secondary importance are fibers to the frontal and parietal cortex and those to the other basal ganglia and the mesencephalon. Irrespective of whether the seizure originates in the lateral cortex or in the amygdalohippocampal region, the systems of seizure propagation seem very comparable. However, the propagation to the diencephalic centers is more rapid and wide in the mesiotemporal origin than in the cortical one, because the cortical focus first evokes the amygdala and hippocampus, through which a secondary propagation takes place to the diencephalon.

Although the number of implanted electrodes are limited in clinical studies, there seem to be many common features between the experimental and clinical studies concerning the systems of seizure propagation. It can be well understood that in spite of considerable differences in clinical and electroencephalographic findings in the early phase of seizures, depending upon the site of the focus, there are fairly common characteristics in the late phase of complex partial seizures.

To better understand the mechanisms of temporal lobe seizures, it may be necessary to elucidate analytically and also synthetically the function of the temporal lobe as an unit.

References

Crandall PH, Walter RD, Rand RW (1963) Clinical applications of studies on stereotactically implanted electrodes in temporal lobe epilepsy. J Neurosurg 21: 827–840

Engel J Jr, Crandall PH, Rausch R (1983) The partial epilepsies. In: Rosenberg RN, Grossman RG et al. (eds) The clinical neurosciences. Churchill Livingstone, New York, pp 1349–1380

Fukuzawa H, Miyasaka M, Ohtaka T (1971) Electroencephalographic and polygraphic study on the phasic structure of psychomotor seizures. Brain Nerve 23: 1169–1180 (in Japanese)

Jasper HH (1964) Some physiological mechanisms involved in epileptic automatisms. Epilepsia 5: 1–20

Mayanagi Y, Walker AE (1974) Experimental temporal lobe epilepsy. Brain 97: 423–446

Neocortical Temporal Seizures

H. G. Wieser and R. U. Müller

Introduction

At the beginning of temporal lobe surgery, the concept prevailed that epileptogenic tissue was rarely limited to a small area. Thus large resective surgery was felt necessary in order to successfully treat patients sufferring from 'psychomotor seizures' of temporal lobe (TL) origin. Then, with growing experience, two major subtypes of what were then called 'temporal lobe seizures' were recognized. One involved mainly the neocortical lateral cortex (this type is extensively described herein) and the other predominantly involved the pararhinencephalic medial structures, i. e., uncus, hippocampus, and amygdala (dealt with in 'Limbic Seizures' by Wieser and Kausel; and by Yaşargil and Wieser, this volume). With the advent and wider application of more sophisticated presurgical evaluation protocols, including chronically implanted depth electrodes and/or subdural strips and grids, there has been a tendency to subclassify further the complex partial seizures (CPSs) into a number of subgroups depending on anatomical, physiological, and clinical criteria (Bancaud et al. 1965; Wieser 1983; Walsh and Delgado-Escueta 1984; Delgado-Escueta and Walsh 1985).

Wieser (1983) - after subjecting his data to an extensive series of statistical electroclinical correlation analyses - concluded that there were four types of temporal lobe CPSs: temporobasal-limbic, temporal-polar, opercular, and posterior temporal-neocortical. While there was a considerable degree of overlap, some features were described that served to distinguish them.

The *temporobasal-limbic* type was the most common. Auras occurred in 80% of patients and included complex memory flashbacks and déjà-experiences as well as the more common epigastric sensations. Consciousness tended to be preserved initially but became maximally impaired when seizures spread to the contralateral hippocampus. An arrest reaction or motionless stare occurred relatively frequently but not always as the earliest sign, and by itself was not a reliable localizing indicator. Oroalimentary automatisms, while frequent, were rarely seen early in the seizures. Complex gestural automatisms were associated with bilateral hippocampal and/or frontobasal-cingulate discharges. Unilateral or bilateral tonic/clonic or postural motor changes were commonly observed during the course of these seizures (Williamson et al. 1987; see also Delgado-Escueta and colleagues' chapter in this volume).

The *temporal-polar* type shared many characteristics with the temporobasal-limbic type. Possible distinguishing features included prominent early autonomic changes and the early appearance of oroalimentary automatisms.

Auditory hallucinations, especially if of elementary nature, were the most prominent feature of the *opercular* type and were considered strong evidence of seizure origin in or near Heschl's gyrus (see Fig. 7). Other features of opercular seizures included focal motor findings (face, upper extremity; see Fig. 10) and focal sensory symptoms (either unilateral or bilateral; bilateral ones being indicative of involvement of the second sensory area).

Temporal Neocortical Seizures

The topic of this chapter, the posterior temporal-neocortical type of TL seizures, is summarized with its main characteristics in Figs. 1 and 2.

Lateralization

Asymmetrical motor phenomena with tonic and/or clonic involvement of the face and upper extremity (thumb, hand) are frequently encountered due to spread and then, of course, are a useful lateralizing sign. Aphasia has a tendency to be more prominent with seizure origin in the language-dominant hemi-

Department of Neurology, University Hospital, Frauenklinikstr. 26, 8091 Zürich, Switzerland

Presurgical Evaluation of Epileptics
Edited by H. G. Wieser and C. E. Elger

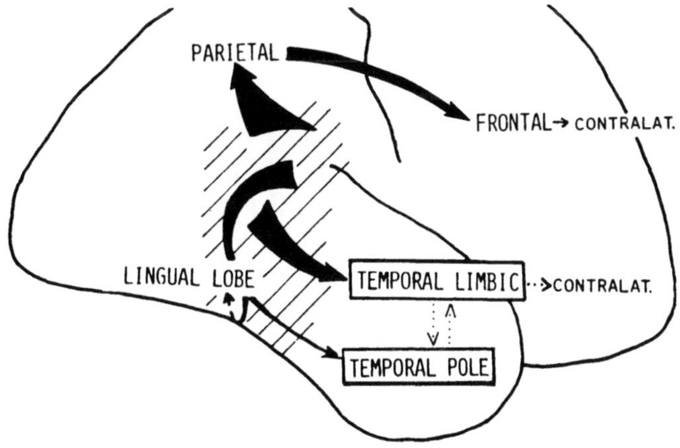

(1. PSYCHIC PHEN. aphasia)
2. BEHAVIOR staring, change of expression
3. PSYCHIC PHEN. vestibular hallucination
4. PSYCHIC PHEN. visual illusion and/or hallucination
5. AUTOMATISM gestic-exploratory, unilateral, upper limb

Fig. 1. Schematic summary with main characteristics of temporal neocortical posterior seizure type (*hatching*, seizure onset area; *arrows* indicate preferential spread). The symptoms are listed in *decreasing order* of their correlation with this seizure type. Auditory phenomena correlating strongly with T1 and regions near Heschl's gyrus are excluded here and were included instead in the 'opercular type' for didactic reasons. Lateralizational aspects are also excluded in this scheme. *Parentheses* indicate that the number of observations is relatively low (from Wieser 1983)

sphere, but this lateralizing aspect is clearly evident only in the postictal phase of the seizure. As noted earlier by Ajmone-Marsan and Ralston (1957), speech reappears usually unimpaired as soon as the ictal episode is over in patients with seizures of this type involving only the "nondominant" hemisphere, whereas impairment of speech persists longer in individuals with posterior neocortical TL seizure onset in the language-dominant hemisphere. Ictal episodes of pure aphasia are observed rarely (see Fig. 3). It has to be emphasized, however, that the establishment of a definite aphasic pattern in the course of such seizures is not always easy. Likewise, the interpretation of language impairments in the immediate postictal period is difficult due to their association with confusion, disorientation, and the possibility of secondary contralateral spread of seizure activity. Speech arrest should not be confounded with aphasia. Its appearance is more frequent but unfortunately of less localizing value.

Parietal signs and symptoms (i.e., features of Gerstmann's syndrome) might be prominent but are also difficult to assess during the ictal episode and, if present, usually can only be detected in the postictal period. The above-mentioned problems, however, can often be eased by careful interactive examination during the seizure episode, as pointed out later in this paper.

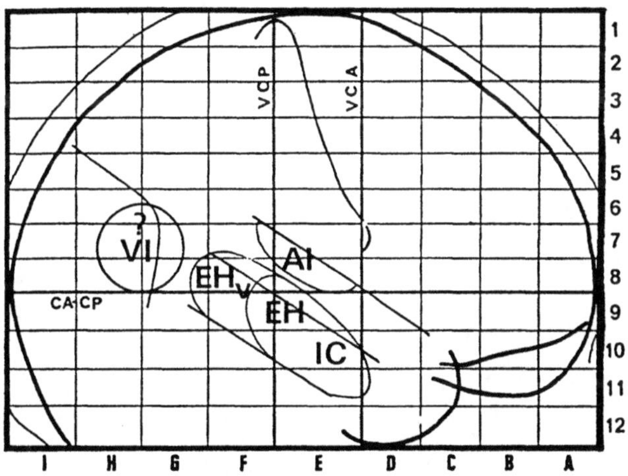

Fig. 2. Summary of the data on "psychical symptomatology." The origin of the seizure discharges during these different manifestations is schematically depicted. *AI*, auditory illusions; *VI*, visual illusions; *IC*, illusions of comparison; *EH*, polymodal experiential hallucinations; *EHv*, visual experiential hallucinations; *VCA*, linea verticalis anterior; *VCP*, linea verticalis posterior; *CA-CP*, bicommissural line (from Bancaud 1981)

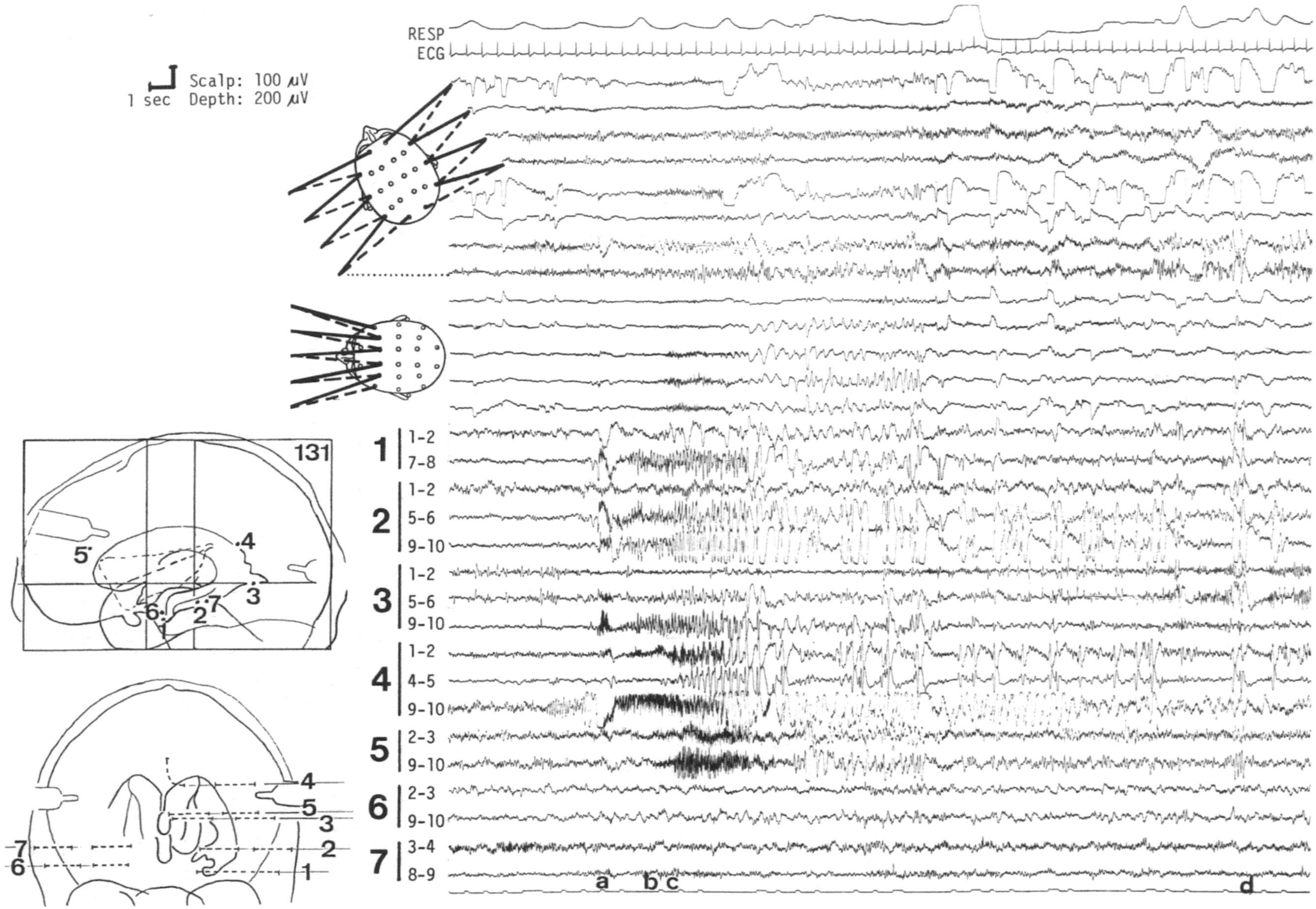

Fig. 3. Combined scalp and depth EEG-recorded aphasic seizure with left lateral posterior seizure onset (4/9–10, close to Wernicke's area). Note early involvement of the posterior cingulate cortex (4/1–2, interhemispherical) and secondary spread to the homolateral frontal cortex (5/9–10). This seizure was accompanied by *(a)* speech arrest and inability to speak, *(b)* flush, and *(c)* tonic contraction of the right corner of the mouth. Language reappeared 30 s after seizure onset *(d)*. (See also Fig. 4.) For electrode positioning *(large numerals)* see *brain map;* each electrode has 10 contacts *(small numerals)*, numbered from inside-out

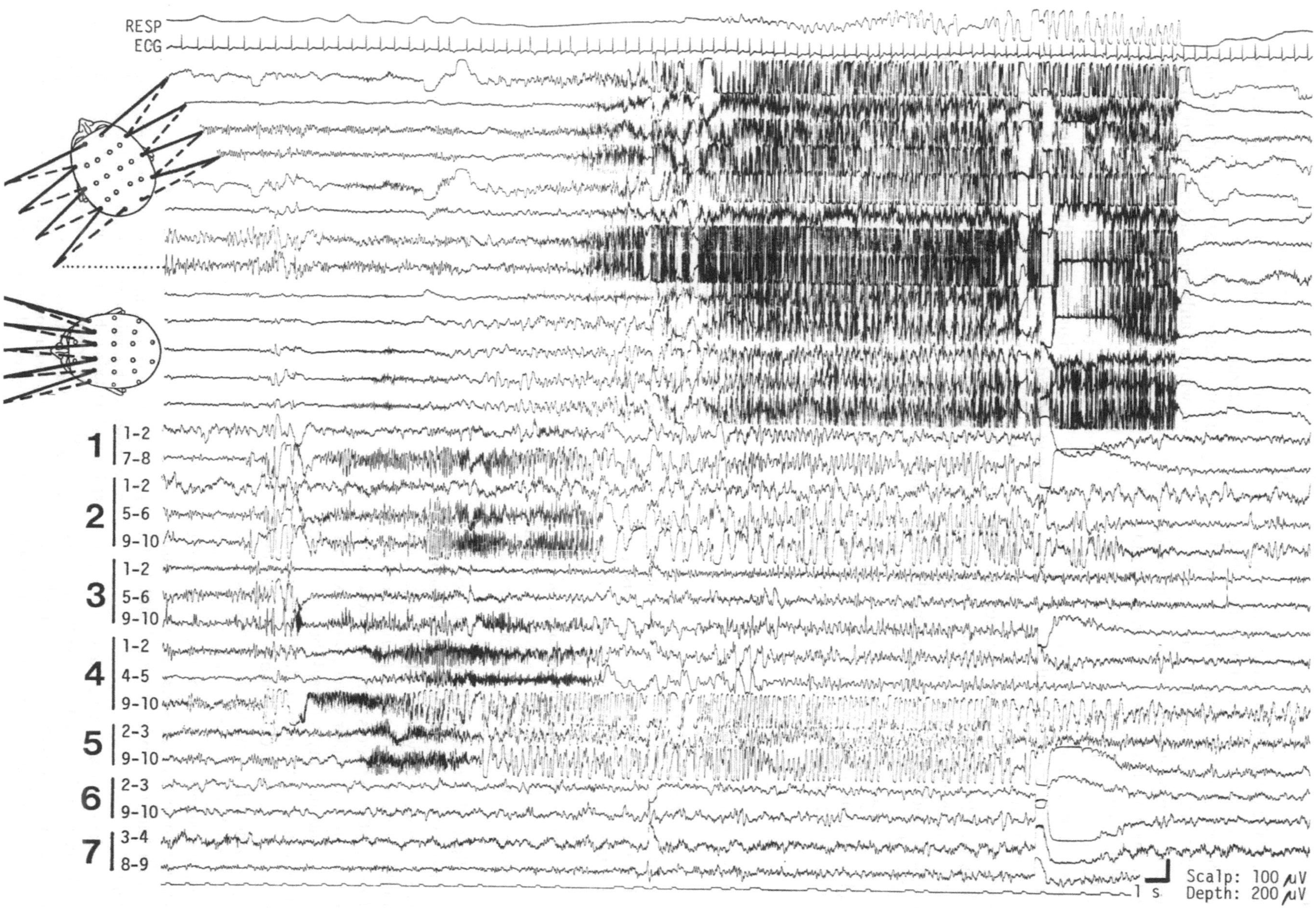

Fig. 4. Compared with the seizure illustrated in Fig. 3, this attack from the same patient shows a clinically identical and electroencephalographically very similar, but prolonged, initial seizure stage. However, it is followed by a second stage with pronounced tonic-clonic motor phenomena, which predominate at the right face and upper extremity. This is indicated by the coinciding muscle activity in the scalp EEG. Aphasia persisted about 2 min into the postictal period (for position of depth electrodes see *brain map* in Fig. 3)

Propagation

With respect to the propagation of temporal neocortical posterior seizures, two main routes have been observed. The more common one is directed to the anterior parts of TL with secondary involvement of homolateral mediobasal-limbic structures. The other is directed to the homolateral parietal cortex (including the posterior cingulate cortex) and then secondarily to the frontal lobe (see Fig. 1).

In an attempt to explain the preferential propagation routes, we may remark that the main sensory areas of the neocortex project to the parietal cortex and frontal cortex in a stepwise manner (Powell 1973). These parietal and frontal regions are interconnected and ultimately project toward the anterior cingulate cortex and to the parahippocampal gyrus (area entorhinalis). This sensory convergence on the parietal regions might serve to explain both the preferential propagation paths and the clinical seizure concomitants of often polymodal experiential hallucinations.

Illustrative Case Reports

Before discussing the specific problems encountered with this posterior neocortical temporal seizure type we would like to give a few examples. The first patient (Figs. 3 and 4) was chosen to exemplify the remarkable intraindividual similarity of seizure onset and early seizure stages and the importance of interactive examination of a patient to detect aphasia within a seizure predominated by secondary motor phenomena.

Figure 3 illustrates an almost pure aphasic episode, and Fig. 4 shows the more elaborated habitual seizure type with secondary motor phenomena. Both attacks in this 31-year-old female suffering from frequent drug-resistant seizures with rare secondary generalization show a similar onset pattern with (a) speech arrest and inability to speak but preserved reagibility to verbal commands, (b) flush and/or pallor, and (c) tonic or tonic-clonic motor phenomena of the right corner of the mouth. If seizures were of the 'stronger type' (as illustrated in Fig. 4), deviation of the eyes and sometimes also of the head to the right occurred. In such seizures the patient obviously suffered from fear and language was disturbed for up to 5 min.

Figure 5 - taken from another patient with similar neocortical posterior TL seizures - serves to illustrate the stereotypical electroencephalographic seizure pattern. This patient with up to 15 seizures daily exhibited a marked verbal memory deficit in association with anomia when interictally tested. The degree of anomia correlated with the frequency and severity of seizures.

Figure 6 illustrates a true vestibular hallucination associated with a fairly well circumscribed, short-lived discharge in the left temporoparietal cortex. (For a more detailed description of this case see Wieser 1982.)

Figure 7 illustrates auditory phenomena associated with seizure origin close to the right Heschl's gyrus and secondary propagation to mediobasal-limbic structures. This case was reported in detail elsewhere (Wieser 1980) because this young female entered a psychomotor status epilepticus with accompanying musical hallucinations, besides other symptoms.

Fig. 5 A–D. Four sections illustrating the stereotyped seizure ▶ onset pattern of this patient with left neocortical posterior TL epilepsy characterized by speech arrest and "absencelike" staring. (Note that only the relevant depth EEG channels are depicted and that the others did not show any ictal signals.) Recording is against a common average reference (intracerebral contacts situated in the white matter)

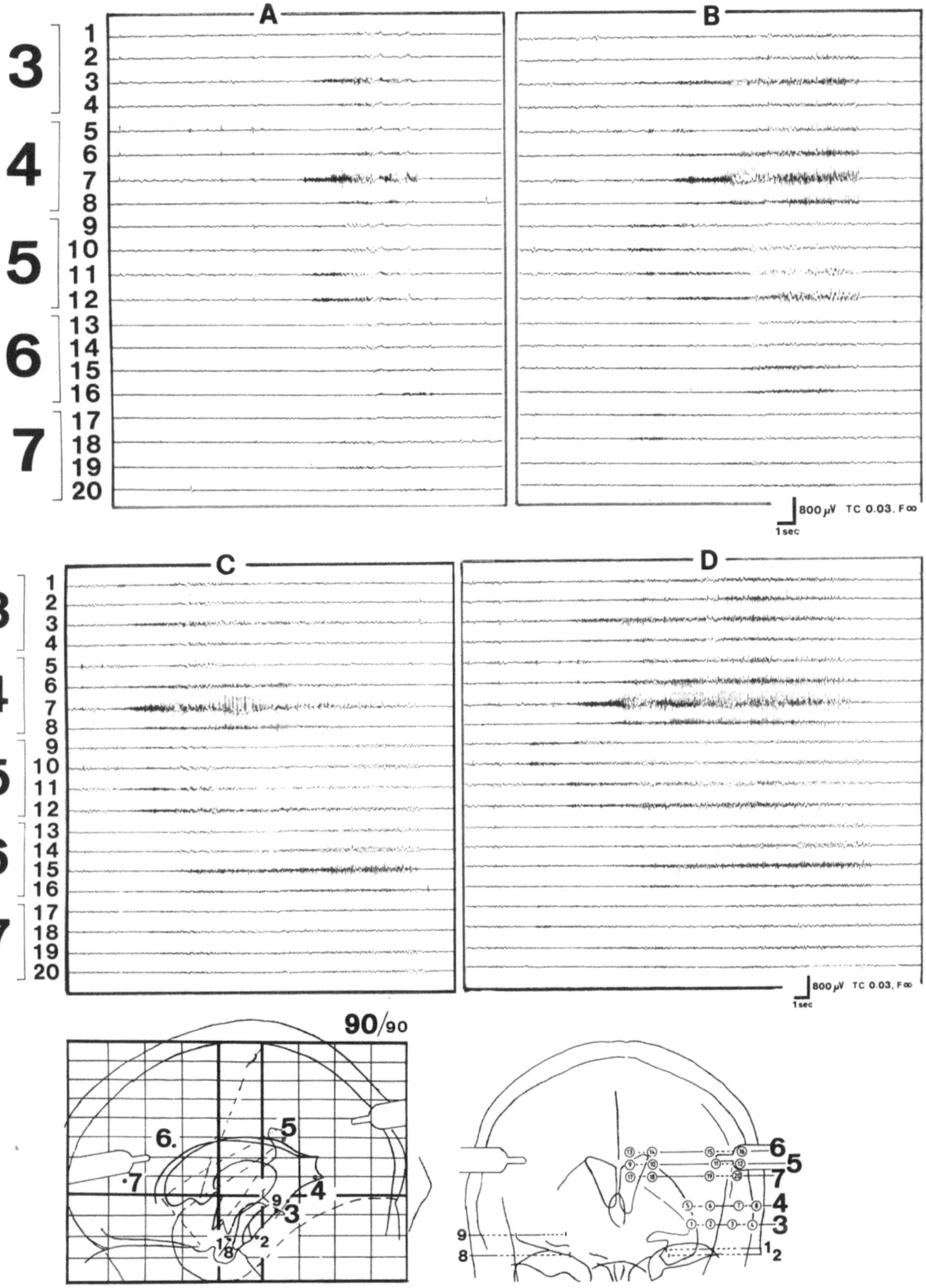
A
B
C
D
3
4
5
6
7
1
2
3
4
5
6
7
8
9
10
11
12
13
14
15
16
17
18
19
20
800 µV TC 0.03, F∞
1 sec
90/90
6.
5
·7
4
9
3
1
8
2
6
5
7
4
3
9
8
1
2

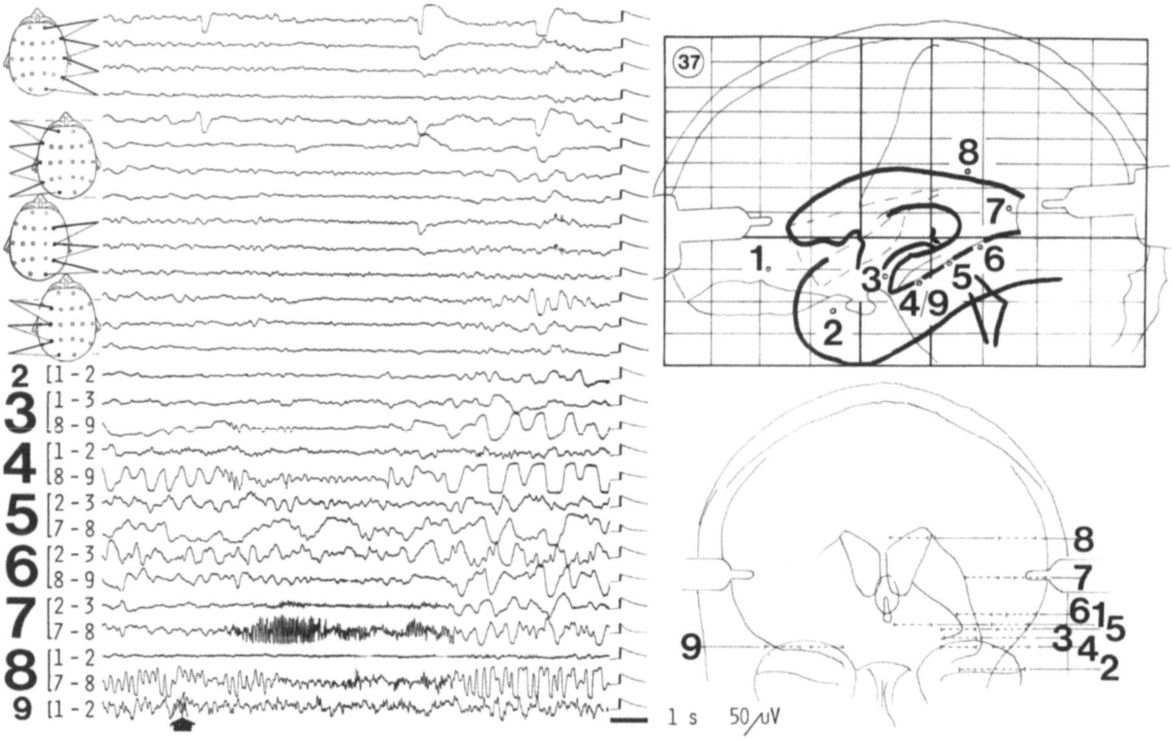

Fig. 6. Combined surface and depth EEG recordings during a 6-s spontaneous epileptic discharge in the left posterior temporal neocortex, accompanied by a vertiginous sensation pulling the patient to the right. When he reported it *(arrow)*, the recorded EEG at the contacts of electrode 7 already showed flat activity, but the high-frequency epileptic discharge was visible only a little bit later. The parietal electrode (8/7-8) is only moderately affected by the discharge, with a latency of about 2 s (from Wieser 1982)

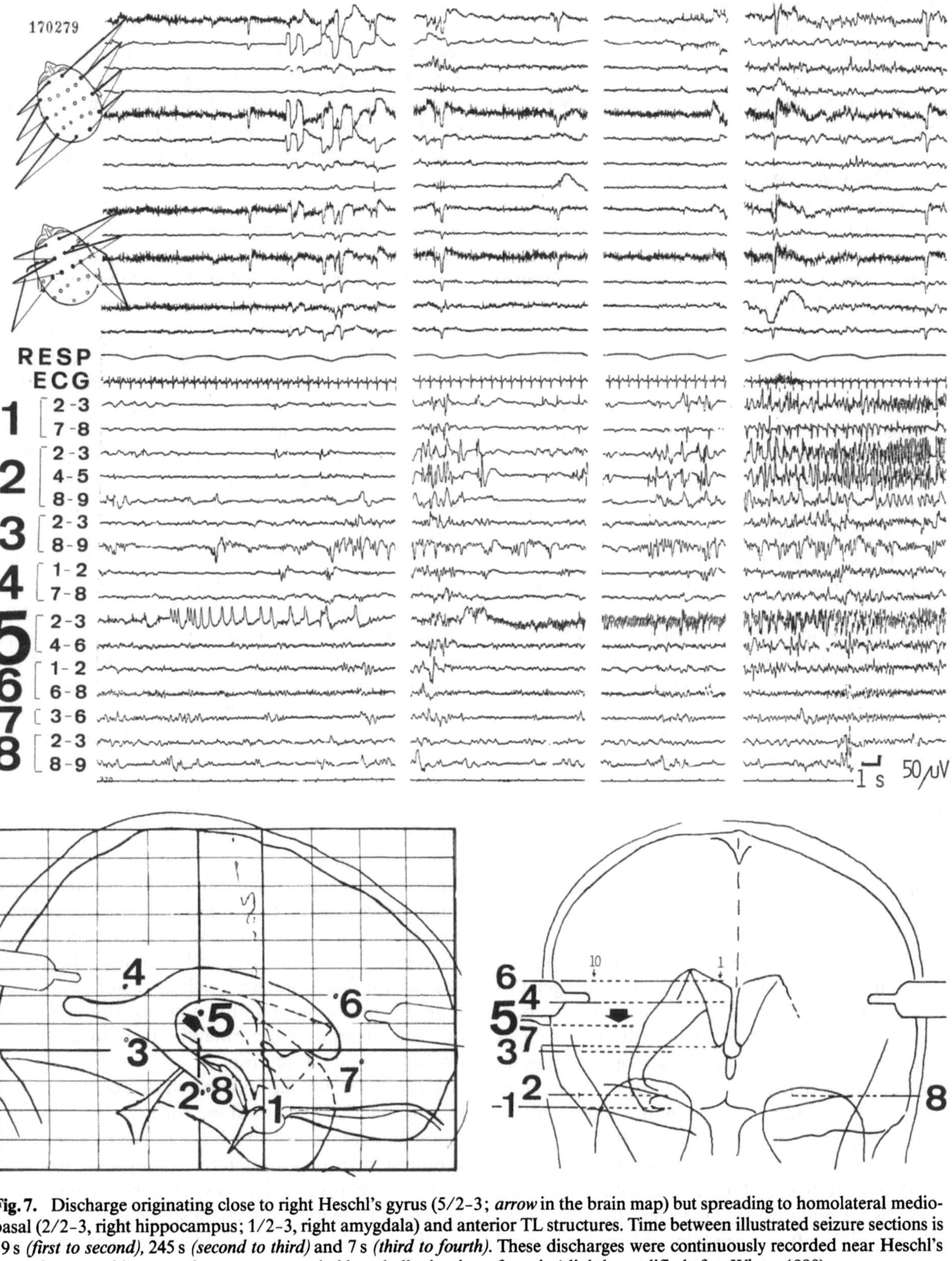

Fig. 7. Discharge originating close to right Heschl's gyrus (5/2-3; *arrow* in the brain map) but spreading to homolateral mediobasal (2/2-3, right hippocampus; 1/2-3, right amygdala) and anterior TL structures. Time between illustrated seizure sections is 19 s *(first to second)*, 245 s *(second to third)* and 7 s *(third to fourth)*. These discharges were continuously recorded near Heschl's gyrus for several hours and were accompanied by a hallucination of music (slightly modified after Wieser 1980)

Specific Problems Encountered with the Lateral Neocortical Posterior Seizure Type of TL Origin

Lateral Temporal Versus Mediobasal-Limbic Seizure Onset

Clinical Semiology. Although most authors recognize that epileptiform discharges of different sectors within the TL will result in specific clinical manifestations, many experts are not convinced that these subtle differences in symptoms justify the creation of a nosologic subclassification. They argue that such an attempt would be based on a relatively small number of - admittedly extensively evaluated - cases, and furthermore, that there is no convincing evidence that "surgical ablation of any of these sectors has a lasting beneficial effect on the seizures" (Ajmone-Marsan 1987). While the first argument is acceptable, at least to a certain degree, the second is no longer valid. The reported results of selective amygdalohippocampectomy (Wieser 1986) clearly favor the possibility of identifying certain subtypes of TL seizures with important surgical consequences.

It is true that localization on basis of ictal behavior alone is rarely sufficient to recommend selective and tailored resective surgery. Localizational hints on the basis of ictal behavior usually refer to involvement of particular functional anatomic systems rather than precise localization to a specific cerebral site (Engel 1987). It is, of course, also true that in many instances the epileptogenic tissue is not limited to a small area.

Imaging Techniques. Large epileptogenic areas are indeed frequently assumed if modern functional imaging techniques, such as positron emission tomography (PET) and single photon emmission computed tomography (SPECT), are consulted (see Engel, this volume; Fig. 8). Since SPECT measures regional cerebral blood flow (rCBF) indirectly, and rCBF is closely linked to local metabolic activity, it is reasonable to assume that SPECT mapping or rCBF alterations reflect underlying brain metabolic disturbances. Regions with diminished CBF may correspond to zones of hypometabolism seen with PET-FDG studies and correlate with underlying pathological brain tissue. However, the zone of diminished rCBF often exceeds the extent of structural lesions noted with CT; a finding which is similar to that with PET-FDG studies (Sperling et al. 1987). This cannot be explained just by the relatively low spatial resolution of the method per se and probably reflects the secondary functional involvement of brain areas anatomically linked to the seizure-inducing structure. Thus, while the lateralizing and regionally localizing potential of these modern functional imaging studies are undisputed, there remain many open questions as to the exact functional and localizational correlation with the seizure-originating sites.

Electrical Data. In comparison with EEG seizure onset localization, interictal spike activity more often leads to the impression that the epileptogenic area consists of a large volume. Moreover, using the interictal spiking, the lateral neocortical temporal cortex seems to be the site of spiking in the majority of cases. Electrocorticography (ECoG) relies essentially on interictal data. Therefore the findings of Rasmussen (1982) are not unexpected: Rasmussen reported that in 100 cases of TL epilepsy seizure-free postoperatively, 46% showed ECoG spiking predominately from the lateral surface, 28% predominately from depth electrodes in the medial TL, and 27% from both. Furthermore, ECoG localization depends on an accurate prior localization. It is particularly well suited for delineating the extent of the epileptogenic cortex as defined by the "spiking area," but - except in rare cases where seizures can be recorded intraoperatively - it is incapable of determining which parts of this abnormal cortex are actually involved in generation of the habitual seizures. On the other hand it is obvious that the sampling problems, inherent in all invasive recording techniques in different manners, are - at least to a certain degree - the cause of quite a large diversity of opinions concerning the question of the prevalence of 'lateral vs mediobasal seizure onset.' Subdural grids and strips might provide better recordings from the lateral neocortex than can be obtained with stereoelectroencephalography (SEEG). The grid technique, however, has several disadvantages, such as requiring craniotomy for insertion, and often inadequate identification of epileptiform abnormalities from medial, particularly limbic, structures.

It must therefore be anticipated that centers which rely on interictal spikes alone for localizing an epileptogenic lesion and which use either ECoG or the subdural strip or grids preferentially diagnose the lateral neocortical epilepsies; centers relying more on ictal information and using stereotactic depth recording would not miss mediobasal seizures. Our work falls in the second group and thus we can confidently discuss lateral neocortical seizure onset type, as mediobasal-limbic structures are almost always adequately monitored by SEEG.

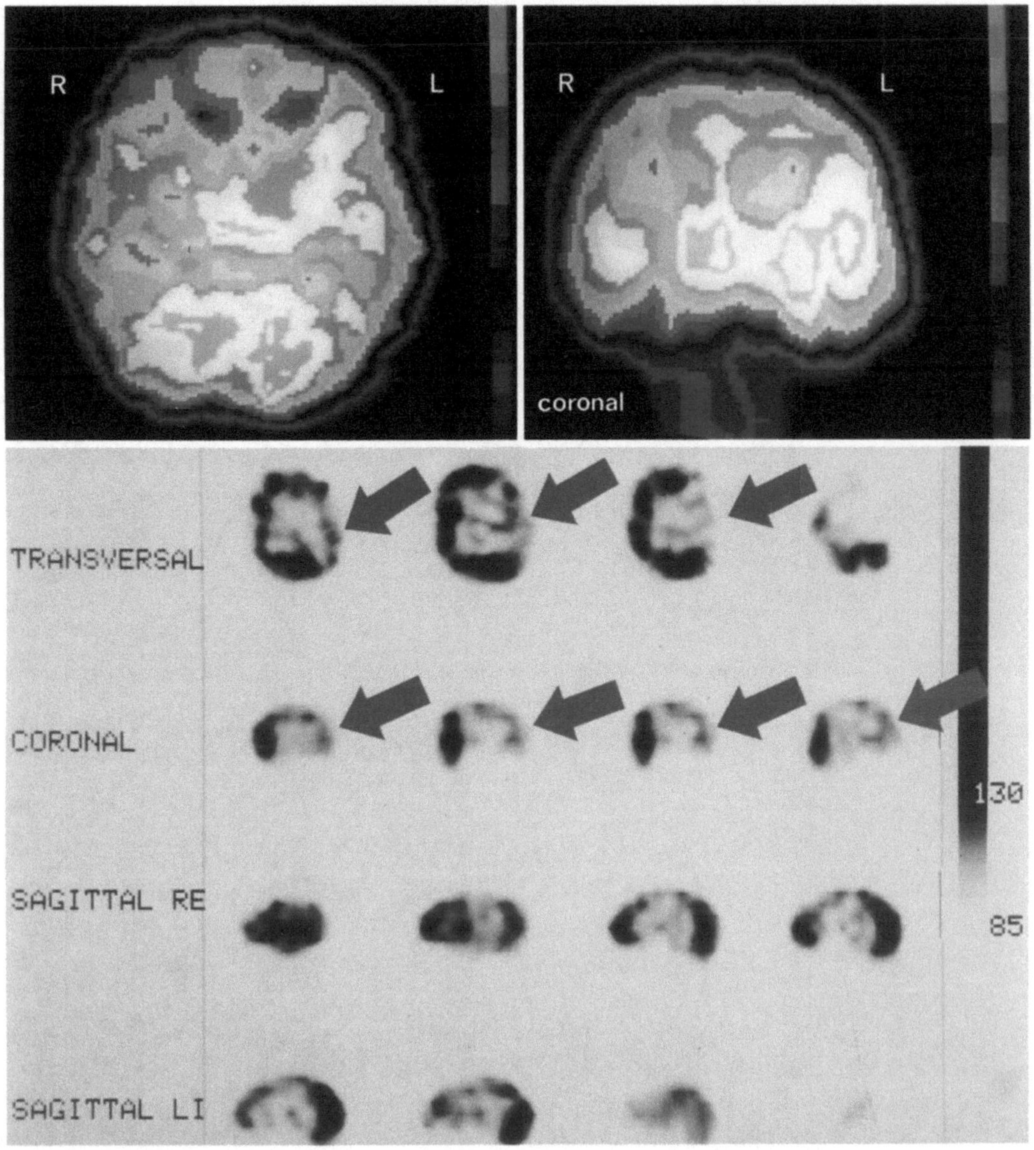

Fig. 8. **a** (HMPAO)-SPECT (800 MBq [^{99m}Tc] hexamethylpropylenamineoxime) of a patient with a well-defined right mediobasal-limbic seizure onset, but widely distributed right TL interictal spiking. Note the large hypoactive area. (Courtesy of Dr. A. Gadže, Zürich) **b** SPECT (^{123}I-N-isopropyl-iodoamphetamine) of a patient with CPS and a large left hemispheric hypoactivity. (Courtesy of Prof. D. Janz, Berlin)

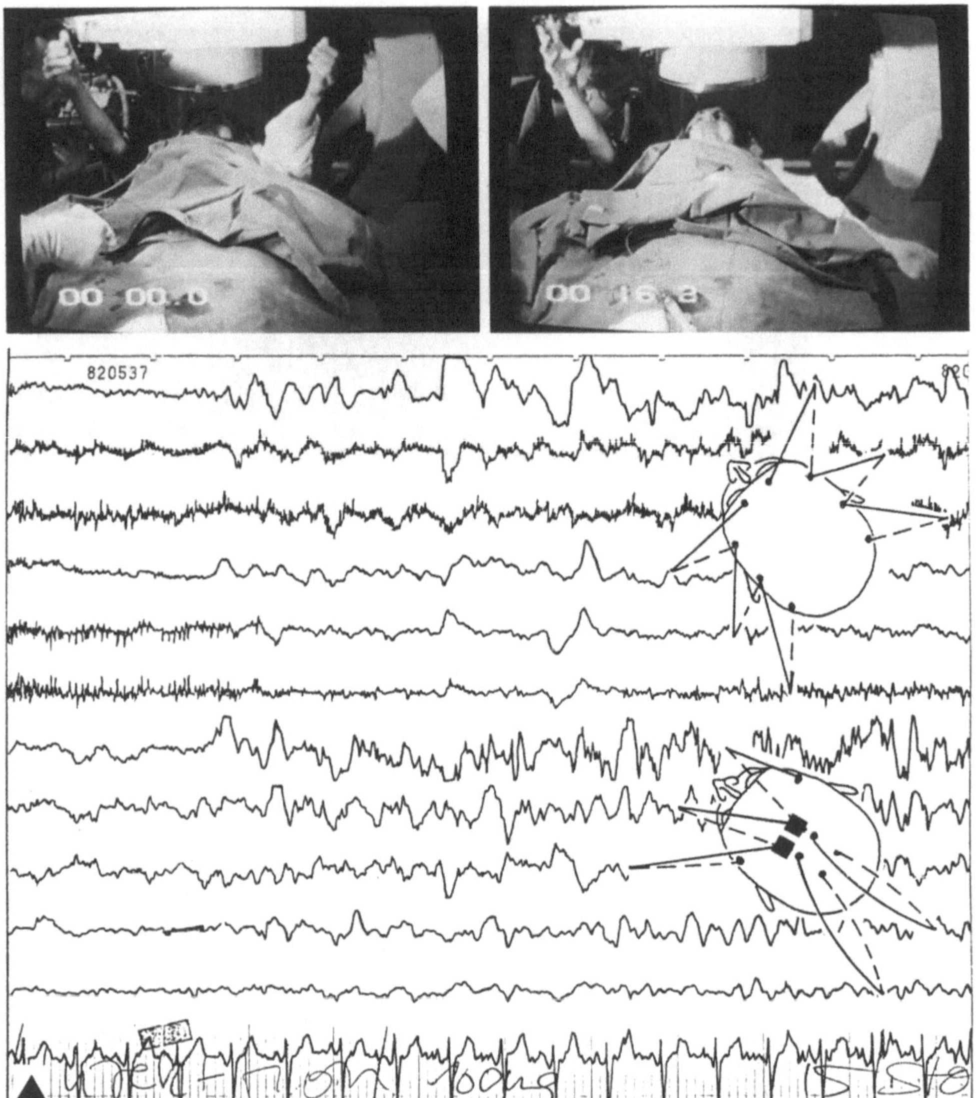

Inject.right
100 mg Amobarb.

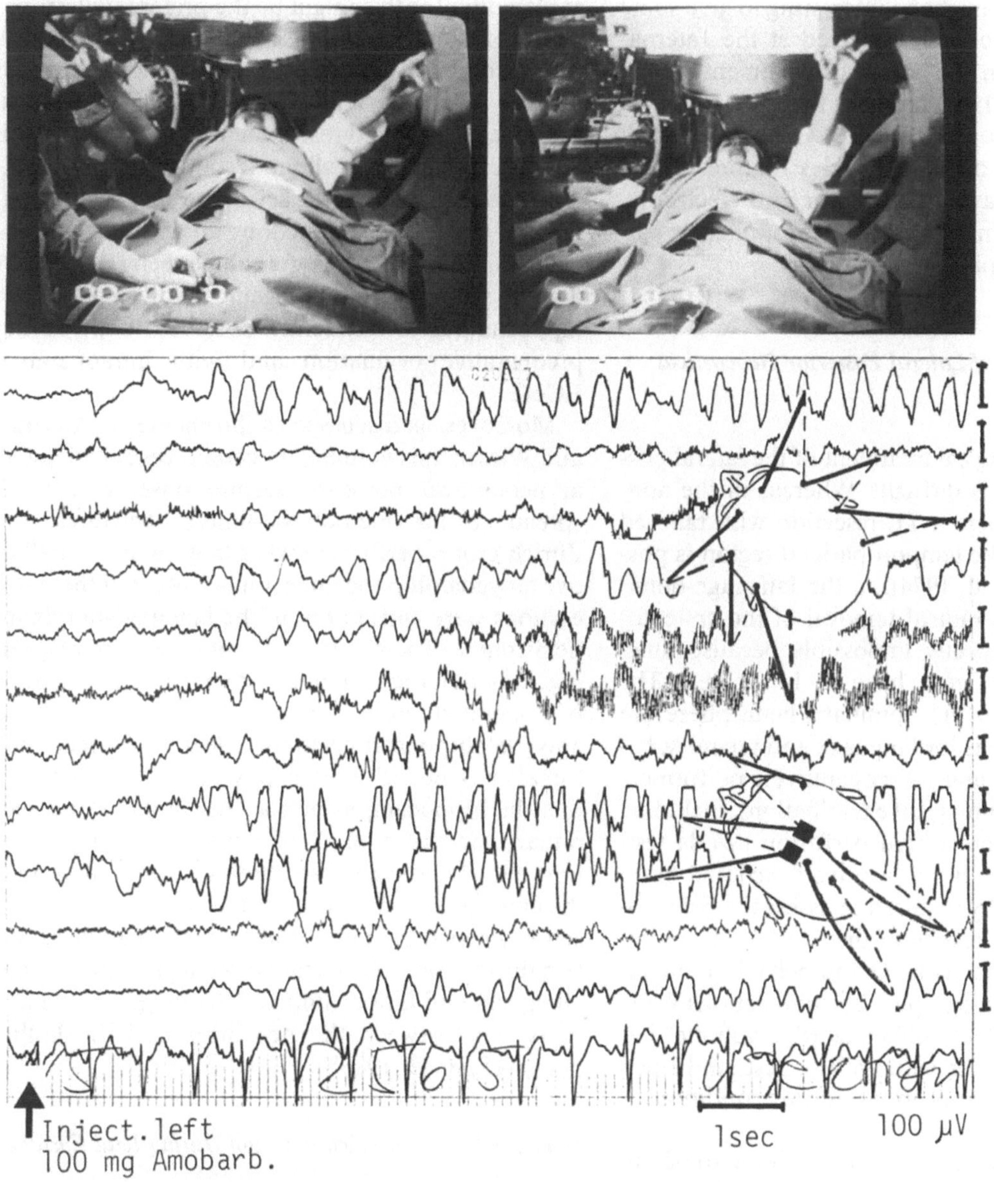

Fig. 9. Illustration of the monitoring situation during the Wada test. The injected amobarbital causes a complete contralateral hemiplegia and slow wave activity in the EEG recorded from scalp and bilaterally inserted foramen ovale electrodes (■; see chapter by Elger and Wieser, this volume). Note the slight difference in the amount of slow wave activity following right intracarotid (p. 262) and left intracarotid (p. 263) amobarbital injection. Angiographic controls did not show any significant flow of the contrast medium to the contralateral hemisphere. Although the neuropsychological results obtained in this patient could be interpreted in favor of the intended right TL resection (global aphasia after left, unimpaired speech after right injection; severe memory loss after left, no significant memory disturbance after right injection of 100 mg amobarbital) in more difficult cases such EEG-differences have to be carefully considered

In this context it might be interesting to give some data which were recently collected at the International Congress on the Surgical Treatment of the Epilepsies. Engel (1987) reported that 84% of the 44 responding centers carry out ECoG in some or all patients; 59% use depth electrodes for presurgical evaluation; 27% have used strip or grid electrodes, but only 2 use them regularly for most of their patients undergoing epilepsy surgery.

Surgical Treatment of Lateral Posterior Neocortical Seizures

Assessment. The surgical treatment of the lateral posterior seizure type is difficult. Whereas in the nondominant hemisphere a TL resection with tailored enlargement into the temporoparietal region is possible (Talairach et al. 1974), in the language-dominant hemisphere a radical removal of the epileptogenic zone is frequently impossible because functionally important areas have to be spared. The functions at risk in the dominant hemisphere are those related to areas for language and memory, besides sensorimotor areas. Preoperatively performed, specific neuropsychological examinations, including the intracarotid amobarbital perfusion (Wada test) and sometimes also intraoperative functional brain mapping, are usually required to assess the risks. Interpretation of the results of the Wada test are improved by simultaneous EEG and behavioral monitoring as well as angiographic documentation (Wada and Rasmussen 1960). Silfvenius (this volume) has discussed in depth the value of this test, and Fig. 9 gives an example of appropriate testing conditions.

Extent of Resection. The other precondition to be fulfilled for surgery to be as radical as possible while avoiding enchroachment into functionally important areas is to employ the 'functional mapping technique' either by using large epidurally placed grids preoperatively (Lüders et al. 1985), or by intraoperative stimulation mapping associated with ECoG (Penfield and Jasper 1954). This is extensively dealt with in the chapter of Ojemann (this volume) and Ojemann and Engel (1987).

Another important question in such cases with clear-cut lateral neocortical seizure onset is whether the resection should be restricted to the lateral TL, or whether the mediobasal TL structures, including the hippocampal formation, should be resected as well. It is our impression that including the mediobasal-limbic structures improves the results, especially with a view towards long-term seizure control. One explanation for this might be the preferential propagation of lateral posterior seizure discharges towards the mediobasal-limbic structures and the often marked secondary epileptogenesis of these regions. Thus we recommend resection of the mediobasal TL in those patients in whom secondary mediobasal-limbic involvement in the seizure event is well documented and memory functions do not depend critically on these structures. As can be seen from Figs. 7 and 10, secondary epileptogenic involvement of the hippocampus is frequently observed, both during preoperative evaluation and with intraoperative ECoG.

Moreover, as discussed in the chapter of Yaşargil and Wieser (this volume), in certain cases with lateral neocortical posterior seizure onset and rapid spread to the mediobasal-limbic structures, the Zürich group performed what they called a "palliative amygdalohippocampectomy." Most of these operations were carried out in the language-dominant hemisphere. The results so far obtained are rewarding, although they are not as good as in the "causally" operated group (Wieser 1986). The selective amygdalohippocampectomy clearly reflects our "smaller is better" philosophy and our hypothesis that the hippocampus acts as an amplifier and secondary pacemaker of TL seizures. Animal degeneration and 2-deoxyglucose studies indeed show that the entorhinal cortex receives the convergence of the sensory neocortex and is the primary projection system during some TL seizure discharges into the dentate gyrus and hippocampus. Therefore, it is the primary link between the neocortex and the limbic system (Spencer 1986).

Need for Interactive Examination During Ictal Episodes

A last point should be emphasized in the context of the evaluation of patients suspected of suffering from this seizure type, namely the indispensible need for interactive examination during ictal episodes. As we are dealing with the highest cortical functions of the temporal and parietal lobes, i.e., with complex, usually polymodal experiential hallucinations and with memory and language in all its permutations of speaking, listening, reading, and writing, even the most sophisticated EEG and behavioral video monitoring cannot substitute for the skillful interactive examination of a patient during the entire seizure episode. If the recording conditions are deprived of this direct interactive examination, essential information required for the lasting benefit of patients who are candidates for surgical resection as a treatment for epilepsy, may be lost.

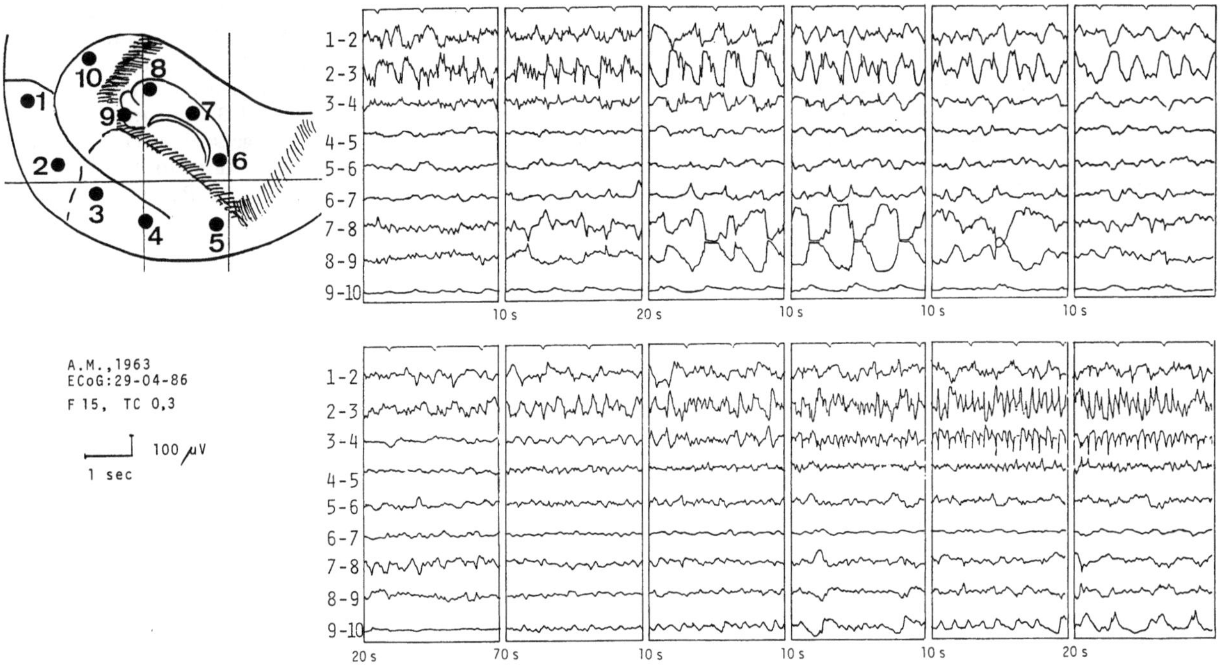

Fig. 10. Electrocorticogram of a right opercular and supra-Sylvian epilepsy case showing the build-up of spontaneous seizure events at supra-Sylvian frontal sites *(electrode 2, upper part; electrode 3, lower part)* associated with clonic motor phenomena of the left face and hand. Note the secondary spread of seizure discharges to the hippocampus *(electrode 8, upper part)*, which shows the build-up of high-amplitude, paroxysmal, slow-wave activity associated with sharp elements. This young male had suffered from CPS with pronounced left tonic-clonic motor phenomena. Despite this symptomatology and prior to his admission in Zürich, surgery had been performed elsewhere, removing parts of the lateral temporal lobe (resection indicated by *hatching*). This operation had been guided by CT and MRI findings which had suggested a low-grade TL astrocytoma. Histologically no tumor was found but a marked neuronal loss, gliosis, and the situation of the patient did not improve following the first operation. At the reoperation (consisting of removal of the insular and supra-Sylvian frontal cortex as well as the remainder of the mediobasal anterior TL) glial-neuronal hamartoma was histologically diagnosed

References

Ajmone-Marsan C (1987) Commentary: clinical characteristics of partial seizures. In: Engel J Jr (ed) Surgical treatment of the epilepsies. Raven, New York, pp 121-127

Ajmone-Marsan C, Ralston BL (1957) The epileptic seizure. Its functional morphology and diagnostic significance. CC Thomas, Springfield

Bancaud J (1981) Epileptic attacks of temporal lobe origin. Jpn J EEG-EMG (Suppl Didactic Lectures): Tenth international Congress of EEG and clinical neurophysiology, September 13-18, Kyoto, pp 61-71

Bancaud J, Talairach J, Bonis A, Schaub G, Szikla G, Morel P, Bordas-Ferer M (1965) La stéréo-électroencéphalographie dans l'épilepsie. Masson, Paris

Delgado-Escueta AV, Walsh GO (1985) Type I complex partial seizures of hippocampal origin: excellent results of anterior temporal lobectomy. Neurology 35: 143-154

Engel J Jr (1987) Approaches to localization of the epileptogenic lesion. In: Engel J Jr (ed) Surgical treatment of the epilepsies. Raven, New York, pp 75-95

Lüders H, Lesser R, Hahn J, Dinner D, Morris H, Harrison M (1985) Global aphasia elicited by stimulation of the dominant fusiform gyrus. J Neurol 232 (suppl): 212

Ojemann GA, Engel J Jr (1987) Acute and chronic intracranial recording and stimulation. In: Engel J Jr (ed) Surgical treatment of the epilepsies. Raven, New York, pp 263-288

Penfield W, Jasper HH (1954) Epilepsy and the functional anatomy of the human brain. Little, Brown, Boston

Powell TPS (1973) Sensory convergence in the cerebral cortex. In: Laitinen LV, Livingstone KE (eds) Surgical approaches in psychiatry. MTP, Lancester, pp 266-281

Rasmussen T (1982) Localizational aspects of epileptic seizure phenomena. In: Thompson RA, Green JR (eds) New perspectives in cerebral localization. Raven Press, New York, pp 177-203

Spencer DD (1987) Postscript summary: should there be a surgical treatment of choice, and if so, how should it be determined? In: Engel J Jr (ed) Surgical treatment of the epilepsies. Raven, New York, pp 477-484

Sperling MR, Sutherling WW, Nuwer MR (1987) New techniques for evaluating patients for epilepsy surgery. In: Engel J Jr (ed) Surgical treatment of the epilepsies. Raven, New York, pp 235-257

Talairach J, Bancaud J, Szikla G, Bonis A, Geier S, Vedrenne C (1974) Approche nouvelle de la neurochirurgie de l'épilepsie. Neurochirurgie 20 (suppl 1): 1-240, Masson, Paris

Wada J, Rasmussen T (1960) Intracarotid injection of sodium amytal for the lateralization of cerebral speech dominance: experimental and clinical observations. J Neurosurg 17: 266-282

Walsh GO, Delgado-Escueta AV (1984) Type II complex partial seizures: poor results of anterior temporal lobectomy. Neurology 34: 1-13

Wieser HG (1980) Temporal lobe or psychomotor status epilepticus. A case report. Electroencephalogr Clin Neurophysiol 48: 558-572
Wieser HG (1982) Zur Frage der lokalisatorischen Bedeutung epileptischer Halluzinationen. In: Karbowski K (ed) Halluzinationen bei Epilepsien und ihre Differentialdiagnose. Huber, Berne, pp 67-92
Wieser HG (1983) Electroclinical features of the psychomotor seizure. G. Fischer/Butterworths, Stuttgart/London
Wieser HG (1986) Selective amygdalohippocampectomy: indications, investigative technique and results. In: Symon L, Brihaye J, Guidetti B, Loew F, Miller JD, Nornes H, Pasztor E, Pertuiset B, Yaşargil MG (eds) Advances and technical standards in neurosurgery 13: 39-133, Springer, Vienna
Williamson PD, Wieser HG, Delgado-Escueta AV (1987) Clinical characteristics of partial seizures. In: Engel J Jr (ed) Surgical treatment of the epilepsies. Raven, New York, pp 101-120

Complex Partial Seizures of Frontal Lobe Origin

A. V. Delgado-Escueta[1, 4], B. E. Swartz[1, 4], H. M. Maldonado[1, 4], G. O. Walsh[1], R. W. Rand[2], and E. Halgren[3, 4]

Introduction

Such as are habituated to their disease have a presentiment when an attack is imminent and run away from men, home, if their house be near, if not, to the most deserted spot...

Hippocrates

Along with this voluminous mental state, there is frequently a "crude sensation" ("warning") of (a) smell or (b) taste; (or, when there is no taste, there may be movements, chewing, tasting, spitting).

J. H. Jackson

As early as 400 years B.C., Hippocrates [26] recognized the warning signs or symptoms (aura) of epilepsy. In 1677, Thomas Willis [59] gave the first description of a psychomotor attack, and in 1822 Prichard [42] first used the term "partial epilepsy." However, it was John Hughlings Jackson who developed the conceptual framework and meaning of epileptic symptomatology and signs. Jackson, a true scholar, completely refounded epileptology and collated bedside observations with anatomical facts. By 1889, he [24, 25] had linked the symptoms and signs of psychomotor seizures to the "sphenoidal lobe" or the medial temporal areas of the uncus. Since then, modern studies, notably by Ajmone-Marsan and Abraham [1-3], Bancaud, Talairach and collaborators in Paris [4-12, 20, 21, 31-33, 48, 53], Wieser and Yaşargil [52, 54, 55] in Zürich, the Spencers et al. [45, 46] and Williamson et al. [56-58] in New Haven, Rasmussen [44], Gloor [22, 23], and Quesney et al. [43] in Montreal, and our Comprehensive Epilepsy Program in Los Angeles [12-15, 34], have characterized a wide range of psychomotor automatisms on cine film or closed-circuit television videotape (CCTV) from simple lip smacking, chewing, pursing lips, and, eye blinking, to running, mimicry, gesticulations, and sometimes bizarre complex bilateral hand and leg movements. These clinical signs have been related to epileptiform discharges recorded by scalp and sphenoidal electroencephalography (EEG) and stereoelectroencephalography (SEEG).

In the following communication, we review complex partial seizures of frontal lobe origin (Tables 1 and 2). In our discussion, we contrast complex seizures of primary hippocampal-amygdalar epilepsy with psychomotor seizures which start from the frontal lobe and temporal neocortex and then secondarily invade the amygdalar-hippocampal regions. In preparing this review, the authors acknowledge that many of the concepts expounded here were long ago the surmise of John Hughlings Jackson [24, 25], Wilder Penfield, Herbert Jasper [37-40], Bancaud and Talairach [4-12, 20, 21, 31, 33, 48, 53] and Ajmone-Marsan and Ralston [1-3]. For these concepts, we do not claim originality. What are original are our observations on psychomotor attacks of extratemporal origin which start from brain regions that had been explored by these eminent predecessors, but whose somatomotor, postural, and contraversive head and eye movements are so subtle that psychomotor automatisms comprise their main event (Type II A complex partial seizures) [12-15, 51]. In addition, we describe gross, frenetic complex motor automatisms of both arms and legs at onset of lapse of consciousness as another form of complex partial seizures of possible frontal lobe origin (Type II B psychomotor attack) [12, 15, 51, 55-57].

Data Base and Methods

Collection of Data and Recording Epileptic Attacks

At the epilepsy unit, three to ten attacks of the habitual seizures of 319 patients were recorded on CCTV and EEG biotelemetry [12-15, 51]. Both split-screen analysis and a digital clock superimposed on videoimages and transcribed in code on the EEG

[1] Comprehensive Epilepsy Program, Neurology Department, UCLA School of Medicine, Los Angeles, CA 90024, USA
[2] Department of Surgery, UCLA Center for the Health Sciences, Los Angeles, CA 90024, USA
[3] Department of Psychiatry, UCLA Brain Research Institute, Los Angeles, CA 90024, USA
[4] VA Southwest Regional Epilepsy Center, Neurology and Research Services, West Los Angeles VA Medical Center (Wadsworth), Los Angeles, CA 90073, USA

Presurgical Evaluation of Epileptics
Edited by H. G. Wieser and C. E. Elger

were used for ascertaining time relations between the clinical attack and electrographic events. Bedside examinations, during at least one attack, by a nurse specialist in epilepsy or one of the authors helped determine the exact onset of alterations of consciousness. In our experience, the clinical signs of impaired consciousness are best demonstrated on CCTV by examining the patients' responses to pinprick, corneal stimulation, and deep pain as well as reaction times to varying command stimuli. Most patients were followed regularly and were familiar with the routine tests performed during prolonged electroencephalographic and videotape-monitoring sessions. They were familiar with methods of responding to the auditory monotone stimuli, such as pushing a response button immediately, delaying response by 3 s, or pushing the response button five times on command. They were also familiar with the test instructing them to turn off flashing lights of various colors, using an off-switch for each color, or reading instructions printed with large letters [12-15, 51].

During periods of prolonged cloudiness most patients were tested for apraxia, written and spoken language, identifying and naming objects, and right and left orientation (one attack at least). Responses to pinprick, corneal stimulation, and deep pain were tested every 1.5 to 3 min. Muscle tone, muscle stretch reflexes, and plantar responses were examined. In some cases amnesia during the attacks was verified when patients were shown the tape-recorded attacks after they had recovered. The duration of attacks was timed from the historically verified onset of altered consciousness to the total disappearance of amnesia.

A 16-channel Grass Model 6 or 8 EEG apparatus was used to record brain waves through 22 scalp electrodes (10-20 system) and two nasopharyngeal or sphenoidal electrodes [12-15]. In clinically indicated cases, multipolar depth electrodes were stereotactically implanted in the amygdala, hippocampus, orbital frontal lobe, cingulate gyrus, or supplementary motor cortex, following the coordinates specified by Talairach and Szikla [48]. Videotape recordings were obtained by mounting multiple cameras directed to the patient and the electroencephalogram. In some patients, black-and-white reproductions of 8-12 EEG channels and of the patient's entire body were displayed on a split-screen TV monitor for on-line viewing [12-15].

Data Analysis

During the past 15 years, all the videotaped attacks were independently analyzed second by second by two or three associates (most recently, H. Maldonado, A. Lee, and B. Swartz) in addition to the two senior epileptologists (G. Walsh and A. V. Delgado-Escueta) [12-15, 51]. Split-screen CCTV-EEG analysis of seizures was performed independently by viewing videotaped recordings of seizures at least ten times independently and without knowledge of the clinical history or electrographic recordings. Automatisms were described according to conventional classifications [1-3, 41]. Ictal electrographic activities recorded through depth and scalp electrodes were studied, also without knowledge of the clinical behavior, by at least three of the authors. To help analyze the clinical and EEG data second for second, the behavioral state of the patient recorded by the TV camera and displayed on the TV monitor (sometimes by split screen with telemetered EEG) was imprinted by a rapid sequence camera [29]. Polygraphic recordings of telemetered EEG were also reproduced by camera, and the digital time code, recorded on both the CCTV and telemetered EEG, was used to line up the clinical and EEG data second for second. Data were analyzed by correlating the onset of behavioral events on the CCTV with the onset of epileptogenic paroxysms in the electroencephalogram. In order to simplify analysis, the constellation of manifestations was divided into six categories:

1. An initial motionless stare during an arrest reaction was considered present when an abrupt fixed gaze occurred at the start of a complex partial seizure accompanied by total unresponsiveness and ictal rhythms in the SEEG or EEG. Complete arrest of motions had to be present. When the patient stopped moving at onset of seizures, but showed chewing, lip smacking, vocalizations, swallowing, or one single discrete movement, we classified the event under categories listed below for purposes of analysis.

2. Oroalimentary automatisms consist of repetitive lip movements, lip smacking, sucking, chewing, tongue movements, throat movements, and deglutition-like movements. They occurred during or after the arrest reaction at onset of seizures or immediately after onset, associated with other stereotyped automatisms. They are associated with total unresponsiveness and SEEG or EEG ictal rhythms.

3. A single discrete movement, involving a particular set of muscles, sometimes seemingly purposeless and in some instances purposeful, should be differentiated from focal movements which are repeti-

tive, localized jerk-like motions, and from perseverative stereotyped automatisms. Examples of a single discrete movement are placing a hand on a lap, putting a newspaper down, moving an audiocassette, pushing a dinner tray away. They should also be differentiated from changes in posture and from bizarre complex automatisms observed in extratemporal epilepsy, i.e., bicycling, bipedal motions or fighting movements. Single, discrete acts may initiate complex partial seizures, precede or occur right at the beginning of an arrest reaction, or follow oroalimentary automatisms. They are associated with abnormal EEG or SEEG changes.

4. Perseverative stereotyped automatisms are a series of stereotyped motions peculiar to each patient, such as repetitive mimicry or gestures, e.g., karate posture, shuffling cards over and over, "pulling in a catch" as in fishing, repeated rubbing of thighs, or running one leg along the shin bone over and over again. Such automatisms occur at the height of ictal epileptiform paroxysms.

5. Reactive automatisms are complex quasipurposeful behavior which the patient would normally display when physically located in a particular environment. These behaviors are different from one attack to the other. They consist of acts, such as drinking a glass of water, going to the bathroom to urinate, or dressing, undressing, etc. They may or may not be accompanied by ictal paroxysms in the SEEG or EEG. Frequently, the surface scalp, sphenoidal, and depth electrodes show slow and low voltage activities. Amnesic responsiveness is usually present.

6. Vocalization is defined as any type of comprehensible or incomprehensible sound, such as intelligent formed speech or mere utterances. They are accompanied by total or partial lack of responsiveness and ictal paroxysms in the SEEG or EEG.

The results of neurologic examination during at least one attack aid in the interpretation of the many attacks recorded on CCTV. Responsiveness was determined by the patient's capability to follow oral commands, to respond orally and accurately to verbal questions, and to respond to pinprick, deep pain, and touch stimulation. It is particularly important to show total unresponsiveness in determining onset of impaired consciousness when the patient lies motionless in bed watching television and electrographic seizures start in the EEG. Amnesic responsiveness may be characterized by the lack of intelligent oral response to simple questions, such as queries about name, place, and time. He may respond by motor acts in response to simple oral commands, e.g., "hold my hand," etc. In some instances, the patient may even give correct answers to complex mathematics and proverbs, but he will be unable to remember his actions when fully recovered from the seizure. The initial motionless stare during an arrest reaction is easily differentiated from the complex bilateral arm and leg automatisms by viewing the motions on CCTV but the examination of the patient is crucial in documenting impaired consciousness [12-15, 29, 51].

Reconstruction of the Psychomotor Attack

The duration of each of the behavioral features recorded on CCTV was measured, and the sequence of these behavioral phenomena was then reconstructed for each attack for each patient and according to their site of origin [29]. To complete the reconstruction of each psychomotor attack, the time of onset of behavioral, SEEG, and EEG changes were entered into an Apple II computer for analysis. Latency between clinical onset and the start of depth discharges was measured. Duration and sequence of behavioral phenomena were related with the onset and spread of electrographic paroxysms. The electrobehavioral features of the psychomotor attack were then compared to the habitual attacks elicited in the history and related to the initial warning signs or symptoms (aura) [29].

Results and Discussion

We evaluated 319 patients with uncontrolled seizures over a span of 15 years. Sixty-five percent were men and 35% were women, and their ages ranged from 22 to 46 years. Uncontrolled seizures consisted of recurring lapses of consciousness with automatisms. Almost all had interictal spikes and sharp waves in nasopharyngeal (NP) or sphenoidal (SP) electrodes during EEG. History further revealed that 82% had one or more secondarily generalized tonic-clonic seizures. Thus, 18% of patients had complex partial seizures only. A total of 1525 seizures of these 319 patients were recorded on CCTV-EEG. These consisted of 1350 complex partial seizures[5], and 74 complex partial seizures that evolved to tonic-clonic seizures (314 patients). In the remaining five patients, no complex partial seizures were captured. Instead, five secondary generalized tonic-clonic sei-

[5] All seizures recorded on CCTV-EEG are available for complete study. The reasons for limiting the number of cases for specific studies are purely of practical nature.

zures and 96 partial elementary seizures were recorded on the CCTV and EEG.

Figure 1 illustrates the age at onset of seizures in 106 patients that were initially studied in collaboration with Kunze, Freedman, Keplinger, and others at the Epilepsy Center at Duke University and Durham VA Medical Center [12, 14] and with Bacsal and Treiman at UCLA and the West Los Angeles VA Southwest Regional Epilepsy Center [13]. Figure 1 also shows that seizures started before the age of 20 years in the majority of patients who underwent anterior temporal lobectomy. This second study was done in collaboration with Walsh [15, 51].

Three major types of complex partial seizures were observed [12-15]. The first type (or Type I) of complex partial seizures turned out to be hippocampal or hippocampal-amygdalar in origin. These seizures are similar to some attacks described under mesial basal limbic [52, 55] or primary rhinencephalic [4, 9] psychomotor seizures. Three phases are present in this first type (Fig. 2). The first phase corresponds to initial akinesis and staring. The patient's eyes show a dazed or wild look and can be transiently proptotic. The patient starts moving during the second phase with stereotyped automatisms like chewing, swallowing, pursing lips, lip smacking and eye blinks. The third phase is the longest part of the attack; although showing purposeful reactions to his environment, the patient still has impaired consciousness and is amnestic. Vocalization can appear during the onset of attacks; however, most patients do not have understandable speech until the third phase.

The second type (Type II) of psychomotor seizures recognized on CCTV turned out to be of extratemporal origin, most commonly frontal lobe [57]. The main events on the CCTV that differentiate extratemporal psychomotor seizures from hippocampal-amygdalar seizures are: (1) the absence of a motionless stare and/or oroalimentary automatisms (lip smacking, chewing) at onset; and (2) the presence of suprasylvian signs, such as contraversive head and eye movements, postures, or partial motor jerks in Type II A complex partial seizures and complex frenetic bilateral arm and leg motor automatisms in Type II B psychomotor attacks. Actual standing up and walking or running away, bipedal bicycling movements, pelvic thrusts, sexual gestures, karate leg and arm movements, and leg thrashing motions occur as de novo automatic behavior during onset of Type II B psychomotor attacks (Fig. 3). The patient is totally unresponsive to superficial pinprick, deep pain, and verbal stimulation during these stereotyped automatisms.

The third type of psychomotor attacks (Type III) are drop attacks followed by confusion and reactive automatisms [12, 13, 36]. They are of unknown origin but are probably frontal lobe. This third form of attack was originally designated "temporal lobe syncope." This is obviously a misnomer since most of these patients have evidence of extratemporal lobe epilepsy [36]. The patient drops to the floor with eyes shut and remains totally unresponsive for 2-3 min. Blood pressure, pupils, and tendon reflexes remain normal. Confusion with partial responsiveness,

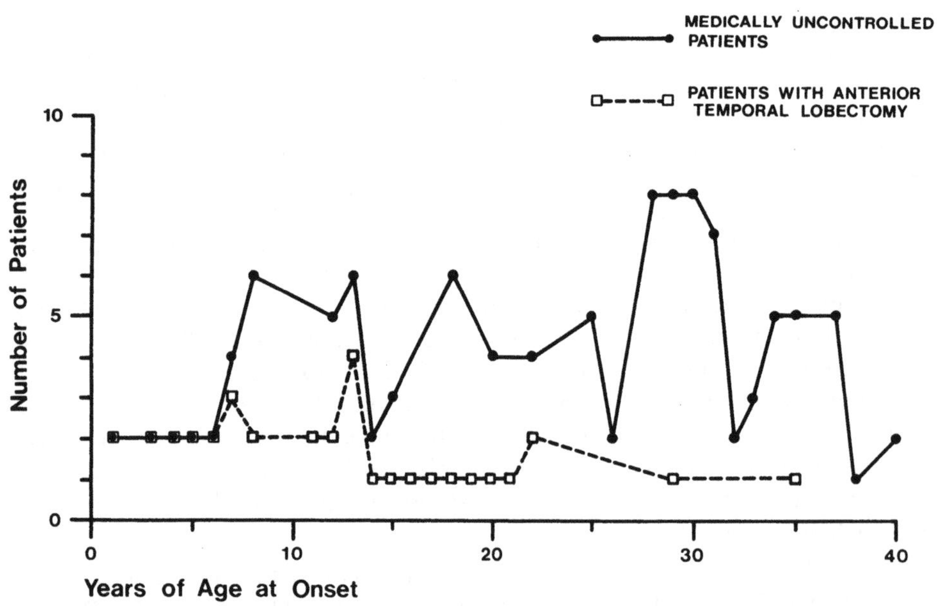

Fig. 1. Complex partial seizures as a function of age at onset. (Reproduced with permission and modified from reference 13)

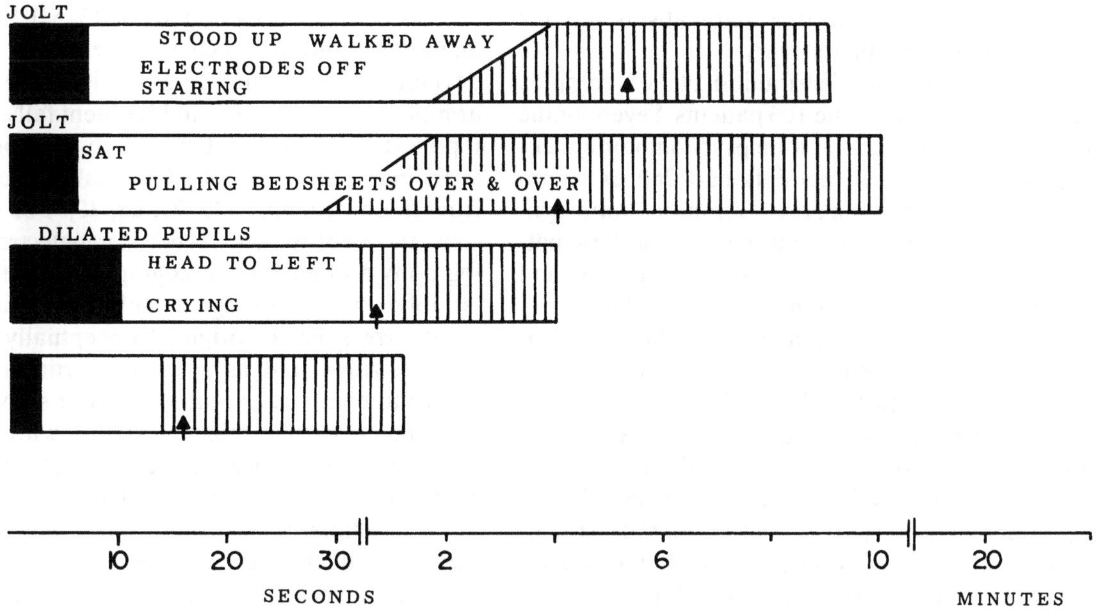

Fig. 2. Onset and sequence of behavior in Type I complex partial seizures from four patients. *Solid* area refers to the initial phase when the patient is motionless, staring, and totally unresponsive to superficial deep pain. *Clear* area refers to the second phase when the patient remains unresponsive with repeated chewing, blinking, swallowing, and other forms of stereotyped automation, and *striped* area refers to the third and longest phase of reactive quasipurposeful automations and partial responses to verbal and painful stimuli. The *arrow* refers to the time when verbal responses appeared as sentences. Perseverative motor behavior peculiar for each patient, and described above, was mixed with the usual stereotyped alimentary automatisms of the second phase. (Reproduced with permission, reference 13)

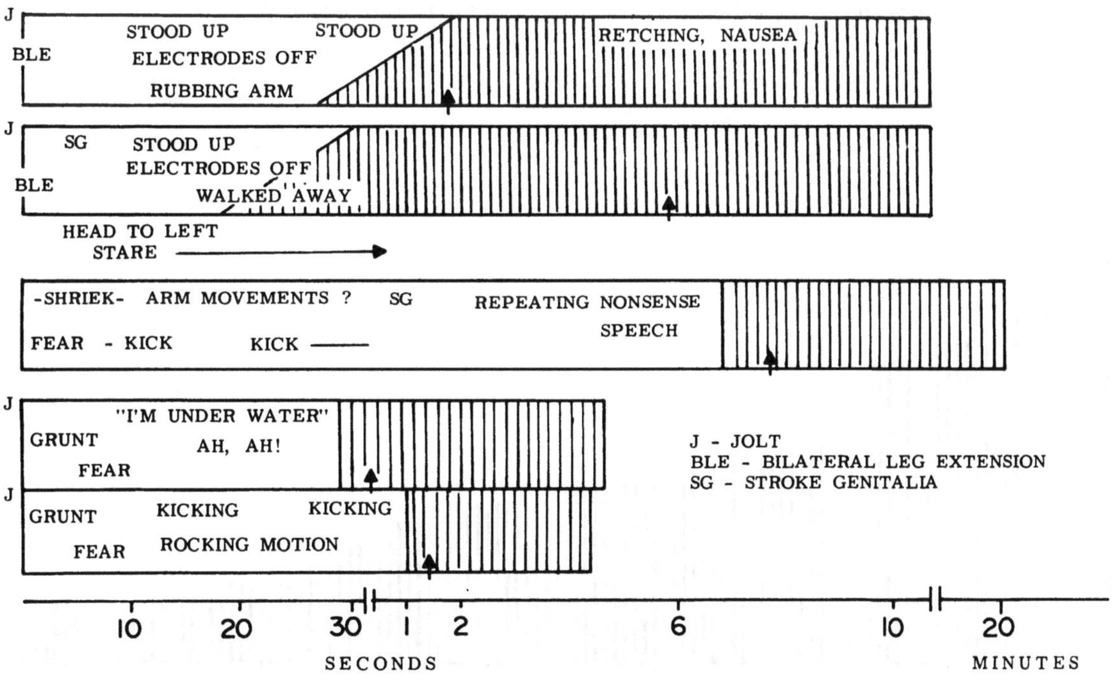

Fig. 3. Five examples of Type II complex partial seizures from four patients. *Clear* area refers to the phase of stereotyped automatisms, and *striped* area refers to the phase of reactive automatisms. During the start of stereotyped automatisms, bilateral leg extension *(BLE)*, apparently purposeless motions of the arms, shrieks, bizarre motions, ambulation appear and can be mixed later in the attacks with oroalimentary automatisms. *Arrow* refers to the time when verbal responses appeared as sentences. (Reproduced with permission, reference 13)

formed speech, and reactive automatisms last for about 2 min before full recovery.

Figure 4 shows the distribution of 905 complex partial seizures among the 106 patients. Seventy-nine patients had 645 temporal lobe psychomotor attacks (Type I). Twenty-seven patients had 260 extratemporal psychomotor seizures. 'Temporal lobe syncope' was noted in six patients [2]. Two of these patients also had Type I attacks, while four also had Type II attacks. 'Temporal lobe syncope' did not occur by itself, i.e., unassociated with either Type I or II attacks. Attention should be given to the fact that 21 patients had both Type I and II attacks.

Surface EEG features frequently focalize or lateralize to the temporal lobe before and during akinesis, staring, and oroalimentary automatisms [12, 13, 15], and our more recent depth electrode studies clearly establish the hippocampal or hippocampal-amygdalar origin of Type I attacks[6] [15] (Fig. 5). Because lateralizing sensory complaints and complex bilateral motor automatisms frequently appeared at the onset of Type II seizures, in the absence of focal temporal signs on the surface EEG and NP/SP electrodes, we first suspected that Type II seizures started outside the temporal lobe [12, 13, 51]. Suprasylvian motor symptoms, such as bilateral leg extensions, deviation of head and eyes, partial tonic or clonic arm or face movements, and frequent falls, had been reported in a few patients with stereoelectroencephalographically verified medial or dorsolateral frontal lobe epilepsy [20, 21, 28, 30]. For these reasons and because our recent stereoelectroencephalography and electrocorticography show their frontal lobe origin, we agree with such studies that emphasize such an origin. Conceptually, some of these seizures start as a simple partial attack and evolve to complex partial seizures or show signs of elementary partial seizures mixed with impaired consciousness at onset of the attack [1-3]. For purposes of objective analysis we had originally called these psychomotor seizures Type II A. Frequently, Type II A attacks are associated with simple partial seizures, such as oculoclonic, contraversive, and postural seizures with preserved consciousness. The remaining problem concerns patients with gross frenetic arm and bipedal movements, bizarre behavior, pelvic thrusts, sexual gestures and automatisms, and running automatisms [13, 50, 51]. We had originally called these attacks Type II B and suspected the medial frontal lobe to play a major role in their clinical expression. The remaining issue is, are *all* of these latter seizures frontal in origin [55-57] or are some frontotemporal?

[6] Ajmone-Marsan and Abraham deserve credit for first suggesting a focal temporal origin for Type I attacks. As stated on *page 19* of their atlas [2], "cases with definite staring are often accompanied by focal temporal lobe activation ..."

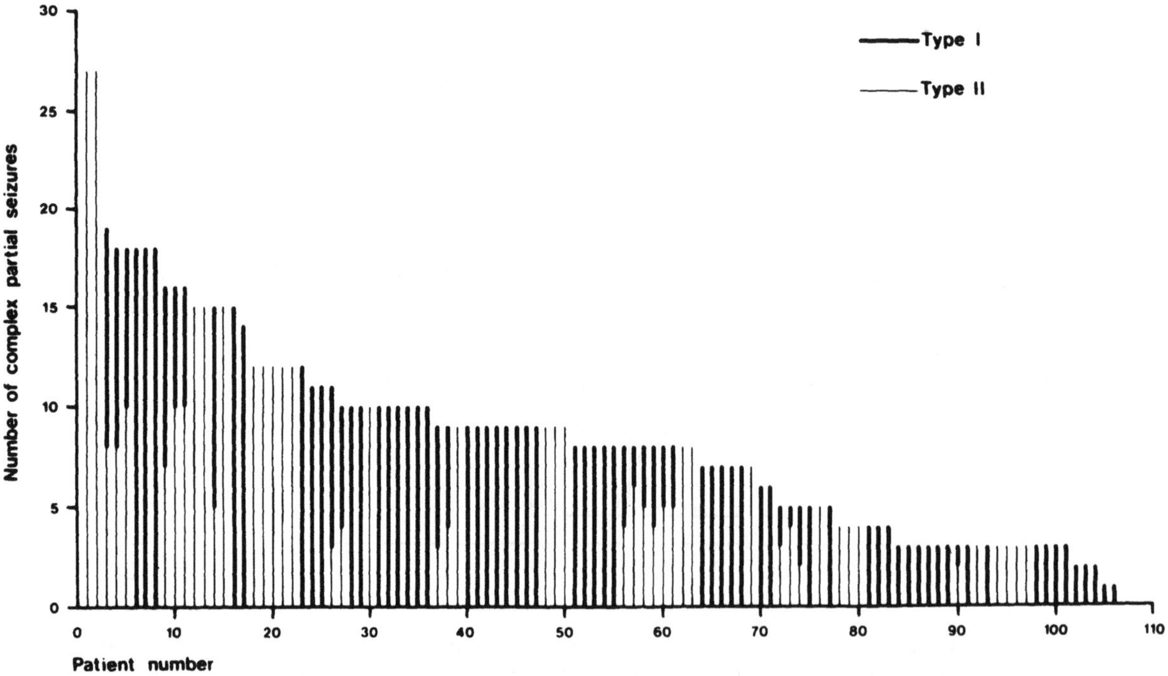

Fig. 4. Distribution of 905 complex partial seizures among 106 patients; the distribution of Type I and Type II attacks is also shown. Seventy-five patients had 3-10 attacks, while 26 patients had 11-27 attacks. In two patients, only one seizure was evaluated while only two attacks were studied in three patients. (Reproduced with permission, reference 13)

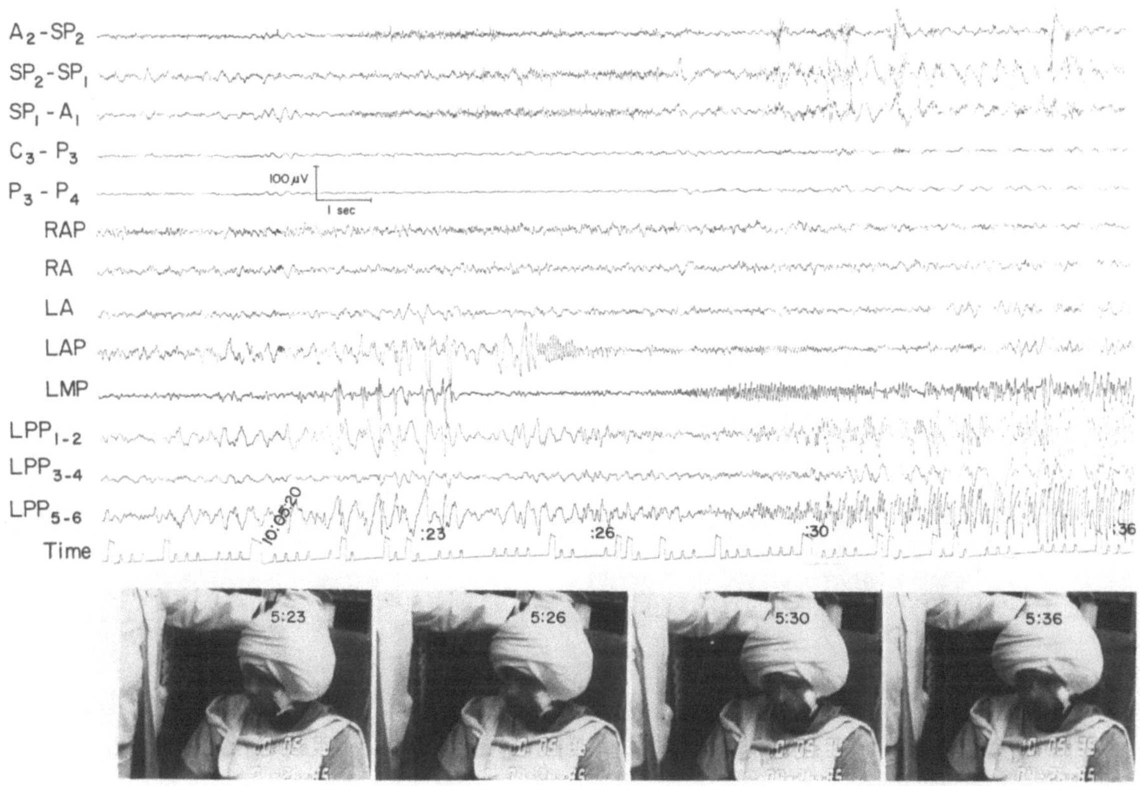

Fig. 5a

Fig. 5a–h. Type I complex partial seizures of hippocampal origin: second-for-second correlations. **a** At 10:05:20, the patient is sitting down conversing with his mother and a nurse. At 10:05:22 a regional burst of polyspikes and waves is seen in the left anterior, mid, and posterior pes hippocampi *(LA, LMP, LPP)*. The patient continues talking. At 10:05:23 a flattening of EEG rhythms is observed on the left mid and posterior pes hippocampi followed at 10:05:26 by fast 22–25 /s low voltage activity in the left mid pes hippocampi. The patient becomes motionless and stares at 05:30 when all the left medial temporal structures are displaying fast activities. The nurse and his mother start to ask him for his name, place, and date. He is unresponsive. He remains motionless and stares from 05:30 to 05:46. **b** At 05:46, chewing and lip smacking appear. He looked down at 05:46 as ictal rhythmic activities involved the right depth structures and the scalp electrodes. **c** At 06:04, the nurse asks for his name and place; she stimulates his face with a pin. The patient does not respond to her questions or to the pinprick stimuli. **d** The bilateral ictal activities continue as the patient remains unresponsive to repeated pinprick stimuli to the face and to simple oral commands. At 06:38, the patient responds to the pinprick by moving his face away from the pin. **e** High amplitude 3/s spikes continue in the left medial temporal structures until 6:52 while the patient remains unresponsive to simple questions. **f–h** The patient responds to questions verbally but remains confused and amnesic until 10:25

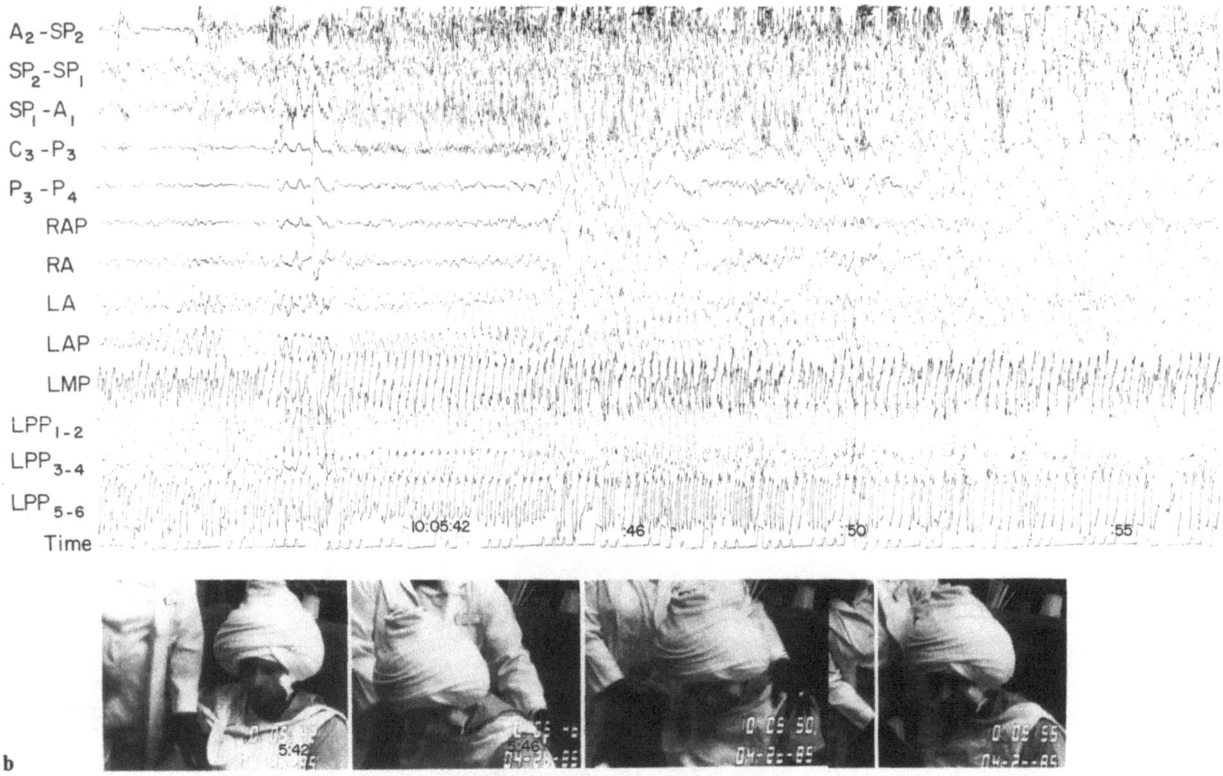
A2-SP2
SP2-SP1
SP1-A1
C3-P3
P3-P4
RAP
RA
LA
LAP
LMP
LPP1-2
LPP3-4
LPP5-6
Time
10:05:42
:46
:50
:55
5:42
5:46

Fig. 5b

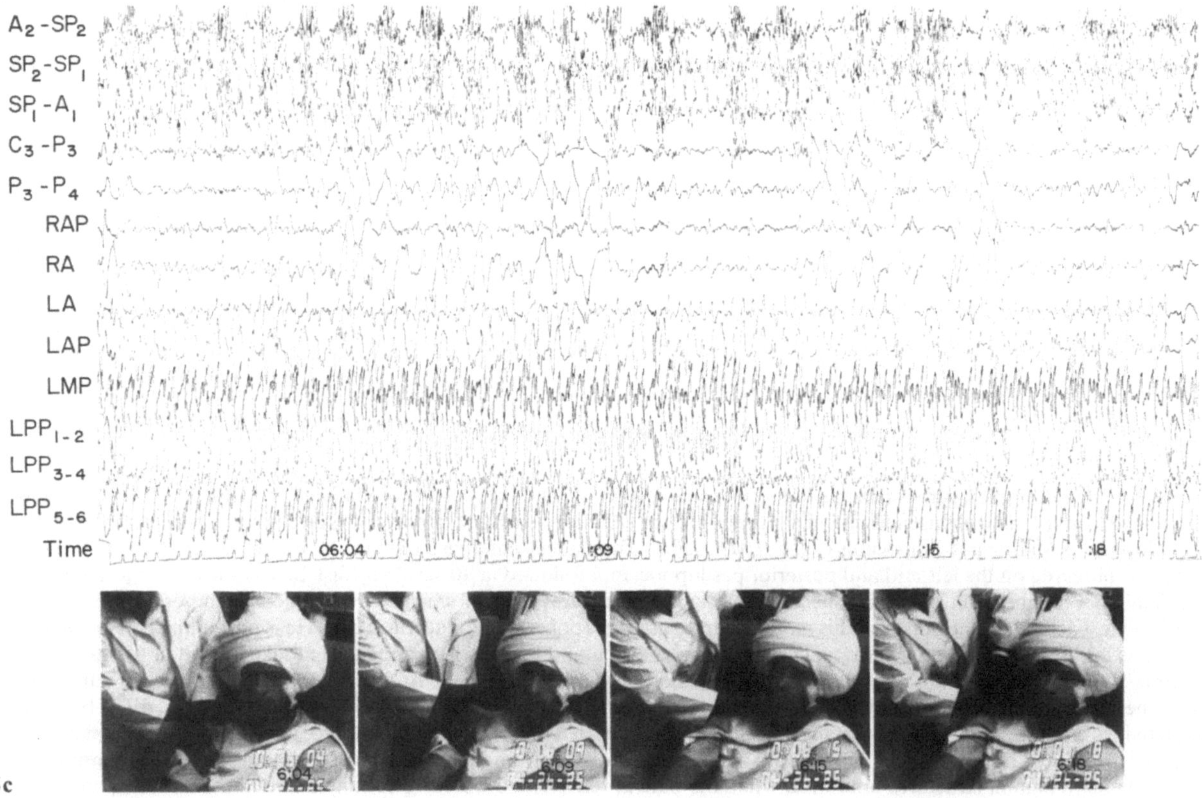
A2-SP2
SP2-SP1
SP1-A1
C3-P3
P3-P4
RAP
RA
LA
LAP
LMP
LPP1-2
LPP3-4
LPP5-6
Time
06:04
:09
:15
:18
6:04
6:09
6:15
6:18

Fig. 5c

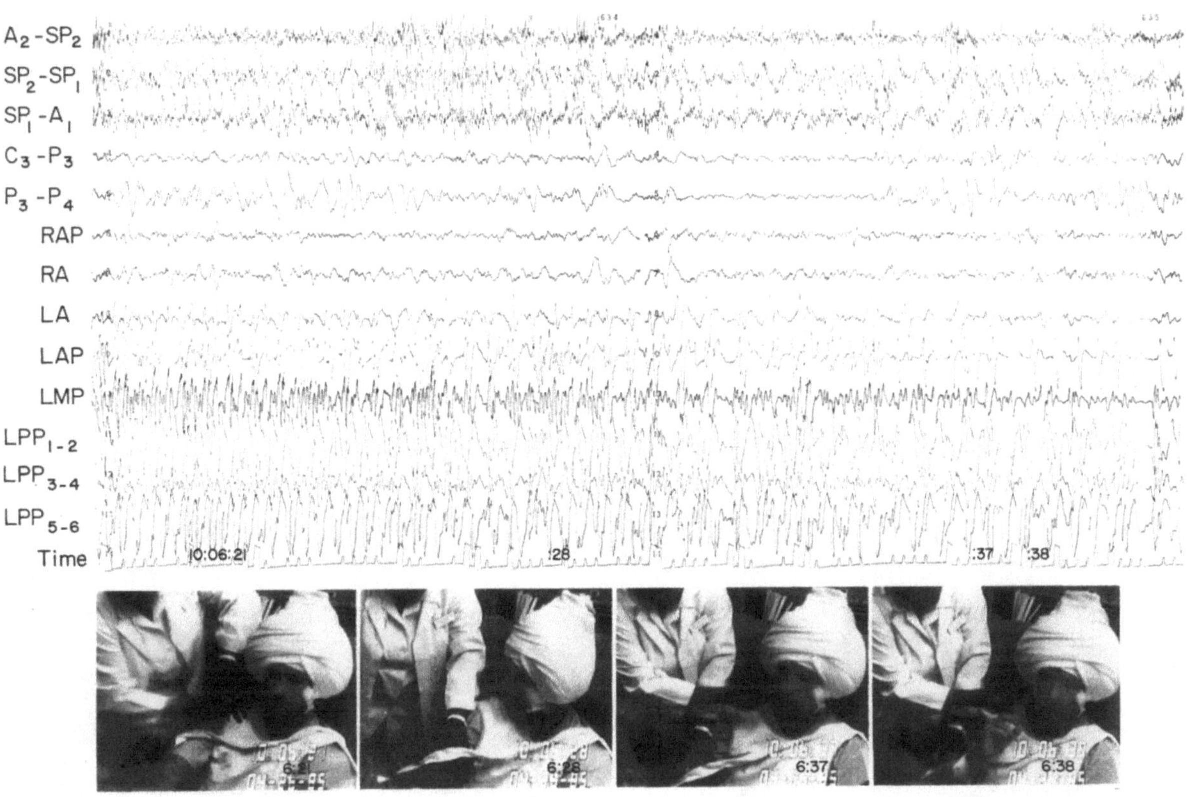
A2-SP2
SP2-SP1
SP1-A1
C3-P3
P3-P4
RAP
RA
LA
LAP
LMP
LPP1-2
LPP3-4
LPP5-6
Time
10:06:21
:28
:37
:38
6:28
6:37
6:38

Fig. 5d

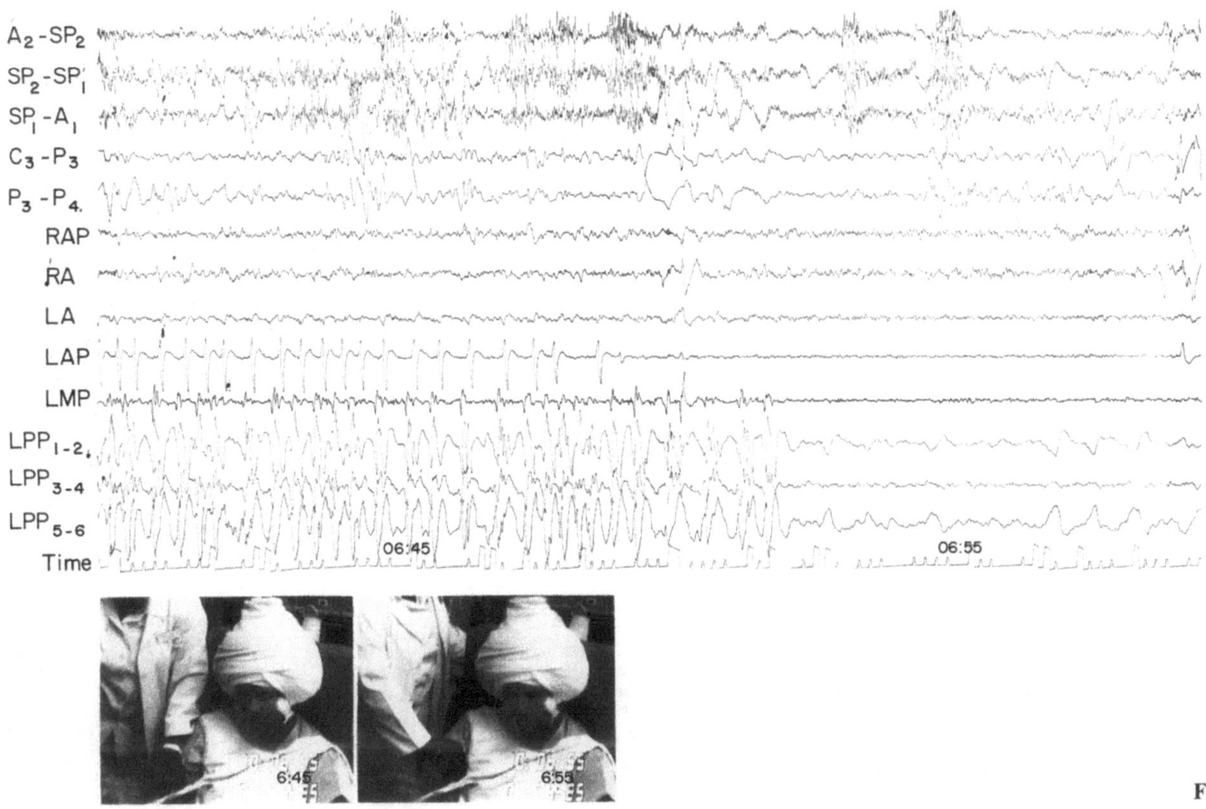
A2-SP2
SP2-SP1
SP1-A1
C3-P3
P3-P4
RAP
RA
LA
LAP
LMP
LPP1-2
LPP3-4
LPP5-6
Time
06:45
06:55
6:45
6:55

Fig. 5e

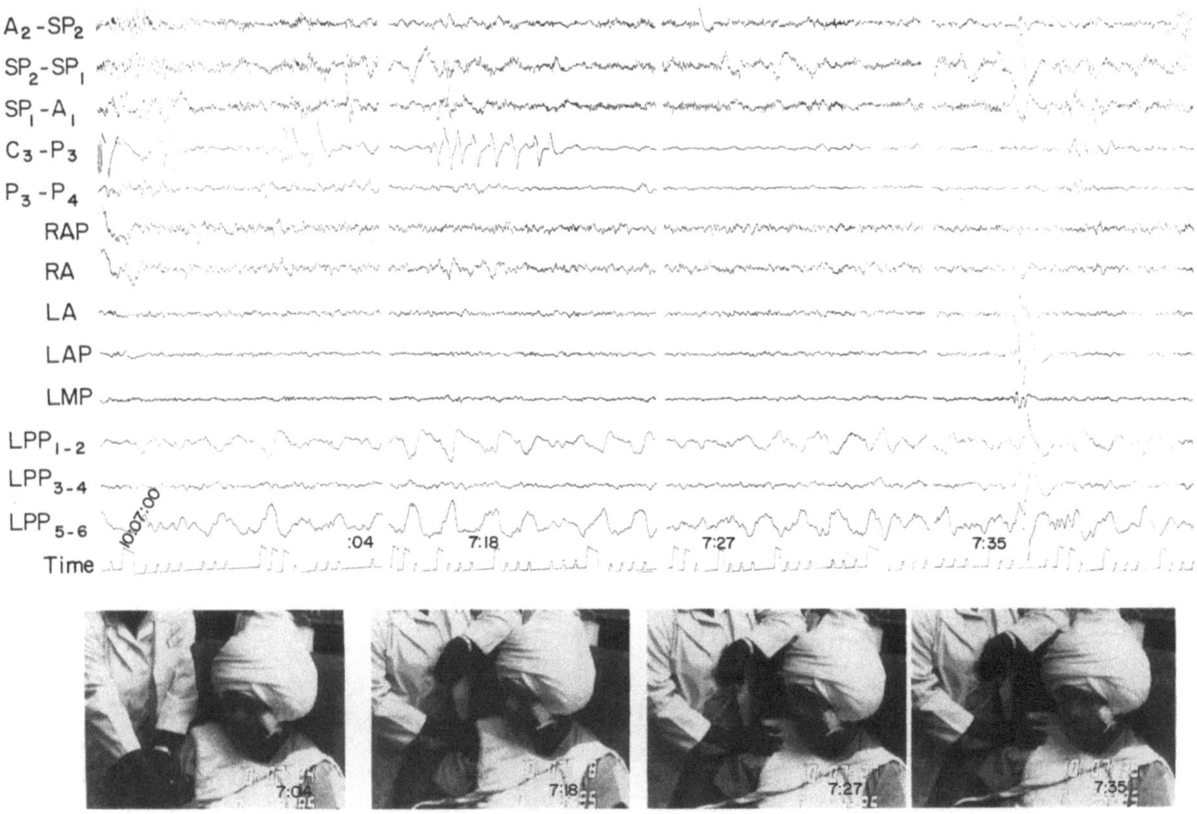

A_2-SP_2
SP_2-SP_1
SP_1-A_1
C_3-P_3
P_3-P_4
RAP
RA
LA
LAP
LMP
LPP_{1-2}
LPP_{3-4}
LPP_{5-6}
Time
10:07:00
:04
7:18
7:27
7:35
7:04
7:18
7:27
7:35

Fig. 5f

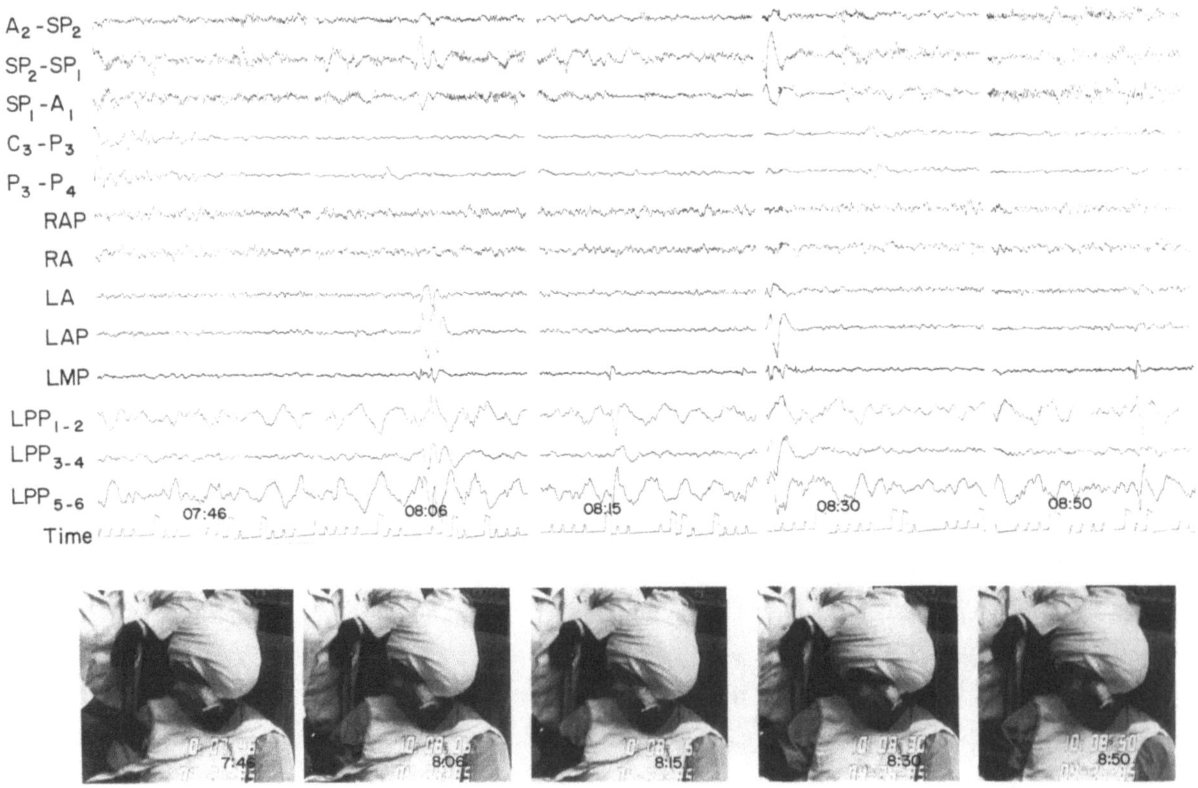

A_2-SP_2
SP_2-SP_1
SP_1-A_1
C_3-P_3
P_3-P_4
RAP
RA
LA
LAP
LMP
LPP_{1-2}
LPP_{3-4}
LPP_{5-6}
Time
07:46
08:06
08:15
08:30
08:50
7:46
8:06
8:15
8:30
8:50

Fig. 5g

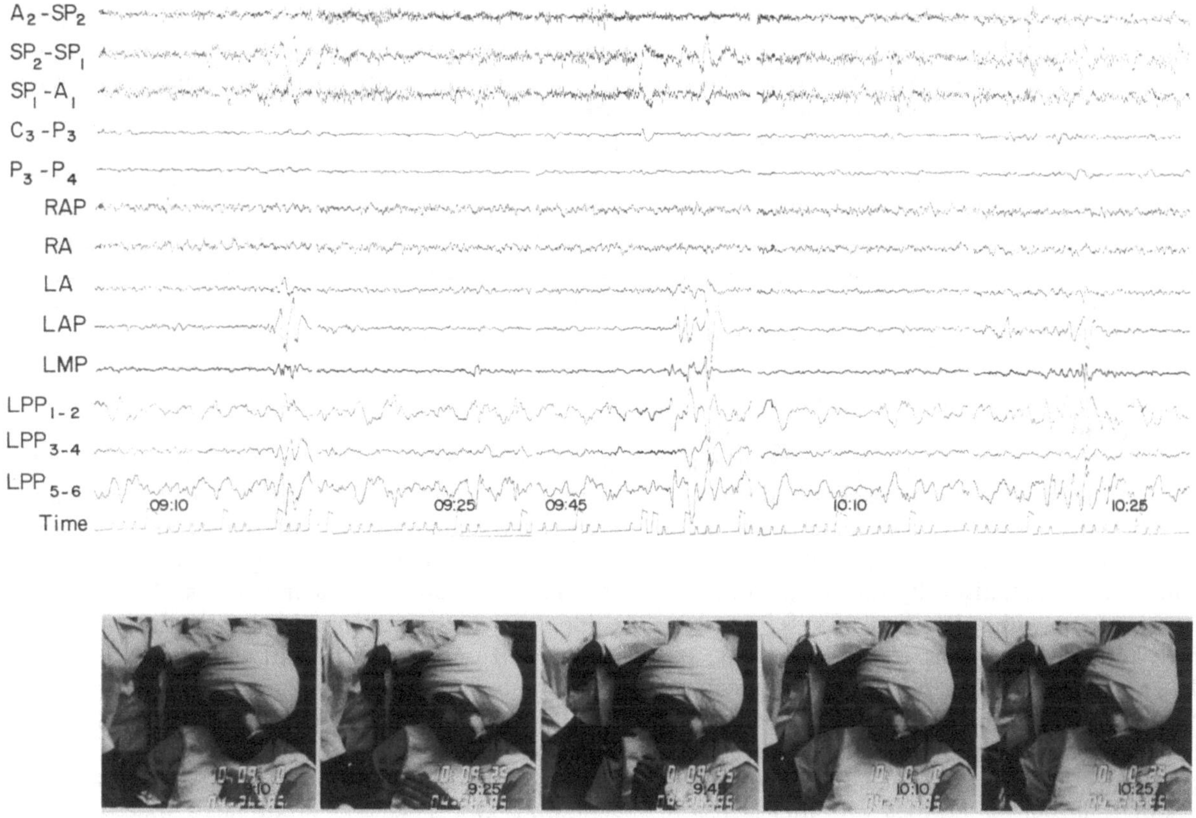
A2-SP2
SP2-SP1
SP1-A1
C3-P3
P3-P4
RAP
RA
LA
LAP
LMP
LPP1-2
LPP3-4
LPP5-6
Time
09:10
09:25
09:45
10:10
10:25

Fig. 5h

Hippocampal and Hippocampal-Amygdalar Epilepsy

For a more thorough discussion, the reader should turn to H.G. Wieser's chapter on limbic seizures (this volume). We discuss temporal lobe psychomotor seizures in this chapter only insofar as is necessary to differentiate them from psychomotor attacks of frontal lobe origin. The fundamental characteristics of complex partial epilepsies of amygdalar-hippocampal origin that allows differentiation from lateral temporal and extratemporal epilepsy are the (a) initial warning symptoms and signs of rising epigastric discomfort, indescribable strange feeling, sense of unreality, fear, loneliness, mood changes, disturbances of memory, déjà vu, déjà vecu, disturbances of ideation, and double consciousness, and (b) automatic behaviors on CCTV, consisting of the arrest reaction with motionless staring and stereotyped oroalimentary automatism (lip smacking and chewing movements) within the first 10 s, followed by reactive automatisms, amnesic responsiveness, and gradual recomposure (Fig. 5). Not all hippocampal or hippocampal-amygdalar seizures, however, are characterized by manifestations of Type I attacks. Most recently, Maldonado et al. [29] studied 18 patients and analyzed the sequence of appearance of automatisms in 68 psychomotor attacks proven by SEEG to be hippocampal amygdalar in origin. Of the hippocampal-amygdalar seizures, 56% showed an arrest reaction during onset, but only 30% had pure motionless staring and the typical features of Type I attacks. Twenty-two percent showed oroalimentary automatisms, while 4% showed one discrete single movement during the arrest reaction. We consider the latter 26% to be variants of Type I psychomotor attacks, since all of these seizures combined motionless staring and/or lip smacking and chewing automatisms and/or single discrete movements during the arrest reaction and within the first 10 s. Most commonly, these combinations of behaviors during onset of a psychomotor attack are associated with hippocampal-amygdalar ictal paroxysms. The problem lies in those 13% of psychomotor attacks of hippocampal-amygdalar origin which start with perseverative stereotyped behavior of a higher order, such as mimicry, repeated gestures, and motions, or the 28% of seizures which show a single discrete movement that precedes or is mixed with perseverative complex motions. At the present time, we are unable to differentiate these latter forms of psychomotor seizures from similar attacks that come from the lateral temporal neocortex or from extratemporal structures like the frontal lobe. Thus, when Type I psychomotor seizures (arrest reaction with motionless staring followed by oral alimentary automatisms, perseveratory and gesticulatory stereotyped automatisms, confusion with reactive automatisms, and amnesia) or variants of Type I psychomotor attacks (as described above) are evident on CCTV, hippocampal or hippocampal-amygdalar origin should be suspected. However, when perseverative stereotyped behavior of a higher order, such as mimicry, gestures, and repeated motions, start a psychomotor attack, the medial temporal, lateral temporal, and extratemporal structures should all be considered possible sites of origin.

Wieser et al. [52, 54], Delgado-Escueta and Walsh [15], Walsh and Delgado-Escueta [51], Maldonado et al. [29], and Swartz et al. [47] also studied the initial warning symptoms and signs (aura) of the psychomotor attack and observed differences between pure hippocampal, combined amygdalar-hippocampal seizures, and pure amygdalar seizures. Pure hippocampal seizures and combined amygdalar-hippocampal seizures commonly start with a strange indescribable feeling and some patients described experiential hallucinations (déjà vu, jamais vu, double consciousness) and interpretative illusions. Pure amygdalar seizures frequently show autonomic symptoms and signs, e.g., pupil, heart, and respiratory rate and pattern changes. Patients with amygdalar seizures commonly describe a rising epigastric discomfort, a sensation of "opening up," and vulnerability with or without nausea. Fear or panic appear when ictal rhythms spread to the hippocampus [47]. Very often, epileptogenic paroxysms involve quickly both amygdalar and hippocampal regions when olfactory and gustatory hallucinations are experienced. It is more common to see psychomotor seizures originate from both hippocampus and amygdala. Pure hippocampal seizures are less frequent than combined hippocampal-amygdalar seizures. Pure amygdalar seizures are rare.

In a recent study, Bossi et al. [11] questioned whether somatomotor manifestations were related to ictal discharges in the temporal lobe or to the spread of paroxysms beyond the temporal lobe. Of 250 patients with temporal lobe epilepsy, 20% had somatomotor manifestations. In particular, early somatomotor manifestations appearing in the first 10 s of the attack were rare. In most cases somatomotor signs appeared late in the attack, most likely due to the spread of ictal discharges to the extratemporal regions, probably the suprasylvian motor cortex. These observations disagree with the observations of Feindel [17-19] and Gloor [22], who produced somatomotor responses to electrical stimulation of the amygdala, while agreeing with previous findings of Bancaud et al. [7-9]. Bancaud et al. showed that local after-discharges in the amygdala and periamygdaloid region do not induce somatomotor signs [8, 9]. These observations by Bossi et al. [11] also agree with previous suggestions of Ajmone-Marsan and Ralston [3] that motor signs in temporal lobe seizures are due to propagation of ictal discharges to extratemporal structures. Our observations on hippocampal-amygdalar seizures are in agreement with reports by Bossi et al. [11], Ajmone-Marsan and Abraham [2], and Ajmone-Marsan and Ralston [3].

The important afferent connections of the hippocampus and the amygdala frequently lead to their involvement by propagating epileptiform paroxysms from the lateral posterior temporal neocortex, operculum, or medial occipital lobe. The clinical symptomatology and signs of individual psychomotor attacks and the initial warning signs and symptoms (aura), however, can distinguish the various sites of origin for the propagating epileptogenic paroxysms. Review of individual attacks reveals vestibular hallucinations, visual illusions, or hallucinations, and unilateral upper limb gesticulatory exploratory automatisms in lateral posterior temporal neocortical epilepsy. Auditory hallucinations, nausea, somesthetic hallucinations, pupillary and facial color changes, feelings of warmth and unilateral head and upper limb tonic posture or clonic jerks are observed in opercular seizures. Simple visual hallucinations (dots, scotomae) and central visual field abnormalities, together with tonic or clonic ocular movements that precede impaired unconsciousness, can signal the start of occipital hippocampal epilepsy.

Frontal Lobe Epilepsy

Most extratemporal psychomotor attacks are frontal lobe in origin. They are classified according to the specific site of the frontal lobe involved. They consist of dorsolateral frontal lobe epilepsies, anterior polar frontal lobe epilepsies, the medial frontal lobe epilepsies, and the frontobasal cingulate epilepsies (Tables 1 and 2). Rarely extratemporal psychomotor seizures start from the parietal lobe, the somatosen-

Table 1. Frontal lobe epilepsies with complex partial seizures: clinical features[a] on CCTV

Frontal Lobe	Clinical Features
Dorsolateral frontal cortex[b] (e.g., intermediate frontal gyrus)	A) Psychomotor attacks: aura usually none; starts with tonic forward extension of head, neck, and trunk as contraversive head or eye movements occur with impaired consciousness; mild clonic facial or arm movements or elevation of one arm above shoulder occur when paroxysms spread to primary cortex or supplementary motor cortex
	B) Sexual automatisms: pelvic thrusts, masturbatory movements, and other sexual automatisms
	C) Pseudoabsence: very brief lapses of consciousness lasting < 10 s with minimal automatisms, sudden in onset and cessation
Medial frontal cortex[b] Supplementary motor	A) Psychomotor attacks: usually no aura and starts with elevation of arm with abduction and flexion of elbow as head turns to look at hand, vocalization, elevation of lips, grimaces, and hissing with impaired consciousness with late appearance of oroalimentary automatisms
	B) Ipsilateral, bilateral, or homolateral motor responses can also start psychomotor attacks
	C) Autonomic seizures: mydriasis, tachycardia, and apnea
	D) Speech arrest
Cingulate cortex	A) Psychomotor attack: starts with sudden fear, mydriasis, epigastric discomfort, leading to impaired consciousness, and usually incontinence. Late appearance of oroalimentary automatisms during impaired consciousness, when paroxysms spread to medial temporal lobe structures; head looks to elevated arm when discharges spread to supplementary motor area
	B) Pseudoabsence: brief impaired consciousness with minimal automatisms but with sphincter incontinence
Frontocingulate[c]	Psychomotor attack: sudden complex motor automatisms involving both legs, arms, and trunk, often bizarre and frenetic, start with impaired consciousness and severe confusion. Pelvic up-down movements, running, bipedal movements similar to bicycling, swimming, karate movements, repeatedly appear as stereotyped from attack to attack mixed with staring, change of facial expression, fear, and oroalimentary automatisms in the middle or late stages of seizure

[a] Denotes a constant electroclinical sequence, characteristic of habitual seizures
[b] Type II A complex partial seizures (Walsh & Delgado-Escueta, 1984)
[c] Type II B complex partial seizures (Walsh & Delgado-Escueta, 1984)

Table 2. Frontal lobe epilepsies with complex partial seizures: electrographic manifestations

Interictal EEG	Ictal EEG/stereo-EEG	Route of spread
Dorsolateral frontal cortex[a] (e.g., intermediate frontal gyrus)		
Focal spikes or sharp waves in the frontal region or normal	- Focal flattening with low voltage 16-24 Hz polyspikes that secondarily generalize or - Generalized flattening most marked in the intermediate frontal gyrus - Generalized flattening is seen on the scalp EEG or bifrontal spikes and sharp waves or rhythmic 2-4 Hz waves	To the supplementary motor cortex, anteriorly to the frontal pole, thalamus and striatum
Medial frontal cortex[a] Supplementary motor (SM)		
1-3 Hz and 4-6 Hz irregular spike-wave complexes (secondary bilateral synchrony) and signs of structural damage	- Focal flattening with low voltage 16-24 Hz polyspikes in the SM cortex increasing amplitude as frequency decreases to 4-8 Hz spkes with secondary generalization - Generalized flattening is seen on scalp EEG	To the contralateral medial hemisphere, anterior thalamus, Ammons' horn
Medial frontal cortex[a] Cingulate gyrus		
1-3 Hz and 4-6 Hz irregular spike-wave complexes (secondary bilateral synchrony) and signs of structural damage	- Focal low voltage 16-24 Hz polyspikes in the cingulate gyrus with secondary generalization and global flattening - Generalized flattening is seen on scalp EEG	To the contralateral medial hemisphere, anterior thalamus, Ammons' horn
Frontobasal-cingulate[b] *psychomotor epilepsy*		
Normal or focal spikes or sharp waves in the frontal region or medial temporal	- Focal flattening with low voltage 16-24 Hz polyspikes with secondary generalization - Generalized flattening is seen on scalp EEG	To the lateral frontal, supplementary motor, parietal and cingulate cortex and temporal limbic lobe

[a] Type II A Complex Partial Seizures
[b] Type II B Complex Partial Seizures

sory cortex, or the occipital lobe and secondarily invade the frontal and/or temporal regions.

Dorsolateral Frontal Lobe Epilepsy with Type II A Complex Partial Seizures. The dorsolateral frontal lobe epilepsies can be subdivided into three groups. One group consists of the epilepsies of the motor cortex and are comprised by partial somatomotor epilepsy, the Bravais-Jacksonian epilepsy and Koshevnikoff's epilepsy syndrome. These forms of epilepsy do not usually manifest with impaired consciousness and psychomotor automatisms and are, thus, not part of our present communication. In the dorsolateral frontal lobe the typical areas which produce psychomotor seizures consist of the intermediate frontal gyrus and its contiguous regions or the anterior frontal pole. Penfield and Jasper [38], Rasmussen [44], Bancaud and Talairach [6, 9], Ajmone-Marsan and Abraham [2], and Ajmone-Marsan and Ralston [3] established our concepts about the dorsolateral frontal lobe epilepsies. Often the interictal EEGs in these patients are normal or spikes and sharp waves may reveal a frontal focus. Attacks recorded with scalp EEG show generalized flattening with low voltage fast rhythms or low voltage asynchronous irregular 0.5- to 3-Hz slow waves [5-9] (Tables 1 and 2).

When the epileptogenic discharges electrically occupy the intermediate frontal gyrus and Brodmann's area 8, quickly spread to area 4 or to Broca's speech area, the initial signs commonly consist of subtle and very brief tonic forward or backward elevation of head, neck and upper trunk, contraversive head and eye movements (cephalotonic or clonic, oculotonic, or oculoclonic), one or two mild facial clonic contractions, and nonfluent dysphasia as consciousness is impaired [2, 3, 5, 9, 51] (Table 1). For purposes of analyses, we had called these psychomotor attacks Type II A. One must emphasize that in these attacks motor signs can be very brief and subtle and the entire clinical gestalt of the seizure onset may be the appearance of impaired consciousness [51]. Repeated viewing of the CCTV however will verify these early motor signs. Paroxysms can frequently spread to the supplementary motor area so that elevation of an abducted extremity with flexed elbows can occur. Late in the sequence of events, oroalimentary automatisms (lip smacking, pursing lips, tongue movements) occur when the amygdalohippocampal complex is invaded by epileptogenic discharges [9, 51].

Penfield and Kristiansen [39], Ajmone-Marsan and Abraham [2], and Ajmone-Marsan and Ralston [3] independently suggested that little clinical manifestations appear when epileptogenic discharges remained confined to the frontal pole. Only after backward spread to posterior frontal regions do head or eyes turning, postures, and unilateral facial tonic contractions appear before evolving to tonic-clonic convulsions. Penfield and Kristiansen [39] and Ajmone-Marsan and Ralston [3] also reported a small number of patients whose seizures started from the anterior frontal region and with impaired consciousness that quickly evolved to tonic-clonic convulsions. Direct rapid symmetrical generalization to the thalamic reticular function or a detour route via the pericallosal-cingulate area accounts for the quick onset of tonic-clonic seizures.

Thus, when epileptogenic discharges start from the anterior polar (frontal) region, seizures start with an initial loss of contact. Adversive and subsequent contraversive movements of head and eyes appear when paroxysms spread backwards to the intermediate frontal gyrus. Axial clonic jerks, falls, autonomic signs, and secondary generalization are also common. Recently, Lüders et al. [27] reported on the presence of a second sensory area localized on the inferior frontal gyrus immediately anterior to the primary area of the motor strip. This report concerned a patient with right frontal lobe epilepsy. Electrical stimulation of this area elicited (a) a paralyzing feeling in the left arm and face, (b) inhibition of rapid alternating movements of left fingers, left hand, and tongue, (c) inability to maintain a strong voluntary muscle contraction of the left hand or tongue, and (d) speech arrest.

In 1965, Bancaud [9] emphasized the meager amount of information on ictal stereo-electrographic recordings of epilepsies from the intermediate frontal lobe. Penfield and Kristiansen [39] and Penfield and Jasper [38] had previously described the clinical manifestations of seizures from the intermediate frontal gyrus based chiefly on ECoG and intraoperative studies. We reiterate Bancaud's statement, since no published work has appeared on ictal SEEG except from the Paris school [9]. More SEEG studies are needed which correlate the onset and spread of electrographic paroxysms from the intermediate frontal gyrus with the sequence of clinical signs during psychomotor attacks recorded on CCTV.

More than one clinical variety of complex partial seizures originate from the intermediate frontal lobe. In addition to Bancaud's 1965 description [9], we have found that epilepsies in this area can also manifest as short attacks 10–20 s in duration, with minimal automatisms suddenly appearing and ending [47]. Farwell and Stuntz [16] recently reported a fronto-parietal astrocytoma causing absence seizures and bilaterally synchronous epileptiform discharges in a 8-year-old child. The seizures were brief staring spells, back arching, eyelid fluttering, lip smacking, and pulling at clothes. The seizures are similar to what we call pseudoabsences or partial seizures with impaired consciousness mainly [47]. We presume that such short seizures occur when paroxysms quickly spread anteriorly to the frontal pole and the deeper subcortical structures of the reticular formation. These can be mistaken for absence attacks but appear more frequently. Figure 6a–c is an example of complex partial seizures originating from the intermediate frontal gyrus. D.B. is 11 years old and has drug-resistant psychomotor attacks that were suspected to be due to Rasmussen's encephalitis. Figure 6a–c illustrates the second-for-second correlation between behavioral and electrographic signs. In this attack contraversive movements of eyes and elevation of arms occur during unconsciousness. She also had short episodes of impaired consciousness with minimal automatic movements, e.g., eyeblinks. In some other seizures, the first sign is a nonfluent dysphasia. The ECoG showed the supplementary motor area to be free of any abnormality. In the intermediate frontal gyrus, the background ECoG was slow, irregular, and arrhythmic while epileptiform sharp waves, spikes, and irregular slow waves were abundant (Fig. 6d). Neuropathological studies were normal, as were routine viral studies in CSF. IgG synthesis of CSF was elevated and in situ hybridization using genomic segments of herpes and measles virus is being carried out.

Medial Frontal Lobe Epilepsies with Type IIA Complex Partial Seizures. Complex partial seizures can also originate from the supplementary motor cortex and the cingulate cortex. In our experience, it can be difficult to differentiate dysphasic seizures that start from the intermediate frontal gyrus to secondarily involve Broca's area, from opercular temporal lobe seizures that involve Broca's areas, or from the speech arrest induced by supplementary motor seizures. Penfield et al. first recognized the existence of a supplementary area in the interhemispheral fissure when stimulation produced vocalization, many different postures and movements, arrest or slowing of voluntary action, autonomic changes, sensations, and speech arrest [38, 39]. Subsequent accounts appeared in a monograph by Ajmone-Marsan and

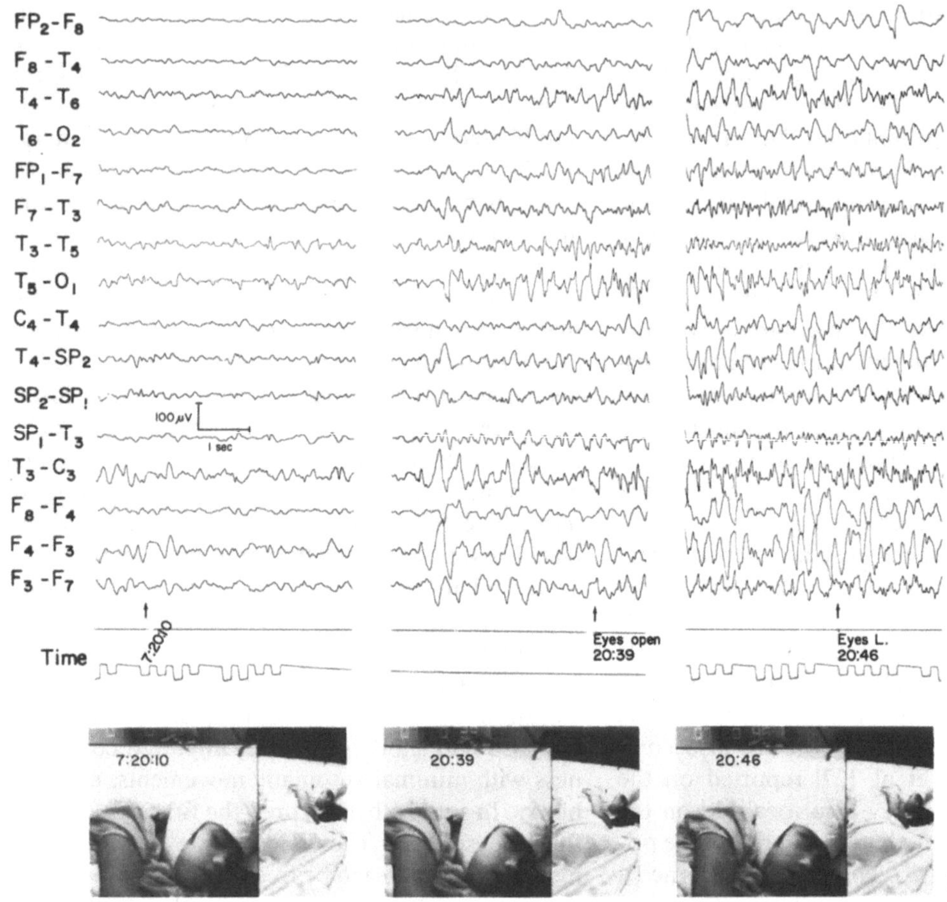

Fig. 6a

Fig. 6a–d. Complex partial seizures with contraversive head and eye movements (Type II A) from the intermediate frontal gyrus. **a** CCTV and scalp-sphenoidal EEG. 3/s high amplitude triangles slow waves appear on the left frontal lobe with rhythmic 9–10/s discharges best observed at F_7, T_3, and SP_1. Repetitive oculoclonic jerks to the left at 20:46 are followed by unconsciousness and **b** contraversive eye movements to the right at 20:52, contraversive head movements to the right at 20:55, elevation of right hand at 21:03, and **c** lingering confusion. **d** Electrocorticography: interictal sharp waves, spikes, decreased fast rhythms and polymorphic 1–2/s waves are almost continuous in the left intermediate frontal gyri (electrodes *2, 3, 4*). The superior frontal gyrus and the supplementary motor gyrus had normal brain waves

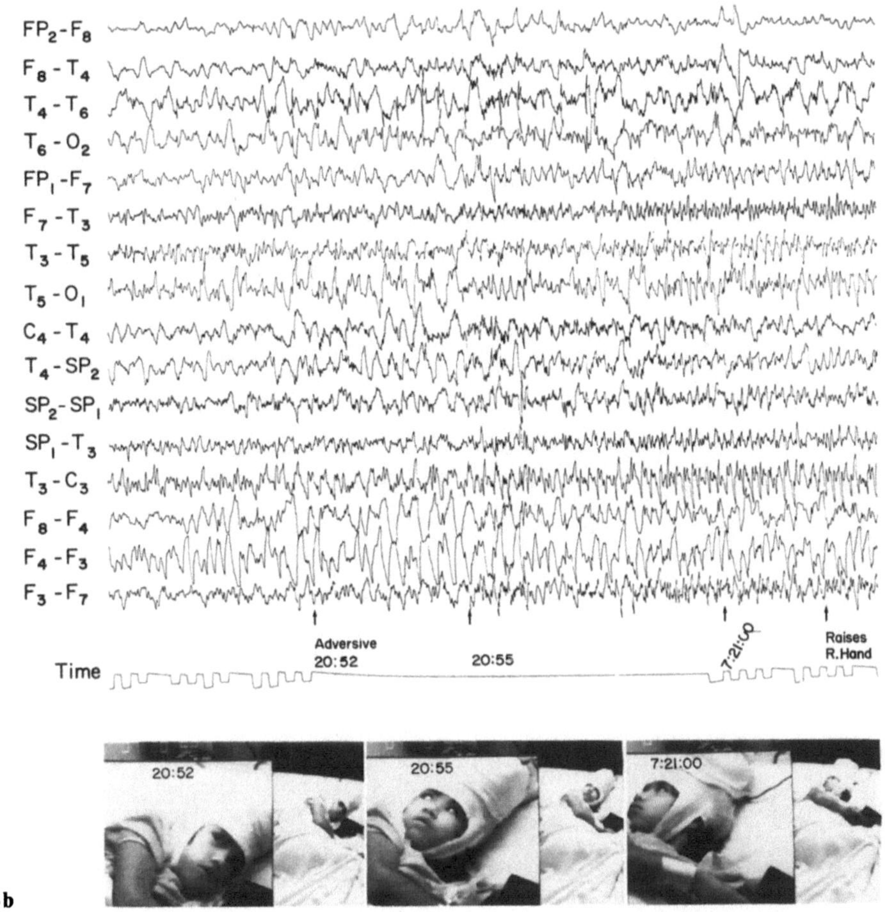

FP2-F8
F8-T4
T4-T6
T6-O2
FP1-F7
F7-T3
T3-T5
T5-O1
C4-T4
T4-SP2
SP2-SP1
SP1-T3
T3-C3
F8-F4
F4-F3
F3-F7
Adversive
20:52
20:55
7:21:00
Raises
R.Hand
Time
20:52
20:55
7:21:00

Fig. 6b

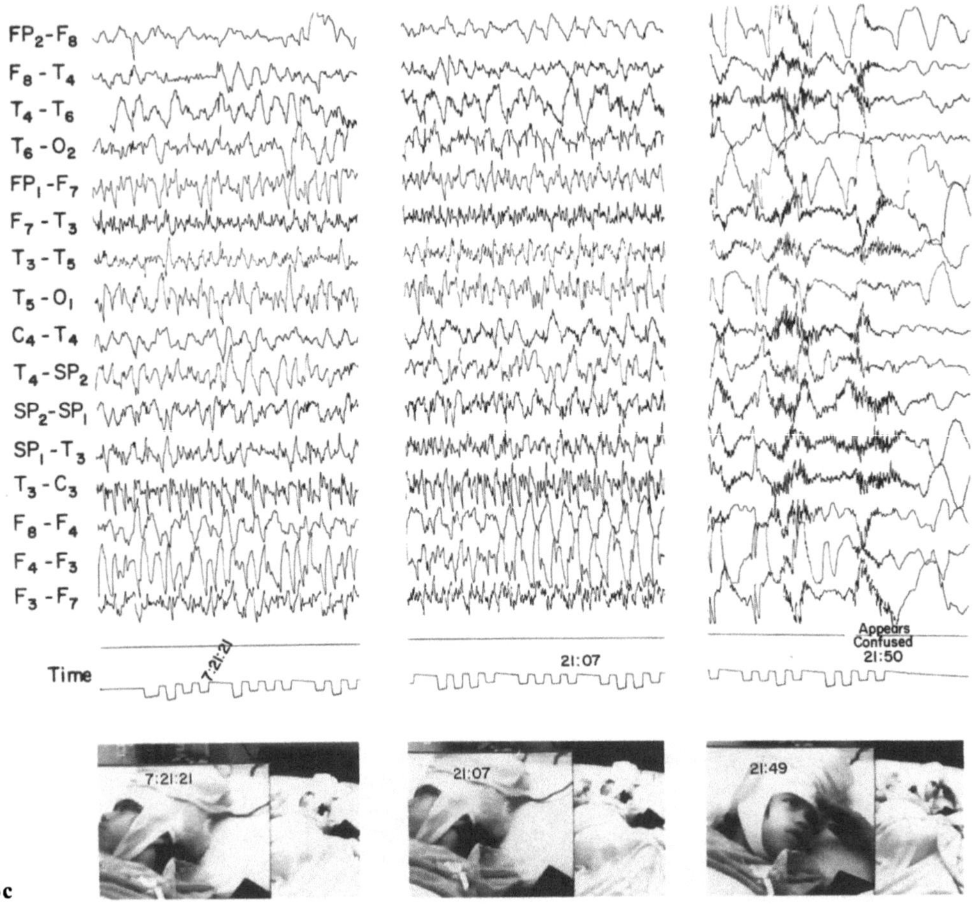
FP2-F8
F8-T4
T4-T6
T6-O2
FP1-F7
F7-T3
T3-T5
T5-O1
C4-T4
T4-SP2
SP2-SP1
SP1-T3
T3-C3
F8-F4
F4-F3
F3-F7
Time
7:21:21
21:07
Appears Confused
21:50
7:21:21
21:07
21:49

Fig. 6c

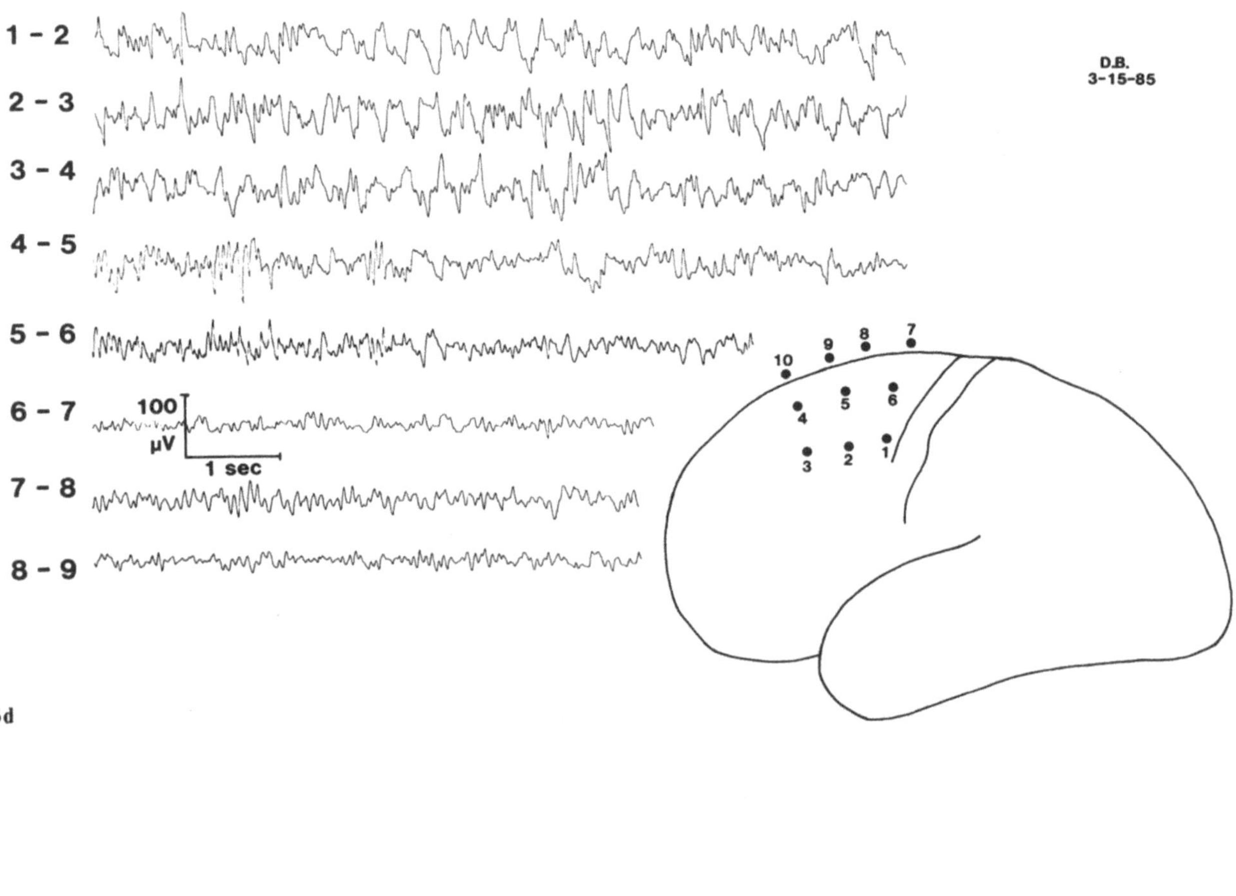
1 - 2
2 - 3
3 - 4
4 - 5
5 - 6
6 - 7
7 - 8
8 - 9
100 μV
1 sec
D.B.
3-15-85
10
9
8
7
4
5
6
3
2
1

Fig. 6d

Ralston [3] and in a seizure atlas by Ajmone-Marsan and Abraham [2]. Penfield and Jasper described the typical posture as "slow movements of the opposite hand outward, backward and somewhat upward while the head and eyes turn toward it (as though the patient were observing what the hand was about to do)" [38]. Ajmone-Marsan and Abraham [2] described similar seizures observed after metrazol stimulation and called the attacks "M2e" - a tonic postural alteration or elevation of one arm which usually begins with flexion of the elbow followed by an abduction of the arm to approximately 90° together with external rotation. These authors hypothesized the origin and route of spread of paroxysms in these seizures [2, 3], but it was Bancaud, Talairach and collaborators in 1965 [9] who proved the actual routes of spread of paroxysms toward the cingulate cortex, the hippocampus, and frontal lobes by SEEG. Symptomatology similar to amygdalar-hippocampal seizures appeared late or in the middle of impaired consciousness. These consisted of autonomic symptoms and signs, oroalimentary automatisms, and even gestural and mimicry automatisms. Partial motor tonic seizures and urinary incontinence are also frequent [51].

In 20% of patients, reported by Ajmone-Marsan and Ralston [3], initial automatisms mixed with postural motions (M2e), focal clonic facial jerks, and head contraversion before evolving quickly to generalized seizures. In another 4%, postural motions and tonic facial contractions preceded and led to automatisms. Ajmone-Marsan and Ralston emphasized the frontal lobe origin of epileptogenic paroxysm and contrasted these clinical sequences to motor phenomena appearing after the onset of automatisms in temporal lobe epilepsy [3].

Six percent of the seizures in the series reported by Ajmone-Marsan and Ralston [3] had attacks which consisted of M2e only or postural motions. Another 4% started with M2e, tonic facial contractions, and automatisms. These were seizures originating from the supplementary motor region. In the experience of these investigators, 45% of the seizures originating from the supplementary motor region showed either an isolated or conscious M2e, while 30% evolved to tonic-clonic convulsions. Eighteen percent showed tonic adversion of head and eyes with a brief disturbance of consciousness followed by convulsions [3]. Our patients are similar to the small percentage of patients (4%) reported by Ajmone-Marsan and Ralston who have automatisms during supplementary motor paroxysms. The M2e posture was less evident on the CCTV in our patients during lapse automatisms [51].

Interestingly, patients of Ajmone-Marsan and Ralston [3] whose seizures were characterized only by a conscious M2e showed "nonconvulsive interictal EEG" while patients whose seizures progressed to unconsciousness and convulsions showed secondary bilateral synchrony on the interictal EEG.

Mazars [30], the group of Bancaud and Talairach [4-11, 31-33, 48] and Tükel and Jasper [49] deserve most of the credit for the delineation of cingulate epilepsy. Like epilepsies from the supplementary motor cortex, the interictal EEG may be normal or show signs of secondary bilateral synchrony (irregular poorly formed spike-wave sequences firing at 1-3/s and 4-6/s) as described by Tükel and Jasper [49]. The ictal EEG shows flattening with rhythmic polyspikes at 16-24/s that quickly generalize. Depth electrode explorations are mandatory if cingulate epilepsy is to be recognized since it is frequently difficult to differentiate this from supplementary motor epilepsy. The problem is that epileptogenic paroxysms commonly spread to other areas very quickly, e.g., from the cingulate cortex to the supplementary motor and vice versa, and the clinical symptoms produced by paroxysms from these two areas may be misleading. In addition, seizure discharges in the cingulate cortex frequently spread to the amygdalar-hippocampal regions and produce autonomic signs and oroalimentary automatisms in the middle and late stages of the psychomotor attacks.

In anterior cingulate epilepsy, Bancaud [5, 9] pointed out the interesting occurrence of sexual and genital events during automatic motions of psychomotor attacks. They reported 15 seizures in six patients whose attacks showed sexual automatisms early in the first 10 s of the attack. Vegetative or autonomic signs and urinary incontinence were also frequently observed in these seizures.

Figure 7a-f illustrates seizures from the supplementary motor area in an adolescent. The second-for-second correlations show the postures briefly occurring, garbled indecent speech, sexual gestures, and body turning during impaired consciousness. Figure 8a, b is the CCTV-EEG correlations of another adolescent with very brief episodes of unconsciousness or "pseudo-absence." Seizures started as fear; he placed his hands on his chest, he looked to the left then to the right, and the facial expression has changed. He was unresponsive for 20-40 s during which time he urinated in his pants. Magnetic resonance imaging showed right frontal and temporal abnormalities.

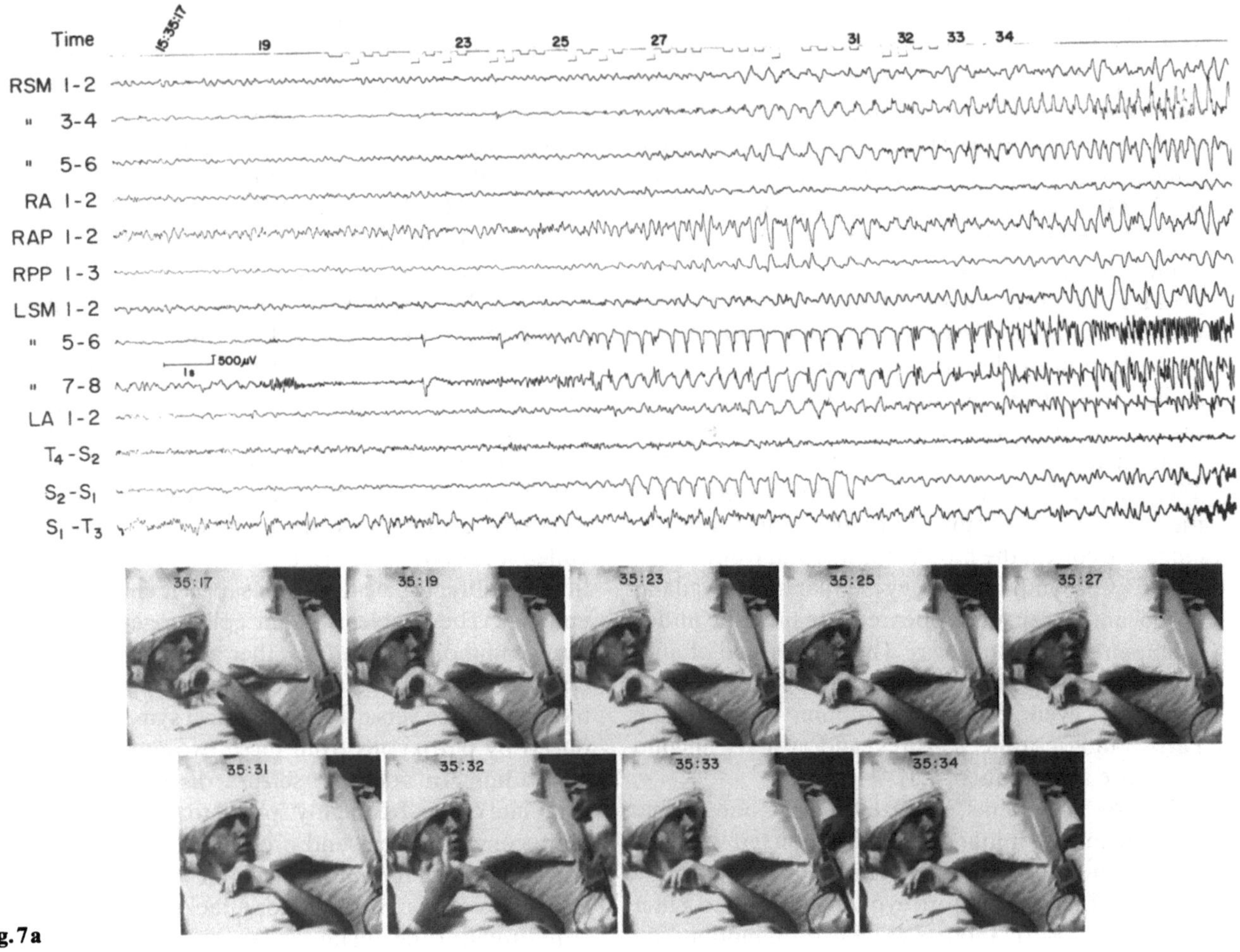

Fig. 7a

Fig. 7a–f. Complex partial seizures with stereotyped obscene gestures associated with electrographic paroxysms from the supplementary motor gyrus. **a** At 15:35:19, ictal 24–26/s rhythmic discharges start in the left supplementary motor regions (*LSM 7–8*). The patient said that he felt funny in his head. By 15:35:25, 2.5/s spikes occupy *LSM 5–6, 7–8* and are reflected at the left sphenoidal electrode (S_1). At this time, the patient mumbled to the nurse unintelligible words. Some words that could be discerned consisted of profane language. At 15:35:32, ictal discharges are clearly evident in the right and left supplementary motor areas and pes hippocampi. **b, c** He extended the right middle finger and continued his profane language. **d** At 15:36:28, head deviated to the right; at 15:36:50 the whole trunk rolled towards the left, eyes turned left and clonic lower facial jerks occurred on the left side. **e** Paroxysms continued as the postural change became fixed, and **f** at 15:37:26, ictal paroxysms stopped

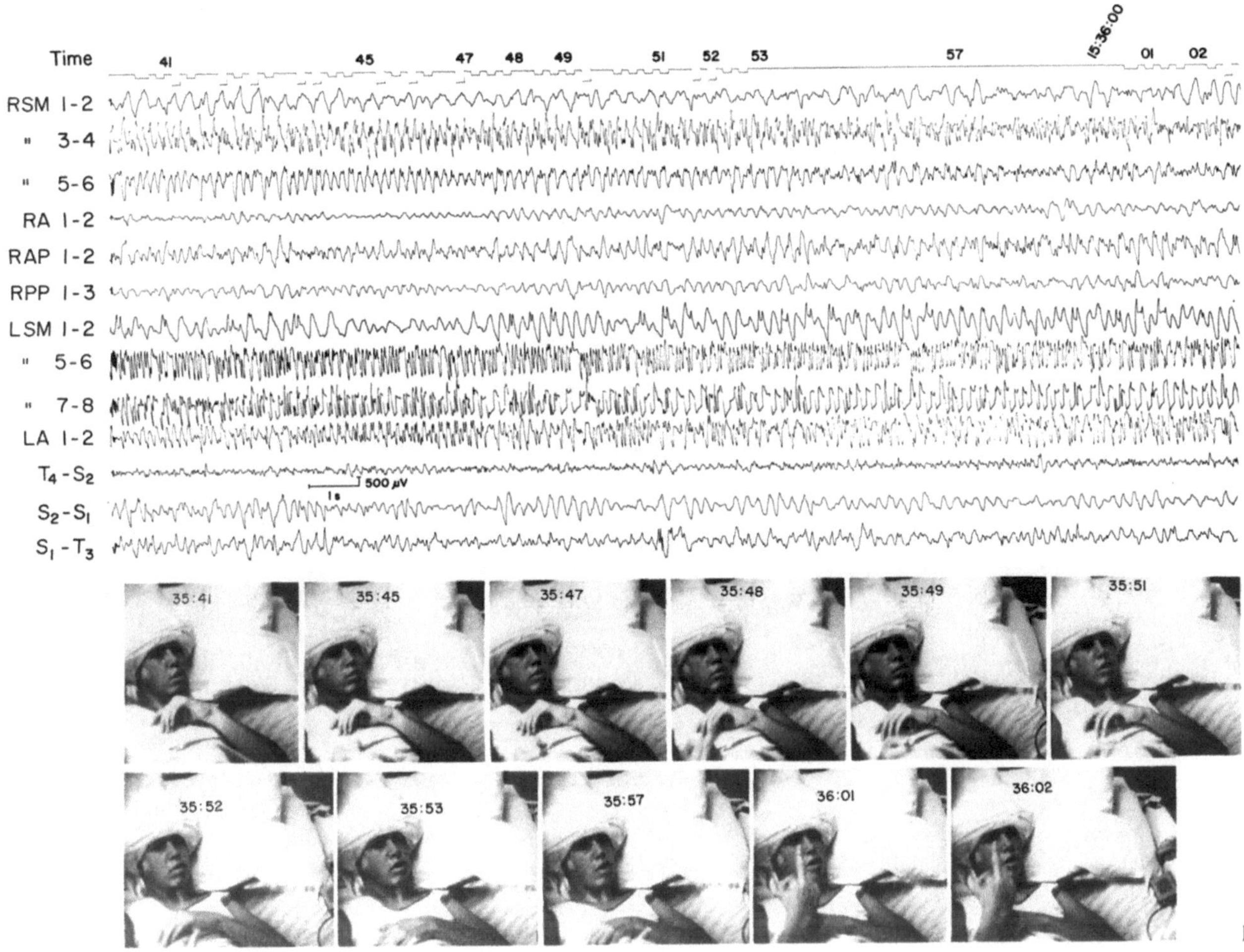
Time
41
45
47
48
49
51
52
53
57
15:36:00
01
02
RSM 1-2
" 3-4
" 5-6
RA 1-2
RAP 1-2
RPP 1-3
LSM 1-2
" 5-6
" 7-8
LA 1-2
T_4 - S_2
S_2 - S_1
S_1 - T_3
500 µV
1 s
35:41
35:45
35:47
35:48
35:49
35:51
35:52
35:53
35:57
36:01
36:02

Fig. 7b

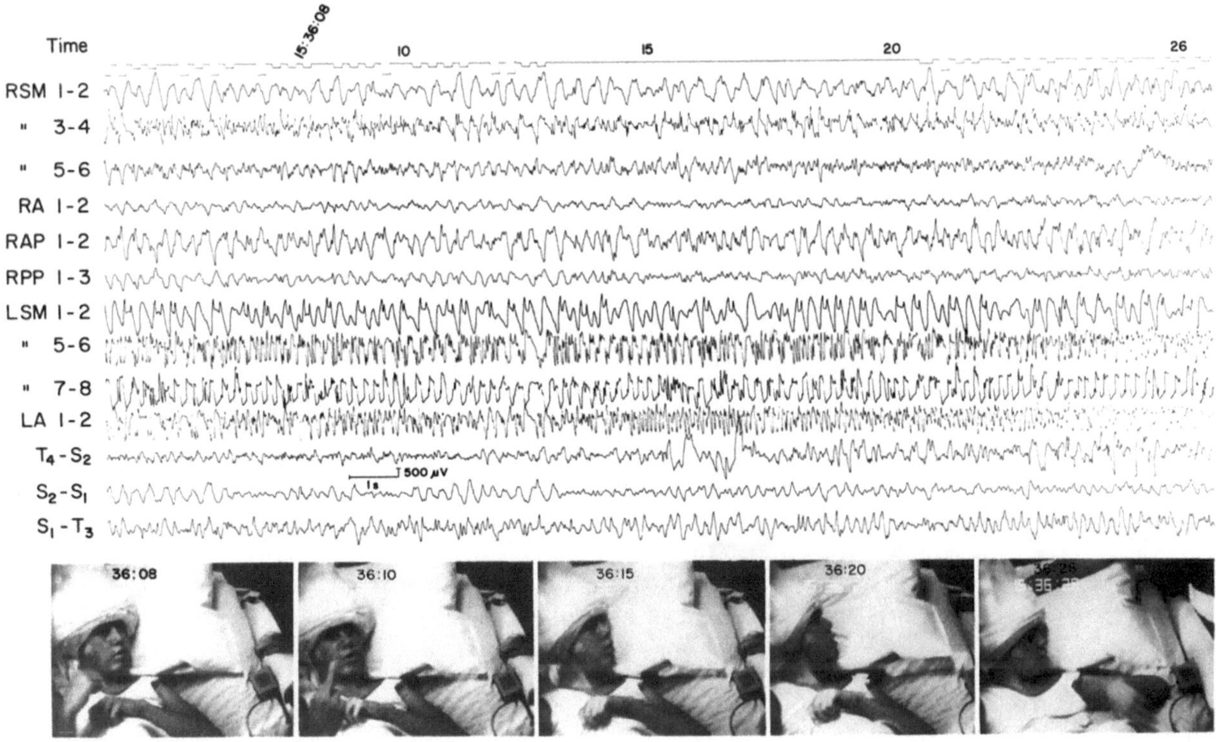
Time
15:36:08
10
15
20
26
RSM 1-2
" 3-4
" 5-6
RA 1-2
RAP 1-2
RPP 1-3
LSM 1-2
" 5-6
" 7-8
LA 1-2
T_4 - S_2
S_2 - S_1
S_1 - T_3
500 µV
1 s
36:08
36:10
36:15
36:20

Fig. 7c

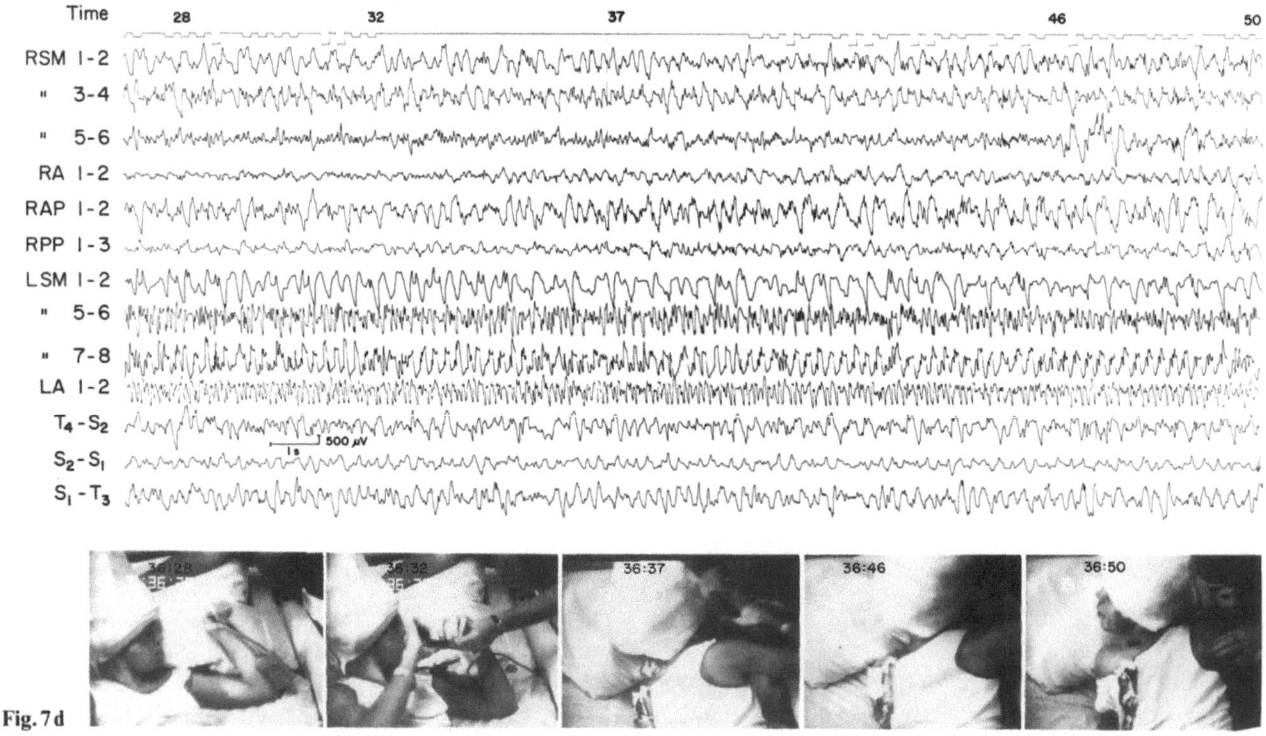
Time
28
32
37
46
50
RSM 1-2
" 3-4
" 5-6
RA 1-2
RAP 1-2
RPP 1-3
LSM 1-2
" 5-6
" 7-8
LA 1-2
T_4-S_2
S_2-S_1
S_1-T_3
500 µV
1 s
36:28
36:32
36:37
36:46
36:50

Fig. 7d

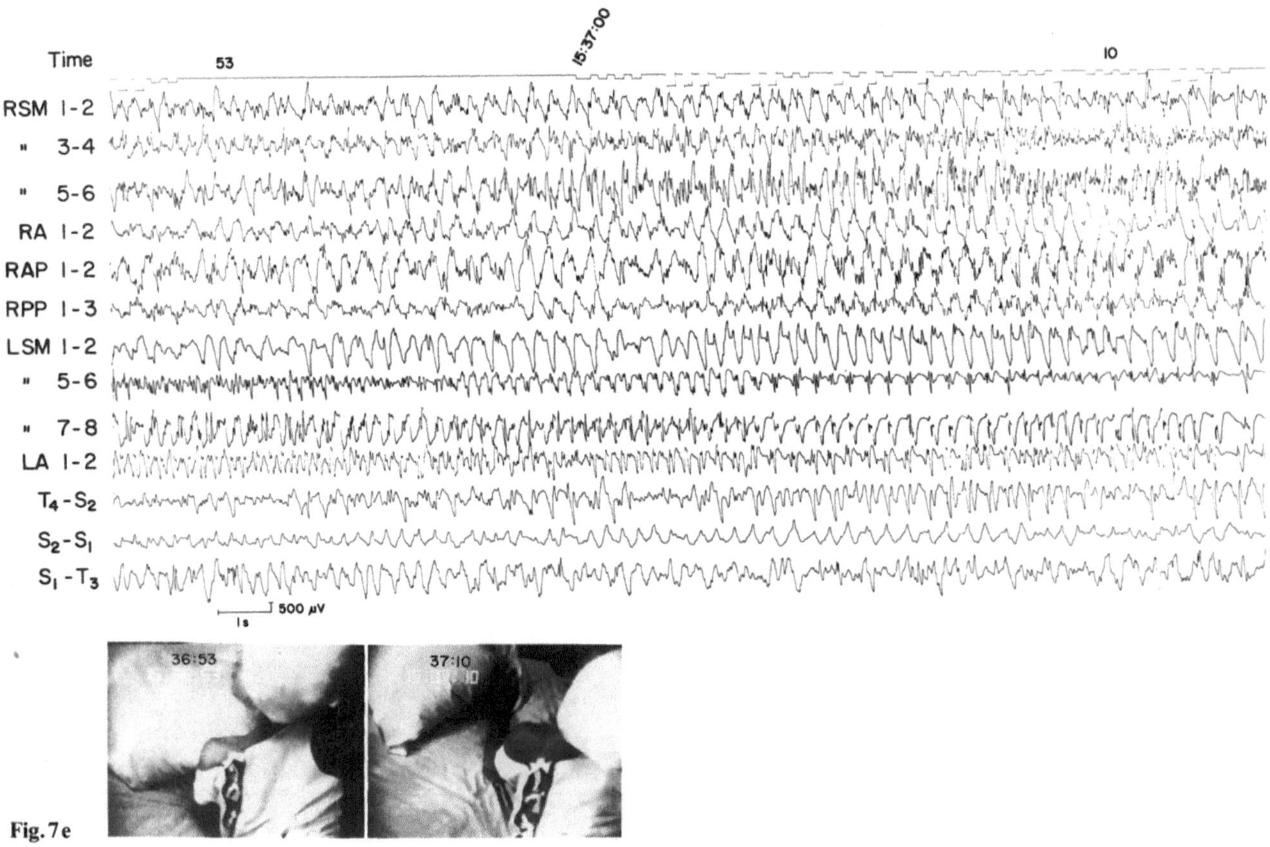
Time
53
15:37:00
10
RSM 1-2
" 3-4
" 5-6
RA 1-2
RAP 1-2
RPP 1-3
LSM 1-2
" 5-6
" 7-8
LA 1-2
T_4-S_2
S_2-S_1
S_1-T_3
500 µV
1 s
36:53
37:10

Fig. 7e

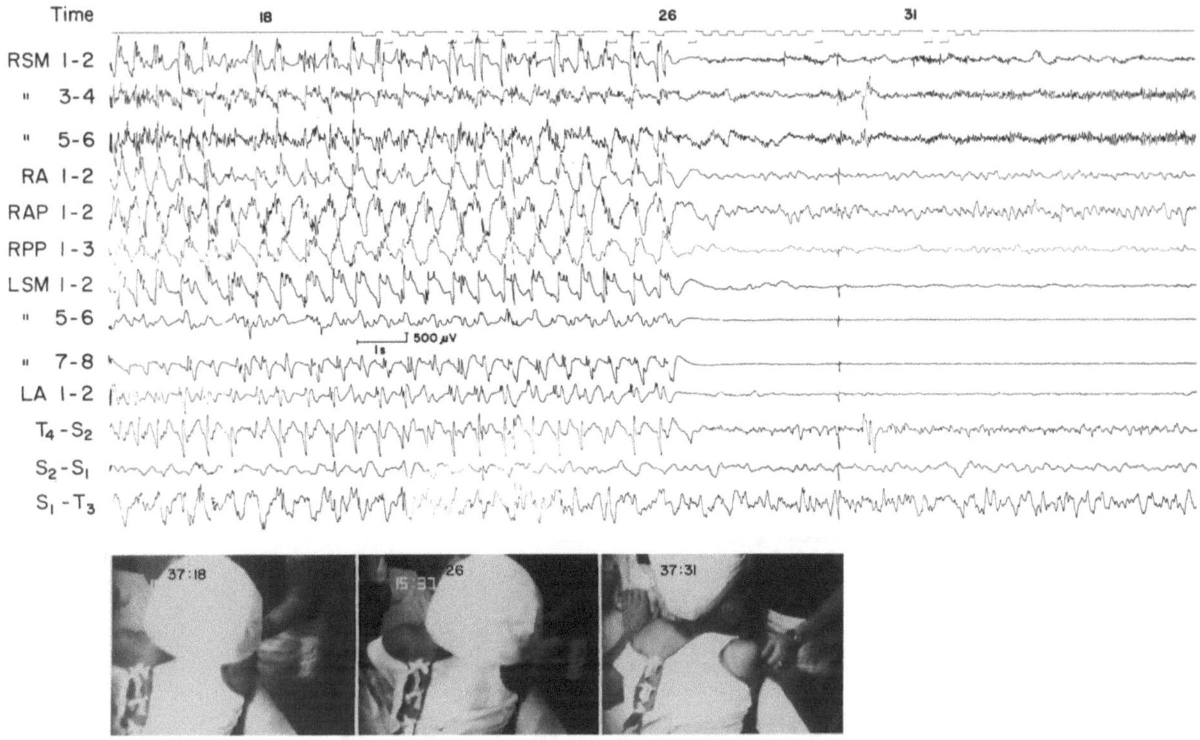
Time
18
26
31
RSM 1-2
" 3-4
" 5-6
RA 1-2
RAP 1-2
RPP 1-3
LSM 1-2
" 5-6
" 7-8
LA 1-2
T4-S2
S2-S1
S1-T3
500 µV
1s
37:18
26
37:31

Fig. 7f

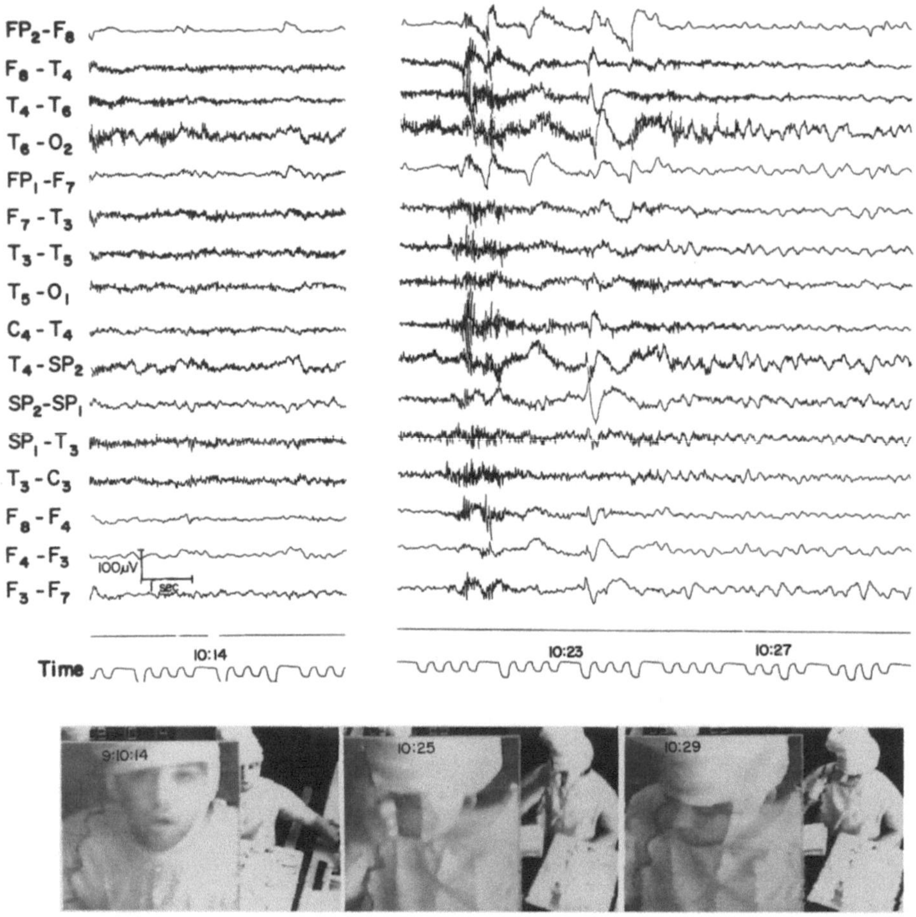

Fig. 8a

Fig. 8a, b. Complex partial seizures with short impairment of consciousness mainly (pseudo-absence) and urinary incontinence. **a** At 10:23, a single sharp wave which phase reverses at T_4, and SP_2 is followed by rhythmic medium voltaged diffuse 4/s slow waves that are best developed at midtemporal, sphenoidal, and central electrodes. At this time, confusion starts, he places his hand over his nose and face. He turns his head from side to side from 10:25 to 10:29. **b** He places his hand back to his nose at 10:35. At 10:48, the patient looks down at his pants, which are wet with urine.
Magnetic resonance imaging revealed increased intensities in the right medial and lateral frontal and temporal lobes

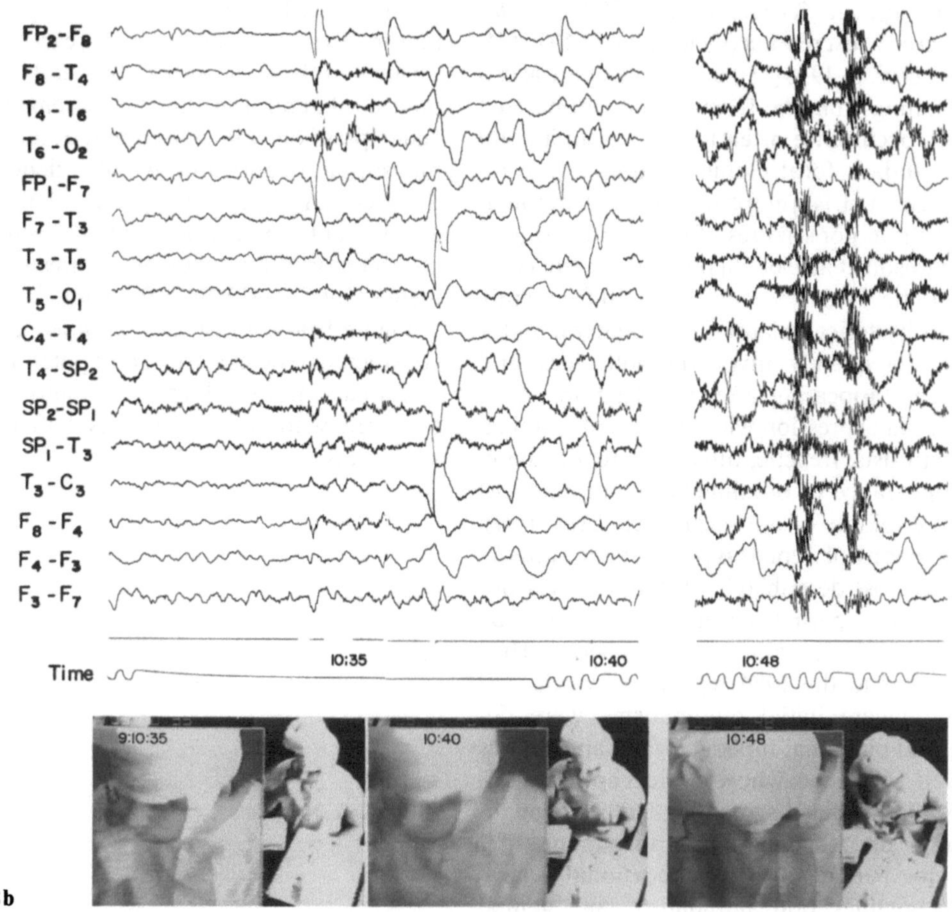
FP2-F8
F8-T4
T4-T6
T6-O2
FP1-F7
F7-T3
T3-T5
T5-O1
C4-T4
T4-SP2
SP2-SP1
SP1-T3
T3-C3
F8-F4
F4-F3
F3-F7
Time
10:35
10:40
10:48
9:10:35
10:40
10:48

Fig. 8b

Frontobasal-Cingulate Seizures or Type II B Complex Partial Seizures. Most reports (Geier et al. in 1977 [21] and Ludwig et al. in 1976 [28]) of complex partial seizures originating from the frontal lobe emphasized the frequent occurrence of early motor activity and focal tonic and/or focal clonic features. More recently, Spencer et al. [46] emphasized sexual automatisms in psychomotor seizures of frontal lobe origin, while Williamson et al. [56-58] pointed out the gradual onset of attacks that were bizarre and frequently mistaken for hysterical seizures, e.g., bilateral motor automatisms frequently involving both legs and arms suddenly appearing after vocalization and a change in facial expression with features of running, fear, pelvic thrusts, frenetic, often bizarre, behavior, and the appearance of oroalimentary automatisms late or in the middle of the seizure [28, 51] (Fig. 9). Our earlier reports of Type II B psychomotor attacks are similar to those described by Williamson as frontal lobe complex partial seizures [56-58] and by Wieser [52] as frontobasal cingulate epilepsy. They are also similar to the 1984 reports by Wada and Purves [50], where initial automatisms were described as oral bimanual-bipedal. Wada and Purves [50] reviewed 14 patients whose clinical episodes, often nocturnal, displayed screaming, yelling, coughing, whining, labored breathing, and bimanual-bipedal activity. Patients also had occasional versive movement, laughter, oral automatism, and genital manipulation during partial or total unresponsiveness. Three patients benefited from anterior two-thirds callosal section while anterior temporal lobectomy in two patients had no effect.

Williamson et al. [56-58] have recently shown ictal rhythmic discharges from the frontal lobe during psychomotor attacks identical to Type II B. In 10 patients, they demonstrated by depth electrodes orbital frontal, medial frontal, diffuse medial-intermediate frontal, diffuse medial frontal, or supplementary motor ictal rhythms [56-58] during such psychomotor attacks. Williamson et al. emphasized, in addition to the bizarre motor and sexual automatisms, the extreme brevity (32-s average with range of 15-70 s) and relative frequency (ten per day average; range 0 to 50 per day) of attacks and their propensity for developing into complex partial status epilepsy.

The interictal EEG is frequently normal or shows rare to infrequent spikes or sharp waves recorded by the sphenoidal nasoethmoidal or orbital electrodes. The ictal scalp EEG shows background low voltage fast rhythms (16-24/s) or irregular 1-4 Hz slowing. SEEG of both frontal and temporal areas is needed to define the exact site of origin and their routes of spread.

Figure 9a, b is an example of Type II B psychomotor seizures. It shows a 38-year-old man whose seizures start as a shriek, bilateral kicking movements, bilateral arm flailing movements, fearful facial expression, and subsequent oroalimentary automatisms and confusion. An oligodendroglioma was detected only on histological examination of the en bloc anterior temporal lobe specimens, but the character and frequency of seizures persist as documented by CCTV-EEG after surgery.

Figure 10a-c illustrates another case originally considered to be hysterical and potentially violent. A scream and frenetic bizarre arm and leg movements occurred with sexual gestures and impaired consciousness.

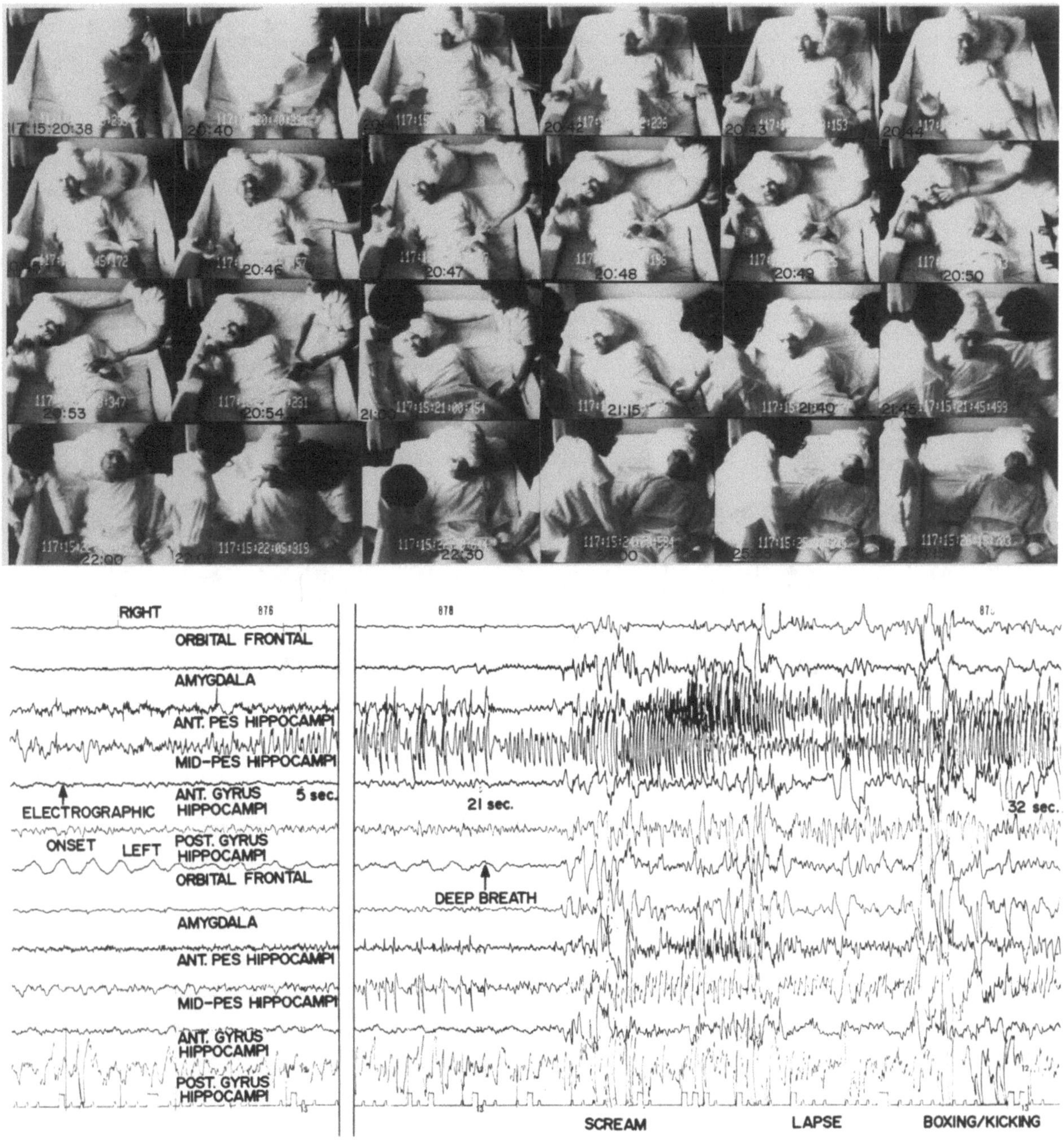

Fig. 9 a, b. Type II B complex partial seizures or frontobasal cingulate psychomotor attacks. **a** The sequence of screaming, impaired awareness, unresponsiveness, hissing, frenetic bilateral arm and leg movements, facial expression of fear, stroking of genitalia, pelvic thrusts, and lip smacking, chewing movements of the mouth followed by confusion. He shrieked at 20:43 and started kicking and flailing his arms from 20:44 to 21:40. At 21:45, he tried to get out of bed. **b** Depth electrographic events during psychomotor attack demonstrating ictal rhythms in the right hippocampal regions. The same pattern of seizures persist after right anterior temporal lobectomy

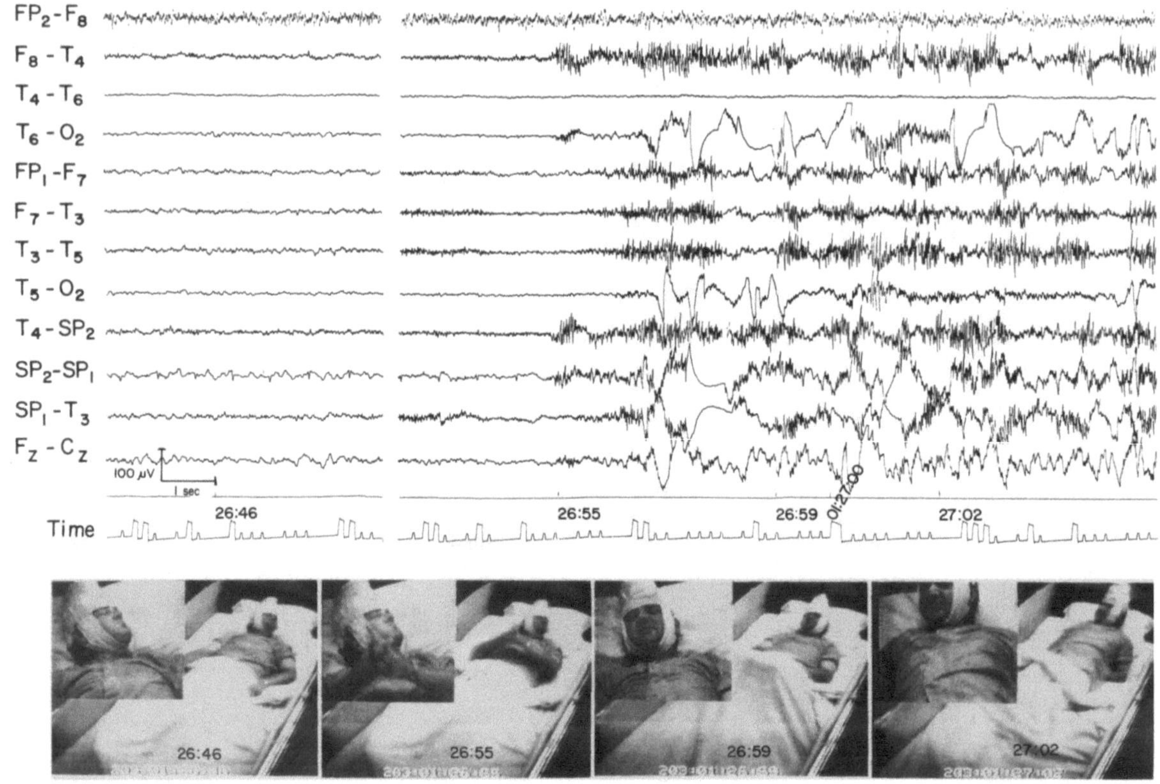

Fig. 10

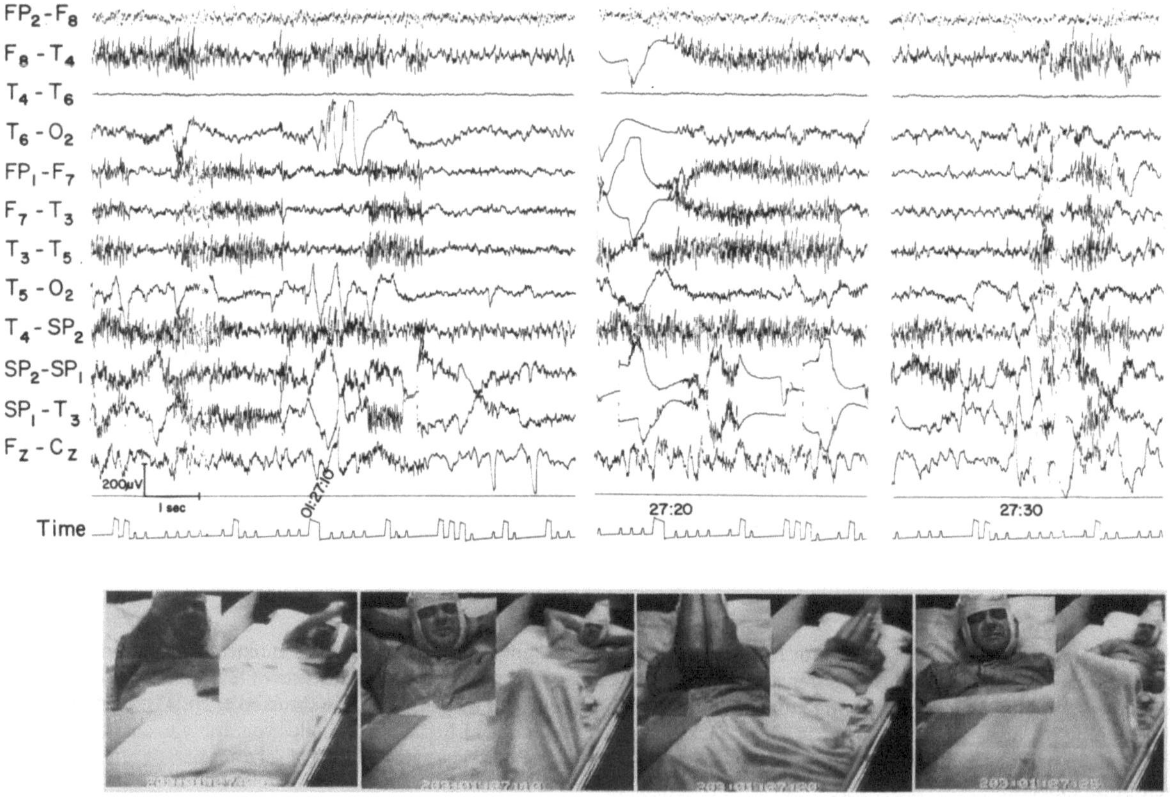

Fig. 10l

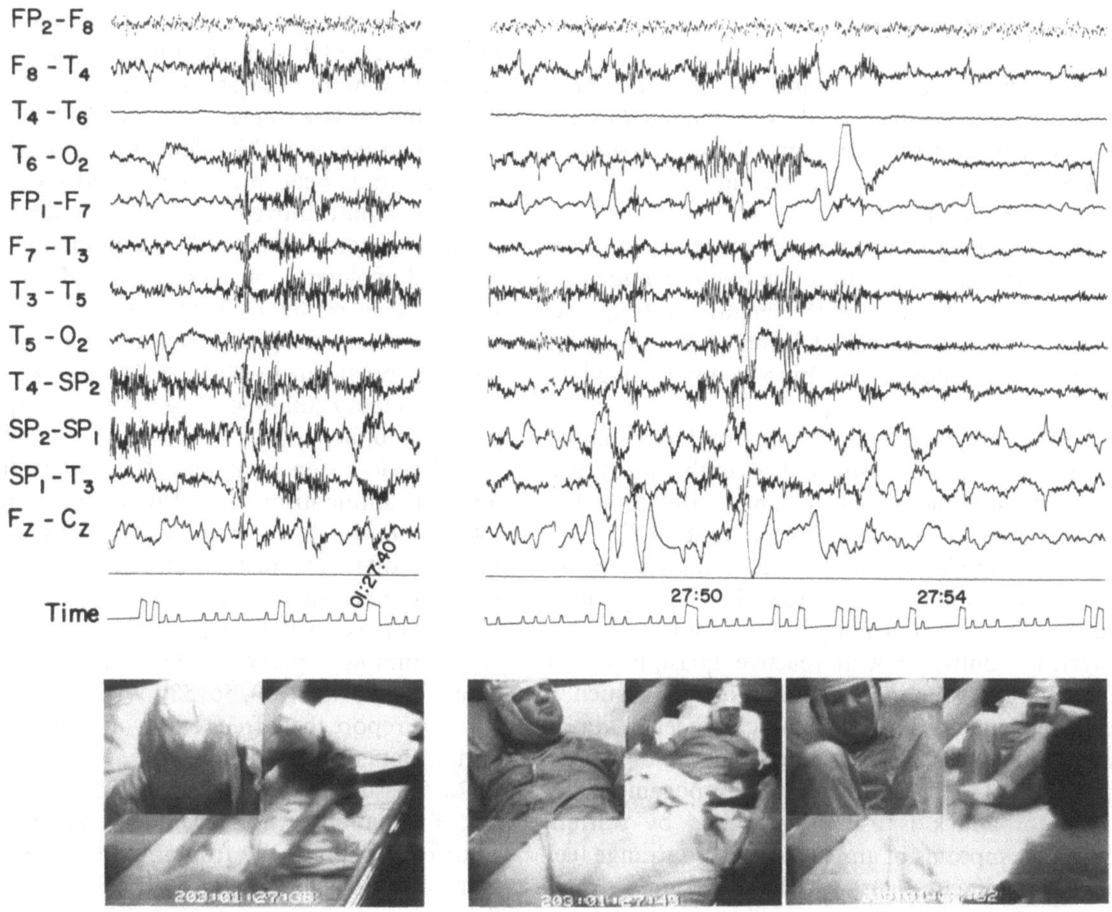

Fig. 10c

Fig. 10a–c. Type II B complex partial seizures. **a** At 26:56, the patient shrieks, "tee tee tee tee", grimaces, looks fearful, extends his lower extremities and feet, places his hands in front, then flexes his knees at 26:59. At 27:02 he places his hands behind his back as he continues to shriek. **b** At 27:08, he extends his upper extremities and hand and at 27:10 he places his hands behind his neck and head. He continues these apparently purposeless movements, grimaces, shrieks, and **c** he attempts to move out of the bed. The scalp EEG is initially low amplitude except for the F_Z-C_Z electrodes where 4–6/s slow waves appear (**a**). **b** Rhythmic 6/s slow waves spread diffusely

Conclusions and Outlook

When a clinical seizure occurs, clinicians frequently have to decide, by the bedside, the meaning of a given behavior and the site from where the behavior comes. We analyzed complex partial seizures mainly from this point of view.

The following summarizes some practical differential points about the aura, the initial conduct, the character of automatisms, the length and frequency of psychomotor seizures of frontal lobe origin and psychomotor seizures of hippocampal-amygdalar origin:

1. Of complex partial seizures of hippocampal-amygdalar origin 56% show at onset the arrest reaction with motionless staring phase, oroalimentary automatisms, or a single discrete motion [29]. The typical seizure lasts 2 min; the motionless stare lasts 5-20 s, and is invariably followed by lip smacking, pursing lips (oroalimentary automatisms), and then lingering confusion with reactive quasipurposeful behavior for 1½ min. Somatomotor signs, such as tonic movements, postures, and clonic jerks, appear late in the attacks as a manifestation of seizure spread [11, 13, 15, 51]. The classic hippocampal psychomotor attack is usually preceded by warning signs or symptoms of an "indescribable strange feeling," fear, olfactory-gustatory hallucinations, experiential hallucinations of déjà vu, jamais vu, or interpretative illusions [5, 9, 24, 52]. When the initial warning signs or symptoms (aura) consist of a rising epigastric discomfort, autonomic symptoms, and signs including borborygmi, belching, pallor, fullness of face, flushing of face, arrest of respiration, pupillary dilation, an amygdalar onset with hippocampal involvement should be considered [47, 52]. A combination of warning signs and symptoms pointing to both amygdalar-hippocampal origin is more common. Signs and symptoms pointing to the lateral posterior temporal, or opercular or occipital cortices during the aura should differentiate the primary hippocampal-amygdalar epilepsies from the secondary involvement of the hippocampus and amygdala by epileptiform paroxysms propagating from the lateral temporal neocortex, operculum or occipital cortex [5, 9, 52].

2. In contrast to the sudden or gradual appearance of an arrest reaction with motionless staring and oroalimentary automatisms seen in amygdalar-hippocampal epilepsy, Type II B complex partial seizures which may originate outside the anterior temporal lobe, such as the frontal basal and cingulate cortex, begin with vocalization, bilateral motor automatisms, bipedal movements, bizarre frenetic arm and leg thrusting behavior, sexual automatisms, running, and fear [50-52]. Oroalimentary automatisms and even a motionless staring phase may appear in the middle or late stages of the attack [2, 3].

3. When postures, M2e fencing positions, contraversive eye or head movements, and partial motor tonic or clonic movements occur at onset of impaired consciousness and psychomotor automatisms, dorsolateral or medial frontal lobe origin should be suspected [2, 3, 37-39]. Sphincter incontinence and sexual automatisms are more common in medial frontal seizures [46, 56-58] like anterior cingulate epilepsy, although they have been reported in opercular epilepsy [5].

4. Pseudo-absence attacks lasting 10-30 s with minimal automatisms are also observed in frontal lobe epilepsies [44]. Such short attacks occur frequently, often in clusters, and may be associated with sphincter incontinence.

5. Complex partial seizures of frontal lobe origin very commonly spread quickly to evolve as a tonic-clonic convulsion [2, 3, 56-58]. Williamson et al. [56, 58] also report their frequent evolution to complex partial status epilepticus.

Rasmussen [44] recently reviewed the characteristics of 40 patients who had undergone cortical excisions of portion of the frontal lobe for the relief of medically refractory epilepsy. The extent of the cortial excisions consisted of (a) excision of the anterior one-third or one-half of the frontal lobe in 18 patients; (b) a more complete frontal lobectomy in nine patients; (c) excision of the frontal parasagittal region in eight patients; and (d) excision limited to the convexity of the frontal lobe in five patients.

The incidence of an aura and the general characteristics of the aura when present were not significantly different among the four categories of frontal lobe excision. The incidence of pseudo-absence episodes and automatisms did not differ among these four categories. Contraversive head movements at onset of generalized seizures appeared to be the most significant observation. It was present in all nine patients with maximal or near maximal partial lobectomy and in half of the remaining 31 patients in the other three categories.

Thirty-six of 40 patients had generalized convulsions. Most patients had more than one type of attack. Pseudo-absences or amnesic episodes were present in 14 patients, automatisms in 12, and both were present in eight. Focal sensorimotor seizures were present in a fourth of patients, while postural seizures occurred in three patients. Seventeen patients had no warning symptoms or signs. And six patients had auras more commonly seen in temporal lobe epilepsy such as confusion or familiarity. Another five patients had an epigastric or chest sensation. Five had a vague head sensation or cephalic aura, and another five had generalized body sensation of some sort. Seven had a somatosensory aura at seizure onset.

From these data, Rasmussen [44] suggested that the spectrum of seizure patterns did not give reliable evidence for localization of the seizure focus. The sole exception, Rasmussen claimed, was initial contraversion of the head. In the patients with contraversive head movements, maximal frontal lobectomy was considered necessary.

Our recent experiences have taught us to be circumspect in differentiating opercular temporal lobe seizures, dorsolateral frontal lobe seizures with spread to Broca's area, and supplementary motor seizures with speech arrest. Even more difficult is the exact delimitation and differentiation of the epileptogenic zones in the cingulate gyrus and the supplementary motor gyrus. We, therefore, agree with Rasmussen's cautious remarks and suggest that SEEG is necessary to clarify the site of origin and route of spread of seizures even when magnetic resonance imaging and positron emission tomography delineate the functionally damaged areas that are potentially epileptogenic.

Because of the precision with which the above subvarieties of complex partial epilepsies can now be identified (see Tables 2 and 3), epileptologists and surgical neurologists now question whether large en bloc lobectomies are necessary to extirpate epileptogenic zones. More precisely, they ask: Should not selective microscopic neurosurgical resections, e.g., partial or total hippocampectomy and amygdalectomy [35, 52, 55], suffice in hippocampal-amygdalar epilepsy or mesiobasal limbic epilepsy? Should not selective and limited cortectomies suffice in well delineated complex partial epilepsies of frontal lobe origin? Selective removal of epileptogenic zones should result in less tissue removed and preservation of brain functions [52, 55].

The recent advent of microneurosurgical techniques has made selective amygdalohippocampectomy possible and the success rate of amygdalohippocampectomies has reached considerable proportions in the hands of Wieser and Yaşargil. Twenty-two out of 27 hippocampectomies resulted in abolition of psychomotor attacks, an 85% success rate [52, 55]. Most recently, they have reported similar success rates in over 100 cases of amygdalohippocampectomies (see chapter by Yaşargil and Wieser, this volume). Similar success rates have also been reported from New Haven, in a modified anterior temporal lobectomy with anterior and posterior hippocampectomy [45].

Because of these recent dramatic developments, our program now advocates selective microscopic neurosurgical resections, such as amygdalohippocampectomies, in amygdalar-hippocampal epilepsies. Anterior temporal lobectomy should be reserved for epilepsies where both lateral and medial anterior temporal lobes contribute to epileptogenesis. In Type II A psychomotor attacks of frontal lobe origin, selective and limited cortectomies should first be tried for well delineated epileptogenic zones. More studies are necessary to delineate the exact origins of Type II B psychomotor attacks.

References

1. Ajmone-Marsan C (1980) Depth electrography and electrocorticography. In: Aminoff MJ (ed) Electrodiagnosis in clinical neurology. Churchill Livingstone, New York, pp 167-196
2. Ajmone-Marsan C, Abraham K (1960) A seizure atlas. Electroencephalogr Clin Neurophysiol 15 (suppl): 1-215
3. Ajmone-Marsan C, Ralston BL (1957) The epileptic seizure: its functional morphology and diagnostic significance. Thomas, Springfield
4. Bancaud J (1980) Surgery of epilepsy based on stereotactic investigations - the plan of the SEEG investigation. Acta Neurochir (Vienna) 30 (suppl): 25-34
5. Bancaud J (1981) Epileptic attacks of temporal origin in man. Jpn J EEG EMG, 61 (Suppl): 71-79
6. Bancaud J, Talairach J (1975) Macro-stereo-electroencephalography in epilepsy. In: Bancaud J (ed) Handbook of EEG and clinical neurophysiology, vol 10. Elsevier, Amsterdam pp 1013-1033
7. Bancaud J, Talairach J, Bresson M, Morel P (1968) Accès épileptiques induits par la stimulation du noyau amygdalien et de la corne d'Ammon. Intérêt de la stimulation dans la détermination des épilepsies temporales chez l'homme. Rev Neurol (Paris) 6: 527-532
8. Bancaud J, Talairach J, Morel P, Bresson M (1966) La corne d'Ammon et le noyau amygdalien: effects cliniques et électriques de leur stimulation chez l'homme. Rev Neurol 3: 329-352
9. Bancaud J, Talairach J, Bonis A, Schaub A, Szikla G, Morel P, Bordas-Ferer M (1965) La stéréo-électroéncephalographie dans l'épilepsie. Masson, Paris
10. Bonis A (1980) Long term results of cortical excision based on stereotactic investigations in severe drug resistant epilepsies. Acta Neurochir (Vienna) 30 (suppl): 55-66
11. Bossi L, Munari C, Stoffels C, Bonis A, Bacia T, Talairach J, Bancaud J (1984) Somatomotor manifestations in temporal lobe seizures. Epilepsia 25 (1): 70-76
12. Delgado-Escueta AV, Kunze U, Waddell G, Boxley J, Nadel A (1977) Lapse of consciousness and automatisms in temporal lobe epilepsy: a videotape analysis. Neurology (Minneap) 27: 144-155
13. Delgado-Escueta AV, Bacsal F Enrile-, Treiman DM (1982) Complex partial seizures on closed-circuit television and EEG: a study of 691 attacks in 79 patients. Ann Neurol 11 (3): 292-300

14. Delgado-Escueta AV, Nashold B, Freedman M, Keplinger C, Waddell G, Miller P, Carwille S (1979) Videotaping epileptic attacks during stereo-electroencephalography. Neurology 29: 473-489
15. Delgado-Escueta AV, Walsh GO (1985) Type I complex partial seizures of hippocampal origin: Excellent results of anterior temporal lobectomy. Neurology 35 (2): 143-154
16. Farwell JR, Stuntz JT (1984) Frontoparietal astrocytoma causing absence seizures and bilaterally synchronous epileptiform discharges. Epilepsia 25 (6): 695-698
17. Feindel W (1961) Response patterns elicited from the amygdala and deep temporal insular cortex. In: Sheer DE (ed) Electrical stimulation of the brain. Texas University Press, Austin
18. Feindel W (1974) Temporal lobe seizures. In: Vinken PJ, Bruyn GW (eds) The Epilepsies. North Holland, Amsterdan, pp 87-106 (Handbook of Clinical Neurology, vol 15)
19. Feindel W, Penfield W (1954) Localization of discharge in temporal lobe automatisms. AMA Arch Neurol Psychiat 72: 605-630
20. Geier S, Bancaud J, Talairach J, et al. (1975) Clinical note: clinical and telestereo-EEG findings in a patient with psychomotor seizures. Epilepsia 16: 119-125
21. Geier S, Bancaud J, Talairach J, et al. (1977) The seizures of frontal lobe epilepsy. Neurology (Minneap) 27: 951-958
22. Gloor P (1975) Physiology of the limbic system. In: Penry JK, Daly DD (eds) Complex partial seizures and their treatment. Raven, New York, pp 27-43
23. Gloor P, Olivier A, Quesney LF, Andermann F, Horowitz S (1982) The role of the limbic system in experiential phenomena of temporal lobe epilepsy. Ann Neurol 12: 129-144
24. Jackson JH (1899) On a particular variety of epilepsy ("Intellectual aura"), one case with symptoms of organic brain disease. Brain 11: 179-207
25. Jackson JH, Stewart Purves (1899) Epileptic attacks with warning sensations of smell and with intellectual aura (dreamy state) in a patient who had symptoms pointing to gross organic disease of the right temporosphenoidal lobe. Brain 22: 534-549
26. Jones WHS (1952) The sacred disease In: Hippocrates. Harvard Univ Press, Cambridge
27. Lüders H, Lesser RP, Dinner DS, Hahn JF, Salanga V, Morris HH (1985) The secondary sensory area in humans: evoked potentials and electrical stimulation studies. Ann Neurol 17 (2): 177-184
28. Ludwig BI, Ajmone-Marsan C, Van Buren J (1976) Depth and direct cortical recording in seizure disorders of extratemporal origin. Neurology (Minneap) 26: 1085-1099
29. Maldonado HM, Delgado-Escueta AV, Rand RW, Walsh GO, Halgren E, Treiman DM, Berman D (1986) Complex partial seizures of hippocampal and amygdalar origin: Epilepsia, submitted 1987
30. Mazars G (1969) Cingulate gyrus epileptogenic foci as an origin for generalized seizures in: Gastaut H, Jasper H, Bancaud J, Waltregny A (eds) The physiopathogenesis of the epilepsies. Thomas, Springfield, pp 186-189
31. Munari C, Bancaud J, Bonis A, Stoffels C, Szikla G, Talairach J (1980) Impairment of consciousness in temporal lobe seizures: a stereo-EEG study. In: Canger R, Angeleri F, Penry JK (eds) Advances in epileptology, XIth Epilepsy international symposium, Raven, New York, pp 111-114
32. Munari C, Bonis A, Stoffels C, Bossi L, Talairach J, Bancaud J (1980) Automatic activities during frontal and temporal lobe seizures: Are they the same? Acta Neurol Scand Suppl 79 (62): 40
33. Munari C, Talairach J, Bonis A, Szikla G, Bancaud J (1980) Differential diagnosis between temporal and "perisylvian" epilepsy in a surgical perspective. Acta Neurochir (Vienna) 30 (Suppl): 97-101
34. Nadel A, Delgado-Escueta AV (1975) Lapse myclonic seizures and automatisms with diffuse 9-Hz rhythms. Epilepsia 16: 511-520
35. Niemeyer P (1958) The transventricular amygdala-hippocampectomy in temporal lobe epilepsy. In: Baldwin M, Bailey P (eds) Temporal lobe epilepsy. Thomas, Springfield, pp 461-482
36. Pazzaglia P, D'Alessandro R, Ambrosetto G, Lugaresi E (1985) Drop attacks: an ominous change in the evolution of partial epilepsy. Neurol 35 (12): 1725-1730
37. Penfield W, Flanigin H (1950) Surgical therapy of temporal lobe seizures. AMA Arch Neurol Psychiatry 64: 491-500
38. Penfield W, Jasper H (1954) Epilepsy and the functional anatomy of the human brain. Little, Brown, Boston
39. Penfield W, Kristiansen K (1951) Epileptic seizure patterns. Thomas, Springfield
40. Penfield WP, Perot P (1963) The brain's record of auditory and visual experience. A final summary and discussion. Brain 86: 595-696
41. Penry JK, Dreifuss FE (1969) Automatisms associated with the absence of petit mal epilepsy. Arch Neurol 21: 142-149
42. Prichard JC (1822) A treatise on diseases of the nervous system. London
43. Quesney LF, Krieger C, Leitner C, Gloor P, Olivier A (1983) Frontal lobe epilepsy: clinical and electrographic presentation. 15th Epilepsy international symposium, Washington DC, Sept 26-30. Abstract p 177
44. Rasmussen T (1983) Characteristics of a pure culture of frontal lobe epilepsy. Epilepsia 24 (4): 482-493
45. Spencer DD, Spencer SS, Mattson RH, Williamson PD, Novelly RA (1984) Access to the posterior medial temporal lobe structures in the surgical treatment of temporal lobe epilepsy. Neurosurg 15: 667-671
46. Spencer SS, Spencer DD, Williamson PD, Mattson RH (1983) Sexual automatisms in complex partial seizures. Neurology (NY) 33 (5): 527-533
47. Swartz BE, Delgado-Escueta AV, Maldonado HM, Walsh GO, Cahan L (1985) Pupil asymmetry, thoughts of helplessness, and fear recorded with depth discharges in amygdalar epilepsy. Epilepsia 26 (5): 530
48. Talairach J, Szikla J, Tournoux P, et al. (1967) Atlas d'anatomie stereotaxique du telencephale. Masson, Paris
49. Tükel K, Jasper H (1952) The electroencephalogram in parasagittal lesions. Electroencephal Clin Neurophysiol 4: 481-494
50. Wada JA, Purves SJ (1984) Oral and bimanual-bipedal activity as ictal manifestations of frontal lobe epilepsy. Epilepsia 15 (5): 668
51. Walsh GO, Delgado-Escueta AV (1984) Type II complex partial seizures: poor results of anterior temporal lobectomy. Neurology (Cleveland) 34: 1-13
52. Wieser HG (1983) Electroclinical features of the psychomotor seizure. Fischer, Stuttgart; Butterworths, New York
53. Wieser HG, Bancaud J, Talairach J, Bonis A, Szikla G (1979) Comparative value of spontaneous and chemically and electrically induced seizures in establishing the lateralization of temporal lobe seizures. Epilepsia 20: 47-59
54. Wieser HG, Meles HP, Bernoulli C, Siegfried J (1980) Clinical and chronotopographic psychomotor seizure patterns

(SEEG study with reference to postoperative results). Acta Neurochir (Vienna) 30 (suppl): 103-112
55. Wieser HG, Yaşargil MG (1982) Selective amygdalohippocampectomy as a surgical treatment of mesiobasal limbic epilepsy. Surg Neurol 17: 445-457
56. Williamson PD, Spencer DD, Spencer SS, Novelly RA, Mattson RH (1983) Complex partial status of frontal lobe origin: a depth electrode study. Epilepsia 24: 260
57. Williamson PD, Spencer DD, Spencer SS, Novelly RA, Mattson RH (1985) Complex partial seizures of frontal lobe origin. Ann Neurol 18 (4): 497-504
58. Williamson PD, Spencer DD, Spencer SS, Novelly RA, Mattson RH (1985) Complex partial status epilepticus: a depth electrode study. Ann Neurol 18 (6): 647-654
59. Willis T (1667) Of the epilepsy or falling sickness. In: An essay of the pathology of the brain and nervous stock in which convulsive diseases are treated of. Dring T, London (1681)

Focal Epilepsies of Nontemporal and Nonfrontal Origin

T. Rasmussen

Introduction

Although neuroanatomists consider the fissure of Rolando to be the dividing line between the frontal and parietal lobes, the effects of cortical stimulation and of discrete excisions of the pre- and postcentral gyri are so different from the frontal cortex in front and the parietal cortex behind that we have considered the pre- and postcentral gyri to be a distinct anatomical and functional brain unit, the sensorimotor or central region. Thus the patients to be considered in this report are those whose epileptogenic cortex involved primarily the central or sensorimotor region, the parietal lobe behind the postcentral gyrus, and the occipital lobe (Table 1).

In 7% of the total MNI surgical seizure series, the principal epileptogenic area involved mainly the central or sensorimotor area, the parietal region was mainly involved in 6%, and the occipital lobe in 1%. The three groups combined contained 320 patients, 14% of the total surgical seizure series.

The etiological classification was approximately the same in these three anatomical regions and the data are combined for ease of presentation (Table 2).

Nontumoral epileptogenic lesions were present in 203 patients (63%) and a variety of neoplasms, other sizeable nodular lesions, and major vascular lesions were present in 117 (37%). This high incidence of tumoral lesions compared with that in the temporal and frontal divisions of our surgical seizure series is to a considerable extent an artifact of the selection of the surgical strategy in many of these cases. The increased safety of operating under local anesthesia resulted in some patients with seizures due to tumors, diagnosed preoperatively and located close to the speech or sensorimotor region, being operated upon with the special surgical seizure techniques, whereas patients with similar lesions further away from critical areas were often operated upon with standard brain tumor surgical technique even though seizures were the patients' main symptom. These patients therefore did not appear in this surgical seizure series. Because of space constraints, this report will consider only the 203 patients with nontumoral epileptogenic lesions, although the patients in the tumoral group are of equal interest from several standpoints.

Among the nontumoral lesions, those due to perinatal brain injury and postnatal head injury and those of unknown etiology were of equal frequency, each 15% or 16% of the total series. The miscellaneous category was next (10%) and included a variety of interesting lesions. The postinflammatory category (5%) and the multiple factor group (2%) were underrepresented compared with the temporal and frontal lobe divisions of this series.

Table 1. Montreal Neurological Institute surgical seizure series: anatomical classification – patients operated upon 1929 through 1980

	No. of patients (%)
Temporal lobe	1210 (56)
Frontal lobe	402 (18)
Central (sensorimotor) region	151 (7)
Parietal lobe	141 (6)
Occipital lobe	30 (1)
Large multilobe lesions	243 (11)
Total	2177

Table 2. Etiology of central, parietal, and occipital epileptogenic lesions

	No. of patients (%)	
Birth trauma or anoxia	51 (16)	203 (63)
Postnatal trauma	49 (15)	
Postinflammatory brain scarring	17 (5)	
Multiple potential etiological factors	6 (2)	
Miscellaneous	33 (10)	
Unknown	47 (15)	
Tumoral and vascular lesions	117 (37)	
Total	320	

Montreal Neurological Institute and Hospital, Department of Neurology and Neurosurgery, McGill University, 3801 University Street, Montréal, Québec H3A 2B4, Canada

Presurgical Evaluation of Epileptics
Edited by H. G. Wieser and C. E. Elger

Preoperative Neurological Status

About 15% of the patients in this group preoperatively had some motor or sensory hemiparesis, minimal or moderate. Another 10% or so had reflex abnormalities indicating damage to the sensorimotor pathways but did not show motor or sensory deficits detectable by routine neurological examinations. Subtle, measurable deficits were sometimes identified by detailed neuropsychological testing, however, in patients whose routine neurological examination showed a normal sensorimotor status (Corkin et al. 1973).

A hemianopsia, usually partial, was often present when the epileptogenic area involved the posterior parietal and/or the occipital region, but was rarely present when the cortical epileptogenic area was restricted to the central region and/or to the parietal parasagittal area.

As in the temporal and frontal lobe subgroups of the MNI surgical seizure series, the majority of the 203 patients in this subgroup had intellectual capacities in the normal range, with only a minority showing significant mental retardation.

Radiological Data

The skull X-ray often showed relative smallness of the involved hemicranium when the brain lesion responsible for the seizures occurred during the first few years of life. Pneumoencephalograms sometimes showed focal areas of atrophy or porencephalic cysts in the posterior half of the hemisphere, but often only showed evidence of moderate generalized hemispheral atrophy, more widespread than the patients' seizure pattern would suggest.

As in the case of frontal and temporal lobe epilepsy the CT scan during the past decade has provided important diagnostic help in identifying focal hypo- and hyperdense epileptogenic brain lesions. Occasionally, however, small discrete nodules such as tuberous sclerosis, small thrombosed vascular malformations, and hamartomas escaped detection even with current CT technology. These lesions are now usually well demonstrated with magnetic resonance imaging techniques, however.

Seizure Patterns

The initial phenomena of the attack patterns in the great majority of the patients of both the central and parietal groups consisted of sensory or motor phenomena involving one arm and/or one side of the face, often associated with or followed by other focal features (Penfield and Kristiansen 1951). When the attacks began with a sensory onset, the sensation was usually described as a tingling numbness or a dead, heavy feeling. In the case of a motor onset, the movements were usually clonic in nature. In some patients, however, the initial motor activity consisted of a tonic posturing of the arm or occasionally consisted of myoclonic-like jerks of the extremity. Secondary generalization of the sensorimotor seizures was relatively frequent and occurred to some extent in most of the patients. Sensory and motor precipitation of the seizures was somewhat more frequent than in the temporal and frontal subgroups of this series.

When the epileptogenic area extended near or into the posterior temporal region, automatisms and perceptual illusions or visual or auditory hallucinatory phenomena were sometimes part of the minor seizure patterns. In the dominant hemisphere, inability to talk during the minor attacks and brief postictal dysphasia were sometimes part of the patients' seizure pattern, particularly when the epileptogenic area was close to either the anterior or the posterior speech areas.

When the epileptogenic area involved mainly the posterior parietal and occipital regions the seizures usually began with unformed visual hallucinations. In a few patients, the initial phenomenon was a graying or blacking out over the whole or part of the visual field.

In about 10% of the patients in this centroparietal-occipital subgroup, the seizure patterns did not give useful localizing information.

EEG Patterns

The epileptiform abnormalities in the EEG were surprisingly varied. Sometimes the spike discharges were well localized to the region indicated by the nature of the seizure onset. In other patients with similar seizure patterns, random spiking involved most of the frontocentral-parietal region even though the patient's seizure pattern suggested the seizures arose in a more restricted cortical area.

Bilaterally synchronous epileptiform discharges, either slow spike-and-wave complexes or multiple spike bursts, were present in a significant number of the patients. These synchronous epileptiform abnormalities occurred in some patients with parasagittal epileptiform lesions, as described by Tükel and Jasper (1952), and in other patients with large epilep-

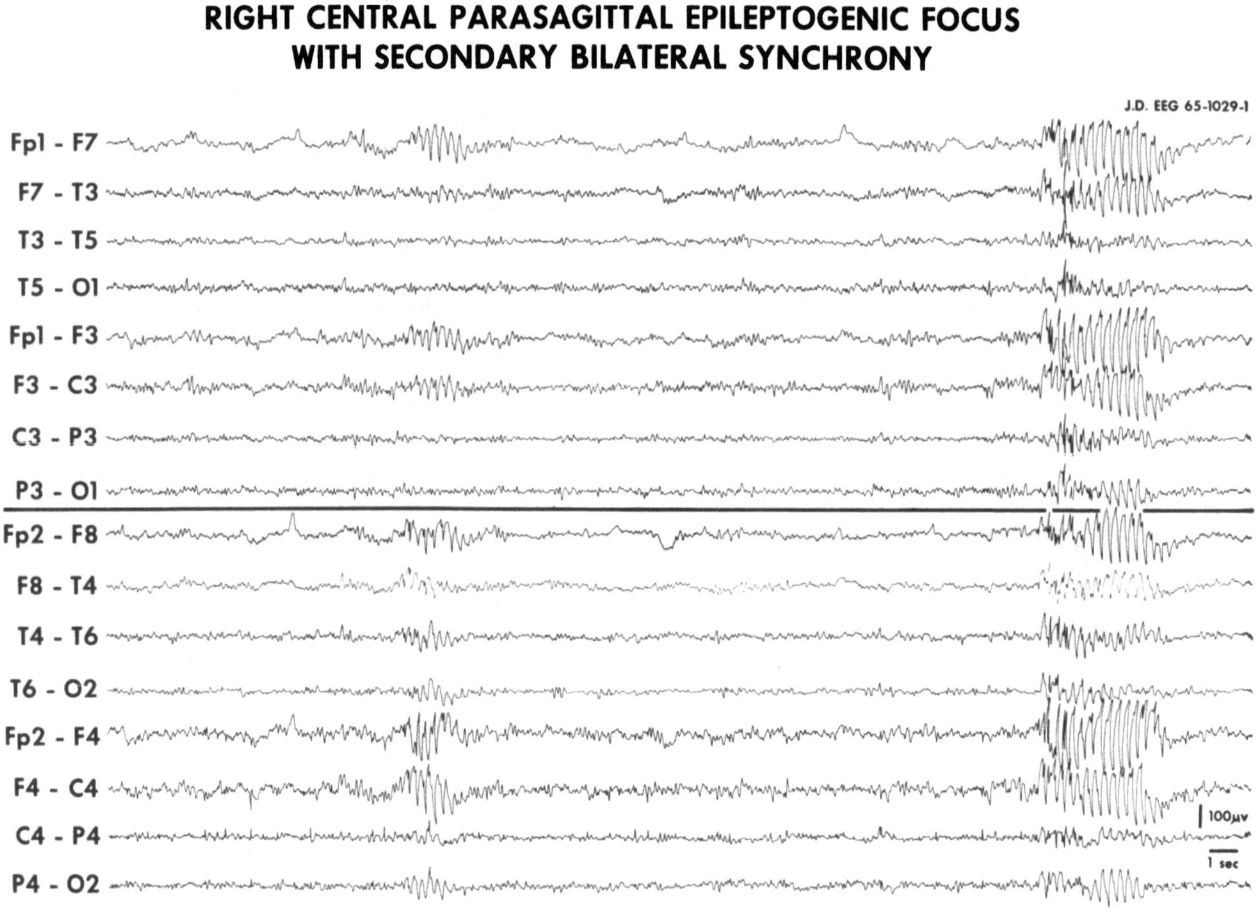

Fig. 1. Patient J. Du. Somatosensorimotor seizures beginning in the left arm. Preoperative EEG

togenic areas involving the convexity of the parieto-occipital region (Fig. 1).

There was often some predominance of the discharges over the side and presumed area of origin of the seizures, as indicated by the patient's seizure pattern, but in some patients the epileptiform discharges were quite symmetrical and gave little or no lateralizing information, even though the X-ray studies gave clear and sometimes spectacular evidence of focal atrophy.

At operation the cortical EEG was nearly as variable, sometimes showing well-restricted spike foci and in other patients showing widespread areas of synchronized epileptiform discharges (Fig. 2). When there was an obvious area of damaged cortex the spiking area nearly always extended into more normal appearing cortex for at least several centimeters.

When the postexcision cortical EEG has shown residual spiking on the margins of the removal, further cortical resection has usually been carried out, unless the spiking area was in or close to speech areas or the motor arm area and more extensive cortical excision was considered to carry unacceptable risk to speech or motor functions.

Patterns of Cortical Excisions

The patterns of the cortical excisions varied widely, being tailored to the individual patient's seizure problem and EEG abnormalities. At one extreme were excisions limited to segments of the pre- and/or postcentral gyrus (Fig. 3).

In other patients the excisions were limited to the parasagittal region or to the mid-convexity of the parietal lobe. At the other extreme were excisions of the entire parietal lobe extending into the upper portion of the posterior temporal region, into the lower central region, or into the occipital lobe (Fig. 4).

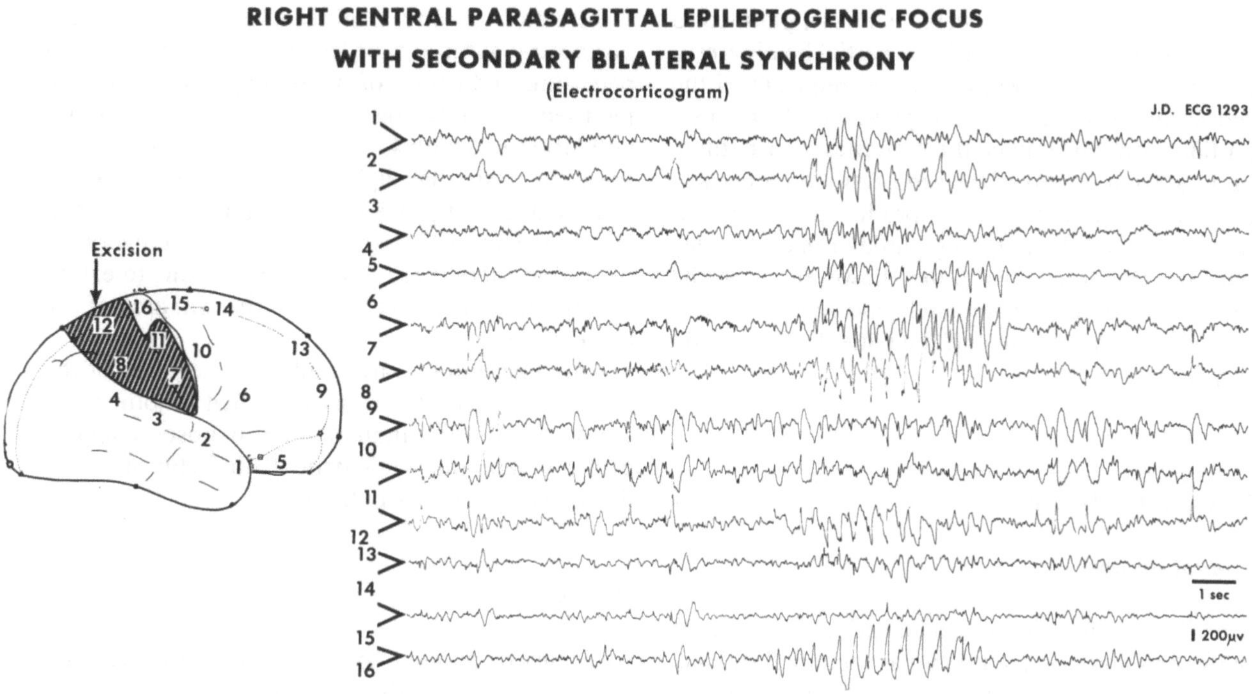

Fig. 2. Patient J. Du. Left frontoparietal preexcision cortical EEG. *Numbers* indicate cortical electrode placements. The *shaded area* indicates subsequent cortical excision of the face and arm area of the postcentral gyrus and adjacent parietal cortex

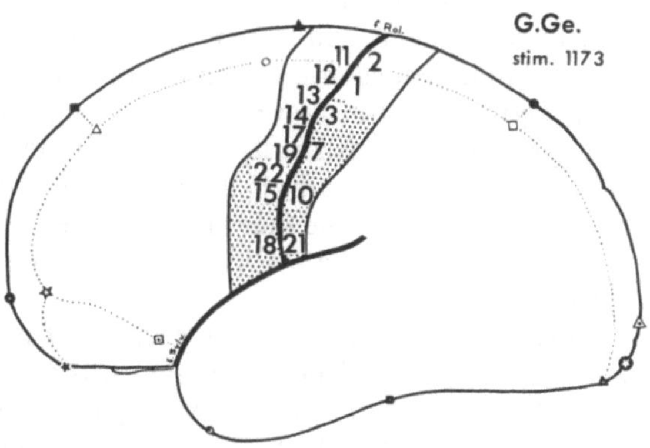

Fig. 3. Patient G. Ge. Somatosensorimotor seizures beginning in the right side of the face. *Numbers* indicate sensory and motor responses to electrical stimulation. The *shaded area* indicates subsequent cortical excision of the face area of the precentral gyrus and face and the lower half of arm area of the postcentral gyrus

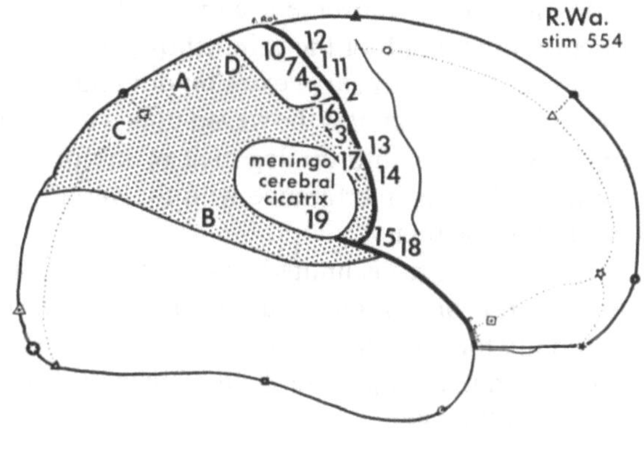

Fig. 4. Patient R. Wa. Somatosensorimotor seizures beginning in the left arm. *Numbers* indicate sensory and motor responses to electrical stimulation. *Letters* indicate sites of maximal epileptic activity in the cortical EEG. The *shaded area* indicates subsequent cortical excision of the face and arm area of the postcentral gyrus and the parietal lobe

The cortical excisions were, as a rule, planned to avoid producing or increasing a partial hemiparesis. In some cases, however, where the frequency of the sensorimotor seizures or the presence of epilepsia partialis continua limited the usefulness of arm function, excision of part of the sensorimotor cortex was carried out, even though this produced some weakness or increased an existing partial arm and/or leg paresis. This was only done, however, after the procedure had been completely and repeatedly discussed with the patient and the patient's family and physician.

Postoperative Deficits

Central Region

When the cortical excision is limited to the face area of the pre- and postcentral gyri as identified by electrical stimulation at threshold intensity, most patients in the early postoperative period show a definite underaction of the lower facial musculature of the contralateral side and a moderate reduction of two-point discrimination over the lips and cheek of the involved side. The motor weakness of the lower face disappears during the next few months in about half of the patients and lessens appreciably in the others. Some reduction in two-point discrimination over the face persists, however, unless the postcentral gyrus resection is small. There is no motor or sensory deficit in the arm if the precentral cortical excision is limited to the face area and does not extend into the arm area, as identified by threshold electrical stimulation.

Most of the patients in whom excision of part or all of the arm area of the precentral gyrus has been carried out have had some preexisting weakness of the arm as a result of the brain damage responsible for the seizures. Excision involving the precentral arm area results in further reduction and slowness of individual finger movement along with a lesser reduction of strength of the grip, elbow, and shoulder movements. The arm area excision does not produce detectable leg weakness, however, unless the cortical excision extends into the leg area of the precentral gyrus or there is interference with the vascular supply to the leg area.

Cortical excisions limited to part or all of the leg area of the precentral gyrus produce reduction of voluntary toe and ankle movement but often there is only a minimal limp present when the patient is seen for late follow-up.

Resection of any significant proportion of the postcentral gyrus arm or leg area always results in a persisting reduction of two-point discrimination, joint sense, and stereognosis in the involved extremity (Corkin et al. 1964, 1970). In the case of the arm area, the resulting slight clumsiness of the hand usually does not prove to be a significant handicap to the patient in the ordinary activities of everyday life unless there is also a motor deficit due to extension of the cortical excision into the arm area of the precentral gyrus.

After excision of part or all of the leg area of the postcentral gyrus, the resulting deficit in joint sense and two-point discrimination rarely poses a significant problem, but a few patients find clumsiness of the lower leg a little troublesome.

Parietal Region

Excisions of parietal cortex behind the postcentral gyrus produce surprisingly little permanent neurologic deficit. In particular the clinical features of the Gerstmann syndrome have not been produced by large parietal cortical excisions of the nondominant hemisphere or by parietal parasagittal excisions above the posterior speech area in the dominant hemisphere. In the early postoperative period there are temporary deficits in the cortical modalities of sensation over the face, arm, or leg or of the entire side, depending on the extent and location of the parietal excision. These deficits disappear, however, when the temporary postoperative disturbances of neighborhood brain function, which we used to attribute to cerebral edema, disappear, unless the cortical excision involves the postcentral gyrus as well as the parietal cortex (Corkin et al. 1964, 1970).

In the early postoperative period following cortical excisions in or near the central region, subtle bilateral sensory deficits in the hands can sometimes be identified by special neuropsychological testing (Corkin et al. 1973). These subtle deficits often disappear during the early postoperative weeks or months. When some bilateral sensory deficit does persist it is not of clinical significance and usually cannot be detected by ordinary neurological examination techniques and is not noticeable by the patient in his or her everyday activities.

Occipital Region

When there is no, or only a minimal visual field defect preoperatively, and when the epileptogenic area

Table 3. Central, parietal, and occipital nontumoral epileptogenic lesions: results of cortical excisions

	No of patients (%)			
Seizure free since discharge	31	63 (34)	105 (56)	186 pts. with follow-up data of 2-51 years (median 18 years)
Became seizure free after some early attacks	32			
Seizure free 3 or more years, then rare or occasional attacks	28	42 (23)		
Marked reduction of seizure tendency	14			
Moderate or less reduction of seizure tendency	81 (44)			
Operative deaths	4			
Died in first two postoperative years	3			
Inadequate follow-up data	10			
Total	203			

is small, the occipital cortical excision can sometimes be carried out with preservation of useful vision in the involved half field. More commonly, however, the epileptogenic area is more widespread over the posterior parietal and occipital region and the preoperative examination demonstrates a significant or complete homonymous hemianopsia. In these patients the cortical excision is done without concern for the calcarine cortex or the optic radiations.

Results of Cortical Excisions on Seizure Tendency

The effectiveness of cortical excisions in reducing the seizure tendency to normal or near normal levels was more often limited by the proximity of the epileptogenic areas to the sensorimotor and speech areas than was the case in the frontal and temporal lobe divisions of this surgical seizure series and in the group with large multilobe epileptogenic lesions. There are follow-up data of 2 or more years in 186 of the 203 patients with nontumoral epileptogenic lesions of the central, parietal, and occipital areas (Table 3). Sixty-three patients (34%) became and remained seizure free. Another 42 patients (23%) have a marked but not quite complete reduction of the seizure tendency. Thus in 105 patients (56%) the cortical resection resulted in a complete or nearly complete reduction of the preoperative medically refractory seizure tendency.

References

Corkin S, Milner B, Rasmussen T (1964) Effects of different cortical lesions on sensory thresholds in man. Trans Am Neurol Assoc 89: 112-116

Corkin S, Milner B, Rasmussen T (1970) Somatosensory thresholds - contrasting effects of postcentral gyrus and posterior parietal lobe excisions. Arch Neurol 22: 41-58

Corkin S, Milner B, Taylor L (1973) Bilateral sensory loss after unilateral cerebral lesion in man. Trans Am Neurol Assoc 98: 25-29

Penfield W, Kristiansen K (1951) Epileptic seizure patterns. Thomas, Springfield

Tükel K, Jasper H (1952) The electroencephalogram in parasagittal lesions. Electroencephalogr Clin Neurophysiol 4: 481-494

Startle Epilepsy with Infantile Brain Damage: The Clinical and Neurophysiological Rationale for Surgical Therapy

P. CHAUVEL[1], J. P. VIGNAL[2], C. LIEGEOIS-CHAUVEL[1], J. P. CHODKIEWICZ[2], J. TALAIRACH[2], and J. BANCAUD[1]

Motor seizures triggered by unexpected acoustic or somatosensitive stimuli have been reported in patients with perinatally acquired hemiplegia (Alajouanine and Gastaut 1955). Up to now, they are generally considered as drug-resistant seizures. Identification of clinical and neurophysiological patterns of startle epilepsies lead us to propose surgical therapy (Bancaud et al. 1967, 1975).

Eighteen patients with severe startle epilepsy underwent an extensive study of electrical and clinical features of their seizures, as evaluation for possible surgical therapy. Most of them presented an infantile hemiplegia as a consequence of a hemispheric perinatal lesion. Neuroradiological data exhibited cranial and cerebral hemiatrophy, predominant in the pericentral area.

Seizures could be triggered by surprising stimuli, such as a sudden noise, or an unexpected tactile and/or proprioceptive (stumbling) stimulus. They were tonic seizures, unilaterally or bilaterally asymmetrical. Axial musculature and proximal parts of the limbs were first involved with adversion of head and eyes. Clonic jerks with partial onset could follow. In some patients, pure clonic seizures were liable to occur. Speech arrest was constant, and complete impairment of consciousness was rare. In every patient, spontaneous seizures with similar symptomatology were present.

Polygraphic recordings demonstrated that startle seizures consisted generally of a two-stage phenomenon, i.e., the startle itself, followed by the spreading seizure. This peculiar startle pattern was asymmetrical, predominant or limited to the hemiplegic side and, with effect of habituation, localized in one muscle. The seizure itself started precisely in that limb (or muscle) first concerned with the startle reflex and propagated to the contralateral corresponding limb as well as to the ipsilateral side. Scalp EEG recordings concomitantly exhibited a high-amplitude evoked spike, culminating at the vertex (the so-called "vertex spike"), followed by flattening and bilateral fast low-voltage discharge.

Stereoelectroencephalography investigations were performed after implantation of depth electrodes in the frontal and anterior parietal lobe (in some cases, in the superior temporal gyrus). The abnormal paroxysmal evoked potential and the low-voltage fast discharge both originated in altered motor areas of the frontal lobe. Precise delineation of the epileptogenic area was based upon SEEG recording of triggered as well as spontaneous seizures, analysis of paroxysmal evoked responses topography, and results of direct electrical stimulations at the same sites. In fact, it was lying in most of the cases in the premotor, supplementary motor, and primary motor areas, with various patterns of extent and overlapping.

Surgical corticectomy was performed in 13 out of the 18 presently reported cases. Detailed outlines of each operation depended on individual delineation of the epileptogenic area by SEEG investigation in each patient. Generally, it consisted of removal of lateral and medial posterior frontal cortex and anterior parietal cortex.

The results of surgical therapy were as follows: eight patients were completely seizure-free (67%); three patients presented an important (>75%) decrease of seizure frequency (25%); one patient was unchanged. No neurological and/or psychological extra deficit was caused by the operation.

The present data strongly suggest that startle epilepsy with infantile hemiplegia is actually a well-defined clinical entity. Its origin and mechanisms lie in an early-lesioned motor area of the cerebral cortex. This is why it represents a good indication for a surgical corticectomy.

The occurrence of an exacerbated startle reflex, localized in one or few muscles whose corresponding sensorimotor cortex is epileptogenic, constitutes the hallmark of this type of startle epilepsies (Chauvel et al. 1982).

[1] Unité de Recherches sur l'Epilepsie, Centre Paul Broca, INSERM U 97, 2 ter rue d'Alésia, 75014 Paris, France and
[2] Service de Neurochirurgie, Hôpital Ste. Anne, 1 rue Cabanis, 75014 Paris, France

Presurgical Evaluation of Epileptics
Edited by H. G. Wieser and C. E. Elger

References

Alajouanine T, Gastaut H (1955) La syncinésie sursaut et l'épilepsie sursaut à déclenchement sensoriel ou sensitif inopiné. 1- les faits anatomocliniques. Rev Neurol 98: 29-41

Bancaud J et al. (1967) Physiopathologie des épilepsies-sursaut (à propos d'une épilepsie de l'AMS). Rev Neurol 117 (3): 441-453

Bancaud J et al. (1975) Hypothèses neurophysiopathologiques sur l'épilepsie-sursaut chez l'homme. Rev Neurol 131 (8): 559-571

Chauvel P, Liegeois C, Bancaud J (1982) Etude neurophysiologique des épilepsies-sursaut: résultats préliminaires. CR Ligue Française contre l'épilepsie 1: 105-110

Symptomatic Partial Epilepsies with Secondary Bilateral Synchrony: Differentiation from Symptomatic Generalized Epilepsies of the Lennox-Gastaut Type

H. Gastaut[1], B. Zifkin[2], A. Maggauda[1], and E. Mariani[1]

Secondary bilateral synchrony (SBS) was first described by Jasper (1951), Jasper et al. (1951), and Tükel and Jasper (1952) as bursts of high-amplitude synchronous slow spike-and-wave complexes which were more or less symmetric over both hemispheres, due to a unilateral epileptogenic lesion of the mesial surface of the frontal or temporal lobe. SBS is thus contrasted with primary bilateral synchrony, which is manifested by more rapid symmetric and bilaterally synchronous spike-and-wave discharges over both hemispheres, due to a generalized epileptogenic process independent of any focal hemispheric lesion. Since then, SBS has been discussed by Gastaut et al. (1968), Niedermeyer (1968, 1972), Gabor and Ajmone-Marsan (1969), Blume and Pillay (1985), Gastaut et al. (1985) and Spencer et al. (1985).

The definition of the Lennox-Gastaut syndrome (LGS) among the secondary generalized epilepsies, electroencephalographically manifested by bursts of slow spike-and-wave complexes resembling those of SBS much more than the typical spike-and-wave complexes of primary generalized epilepsy, has made it imperative to describe the electroclinical features of the symptomatic partial epilepsies with SBS so that they may be better distinguished from those of LGS.

Material and Method

We studied 33 patients with symptomatic partial epilepsy and SBS confirmed by EEG. The EEG features necessary for this diagnosis were the presence of focal epileptiform discharges followed by bursts of bilaterally synchronous spike-and-wave complexes. The causal relationship between these two features was considered as adequately shown when the discharges of the primary focus were followed within 100–200 ms by the bilaterally synchronous spike-and-wave complexes. This relationship may be difficult to determine if the primary focus gives rise only to occasional and brief bilateral spike-and-wave bursts, or if the bilateral spike-and-wave activity is so abundant and widespread as to mask the primary focus. In the first instance, hyperventilation may be useful to evoke rhythmic repetition of the original focal discharge, and thus to enhance the bilaterally synchronous activity. When the bilaterally synchronous spike-and-wave activity may be obscuring the primary focus, a single oral dose of 1 mg/kg clobazam is often useful, attenuating the SBS within 20–40 min, and reducing or eliminating the bilateral spike-and-wave complexes before affecting the primary focus.

To distinguish this syndrome from symptomatic partial epilepsy secondary to a cerebral lesion but without SBS, and from the classic LGS, we also reviewed two control groups consisting of the 33 most recent cases of partial epilepsy secondary to a cerebral lesion, without SBS (group C 1); and the 33 most recent cases of LGS (group C 2). Each of the 99 subjects making up the three groups had several EEGs and at least one CT scan. Each of them had been examined by the same person (HG), who also interpreted all the EEGs.

Case Reports

Case 1: A. D. This 14-year-old girl with an unremarkable family history had a 95-min left-sided hemiclonic convulsion with fever at the age of 6 years, with a residual left hemiplegia. Her intelligence was not affected. At 10 years she developed brief seizures occurring several times weekly, with complex symptomatology, and mouthing and swallowing automatisms. Because of her EEG, these were erroneously thought to be absences and were unsuccessfully

[1] WHO Collaborating Centre for Teaching and Research in Neurology, Institut de Recherches Neurologiques, Faculté de Médecine, 13385 Marseille, France
[2] New York State Institute for Basic Research in Developmental Disabilities, Staten Island, New York, and Downstate Medical Center, Brooklyn, NY 11203, USA

Presurgical Evaluation of Epileptics
Edited by H. G. Wieser and C. E. Elger

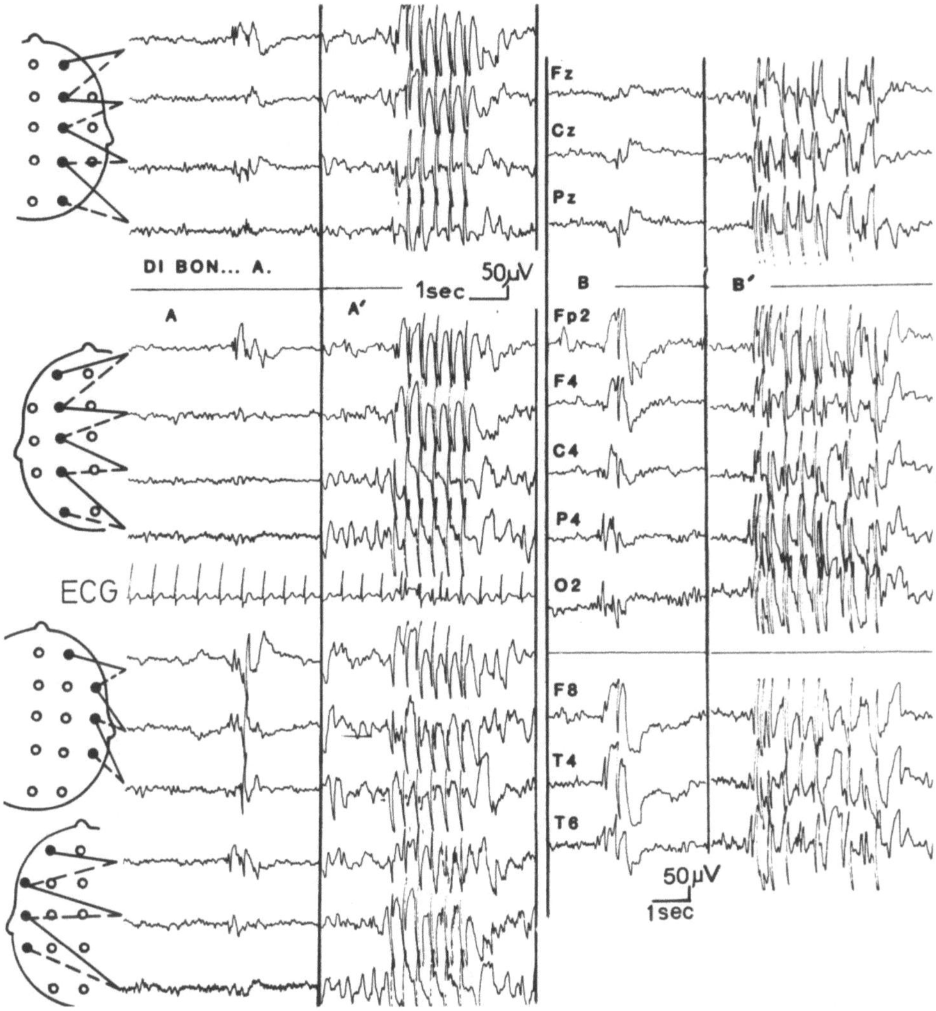

Fig. 1. *Case 1*. Bipolar *(A, A')* and average reference *(B, B')* EEGs show at *A* and *B* isolated focal spike-and-wave activity over the right temporal region at times with frontal spread. During hyperventilation *(A'* and *B')* bursts of generalized 3/s spike-and-wave complexes are seen without clinical accompaniment

treated with valproate. Although the EEG showed generalized 3/s spike-and-wave complexes, these were related by SBS to a right temporal spike focus (Fig. 1). The CT scan showed right hemisphere atrophy characteristic of the HHE syndrome (Gastaut et al. 1960, 1979). The addition of clobazam suppressed the seizures for only 2 months. Since replacing valproate with carbamazepine, she has been seizure free for the past 3 years.

Case 2: S. B. This 10-year-old girl was one of twins. Her father had primary generalized epilepsy. At 6 months, a congenital left hemiplegia was diagnosed. Her IQ is 87 and her mental status unremarkable. At 8 years, left hemiclonic seizures began to occur monthly, and complex seizures with oral and gestural automatisms also occurred several times weekly. These were resistant to valproate and diazepam and she was referred with a diagnosis of LGS because of her EEG. This showed diffuse spike-and-wave complexes, but these were related through SBS to a right frontal spike focus which became particularly visible after the diffuse spike-and-wave activity had been suppressed by oral clobazam (Fig. 2). The CT scan showed a right anterior Sylvian infarct. Clobazam was substituted for diazepam and both the seizures and SBS disappeared, but only for 3 months.

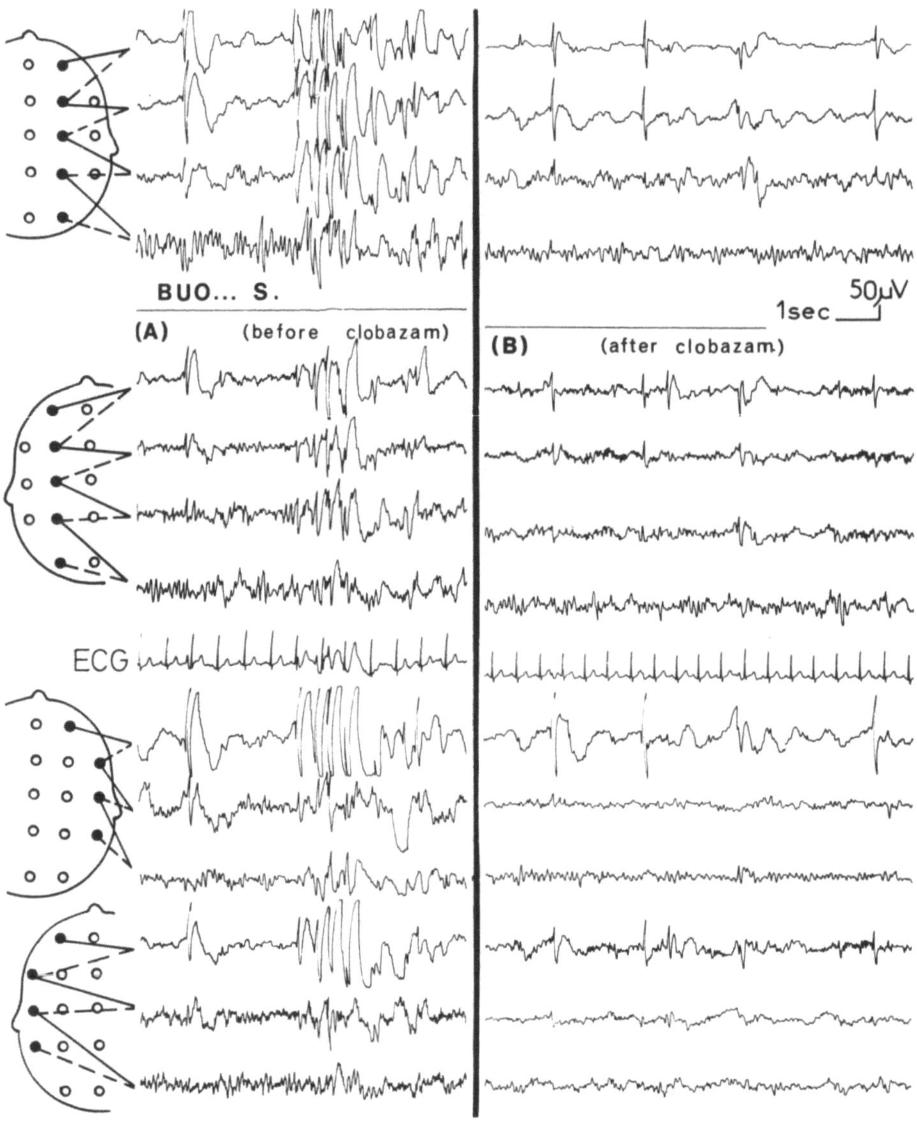

Fig. 2. *Case 2.* The EEG prior to oral clobazam *(A)* shows high-amplitude right frontal spike-and-wave activity which, when isolated, spreads over the left frontal region. When repetitive, the spike-and-wave activity tends to spread over both hemispheres. After 1 mg/kg oral clobazam *(B)* the secondary bilateral synchrony is suppressed, leaving the focal spikes

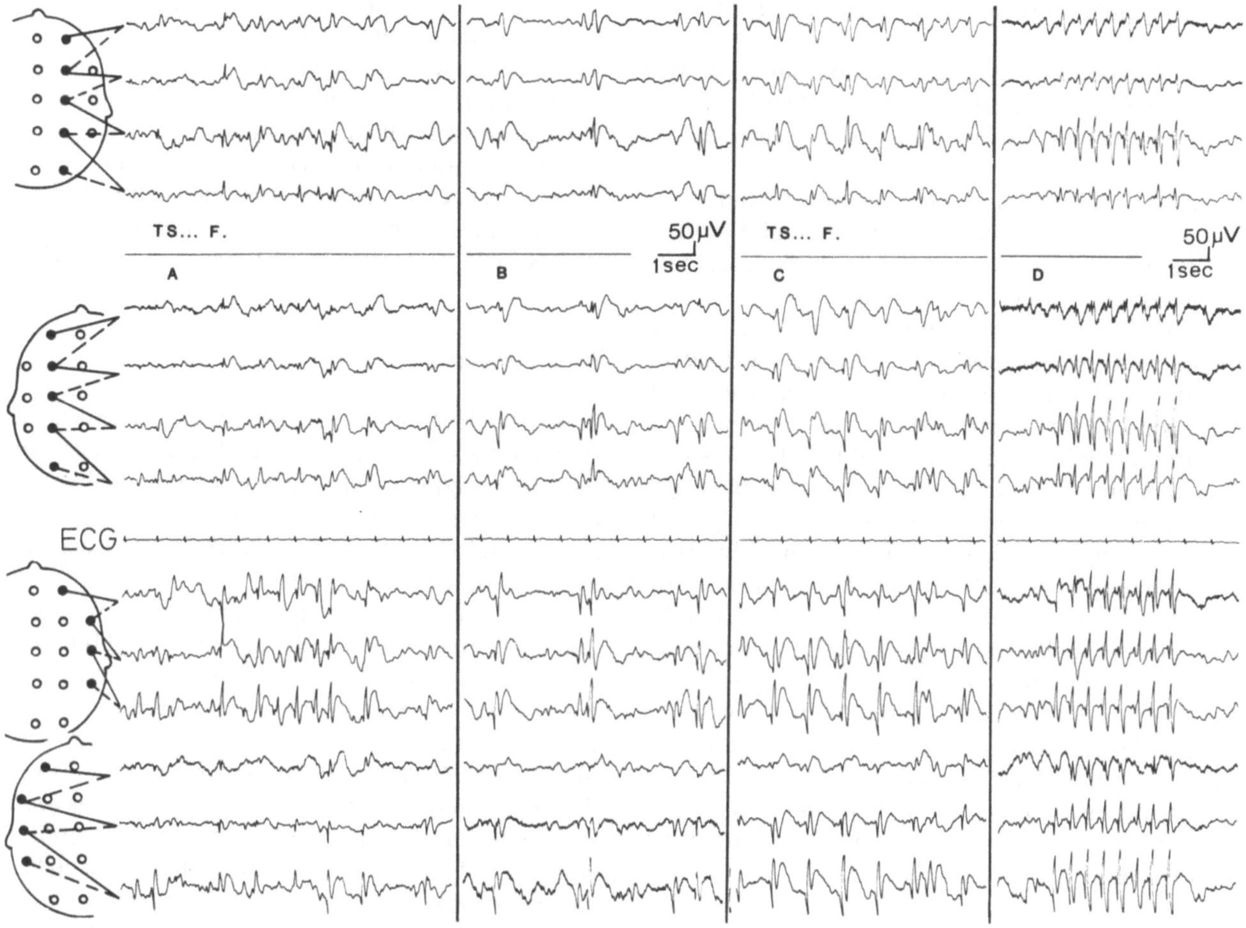

Fig. 3. *Case 3.* Independent spike foci over the right midtemporal and left posterior temporal regions *(A)*. Only when the right-sided spikes repeat rhythmically between 0.5 and 3/s *(B–D)* they are followed by left-sided spikes and then bilaterally synchronous and symmetric spike-and-wave activity. This pattern successively resembles that of subacute sclerosing panencephalitis *(B)*, LGS *(C)*, and primary generalized epilepsy *(D)*

Case 3: F.T. This 17-year-old woman has a brother with primary generalized epilepsy. Following a head injury at 13 years, she was comatose for 8 days, but there were no residual neurologic or psychiatric findings. At 16 years, tonic seizures and increasing difficulties at school developed. Because her EEG at the time showed bursts of generalized slow spike-and-wave complexes, a diagnosis of post-traumatic LGS was made, and valproate and phenobarbital were prescribed. In fact, these slow spike-and-wave complexes are preceded by right temporal focal spikes with SBS (Fig. 3). The CT scan is normal. Her medications were changed to carbamazepine and clobazam. She has been seizure free for the past 3 months and her school performance has improved. The EEG paroxysmal activity has also disappeared.

Results

Age and Sex of Patients

Age at the most recent follow-up visit was from 5 to 25 years (mean, 16.5 years; SD, 4.88 years). Seventeen (51.5%) were females, and 17 males (48.5%).

Etiologic Factors

A family history of epilepsy, a family history of febrile convulsions, or a personal history of febrile convulsions suggested an underlying epileptic predisposition in 13 cases (39%). In 27 cases, the history suggested a cerebral lesion, usually seen on the CT scan:

- Congenital Sylvian infarct seen on the CT scan: six cases.
- Prolonged seizures with fever and an HHE syndrome: four cases, three confirmed by CT scan.
- Head trauma: six cases, three with lesions seen on CT scan.
- Arachnoid cyst: one case, seen on CT scan.
- Subdural empyema: one case, seen on CT scan.
- Perinatal and/or neonatal disturbances: nine cases, five with CT scan lesions.

CT scan lesions were seen in 19 patients. The lesions were usually large and involved most frequently the mid-portion of one hemisphere. They were temporal or centrotemporal in 11 cases, frontal in 5, centroparietal in 2, and occipital in 1. Lesions were much more often superficial or deep than parasagittal. The mesial surface was involved in only three cases. Combining clinical, CT scan, and EEG data, one hemisphere could always be implicated, the right in 19 cases (58%) and the left in 14 cases (42%). In four cases, bilateral temporal EEG foci were seen, but with lateralized predominance over the hemisphere from which the SBS arose. In the 26 cases for which data are available, the lesion was acquired between birth and 15 years of age (mean 4 years and 5 months), including those 16 cases (48.5% of the total) in whom the lesion was acquired during or before birth.

Interictal Symptoms

The neurologic examination was normal in 22 cases (66.6%). Hemiparesis or hemiplegia was found in ten cases, quadriplegia with lateralized predominance in one case, and hemianopia in one. Intelligence was normal or borderline in 21 patients (64%). Mild retardation was present in four (12%), moderate retardation in five (15%), and profound retardation in three (9%). In seven cases, transient slowing of mental functions was observed with the appearance of SBS, resolving with its disappearance.

Ictal Symptoms

Seizures began over an age range from 3 to 21 years (mean age 11 years, 7 months). In the 26 cases for which data were available, there was a delay between acquisition of a lesion and seizure onset ranging from 6 months to 19 years (mean delay 7 years, 6 months).

The clinical seizure types were:

- Partial seizures with complex symptomatology: 12 patients
- Partial seizures with adversion: 2 patients
- Tonic-clonic seizures: 11 patients
- Tonic seizures: 7 patients
- Clonic seizures: 2 patients
- Hemiclonic seizures: 9 patients
- Tonic or atonic astatic seizures: 7 patients

Seizures occurred several times per year in 3 (9%), several times monthly in 15 (45%), several times weekly in 8 (24%), and several times daily in 7 (21%).

EEG Findings

The focal EEG abnormality consists of spikes or spike-and-wave complexes which are localized or relatively predominant over the temporal (20 cases, 61%), frontal (7 cases, 21%), central (5 cases, 15%), or occipital (1 case) regions. Normal EEG activity may also be altered regionally.

The SBS is shown by high-amplitude slow spike-and-wave complexes which are generalized but predominant over both anterior quadrants. These spike-and-wave complexes are bilaterally synchronous but not necessarily symmetric. Bilateral symmetry was observed in 18 cases (55%), higher amplitude over the hemisphere containing the spike focus was seen in 9 cases (27%) and predominance over the other side in 5 (15%). These occur only rarely as isolated spike-and-wave complexes. They are typically rhythmic and repetitive, either spontaneously or with hyperventilation. Frequency is between 0.5 and 3/s, most often 2/s. Thus, they usually appear similar to the slow spike-and-wave complexes of LGS rather than to the typical spike-and-wave complexes of a primary generalized epilepsy.

Secondary bilateral synchrony is considerably activated by hyperventilation and, to a lesser degree, by slow wave sleep, which in five cases with tonic and/or astatic seizures also evoked brief subclinical bursts of epileptic recruiting rhythm. It can be transiently reduced or eliminated by a single dose of clobazam. Between bursts, slowing of background activity was seen in 21 cases (64%).

Drug Treatment

Drugs which are useful in treating partial epilepsies, such as phenobarbital, phenytoin, and carbamazepine, are often effective. Benzodiazepines have been useful adjuvants, but their effects may be transient. Abolition of seizures was observed in seven cases (21%) and a 50% or greater reduction in seizures in ten others (30%). In most cases, improvement in the

EEG parallels clinical improvement, and the SBS has more often disappeared than has the associated focal abnormality.

Discussion

Etiology of Symptomatic Partial Epilepsies with SBS

Symptomatic partial epilepsies with SBS are clearly due to a lesion of the mid-portion of one hemisphere in almost all cases. The lesions have been atrophic, post-ischemic, or post-traumatic. They are congenital in half of the cases. The CT scans show that these lesions are usually much more widespread than the focal EEG spikes which are related to them and that they usually involve the superficial and deep temporal and central regions, and not only the mesial surface of the frontal or temporal lobes as had originally been supposed (Jasper 1951; Jasper et al. 1951; Tükel and Jasper 1952).

The location of the epileptogenic lesion is responsible for the interictal and ictal symptoms as well as for the location of the EEG focus:

1. Thus, 24% of these patients had complex partial seizures, 18% had hemiclonic and 10% adversive seizures with or without secondary generalization. This proportion is similar to those found by Gabor and Ajmone-Marsan (1969), who noted 14% with complex partial seizures and 25% with simple partial seizures with sensorimotor symptoms; and by Niedermeyer (1972), who found 23% with psychomotor seizures and 23% with focal motor, hemiconvulsive, or adversive seizures.

2. Interictal symptoms consist almost exclusively of hemiplegia or hemiparesis with or without associated motor or sensory signs in 33% of cases. Mental retardation was found in 36%. Gabor and Ajmone-Marsan (1969) and Niedermeyer (1972) found 20% and 12% respectively with hemiplegia, and 17% and 12% respectively with mental retardation.

3. The EEG focus is localized to, or predominates over, the anterior part of one hemisphere. It was found over the temporal region in 61%, the frontal region in 21%, and the central region in 15%. Only one case of an occipital focus was found. Gabor and Ajmone-Marsan reported temporal foci in 34% of cases, frontal in 34%, central or frontocentral in 26%, and occipital in a single case. Niedermeyer found a temporal focus in 31% of cases, frontal in 38%, frontotemporal in 15%, central or frontocentral in 8%, and parietal in 8%.

Although the symptomatic partial epilepsies with SBS are clearly due to a cerebral lesion, it is equally clear that a constitutional epileptic predisposition must play a role in the genesis of the seizures and the SBS. This is shown by:

1. The high proportion (39%) of patients with a personal or family history of seizures, similar to that noted for the primary epilepsies, generalized or partial, which are a function of a genetically transmitted epileptic predisposition (Gastaut 1982). 2. The development of seizures and SBS exclusively during childhood and adolescence, that is, at the age when the epileptic predisposition is most marked and when the primary generalized and partial epilepsies related to it also appear (Gastaut 1982). 3. The secondary generalization to tonic-clonic seizures of the partial attacks, which is more frequent in partial epilepsy with SBS (22%) than in the partial epilepsies without SBS (12%), as had also been noted by Gabor and Ajmone-Marsan (1969).

It thus appears that the etiology of the partial epilepsies with SBS is complex, including an organic factor represented by the focal cerebral lesion, and a constitutional factor, which is the epileptic predisposition. This mixed etiology helps us to understand the differences which separate the symptomatic partial epilepsies with SBS from those without SBS, and from LGS.

Differentiation Between Symptomatic Partial Epilepsies With and Without SBS

Significant differences ($P < 0.05$) between the study group and the control group C 1 with secondary partial epilepsy without SBS were found for the following characteristics:

1. Unlike the secondary partial epilepsies without SBS, they begin in childhood and adolescence and not at any age.

2. They much more frequently involve a constitutional epileptic predisposition associated with the organic factor of a localized cerebral lesion. A family history of epilepsy was found in 39% of cases compared with 13% in the control group C 1. The focal cerebral lesion representing the organic factor, contrary to Tükel and Jasper (1952), has never been a tumor.

3. The seizures differ in the two groups. With SBS, two or three different seizure types are more frequently associated in the same patient. This was observed in 45% of cases compared with 18% of the control group. Generalized convulsions were also more frequently seen than were seizures which remain partial or hemiclonic. These were seen in 55%, 27%, and 18% of the study group respectively, compared with 21%, 67%, and 9% of the control group.

4. In SBS, by definition, the EEG shows bursts of bilaterally synchronous slow spike-and-wave complexes which are not present in cases of partial epilepsies without SBS.

Nonsignificant trends were observed for several other possible differentiating features. A larger proportion of patients with SBS were hemiplegic or mentally retarded, and more SBS patients had more frequent seizures. More also had astatic seizures. These were similar to those described as temporal lobe syncope by Landolt (1960) and Caffi (1973), as temporal drop attacks (Delgado-Escueta 1979) or astatic seizures (Pazzaglia et al. 1980, 1985). A larger proportion of SBS patients had frontal EEG foci than were found in group C1, while control patients had more central, parietal, and occipital foci. Seizures also seemed more difficult to control in patients with SBS.

Comparison of Symptomatic Partial Epilepsies with SBS and LGS

Lennox-Gastaut syndrome is characterized by frequent seizures, multiple seizure types including at least two of atypical absences, tonic or atonic astatic seizures or tonic seizures; arrest or retardation of psychomotor development, and diffuse slow spike-and-wave complexes (Gastaut et al. 1966; Beaumanoir 1982; Gastaut 1982).

In the study group, the organic factor consists of a focal cerebral lesion, while the organic factor in LGS is diffuse, shown on CT scan by diffuse cerebral atrophy which may predominate over both frontotemporal regions, as seen in 57% of group C2, and in 54% of our previously reported cases (Gastaut et al. 1980).

Although it may be difficult to differentiate the clinical pattern from that of the partial epilepsies with SBS, significant differences were found for the following characteristics:

1. Partial epilepsies with SBS begin in later childhood and adolescence (mean age of onset 11 years and 7 months) rather than in early and mid-childhood (mean age of 6 years and 7 months) in group C2 with LGS.

2. Focal neurologic signs are much more frequently seen in the study group (33% had hemiplegias compared with 9% in the control group) while obvious mental retardation was less frequently seen (36% compared with 87%).

3. Comparison of the seizures of the partial epilepsies with SBS with those of LGS shows that:

a) The seizures of SBS are less frequent, occurring more than several times weekly in 45% of cases compared with 83% of LGS patients. Multiple seizure types are less often associated in the same patient. This was observed in 45% of cases with SBS compared with 85% of those with LGS.
b) Partial seizures were more frequently seen in SBS. They accounted for 28% of seizures but were not observed in group C2.
c) Tonic and tonic or atonic astatic seizures were less frequently seen in SBS, each representing 14% of cases while they accounted for 38% and 22% respectively of cases in group C2.
d) Atypical absences and episodes of absence status were not seen in SBS. Atypical absences were frequent and absence status occurred in 27% of the LGS cases.

4. Comparison of the EEG patterns shows that:

a) Bilateral bursts of slow spike-and-wave activity are seen in all cases with SBS, as in LGS. However, the spike-and-wave complexes of SBS are more often bilaterally synchronous than those of LGS, which are more diffuse. This activity also more often showed a lateralized asymmetry in SBS (52% of cases compared with 23% in the control group).
b) An epileptic focus was seen in all cases of SBS and not at all in LGS. In the latter, a localized predominance of slow spike-and-wave activity may be seen. This may remain stable or may shift, but is not a true epileptic focus.
c) Bursts of epileptic recruiting rhythm in slow wave sleep were rare, seen in only 15% of cases, all of whom had tonic or astatic seizures. Recruiting rhythm was seen in 79% of cases in group C2.

Nonsignificant trends were observed for some possible differentiating features. Evidence of a constitutional epileptic predisposition was more often seen in the study group. A larger proportion of patients with SBS had hemiclonic seizures. Drug treatment also seemed somewhat more effective for these patients. Benzodiazepines, including clobazam, seemed about equally effective for the two disorders.

Differential Diagnosis and Classification of Symptomatic Partial Epilepsies with SBS

The number of nonsignificant differences noted between the symptomatic partial epilepsies with and without SBS suggests that it may be difficult to distinguish the two on clinical and CT grounds alone. Their EEGs, however, are dramatically different, perhaps reflecting a greater role for the epileptic predisposition in cases with SBS, and allowing differential diagnosis.

The data presented suggest that the electroclinical findings in the partial epilepsies with SBS are more like those of LGS than those of the partial epilepsies without SBS. Often, symptomatic partial epilepsies with SBS are misdiagnosed as the much better known LGS. In a retrospective study, Beaumanoir (1982) reported misdiagnosis in 37% of 103 patients said to have LGS. Similarly, we noted in the present study that 17 of the 33 patients (51.5%) with partial epilepsies and SBS had been referred with a diagnosis of LGS.

The greatest difficulty in differentiating these two syndromes occurs in cases in whom a partial epilepsy with SBS is manifested by the most characteristic seizure type of LGS, frequent astatic attacks. This has been discussed by Pazzaglia (1984; Pazzaglia et al. 1980), who studied partial epilepsies with frequent astatic seizures. In our opinion, these should be considered examples of partial epilepsies with SBS in the great majority of cases. Although associated with retardation and EEGs showing bilaterally synchronous slow spike-and-wave complexes with bursts of epileptic recruiting rhythm during sleep, they are also characterized by childhood- or adolescence-onset complex partial seizures with frontal or temporal EEG epileptic foci. Pazzaglia has also made these points and has emphasized that these epilepsies should not be confused with the late-onset LGS as suggested by Lipinski (1977). Positron emission tomography results in patients thought clinically to have LGS (Gur et al. 1982; Chugani et al. 1987) suggest that this method may also be useful in distinguishing the two syndromes.

It is also possible that the natural history of the epilepsies with SBS is different from that of LGS, either due to the effects of drug treatment to which the first may be more responsive or due to the natural reduction in the epileptic predisposition after adolescence (Gastaut 1982). We have only been able to study the short-term course and have seen that in about ten cases antiepileptic drugs can suppress the astatic seizures, psychomotor slowing, and the bilaterally synchronous spike-and-wave activity. Moreover, since the syndrome was of recent onset in many cases, it appears that the symptoms in many cases of partial epilepsies with SBS resembling LGS can be transitory, while chronicity and intractability are hallmarks of LGS.

The superficial clinical similarities between the partial epilepsies with SBS and LGS should not influence the classification of these seizure disorders of childhood and adolescence. Both belong in the category of secondary, symptomatic, or lesional epilepsies. The partial epilepsies with secondary bilateral synchrony belong among the secondary or symptomatic partial epilepsies, while LGS is part of the secondary or symptomatic generalized epilepsies as categorized in the present international classification of the epilepsies (Dreifuss et al. 1985; Gastaut and Zifkin 1985).

Intractable astatic seizures or drop attacks are a disabling seizure type noted both with LGS and in symptomatic partial epilepsies with SBS. Review of the reports of section of the corpus callosum for intractable epilepsy shows that there is agreement among many authors that drop attacks often respond dramatically to this operation. The clinical, radiologic, and EEG data show that the best results were obtained in patients with evidence of gross hemisphere lesions, who were not severely retarded. Their EEGs showed focal, lateralized multifocal, or hemispheric epileptic discharges either associated with or evolving to diffuse spike-and-wave activity interictally or ictally (Wilson et al. 1975, 1977, 1978, 1982; Amacher 1976; Avila et al. 1980; Huck et al. 1980; Bouvier et al. 1983; Geoffroy et al. 1983; Harbaugh et al. 1983; Rayport et al. 1983; Roberts et al. 1983; Blume 1984; Gates et al. 1984; Spencer et al. 1985).

They were similar to the cases of partial epilepsies with SBS reported here. Disappointing results in patients with drop attacks were noted in severely retarded microcephalic patients who did not fit this clinical pattern. Good results also seemed less likely in other cases with severe retardation or apparently typical LGS. Gastaut et al. (1985) noted failure of anterior callosotomy for seizure control in three subjects with LGS, but benefit in two patients with symptomatic partial epilepsy with a frontal focus, SBS, and falling spells.

The distinction between partial epilepsies with SBS and LGS is thus not of theoretical importance alone. Although the medical treatment and response to anticonvulsants are similar, patients with SBS may profit from callosal section, with relief of drop attacks. Those with LGS seem less likely to benefit. As callosal section in man has become a procedure with predictably low morbidity and mortality, heightened awareness of this syndrome, with early clinical and EEG diagnosis, should lead to more examples of successful surgical treatment, selecting those patients most likely to benefit.

References

Amacher AL (1976) Midline commissurotomy for the treatment of some cases of intractable epilepsy. Child's Brain 2: 54-58

Avila JO, Radvany J, Huck FR, Pires de Camargo CH, Marino R Jr, Ragazzo PC, Riva D (1980) Anterior callosotomy as a substitute for hemispherectomy. Acta Neurochir [Suppl] 30: 137-143
Beaumanoir A (1982) The Lennox-Gastaut syndrome: a personal study. In: Broughton R (ed) Henri Gastaut and the Marseille school's contribution to the neurosciences. Electroencephalogr Clin Neurophysiol [Suppl] 35: 85-99
Blume WT (1984) Corpus callosum section for seizure control: rationale and review of experimental and clinical data. Cleve Clin Q 51: 319-332
Blume WT, Pillay N (1985). Electrographic and clinical correlates of secondary bilateral synchrony. Epilepsia 26: 636-641
Bouvier G, Mercier C, St-Hilaire J-M, Giard N, Labrecque R, Beique RA (1983) Anterior callosotomy and chronic depth electrode recording in the surgical management of some intractable seizures. Appl Neurophysiol 46: 308-316
Caffi J (1973) Zur Frage klinischer Anfallformen bei psychomotorischer Epilepsie. Schweiz Med Wochenschr 103: 469-475
Chugani HT, Mazziota, JC, Engel J, Phelps ME (1987). The Lennox-Gastaut syndrome: Metabolic subtypes determined by 2-deoxy-2 [^{18}F] fluoro-D-glucose positron emission tomography. Ann Neurol 21: 4-13
Delgado-Escueta A (1979) Epileptogenic paroxysms: modern approaches and clinical correlations. Neurology 29: 1014-1022
Dreifuss F et al. (1985). Proposal for classification of epilepsies and epileptic syndromes. Epilepsia 26: 268-278.
Gabor A, Ajmone-Marsan C (1969) Co-existence of focal and bilateral diffuse paroxysmal discharges in epileptics. Epilepsia 10: 453-472
Gastaut H (1982) "Benign" or "functional" (versus "organic") epilepsies in different stages of life. In: Broughton R (ed) Henri Gastaut and the Marseille school's contribution to the neurosciences. Electroencephalogr Clin Neurophysiol [Suppl] 35: 17-44
Gastaut H, Zifkin BG (1985). Classification of the epilepsies, J. Clin. Neurophysiol. 2: 313-326.
Gastaut H, Poirier F, Payan H, Salamon G, Toga M, Vigouroux M (1960) HHE syndrome: hemiconvulsions, hemiplegia, epilepsy. Epilepsia 1: 418-447
Gastaut H, Roger J, Soulayrol R, Tassinari CA, Regis H, Dravet C, Bernard R, Pinsard N, Saint-Jean M (1966) Childhood epileptic encephalopathy with diffuse slow spike-waves (otherwise known as 'petit mal variant') or Lennox's syndrome. Epilepsia 7: 139-179
Gastaut H, Mouren P, Paillas J (1968) A propos de la "bisynchronie secondaire" en électroencéphalographie: paroxysmes bilateraux, synchrones et symétriques révélateurs d'un abcès temporal. Rev Neurol 119: 295-298
Gastaut H, Pinsard N, Gastaut JL, Regis H, Michel B (1979) Computerized tomography of acute hemiplegia in children. In: Goldstein M, Bolis L, Fieschi C, Gorini S, Millikan CH (eds) Advances in neurology, vol 25. Raven, New York, pp 329-337
Gastaut H, Pinsard N, Genton P (1980) Electroclinical correlations of CT scans in secondary generalized epilepsies. In: Canger R, Angeleri F, Penry JK (eds) Advances in epileptology, XIth epilepsy international symposium. Raven, New York, pp 45-52
Gastaut H, Magaudda A, Mariani E (1985) Corrispondenza anatomo-clinica del fenomeno elettroencefalografico della sincronia bilaterale secondaria. Rivista Ital EEG Neurofis
Gates JR, Leppik IE, Yap J, Gumnit RJ (1984) Corpus callosotomy: clinical and electroencephalographic effects. Epilepsia 25: 308-316
Geoffroy G, Lassonde M, De Lisle F, Décarie M (1983) Corpus callosotomy for control of intractable epilepsy in children. Neurology 33: 891-897
Gur RC, Sussman NM, Alavi A, Gur RE, Rosen AD, O'Connor M, Goldberg HI, Greenberg JH, Reivich M (1982) Positron emission tomography in two cases of childhood epileptic encephalopathy (Lennox-Gastaut syndrome). Neurology 32: 1191-1194
Harbaugh RE, Wilson DH, Reeves AG, Gazzaniga MS (1983) Forebrain commissurotomy for epilepsy. Acta Neurochir 68: 263-275
Huck FR, Radvany J, Avila SD, Pires de Camargo CH, Ragazzo PC, Riva D, Arlant P (1980) Anterior callosotomy in epileptics with multiform seizures and bilateral synchronous spike and wave EEG pattern. Acta Neurochir [Suppl] 30: 127-135
Jasper H (1951) Etude anatomo-physiologique des épilepsies. In: Fischgold H (ed) Compte-rendus du 2ème congrès international d'EEG, Paris, 1949. Masson, Paris, pp 99-111
Jasper H, Pertuiset B, Flanigin H (1951) EEG and cortical electrograms in patients with temporal lobe seizures. Arch Neurol Psychiatry 65: 272-290
Landolt H (1960) Die Temporallappenepilepsie und ihre Psychopathologie. Karger, Basel
Lipinski C (1977) Epilepsies with astatic seizures of late onset. Epilepsia 18: 13-20
Niedermeyer E (1968) The occurrence of generalized (centrencephalic) and focal seizure patterns in the same patient. Johns Hopkins Med J 122: 11-25
Niedermeyer E (1972) The generalized epilepsies. Thomas, Springfield, pp 74-86
Pazzaglia P, D'Allessandro E, Ambrosetto G (1980) Le crise epilettiche astatiche. Progressi in epilettologia, Lega Ital contro l'epilepsia, vol 29-30. Milano, pp 179-182
Pazzaglia P, D'Alessandro R, Ambrosetto G, Lugaresi E (1985). Drop attacks: an ominous change in the evolution of partial epilepsy Neurology 35: 1725-1730.
Rayport M, Ferguson S, Corrie WS (1983) Outcomes and indications of corpus callosum section for intractable seizure control. Appl Neurophysiol 46: 47-51
Roberts DW, Allen CD, Allen AH, Reeves AG (1983) Depth electrode recording in patients undergoing corpus callosotomy for intractable epilepsy. Appl Neurophysiol 46: 26-32
Spencer SS, Spencer DD, Williamson PD, Mattson RH (1985), Effect of corpus callosum section on secondary bilaterally synchronous interictal EEG discharges. Neurology 35: 1089-1094
Tükel K, Jasper H (1952) The electroencephalogram in parasagittal lesions. Electroencephalogr Clin Neurophysiol 4: 481-494
Wilson DH, Culver C, Waddington M, Gazzaniga MS (1975) Disconnection of the cerebral hemispheres. Neurology 25: 1149-1153
Wilson DH, Reeves AG, Gazzaniga MS, Culver C (1977) Cerebral commissurotomy for control of intractable seizures. Neurology 27: 708-715
Wilson DH, Reeves AG, Gazzaniga MS (1978) Division of the corpus callosum for uncontrollable epilepsy. Neurology 28: 649-653
Wilson DH, Reeves AG, Gazzaniga MS (1982) "Central" commissurotomy for intractable generalized epilepsy: series two. Neurology 32: 687-697

Editors' Commentary 7

Inference from the clinical features of seizures as to the most likely seizure onset area is one of the central problems of this book. It involves correlating epileptic symptoms and signs with certain cerebral localizations. Sound localizational judgements, often based merely on the description of clinical seizure patterns, are first expected from the referring doctor, who has to decide whether he should send a patient to a special epilepsy surgical center. Frequently, the seizure patterns are the most decisive single factor for acceptance of a patient as a possible candidate for epilepsy surgery. Most effort during the presurgical evaluation is directed towards discovering whether the electroclinical accompaniments of the patient's habitual seizures point to a single, restricted, and surgically accessible brain region. Furthermore, the clinical seizure pattern is often instrumental in making the final decisions when intracranial electrode placement is planned.

These chapters deal with the anatomoelectroclinical correlations of certain seizure types. They differ insofar as the authors validated differently their localizational conclusions by means of successfully performed resections, and placed different emphasis on lesional and/or electroencephalographic (surface and/or depth EEG) data. Furthermore, there exist methodological differences in terms of ictal clinical data aquisition (not all studies make use of simultaneously monitored and interactively examined behavior) and in terms of precision of synchronization of electrical with clinical data.

At the symposium there was general agreement that a localization-related subclassification of partial seizures is possible. However, much discussion arose as to what degree reliable "small compartmentalized localizational conclusions" are methodologically and conceptually possible. It was cautioned that a single ictal sign or symptom should not be overestimated in terms of its localizing and/or lateralizing value. It was underlined that the final weight of certain ictal clinical features depends largely on the context and mode of their appearance, and that in many instances the distribution of the early epileptic discharge, measured by appropriate electrophysiological techniques, is decisive for surgical evaluation.

It cannot be denied that some of the proposed seizure "subtypes" have created considerable controversy and need validation by further studies with more data and in more centers. On the other hand, although at present there remain substantial gaps in our understanding of partial epilepsies, especially of extratemporal and of frontal lobe origin, progress has been made and is being realized for certain clinical seizure syndromes. The recognition of certain distinctive seizure types in turn has sharpened observation and improved documentation. Therefore more precise chronotopographical and electroclinical seizure analyses can be expected in the near future which hopefully will allow confirmation of the significance of identified clinical as well as electrical seizure types, as discussed in the foregoing chapters.

The Surgical Approach: Preconditions, Rationale, Efficacy

Drug-Resistant Partial Epilepsy: Clinical and Pharmacological Criteria

D. SCHMIDT

The second of Walker's (1974) widely accepted criteria for selection of patients for surgical evaluation requires "that the epileptic attacks cannot be controlled by adequate trials of antiepileptic drugs." More recently, Delgado-Escueta and Walsh (1983) recommended that complex partial seizures and secondarily tonic-clonic seizures should be "truly intractable" to drug therapy in surgical candidates.

What constitutes an adequate trial of medical treatment? McNaughton and Rasmussen (1975) very clearly suggested that medical treatment in any patient should not be considered adequate unless the antiepileptic drugs have been used in the maximum doses that can be tolerated. The definition of intractable or drug-resistant epilepsy certainly implies the failure of adequate medical treatment with the optimum choice of drugs, the most promising drug regimen, and the most effective daily dose (Schmidt and Morselli 1986).

Choice of Drug(s)

For partial seizures and secondarily generalized tonic-clonic seizures, four primary agents are available. These are: carbamazepine, phenytoin, phenobarbital, and primidone (Table 1). All four agents are similarly effective for generalized tonic-clonic seizures, while carbamazepine is more effective than phenobarbital or primidone for partial seizures (Mattson et al. 1985). Any adequate medical treatment starts with a single drug. The dosage is gradually increased until complete seizure control is reached. Enough time should be allowed to reach steady-state concentrations of the individual drug (Table 2). It seems appropriate to adjust dosage individually on clinical grounds, starting with a low plasma concentration in patients with a low seizure frequency and to increase the dosage if necessary until clinical toxicity prevents an increment (see daily dose). If single-drug therapy fails to produce complete control even though the daily dose cannot be increased because of overt clinical toxicity, epilepsy is intractable or resistant to the individual drug (Table 3). An alternative primary agent can be employed successfully

Department of Neurology, Klinikum Großhadern, Ludwig-Maximilian University, 8000 Munich 70, Federal Republic of Germany

Table 1. Choice of antiepileptic drugs for treatment of partial seizures and secondarily generalized tonic-clonic seizures

Primary drugs	Carbamazepine, phenytoin, phenobarbital, primidone
Secondary drugs	Valproate, methsuximide, clobazam, clonazepam, clorazepate, oxazepam, acetazolamide, sulthiame
Tertiary drugs	Mephenytoin, phenacemide, ethotoin, bromides
Experimental drugs	

Table 2. Relevant clinical pharmacokinetics of primary antiepileptic drugs when used in single-drug therapy in adults

Drug	Mean daily dose (mg/kg)	Distribution of daily dose	Time before steady state (days)
Carbamazepine	15-20	1-4	3- 6
Phenytoin	5	1-2	5-10
Primidone	10	3-4	1- 5
Phenobarbital	2- 4	1-2	14-21
Valproate	20	1-3	2- 4
Ethosuximide	15-30	2-3	12

Table 3. Recommended drug regimen(s) for patients with uncontrolled partial seizures and secondary generalized tonic-clonic seizures

Single-drug therapy with primary agent
- if unsuccessful, change to →

Alternative single-drug therapy with another primary agent
- if unsuccessful, introduce →

Two-drug therapy with primary agents or secondary agent(s)
- if unsuccessful, return to →

Single-drug therapy with best individual tolerance

Presurgical Evaluation of Epileptics
Edited by H.G.Wieser and C.E.Elger

when single-drug therapy failed, leading to complete control in 12% of the patients (Schmidt and Richter 1985).

Only when successive single-drug therapies with maximum doses have failed is a second drug added. The efficacy and toxicity of adding a second drug in chronic partial epilepsy has been studied carefully in recent years both in Europe and the United States (Schmidt 1982; Mattson et al. 1985). The results were similar in both trials: 11%–13% will become completely controlled when a second drug is added, with the risk of increased drug toxicity. Unfortunately, we have no means of predicting which patients can be treated successfully with two-drug therapy.

If primary agents fail it may be worthwhile considering adding a secondary drug, e.g., because phenytoin and carbamazepine may differ in their mechanism of action from the GABAergic benzodiazepines, twenty percent of patients with chronic complex partial seizures resistant to primary agents became free of seizures when clobazam was added (Schmidt et al. 1986a). Unfortunately, tolerance to the antiepileptic effect and mild sedation compromise the use of clobazam and other benzodiazepines, e.g., clonazepam. Clonazepam may be useful as a second drug in some patients. Tolerance, sedation, and withdrawal effects, however, reduce its clinical value. Intermittent use of clobazam (Oxley 1986) or clonazepam (Sher 1985) in chronic epilepsy may be valuable. Adding valproate may improve generalized seizures (Chayasirisobhon and Russell 1983) but produces only an equivocal response in chronic complex partial seizures.

Methsuximide, as a second drug, leads to complete seizure control in some patients with chronic complex partial seizures (Browne et al. 1983). In the future, new antiepileptic drugs may be available such as GABAergic agents which have been shown to be effective even in many patients with previously unresponsive complex partial seizures (Browne et al. 1984). A number of tertiary drugs (Table 1) should be employed in epilepsy centers only because of the severe side effects of these agents (Schmidt 1985).

Daily Dose

Any assessment of the daily dose required, before the epilepsy can be called resistant, relies on the improvement of efficacy with increasing daily dose or rising plasma concentration. Seizure frequency usually declines as the plasma concentration of phenytoin, phenobarbital, or carbamazepine increases. The plasma concentrations when seizures cease vary

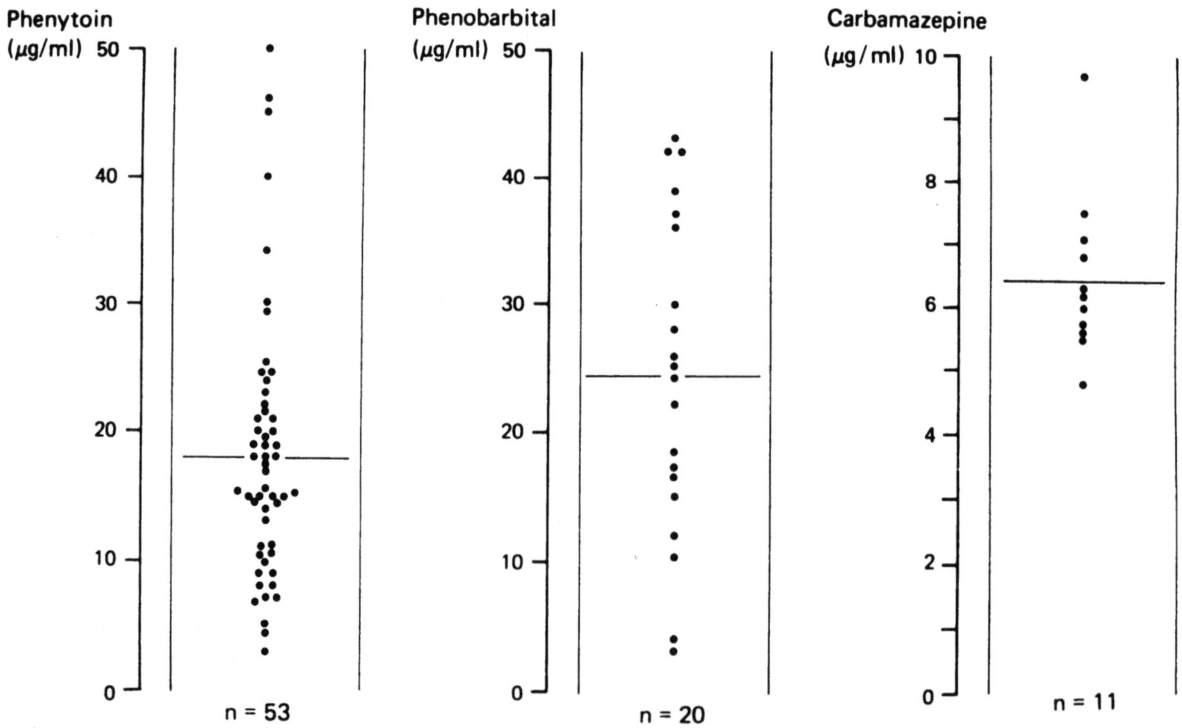

Fig. 1. Effective plasma concentrations of phenytoin, phenobarbital, and carbamazepine when given during single-drug therapy. Each *point* represents the lowest individual plasma concentration of one patient associated with complete seizure control for 1 year. None of the patients ($n = 84$) had signs of overt clinical toxicity. (Schmidt and Haenel 1984)

Table 4. Therapeutic plasma concentrations of primary agents when used during single-drug therapy for specific types of seizures (Schmidt et al. 1986b)

Classification of seizure	Phenytoin (μg/ml)	*(n)*	Phenobarbital (μg/ml)	*(n)*	Carbamazepine (μg/ml)	*(n)*
Generalized tonic-clonic seizures	14	28	18	10	5.5	2
Partial seizures with or without secondary generalization	23**	25	38*	6	7.0*	7

* $P<0.01$; ** $P<0.02$

severalfold among patients (Fig. 1). Complete cessation of seizures was observed at plasma concentrations of 3-50 μg/ml phenytoin, 3-43 μg/ml phenobarbital, and 4.8-9.7 μg/ml carbamazepine.

Patients with partial seizures (Table 4), and a high number of seizures prior to the drug treatment, require plasma concentrations in the upper range (Schmidt and Haenel 1984). Thirty percent of the plasma concentrations of more than 20 μg/ml phenytoin were required for seizure control and were tolerated without overt clinical toxicity. Therefore, it seems appropriate to adjust dosage individually on clinical grounds, starting with a low plasma concentration in patients with a low seizure frequency and increasing the dosage if necessary until overt clinical toxicity prevents an increment, rather than adhering strictly to published therapeutic ranges.

When seizures cannot be adequately controlled, a number of factors can be responsible (Table 5) which require careful assessment by experienced specialists for epilepsy.

The potential long-term toxicity of drug therapy in patients requiring high plasma concentrations for complete control has been a growing concern (Schmidt 1985). In a recent study on the correlation of phenytoin toxicity and the plasma concentration, the highest individual plasma concentration not associated with overt clinical side effects was 33 μg/ml, with a mean of 19 μg/ml. The lowest individual toxic plasma concentrations associated with neurological drug toxicity during single-drug therapy ranged from 7 to 49 μg/ml, with a mean of 25 μg/ml. Unfortunately, patients with clinical drug toxicity at plasma concentrations of 25 μg/ml or below more often had partial epilepsy and a preexisting neurological deficit (Schmidt and Lempert 1987). Forty-five percent tolerated plasma concentrations of 20-33 μg/ml phenytoin during single-drug therapy without any clinical side effects. From a clinical perspective any increase above 33 μg/ml phenytoin requires close clinical supervision and frequent antiepileptic drug monitoring because of an increased risk of drug toxicity.

Table 5. Factors responsible for inadequate control of seizures

Noncompliance
Inadequate drug dosage
Improper choice of drug
Inappropriate combination of drugs
Drug interactions
Precipitating factors, e.g., stressful situations, sleep deprivation, alcohol, menstruation
Erroneous classification of epileptic seizures, e.g., complex partial seizures are mistaken for absence seizures and vice versa
Misdiagnosed nonepileptic seizures, e.g., hysterical seizures, myoclonus, extrapyramidal seizures
Unrecognized progressive neurological lesions, e.g., astrocytoma, oligodendroglioma, hamartomatous disorders

Awareness of the limits of toxic plasma concentrations will improve the early detection of clinical side effects and thus prevent chronically sustained intoxications which may carry the risk of irreversible cerebellar damage if uncorrected (Höglmeier and Wenzel 1969). Furthermore, we are just beginning to explore the neuropsychological impairment associated with chronic drug therapy and other features of chronic toxicity (Hirtz and Nelson 1985; Schmidt 1985).

Even successful epilepsy surgery will not allow the immediate discontinuation of drug treatment in most cases, but will reduce the epileptogenic activity and gradually allow complete seizure control with lower plasma concentration or even without drug therapy in a considerable proportion of patients.

If insufficient reduction of seizure activity, despite adequate medical treatment and drug toxicity or both, impairs the pursuit of economic and social activity and personal life, drug resistance of the epilepsy can be assumed, and presurgical evaluation should be considered.

References

Browne TR, Feldman RG, Buchanan RA, Allen NC, Fawcett-Vickers L, Szabo GK, Mattson GF, Norman SE, Greenblatt DJ (1983) Methsuximide for complex partial seizures: efficacy, toxicity, clinical pharmacology, and drug interactions. Neurology 33: 414-418

Browne TR, Mattson RH, Napoliello MJ, Penry JK, Smith DB, Treiman DM, Wilder BJ (1984) Multicenter single-blind study of gamma-vinyl-GABA (GVG) for refractory complex partial seizures (CPS). Neurology 34 [Suppl 1]: 265

Chayasirisobhon S, Russell M (1983) Valproic acid and intractable seizures in severely brain-damaged patients. Neurology 33: 99-101

Delgado-Escueta AV, Walsh GO (1983) The selection process for surgery of intractable complex partial seizures: surface EEG and depth electrography. In: Ward AA Jr, Penry JK, Purpura D (eds) Epilepsy. Raven, New York, pp 295-326

Hirtz DG, Nelson KB (1985) Cognitive effects of antiepileptic drugs. In: Pedley AT, Meldrum BS (eds) Recent advances in epilepsy, no 2. Livingstone, Edinburgh, pp 161-182

Höglmeier H, Wenzel U (1969) Zerebellarer Dauerschaden durch vorübergehende Hydantoinüberdosierung. Dtsch Med Wochenschr 94: 1330-1332

Mattson RH, Cramer JA, Collins JF, Smith DB, Delgado-Escueta AV, Browne TR, Williamson PD, Treiman DM, McNamara JO, McCutchen CB, Homan RW, Crill WE, Lubozynski MF, Rosenthal NP, Mayersdorf A (1985) Comparison of carbamazepine, phenobarbital, phenytoin, and primidone in partial and secondarily generalized tonic-clonic seizures. N Engl J Med 313: 145-151

McNaughton FL, Rasmussen T (1975) Criteria for selection of patients for neurosurgical treatment. In: Purpura DP, Penry JK, Walter RD (eds) Advances in neurology, vol 8, neurosurgical management of the epilepsies. Raven, New York, pp 37-48

Oxley J (1986) Intermittent treatment of uncontrolled epilepsy. In: Schmidt D, Morselli PL (eds) Intractable epilepsy. Raven, New York, pp 203-207

Schmidt D (1982) Two antiepileptic drugs for intractable epilepsy with complex-partial seizures. J Neurol Neurosurg Psychiat 45: 1119-1124

Schmidt D (1985) Adverse effects. In: Frey HH, Janz D (eds) Antiepileptic drugs. Handbook of experimental pharmacology, chapter 27. Springer, Berlin Heidelberg New York Tokyo, pp 791-818

Schmidt D, Haenel F (1984) Therapeutic plasma levels of phenytoin, phenobarbital, and carbamazepine: individual variation in relation to seizure frequency and type. Neurology 34: 1252-1255

Schmidt D, Lempert T (1987) Phenytoin toxicity: relation to plasma concentration. Arch Neurol (to be published)

Schmidt D, Morselli PL (eds) (1986) Intractable epilepsy. Raven, New York

Schmidt D, Richter K (1985) Alternative single antiepileptic drug therapy for refractory epilepsy. Ann Neurol 151: 85-87

Schmidt D, Rohde M, Wolf P, Roeder-Wanner U (1986a) Clobazam for refractory focal epilepsy. Arch Neurol 43: 824-826

Schmidt D, Einicke I, Haenel F (1986b) The influence of seizure type on the efficacy of plasma concentrations of phenytoin, phenobarbital and carbamazepine. Arch Neurol 43: 263-265

Sher PK (1985) Alternate-day clonazepam treatment of intractable seizures. Arch Neurol 42: 787-788

Walker AE (1974) Surgery for epilepsy. In: Magnus O, Lorentz de Haas AM (eds) Handbook of clinical neurology. The epilepsies, vol 15. North Holland, Amsterdam, pp 739-758

Preoperative Indicators for Optimal Surgical Outcome for Temporal Lobe Epilepsy*

P. H. CRANDALL[1, 5], R. RAUSCH[2, 3], and J. ENGEL, Jr.,[3, 4, 5, 6]

This title must be defined at the outset because it is obvious that under the circumstances of our present knowledge a firm model cannot be constructed for the surgical treatment of a clinical condition with so many variables. Rather we are pointing out some factors observed during a 25-year experience at the UCLA Medical Center which we hope will be confirmed by others and prove to be clinically useful. The concepts of psychomotor epilepsy, temporal lobe epilepsy, and partial complex epilepsy are not the same. The main components of the clinical features such as autonomic auras, staring states, clouding consciousness, and automatisms are poorly defined and overlapping in regard to topographical origin in the brain. Therefore, we are referring to temporal lobe epilepsy as defined by depth electrode studies during spontaneous seizures with an electrographic origin and propagation in the limbic system, with microscopic pathology in the removed temporal lobe and with successful alleviation of seizures for follow-up periods of 2-25 years. There are also unsuccessful cases which illustrate the points to be made.

It is well known that neuropsychological and neuropsychiatric disorders are commonly encountered with such patients. These conditions can also be oriented with regard to right- and left-sided epileptogenesis and surgery which should be entered into considerations of prognosis. Added to this are psychosocial problems which are often more difficult to manage than the problem of seizure control. Not all of these can be legitimate surgical goals but at least the avoidance of significant cognitive deficits and psychological handicaps is important as well as seizure relief.

Security of Clinical Diagnosis

Clinical Features

Most clinicians recall that the term "psychomotor epilepsy" was coined by Gibbs (Gibbs et al. 1938). It was later associated with "low temporal spiking" in the EEG, enhanced by sleep (Gibbs and Gibbs 1947, 1952). Also the psychiatric aspects were described quite early (Gibbs 1951). Despite considerable clinical study of auras and "signal symptoms" of psychomotor epilepsy, no special significance could be found from these per se with regard to the location of the epileptic process as had been the case in elementary seizures of the Jacksonian type, aphasic attacks, and auditory or visual hallucinations in the primary cortices. A large body of information was collected concerning the topographical organization of the human brain by Penfield and his colleagues at the Montreal Neurological Institute, 1927-1974. Perhaps the best-known summary appeared in *Epilepsy and the Functional Anatomy of the Human Brain* in 1954.

Stereoelectroencephalography (SEEG) was introduced by Talairach, Bancaud, and associates in Paris (1958-1974) and provided accurate access to superficial and deep cortical fields as well as to subcortical structures to the level of the diencephalon. This provided the first opportunity to make firm anatomoelectroclinical relationships. To the enormous amount of data which could be collected, Wieser

[1] Division of Neurosurgery, University of California at Los Angeles, School of Medicine, 760 Westwood Plaza, Los Angeles, CA 90024, USA
[2] Department of Psychiatry and Biobehavioral Sciences, University of California at Los Angeles, School of Medicine, 760 Westwood Plaza, Los Angeles, CA 90024, USA
[3] Department of Neurology, University of California at Los Angeles, School of Medicine, 760 Westwood Plaza, Los Angeles, CA 90024, USA
[4] Department of Anatomy, University of California at Los Angeles, School of Medicine, 760 Westwood Plaza, Los Angeles, CA 90024, USA
[5] Brain Research Institute, University of California at Los Angeles, School of Medicine, 760 Westwood Plaza, Los Angeles, CA 90024, USA
[6] Laboratory of Nuclear Medicine and Biophysics, University of California at Los Angeles, School of Medicine, 760 Westwood Plaza, Los Angeles, CA 90024, USA

* Supported by US PHS Grant NS 02808.

Presurgical Evaluation of Epileptics
Edited by H. G. Wieser and C. E. Elger

(1983) brought the power of the computer, especially cluster analysis, so that he could analyze electrographic seizures and behavior in 10-s epochs to correlate epileptic activity to structure and manifestation. He took care to analyze seizures of spontaneous origin mainly.

Wieser found that there were five types of psychomotor epilepsy and that SEEG revealed a high electroclinical correlation during attacks. The sequence of involvement of structures was indicative of certain circuits being operative for each type, although any single symptom or sign remains of doubtful localizing value. If we look for specific symptomatology in the five types, the results are not entirely conclusive and might not be reliable without the SEEG evidence. However, there appear to be prominent fear and unilateral face and mouth contractions and oroalimentary automatisms in the unilateral temporal mesiobasal type and amnesia in the bilateral. Temporal neocortical posterior may be characterized by aphasia and vestibular and visual hallucinations; opercular by auditory and somesthetic hallucinations. Frontobasal-cingulate types involve complex automatisms including verbal automatisms.

In conclusion, from symptomatology and behavioral information, one may classify the seizure in the general class of partial complex epilepsy, but for surgical treatment additional EEG analysis is essential.

Electroencephalographic Data

The EEG is an essential instrument for the confirmation of the diagnosis. However, considerable variations in time and space in the scalp "EEG focus" were also observed. Even when predominantly unilateral, on repeated examinations the spiking might shift to frontal and temporal, and diffuse in anteroposterior location. There was no particular correlation of location of temporal spiking to any clinical constellation of symptoms. Further, there was a large fraction of patients displaying bitemporal, independent discharges, which seemed to be the case when more extensive, prolonged or repeated examinations were made. Approximately 60%–70% of patients with psychomotor epilepsy display interictal paroxysmal EEG changes in the waking state (Gastaut 1950). Sleep recordings (Gibbs and Gibbs 1947), as well as the use of pharyngeal and sphenoidal electrodes, increase the incidence of positive EEG findings (Rovit et al. 1961). The localizing evidence of an interictal spike focus at the scalp can be questionable or even misleading for surgical purposes. It may be only a part of a deeply localized, more extended focus or represent a projection of a distant primary focus which propagates to the surface (Bancaud et al. 1969, 1975). Jasper and colleagues (1951) showed that when interictal spikes were strictly unilateral the temporal lobe surgery was successful in 67% of cases as compared with only 21% if there were asynchronous discharges in both temporal regions. This was further amplified by Bloom et al. (1960). It was commonly concluded that this failure of surgery associated with bitemporal interictal discharges might have been due to bilateral seizure foci or secondary seizure foci development. Falconer adopted an arbitrary rule that interictal abnormalities must be predominantly on one side by a ratio of 4:1 for anterior temporal surgery to be considered. He wrote in 1967 "the selection of patients has been rigorous, and we estimate that only about one in nine patients (11%) referred for surgery fulfill (our) criteria." The presence of bilateral interictal temporal lobe spikes even in this patient population did not correlate with a poor outcome (Engel et al. 1975).

Studies of these perplexing bitemporal interictal abnormalities by means of stereotactically implanted depth electrodes prompted the beginning of the surgery program at the University of California, Los Angeles, in 1960. The methodology has been reported (Crandall et al. 1963). Basically this consisted of utilizing electrodes implanted into the limbic structures medial to the temporal lobes, bilaterally and symmetrically in all patients for purposes of recording over several weeks. These sites were chosen because of the bitemporal, interictal discharges, because the hippocampal formation is the most seizure sensitive region of the brain, and the demonstration that the pathological substrate of temporal lobe epilepsy involved the medial temporal lobe or lobes.

The Ictal EEG Signal

The remarkable finding in all series utilizing SEEG is that an ictal EEG focus at one site can be found in the majority of patients with partial complex epilepsy (at UCLA in over 70%), even with EEG evidence of bilateral independent temporal epileptiform abnormalities. This has been confirmed by Gloor et al. (1980).

In the words of Talairach, Bancaud, and colleagues (1974) "One can say that the appearance of a discharge of spikes of high frequency, low voltage and rapid propagation constitutes for most partial epilepsies an essential index of localization of the source . . .". However, in their view one must not only define the area of onset of the ictal discharges but also the pathways of diffusion of the discharges.

A study by a member of our group (Lieb et al. 1981) involved long-term observations of 52 epileptics following anterior temporal lobectomy using multivariate analysis. Variables found to correlate with good surgical outcome included: (a) a high proportion of ictal onsets (surface, deep, and surface/deep) from the side chosen for lobectomy and (b) a high proportion of deep focal onsets. Variables found to correlate with poor surgical outcome included: (a) a high proportion of bilaterally synchronous onsets of any type (surface, deep, and surface/deep) and variable onset locations. Other data suggested a weak but significant correlation between the occurrence of unilateral surface or surface/deep onsets from either hemisphere and poor outcome. Bilateral independent interictal activity did not predict outcome.

Additional Diagnostic Tests

A number of other diagnostic tests have been added to the diagnostic criteria of ictal SEEG findings since 1977. Table 1 summarizes the testing in two phases. Most of these tests are probes for possible areas of intracerebral lesions associated with this type of epilepsy and termed tests of functional deficit. A detailed account of this protocol has been published as well as a description of the surgical methods utilized (Engel et al. 1983).

Table 1. Current clinical evaluation scheme

Phase I
Scalp/sphenoidal interictal EEG
Video, scalp/sphenoidal telemetry
Neuropsychologic testing
Neuroradiologic testing (CT scan, PEG,[a] angiography)
PET scan of cerebral glucose metabolism
Magnetic resonance imaging
Phase II
Video and SEEG telemetry
Afterdischarge thresholds
Intracarotid amobarbital (Wada test)
Thiopental study[b]

[a] PEG is done only if phase II is recommended

[b] If the initial phase I studies suggest that the patient may meet the criteria for lobectomy without implantation described in the text, the Wada test and thiopental study are done during phase I with scalp and sphenoidal leads in place rather than with depth electrodes

Prognosis Related to Neuropathological Substrate

The importance of the presence of a neuropathological substrate for temporal lobe epilepsy was first disclosed by Falconer and his associates after he had developed the en bloc resection of the anterior temporal lobe (Falconer et al. 1955; Cavanagh and Meyer 1956). He noted that 90% of patients in which the operative specimen had the most common type of pathology, mesial temporal sclerosis, were seizure free. Efforts to detect this preoperatively were not very successful. More recently it was stated by Bancaud (1980), "The surgical results are clear in our population and affirm that there is a close relationship between lesion and epileptogenic area. In each patient cured by cortical resection, there were cellular alterations in the surgical specimen, while the absence of a lesion involved a surgical failure. However, our experience has shown that the excision of the lesion was far from being complete in patients cured of epilepsy; therefore, other criteria are needed to define an epileptogenic zone." Tests of epileptogenic excitability are still required.

Positron computed tomography (PET) of fluorodexyglucose, containing the positron emitter fluorine-18 (^{18}F), labels metabolically active elements in the brain and produces an image with PET that defines patterns of local cerebral metabolic rate for glucose (Engel et al. 1982a, b, c).

Seventy percent of patients with partial complex seizures in our series have demonstrated interictal zones of hypometabolism on FDG scans. The localization of hypometabolism correlates well with the site of predominant epileptiform EEG abnormalities and with the presence of a structural lesion found on pathologic examination of the resected lobe. The area of hypometabolism is usually much larger than the microscopically identified abnormality.

In 1979, criteria were developed to omit SEEG telemetry in some patients: (1) clear-cut scalp EEG findings of localized seizure onset in mesial temporal lobe (via sphenoidal leads) with frequent interictal spikes in the same area; (2) focal hypometabolism of the same temporal lobe on PET scanning; and (3) two additional confirmatory tests of local cerebral dysfunction or nonspecific structural changes in radiologic examination, such as dilated temporal horn or calcification. During the past 5 years, about one-third of patients undergoing temporal lobectomy have met these criteria.

Results of Temporal Lobectomy for the Control of Seizures

Patients with more than 2 years postoperative follow-up are presented below. Class 1 refers to those patients who are seizure-free with or without auras; class 2, one seizure or less per year with or without rare generalized attacks or nocturnal seizures only; class 3, worthwhile reduction of seizures (not more than one per month); and class 4, no improvement. Over three-fourths of patients referred with a diagnosis of complex partial epilepsy come to surgery at UCLA. The improvement in the results seen in Table 2 reflects the improvement in diagnostic methods, as a standardized anterior temporal lobectomy was used in the entire series.

Table 2. Summary of results of surgery (105 temporal lobectomy patients only)

Seizure class	Patient group: Pre-1977 with SEEG only				Post-1977 with SEEG and focal dysfunction tests				Post-1977 without SEEG			
1	26	44	51	83%	21	62	71	97%	9	75	92	100%
2	4				3				2			
3	19				9				1			
4	10				1				0			

Prediction of Psychosocial Functioning Following Temporal Lobe Surgery for Seizures

A major, if not the most important, drawback of having a seizure disorder is its adverse effect on psychosocial functioning. Work and school performance, family and nonfamily relationships, and personal satisfaction may all be affected. These indirect effects of epilepsy are a major consideration in evaluating the appropriateness of surgical intervention for an intractable seizure disorder. It is, therefore, assumed that successful surgical control of seizures would result in a significant change in the psychosocial aspects of the patient's life.

Seizure Control Predicts Psychosocial Change

The results of a previous UCLA study of patients who underwent temporal lobe surgery for intractable seizures confirmed the relation of seizure control and psychosocial change. Specifically, we found that improvement in psychosocial status was directly related to the extent that seizures were relieved by temporal lobe surgery (Rausch and Crandall 1982). Our data also contained prognostic information that patients with higher preoperative functioning level were more likely to have their seizures controlled by surgery and thus were more likely to improve psychosocially after surgery.

Psychosocial functioning was measured as follows:

Degree of Dependency. A rating of 1-6, with the lowest score (1) representing "self-sufficient, able to care for others where necessary" and the highest score (6) indicating "personal needs only or unable to care for self".

Work, Housework, or School Performance. A rating of 1-5, with (1) representing "works steadily and efficiently with no limitations or able to run home efficiently" and (5) indicating "unable to work, dependent at home on others."

Family Relationships. A rating of 1-5 with (1) representing "social adjustment normal" and (5) representing "unable to live with family as a direct consequence of epilepsy or personality difficulties."

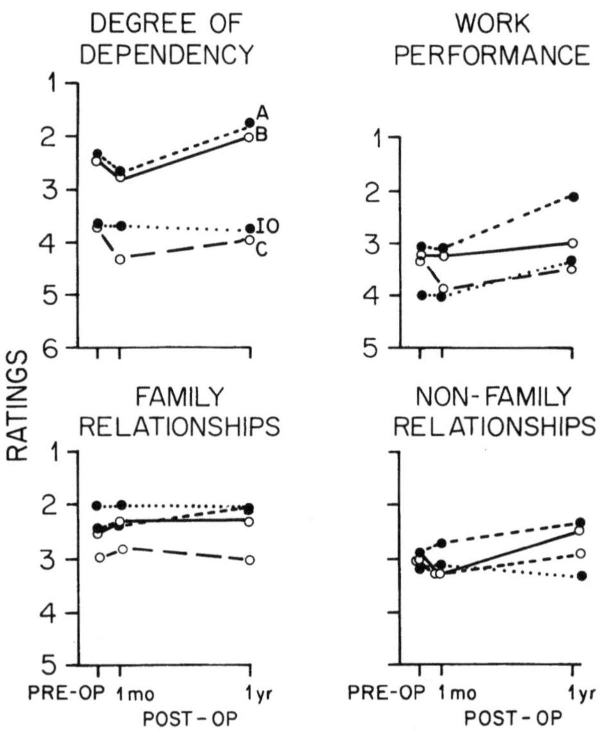

Fig. 1. Mean psychosocial ratings before surgery and at the 1-month and 1-year postoperative evaluations. Group A, seizures controlled by surgery; group B, worthwhile reduction of seizures (>75% reduction); group IO, implant only (without lobectomy); group C, little or no relief after surgery. For ratings see text

Nonfamily Relationships. A rating of 1-5 with (1) representing "relations normal" and (5) indicating "unable to make relationships outside of the family."

Figure 1 shows the mean psychosocial ratings before surgery, and at the 1-month and the 1-year evaluation, for 32 patients grouped as to surgical seizure control. As can be seen in this figure, patients whose seizures were controlled by surgery (group A) function better even before surgery than those whose seizures were not controlled (group C). Also, it was *only* the patients whose seizures were controlled (group A) that improved significantly on the psychosocial ratings by 1 year postoperatively. The group A patients showed improvement on the following scales; degree of dependence; work/school performance; and nonfamily relationships (all $Ps < 0.05$). Family relationships did not significantly change and reflect the difficulty frequently observed in attempts to modify family dynamics of the patients.

Prediction of Psychosocial Change in a "Success Group"

In the above study, we showed that extent of seizure control was a major determinant of psychosocial improvement following surgical treatment. The following study extends these findings to show that within a group of patients whose seizures were "successfully" surgically treated, there are clear indices available before the surgery that can identify which of the surgically "successful" patients will most likely benefit psychosocially.

Method

Patient Group. The patients who constituted our successfully treated group were 41 consecutive "protocol" patients. Since the introduction of our multicriteria protocol in 1977, *all* but one of the patients in the series, both with and without SEEG, had significant seizure reduction with surgery. Patients in this series, who have had at least a 2-year postoperative evaluation of their seizure condition, are included in the present study.[7]

Predictive Measures. An initial study was undertaken to determine from a larger set of potential predictive variables, i.e., those variables which would most likely distinguish between patients who showed the most improvement psychosocially and those patients who showed the least improvement or worsened following surgical control of their seizures. Patient selection for the two extreme outcome groups, the most and least improved groups, was based upon their performance score on the four psychosocial ratings from Rausch and Crandall (1982), as described earlier. Twenty of the 41 patients fell in the extreme groups, i.e., three of the four scales (75%) changed in the same direction, either to the better or worse. The list of predictive variables initially entered into these evaluations is shown in Table 3.

Table 3. Dichotomized variables in the study of characteristics of extreme surgical outcome groups

Clinical parameters
Sex
Age at surgery[a]
Age at onset
Seizure duration
Psychological parameters
Preoperative family support[a]
MMPI's clinical scales (scales 1-10)
Degree of psychopathology[a] (# of MMPI scales $\geqslant 70$)
Ego strength (MMPI subscale)
MacAndrew scale[a] (MMPI subscale)
Goldberg scale (MMPI subscale)
Cognitive parameters
WAIS full scale
Verbal memory scores
Nonverbal memory scores

[a] Variables significantly differentiating ($P < 0.05$) most psychosocially from least psychosocially improved patient groups

The five variables which emerged as significantly differentiating these extreme groups ($P < 0.05$) are defined below. The italicized dichotomy was associated with the group with the best outcome.

1. Age at surgery: (*< 30 years* vs. $\geqslant$30 years).
2. Preoperative good family support (*Yes* vs. No):
Appropriate relative (e.g., spouse, if patient is married or parents and siblings if patient is single) provides the patient with emotional security. Is supportive in stressful/critical situations. Helps the patient follow medical instructions including the taking of medicines and getting to appointments if the patient requires this assistance. The relative does not deprecate the patient's self-esteem but rather provides positive reassurance.
3. Degree of psychopathology: (*< 3* vs. > 3 MMPI Clinical Scales $\geqslant$2 standard deviations above mean, i.e., *T*-score $\geqslant$70).
4. MacAndrew Scale (*< 24* vs. > 24):
A subscale of the MMPI useful as a measure of an addictive personality. This measure has been shown to relate to personality characteristics of alcoholics.

[7] Several patients in this series have since begun having seizures; however, since they were "successful" patients at the time of this evaluation, they are included in the present analyses.

These results show that a relatively young age at surgery, strong family support, and stable personality characteristics characterize patients with the best psychosocial outcome following surgery (i.e., those patients who demonstrated the most positive change in their psychosocial living). These findings were a confirmation of our clinical impression of the factors influencing psychosocial change. These four preoperative predictors were then entered into more detailed analyses of the psychological outcome of all 41 patients, not just the two extreme groups. In addition to the four predictor variables identified above, the following two direct consequences of surgery were included in our more extensive analyses:

5. *Extent of Seizure Control.* While all patients received significant reduction of seizure frequency, patients could be further categorized in a finer scale as to extent of control (as reported in results of surgery). Seizure control scores were dichotomized as 1 vs. 2 or 3.

6. *Cognitive Decrease.* Following unilateral anterior temporal lobe resection, some selective loss in memory functioning is expected to occur (Milner 1958; Blakemore and Falconer 1967; Rausch and Crandall 1982). A decrease in verbal memory is associated with the left or language-dominant resection, and decrease in nonverbal memory is associated with right temporal lobe resection. For the present study, a significant cognitive decrease is defined as "one in which scores on the appropriate memory tests dropped considerably (greater than one standard deviation) and the patient or family reported associated difficulty in this area."

Outcome Measure. Two primary psychosocial outcome variables were investigated; (a) *change* in psychosocial functioning as compared with the preoperative functioning level and (b) the *attainment* postoperatively of the *highest levels* of psychosocial functioning. An overall psychosocial score which represents the mean of five subscales of psychosocial areas was used to assess the patients preoperatively and also at the most recent annual postoperative follow-up check. Each scale has a rating of 1-5, with 1-2 representing the highest levels of functioning and 5 the lowest. The first four scales, *Degree of Dependency, Work Performance, Family Relations,* and *Nonfamily Relations,* are modified from the scales used in the previous above-mentioned studies. A fifth scale, *Personal Satisfaction,* was added, which similarly ranged from a high of (1) satisfied or reasonably contented to (5) suicidally discontented.

Results

The preoperative and postoperative rating scores of our 41 consecutive patients are shown in Fig. 2. Variables depicted in Fig. 2 include the two surgical result variables, seizure control and cognitive decrease, as well as two of the predictor variables,

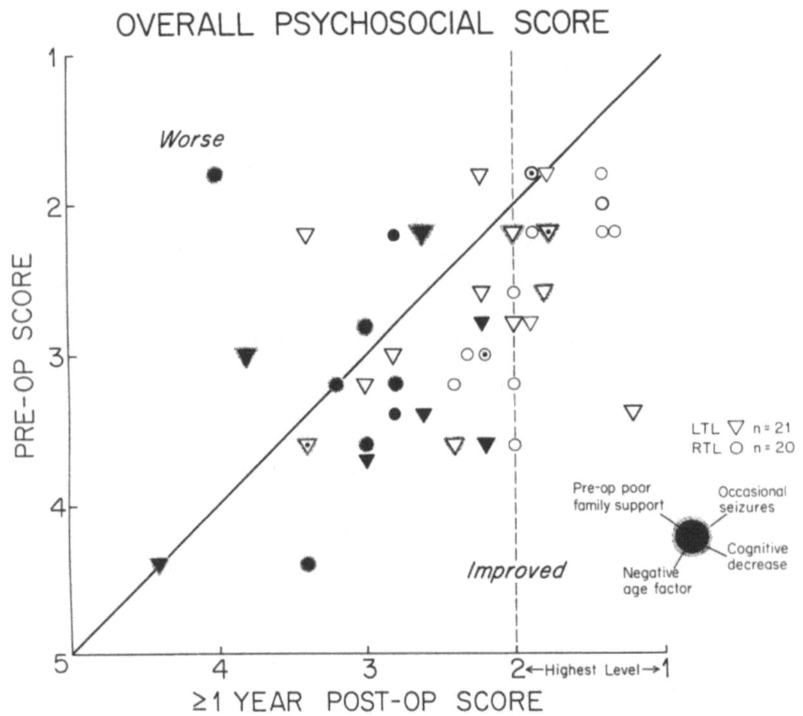

Fig. 2. Preoperative and postoperative rating scores of 41 consecutive patients

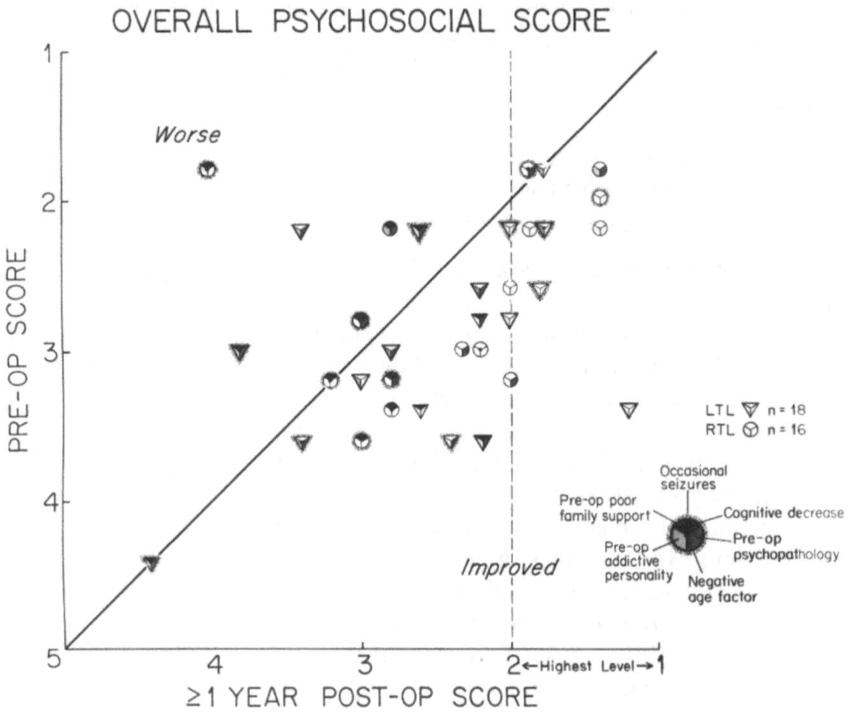

Fig. 3. Preoperative and postoperative rating scores of the subgroup of 34 patients with the two predictive personality variables "degree of psychopathology" and "addictiveness"

preoperative family support and the age at surgery factor. Thirty-four of the 41 patients also had available the two predictive personality variables degree of psychopathology and addictiveness; the complete data of this smaller subgroup are shown in Fig. 3.

The figures are illustrated such that if a factor is indicated it occurs in the negative, e.g., an older age at surgery would be indicated by shadowing, while younger age would be represented by no coloring or marking of the appropriate area. Left and right temporal lobe patients are also identified by different symbols. Patients who *improve* psychosocially after surgery are shown *below* and *on* the diagonal line. (We chose to include those three patients who remain unchanged as measured by our psychosocial rating in our "improved" group. Their seizure frequency and medication levels have reduced since surgery; and although no change in our gross psychosocial rating was depicted, it is assumed that if everything else remains the same, with their positive seizure control, some improvement in daily life-style would have occurred.)

Patients who, despite improved seizure control, received a worse psychosocial score after surgery are shown *above* the line. In addition, those patients who, regardless of their preoperative functioning level, *attained the highest levels* of psychosocial functioning after surgery are shown on and to the *right* of the vertical dashed line.

Visual Analyses. The main visual impressions obtained from these figures are:

1. A positive but less than perfect correlation exists between preoperative and postoperative psychosocial ratings.

2. The clearer a patient's symbol, the more likely he or she is both (a) to improve (fall on or below the diagonal line) and (b) to function at the highest levels postoperatively (on or to the right of the vertical line).

3. The larger the number of areas shaded, the more likely a patient (a) will not improve and (b) will fall within the lowest area of psychosocial functioning.

4. Despite more than one negative indicator, several patients improve dramatically after surgery. The unexpectedly improved patients tend to be younger at the time of surgery than other comparable patients who did not do as well.

Conditional Probabilities. Predictions of a *"good" outcome* are indicated by the conditional probabilities in Table 5. Estimates of the probability of (1) psychosocial improvement and (2) the highest levels of postoperative functioning are provided separately. Estimates are based on the percentage of patients who improve psychosocially (or who attained the highest levels of functioning) who also had a variable present in the positive direction. For example:

$$\text{Probability of improvement} = \frac{\text{number of patients with positive signs who improve}}{\text{number of patients with positive signs}}$$

Conditional probabilities are shown with 68% confidence limits or + one standard deviation.

The conditional probabilities for improvement in psychosocial status after surgery are shown by the first column of Table 4. Given the fact that a seizure patient has met our diagnostic criteria for surgery, there is an estimated 83% chance of improvement postoperatively in overall psychosocial functioning. The percentage of patients who had good indicators and who improved are also shown selectively for each of the predictor variables. Of the preoperative variables only a younger age at surgery did not improve our predictive power. There were three patients in our study who had all factors in the positive direction and all of these patients improved after surgery. There were eight patients who had all four preoperative predictors in the positive direction and all of these patients improved after surgery. As a matter of fact, the combination of any two of the three contributing preoperative predictors (no or mild psychopathology, good family support, and nonaddictive personality) was 100% in predicting improvement after surgery.

The second column of Table 4 shows the *conditional probabilities* of *attaining* the highest functioning level. The individual variables are not as effective in predicting achievement of the highest functioning level, with probabilities averaging at 50/50 chance level. However, if all preoperative positive variables are present, the estimated probability increased significantly, up to 1.00. Analyses of any two combinations of "good" predictors revealed the highest estimates resulted from the presence of "good family support" in conjunction with either of the personality factors (both 83%).

Table 5 provides the estimates for a negative *change in postoperative functioning*. Unlike predicting good postoperative functioning, given "good" preoperative indicators, it is much more difficult to predict who will be worse following treatment, in the light of "poor" indicators. As a matter of fact, given the fact of surgery, patients are not likely to worsen psychosocially after surgery ($P=0.17$ or 1 P of improvement). For any given "poor" predictor, the highest estimate is still below chance. Combinations of any two poor predictors were reviewed and the following two combinations gave the highest probabilities of failure:

Poor family support with significant cognitive decrease: 0.80 ± 0.18
Poor family support with addictive personality: 0.75 ± 0.22

If patients had three or more "poor" predictor factors of any combination, the estimate for worsening of their functioning level increased significantly, with the conditional probabilities ranging from 0.50–1.00. Thus, the presence of two or more poor predictors is required to raise the probability of worsening after surgery.

Preoperative prediction of less than the highest levels of functioning given poor predictors is shown in the second column of Table 5. Interestingly, given the presence of any one "poor" indicator, it can be predicted that these patients are *not* likely to function at

Table 4. Predictors of positive psychosocial outcome

	(n)	Estimate of probability + 1 SD	
		Improved/ same	Highest level
If surgery	41	0.83 ± 0.06	0.37 ± 0.08
No or mild cognitive decrease	25	0.96 ± 0.04	0.36 ± 0.10
Excellent seizure control	27	0.89 ± 0.06	0.44 ± 0.10
No or mild psychopathology[a]	17	0.94 ± 0.06	0.59 ± 0.12
Good family support[a]	26	0.92 ± 0.05	0.58 ± 0.10
Nonaddictive personality[a]	22	0.91 ± 0.06	0.45 ± 0.11
Young age at surgery[a]	32	0.84 ± 0.07	0.41 ± 0.09
All preoperative variables[a]	8	1.00	1.00
All variables	3	1.00	1.00

[a] Preoperative variables

Table 5. Predictors of psychosocial outcome

	n	Estimate of probability + 1 SD	
		Worse	Less than highest level
If surgery	41	0.17 ± 0.06	0.63 ± 0.08
Cognitive decrease	16	0.38 ± 0.12	0.63 ± 0.12
Occasional seizures	14	0.29 ± 0.12	0.79 ± 0.11
Psychopathology[a]	17	0.29 ± 0.12	0.82 ± 0.09
Poor family support[a]	15	0.33 ± 0.12	1.00
Addictive personality[a]	12	0.33 ± 0.12	0.75 ± 0.13
Older age at surgery[a]	9	0.22 ± 0.14	0.78 ± 0.14
All preoperative variables[a]	2	0.50 ± 0.35	1.00
All variables	1	1.00	1.00

[a] Preoperative variables

the highest level postoperatively. The best predictor of less than optimal outcome is *poor family support*. Not one patient with poor family support, despite the presence of other positive signs, is functioning optimally after surgery. The presence of more than *one* poor predictor insured the failure of a patient from reaching the highest level of psychosocial functioning.

Summary

New data are presented to show that within a group of "successfully" surgically treated patients, indices are present before surgery which can provide a high estimate of individual postoperative psychosocial functioning.

An initial study revealed four factors available preoperatively which significantly differentiated two extreme groups, i.e., a group of patients who improved the most and a group who improved the least (or worsened) psychosocially after surgery. Characteristics of the best outcome group were:

1. Younger than 30 years of age at surgery
2. Strong family support
3. None or mild psychopathology
4. Absence of an addictive personality

These four predictor variables plus two surgical outcome variables (seizure control and cognitive decrease) were used to calculate conditional probabilities for postoperative psychosocial functioning for *all* patients (not just the extreme groups). Both change in and optimal functioning in the psychosocial area were investigated.

Improvement in psychosocial functioning is likely (83%) just given the fact a patient meets our surgical criteria and undergoes surgery. The presence of the following "good" preoperative indicators, strong family support, and stable personality characteristics, increased the probability of improvement. The presence of any two combinations of "good" predictors increased the probability to 1.00.

A *decrease* in functioning is not likely (0.17) with surgery. Even in the presence of a "poor" indicator, the probability is still below chance that a patient would worsen after surgery, i.e., the patient is likely to improve. However, as the number of "poor" indicators increases for a given patient, the probability of worsening becomes increasingly higher.

Prediction of the attainment of the highest functioning level postoperatively is difficult for any given predictor. However, good family support in conjunction with stable personality characteristics increased the probability of this achievement up to 83%.

Given the presence of a "poor" predictor, it is unlikely a patient will achieve the ultimate in psychosocial function. Interestingly, it is the presence of "poor family support" which weighs the heaviest in preventing this optimal achievement.

In conclusion, a number of variables have emerged which if present increase the probability of a person improving psychosocially following surgical control of intractable seizure condition. These include good family support, no or mild psychopathology, a nonaddictive personality, and a relatively young age at surgery. While all of these variables also play a role in determining whether or not a patient achieves the highest level of psychosocial functioning after surgery, it is good family support which plays the most critical role. Good family support has proven to be a necessary condition for the attainment of optimal psychosocial function.

References

Bancaud J (1980) Topographic relationship between cerebral lesions and seizure discharges. In: Canger R, Angeleri F, Penry JK (eds) Advances in epileptology: XIth epilepsy international symposium. Raven, New York, pp 102–109

Bancaud J, Talairach J, Bonis A, Bordas-Ferrer M (1969) Anomalies EEG intercritiques bitemporales asynchrones dans les épilepsies unilaterales (à propos d'une épilepsie d'origin tumorale). Rev Neurol 121/3: 369–379

Bancaud J, Riber MF, Chagot D (1975) Origine comparée des paroxysmes de pointes "infracliniques" et des crises électrocliniques spontanées dans l'épilepsie. Rev Electroencephalogr Neurophysiol 5: 63–66

Blakemore CB, Falconer MA (1967) Long-term effects of anterior temporal lobectomy on certain cognitive functions. J Neurol Neurosurg Psychiatry 30: 364–367

Bloom D, Jasper HH, Rasmussen T (1960) Surgical therapy in patients with temporal lobe seizures and bilateral EEG abnormality. Epilepsia 1: 351–365

Cavanagh JP, Meyer A (1956) Aetiological aspects of Ammon's horn sclerosis associated with temporal lobe epilepsy. Br Med J 2: 1403–1405

Crandall PH, Walter RD (1971) The ictal electroencephalographic signal identifying limbic system seizure foci. Proc Am Assoc Neurol Surg 1: 1–9

Crandall PH, Walter RD, Rand RW (1963) Clinical application of studies of stereotactically implanted electrodes in temporal lobe epilepsy. J Neurosurg 21: 827–840

Crandall PH, Engel J Jr, Rausch R (1983) Indications for depth electrode recordings in complex partial epilepsy and subsequent surgical results. In: Rose RC (ed) Progress in epilepsy, Pitman Medical, London, pp 507–526

Engel J Jr, Driver MV, Falconer MA (1975) Electrophysiological correlates of pathology and surgical results in temporal lobe epilepsy. Brain 98: 129–156

Engel J Jr, Kuhl DE, Phelps ME, Crandall PH (1982a) Comparative localization of the epileptic focus in partial epilepsy by PCT and EEG. Ann Neurol 12: 529–537

Engel J Jr, Brown WJ, Kuhl DE, et al. (1982b) Pathological findings underlying focal temporal lobe hypometabolism in partial epilepsy. Ann Neurol 12: 518–528

Engel J Jr, Kuhl DE, Phelps ME, Mazziotta JC (1982c) Interictal cerebral glucose metabolism in partial epilepsy and its relation to EEG changes. Ann Neurol 12: 510-517

Engel J Jr, Crandall PH, Rausch R (1983) The partial epilepsies. In: Rosenberg RN (ed) The clinical neurosciences, vol 2. Churchill, New York, chap 15, pp 1349-1380

Falconer MA (1967) Surgical treatment of temporal lobe epilepsy. New Z Med J 66: 539-544

Falconer MA, Hill D, Meyer W, Mitchell W, Pond DA (1955) Treatment of temporal lobe epilepsy by temporal lobectomy: a survey of findings and results. Lancet 1: 827-835

Gastaut H (1950) Etude électro-encéphalographique du lobe temporal. Rev Otoneuroophthalmol 22: 301-328

Gibbs EL, Gibbs FA (1947) Diagnostic and localizing value of electroencephalographic studies in sleep. Res Public Assoc Res Nerv Ment Dis 26: 366-376

Gibbs FA (1951) Ictal and nonictal psychiatric disorders in temporal lobe epilepsy. J Nerv Ment Dis 113: 522-528

Gibbs FA, Gibbs EL (1952) Atlas of electroencephalography, vol 2, epilepsy. Addison-Wesley, Reading

Gibbs FA, Gibbs EL, Lennox WG (1938) Cerebral dysrhythmias of epilepsy. Measures for their control. Arch Neurol Psychiatry 39: 298-314

Gloor P, Olivier A, Ives J (1980) Prolonged seizure monitoring with stereotactically implanted depth electrodes in patients with bilateral interictal epileptic foci: how bilateral is bitemporal epilepsy? In: Penry JK, Wada J (eds) Advances in epileptology, Xth epilepsy international symposium. Raven, New York, pp 83-88

Jasper HH, Pertuiset B, Flanigin H (1951) EEG and cortical electrograms in patients with temporal lobe seizures. Arch Neurol Psychiatry 65: 272-290

Lieb JP, Engel J Jr, Gevins A, Crandall PH (1981) Surface and deep EEG correlates of surgical outcome in temporal lobe epilepsy. Epilepsia 22: 515-538

Milner B (1958) Psychological defects produced by temporal lobe excision. Res Public Assoc Res Nerv Ment Dis 36: 244-257

Penfield W, Jasper H (1954) Epilepsy and the functional anatomy of the human brain. Little, Brown, Boston, 896 pp

Rausch R, Crandall PH (1982) Psychological status related to surgical control of temporal lobe seizures. Epilepsia 23: 191-202

Rovit RL, Gloor P, Rasmussen T (1961) Sphenoidal electrodes in the electrographic study of patients with temporal lobe epilepsy - an evaluation. J Neurosurg 18: 151-158

Talairach J, Bancaud J (1974) Stereotactic exploration and therapy in epilepsy. In: Vinken PJ, Bruyer GW (eds) Handbook of clinical neurology, vol 15. North-Holland, Amsterdam, pp 758-782

Talairach J, Szikla G (1967) Atlas d'anatomie stéréotaxique du telencéphale. Masson, Paris, 326 pp

Talairach J, Szikla G (1979) Stereotactic neuroradiological concepts applied to surgical removal of cortical epileptogenic areas. In: Rasmussen T, Marino R (eds) Functional neurosurgery. Raven, New York, pp 219-242

Talairach J, David M, Tournoux P (1958) L'exploration chirurgicale stéréotaxique du lobe temporal dans l'épilepsie temporale. Masson, Paris, 136 pp

Talairach J, Bancaud J, Szikla GA, Bonis A, Geier S, Vedrenne C, et al. (1974) Approche nouvelle de la neurochirurgie de l'épilepsie. Methodologie stéréotaxique et résultats therapeutiques. Neurochirurgie 20, Suppl 1, 240 pp

Wieser HG (1983) Electroclinical features of the psychomotor seizure. Fischer/Butterworths, Stuttgart/London, 242 pp

The Preoperative Electroclinical Evaluation Determines Postoperative Seizure Control after Temporal Lobectomy

M. Rayport, W. S. Corrie, and S. M. Ferguson

Introduction

We have tested the hypothesis that the outcome of temporal corticectomy for seizure control is a function of the clinical and electrophysiological methodology employed in case selection in 25 consecutive cases of temporolimbic seizures from a study population of 68 sequential patients with medically intractable epilepsy who had received complete inhospital neurological, neuropsychiatric, radiological, extracranial EEG, and stereoelectroencephalographic (SEEG) assessment. The relatively low published rates of complete seizure control (30%–40% seizure-free) after temporal lobectomy (Jensen 1975; Rasmussen 1975) indicate the limitations of seizure control consequent to reliance on interictal extracranial EEG recording for selection of surgical candidates. Our methodology comprised three aspects: surgical philosophy, surgical selection criteria, and indications for SEEG.

Surgical Philosophy. The availability of multitechnical diagnostic capabilities for electrophysiological detection and localization of the epileptogenic cortex, including extracranial EEG, chronic subdural recording, acute or chronic SEEG, and electrical brain stimulation, in the same surgical epilepsy unit avoids the biasing influence of strong unitechnical preferences that limit the preoperative data base, e.g., to interictal recording and/or one particular level of electrode placement, extracranial, epidural, subdural, or intracerebral, for all cases. Insofar as the objective of neurosurgical treatment is the elimination of seizures and not of interictal spikes, preoperative electrographic localization for surgical case selection must rely primarily on EEG recording of the patient's spontaneous habitual seizures. In so far as epilepsy pervades both the brain and the life of the afflicted person, a multidisciplinary clinical management team including behavioral specialists is more effective in helping such persons and in research on epilepsy. Each patient would be an individual case study in brain pathophysiology, pathomorphology, and behavioral pathology.

Criteria of Clinical and EEG Concordance for Seizure Control Surgery. Patients who met the following criteria of electroclinical concordance were eligible for temporal lobectomy with ECoG: (1) stable unifocal seizure pattern; (2) concordant unifocal ictal EEG discharge starts before or with the spontaneous clinical seizures in split-screen closed-circuit television recordings; (3) perictal cognitive testing concords with putative EEG localization; (4) neuropsychiatric, neuropsychological, and radiological evaluations confirm localization, or are nonlocalizing.

Indications for SEEG. (1) stable unifocal seizure pattern; (2) criteria of EEG/clinical concordance are not met; (3) observed data do not permit localization of seizure origin.

Results and Conclusion

The results are summarized in Table 1. Of the 25 consecutive patients, 14 met the criteria of clinico-EEG concordance and 11 were found to be operable by SEEG. Durations of follow-up ranged from 13 to 183 months (mean 81 months). Twenty-two patients (88%) have been seizure-free since surgery, and 1 became seizure-free after some early attacks. Of the seizure-free group 3 have been followed for 19, 14,

Table 1. Postoperative seizure control in 25 consecutive cases of intractable temporolimbic complex partial seizures

Seizure-free since surgery	19 (76%)	22 (88%)	23 (92%)
Auras only	3 (12%)		
Seizure-free after some early attacks	1 (4%)		
Moderate reduction (50%)	1 (4%)		
Unimproved	1 (4%)		

Department of Neurological Surgery, Medical College of Ohio, Toledo, OH 43699-0008, USA

Presurgical Evaluation of Epileptics
Edited by H. G. Wieser and C. E. Elger

and 13 months. Removal of these 3 cases from the seizure-free group would leave 20 (80%) who have been seizure-free for more than 2 years. We believe that these results support the hypothesis that case selection for surgical control of temporolimbic partial seizures on the basis of concordant clinical and electrophysiological localization of the *ictal* mechanism results in a high probability of postoperative cessation of seizures.

References

Jensen I (1975) Temporal lobe surgery around the world. Results, complications, and mortality. Acta Neurol Scand 52: 354–373

Rasmussen T (1975) Surgical treatment of patients with complex partial seizures. In: Penry JK, Daly DD (eds) Complex partial seizures and their treatment. Advances in neurology, vol. 11. Raven, New York, pp 415–442

High-Level Drug Therapy Versus Surgical Treatment: Long-Term Outcome

I. Jensen

The main objective of treatment of any disease is to cure the patient. In epilepsy the primary aim has been to render the patient seizure-free. However, within the past decades it has become increasingly evident that the epileptic seizures constitute only a part of the total disability. Displaying both physical and psychic symptoms, epilepsy more than any other disease has been open to interpretation both as a physiological process and as the effect of "spiritual" influences. Through the centuries, this unfortunate group has been subjected to a variety of "therapies."

The medieval physican ordered regimen, drugs, cauterization, and trephining in continuation of ancient traditions. In the seventeenth century, castration was performed in cases of supposed sympathetic epilepsy originating from the testicles to stop masturbation and thereby cure the epilepsy. Double-blind trials had not yet been invented. Diets, both oral and external water-therapies, and bleeding were later practiced, also in vain. The pharmacological arsenal in the late nineteenth century included mistletoe, preparations of silver nitrate given internally, turpentine, indigo, belladonna, and inhalation of chloroform, among others [66].

In the twentieth century, a more enlightened attitude has developed. On the basis of clinical observations and neurophysiological examinations, epilepsy is no longer considered one disease, but a collection of symptoms which may originate from various parts of the brain, presenting different etiologies and prognoses and, as has become evident during the past decades, also requiring a highly specific pharmacotherapy [56]. Phenobarbital was introduced in 1912 and phenytoin in 1938; both drugs have been and are still widely used anticonvulsants, with the advantages of being effective in controlling various types of seizures, as well as being inexpensive. However, both drugs have demonstrated side effects of such a degree that, if a pharmaceutical firm would apply today for their introduction onto the market, it would probably not be allowed.

Carbamazepine has for the past 25 years proved its efficacy as monotherapy in most forms of partial seizures, particularly those of the complex partial variety, but it might also be effective in the treatment of generalized, and especially grand mal, seizures. Serious side effects are infrequent, and the milder toxic side effects are usually dose related. The serum level is directly related to the dose taken, thus facilitating the optimal therapeutic medication.

Valproic acid has been used for some 10 years and is potent in the treatment of generalized seizures; there is a low frequency of side effects, but a small percentage of patients experience an intolerable marked weight gain, which causes the drug to be discontinued. The most disturbing side effect is, however, an unheralded liver failure which, though extremely rare, is frequently fatal [16, 55].

Carbamazepine is the primary drug of choice in the treatment of complex partial seizures, while valproic acid is preferred in generalized seizures.

The efficacy of drug treatment is frequently assessed on the basis of a reduction of seizure frequency and/or severity without toxic side effects. However, the value of the treatment can hardly be determined without repeated observations and psychological evaluations of possible cognitive and other drug-related problems, these being extremely marked during treatment with phenobarbital and to a lesser degree with phenytoin. The impairments cannot be dismissed as "natural and inevitable" in epileptic patients [12, 67].

Epilepsy affects approximately 1 in every 200 persons. For every patient with generalized epilepsy, at least two suffer from partial seizures. Thus, in a population of 100000 epileptics, upward of 60000 will have partial seizures. Since detailed studies indicate failure of seizure control in 30% or more of epileptic patients, there are about 20% with uncontrolled partial seizures. The number of those who might be helped by excision of an epileptogenic focus has

Rigshospitalet, University of Copenhagen, Blegdamsvej 9, 2100 Copenhagen, Denmark, and Vejlesøevej 122, 2840 Holte, Denmark

Presurgical Evaluation of Epileptics
Edited by H. G. Wieser and C. E. Elger

been variously cited as 4.5%–10% of the total group of epileptic patients [62].

Because a patient has not been controlled with an adequate amount of a specific anticonvulsant in the past, does this mean that a retrial will be unsuccessful? A second delicate question would be the advisability of a trial of less commonly used or still experimental anticonvulsants. Should surgical treatment be delayed until an adequate trial has been conducted? Again, only a clinical and somewhat arbitrary decision can be made, based upon the low probability that such a drug will make a significant difference considering the patient's past anticonvulsant history. Balanced against the time necessary for a search for a new drug regimen for the patient, is the matter of postponing the decision for surgery [72]. Statistics indicate clearly that the surgical outcome is partly influenced by the length of the seizure history and that further attempts at medical management beyond a point will only prolong the disability of the patient, incurring the risks of side effects, and making rehabilitation difficult [3, 5, 12, 14, 31, 32, 37, 43, 48, 54, 59, 65].

At the present state of our knowledge, neurosurgical treatment of partial epilepsy is dependent on the demonstration of the patient's epileptic seizures originating in a circumscribed area of the brain which can be removed without exposing the patient to the risk of intolerable neurological deficits [23].

The reliable localization of an epileptogenic cerebral lesion continues to be a controversial issue. Neuroradiological procedures, including CT scans, can usually identify mass lesions, such as tumors, major vascular malformations, and localized or general atrophies; removal of tumors and malformations produces a high rate of seizure remission, but most often such abnormalities are absent in a population of patients with drug-resistant complex partial seizures.

To establish that a patient is suffering from drug-resistant epilepsy of the complex partial seizure type, it is necessary that the drugs have been administered in the maximal permissible doses, within toxic limits, with assiduous control of the serum levels. Preferably monotherapy, but also adjunctive medication, must have been already tried in vain [16, 55, 67].

In spite of the obvious difficulties, an attempt should be made to secure from observers as accurate a description of the seizures as possible. It is the practice at the Montreal Neurological Institute and at UCLA gradually to reduce the anticonvulsant medication and/or utilize sleep deprivation in an effort to increase the probability of observing a seizure [8, 12, 19, 23, 35, 72]. Other methods of recording the clinical aspects of the seizures require basically some type of pictorial preservation. Most centers working with these methods have found that videotape is the most practical method of recording seizures, because of the ease in editing and its economy. A further increase in the yield of useful information can be obtained by the split-screen technique of recording the surface EEG along with the clinical behavior of the patient.

Demonstration in the EEG of a focal onset of ictal discharge in the record of a spontaneous seizure most closely approximates the ideal requirement for an unequivocal identification of the site of origin of an attack. The second-best diagnostic and localized evidence is represented by unequivocal and localized interictal epileptiform discharge in the EEG [23, 35].

Other clinical and paraclinical data – including neurological signs, radiological changes, and the psychological test pattern – often support the localizing evidence obtained from EEG recordings, but their reliability as indicators of the area of seizure onset is not great. Nasopharyngeal electrodes can be easily and quickly applied by a skilled technician, without undue discomfort or risk to the patient. Sphenoidal electrodes, if properly inserted, provide high-quality, high-resolution records with very little artifact from a region which, in most patients suffering from temporal lobe epilepsy, represents their major epileptogenic area [23, 48].

Depth electrode studies have proved invaluable, if not essential, in the evaluation of those patients in whom serial scalp and sphenoidal recordings have demonstrated bilateral, independent, surface spiking, approaching a parity between right and left temporal areas. The implantation of stereotactic depth electrodes also affords the opportunity to stimulate electrically the various contact points within the brain in an attempt to reproduce the patient's habitual seizures. Chronic depth electrode studies are also essential in planning for surgical treatment outside the limbic system [14, 19, 23, 26, 34, 35, 43, 48, 62, 64, 72, 75].

Electrocorticography associated with electrical stimulation is considered indispensible at the Montreal Neurological Institute and at the University of Washington, Seattle [23, 52, 53]. If the surgical procedure elected is a maximal en bloc temporal lobectomy, it could be argued that electrocorticography at the time of the operation would be superfluous [12, 20, 33, 72]. Outside the temporal lobes, electrocorticography combined with electrical stimulations at the time of surgery continues to be the basic technique for determining the area of resection.

The first recorded application of temporal lobe resection as a treatment in itself, and not associated with the removal of a tumor, scar, or abscess, was by Burckhardt in 1891, when he tried unsuccessfully to relieve two schizophrenic patients of their auditory hallucinations [7]. The first operation for epilepsy performed under the guidance of electrocortigraphy was by Percival Bailey in 1947 [2]. The operative procedures applied by various neurosurgical clinics follow either the technique by Wilder Penfield [50], according to which the operation is performed with the patient under local anesthesia and under guidance of electrocortical stimulation and recording or, alternatively, that of Murray Falconer [20], in which as much of the affected temporal lobe as possible is removed en bloc under general anesthesia [30].

According to even the earliest reports, which included only limited follow-up periods, the final outcome was favorable, as judged by abolition of seizures or a marked reduction in seizure frequency. So far, surveys covering the results of more than 5000 operations for temporal lobe epilepsy have been published: 4800 lobectomies (Tables 1, 2 with accompanying references), 350 stereotactic lesions [3, 26, 38, 39, 43, 45, 60, 61, 71, 75], and 103 amygdalohippocampectomies [76, 77]. These are minimum figures, as it is virtually impossible to obtain results from various centers in eastern Europe including the USSR, as well as from China and Japan.

The overall impression is that a unilateral anterior temporal lobectomy is, for the time being, the optimal treatment for drug-resistant temporal lobe epilepsy, the elegant stereotactic procedures not having fulfilled their promising early results. The period of observation for the amygdalohippocampectomies is still too short, but quite possibly these operations which leave the lateral cortex intact will turn out to be the intervention of choice. The results of the most recent survey of this series are given by Yaşargil and Wieser in this volume.

The chances of a favorable outcome regarding partial or complete relief from seizures have been increased during the years. This is primarily due to improved indications for operations, based upon the results from the first surveys, and secondly due to the institution of more refined preoperative examinations, i.e., sphenoidal leads, depth electrode recordings and stimulations, telemetry, and videotape recordings. An improved surgical technique as well as better anesthesiology service have reduced the risks for complications to a minimum, just as there now is insignificant operative or immediate postoperative mortality risk. The various centers have changed the definitions of their postoperative results during the years. But by and large, it can be expected that 50%-60% of all operated on patients will be free from any seizures, that a total of at least 80% will have rare or only few seizures, and that only 5%-10% will belong to the unfortunate group of "no change." The UCLA group, in their last survey of patients, had only one who did not belong to the successful groups [8].

While the main objective of treatment is to render the patients seizure-free, it must be remembered that the seizures are only one part of the patient's total disability. Psychological, social, educational, and vocational limitations are also imposed on those with epilepsy [1, 4, 9, 12, 19, 22, 31, 36, 37, 42, 48, 54, 56, 57, 65, 70]. The time has come, as already stated by Crandall 10 years ago [12], to establish common criteria by which to evaluate medical and surgical treatment of epilepsy, associated social services, and employment standards. For example, the efficacy of drug treatment is frequently assessed on the basis of reduction of seizure frequency and/or severity without toxic side effects. However, the value of treatment can hardly be determined without repeated psychological evaluation of possible drug-related cognitive problems and observations concerning social relationships. These impairments cannot be dismissed as "natural or inevitable" in epileptic patients. Similarly, a high success of surgical treatment must be based on an abolition of, or a reduction in, the frequency of seizures, without cognitive or neurological deficits and without aggravation of psychiatric symptoms [54].

Cognitive deficits can result from temporal lobe ablations, tumors, and congenital malformations. However, they occur commonly from epileptic disruption of function. Thus the definite postoperative improvement in certain cognitive functions, especially observed in young children, followed relief from seizures. The most marked changes in intellectual scores following surgery occur in patients whose seizures are surgically controlled. The patients with left-sided operations generally improve in performance/perceptual scores, while the patients with right-sided operations show increased verbal scores. Patients with poor post-lobectomy seizure relief tend to have lower presurgical intellectual scores and a further drop in these following lobectomy, as compared with the successfully operated patients [36]. It can also be concluded that patients whose seizures are relieved improve in degree of independence, working performance, and nonfamily relationships, whereas patients whose seizures are not controlled by surgery show no significant change in psychosocial measurement. The UCLA comparative investi-

Table 1. Results of temporal lobe resection for seizures (indicated in percentages)

Country	Period	Total *N*	Follow-up *N*	Success A	Success B	Success C	Worthwhile improvement D	Worthwhile improvement E	Failure/ mortality/ no follow-up	References
Canada	1928–1980	1034	914	36	13		13	36	2	[48, 53]
Denmark	1960–1984	124	74	61	—	20	—	7	12	[30, 32, 33]
England, Guy's-Maudsley	1951–1970	255	171	42	—	19	—	22	16	[20, 65]
France, la Salpetrière	1949–1956	110	110	29	—	36	—	18	17	[25]
la Timone	1949–1956	50	50	40	—	—	—	60	—	[49]
Sainte-Anne	1957–1973	68	68	62	—	19	—	—	19	[64]
Hungary	1951–1975	67	67	50	—	—	46	—	4	[29]
Norway	1952–1973	35	35	54	—	23	—	9	14	[30]
Poland	1957–1979	262	262	49	—	16	—	18	16	[63]
Switzerland	1975–1984	48	48	52	—	4	25	—	19	[75]
USSR		130?	69	54	—	22	—	12	13	[4]
USA, Washington DC	1946–1956	31	31	45	—	36	—	10	10	[44]
Chicago	1948–1951	72	36	28	—	—	—	44	28	[2]
Phoenix	1948–1965	78	78	30	—	26	—	9	36	[24]
Columbia, New York	1951–1977	45	25	40	—	—	36	—	24	[11]
NINDB, Maryland	1954–1969	76	76	32	18	—	21	9	20	[70]
UCLA	1959–1972	38	38	—	—	71	—	18	11	[8, 12]
	1972–1976	106	106	44	39	—	—	17	—	[12]
	1977–1982[a]			62	35	—	—	3	—	[12]
Johns Hopkins	1970–1982	27	23	40	—	—	—	35	25	[22]
Seattle	1976–1980	31	23	61	—	—	—	17	22	[68]
India, Velore	–1969	24	24	33	—	—	46	—	21	[41]
Australia	1949–1972	46	46	41	—	—	27	16	15	[15]

A completely seizure-free, apart from perhaps a single seizure in the immediate postoperative period; *B* isolated or occasional attacks, after being free from seizures for several years; *C* maximal two to three seizures a year. *D* only 1%–10% as many attacks as preoperatively; *E* at least 50% reduction in seizure frequency

[a] Temporal lobectomy with SEEG telemetry

Table 2. Results of temporal lobe resection for seizures (surveys with less elaborate data)

Country	Period	Total *N*	Follow-up *N*	Success A	Success B	Success C	Worthwhile improvement D	Worthwhile improvement E	Failure/ mortality/ no follow-up	References
Czechoslovakia,										
Bratislava	ca. 1959–1970	44	44	—	—	—	57	21	21	[10, 30]
Prague		286								[71]
England, London	–1967?	62	62	45	—	—	—	23	32	[13, 30]
Hungary, Budapest	1955–1958	60?	34	24	—	—	—	33	44	[21, 30]
Italy	ca. 1951–1952	28	13	46	—	—	—	23	31	[40, 30]
The Netherlands	1949–1969	45	45	40	—	—	29	11	20	[17, 30]
Poland		35	35	54	—	—	23	9	14	[28, 30]
Sweden	1951–1967	154	154	47	—	—	—	24	29	[30, 47]
Switzerland	ca. 1957–1966	35	35	32	—	—	21	—	47	[30, 73]
West Germany	1960–1969	19	19	58	—	—	—	—	42	[18, 30]
USA, Minnesota	1949–1953	50	25	48	—	—	—	—	52	[6, 30]
New York	ca. 1959–1962	21	18	28	—	—	39	17	17	[27, 30]
San Francisco	–1968	10?	10	60	—	—	—	20	20	[30, 46]
Yale	ca. 1972–1982	26?	26							[1]
Brazil, Rio de Janeiro	1956–1957	37	10	50	—	—	—	—	50	[30, 51]
Cuba	1950–1954	18	18	39	—	—	22	17	22	[30, 58]
China, Shanghai	1959–1964	8	8	38	—	—	12	25	25	[30, 58]
Israel	1955–1963	11	9	44	—	—	33	—	22	[30, 74]

gations have shown that, in this regard, nonoperated patients will demonstrate the same impairments as the surgically unsatisfactory group [54].

Patients with seizures in early childhood and adolescence have an additional problem in that adequate personality development has not taken place during the various phases of maturation. Nevertheless, seizure relief in adolescents or young adults makes it possible for them to reestablish identity and to resume maturation in social relationships [12, 31]. In the younger age groups, i.e., below 25 years of age, relief from seizures is associated with improvement in all psychiatric states, even in patients with more or less overt psychoses, thus facilitating rehabilitation. In the older age group, with long duration of epilepsy, frequent psychiatric problems are encountered in rehabilitation [12, 14, 57, 70].

Lindsay, Ounsted, and collaborators have followed 100 children with epilepsy since 1948. In nine of these, a temporal lobectomy has been performed. They concluded:

In patients who deteriorate gravely between the ages of five and 15 years surgical considerations become urgent early in the biography, and should not be delayed. All too often in our wider experience we see children who for many years have been treated with combinations of the so-called anticonvulsants, without gaining any true relief, but there are signs that this practice is coming to an end. When operations were performed in this series, and in other children we have studied, we saw quite remarkable reversal of social, intellectual, and characterological handicap. This suggests to us that the adolescent brain retains powers of recovery which are worthy of further studies [37].

Several other studies, including the Danish surveys, support this recommendation of early surgery [4, 12, 14, 31, 32, 43, 48, 54, 59, 69, 70].

Conclusion

The ultimate therapeutic aim in epilepsy is, of course, the correction of the basic abnormality of neuronal mechanisms responsible for the seizures. However, until our understanding of these abnormalities permits the development of truly definitive medical therapy, we have to acquiesce in what is currently designated as the optimal treatment. Patients with epilepsy should be examined to diagnose their specific type of epilepsy and to determine a possible causal treatment of the disorder. The adequate treatment should thereafter be instituted, be it surgery or medication. If, in patients with complex partial seizures, the medical treatment fails to succeed within a few years in spite of optimal, maximal, nontoxic medication with frequent serum level controls, it might today be considered unethical to withhold from these patients an intensive presurgical evaluation. This might eventually lead to a surgical procedure, such as unilateral anterior temporal lobectomy.

References

1. Augustine EA, Novelly RA, Mattson RH, Glaser GH, Williamsson PD, Spencer DD, Spencer SS (1984) Occupational adjustment following neurosurgical treatment of epilepsy. Ann Neurol 15: 68-72
2. Bailey P (1961) Surgical treatment of psychomotor epilepsy. South Med J 54: 299-301
3. Barcia-Salorio L, Broseta J (1976) Stereotactic fornicotomy in temporal lobe epilepsy: indications and long-term results. Acta Neurochirurgica [Suppl] 23: 167-175
4. Bein BN, Boreiko VB (1975) The dynamics of temporal lobe epilepsy manifestations and the social-occupational prognosis in children following surgical removal of the epileptic focus. ZH Nevropatol Psikhiatr 75: 512-517
5. Birri R, Perret E, Wieser HG (1982) Der Einfluß verschiedener Temporallappenoperationen auf das Gedächtnis bei Epileptikern. Nervenarzt 53: 144-149
6. Brown IA, French LA, Ogle WS, Jahnson S (1956) Temporal lobe epilepsy; its clinical manifestations and surgical treatment. Medicine 35: 425-459
7. Burckhardt G (1891) Über Rindenexcisionen, als Beitrag zur operativen Therapie der Psychosen. Zeitschrift für Psychiatrie und psychisch-gerichtliche Medizin 47: 463-548
8. Cahan LD, Sutherling W, McCullough MA, Rausch R, Engel J Jr, Crandall PH (1984) Review of 20-year UCLA experience with surgery for epilepsy. Cleve Clin Q 51: 313-318
9. Cherlow DG, Serafetinides EA (1976) Speech and memory assessment in psychomotor epilepsy. Cortex 12: 21-26
10. Cigánek L, Likavcanova E (1972) Long-term results of the surgical treatment of temporal epilepsies. Bratisl. lek. Listy 58: 525-531
11. Cogen PH, Antunes JL, Correll JW (1979) Reproductive function in temporal lobe epilepsy: the effect of temporal lobectomy. Surg Neurol 12: 243-246
12. Crandall PH (1975) Postoperative management and criteria for evaluation. Adv Neurol 8: 265-279
13. Currie S, Heathfield KWG, Henson RA, Scott DF (1972) Clinical course and prognosis of temporal lobe epilepsy. Brain 94: 173-190
14. Davidson S, Falconer M (1975) Outcome of surgery in 40 children with temporal lobe epilepsy. Lancet 1: 1260-1263
15. Davis E (1974) Temporal lobectomy for epilepsy - a follow-up. Proc Aust Ass Neurol 11: 245-246
16. Delgado-Escueta AV, Treimann DM, Walsh GO (1983) The treatable epilepsies. N Engl J Med 308 (26): 1576-1784
17. de Vet AC (1972) Temporal epilepsy, c. q. psychomotor epilepsy. Schweiz Arch Neurol Neurochir Psychiat 111/2: 453-461
18. Dill R, Gullota F (1970) Pathomorphologische Befunde bei Temporallappenepilepsien. Schweiz Arch Neurol Psychiat 106: 241-255

19. Engel J, Rausch R, Lieb JP, Kuhl DE, Crandall PH (1981) Correlation of criteria used for localizing epileptic foci in patients considered for surgical therapy of epilepsy. Ann Neurol 9: 215-224
20. Falconer M, Serafetinides EA (1963) A follow-up study of surgery in temporal lobe epilepsy. J Neurol Neurosurg Psychiat 26: 154-165
21. Fenyes I, Zoltan L, Fenyes G (1961) Temporal epilepsies with deep-seated epileptogenic lesions. Arch Neurol (Chic) 4: 559-571
22. Fisher RS, Uematsu S (1982) Surgical therapy of complex partial seizures. Johns Hopkins Med J 151(6): 332-343
23. Gloor P (1975) Contributions of electroencephalography and electrocorticography to the neurosurgical treatment of the epilepsies. Adv Neurol 8: 59-105
24. Green JR (1967) Temporal lobectomy, with special reference to selection of epileptic patients. J Neurosurg 26: 584-593
25. Guillaume J, Mazard G (1956) Indication et résultats du traitement chirurgical des épilepsies temporales. Sem Hop Paris 32/2: 2013-2018
26. Hood TW, Siegfried J, Wieser HG (1983) The role of stereotactic amygdalotomy in the treatment of temporal lobe epilepsy associated with behavioral disorders.Appl Neurophysiol 46: 19-25
27. Horowitz MJ, Cohen FM, Skolnikoff AA, Saunders FA (1970) Psychomotor epilepsy: rehabilitation after surgical treatment. J Nerv Ment Dis 150: 273-290
28. Huber Z (1967) Effectiveness of anterior temporal lobectomy in the treatment of temporal epilepsy. Neurol Neurochir Pol 1: 379-383
29. Hullay J, Gombi R, Velok G (1976) Surgical and stereotactic attempts in intractable epilepsy. Acta Med Acad Scient Hung 33 (2): 119-124
30. Jensen I (1975) Temporal lobe surgery around the world. Results, complications, and mortality. Acta Neurol Scandinav 52: 354-373
31. Jensen I (1976) Temporal lobe epilepsy. Social conditions and rehabilitation after surgery. Acta Neurol Scandinav 54: 22-44
32. Jensen I (1977) Temporal lobe epilepsy. With special reference to surgical results, neuropathology, and social conditions. Dan Med Bull 24: 235-249
33. Jensen I, Værnet K (1977) Temporal lobe epilepsy. Follow-up investigation of 74 temporal lobe resected patients. Acta Neurochir (Wien) 37: 173-200
34. Laws ER Jr, Niedermeyer E, Walker AE (1970) Diagnostic significance of scalp and depth EEG findings in patients with temporal and frontal lobe epilepsy. Johns Hopkins Med J 126: 146-153
35. Lieb JP, Engel J Jr, Gevins A, Crandall PH (1981) Surface and deep EEG correlates of surgical outcome in temporal lobe epilepsy. Epilepsia 22: 515-538
36. Lieb JP, Rausch R, Engel J Jr, Brown WJ, Crandall PH (1982) Changes in intelligence following temporal lobectomy: relationship to EEG activity, seizure relief, and pathology. Epilepsia 23: 1-13
37. Lindsay J, Ounsted C, Richards P (1984) Long-term outcome in children with temporal lobe seizures. V: indications and contra-indications for neurosurgery. Dev Med Child Neurol 26: 25-32
38. Łuczywek E, Mempel E (1980) Memory and learning in epileptic patients treated by amygdalotomy and anterior hippocampotomy. Acta Neurochirurgica [Suppl] 30: 169-175
39. Marossero F, Ravagnati L. Sironi VA, Miserocchi G, Franzini A, Ettore G, Cabrini GP (1980) Late results of stereotactic radiofrequency lesions in epilepsy. Acta Neurochirurgica [Suppl] 30: 145-149
40. Maspes PE, Marossero F (1953) Survey of 28 patients operated for temporal lobe epilepsy. Rev Neurol 88: 578-580
41. Mathai KV, Chandy J (1970) Surgical treatment of temporal lobe seizures. Neurology (Bombay) 18: 158-164
42. Mempel E, Witkiewicz B, Łuczywek E (1981) Temporal lobe epilepsy during the growth stage treated by amygdalotomy and hippocampotomy. Neurol Neurochir Pol 15: 611-615
43. Mempel E, Witkiewicz B, Stadnicki R, Łuczywek E, Kuciński L, Pawłowski G, Nowak J (1980) The effect of medial amygdalotomy and anterior hippocampotomy on behaviour and seizures in epileptic patients. Acta Neurochirurgica [Suppl] 30: 161-167
44. Morris AA (1956) Temporal lobectomy with removal of uncus, hippocampus, and amygdala. Arch Neurol Psychiat (Chic) 76: 479-496
45. Mundinger F, Becker P, Groebner AE, Bachschmid D (1976) Late results of stereotactic surgery of epilepsy, predominantly temporal lope type. Acta Neurochirurgica [Suppl] 23: 177-182
46. Niemeyer P (1958) The transventricular amygdala-hippocampectomy in temporal lobe epilepsy. In: Baldwin M, Bailey P (eds) Temporal lobe epilepsy. Thomas, Springfield, pp 461-482
47. Olivecrona H, Widen L (1969) Kirurgisk behandling av temporallobsepilepsi. Svenska Läk-Tidn 66: 2681-2689
48. Olivier A (1983) Surgical management of complex partial seizures. Prog Clin Biol Res 124: 309-324
49. Paillas JE, Gastaut H, Vigoroux R, Naquet R (1964) Corrélations anatomo-électro-cliniques dans l'épilepsie temporale; àpropos des résultats obtenus chez 38 opérés. Neurochirurgia (Paris) 10: 477-505
50. Penfield W, Jasper H (1954) Epilepsy and the functional anatomy of the human brain. Little, Brown, Boston
51. Picaza JA, Gumá (1956) Experience with the surgical treatment of psychomotor epilepsy. Arch Neurol Psychiat (Chic) 75: 57-61
52. Rasmussen T (1980) Surgical aspects of temporal lobe epilepsy. Results and problems. Acta Neurochirurgica [Suppl] 30: 13-24
53. Rasmussen T (1983) Surgical treatment of complex partial seizures. Results, lessons, and problems. Epilepsia 24 [Suppl 1]: S 65-S 76
54. Rausch R, Crandall PH (1982) Psychological status related to surgical control of temporal lobe seizures. Epilepsia 23: 191-202
55. Reynolds EH (1983) Mental effects of antiepileptic medication: a review. Epilepsia 24 [Suppl 2]: 85-95
56. Rodin EA (1968) The prognosis of patients with epilepsy. Charles C Thomas, Springfield
57. Serafetinides EA, Cherlow DG (1976) Psychiatric test assessment of patients with psychomotor epilepsy. Am J Psychiatry 133: 844-847
58. Shih Yü-Ch'üan, Yin Hsiao-Tung, Ch'ü Chih-P'ing, Ch'en Han-Pai, Huang Suei-Jen (1966) Experience in anterior temporal lobectomy for temporal lobe epilepsy. China med J 85: 37-43
59. Sigua OA, Chkhenkeli SA (1980) Results of surgical treatment of severe forms of temporal epilepsy. Zh Vopr Neirokir 4: 3-10
60. Skriabin VV, Bein BN, Ivanov EV (1976) Stereotactic surgery of temporal lobe epilepsy. Zh Vopr Neirokir 4: 10-16

61. Sonnen AEH, Manen JV, van Dijk B (1976) Results of amygdalotomy and fornicotomy in temporal lobe epilepsy and behaviour disorders. Acta Neurochirurgica [Suppl] 23: 215-219
62. Spencer SS (1981) Depth electroencephalography in selection of refractory epilepsy for surgery. Ann Neurol 9: 207-214
63. Stępien L, Bacia T, Bidziński J, Wisáwski J (1981) Late results of operation in temporal lobe epilepsy in adults and children. Neurosurg Rev 4: 61-69
64. Talairach J, Bancaud J (1973) Stereotaxic approach to epilepsy. Prog Neurol Surg 5: 297-354
65. Taylor DC, Falconer M (1968) Clinical, socio-economic and psychological changes after temporal lobectomy for epilepsy. Brit J Psychiatry 114: 1247-1261
66. Temkin O (1971) The falling sickness. Johns Hopkins Press, Baltimore/London
67. Trimble MR, Thompson PJ (1983) Anticonvulsant drugs, cognitive function, and behavior. Epilepsia 24 [Suppl 1]: 55-64
68. Turner DA, Wyler AR (1981) Temporal lobectomy for epilepsy: mesial temporal herniation as an operative prognostic finding. Epilepsia 22: 623-629
69. Værnet K (1983) Temporal lobectomy in children and young adults. In: Parsonage M et al. (eds) Advances in epileptology: XIV epilepsy international symposium Raven, New York, pp 255-261
70. Van Buren JM, Ajmone-Marsan C, Mutsuga N, Sadowsky D (1975) Surgery of temporal lobe epilepsy. Adv Neurol 8: 155-196
71. Vladyka V (1978) Surgical treatment of epilepsy and its application in temporal epilepsy. Cesk Neurol Neurochir 41: 95-106
72. Walter RD (1975) Principles of clinical investigation of surgical candidates. Adv Neurol 8: 49-58
73. Weber G (1967) Chirurgische Therapie der fokalen Epilepsie. Ther Umsch 24: 517-521
74. Weiss AA (1965) Psychodiagnostic follow-up of eight cases of temporal lobectomy. Israel Ann Psych rel Disc 3: 259-266
75. Wieser HG (1985) Derzeitige Möglichkeiten der operativen Epilepsiebehandlung. Nervenarzt 56: 404-409
76. Wieser HG (1976) Selective amygdalohippocampectomy: indications, investigative technique and results. In: Symon L et al. (eds) Advances and technical standards in neurosurgery, vol 13: 39-133
77. Wieser HG, Yaşargil MG (1982) Die "selektive Amygdala-Hippokampektomie" als chirurgische Behandlungsmethode der mediobasal-limbischen Epilepsien. Neurochirurgia 25: 39-50
78. Wieser HG, Meles HP, Bernoulli C, Siegfried J (1980) Clinical and chronotopographic psychomotor seizure patterns (SEEG study with reference to postoperative results). Acta Neurochir [Suppl] 30: 103-112

Cortical Resection for Multilobe Epileptogenic Lesions

T. Rasmussen

Introduction

This report concerns those patients, operated upon for medically refractory seizures, whose epileptogenic area extended well beyond one of the standard anatomical brain divisions, temporal, frontal, central, parietal, or occipital. In the Montreal Neurological Institute (MNI) surgical seizure series, as of the end of 1980, these patients, 243 in number, constituted 11% of the total series of 2177 patients (Table 1).

Many of the patients in this group exhibited large areas of brain atrophy, porencephalic cysts, etc., easily demonstrable in the pneumoencephalogram or CT scan. The others usually showed diffuse enlargement of the ventricle of the involved hemisphere. When the brain damage had occurred early in life the skull X-ray nearly always showed relative smallness of the hemicranium on the involved side.

Perinatal brain injury was responsible for the brain damage and resulting seizure tendency in 44%, with postinflammatory brain scarring responsible for 23% (Table 2). Next in frequency was postnatal brain trauma (14%), with a variety of miscellaneous lesions responsible for 10%. There was more than one potential etiological factor in 4% and no obvious cause in 5%.

Nearly half of the patients were first operated upon in childhood with an additional 36% operated upon in the decade of 16-25 years of age. Fifteen percent have had one or more reoperations. In these cases the follow-up duration and seizure status are listed after the last operation.

This multilobe surgical seizure group is a heterogeneous one from several standpoints and it is convenient to separate the discussion into five subdivisions:

1. Hemispherectomy candidates - patients with maximal or near maximal hemiplegia and hemianopsia, most commonly so-called infantile hemiplegia
2. Subtotal hemispherectomy candidates - patients with similar extensive brain lesions but not considered suitable for hemispherectomy because the hemiphlegia was not maximal, and/or because there was minimal or no hemianopsia
3. Patients in whom only about a half or a third of the hemisphere was removed
4. Patients with the syndrome of "chronic encephalitis," consisting of focal seizures and slowly progressive neurologic deficits. This small, easily diagnosed group of patients often poses particularly difficult therapeutic problems
5. Early pressure complications of large brain excisions

Table 1. Montreal Neurological Institute surgical seizure series: Anatomical classification - patients operated upon 1929 through 1980

	No. of patients (%)
Temporal lobe	1210 (56)
Frontal lobe	402 (18)
Central (sensorimotor) region	151 (7)
Parietal lobe	141 (6)
Occipital lobe	30 (1)
Large multilobe lesions	243 (11)
Total	2177

Table 2. Multilobe epileptogenic lesions - Etiology (neoplasms excluded)

	No. of patients (%)
Birth trauma or anoxia	106 (44)
Postnatal trauma	33 (14)
Postinflammatory brain scarring	58 (23)
Multiple potential epileptogenic factors	9 (4)
Miscellaneous	24 (10)
Unknown	13 (5)
Total	243

Montreal Neurological Institute and Hospital, Department of Neurology and Neurosurgery, 3801 University Street, Montréal, Québec H3A 2B4, Canada

Presurgical Evaluation of Epileptics
Edited by H. G. Wieser and C. E. Elger

Hemispherectomy Candidates

Since Krynauw's 1950 report on hemispherectomy for infantile hemiplegia and medically refractory seizures, this operation quickly became recognized as a highly effective procedure for this small, well-defined patient population (McKissock 1953; Ransohoff 1954; Gros and Vlahovitch 1955; Laine and Gros 1956; White 1961). In the mid 1960s, however, it became apparent that serious late complications appeared in one-fourth to one-third of hemispherectomized patients who were followed for 4 years or more (Laine et al. 1964). Our experience has been typical. By 1973 we had recognized nine patients with serious late complications. Since then 2 more have occurred, so to date 11 patients have developed this complication in our series of 31 patients who have had an anatomically complete hemispherectomy carried out in one or in several stages.

The usual cause of this late, often fatal, complication was soon recognized to be the result of gradual seepage of red blood cells into the hemispherectomy cavity (Ulrich et al. 1965; Oppenheimer and Griffith 1966; Griffith 1967; Till 1967; Brett 1969; Falconer and Wilson 1969; Wilson 1970; Rasmussen 1973) causing the syndrome of superficial cerebral hemosiderosis, first described by Noetzel in 1940 and produced experimentally in dogs by Iwanowski and Olszewki in 1960. As a result, relatively few hemispherectomies have been carried out in the years since the early 1970s.

A review of the lengthy MNI experience with subtotal hemispherectomies, in which one-fifth to one-third of the damaged and epileptogenic hemisphere was preserved, however, revealed that the late complication of superficial cerebral hemosiderosis had not appeared in this group of patients. Therefore during the period from 1968 to 1974 a subtotal hemispherectomy was routinely carried out in all patients who previously would have been considered suitable candidates for a complete hemispherectomy. We preserved the least epileptogenic one-fourth or one-third of the hemisphere, usually the frontal or the occipital pole. This subtotal hemispherectomy has satisfactorily protected the patients from the late complication of superficial cerebral hemosiderosis (Rasmussen 1973), but at the cost of a moderate reduction in the percentage of patients experiencing a complete or nearly complete reduction of the seizure tendency.

We then realized that disconnecting the preserved segment of the hemisphere from the corpus callosum and the upper brain stem would yield the superior reduction of seizure tendency of a complete hemispherectomy and, at the same time, safeguard against the late complication of superficial cerebral hemosiderosis, as was demonstrated by the subtotal hemispherectomy series (Rasmussen 1983), and in a still larger series of patients in whom about half of the hemisphere had been removed, and had also been followed for long periods. This functional hemispherectomy has now been carried out in 14 patients during the period 1974 through 1983 (Table 3).

Ten patients are seizure free and one patient has had only a single nocturnal seizure in the third postoperative year. Thus 11 patients, 79%, have had a complete or nearly complete reduction of the preoperative medically refractory seizure tendency. Of the remaining three, one has less than 1% of the preoperative seizure rate and severity and the other two about 5%-20% of the preoperative seizure situation. In addition, 13 of these 14 patients have had a significant improvement in behavior and social status. The one remaining patient, severely retarded, is unchanged in this respect. The effectiveness of this

Table 3. Late intracranial pressure complications of large brain excisions

Operations	No. of patients	Late pressure complication	Interval between operation and complication	Duration of follow-up
Anatomical complete hemispherectomy (1 stage 24 patients) (2, 3, or 4 stages 7 patients)	31	11 patients (35%)	4 ½-24 years (median 9 years)	7-33 years (median 20 years)
Functional complete but anatomical subtotal hemispherectomy	14	1 patient } (2.6%)	3 years	2-10 years (median 4 ½ years)
Subtotal hemispherectomy, one-fifth to one-third of hemisphere preserved	57	1 patient } (2.6%)	10 years	4-34 years (median 29 years)
Hemi-hemispherectomy, about one-half of hemisphere preserved	71	1 patient (1.4%)	13 years	4-39 years (median 12 years)

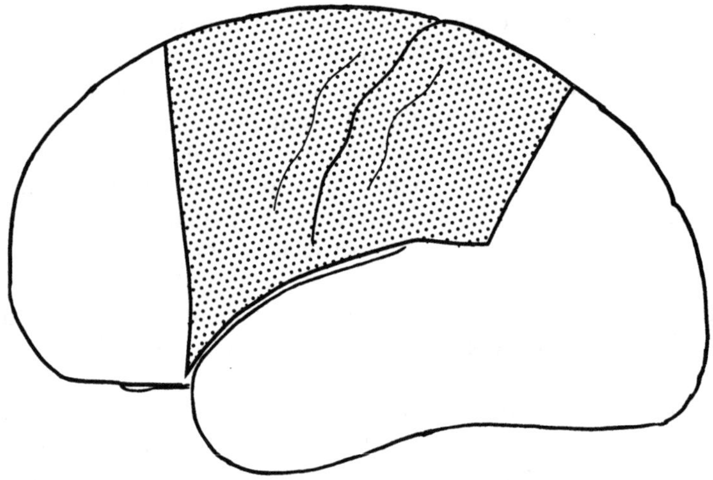

Fig. 1. First-stage excision for functional hemispherectomy

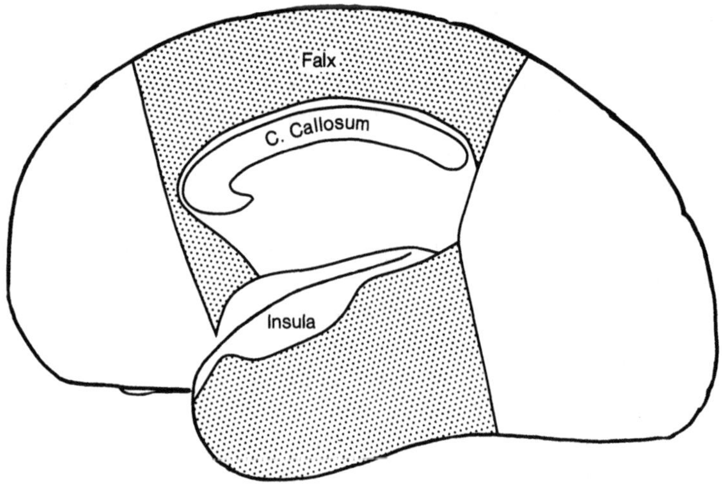

Fig. 2. Final stage of functional hemispherectomy. Anterior frontal and posterior parietal and occipital regions preserved but disconnected from the upper brain stem and the corpus callosum

functional hemispherectomy is thus similar to that reported in the extensive hemispherectomy literature from many parts of the world (White 1961).

One patient needed a shunt 3 years postoperatively because of inadequate CSF absorptive mechanisms, but none of the 14 has developed superficial cerebral hemosiderosis. Although the follow-up is short in relation to this late complication, the long-term story should not be different from our experience with our series of 57 patients with subtotal hemispherectomy and 71 patients with a hemi-hemispherectomy, followed for 4 years or longer.

This operation, which we have called a functional hemispherectomy, consists of first excising the central part of the hemisphere above the fissure of Sylvius extending anteriorly to a level just in front of the genu of the corpus callosum and extending posteriorly to a level just behind the splenium of the corpus callosum (Fig. 1).

The frontal white matter is then sectioned in front of the corpus callosum down to the medial pia-arachnoid layer lying on the falx and the parietal white matter similarly sectioned behind the corpus callosum. The remaining anterior frontal and posterior parietal and occipital regions are therefore disconnected from the corpus callosum and from the upper brain stem. Finally, the temporal lobe including the amygdala and hippocampus is then excised (Fig. 2).

Our hemispherectomy series, including both the anatomically complete and functional hemispherectomy cases just described, includes 41 patients who survived the first 2 postoperative years (Table 4). Twenty-one patients (51%) have become and remained seizure free, and 83% have had a complete or nearly complete reduction of seizure tendency. The remaining seven patients have all had a significant reduction in seizure tendency, but not sufficient

Table 4. Results of hemispherectomy for seizures, operated 1952 through 1983

Seizure free since discharge	18 patients	21 patients (51%)	34 patients (83%)	41 patients with follow-up data of 2-33 years (median 15 years)
Became seizure free after some early attacks	3 patients			
Free 3 or more years, then rare or occasional attacks	12 patients	13 patients (32%)		
Marked reduction of seizure tendency	1 patient			
Moderate or less reduction of seizure tendency	7 patients (17%)			
Died in first postoperative year	2 patients			
Total	43 patients			

Table 5. Results of subtotal hemispherectomy for seizures (one-fifth to one-third of hemisphere preserved)

Seizure free since discharge	15 patients	26 patients (45%)	40 patients (69%)	58 patients with follow-up data of 2-34 years (median 29 years)
Became seizure free after some early attacks	11 patients			
Free 3 or more years, then rare or occasional attacks	11 patients	14 patients (24%)		
Marked reduction of seizure tendency	3 patients			
Moderate or less reduction of seizure tendency	18 patients (31%)			
Postoperative deaths	2 patients			
No follow-up data	2 patients			
Total	62 patients			

to qualify them for one of the preceding groups as we have defined them.

In summary, we believe this small, well-defined patient population of hemispherectomy candidates no longer poses difficult diagnostic or therapeutic problems.

Subtotal Hemispherectomy Candidates

This larger group of 62 patients with extensive hemispheral brain damage presents more difficult diagnostic and therapeutic problems. These patients usually exhibit several different seizure patterns and show widespread epileptiform discharges in the EEG. They were not considered to be good candidates for total hemispherectomy, however, because the hemiparesis was not maximal and/or because a useful visual field was still present. Often the seizure investigation only enabled one to say that the maximum epileptogenicity involved mainly the anterior or the posterior half of the involved hemisphere. The radiological findings sometimes helped considerably to decide whether the craniotomy should expose mainly the anterior or the posterior half of the hemisphere, but often the pneumoencephalogram or the CT scan showed only diffuse enlargement of the entire ventricle in the involved hemisphere.

At operation the cortical EEG sometimes defined the limits of the epileptogenic cortex more satisfactorily, but often the epileptiform discharges were present and often synchronized over most of the exposed damaged and adjacent cortex. The surgical plan has been to remove as much of the epileptogenic cortex, as defined by the cortical EEG and by consideration of the patient's seizure pattern, as is necessary without incurring too great a risk of producing a neurologic deficit, or of significantly increasing an existing deficit.

A considerable variety of cortical excisions has been carried out in this group of 62 patients. When a less than maximal hemiparesis was present, the arm and leg area of the sensorimotor gyri was preserved along with the posterior parietotemporal and occipital region if a useful visual field was present. If there was a maximal or near maximal hemiparesis but a preserved visual field, the anterior two-thirds of the hemisphere was removed, preserving the occipital and posterior parietotemporal region. As just mentioned, during the late 1960s and early 1970s the least epileptogenic one-fourth or one-fifth of the hemisphere, usually the frontal or occipital pole, was preserved to protect against the late development of superficial cerebral hemosiderosis, even though the patient was a good hemispherectomy candidate with a maximal hemiplegia and a complete homonymous hemianopsia.

Fifty-eight of these 62 patients have been followed for 2 to 34 years (Table 5). Twenty-six (45%) have become and remained seizure free, and another 14 patients have had a marked reduction in seizure tendency. Thus, the subtotal hemispherectomy has resulted in a complete or nearly complete reduction of the medically refractory seizure tendency in 69%, compared with 83% of the hemispherectomy series.

Patients in Whom About a Half or a Third of the Damaged Hemisphere Was Removed

There were 144 patients in this category. Nearly all had some degree of hemiparesis, usually minimal or moderate. About 15% had a partial or complete homonymous hemianopsia. A few had speech functions still present, or suspected of being still present, in a damaged left cerebral hemisphere.

A still greater variety of cortical excisions have been carried out in this group, being tailored in each instance on the basis of the seizure patterns, the preoperative and operative EEG studies and the nature and severity of the patients' preoperative neurologic deficit. Only 40% have experienced a complete or nearly complete reduction of seizure tendency, compared with 68% in the subtotal hemispherectomy series and 83% in the hemispherectomy series.

In many of the patients in this group, and in the preceding subtotal hemispherectomy series, part of the area of obviously damaged cortex was left in situ for one ore more reasons.

1. In about a third of these patients the judgment was made, on the basis of the patients' attack patterns, the preoperative EEG studies and the pre- and postexcision cortical EEG, that the epileptogenic area removal was adequate and the remaining obviously damaged cortex could be left in situ without lessening the likelihood of a satisfactory reduction of the seizure tendency. Follow-up analyses over the years have indicated that this decision was correct in about half of the patients and a complete or nearly complete reduction of the seizure tendency was achieved. In the remaining half, however, a less satisfactory seizure reduction occurred, and in retrospect a more complete excision of the damaged cortex would often have been wise.

2. In about one-fourth of the patients in this group the cortical excision was limited primarily to minimize the risk of increasing a partial hemiparesis. In many of these patients a more extensive excision might have been unwise since a significant increase in hemiplegia is seldom a good trade for a possibility, but not a certainty, of improved reduction of seizures. In some of these patients, however, a more complete removal could have been carried out without too great a risk of increasing the hemiparesis, particularly when the hemiparesis was moderate rather than slight.

3. In another 20% of the patients in this subgroup the cortical excision was limited primarily to avoid producing a hemianopsia. Most of these patients, however, did achieve a reasonably satisfactory reduction of the seizure tendency.

4. The cortical excision was limited primarily to avoid risk of producing a dysphasia in about 10%, and in these patients, most of them children, a more extensive excision would not have been wise until subsequent testing had proved that the speech functions were actually not in the damaged left hemisphere (Wada and Rasmussen 1960; Milner et al. 1964; Branch et al. 1964).

5. In the remaining 10% of this group the persisting area of cortex was left in situ prior to the development of the technique of functional hemispherectomy, described in the first section of this report, to protect against the late development of superficial cerebral hemosiderosis. Two-thirds of these patients became seizure free. Some of the remaining third would doubtless have benefited more significantly if the remaining cortical area had been disconnected as in the functional hemispherectomy technique.

In summary, one gets the strong impression in reviewing the data in these patients, where the seizure tendency is associated with a large area of obviously damaged brain tissue, that neither the attack patterns nor the preoperative and operative EEG studies give the surgeon as good guidance as to the extent of the cortical excision required to produce a satisfactory reduction of the seizure tendency, as is the case with more restricted and less severely damaged epileptogenic cortical areas. It would seem wise, in these multilobe seizure cases with large areas of obviously damaged brain tissue, to carry out as complete a removal of the obviously damaged cortex as can be done without too great a risk of significantly increasing the patients' neurologic deficits, even though the clinical and EEG evidence suggest that a partial removal of the damaged area would be adequate.

Patients with "Chronic Encephalitis"

Patients with this syndrome, which we first recognized in the 1950s (Rasmussen et al. 1958), constitute only about 1% of the patients admitted to the MNI each year for investigation of epilepsy. The syndrome is relatively stereotyped and is easily diagnosed, but the patients usually present urgent and difficult therapeutic problems. This small series, 38 in number at the end of 1980, is of special interest for several reasons, however.

In the patients with this syndrome, the seizures typically begin in childhood between 2 and 15 years of age. Over half of these patients had some mild illness shortly before the onset, such as a mild flu, gastrointestinal upset, etc. The seizures are often initially generalized but soon show an increasing proportion of unilateral sensorimotor patterns, often developing into epilepsia partialis continua, involving the face and/or arm of one side and sometimes extending to the leg. With evidence of lateralization of the seizure tendency, a slowly progressive hemiparesis develops. If the dominant hemisphere is involved and the onset of seizures occurs after age 3, 4, or 5 years, there is progressive dysphasia as well. A slowly progressive mental retardation is also common. The neurologic deterioration is slowly progressive over periods up to 10 or more years. The median duration of the progressive neurologic deterioration in this series was 4 years, as accurately as we could estimate it from all available evidence. Serial pneumoencephalograms and CT scans give dramatic evidence of the progression of the brain atrophy, slowly advancing year by year.

The severity of the seizure tendency nearly always makes surgical therapy essential. When the patients were studied after a maximal or near maximal hemiplegia was already present, an anatomic or, since 1976, a functional hemispherectomy has been carried out with gratifying results (Rasmussen 1978, 1983).

When the patient has been studied earlier in the course of the illness and only a slight or moderate hemiparesis is present, or if speech functions are probably still present in the involved hemisphere, the surgical decision is more difficult. It has been our practice in such cases to carry out as large a cortical excision of the most epileptic region of the brain as possible without increasing the hemiparesis. This has usually consisted of a frontal and temporal lobectomy, sparing the speech areas if the dominant hemisphere was involved, and sometimes including the lower portions of the pre- and postcentral gyri when epilepsia partialis continua was present and involved mainly the face and arm. This has usually produced a considerable reduction of the seizure tendency, stopping the epilepsia partialis continua, but has not produced the complete or near complete reduction of seizure tendency that often follows the more radical cortical excision that can be carried out once the disease process has become arrested and the neurologic deficit stabilized. Reoperation has been carried out in six of these patients, with significant benefit, and has been recommended in several others.

Microscopic examination of the surgical specimen shows perivascular lymphocytic cuffs and glial nodules in areas where the brain tissue is relatively well preserved. In later stages of the disease there are areas of microcystic degeneration with marked neuronal fallout and no evidence of the inflammatory elements.

The nature of the infectious process remains mysterious, however. Standard viral studies have been carried out in the last 25 patients in the series, and have been negative or have shown positive reactions to herpes simplex and measles virus in low dilutions, which is probably of no clinical significance. Limited electron microscopic searches for virus particles have been carried out in some of the recent patients but so far none have been seen. Studies at the Slow and Latent Virus Laboratory of the National Institutes of Health at Bethesda were carried out in six patients operated upon between 1966 and 1971 and all were negative. Another attempt with new culture techniques is currently under way by this laboratory, however.

Our experience indicates that cortical excision neither exacerbates nor cures the disease process, which does become arrested sooner or later. Most patients, unfortunately, are ultimately left with a moderate or severe hemiparesis, some dysphasia if the dominant hemisphere is involved too late in childhood to permit good transfer of speech functions to the other hemisphere, and some degree of mental retardation.

These patients came to light because of the severity of the seizure tendency and because of its lateralization to one cerebral hemisphere. The importance of those few patients, in my opinion, rests on two factors: (1) the possibility that unrecognized, less-extensive, and less-prolonged invasion of the brain by an agent like this might be relatively common and might account for some of the areas of gliosed or partially destroyed brain tissue that we encounter in operations for focal epilepsy for which there has been no obvious etiology and (2) the possibility that developments in the viral field may some day permit us to make an early etiologic diagnosis of mild childhood viral diseases, upper respiratory infections, etc., and identify those that may invade the brain in a silent fashion. Ultimately this might lead to effective therapy before significant brain damage has occurred. If and when this happy development comes about it may result in the possibility of preventing some cases of mental retardation, multifocal or generalized epilepsy, and neurologic deficits for which there is no recognized etiology at present.

Table 6. Early intracranial pressure complications of large brain excisions

Operations	No. of patients	Early pressure complication
Anatomical complete hemispherectomy (1 stage 24 patients) (2, 3, or 4 stages 7 patients)	31	3 patients (9%)
Functional complete but anatomical subtotal hemispherectomy	14	1 patient } (8.1%)
Subtotal hemispherectomy, one-fifth to one-third of the hemisphere preserved	60	5 patients } (8.1%)
Hemi-hemispherectomy, about one-half of the hemisphere preserved	78	2 patients (2.6%)

Early Pressure Complications of Large Brain Excisions

Although a low-grade increase in intracranial pressure is commonly present for a week or 10 days in nearly all patients who have had large brain removals carried out, this usually resolves spontaneously during the next few days and does not require specific treatment. In a few patients, however, this increase in pressure persists beyond the usual postoperative period as a result of inadequate absorption of CSF, secondary to the extensiveness of the removal of pia-arachnoid absorbing surfaces (Table 6).

This is particularly apt to occur in patients in whom the original brain insult has produced some chronic fibrosis of the leptomeninges, as is apt to occur after head injury with extensive subarachnoid hemorrhage or following meningoencephalitis. This early pressure complication has occurred in 3 (9%) of our 31 patients with an anatomically complete hemispherectomy, one of the 14 patients with the functional hemispherectomy, 5 of 60 patients with subtotal hemispherectomy (8.1% of these 74 patients), and 2 (2.6%) of 78 patients with removal of about one-half of the hemisphere. If this early postoperative increased pressure persists beyond the third or fourth postoperative week, we have learned that a shunt should not be unduly delayed even though the increased intracranial pressure seems minimal. With prompt shunting, this early pressure complication has not had a deleterious effect on the long-term results.

References

Branch C, Milner B, Rasmussen T (1964) Intracarotid sodium amytal for the lateralization of cerebral speech dominance: observations on 123 patients. J Neurosurg 21: 399–405
Brett E (1969) Second thoughts on hemispherectomy in infantile hemiplegia. Dev Med Child Neurol 11: 374–376
Falconer MA, Wilson PJE (1969) Complications related to delayed hemorrhage after hemispherectomy. J Neurosurg 30: 413–426
Griffith HB (1967) Cerebral hemispherectomy for infantile hemiplegia in the light of late results. Ann R Coll Surg Engl 41: 183–201
Gros C, Vlahovitch B (1955) L'hémisphérectomie cérébrale. Imprimerie Causse, Graille and Castelnau, Montpellier
Iwanowski I, Olszewski J (1960) The effects of subarachnoid injections of iron containing substances on the central nervous system. J Neuropathol Exp Neurol 19: 433–448
Krynauw RA (1950) Infantile hemiplegia treated by removing one cerebral hemisphere. J Neurol Neurosurg Psychiat 13: 243–267
Laine E, Gros C (1956) L'hémisphérectomie. Masson, Paris
Laine E, Pruvet P, Osson D (1964) Résultats éloignés de l'hémisphérectomie dans les cas d'hémiatrophie cérébrale infantile génératrice d'épilepsie. Neuro-Chirurg 10: 507–522
McKissock W (1953) Infantile hemiplegia. Proc R Soc Med 46: 431–434
Milner B, Branch C, Rasmussen T (1964) Observations on cerebral dominance. In: de Reuck AVS, O'Connor M (eds) Disorders of language (Ciba foundation symposium). Churchill, London

Noetzel H (1940) Diffusion von Blutfarbstoff in der inneren Randzone und äußeren Oberfläche des Zentralnervensystems bei subarachnoidaler Blutung. Arch Psychiat 111: 129-138

Oppenheimer DR, Griffith HB (1966) Persistent intracranial bleeding as a complication of hemispherectomy. J Neurol Neurosurg Psychiat 29: 229-240

Ransohoff JC (1954) Hemispherectomy in the treatment of convulsive seizures associated with infantile hemiplegia. Res Publ Assoc Res Nerv Ment Dis 34: 176-195

Rasmussen T (1973) Postoperative superficial hemosiderosis of the brain, its diagnosis, treatment and prevention. Trans Am Neurol Assoc 98: 133-137

Rasmussen T (1978) Further observations on the syndrome of chronic encephalitis and epilepsy. Appl Neurophysiol 41: 1-12

Rasmussen T (1983) Hemispherectomy for seizures revisited. Can J Neurol Soc 10: 71-78

Rasmussen T, Olszewski J, Lloyd-Smith D (1958) Focal seizures due to chronic localized encephalitis. Neurology 8: 435-445

Till K (1967) Hemispherectomy for infantile hemiplegia. Dev Med Child Neurol 9: 773-774

Ulrich J, Isler W, Vassali L (1965) L'effet d'hémorragies leptoméningées répétées sur le système nerveux (la sidérose marginale du système nerveux central). Rev Neurol 112: 466-471

Wada J, Rasmussen T (1960) Intracarotid injection of sodium amytal for the lateralization of cerebral speech dominance: experimental and clinical observations. J Neurosurg 17: 266-282

White HH (1961) Cerebral hemispherectomy in the treatment of infantile hemiplegia: review of the literature and report of 2 cases. Confinia Neurol 21: 1-50

Wilson PJE (1970) Cerebral hemispherectomy for infantile hemiplegia. A report of 50 cases. Brain 93: 147-180

Selective Microsurgical Resections*

M. G. Yaşargil[1] and H. G. Wieser[2]

Introduction

Selective resective surgical therapy offers a real chance of cure only in those patients in whom a well-defined and narrowly restricted "epileptogenic focus," i. e., seizure onset locus, can be identified and precisely localized. As there is a spectrum of epileptic disorders which ranges from simple partial phenomena (due to small well-defined epileptogenic lesions), through bilaterally independent and multifocal disorders, to the so-called secondary generalized epilepsies in which the cerebral involvement is so diffuse that seizures manifest as generalized from the onset, a priori only a small proportion of the general seizure population may be expected to meet the criteria for selective surgical interventions.

In those patients, however, in whom the preoperative evaluation can demonstrate unequivocally such a focal seizure onset, the improved localization precision has to be translated into adequate surgical procedures. These have to be as radical as necessary, yet at the same time as careful and selective as possible, i. e., they should not remove or damage healthy or only secondarily recruited brain.

The need for selective removal of the mesiobasal limbic structures has derived from refined presurgical evaluation of patients suffering from drug-resistant complex partial seizures. As discussed in another chapter of this volume (by Wieser), a strictly focal unilateral mesiobasal seizure focus involving the hippocampal formation and the amygdala was demonstrated in a considerable proportion of such patients.

The development in Zurich of selective amygdalohippocampectomy took advantage of several previously developed refinements in aneurysm surgery using the so-called pterional approach and the experience that various tumors and arteriovenous malformations (AVM) confined to the mesiobasal temporal lobe (TL) could be effectively extirpated without recourse to lobectomy. Moreover, we had previously noticed that local removal of such mesiobasal tumors and AVMs had been attended in a large proportion of patients by cessation of preoperatively present seizures if the amygdala and/or hippocampal formation including the parahippocampal gyrus were resected simultaneously with the lesion.

The principal objectives in this chapter are to briefly review the surgical technique of selective amygdalohippocampectomy and the subsequent results. A more detailed description of the operative anatomy and the surgical technique, as well as of the indications, investigative techniques, and the results has been published recently (Yaşargil et al., 1985; Wieser 1986). We use the term "amygdalohippocampectomy" to denote the removal of the amygdala, the hippocampus, and the parahippocampal gyrus, extending in the coronal plane from the fimbria to the collateral sulcus. In the anteroposterior plane, the posterior transsection of the parahippocampal gyrus is at the level of the bifurcation of the P-2 segment to form the P-3 segments. This is at the level of the lateral geniculate body where the fimbria ascends to the splenium to form the crus of the fornix. Thus, the en bloc resected specimen of the hippocampus and parahippocampal gyrus measures approximately 4 cm in length, 1.5 cm in width, and 2 cm in depth.

[1] Department of Neurosurgery, University Hospital, Rämistr. 100, 8091 Zürich, Switzerland
[2] Department of Neurology, University Hospital, Frauenklinikstr. 26, 8091 Zürich, Switzerland

* An extensive version of this paper has been published in: Advances and Technical Standards in Neurosurgery, Vol. 12 (1985) and Vol. 13 (1986), (L. Symon et al., eds.), Springer-Verlag, Vienna - New York.

Presurgical Evaluation of Epileptics
Edited by H. G. Wieser and C. E. Elger

Brief Description of the Operative Technique

Craniotomy, Opening the Sylvian Fissure, Inspection and Mobilization of the Middle Cerebral Artery and Its Branches, Cortical Incision, Opening the Temporal Horn, Removal of Amygdala

In principal, for the removal of the amygdala and hippocampal formation three approaches could be used: either the lateral "transventricular" (Niemeyer 1958), the "subtemporal," or via the anterior portion of the superior temporal sulcus. As the lateral transcortical, or sulcal, approach unavoidably produces cortical injury, and the subtemporal approach requires excessive retraction, they did not satisfy our needs and the *trans-Sylvian route* was adapted for our specific aims.

The craniotomy for this operation is a variation of the now standard interfascial pterional craniotomy (Yaşargil 1984) with the skin incision placed 2 cm more posteriorly, thus enlarging the craniotomy over the TL and exposing the anterior third of the superior temporal gyrus (Fig. 1).

After CSF drainage the posterolateral orbital roof and posterior ridge of the greater wing of the sphenoid are completely flattened down to the anterior clinoid process using a high speed electric drill. The dura is opened in a semicircular fashion above the Sylvian fissure and the dural flap is reflected over the sphenoidal ridge (Fig. 2a). The next step is to demonstrate the carotid cistern and its opening between the optic nerve and internal carotid artery (ICA). This maneuver, together with the opening of the arachnoid over and lateral to the ICA and A-1 segment, results in release of a considerable amount of CFS. This minimizes the need for retraction and allows the checking of size, position, and variations of the posterior communicating artery (PcoA), anterior choroidal artery (AchoA), uncal artery, and oculomotor nerve (Fig. 2b).

The Sylvian fissure is then gently opened from the carotid bifurcation to the middle cerebral artery bifurcation and some 1.5-2.0 cm beyond, exposing approximately the anterior one third of the insula and 1-2 cm of the M-2 segment. With the arachnoid between the temporal and fronto-orbital areas now well opened, one may retract the frontal lobe a little more medially and inspect the lateral aspect of the ICA more fully. The position of the uncus and parahippocampal gyrus should be noted with particular reference as to whether there is deep medial herniation. The lateral branches of the M-1 segment are now studied carefully and the temporopolar, anterior temporal, and middle temporal arteries are identified. At the peak of the limen insulae the curve of the inferior trunk of the M-2 segment is followed as it runs to the inferior part of the circular sulcus of the insula.

A cortical incision 1-2 cm in length is then made lateral to M-1 and anteromedial to M-2, into the inferior part of the circular sulcus of the insula opening the anterior portion of the uncinate fasciculus. The incision is placed where the perforating branches to the insula have been coagulated and lies between the temporopolar and anterior temporal arteries (see Fig. 2b). The amygdala is found some few millimeter below the cortical surface. It is advisable to advance first in the direction of the inferior horn and to open into it. This will provide a clearer orientation as to the size, direction, and extent of the amygdaloid body. If uncertainty exists as to the location of the inferior horn, the amygdala may first be removed piecemeal both by microrongeur (to provide histological specimens) and by gentle suction. Removal of the amygdala in the medial/mesiobasal direction proceeds with great caution until the optic tract, lying medially, has been identified.

The parts of the amygdala lying medially close to the claustrum, putamen and pallidum must remain unresected. Following the subpial resection of the most anterior part of the parahippocampal gyrus, the transparent curtain of pial and arachnoid membranes, adjacent to the lateral part of the carotid cistern and the anterior part of the ambient cistern, may readily be identified antero-inferiorly/infero-anteriorly. Upon opening the pia one identifies the uncal and anterior choroidal arteries entering the choroidal fissure from the crural cistern. Medial to this lies the optic tract and laterally the vein of Rosenthal. The cerebral peduncle, P-2 segments, and third cranial nerve may also be seen within the peduncular portion of the ambient cistern. Furthermore, through the delicate arachnoid membranes and fibers between the third cranial nerve and posterior communicating artery, the P-2 and P-1 segments, with their branches, may now be identified within the interpeduncular cistern.

One can now appreciate how the combined subfrontal and insula/temporal approach enables the surgeon to see not only the course of the anterior choroidal artery but also the optic tract. This is particularly important in those cases demonstrating deep mesial herniation.

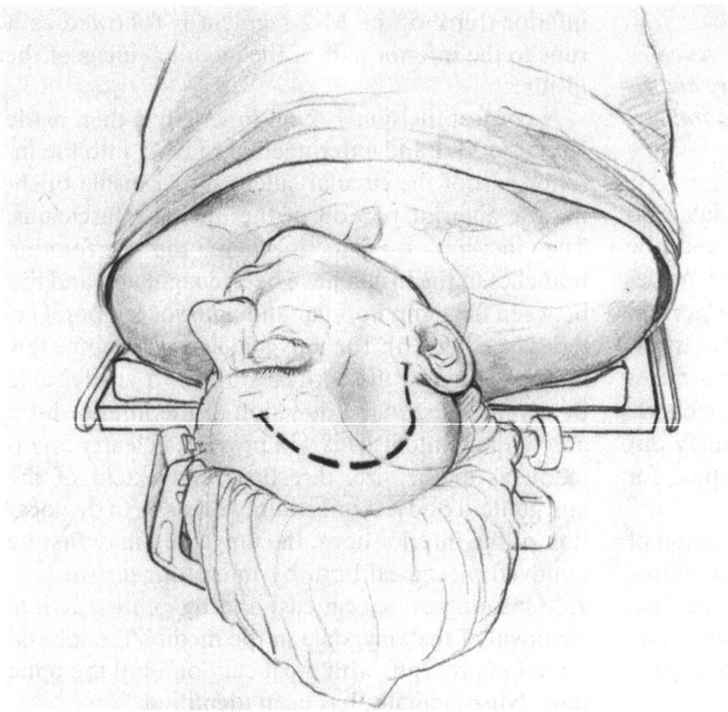

a

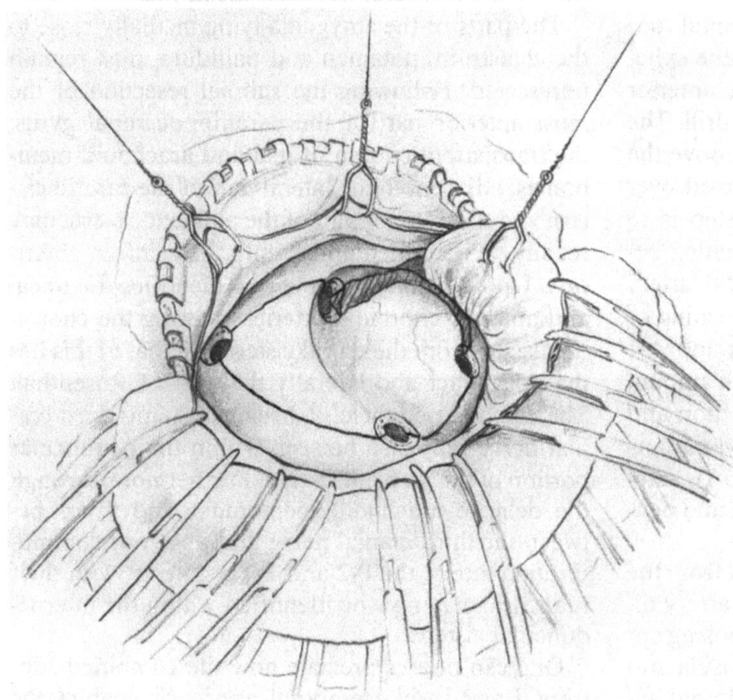

b

Fig. 1. **a** Position of the head and location of skin incision. **b** Enlargement of standard pterional craniotomy exposing the superior temporal surface. (From Yaşargil et al. 1985)

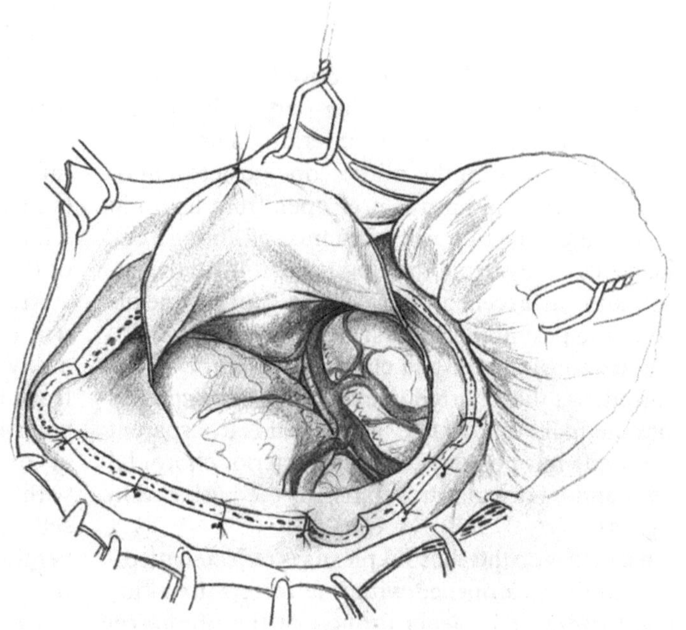

a

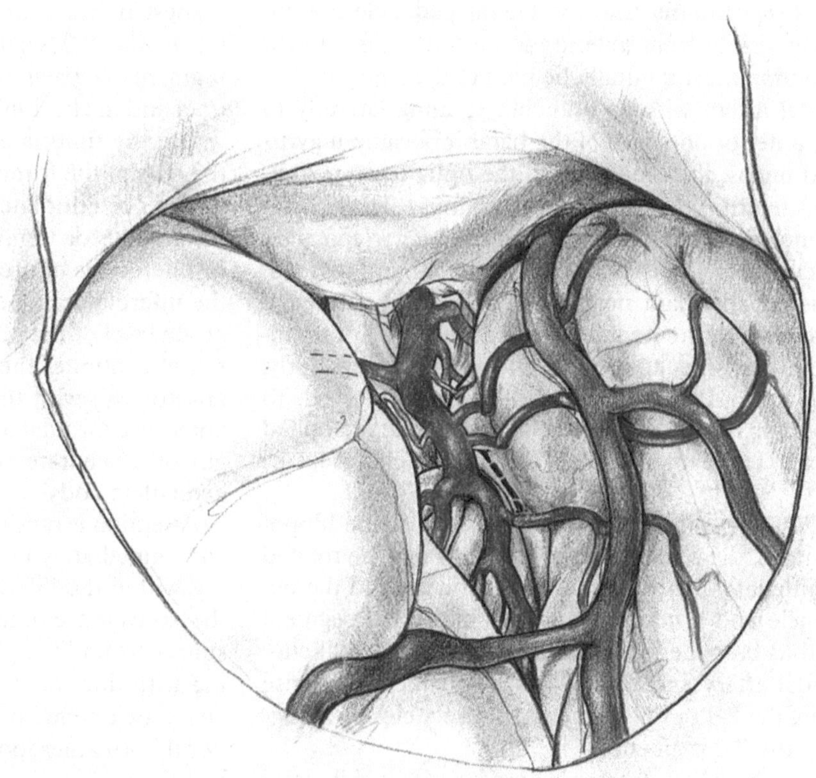

b

Fig. 2. **a** Dural opening and exposure of the proximal Sylvian structures. **b** Visualization of the ICA, PcoA, anterior choroidal artery, M-1 and M-2 segments with temporal branches. Dotted line indicates the place of incision. (From Yaşargil et al. 1985)

Removal of the Hippocampus and Parahippocampal Gyrus

Once the amygdala has been removed, the next stage is somewhat simpler. The earlier part of the dissection has been directed antero-inferiorly and the surgeon must now turn the microscope postero-inferiorly. The temporal horn is now open from its tip in an occipital direction for a distance of about 2 cm. This gives access from the tip of the horn to the trigone and an excellent view of the choroid plexus and of the pes hippocampus.

To demonstrate the tela choroidea over the sulcus choroideus, the choroid plexus may be displaced from medially to laterally. Through this transparent membrane it is possible to see the anterior choroidal artery and ventricular tributaries of the basilar vein (Fig. 3 a).

In a next step the choroid plexus is reflected medially and the tela opened with fine forceps or a fine dissector until the taenia fornicis of the fimbra on the hippocampus and the lateral peduncle are entirely visible from anterior to posterior rims. Great attention must naturally be paid to the anterior choroidal artery with its branches running laterally to the anterior one third of the parahippocampal gyrus and uncus, then medially to the optic tract, peduncle, internal capsule and thalamus. The lateral branches of the anterior choroidal artery (so-called uncus artery) must now be coagulated and divided, whereas the main stem of the artery and its medial branches to the peduncle, optic tract, pallidum, internal capsule, thalamus, lateral geniculate body, and choroid plexus must at all cost be preserved. To avoid spasm of these vessels, papaverine is applied locally to the main trunk of the anterior choroidal artery.

After opening the choroidal fissure of the hippocampus, the parahippocampal gyrus may be rotated a little laterally to permit inspection around the peduncle and visualization of the entire P-2 segment and its branches. Ultimately the posteromedial choroidal artery and collicular artery, which may arise from the P-1 or P-2 segments, run very closely alongside the P-2 segment.

The so-called Ammon's horn arteries which enter the hippocampal sulcus originate partially from the anterior choroidal artery, but mainly from the P-2 segment, just proximal to the origin of the posterolateral choroidal artery. The Ammon's horn arteries must be coagulated and divided. Transverse section of the hippocampus is then carried out at the level of the posterior rim of the peduncle as the P-2 bifurcates to form the inferolateral and superomedial trunks. From the inferolateral trunk arise the entire inferotemporal branches, (antero-inferior, inferomedial, inferoposterior, and infero-temporo-occipital). The branches for the parahippocampal gyrus arise within the collateral sulcus from the inferolateral trunk or in some cases from the infero-anterior and inferomedial temporal arteries. The second trunk (superomedial) can make deep or superficial loops within the ambient cistern. This is the parent vessel of the calcarine and parieto-occipital arteries.

Transsecting the hippocampus at this level, one will encounter small hippocampal veins which are easily coagulated and divided. There will also be from one to three larger cortical veins coursing to veins in the collateral sulcus and thence to the medial cortical vein. These need to be eliminated together with a few parahippocampal veins in the sulcus itself. The hippocampal veins drain to the inferior ventricular vein, thence to the basilar vein; the inferior ventricular vein must also be coagulated and divided in a proper distance to the basilar vein. At the highest point of the posterior rim of the peduncle where the P-2 segment bifurcates to form the P-3 segments, one will find the terminal part of the optic tract and the beginning of the lateral geniculate body where the fimbria ascends to the splenium to form the crus of the fornix.

The posterior lateral choroidal artery may arise in the middle or terminal portion of the P-2 segment just before its bifurcation, or it may arise from either the inferolateral or superomedial trunk. This occurs at the level of the posterior rim of the peduncle. The vessel continues then to enter the choroid plexus to anastomose with the choroid plexus branch of the anterior choroidal artery. Also at this level is the origin of a separate vessel which supplies the lateral geniculate body.

Attention is now directed toward an unnamed sulcus immediately lateral to the hippocampus and lying within the inferior horn. This is an extension of the so-called eminentia collateralis, a semicircular sulcus around the pes hippocampus. Gentle spreading with fine foreceps or a dissector allows this sulcus to be opened from the anterior floor of the temporal horn anteroposteriorly or postero-anteriorly in the direction of the peduncle where the rhinal sulcus and its posterior extension, the collateral sulcus, are found (Fig. 3 b).

This unnamed semicircular sulcus around the pes hippocampus and the bottom of the temporal horn is separated only by some 1-3 mm from the rhinal and collateral sulci. It is very easily opened in a fashion similar to the opening of the choroidal fissure medial to the pes hippocampus.

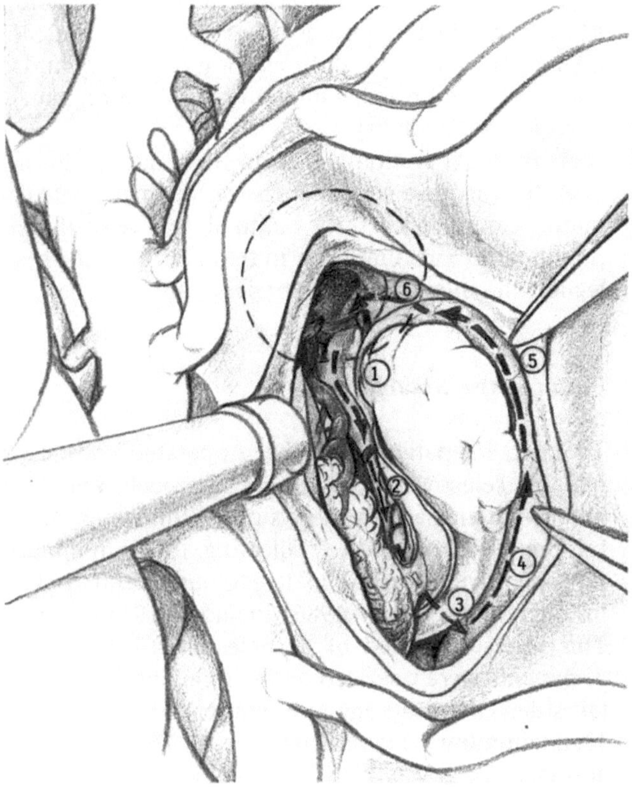

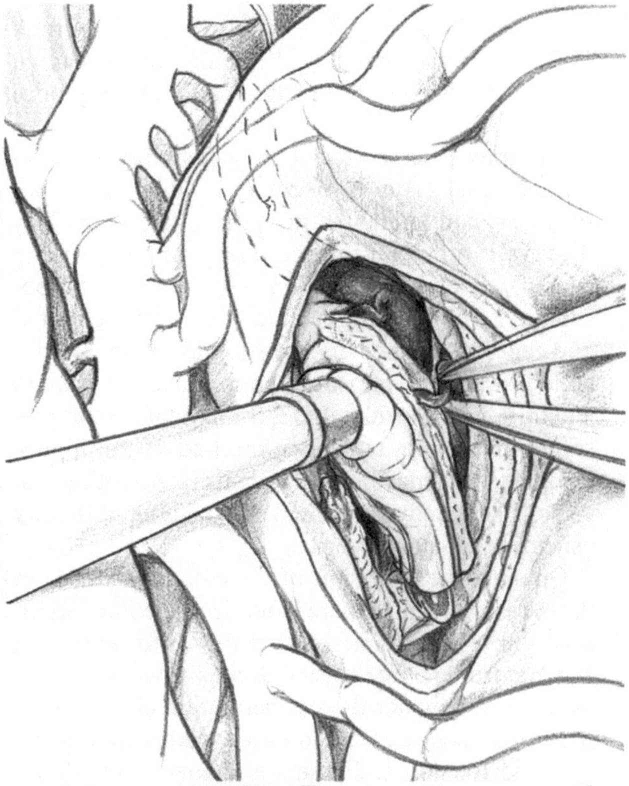

b

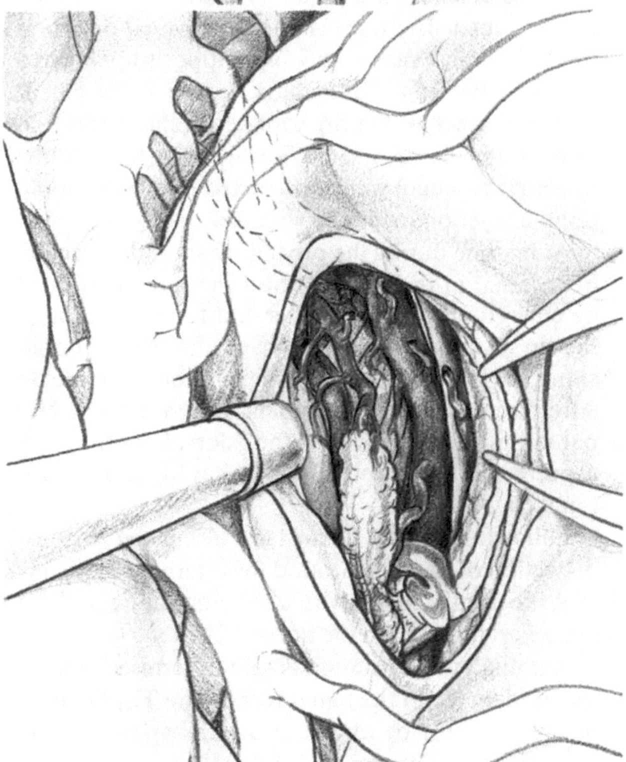

c

Fig. 3. **a** The amygdala is removed *(dotted lines)*, exposure of the entire temporal horn with demonstration of the anterior choroidal artery, basilar vein of Rosenthal, and optic tract. The direction of the *arrows* indicates dissection steps *(1-6)* around the hippocampus. **b** The incision along the lateral hippocampus leads to the rhinal sulcus and collateral sulcus. P-2 segment and branches are visible. The artery to the parahippocampal gyrus is coagulated and thereafter divided. **c** Following amygdalohippocampectomy the important structures surrounding the lateral peduncle are visible, i.e., optic tract, basilar vein, anterior and posterior choroidal arteries, P-2 segment, and choroid plexus. (From Yaşargil et al. 1985)

In this lateral opening are found the several temporal vessels arising from the inferolateral trunk (P-3), not simply crossing the rhinal and collateral sulci but entering, bifurcating, and leaving once again.

One may observe which branches supply the parahippocampal gyrus, fusiform gyrus, inferior temporal gyrus, and lateral temporo-occipital gyrus. Only the vessels supplying the parahippocampal gyrus are coagulated and divided (Fig. 3b). The parahippocampal gyrus can then be readily reflected, elevated and removal en bloc via the subpial plane. The resected specimen measures approximately 4 cm in length, 1.5 cm in width and 2 cm in depth. In the resected area, there remain in the bed of the hippocampal gyrus small bleeding points from exiting pial veins. These may be coagulated without difficulty using fine bipolar forceps.

Opening the extension of the collateral sulcus in those cases in which there is no deep medial herniation, one will ultimately reach the tentorium some 2–5 mm from its free edge and in its anterior half. In those cases in which there is herniation of the mesial structures, one must be more concerned about the line of dissection and potential damage to underlying structures. In these instances, provided one stays in the subpial plane, the parahippocampal gyrus may still be removed en bloc, as the P-2 segment with its branches, the superior cerebellar artery, the third cranial nerve, and the fourth nerve (lying below the tentorial edge) will be protected by the pia and a double layer of arachnoid (Fig. 3c).

At no time during the procedure should retractors be inserted in the small cortical incision. This must be entered only by the sucker and forceps although the tip of the sucker placed over a moist cottonoid sponge may be used as a gentle temporary retractor. After removal of the amygdala, hippocampus, and parahippocampal gyrus, the dorsal and polar aspects of the temporal lobe should appear untouched.

Following careful hemostasis along the lines of dissection of M-1 and around the internal carotid artery, the dura is closed with a running suture and the bone flap replaced in the normal fashion.

Although this operation is safe in terms of inflicting no damage to the remainder of the TL, the neurosurgeon must be aware of a number of specific problems (see Yaşargil et al. 1985) during this procedure. They arise mainly from frequently encountered variations of the form and distribution of the vessels and their supply. Moreover, the size and form of the hippocampus and parahippocampal gyrus can vary considerably, and their deep temporal herniation is a frequent finding.

Thus, it must be emphasized that considerable microsurgical experience is necessary before undertaking this extremely confined form of surgery and, above all, a thorough familiarity with the regional anatomy is mandatory.

As the next section will show, this microsurgical operation, using a cortical incision which is at maximum 2 cm in length, turns out to be very rewarding, provided the preconditions in terms of patients' selection and microsurgical experience are fully met.

Postoperative Results

To date[3], 163 patients have been operated upon using this selective microsurgical approach, and another 3 patients are awaiting surgery following careful preoperative studies. Allowing for a minimal postoperative follow-up of 1 year, our recently published report (Wieser 1986) included 103 patients. This series consisted of 62 males and 41 females; 51 patients were operated on the right and 52 on the left side. The mean age at operation was 30 (SD ± 13.4, minimum 1, maximum 67) years. The mean age at seizure onset was 20 (SD ± 16, minimum 0.3, maximum 59) years. The mean follow-up at the time of writing the above-mentioned article was 38.7 (SD ± 21.2, minimum 12, maximum 118) months. There were 54 nontumoral and 49 tumor cases. Within the tumor group 34 patients had benign (astrocytoma 17, ganglioglioma 9, oligodendroglioma 6, meningioma 2) and 15 malignant tumors. In the nontumoral group 29 cases were found to have marked hippocampal sclerosis (three times associated with old hemorrhage, once with a marked scar); 15 cases had a vascular pathology (AVM 13, venous angioma 1, aneurysm 1); and 3 patients had tuberous sclerosis. Thirty-one nontumoral cases had been evaluated by stereo-EEG and the remaining by 'foramen ovale' or sphenoidal electrode recordings in almost all instances.

General Clinical Outcome

The general clinical outcome as rated by the patient and the relatives, as well as by neurologists, neurosurgeons, and neuropyschologists, was graded into four classes. As complications related to diagnostic procedures and/or surgery were fully integrated into this assessment, they are discussed first.

Ten patients of this series have died. Two of them are within the nontumoral group and had accidentally been inoculated with Creutzfeldt-Jakob disease

[3] April 1987

during presurgical evaluation. The circumstances of this extraordinary complication have been reported elsewhere (Bernoulli et al. 1977). Seven patients died as a consequence of having recurrent malignant tumor, and 1 for unknown reason. Excluding the malignant tumors, the remaining 88 patients were markedly improved in 64%, improved in 28%, unchanged in 5%, and worsened in 3% of the cases. The latter number is a consequence of the two patients with Creutzfeldt-Jakob disease.

Postoperative Seizure Control

With respect to seizure control, the results[4] of this entire series ($n = 103$) are the following: seizure-free 62%; persistent rare seizures (less than 1/year) 5%; worthwhile improvement 26%; no worthwhile improvement 7%. Disregarding malignant tumors and "palliatively" operated nontumoral cases, 75% are seizure-free, 4% have rare seizures, 16% have worthwhile, and 4% have no worthwhile improvement ($n = 73$). In the nontumoral group, from those patients who were operated because of an unequivocally and strictly focal seizure onset demonstrated by SEEG within the resected structures ("causally" operated, $n = 16$), as many as 88% are seizure-free and the remaining 12% show worthwhile improvement, i.e., there is no single patient who has not benefitted from this operation.

These data, however, must be seen on the basis of the actual anticonvulsant drug regimen, as given elsewhere (Wieser 1986). Here it may suffice to mention that only 12% of all patients actually have a more or less unchanged drug regimen compared to the preoperative level. In 88% of the patients, the anticonvulsants are reduced compared to the preoperative level. Nineteen percent of these, however, are still taking two drugs. Eighteen percent of the previously drug-resistant seizure patients are already completely off drugs and remain seizure-free.

Postoperative Neuropsychological Outcome

As far as specific learning and memory performances are concerned, these findings are described in the chapter by Nadig and Wieser, in this volume, and by Wieser (1986).

In those patients in whom a pre-/postoperative comparison was made with the same type of tests ($n = 17$), the findings can be summarized as follows. (a) In general, a postoperative improvement of the learning performance was noted. (b) This postoperative improvement was more marked for material specific to the contralateral, nonoperated side. (c) There was a significant relationship between seizure outcome and postoperative learning and memory performance, insofar as patients with complete postoperative seizure control showed better results than those with incomplete control. (d) Concerning memory (delayed recall), the overall trend is also that of a slight postoperative improvement. This is, however, not significant regarding the performance for test-material specific to the operated hemisphere and is also - compared to the learning results - less marked regarding the contralateral, nonoperated side.

Postoperative Visual Field Examinations

Pre-/postoperative comparisons of visual fields, as carried out by routine Goldmann perimeter examination and since 1983 by the OCTOPUS, did not show any additive postoperative visual field defects.

Concluding Remarks

From the evidence of this series, we conclude that in patients with strictly localized drug-resistant, recurrent mesiobasal limbic seizure onset the described selective amygdalohippocampectomy is to be preferred to the "standard" anterior temporal lobectomy or to other recently described operations with access to the posterior medial temporal lobe structures (Spencer et al. 1984).

Moreover, this selective amygdalohippocampectomy may also be seriously considered in those cases in which seizure onset is outside the limits of a safe anterior temporal lobectomy on the speech dominant side, and in which a secondary "pacemaker" role for anterior limbic temporal structures in seizure generation can be ascertained. Although the results of such palliative operations are not as good as for causal operations and palliatively operated patients must be expected to continue to experience auras postsurgically, a considerable proportion of patients with such a seizure constellation can be helped significantly. Otherwise, they would probably be judged inoperable. It is abvious that such palliative operations with an a priori lower success rate can be recommended only if the intended surgery is safe and will not introduce any significant functional deficits.

[4] Results are given according to the recommendations of the "International Temporal Lobe Club" (Meeting at New York; J. Engel Jr., Dec. 1981, pers. comm.).

In conclusion, we wish the amygdalohippocampectomy described above to be regarded as neither more nor less than a first small step towards our goal of highly tailored "selective" but still effective epilepsy surgery. It should be emphasized that this operation reflects both progress in presurgical evaluation of seizure patients and improved microsurgical technique and experience. As this operation can help only a relatively small subgroup of the large seizure population in whom surgical therapy has to be seriously considered, similar efforts should be directed to other seizure subtypes with the accompanying aim of studying the possibility of more selective interventions. It is our guess that this will turn out to be even much more difficult, but not impossible.

It is, however, expected that the findings in many patients will continue to preclude any selective resective surgery and - if surgery still has to be contemplated in such cases - only large traditional resections can offer a real chance to ease the lot of these patients.

Acknowledgements. We wish to thank Dr. P.J. Teddy and P. Roth, as well as the editor-in-chief, Prof. Dr. L. Symon, and the Springer-Verlag, Vienna, for their generous permission to reproduce parts of this article and the figures, which were originally published in Vol. 12 of *Advances and Technical Standards in Neurosurgery.*

References

Bernoulli C, Siegfried J, Baumgartner G, Regli F, Rabinowicz T, Gajdusek DC, Gibbs CJ Jr (1977) Danger of accidental person-to-person transmission of Creutzfeldt-Jakob disease by surgery. Lancet 1: 478-479, 569

Niemeyer P (1958) The transventricular amygdalohippocampectomy in temporal lobe epilepsy. In: Baldwin M, Bailey P (eds) Temporal lobe epilepsy. Springfield, Thomas, pp 461-482

Spencer DD, Spencer SS, Mattson RH, Williamson PD, Novelly RA (1984) Access to the posterior medial temporal lobe structures in the surgical treatment of temporal lobe epilepsy. Neurosurgery 15: 667-671

Wieser HG (1986) Selective amygdalohippocampectomy: indications, investigative technique and results. In: Symon L et al. (eds) Advances and technical standards in neurosurgery, vol 13. Springer, Vienna New York, pp 39-133

Wieser HG, Yaşargil MG (1982) Selective amygdalohippocampectomy as a surgical treatment of mesiobasal limbic epilepsy. Surg Neurol 17: 445-457

Yaşargil MG (1984) Microneurosurgery, vol 1. Thieme, Stuttgart

Yaşargil MG, Teddy PJ, Roth P (1985) Selective amygdalohippocampectomy: operative anatomy and surgical technique. In: Symon L et al. (eds) Advances and technical standards in neurosurgery, vol 12. Springer, Vienna New York, pp 93-123

Stereotactic Surgery for Epilepsy

L. Ravagnati

Different stereotactic approaches have been devised for the treatment of medically refractory epilepsies, since Spiegel et al. (1958) and Wycis et al. (1966) proposed pallidotomy for convulsive disorders and Hassler and Riechert (1957) introduced fornicotomy for temporal lobe epilepsy. Based upon the premise that epileptic seizures are due to the firing of hyperexcitable neurons whose abnormal discharges propagate along preferential pathways, various procedures have been designed to (1) destroy the localized centers of the seizure onset; (2) interrupt the pathways of propagation; and (3) lessen the neuronal excitability, both by decreasing facilitory impulses and by increasing inhibitory influences (Ramamurthi and Kalyanaraman 1982; Walker 1982). Stereotactic surgery has been used mainly in those cases that do not meet the criteria for classical open surgery with cortical resection. Such cases are estimated to represent more than 60% of all medically intractable epilepsies (Bouchard 1976; Cigánek et al. 1976).

This paper will present the data collected at the Institute of Neurosurgery of the University of Milan, and evaluate the published data, which have been recently reviewed by different authors (Jinnai and Mukawa 1973; Ojemann and Ward 1975; Ramamurthi and Kalyanaraman 1982; Walker 1982).

Materials and Method

Our results in stereotactic surgery for epilepsy are based upon radiofrequency lesions in 37 patients, with a follow-up of 2–13 years in 30 of them (Marossero et al. 1980). They are part of a series of 165 epileptic patients who have undergone surgical procedures for diagnostic and therapeutic purposes, out of 1700 patients admitted for nontumoral epilepsy from 1958 to 1980 (Ravagnati et al. 1984). Depth EEG studies were done in 114 cases; 24 of them were judged unsuitable for open or stereotactic surgery and 5 refused further interventions. The other surgical procedures consisted of lobectomy or topectomy (93 cases, including 11 previously treated by stereotactic approach) and hemispherectomy (17 cases).

With reference to the current literature, the data presented have been found to be nonhomogeneous in many aspects, such as criteria for patient selection, seizure classification, age grouping, outcome categories, and follow-up studies. The data have been classified according to the surgical rationale and targets used in similar types of epilepsies. As different authors use different outcome groups, we have tried to fit the surgical results into three broad categories: group A includes seizure-free patients and patients with occasional seizures only; group B includes patients who had a worthwhile improvement, with a significant reduction in seizure frequency; group C includes both patients who had only a transitory or mild improvement, and failures. Other parameters, such as social factors, behavior, and employment, have not been considered in the outcome, even in cases where the indications for surgery (amygdalotomy, hippocampotomy) were more for behavioral disturbances than for epileptic seizures. Regarding the follow-up, out of 51 series, with a total of 1098 patients, the range of follow-up is short, being less or a little more than 1 year only in three series (Laitinen 1967; Ojemann and Ward 1975; Talairach and Szikla 1965) and not indicated in ten (Bouvier et al. 1980; Chiorino et al. 1968; Flanigin and Nashold 1976; Ganglberger 1984; Gillingham and Campell 1980; Hori et al. 1968; Nesterov and Kulikov 1973; Ramamurthi et al. 1974; Schaltenbrand et al. 1966; Šramka et al. 1972). In rare cases, series with long follow-ups also include a few patients with inadequate observation (Balasubramaniam and Kanaka 1976; Barcia-Salorio and Broseta 1976; Cigánek et al. 1976; Laitinen 1967; Schwab et al. 1965; Spiegel et al. 1958). It can be calculated that about 250 patients have inadequate or nonspecified follow-up and that about 750 have a mean follow-up of 3–5 years.

Institute of Neurosurgery, University of Milan, Policlinico, Via. F. Sforza 35, Milano 20122, Italy

Presurgical Evaluation of Epileptics
Edited by H. G. Wieser and C. E. Elger

Results

Stereotactic Destruction of the Epileptogenic Zone

In the surgical treatment of temporal lobe epilepsy, lesions have been placed in the areas of seizure onset, such as amygdala and hippocampal formation, as determined by depth EEG. The lesions were unilateral, except for a few cases of bilateral lesions of the amygdala in bitemporal epilepsy (Table 1).

Stereotactic Interruption of Pathways of Seizure Discharge

This type of approach has been designed to prevent the ictal spread to distant neuronal aggregates, and thus prevent the clinical exteriorization of the seizure in cases of multiple or diffuse foci and generalized seizures.

Hippocampal-Amygdaloid Complex. Table 2 summarizes the combination of target structures in the limbic system and limbic-thalamic connections, as used mainly in temporal lobe epilepsy with complex partial and secondary generalized seizures (Bouchard 1976; Ganglberger 1984; Mundinger et al. 1976; Narabayashi 1979).

The results of lesioning multiple target points are shown in Table 3.

This approach was exclusively directed to interrupting pathways of spread, except in some cases of Adams and Rutkin (1969), where combined lesions of both the areas of seizure onset and their efferent connections were done.

Instead of lesioning multiple limbic pathways, other authors have limited their target points to the medial or lateral parts of the amygdala, in cases of psychomotor seizures (Table 4) and grand mal seizures (Table 5).

Striatum and Thalamus. Lesions of these neuronal relay aggregates, which were supposed to activate, facilitate, or propagate epileptic discharges, have been performed mainly in primary and secondary generalized epilepsies, and in a few cases of partial epilepsy.

The results of stereotactic lesions in pallidum (Table 6), putamen (Table 7), nucleus ventralis anterior of the thalamus (Table 8), nucleus ventralis lateralis (Table 9), nucleus centromedianus (Table 10), and internal medullary lamina (Table 11) have been rather disappointing.

Fields of Forel. Lesions in the fields of Forel produced good results in more than 30% of cases, characterized by generalized and hemiconvulsive seizures (Table 12). In the opinion of Jinnai et al. (1976), this target is indicated in cases of grand mal and neocortical motor seizures and not for limbic and myoclonic epilepsy.

Internal Capsule. Stereotactic interruption of the corticospinal bundle at the level of the internal capsule has been performed in cases of epilepsia partalis continua, with good seizure control but also with obvious motor deficits.

A "premotor bundle" located in the medial side of the internal capsule, where motor responses were not evoked by stimulation, has been another target in cases of grand mal seizures, with uncertain results (Table 13).

Abbreviations used in the tables

Targets

AC, anterior commissure
AMYG, amygdala
CL, nucleus centrolateralis thalami
CM, nucleus centromedianus thalami
DM, nucleus dorsalis medialis thalami
F.UNCIN, fasciculus uncinatus
FO, fornix
HIPP, hippocampus
HYPOTH, hypothalamus
INT.MED.LAM, internal medullary lamina of thalamus
INTRALAM.NCL, intralaminar nuclei of thalamus
PAL, pallidum
VA, nucleus ventralis anterior thalami
VL, nucleus ventralis lateralis thalami
VOA, nucleus ventro-oralis anterior thalami

BILAT, bilateral
CONTRALAT, contralateral
UNILAT, unilateral

Seizure type

CENTRENC, centrencephalic epilepsy
CPS, complex partial seizures
GENER, generalized seizures
GM, grand mal
HEMIFOC.CONV, hemiconvulsions and focal convulsions
MYOCL, myoclonus epilepsy, myoclonus
PART.CONT, epilepsia partialis continua
PART.MOTOR, partial motor seizures
PART.SENS, partial sensory seizures
PM, petit mal
PRIM.GENER, primary generalized seizures
PSYCHOMOT, psychomotor seizures
sf, seizure-free
SEC.GENER, secondary generalized seizures
TEMP EEG, temporal EEG abnormalities
TLE, temporal lobe epilepsy

Table 1. Results of stereotactic lesioning in medial temporal epileptogenic areas (amygdala - hippocampus). (Few cases of Flanigin's series had thalamic lesions as well)

Author	Number of patients	Depth-EEG findings	Results		
			Seizure-free or occasional seizures A	Worthwhile improvement B	Failure C
Hood et al. (1983)	4	UNILAT AMYG	2 sf	1	1
Marossero et al. (1980)	19	UNILAT BILAT	5 1	1 5	3 4
Bouvier et al. (1980)	5	UNILAT	-	5	-
Flanigin and Nashold (1976)	16		4	6	6
Talairach and Szikla (1965)	14	UNILAT BILAT	9 sf	2	3
Total %	58		21 35%	20 35%	17 30%

Other Targets. Stereotactic lesions of portions of the corpus callosum, with combined interruption of the fornix, have been performed by Schaltenbrand et al. (1966) in three cases: two of them were ameliorated.

Upper mesencephalic reticulotomy was performed by Sano (1974) in six cases: two patients with generalized epilepsy were reported to be seizure free.

Combined Approach. A combined approach, consisting of multiple lesions in different structures of a single patient, is adopted in some eastern European centers. Depth-EEG investigations are followed by stereotactic lesions placed in the structures where epileptogenic discharges were recorded. The results of this type of approach are shown in Table 14.

Standardized Stepwise Approach. A stepwise type of stereotactic approach is reported from India by other authors. They begin first with a bilateral amygdalotomy which, in case of failure, is followed by unilateral hypothalamotomy and finally by thalamic lesions (Table 15).

Approaches Designed to Decrease the Neuronal Excitability

The posterior hypothalamus has been proposed as a stereotactic target, on the assumption that lesions of the "ergotropic zone" of the hypothalamus could decrease the excitability of the cortex and subcortical structures (Table 16).

Others authors have attempted to decrease the neuronal excitability by stimulating the nucleus caudatus (Table 17).

Mild improvements have been reported following bilateral stereotactic lesions of the nucleus dentatus in seven patients with dyskinetic syndromes and epileptic seizures, and in three patients with only epilepsy, generalized in two and partial in one (Fraioli and Guidetti 1975).

Besides the stereotactic approaches, other procedures have been proposed to decrease the cortical excitability. It is claimed that chronic cerebellar stimulation is followed by good results in primary generalized seizures, as shown in Table 18.

Table 2. Targets used by different authors and schools for interruption of activating and reverberating circuits of the hippocampal - amygdaloid complex

Fornix - anterior commissure	Hassler and Riechert (1957) Umbach (1966)
Amygdala	Narabayashi (1979)
Fornix - anterior commissure-medial amygdala	Freiburg occasionally (Mundinger et al. 1976) Berlin routinely (Umbach 1966)
Fornix - anterior commissure-basolateral amygdala	Berlin (Kim 1971)
Fornico - amygdalotomy - VA thalamotomy	Bouchard (1976)
Amygdalo-mediothalamic pathway (DM)	Ganglberger (1984)

Table 3. Results of stereotactic interruption of hippocampal-amygdaloid pathways

Author	Number of patients	Seizure type	Lesion (main target: hippocampal-amygdaloid pathways)	Results		
				A	B	C
Umbach (1966)	18	TLE+GM	FO (5 BILAT)+AC	4 sf GM free: 11/13		
Bouchard (1976)	44	½ TLE+GM ½ GM MULTIFOCAL	FO+AC+AMYG +VA 50% BILAT	17 (7 sf)	17	10
Mundinger et al. (1976)	29 4	PSYCHOMOT PRIM.GENER	FO (3 BILAT)+AC +AMYG.(BILAT)	6 0	13 2	10 2
Ganglberger (1984)	10	TLE+GM	FO+AMYG+DM	5 (4 sf)	4	1
Schaltenbrand et al. (1966)	5	PARTIAL: (PSYCHOMOT+ MOTOR)	AC FO (3) PAL (1)	-	3	2
Orthner and Lohman (1966)	6	TLE	FO AC (2) PAL (2)	-	5	1
Adams and Rutkin (1969)	16	PSYCHOMOT+GM	AMYG-HIPP AC PAL (1)	1 sf	10	5
Barcia-Salorio and Broseta (1976)	42	UNILAT TLE+GM	FO+AC (UNILAT) 3 AMYG (CONTRALAT)	13 sf	20	9
Marossero et al. (1980)	9	CPS	AMYG FO F. UNCIN	-	1	8
Total %	183			46 25%	(75) (45%)	(48) (30%)

Table 4. Results of amygdalotomy in patients with complex partial seizures and temporal lobe epilepsy

Author	Number of patients	Seizure type	Lesion (main target: amygdala)	Results		
				A	B	C
Sonnen et al. (1976)	8	PSYCHOMOT+GM	BILAT (7) UNILAT FO (3)	1	1	6
Ramamurthi et al. (1974)	31	TLE	UNILAT BILAT	10 sf 4 sf	7 3	4 3
Vaernet (1972)	45	UNILAT TLE BILAT TLE BILAT TLE	UNILAT UNILAT BILAT	5 sf 1 sf 2 sf	10 5 2	12 4 4
Narabayashi and Mizutani (1970)	3	PSYCHOMOT	UNILAT	-	-	3
Chitanondh (1969)	31	PSYCHOMOT	UNILAT (26) BILAT (5)	3 sf	7	21
Wycis (1969)	2	PSYCHOMOT	UNILAT	-	-	2
Schaltenbrand et al. (1966)	6	PARTIAL (MOTOR, PSYCHOMOT)	UNILAT	-	-	6
Schwab et al. (1965)	10	PSYCHOMOT GM (1)	UNILAT	3	4	3
Total %	136			29 21%	39 29%	68 50%

Table 5. Results of amygdalotomy in patients with primary generalized seizures

Author	Number of patients	Seizure type	Lesion (main target: amygdala)	Results		
				A	B	C
Narabayashi and Mizutani (1970)	25	GM GM + PSYCHOMOT	UNILAT ⅓ BILAT	9 sf	8	8
Diemath (1972)	2	GM	UNILAT AMYG CONTRALAT DM	2 sf	-	-
Heimburger et al. (1978)	49	GM PSYCHOMOT PM	⅔ BILAT	-	21	28
Narabayashi (1979)	18	GM (12) PSYCHOMOT (2) MIXED (4)	BILAT (11)	1 sf	about 40%	(? 10%)
Total	94			12	(36)	(? 46)
%				13%	(38%)	(? 49%)

Table 6. Results of stereotactic lesioning in the globus pallidus

Author	Number of patients	Seizure type	Lesion (main target: pallidum)	Results		
				A	B	C
Spiegel et al. (1958) Wycis et al. (1966)	14	SALAAM Seizures GM PSYCHOMOT	PAL (2 BILAT) + AMYG (8)	3 sf	6	5
Schaltenbrand et al. (1966)	4	PARTIAL	FO + AC	1	1	2
Orthner and Lohman (1966)	2	PARTIAL	+ FO + AC	-	1	1
Gillingham and Campbell (1980)	14	PARTIAL SEC.GENER	11 BILAT.	-	12	2
Total	34			4	20	10
%				12%	59%	29%

Table 7. Results of stereotactic lesioning in the putamen in patients with generalized seizures

Author	Number of patients	Lesion (main target: putamen)	Results
Dierssen (1966)	11		5 sf
Hori et al. (1968)	68	BILAT	20% sf 44% improved
Total	80		

Table 8. Results of stereotactic lesioning in the nucleus ventralis anterior of the thalamus (VA)

Author	Number of patients	Seizure type	Lesion (main target: VA)	Results		
				A	B	C
Talairach (1952)	2	GM	VA BILAT	–	–	2
Mullan et al. (1967)	3	GM (2) PART.SENS (1)	VA+VL	–	2	1
Pertuiset et al. (1969)	12	GM	VA+CM 5 BILAT	2 sf	8	2
Šramka et al. (1972)	4			–	4	–
Ojemann and Ward (1975)	3	GM (1) PSYCHOMOT (1) PARTIAL (1)	UNILAT	–	–	3
Total %	24			2 8%	14 58%	8 33%

Table 9. Results of stereotactic lesioning in the nucleus ventralis lateralis of the thalamus (VL)

Author	Number of patients	Seizure type	Lesion (main target: VL)	Results		
				A	B	C
Mullan et al. (1967)	3		combined lesions	–	1	2
Nesterov and Kulikov (1973)	55	PART.CONT	VL+CM	–	45	10
Jelsma et al. (1973)	10	PARTIAL or PRIM.GENER	BILAT	–	5	5
Ciganek et al. (1976)	2	JACKSONIAN	VOA	–	2	–
Total %	70			– –	53 76%	17 24%

Table 10. Results of stereotactic lesioning in the nucleus centromedianus of the thalamus (CM)

Author	Number of patients	Seizure type	Lesion (main target: CM)	Results		
				A	B	C
Ramamurthi et al. (1974)	6	MYOCL SALAAM seizures	CM	3 sf	1	2
Sano (1974)	5	GENER	CM+CL INTRA- LAM.NCL	–	2	3
Nesterov and Kulikov (1973) Mullan et al. (1967) Pertuiset et al. (1969) Koshino et al. (1975)		GENER	combined lesions		?	

Table 11. Results of stereotactic lesions in the internal medullary lamina of the thalamus

Author	Number of patients	Seizure type	Results		
			A	B	C
Hassler and Riechert (1957)	3	ABSENCE	-	-	3
Spiegel et al. (1962)	6	GM ABSENCE	-	3	3
Orthner and Lohman (1966)	4		-	-	4
Total	13		-	3	10

Table 12. Results of stereotactic lesions placed in the fields of Forel

Author	Number of patients	Seizure type	Lesion (main target: fields of Forel)	Results		
				A	B	C
Jinnai et al. (1976)	46	GM, HEMIFOC.CONV PSYCHOMOT	BILAT UNILAT	5 15	6 4	4 12
Bidzinski and Ostrowski (1981)	7	GENERAL.EEG	UNILAT	1	3	3
Marossero et al. (1980)	2	GM	BILAT	-	-	2
Ramani et al. (1980)	6	GM - MYOCL - PSYCHOMOT	BILAT	-	4	2
Yoshii (1978)	15	GM - others	UNILAT BILAT	7	5	3
Koshino et al. (1975)	13	GM (7) Hemiconvulsions (6)		2	1	10
Chiorino et al. (1968)	2	GM	UNILAT	-	2	-
Mullan et al. (1968)	5	TLE PART.MOTOR SEC. GENER	UNILAT	1	1	3
Laitinen (1967)	7	GM - MYOCL	Subventral thalamotomy	-	5	2
Total %	103			31 30%	31 30%	41 40%

Table 13. Results of stereotactic lesioning in the internal capsule

Author	Number of patients	Seizure type	Lesion (main target: internal capsule)	Results		
				A	B	C
Jelsma et al. (1973)	3	PART.CONT	internal capsule	-	3	-
Schaltenbrand et al. (1966)	1	PART.MOTOR	internal capsule +PAL	1	-	-
Kalyanaraman and Ramamurthi (1970)	7	GM	PREMOTOR BUNDLE	-	3	4
Total	11			1	6	4

Table 14. Results of combined, multiple stereotactic lesioning in various target areas ("combined approach")

Author	Number of patients	Seizure type	Targets: "combined approach"	Results		
				A	B	C
Cigánek et al. (1976)	61	Multiform types, often generalized, including: TLE, MYOCL (6), JACKSONIAN (2), PRIM.GEN (1)	12 different thalamic ncl., AMYG, HIPP, HYPOTH, Ncl. ruber, putamen, PAL, Ncl. caudatus, AC, fields of Forel, FO, medial frontal cortex	15 (25%)	24 (39%)	22 (36%)
Mempel et al. (1980)	70	simple partial, CPS, GENER, TEMP.EEG	AMYG, HIPP	8 sf (11%)	52 (74%)	10 (14%)

Table 15. Results of stereotactic lesioning using the "standardized stepwise approach"

Author	Number of patients	Seizure type	Targets ("standardized stepwise approach")	Results		
				A	B	C
Balasubramanian and Kanaka (1976)	85	GM	BILAT AMYG ↓			
	15	GM + FOCAL	UNILAT HYPOTH ↓	31 sf	29	40
	5	TLE	INT.MED.LAM ↓	3 sf	1	1
	1	ABSENCE	BILAT CM (5)	1	–	–
	1	MYOCL.		1 sf	–	–
Total	107			36	30	41
%				34%	28%	38%

Table 16. Results of stereotactic lesioning in the "ergotropic zone" of the posteromedial hypothalamus

Author	Number of patients	Seizure type	Results		
			A	B	C
Sano et al. (1970)	22	GM, ABSENCE, PSYCHOMOT	1	6	15

Table 17. Results of therapeutic temporary electrical stimulation of the nucleus caudatus

Author	Number of patients	Seizure type		Results		
				A	B	C
Drlickova et al. (1984)	13	GM TLE TLE + GM CENTRENC	5 3 4 1	2 (1 sf)	7	4

Table 18. Results of chronic cerebellar stimulation

Author	Number of patients	Seizure type	Results		
			A	B	C
Davis et al. (1984)	32	GM PM MYOCL PSYCHO-MOT	18 57%	9 28%	5 15%
Other authors (Davis et al. 1984)	55				
Total	87		61 (70%) BENEFITED		

Complications

Table 19 shows the incidence of complications due to hemorrhage and other vascular lesions (thrombosis, vasospasm) following stereotactic surgery, derived from the analysis of more than 20000 operations for epilepsy and extrapyramidal disorders (Crevier 1974).

Besides these mechanical complications, other complications are reported to be due to inaccurate localization of the targets and misplaced stereotactic lesions. Their incidence varies in different series. The neurological deficits are reported to clear within a few months in the majority of cases.

One should mention a last category including the complications following correctly placed lesions, due to functional imbalance. Temporary confusion, brief memory impairment, and confabulation were reported in two cases following bilateral fornicotomy and commissurotomy (Mundinger et al. 1976). Memory disturbances following lesions in both amygdala and hippocampus in the dominant hemisphere are also reported in one case (Flanigin and Nashold 1976). Following bilateral amygdalotomy, no evidence of Kluver-Bucy syndrome was observed. Bilateral lesions in the fields of Forel are reported to be followed by mild to moderate cognitive impairment in some cases (Ramani et al. 1980).

Table 19. Occurrence of complications in large series of stereotactic surgery (Crevier 1974)

	%
Average incidence of hematomas	1-3
Deaths secondary to hematomas	0.5-1.5
Motor deficits secondary to hematomas	0.3-1
Motor deficits secondary to a vascular trauma (other than hematomas)	1.4

Discussion

Despite theoretical and experimental propositions underlying various rationales for target determination, there are as yet no standardized stereotactic procedures for the different classes of epilepsy. The most successful targets appear to be those related to mesial temporal structures, with regard to both the area of seizure onset as determined by depth EEG, and the propagation pathways for cases of temporal lobe epilepsy with bitemporal foci unsuitable for unilateral resection. Thirty percent of the patients have been rendered seizure-free or left with occasional seizures only, and another 40% had a worthwhile improvement. Asymmetrical stereotactic lesioning of both temporal lobes may thus be proposed as a method for treating both complex partial and secondary generalized seizures in bitemporal epilepsy.

In primary generalized seizures as well as in multifocal epilepsies, lesions of the thalamus, which has been considered a relay of the pathways of seizure discharge from the cortex as well as a pacemaker in generalized epilepsy, have produced disappointing results. For grand mal seizures, whether of primary or secondary generalized type, the fields of Forel can be considered a promising target, but further evaluation of its efficacy and safeness is needed (Ramani et al. 1980).

At the present time, depth EEG investigations to identify the sites of origin and the pathways of seizure discharges seem to be the most useful preliminary step in determining the stereotactic targets in a particular patient.

Stimulation techniques, designed to lessen the excitability of the cortex and subcortical structures, might constitute an alternative way of treating otherwise intractable primary generalized epilepsy. It should be said that a number of procedures, designed to interrupt the pathways of seizure discharges, might attain their results by interrupting activating systems, thus decreasing the neuronal excitability.

The incidence of complications in stereotactic surgery has been proven to be very low and could be further reduced by routine use of stereotactic angiography to prevent vascular accidents, by careful search for anatomical lesions in temporal lobes before lesioning targets of the hippocampal formations, and by using neurophysiological tests to identify the target structures (Maspes et al. 1972).

The results obtained by stereotactic surgical treatment for epilepsy are not comparable to those of resective surgery. The generous admission to stereotactic procedures of otherwise intractable patients, as compared with the strict selection of cases for resective surgery, is responsible for the differences in outcome. A protocol improved with regard to selection of patients, the targets for different types of epilepsy, and the follow-up studies could ameliorate the final results.

Acknowledgments. The author is greatly indebted to Rossana Bianchi; without her editorial assistance this work would not have been possible. He is also grateful to Prof. F. Marossero, Prof. L. Infuso, Prof. G. Ettorre, Prof. G. P. Cabrini, Dr. V. A. Sironi for their suggestions.

References

Adams J, Rutkin B (1969) Treatment of temporal lobe epilepsy by stereotactic surgery. Confin Neurol 31: 80-85

Balasubramaniam V, Kanaka TS (1976) Stereotactic surgery of the limbic system in epilepsy. Acta Neurochir [Suppl] 23: 225-234

Barcia-Salorio JL, Broseta J (1976) Stereotactic fornicotomy in temporal lobe epilepsy: indications and long-term results. Acta Neurochir [Suppl] 23: 167-175

Bidzinski J, Ostrowski K (1981) Destruction of Forel's field in the surgical treatment of epilepsy. Neurol Neurochir Pol 15: 49-52

Bouchard G (1976) Basic targets and the different epilepsies. Acta Neurochir [Suppl] 23: 193-199

Bouvier G, Saint-Hilaire JM, Maltais R, Bélique R, Desrochers P (1980) Stereotactic lesions in primary epilepsy of the limbic system. Acta Neurochir [Suppl] 30: 151-159

Chiorino R, Donaso P, Diaz G, Aranda L, Asenjo A, Garcia C (1968) Stereotaxic surgery in the treatment of epilepsy. Description of a new technique: campotomy. Neurochirurgia 26: 143-147

Chitanondh H, Laksanavicharn U (1969) Stereotaxic amygdalotomy in the treatment of temporal automatism. Excerpta Medica Int Congress Series 193: 9

Cigánek L, Šramka M, Nadvornik P, Fritz G (1976) Effects of stereotactic operations in the treatment of epilepsies: neurological aspects. Acta Neurochir [Suppl] 23: 201-204

Crevier PH (1974) Post-stereotaxic intracranial hematomas. Acta Neurochir [Suppl] 21: 71-73

Davis R, Gray E, Engle H, Dusnak A (1984) The reduction of seizures in cerebral palsy and epileptic patients using chronic cerebellar stimulation. Acta Neurochir [Suppl] 33: 161-167

Diemath H (1972) Discussion. In: Vaernet K (ed): Stereotaxic amygdalotomy in temporal lobe epilepsy. Confin Neurol 34: 182

Dierssen G (1966) Discussion. In: Wycis H, Baird H, Spiegel E (eds) Long range results following pallidotomy and pallidoamygdalotomy in certain types of convulsive disorders. Confin Neurol 27: 114-120

Drlickova V, Šramka M, Nadvornik P (1984) Caudate nucleus stimulation during epilepsy treatment. Acta Neurochir [Suppl] 33: 155-160

Flanigin HF, Nashold BS (1976) Stereotactic lesions of the amygdala and hippocampus in epilepsy. Acta Neurochir [Suppl] 23: 235-240

Fraioli B, Guidetti B (1975) Effects of stereotactic lesions of the dentate nucleus of the cerebellum in man. Appl Neurophysiol 38: 81-90

Ganglberger JA (1984) Additional new approach in treatment of temporal lobe epilepsy. Acta Neurochir [Suppl] 33: 149-154

Gillingham FJ, Campbell D (1980) Surgical interruption of the conduction pathways for the control of intractable epilepsy. Acta Neurochir [Suppl] 30: 67-74

Hassler R, Riechert T (1957) Über einen Fall von doppelseitiger Fornicotomie bei sogenannter temporaler Epilepsie. Acta Neurochir 5: 330-340

Heimburger RF, Small IF, Small JG, Milstein V, Moore D (1978) Stereotactic amygdalotomy for convulsive and behavioral disorders. Appl Neurophysiol 41: 43-51

Hood TW, Siegfried J, Wieser HG (1983) The role of stereotactic amygdalotomy in the treatment of temporal lobe epilepsy associated with behavioral disorders. Appl Neurophysiol 46: 19-25

Hori Y, Terada C, Miyamoto S (1968) The effect of stereotaxic putamectomy for epileptic seizures. Neurol Med Chir (Tokyo) 10: 321-323

Jelsma R, Bertrand C, Martinez S, Molina-Negro P (1973) Stereotaxic treatment of frontal-lobe and centrencephalic epilepsy. J Neurosurg 39: 42-51

Jinnai D, Mukawa J (1973) Surgery for epilepsy. In: Krayenbühl H, Maspes PE, Sweet WH (eds) Progress in neurological surgery, vol 5. Karger, Basel, pp 229-296

Jinnai D, Mukawa J, Kobayashi K (1976) Forel-H-tomy for the treatment of intractable epilepsy. Acta Neurochir [Suppl] 23: 159-166

Kalyanaraman S, Ramamurthi B (1970) Stereotaxic surgery for generalized epilepsy. Neurol India 18 [Suppl 1]: 34-41

Kim YK (1971) Effects of basolateral amygdalotomy. In: Umbach W (ed) Special topics in stereotaxis. Hippokrates, Stuttgart, pp 156-167

Koshino K, Nakano M, Miki M, Matsumura H (1975) Stereotaxic operation for the control of infantile epilepsy and associated behavioral disorder. Confinia Neurol 37: 223-231

Laitinen LV (1967) Thalamotomy in progressive myoclonus epilepsy. Acta Neurol Scand [Suppl] 31: 170-171

Marossero F, Ravagnati L, Sironi VA, Miserocchi G, Franzini A, Ettorre G, Cabrini GP (1980) Late results of stereotactic radiofrequency lesions in epilepsy. Acta Neurochir [Suppl] 30: 145-149

Maspes PE, Marossero F, Cabrini GP (1972) Neurophysiological tests for stereotactic identification of rhinencephalic targets. Schweizer Archiv Neurol Neurochir Psychiat 111: 331-340

Mempel E, Witkiewicz B, Stadnicki R, Luczywek E, Kucinski L, Pawlowski G, Nowak J (1980) The effect of medial amygdalotomy and anterior hippocampotomy on behavior and seizures in epileptic patients. Acta Neurochir [Suppl] 30: 161-168

Mullan S, Vailati G, Karasick J, Mailis M (1967) Thalamic lesions for the control of epilepsy. Arch Neurol 16: 277-285

Mundinger F, Becker P, Gröbner E, Bachschmid G (1976) Late results of stereotactic surgery of epilepsy of predominantly temporal lobe type. Acta Neurochir [Suppl] 23: 177-182

Narabayashi H (1979) Long range results of medial amygdalotomy on epileptic traits in adult patients. In: Rasmussen T, Marino R (eds) Functional neurosurgery. Raven, New York, pp 243-252

Narabayashi H, Mizutani T (1970) Epileptic seizures and stereotaxic amygdalotomy. Confin Neurol 32: 289-297

Nesterov L, Kulikov A (1973) Stereotactic operations on the subcortical structures of the brain in the treatment of Kozhevnikov's epilepsy. Excerpta Medica Int Congress Series 293: 21

Ojemann GA, Ward AA Jr (1975) Stereotactic and other procedures for epilepsy. In: Purpura DP, Penry JK, Walter RD (eds) Advances in neurology, vol 8. Raven, New York, pp 241-263

Orthner H, Lohman R (1966) Experience with stereotaxic interventions in epilepsy. Dtsch Med Wochenschr 91: 984-991

Pertuiset B, Hirsch J, Sachs M, Landau-Ferey J (1969) Selective stereotaxic thalamotomy in "grand mal" epilepsy. Excerpta Medica Int Congress Series 193: 72

Ramamurthi B, Kalyanaraman S (1982) Stereotaxic targets for epilepsy. In: Schaltenbrand G, Walker AE (eds) Stereotaxy of the human brain. Thieme, Stuttgart, pp 653-660

Ramamurthi B, Kalyanaraman S, Sayeed ZA, Dharmapal N (1974) Surgical treatment of epilepsy. In: Harris P, Mawdsley C (eds) Epilepsy. Churchill Livingstone, London, pp 203-208

Ramani SV, Yap JC, Gumnit RJ (1980) Stereotactic fields of Forel interruption for intractable epilepsy. Appl Neurophysiol 43: 104-108

Ravagnati L, Sironi VA, Ettorre G, Marossero F (1984) Surgical management of partial epilepsies: a critical evaluation of electrical, CT scan, and pharmacological studies of the Milan series. Acta Neurochir [Suppl] 33: 139-144

Sano K (1974) Indications for and results of making central lesions in the brain for epileptic patients. In: Harris P, Mawdsley C (eds) Epilepsy. Churchill Livingstone, London, pp 215-221

Sano K, Mayanaji Y, Sekino H, Ogashiwa M, Ishijima B (1970) Results of stimulation and destruction of the posterior hypothalamus in man. J Neurosurg 33: 689-707

Schaltenbrand G, Spuler H, Nadjimi M, Hopf H, Wahren W (1966) Die stereotaktische Behandlung der Epilepsien. Confin Neurol 27: 111-113

Schwab R, Sweet W, Mark V, Kjellberg R, Ervin F (1965) Treatment of intractable temporal lobe epilepsy by stereotactic amygdala lesions. Tr Am Neurol Assoc 90: 12-19

Sonnen AEH, Manen J, Dijk B (1976) Results of amygdalotomy and fornicotomy in temporal lobe epilepsy and behaviour disorders. Acta Neurochir [Suppl] 23: 215-220

Spiegel E, Wycis H, Baird H (1958) Long-range effects of electro-pallido-ansotomy in extrapyramidal and convulsive disorders. Neurology 8: 734-740

Spiegel E, Wycis H, Baird H (1962) Subcortical mechanisms in convulsive disorders. In: Spiegel E, Wycis H (eds) Stereoencephalotomy, part II. Grune and Stratton, New York, pp 447-480

Šramka M, Lojka J, Nadvornik P, Cigánek L (1972) Relation of the anterior thalamic reticulum to epilepsy. In: Fusek I, Kunc Z (eds) Present limits of neurosurgery. Excerpta Medica, Amsterdam

Talairach J (1952) Destruction du noyau ventral anterior thalamique dans le traitement des maladies mentales. Rev Neurol 8: 352-357

Talairach J, Szikla G (1965) Destruction partielle amygdalo-hippocampique par l'Yttrium^{90} dans le traitement de certaines épilepsies à expression rhinencephalique. Neuro-Chirurgie 11: 233-240

Umbach W (1966) Long term results of fornicotomy for temporal epilepsy. Confin Neurol 27: 121-123

Vaernet K (1972) Stereotaxic amygdalotomy in temporal lobe epilepsy. Confin Neurol 34: 176-180

Walker AE (1982) General principles of stereotaxic surgery for epilepsy. In: Schaltenbrand G, Walker AE (eds) Stereotaxy of the human brain. Thieme, Stuttgart, pp 645-652

Wycis H (1969) Discussion. In: Adams J, Rutkin B (eds) Treatment of temporal lobe epilepsy by stereotaxic surgery. Confin Neurol 31: 80-85

Wycis H, Baird H, Spiegel E (1966) Long range results following pallidotomy and pallidoamygdalotomy in certain types of convulsive disorders. Confin Neurol 27: 114-120

Yoshii N (1978) Follow-up study of epileptic patients following Forel-H-tomy. Appl Neurophysiol 40: 1-12

Consequences for the Present Practice of Epilepsy Therapy in Europe

D. Janz

The treatment of epilepsy is essentially pharmacological treatment. Surgery can, however, be considered under the following circumstances:

- When the seizures are pharmacoresistant
- When they stem from an operable focus
- When the type and frequency render them personally and socially intolerable

This is nothing new. Surgical treatment of epilepsy has existed for 100 years now, as is testified by the names Victor Horsley, Fedor Krause, Otfried Foerster, Wilder Penfield, and Murray Falconer, to mention but a few pioneers in this field. This relatively modest branch on the tree of epilepsy treatment has begun to sprout and bear fruit again in the past few years. In my opinion there are two main reasons for this. On the one hand, it is because the enormous progress in antiepileptic drug therapy in the past 2 decades has practically benefited only generalized epilepsies and, on the other hand, because focal epilepsies have benefited more from the recent rapid development of diagnostic and microsurgical techniques.

How great is the proportion of people suffering from epilepsy who can be significantly helped by surgery? Based on the number of new cases only, we can expect the aforementioned criteria to apply to at least 1.7/100000, i.e., 1000 people per year in Germany. We can assume that at least half of the total number of new cases (49/100000 according to Woodbury 1978), or as many as two-thirds according to the study by Hauser and Kurland (1975), will suffer from focal seizures. About 60% of these will suffer from complex focal seizures and at most half of these 60% will achieve remission under monotherapy. The addition of a second drug reduces seizures by more than 75% in only a further 13% (Schmidt 1982) or completely in 11% (Mattson et al. 1985). According to conservative estimates, we must therefore expect that, in one-third of patients with complex focal seizures, drug treatment will definitely not be satisfactory and that one-third of these will also fulfil the other two criteria, i.e., they will have an identifiable operable focus and be under severe personal and social strain. This gives us an annual increase of 1.7/100000 or in Germany of 1000 patients with temporal lobe epilepsy who are candidates for surgical treatment. The number becomes correspondingly greater if we base the calculation on the total number of epilepsy cases instead of on new cases only, and if we also take into consideration other possible indications for different forms of surgery such as corpus callosotomy, hemispherectomy, or other procedures. However, I will limit myself to the indications for local resection of the epileptic focus.

In the past, there were difficulties with the criteria for drug resistance and the methods for identification of operable foci, but the Zurich Symposium has convinced us that greater precision has become possible in both respects. This has considerably increased the prospects of success with surgical treatment of focal epilepsies, particularly the most frequent form, i.e., temporal lobe epilepsy.

Thus - to summarize the results again briefly - it has now become possible to give a scientifically sound definition of drug resistance using pharmacological and instrumental means of monitoring therapy. Well-tried means of monitoring therapy, which should be routine in determining drug resistance, are seizure charts, compliance control with the aid of dispensers, video-supported simultaneous long-term observation of seizures and of the electrical activity of the brain, monitoring of the plasma concentrations of antiepileptic drugs, and observation of signs of intoxication.

According to Schmidt's definition (1982, 1984), there is drug resistance "when despite successive monotherapy or combination therapy with antiepileptic drugs of first choice for focal seizures, i.e., phenytoin, carbamazepine, primidone or phenobarbital, and despite high plasma concentrations and signs of beginning intoxication epileptic seizures still persist".

Department of Neurology, Free University of Berlin, Klinikum Charlottenburg, Spandauer Damm 130, 1000 Berlin 19, Federal Republic of Germany

Presurgical Evaluation of Epileptics
Edited by H.G.Wieser and C.E.Elger

With regard to the methods for identification of morphological and functional epileptogenic regions, I would like to distinguish between methods which are already established and those which are still at the stage of development and testing. The former group encompasses all noninvasive and invasive EEG recordings, which are only relevant for focus determination if they are made over fairly long periods and during several seizures. It also includes computer tomography, magnetic resonance imaging, and positron emission tomography. Methods which I have classed as still being on trial are the single photon emission computer tomography (SPECT) method, which we have found very useful (Inoue et al. 1985) and which may one day be a considerably cheaper alternative to the PET method; magnetic resonance spectroscopy, although I know nothing of its use in epilepsy as yet; and magnetencephalography which may one day - so we at least hope - replace the invasive EEG methods.

As we have learned, the selection of suitable candidates for surgery is an arduous procedure which demands a thorough evaluation of all methods of therapy control on the one hand and focus determination on the other hand. In addition, neuropsychological tests and psychological examinations are necessary and social criteria of prognosis should be taken into consideration. After all, the operation is not an end in itself but is intended to improve the quality of life, even though pharmacotherapeutic, psychological, and social care will continue to be necessary for a long time. Selection, operation, and aftercare therefore call for close cooperation of a group of highly specialized experts of various disciplines, such as does not yet exist in an ideal form anywhere in Europe. There are groups which, despite excellent cooperation between epileptologists and surgeons, lack the facilities for depth electroencephalography and the resultant tailored or selective surgical procedures. Other groups lack the facilities for exhausting all forms of conventional pharmacotherapy and providing final proof of drug resistance - to name but two characteristic deficiencies. Although it might be desirable to welcome and promote every local initiative in order to meet at least the most basic needs and also to encourage competition and the exchange of ideas, on the other hand the disadvantages of teams which are inadequately qualified and lack technical facilities should not be underestimated. We know of cases where patients were referred directly to a neurosurgeon and operated on, on the basis of isolated CT or MRT findings instead of a preoperative team diagnosis, and where a focus and seizures still persist. Modern techniques seduce us into assuming that what we see morphologically is in fact identical with the epileptogenic region and thus into overhasty action. We know of cases where pharmacotherapy was not conscientiously kept up, or was even altered, while the patient was in the care of the neurosurgeon.

The German Epilepsy Board (Epilepsie-Kuratorium) considers that the Federal Republic of Germany and West Berlin urgently need at least two centers which would concentrate on preoperative diagnostic procedures and surgical treatment of epilepsy (Janz et al. 1985). Correspondingly Europe would need at least six to eight such centers, having at their disposal all the facilities for pre- and postoperative screening which I have listed here in a plan which we aspire to realize in Berlin (Fig. 1). Concerning the organizational structure of such centers the German Epilepsy Board considers that the following arrangement would be ideal:

Units for preoperative diagnostic procedures in epilepsy should be integrated in a neurological and neurosurgical department and cooperate with physi-

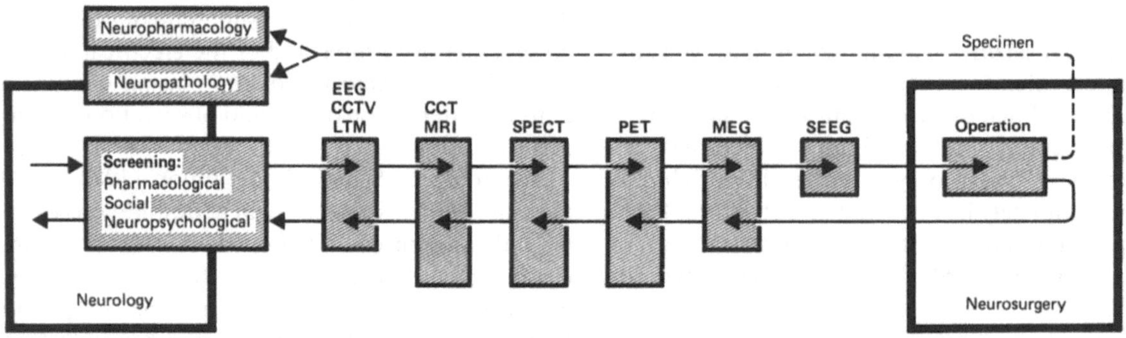

Fig. 1. Pre- and postoperative diagnostic techniques for surgical treatment in focal epilepsy. *CCT* Cranial Computerized Tomography; *CCTV* Closed Circuit Television; *EEG* Electroencephalography; *LTM* Long Term EEG Monitoring; *MEG* Magneto-Encephalography; *MRI* Magnetic Resonance Imaging; *PET* Positron Emission Tomography; *SPECT* Single Photon Emission Tomography; *SEEG* Stereo-Electroencephalography

cians specializing in conservative and surgical treatment of epilepsy to form a working group. This will ensure that the determination of drug resistance, indications for diagnosis and surgery, and aftercare all follow a uniform concept of treatment.

The working groups described here are expensive and are thus only justified if at the same time they are research groups. Natural topics should be:

- Research into the neurological, neurochemical, and anatomical causes of drug resistance
- The development of noninvasive procedures for defining epileptogenic functional disorders
- Comparison of noninvasive and invasive diagnostic procedures
- The development of specific surgical procedures for the removal or influencing of epileptogenic foci

Such centers should collaborate in comparing different diagnostic and therapeutic methods with the aim of optimizing them. This would then lead to proposals for economical use of these methods. The evaluation of appropriate methods is a scientific task which is only feasible at suitably equipped centers.

Such centers should collaborate in comparing different diagnostic and therapeutic methods with the aim of optimizing them. This would then lead to proposals for economical use of these methods. The evaluation of appropriate methods is a scientific task which is only feasible at suitably equipped centers.

References

Hauser WA, Kurland LT (1975) Epidemiology of epilepsy in Rochester, Minnesota, 1935-1967. Epilepsia 16: 1-66

Inoue Y, Wolf P, Hedde JP, Meencke HJ (1985) SPECT with Xenon 133 in the diagnosis of temporal lobe epilepsy. XIIIth World congress of neurology, Hamburg

Janz D, Penin H, Scheidemann KF, Thorbecke R (1985) Epilepsie-Bericht 85. Rheinland, Köln

Mattson RH, et al. (1985) Comparison of carbamazepine, phenobarbital, phenytoin, and primidone in partial and secondarily generalized tonic-clonic seizures. N Engl J Med 313: 145-151

Schmidt D (1982) Two antiepileptic drugs for intractable epilepsy with complex partial seizures. J Neurol Neurosurg Psychiatry 45: 1119-1124

Schmidt D (1984) Behandlung der Epilepsien, 2nd edn. Thieme, Stuttgart

Woodbury LA (1978) Incidence and prevalence of seizure disorders including the epilepsies in the United States of America. In: The commission for the control of epilepsies and its consequences, vol IV. DHEW publication (NIH) 78-276, pp 24-77

Editors' Commentary 8

The chapters grouped within this last topic deal with several aspects such as: identification of candidates for, and goals of, epilepsy surgery; predictors of a good outcome; and several surgical approaches per se. We intended to present data from which 'consequences for the present practice of epilepsy therapy in Europe' could be derived. This was discussed in a final panel (moderator: Engel; panelists: Crandall, Creutzfeldt, Gloor, Janz, Lopes da Silva, Ojemann, Petsche, Rassmussen, Wyler).

Who Should Be Referred for Surgery?

Historically, surgery for epilepsy has been a last resort. For various reasons it has now become an option which should be considered along side the choice of medical treatment. The discussion confirmed the increasing awareness that treating epilepsy is highly rewarding, and seizure control is most decisive for the quality of life. The participants agreed to recommend that patients whose seizures were not brought under control by a neurologist after 9-12 months should be referred to a specialized epilepsy center where surgical options could be considered. There is no simple definition of 'intractability,' and intractability alone is not an indication for surgery. The criteria for referral depend largely on the impact of the seizures on the patient's life.

Among the relative contraindications for surgical treatment, serious psychopathology and actual psychosis, but not the mere history of psychotic problems, were generally accepted. Surgery is not performed routinely in infants and young children, because many seizure problems in this age group are not amenable to surgery, and cooperation for presurgical diagnostic evaluation is often difficult to obtain. However, there are strong arguments for early surgery: (1) the longer the duration of the epilepsy the harder it is to treat it, possibly because of "kindling"; (2) the major social and emotional problems of epileptic patients develop during adolescence; (3) patients over 40 who have had epilepsy for most of their lives often remain disabled even when surgery succeeds in abolishing their seizures; (4) plasticity of brain functions is progressively lost with advancing age.

On the other hand, the discussion showed that even patients who suffer from substantial neurological or psychosocial problems can benefit from surgery in a major way. It was agreed that the question of social rehabilitation after surgery deserves much more attention.

How Should the Surgical Treatment of Choice Be Determined?

When we were preparing the Zürich meeting we knew that Dr. Crandall was working on an extensive review of resective surgery for the first international meeting on surgery for medically intractable epilepsy (Palm Desert). This review (Crandall 1987), complete with integrative summaries (Spencer 1987) and comments (Rasmussen 1987), will soon be available. Moreover, we knew that the complications of epilepsy surgery, which could not be sufficiently dealt with at the Zürich meeting, were to be precisely listed by Van Buren (1987) for various diagnostic and therapeutic surgical approaches in the same Palm Desert proceedings. Therefore, we found it justified to confine discussions at the Zürich meeting to assessing the efficacy of the more widely accepted surgical approaches. The majority of neurosurgeons presented and discussed resective surgery ranging from large multilobe resections to restricted microsurgical resections. Obviously, defining the extent of the epileptogenic cortex remains the major issue even in such cases where the questions of lateralization and localization have been solved with a high degree of reliability. It was proposed that in certain patients without obvious lesions it might become necessary to apply the whole battery of the various neurophysiological recording techniques in a stepwise and graded fashion in order to define as precisely as possible the extent of the resection.

While alternative, stereotactic, surgical approaches to epilepsy were included, other possibili-

Presurgical Evaluation of Epileptics
Edited by H. G. Wieser and C. E. Elger

ties, such as corpus callosotomy, cerebellar stimulation, and subpial incisions were not addressed (for a recent review see Rasmussen 1987). The majority agreed that the future role of many of these latter procedures was ill-defined. It became clear that the outcome has to be accurately predetermined, and it was anticipated that this issue would be discussed in depth at the Palm Desert Conference (see Hauser 1987).

The discussion showed that the other main goal of surgical presentations was also reached, namely, defining the role of electrophysiological and functional tests in the surgeon's decision to operate. It became apparent in what way the presurgical evaluation is influencing referral patterns and how these in turn influence surgical approaches and their results. Probably for these reasons there was a strong inclination among participants to avoid conclusions cautiously as to what should be "the best approach". It was evident that the best investigative procedures and surgical techniques will remain the ones which are well suited to individual patients and their specific problems. The discussion furthermore showed that many of the differences in surgical approaches did not so much exhibit different interpretations or "philosophies" of the various groups but rather reflected the concentration of their efforts on certain types of epilepsy which they could handle best with their available resources.

Although there was tacit agreement that a certain "subspecialization" of each center with given individual resources and talents might even be desirable with a view towards optimal treatment of specific epilepsy types, most participants made a plea for having at their disposal the whole range of diagnostic (including neuropathology) and therapeutic facilities. Optimal and minimal requirements were formulated by Janz, and compared to the actual situation in German-speaking countries.

The panel discussion emphasized the following tasks:

1. To identify areas of general agreement and controversy
2. To pinpoint possible causes of existing disagreement and to show how these can be rationalized
3. To analyze the role of epilepsy surgery from a medical and scientific perspective, and to predict future trends

The moderator's view that, at present, epilepsy surgery is strongly underutilized, directed the discussion towards identification of the possible reasons. The following arguments were raised.

First, successful epilepsy surgery programs require a specialized team and should therefore be established at academic centers. The need to have a large, intensely collaborating team which could concentrate fully on this work is actually often prohibitive, not only because of the financial aspect but also because it may bring the ultimately responsible and performing neurosurgeon to the "passive end." The neurosurgeon is then in danger of relinquishing the more rewarding theoretical aspects of epilepsy surgery to his neurophysiological and neurological colleagues. A strong participation in the theoretical aspects of epilepsy surgery was found to be most important in order to motivate young neurosurgeons in this field. A certain minimal size of centers, guaranteeing a reasonable number of epilepsy operations per year, was thought to be another important condition for collecting sufficient experience, and for survival as an "epilepsy surgeon."

Second, few speakers emphasized the urgent need for better education of colleagues, referring physicians, and also administrations. Examples showed how certain educational programs had indeed had a beneficial influence on the care of epileptic patients, and also on the cost/benefit aspect of existing surgical centers. A strong point for earlier referral was made.

Third, more intense and worldwide collaboration was asked for, especially with the aim to collect enough data to answer pending questions and to direct future trends in the most fruitful directions.

The following possible causes were identified for the renewed interest in and increasing acceptance of epilepsy surgery: (1) increased knowledge, (2) more sophisticated and better methods to study "more complicated" cases, (3) the possibilities to operate on more simple cases without resorting to invasive studies, and - probably most important - (4) better results with fewer postoperative deficits.

References

Crandall PH (1987) Cortical resections. In: Engel J Jr (ed) Surgical treatment of the epilepsies. Raven, New York, pp 377-404

Hauser A (1987) Postscript: how should outcome be determined and reported? In: Engel J Jr (ed) Surgical treatment of the epilepsies. Raven, New York, pp 573-582

Rasmussen T (1987) Commentary: extratemporal cortical excisions and hemispherectomy. In: Engel J Jr (ed) Surgical treatment of the epilepsies. Raven, New York, pp 417-424

Spencer DD (1987) Integrative summary: should there be a surgical treatment of choice, and, if so, how should it be determined? In: Engel J Jr (ed) Surgical treatment of the epilepsies. Raven, New York, pp 477-484

Van Buren JM (1987) Complications of surgical procedures in the diagnosis and treatment of epilepsy. In: Engel J Jr (ed) Surgical treatment of the epilepsies. Raven, New York, pp 465-476

Concluding Remarks

H. Petsche

It is a great pleasure for me to have been asked to round off this gathering with some closing remarks. This gives me an opportunity to say a few words that I feel have to be said. At the very first, I think you all agree with me that this gathering was of great importance for epileptology. The reasons for this success, I think, are due mainly to the outstanding scientific work that has been done here - otherwise it would have been impossible to gather, in one place, so many qualified workers in this field in the middle of a season jammed with congresses and symposia. This centre for the treatment of grave epilepsies, which grew slowly and finally obtained a worldwide reputation, had its origins in the work of Prof. Rudi Hess who, from the beginning of his scientific career, fostered the idea of a treatment of epilepsies by neurosurgery. Professor Hess was among the first to supervise surgical interventions on the brain with his EEG machine. Therefore, it is only logical that the final success of this department, its great reputation, and with this also the success of our meeting are ultimately owed to him.

During these 3 days, an incredible abundance of methods, facts, and perspectives with respect to epilepsy has been presented to us. To me, two aspects seemed to be of particular importance. First, the recognition that even apparently therapy-resistant cases of epilepsy, which a decade ago would never have been considered for surgical attention, are now often amenable to treatment through the utilization of highly sophisticated diagnostic tools and operative methods. Second, the wealth of essential insights into the behavior of large neuronal populations in man has brought a deepened knowledge of the mechanisms of spreading and by that also allowed for a better understanding of the clinical phenomena accompanying a seizure. In this context I may mention Wieser's book on psychomotor seizures which grew out of his experience obtained here, and which, for the first time, helps us to understand the structural background of a great deal of the clinical symptoms of such seizures.

Any attempt to summarize the many results supplied at this gathering seems futile. Nevertheless, the last word on such an event should not only be a honorary address. Thus I should like to glance at a few tendencies that have seemed to emerge from this stream of ideas with which I have been confronted.

To my mind, one of these seemed to be the following: the epileptic seizure is a spatio-temporal and most complex electrical reaction of the brain to cerebral lesions, the number and distribution of which may vary within wide ranges: a restricted scar may be equally sufficient to give rise to a seizure as would a diffuse, generalized deterioration of, say, metabolic origin. In the majority of cases, the morphological topography of the underlying lesions simply defies exact delineation. One main avenue of progress that became apparent in this conference seems to be the improvement of the knowledge of the topography of brain lesions by using depth recordings with many electrodes. Fortunately, however, successful therapy does not only depend on an exact knowledge of this topography but also on knowledge of the intracerebral ways of spreading of the epileptic process. The experience obtained by multiple depth recordings, as we have seen, has brought us new and essential information about the road maps of the epileptic process.

Up to a short time ago, "centrencephalic thinking," as I am used to calling it, originating from Penfield's ideas on the generalization of a seizure, predominated in epileptology. This idea was nourished by the observations of the EEG during generalized seizures, particularly of the petit mal type, which suggested the presence of a common subcortical pacemaker.

Depth recordings have, however, demonstrated that no common subcortical pacemaker does in fact exist: deep brain structures involved in a seizure usually exhibit electrical patterns of great complexity both in space and time. The depth EEG has thus

Institute of Neurophysiology, University of Vienna, and Brain Research Institute, Austrian Academy of Sciences, Währinger Str. 17, 1090 Vienna, Austria

Presurgical Evaluation of Epileptics
Edited by H. G. Wieser and C. E. Elger

brought more and more evidence that the EEG recorded from the scalp is unsuited for conclusions regarding possible phenomena in subcortical structures. Not even cerebral events in the more deeply located parts of the neocortex, as in the temporal lobe, can always be reliably detected by the surface EEG.

Another surprising finding often seen with depth electrodes is that even closely adjacent recording sites may produce quite different electrical activities. These events need not take place in different cortical structures, since even in homogeneous cortical regions, such as in the nongyrated striatal cortex of a rabbit, voltage gradients of up to several millivolts per millimeter are commonly found during seizures. These findings raise doubts about whether the concept of a "primary generalized epilepsy" is still meaningful, since all generalized seizures seem to start from one or perhaps even several lesions in the brain.

One may argue that this is of purely academic interest. I do not think so. For as soon as we dismiss the idea of a primary generalized epilepsy from our minds, we necessarily must ask ourselves in any drug-resistant case with generalized pathological EEG without any lateral accentuation, if this case might not be given relief by neurosurgical intervention of some type.

As I am a neurologist and an electroencephalographer, and neurophysiology also being very dear to me, I should like to stress a last point: in 1870, Hughlings Jackson wrote: "A convulsion is but a symptom, and implies only that there is an occasional, an excessive, and a disorderly discharge of nerve tissue on muscles." He used the word "discharge" without even knowing the very nature of it. It was perhaps his statement of "disorderly discharges" underlying convulsions that has led to the still widely used analogy between a seizure and a thunderstorm in the brain, whereby "thunderstorm" is conceived as an equivalent for "electrical chaos."

The results presented during these days, as well as the neurophysiological observations on epilepsy models, however, have made it quite clear that the epileptic seizure is far from being electrical chaos in the brain, but rather a superimposition of phenomena of high degrees of order, determined by the structural and functional properties of nerve tissue. I see this aspect from the stance of the neurophysiologist and am heartened by this. For it demonstrates once more that it is sensible to search for the principles of the origin and spread of excitation: by this we may not only obtain more insights into the essence of the epileptic seizure and find new keys to therapeutic strategies, but may also learn to understand better the properties of neuronal populations and their structural dependencies.

I learnt a great deal in these few days and should like to express my sincerest gratitude to all participants for this. At the same time, and on behalf of all participants, my thanks go to the organizers of this meeting, Drs. Wieser and Elger, as well as to all those anonymous ladies and gentlemen who have contributed to this obvious success, for their efforts and their splendid hospitality which have made this event a source of great enrichment to all of us.

Subject Index

MIX
Papier aus verantwortungsvollen Quellen
Paper from responsible sources
FSC® C105338

If you have any concerns about our products,
you can contact us on
ProductSafety@springernature.com

In case Publisher is established outside the EU,
the EU authorized representative is:
Springer Nature Customer Service Center GmbH
Europaplatz 3, 69115 Heidelberg, Germany

Printed by Libri Plureos GmbH
in Hamburg, Germany